DATE DUE

Methods in Enzymology

Volume 316
VERTEBRATE PHOTOTRANSDUCTION AND THE
VISUAL CYCLE
Part B

METHODS IN ENZYMOLOGY

EDITORS-IN-CHIEF

John N. Abelson   Melvin I. Simon

DIVISION OF BIOLOGY
CALIFORNIA INSTITUTE OF TECHNOLOGY
PASADENA, CALIFORNIA

FOUNDING EDITORS

Sidney P. Colowick and Nathan O. Kaplan

Methods in Enzymology

Volume 316

Vertebrate Phototransduction and the Visual Cycle

Part B

EDITED BY

Krzysztof Palczewski

UNIVERSITY OF WASHINGTON SCHOOL OF MEDICINE
SEATTLE, WASHINGTON

ACADEMIC PRESS

San Diego   London   Boston   New York   Sydney   Tokyo   Toronto

This book is printed on acid-free paper.

Copyright © 2000 by ACADEMIC PRESS

All Rights Reserved.
No part of this publication may be reproduced or transmitted in any form or by any means, electronic or mechanical, including photocopy, recording, or any information storage and retrieval system, without permission in writing from the Publisher.

The appearance of the code at the bottom of the first page of a chapter in this book indicates the Publisher's consent that copies of the chapter may be made for personal or internal use of specific clients. This consent is given on the condition, however, that the copier pay the stated per copy fee through the Copyright Clearance Center, Inc. (222 Rosewood Drive, Danvers, Massachusetts 01923) for copying beyond that permitted by Sections 107 or 108 of the U.S. Copyright Law. This consent does not extend to other kinds of copying, such as copying for general distribution, for advertising or promotional purposes, for creating new collective works, or for resale. Copy fees for pre-1999 chapters are as shown on the chapter title pages. If no fee code appears on the chapter title page, the copy fee is the same as for current chapters.
0076-6879/00 $30.00

Academic Press
A Harcourt Science and Technology Company
525 B Street, Suite 1900, San Diego, California 92101-4495, USA
http://www.academicpress.com

Academic Press Limited
24-28 Oval Road, London NW1 7DX, UK
http://www.hbuk.co.uk/ap/

International Standard Book Number: 0-12-182217-6

PRINTED IN THE UNITED STATES OF AMERICA
00 01 02 03 04 05 06 MM 9 8 7 6 5 4 3 2 1

Table of Contents

CONTRIBUTORS TO VOLUME 316 xi

PREFACE. xvii

VOLUMES IN SERIES xix

Section I. Photoreceptor Proteins

1. Inhibition of Rhodopsin Phosphorylation by S-Modulins: Purification, Reconstitution, and Assays — SATORU KAWAMURA — 3

2. Analysis of Protein–Protein Interactions in Phototransduction Cascade Using Surface Plasmon Resonance — DAULET K. SATPAEV AND VLADLEN Z. SLEPAK — 20

3. Expression of Phototransduction Proteins in *Xenopus* Oocytes — BARRY E. KNOX, ROBERT B. BARLOW, DEBRA A. THOMPSON, RICHARD SWANSON, AND ENRICO NASI — 41

4. *Xenopus* Rod Photoreceptor: Model for Expression of Retinal Genes — SUCHITRA BATNI, SHOBANA S. MANI, CHARISSE SCHLUETER, MING JI, AND BARRY E. KNOX — 50

5. Fusion between Retinal Rod Outer Segment Membranes and Model Membranes: Functional Assays and Role for Peripherin/rds — KATHLEEN BOESZE-BATTAGLIA — 65

6. Bovine Retinal Nucleoside Diphosphate Kinase: Biochemistry and Molecular Cloning — NAJMOUTIN G. ABDULAEV, DMITRI L. KAKUEV, AND KEVIN D. RIDGE — 87

Section II. Calcium-Binding Proteins and Calcium Measurements in Photoreceptor Cells

7. Calcium-Binding Proteins and Their Assessment in Ocular Diseases — ARTHUR S. POLANS, RICARDO L. GEE, TERESA M. WALKER, AND PAUL R. VAN GINKEL — 103

8. Molecular Structure of Membrane-Targeting Calcium Sensors in Vision: Recoverin and Guanylate Cyclase-Activating Protein 2	JAMES B. AMES, MITSUHIKO IKURA, AND LUBERT STRYER	121
9. Measurement of Light-Evoked Changes in Cytoplasmic Calcium in Functionally Intact Isolated Rod Outer Segments	PETER B. DETWILER AND MARK P. GRAY-KELLER	133
10. Laser Spot Confocal Technique to Measure Cytoplasmic Calcium Concentration in Photoreceptors	HUGH R. MATTHEWS AND GORDON L. FAIN	146

Section III. Phototransduction

11. Functional Study of Rhodopsin Phosphorylation *in Vivo*	ANA MENDEZ, NATALIJA V. KRASNOPEROVA, JANIS LEM, AND JEANNIE CHEN	167
12. Mechanisms of Single-Photon Detection in Rod Photoreceptors	FRED RIEKE	186
13. Electroretinographic Determination of Human Rod Flash Response *in Vivo*	DAVID R. PEPPERBERG, DAVID G. BIRCH, AND DONALD C. HOOD	202
14. Electrophysiological Methods for Measurement of Activation of Phototransduction by Bleached Visual Pigment in Salamander Photoreceptors	M. CARTER CORNWALL, G. J. JONES, V. J. KEFALOV, G. L. FAIN, AND H. R. MATTHEWS	224
15. Exploring Kinetics of Visual Transduction with Time-Resolved Microcalorimetry	T. MINH VUONG	253
16. Use of α-Toxin-Permeabilized Photoreceptors in *in Vitro* Phototransduction Studies	ANNIE E. OTTO-BRUC, J. PRESTON VAN HOOSER, AND ROBERT N. FARISS	269
17. Photoreceptors in Pineal Gland and Brain: Cloning, Localization, and Overexpression	TOSHIYUKI OKANO AND YOSHITAKA FUKADA	278
18. Cultured Amphibian Melanophores: A Model System to Study Melanopsin Photobiology	MARK D. ROLLAG, IGNACIO PROVENCIO, DAVID SUGDEN, AND CARLA B. GREEN	291

Section IV. Enzymes of the Visual Cycle

19. High-Performance Liquid Chromatography Analysis of Visual Cycle Retinoids	GREGORY G. GARWIN AND JOHN C. SAARI	313

20. Quantitative Measurements of Isomerohydrolase Activity	ANETTE WINSTON AND ROBERT R. RANDO	324
21. Multienzyme Analysis of Visual Cycle	HARTMUT STECHER AND KRZYSZTOF PALCZEWSKI	330
22. Analyzing Membrane Topology of 11-*cis*-Retinol Dehydrogenase	ANDRAS SIMON, ANNA ROMERT, AND ULF ERIKSSON	344
23. Phase Partition and High-Performance Liquid Chromatography Assays of Retinoid Dehydrogenases	JOHN C. SAARI, GREGORY G. GARWIN, FRANÇOISE HAESELEER, GEENG-FU JANG, AND KRZYSZTOF PALCZEWSKI	359
24. Short-Chain Dehydrogenases/Reductases in Retina	FRANÇOISE HAESELEER AND KRZYSZTOF PALCZEWSKI	372
25. Substrate Specificities of Retinyl Ester Hydrolases in Retinal Pigment Epithelium	ANDREW T. C. TSIN, NATHAN L. MATA, JENNIFER A. RAY, AND ELIA T. VILLAZANA	384
26. Molecular Characterization of Lecithin-Retinol Acyltransferase	ALBERTO RUIZ AND DEAN BOK	400
27. Analysis of Chromophore of RGR: Retinal G-Protein–Coupled Receptor from Pigment Epithelium	WENSHAN HAO, PU CHEN, AND HENRY K. W. FONG	413

Section V. Posttranslational and Chemical Modifications

28. Low Temperature Photoaffinity Labeling of Rhodopsin and Intermediates along Transduction Path	MARIA L. SOUTO, BABAK BORHAN, AND KOJI NAKANISHI	423
29. Structural Analysis of Protein Prenyl Groups and Associated C-Terminal Modifications	MARK E. WHITTEN, KOHEI YOKOYAMA, DAVID SCHIELTZ, FARIDEH GHOMASHCHI, DEREK LAM, JOHN R. YATES III, KRZYSZTOF PALCZEWSKI, AND MICHAEL H. GELB	436
30. Isoprenylation/Methylation and Transducin Function	CRAIG A. PARISH AND ROBERT R. RANDO	451
31. Functional Analysis of Farnesylation and Methylation of Transducin	TAKAHIKO MATSUDA AND YOSHITAKA FUKADA	465
32. Identification of Phosphorylation Sites within Vertebrate and Invertebrate Rhodopsin	HIROSHI OHGURO	482

33. Ocular Proteomics: Cataloging Photoreceptor Proteins by Two-Dimensional Gel Electrophoresis and Mass Spectrometry — Hiroyuki Matsumoto and Naoka Komori — 492

Section VI. Analysis of Animal Models of Retinal Diseases

34. Animal Models for Inherited Retinal Diseases — Jeanne Frederick, J. Darin Bronson, and Wolfgang Baehr — 515

35. Avian Models of Inherited Retinal Disease — Susan L. Semple-Rowland and Nancy R. Lee — 526

36. Biochemical Analysis of Phototransduction and Visual Cycle in Zebrafish Larvae — Michael R. Taylor, Heather A. Van Epps, Matthew J. Kennedy, John C. Saari, James B. Hurley, and Susan E. Brockerhoff — 536

37. Genetic Models to Study Guanylyl Cyclase Function — Susan W. Robinson and David L. Garbers — 558

38. Analysis of Visual Cycle in Normal and Transgenic Mice — J. Preston Van Hooser, Gregory G. Garwin, and John C. Saari — 565

Section VII. Inherited Retinal Disease: From the Defective Gene to Its Function and Repair

39. *In Situ* Hybridization Studies of Retinal Neurons — Linda K. Barthel and Pamela A. Raymond — 579

40. Cloning and Characterization of Retinal Transcription Factors, Using Target Site–Based Methodology — Shiming Chen and Donald J. Zack — 590

41. *In Vivo* Assessment of Photoreceptor Function in Human Diseases Caused by Photoreceptor-Specific Gene Mutations — Artur V. Cideciyan — 611

42. Spectral Sensitivities of Human Cone Visual Pigments Determined *in Vivo* and *in Vitro* — Andrew Stockman, Lindsay T. Sharpe, Shannath Merbs, and Jeremy Nathans — 626

43. Molecular Analysis of Human Red/Green Visual Pigment Gene Locus: Relationship to Color Vision — Samir S. Deeb, Takaaki Hayashi, Joris Winderickx, and Tomohiko Yamaguchi — 651

44. Expression and Characterization of Peripherin/rds-rom-1 Complexes and Mutants Implicated in Retinal Degenerative Diseases	ANDREW F. X. GOLDBERG AND ROBERT S. MOLDAY	671
45. Isolation of Retinal Proteins That Interact with Retinitis Pigmentosa GTPase Regulator by Interaction Trap Screen in Yeast	RONALD ROEPMAN, DIANA SCHICK, AND PAULO A. FERREIRA	688
46. Genetic Analysis of RPE65: From Human Disease to Mouse Model	T. MICHAEL REDMOND AND CHRISTIAN P. HAMEL	705
47. Construction of Encapsidated Gutted Adenovirus Minichromosomes and Their Application to Rescue of Photoreceptor Degeneration	RAJENDRA KUMAR-SINGH, CLYDE K. YAMASHITA, KEN TRAN, AND DEBORA B. FARBER	724
48. Production and Purification of Recombinant Adeno-Associated Virus	WILLIAM W. HAUSWIRTH, ALFRED S. LEWIN, SERGEI ZOLOTUKHIN, AND NICHOLAS MUZYCZKA	743
49. Ribozymes in Treatment of Inherited Retinal Disease	LYNN C. SHAW, PATRICK O. WHALEN, KIMBERLY A. DRENSER, WEIMING YAN, WILLIAM W. HAUSWIRTH, AND ALFRED S. LEWIN	761
50. Cross-Species Comparison of *in Vivo* Reporter Gene Expression after Recombinant Adeno-Associated Virus–Mediated Retinal Transduction	JEAN BENNETT, VIBHA ANAND, GREGORY M. ACLAND, AND ALBERT M. MAGUIRE	777

AUTHOR INDEX 791

SUBJECT INDEX 827

Contributors to Volume 316

Article numbers are in parentheses following the names of contributors.
Affiliations listed are current.

NAJMOUTIN G. ABDULAEV (6), *Center for Advanced Research in Biotechnology, National Institute of Standards and Technology and University of Maryland Biotechnology Institute, Rockville, Maryland 20850*

GREGORY M. ACLAND (50), *James A. Baker Institute, Cornell University, Ithaca, New York 14853-6401*

JAMES B. AMES (8), *Center for Advanced Research in Biotechnology, University of Maryland Biotechnology Institute, Rockville, Maryland 20850*

VIBHA ANAND (50), *F. M. Kirby Center, Scheie Eye Institute, University of Pennsylvania School of Medicine, Philadelphia, Pennsylvania 19104-6069*

WOLFGANG BAEHR (34), *Moran Eye Center, University of Utah, Salt Lake City, Utah 84132*

ROBERT B. BARLOW (3), *Department of Ophthalmology, State University of New York Health Science Center, Syracuse, New York 13210*

LINDA K. BARTHEL (39), *Department of Anatomy and Cell Biology, University of Michigan Medical School, Ann Arbor, Michigan 48109-0616*

SUCHITRA BATNI (4), *Neuroscience Research Institute, University of California, Santa Barbara, California 93117*

JEAN BENNETT (50), *F. M. Kirby Center, Scheie Eye Institute, University of Pennsylvania School of Medicine, Philadelphia, Pennsylvania 19104-6069*

DAVID G. BIRCH (13), *Retina Foundation of the Southwest, and Department of Ophthalmology, University of Texas Southwestern Medical School, Dallas, Texas 75231*

KATHLEEN BOESZE-BATTAGLIA (5), *Department of Molecular Biology, School of Osteopathic Medicine, University of Medicine and Dentistry of New Jersey, Stratford, New Jersey 08084*

DEAN BOK (26), *Jules Stein Eye Institute, University of California, Los Angeles, California 90095*

BABAK BORHAN (28), *Department of Chemistry, Michigan State University, East Lansing, Michigan 48824*

SUSAN E. BROCKERHOFF (36), *Department of Biochemistry, University of Washington School of Medicine, Seattle, Washington 98195*

J. DARIN BRONSON (34), *Moran Eye Center, University of Utah, Salt Lake City, Utah 84132*

JEANNIE CHEN (11), *Departments of Ophthalmology and Cell and Neurobiology, Mary D. Allen Laboratory for Vision Research, Doheny Eye Institute, University of Southern California School of Medicine, Los Angeles, California 90033*

PU CHEN (27), *Department of Craniofacial Biology, University of Southern California School of Dentistry, Los Angeles, California 90033*

SHIMING CHEN (40), *Department of Ophthalmology and Visual Science, Washington University School of Medicine, St. Louis, Missouri 63110*

ARTUR V. CIDECIYAN (41), *Scheie Eye Institute, University of Pennsylvania, Philadelphia, Pennsylvania 19104*

M. CARTER CORNWALL (14), *Department of Physiology, Boston University School of Medicine, Boston, Massachusetts 02118*

SAMIR S. DEEB (43), *Departments of Medicine and Genetics, University of Washington, Seattle, Washington 98196*

PETER B. DETWILER (9), *Department of Physiology and Biophysics, University of Washington, Seattle, Washington 98195*

KIMBERLY A. DRENSER (49), *Department of Molecular Genetics and Microbiology, University of Florida College of Medicine, Gainesville, Florida 32610*

ULF ERIKSSON (22), *Ludwig Institute for Cancer Research, S-17177 Stockholm, Sweden*

G. L. FAIN (10, 14), *Departments of Physiological Science and Ophthalmology, University of California, Los Angeles, California 90095*

DEBORA B. FARBER (47), *Jules Stein Eye Institute, University of California School of Medicine, Los Angeles, California 90095*

ROBERT N. FARISS (16), *Department of Ophthalmology, University of Washington School of Medicine, Seattle, Washington 98195-6485*

PAULO A. FERREIRA (45), *Department of Pharmacology and Toxicology, Medical College of Wisconsin, Milwaukee, Wisconsin 53226*

HENRY K. W. FONG (27), *Doheny Eye Institute, and Departments of Ophthalmology and Molecular Microbiology, University of Southern California School of Medicine, Los Angeles, California 90033*

JEANNE FREDERICK (34), *Moran Eye Center, University of Utah, Salt Lake City, Utah 84132*

YOSHITAKA FUKADA (17, 31), *Department of Biophysics and Biochemistry, Graduate School of Science, University of Tokyo, Bunkyo-ku, Tokyo 113-0033, Japan*

DAVID L. GARBERS (37), *Howard Hughes Medical Institute and Department of Pharmacology, University of Texas Southwestern Medical Center, Dallas, Texas 75235-9050*

GREGORY G. GARWIN (19, 23, 38), *Department of Ophthalmology, University of Washington School of Medicine, Seattle, Washington 98195-6485*

RICARDO L. GEE (7), *Department of Ophthalmology and Visual Sciences, Madison Medical School, University of Wisconsin, Madison, Wisconsin 53792*

MICHAEL H. GELB (29), *Departments of Chemistry and Biochemistry, University of Washington, Seattle, Washington 98195*

FARIDEH GHOMASHCHI (29), *Departments of Chemistry and Biochemistry, University of Washington, Seattle, Washington 98195*

ANDREW F. X. GOLDBERG (44), *Eye Research Institute, Oakland University, Rochester, Michigan 48309*

MARK P. GRAY-KELLER (9), *Department of Physiology and Biophysics, University of Washington, Seattle, Washington 98195*

CARLA B. GREEN (18), *Department of Biology, University of Virginia, Charlottesville, Virginia 22903*

FRANÇOISE HAESELEER (23, 24), *Department of Ophthalmology, University of Washington School of Medicine, Seattle, Washington 98195-6485*

CHRISTIAN P. HAMEL (46), *INSERM Unité 254, Neurobiologie de l'Audition, F-34090 Montpellier, France*

WENSHAN HAO (27), *Division of Biology, California Institute of Technology, Pasadena, California 91125*

WILLIAM W. HAUSWIRTH (48, 49), *Departments of Ophthalmology, and Molecular Genetics and Microbiology, University of Florida College of Medicine, Gainesville, Florida 32610*

TAKAAKI HAYASHI (43), *Departments of Medicine and Genetics, University of Washington, Seattle, Washington 98196*

DONALD C. HOOD (13), *Department of Psychology, Columbia University, New York, New York 10027*

JAMES B. HURLEY (36), *Howard Hughes Medical Institute and Department of Biochemistry, University of Washington School of Medicine, Seattle, Washington 98195*

MITSUHIKO IKURA (8), *Division of Molecular and Structural Biology, Ontario Cancer Institute, University of Toronto, Toronto, Ontario 25G 2M9, Canada*

GEENG-FU JANG (23), *Department of Ophthalmology, University of Washington School of Medicine, Seattle, Washington 98195-6485*

MING JI (4), *Department of Biochemistry and Molecular Biology, State University of New York Health Science Center, Syracuse, New York 13210*

G. J. JONES (14), *Department of Physiology, Boston University School of Medicine, Boston, Massachusetts 02118*

DMITRI L. KAKUEV (6), *Shemyakin and Ovchinnikov Institute of Bioorganic Chemistry, Russian Academy of Sciences, Moscow, Russia*

SATORU KAWAMURA (1), *Department of Biology, Graduate School of Science, Osaka University, Toyonaka, Osaka 560-0043, Japan*

V. J. KEFALOV (14), *Department of Physiology, Boston University School of Medicine, Boston, Massachusetts 02118*

MATTHEW J. KENNEDY (36), *Department of Biochemistry, University of Washington School of Medicine, Seattle, Washington 98195*

BARRY E. KNOX (3, 4), *Departments of Biochemistry and Molecular Biology, and Ophthalmology, State University of New York Health Science Center, Syracuse, New York 13210*

NAOKA KOMORI (33), *Department of Biochemistry and Molecular Biology, University of Oklahoma Health Sciences Center, Oklahoma City, Oklahoma 73190*

NATALIJA V. KRASNOPEROVA (11), *Departments of Ophthalmology and Genetics, New England Medical Center, Tufts University School of Medicine, Boston, Massachusetts 02111*

RAJENDRA KUMAR-SINGH (47), *Jules Stein Eye Institute, University of California School of Medicine, Los Angeles, California 90095*

DEREK LAM (29), *Departments of Chemistry and Biochemistry, University of Washington, Seattle, Washington 98195*

NANCY R. LEE (35), *University of Florida Brain Institute, Gainesville, Florida 32610-0244*

JANIS LEM (11), *Departments of Ophthalmology and Genetics, New England Medical Center, Tufts University School of Medicine, Boston, Massachusetts 02111*

ALFRED S. LEWIN (48, 49), *Department of Molecular Genetics and Microbiology, Gene Therapy Center, University of Florida College of Medicine, Gainesville, Florida 32610*

ALBERT M. MAGUIRE (50), *F. M. Kirby Center, Scheie Eye Institute, University of Pennsylvania School of Medicine, Philadelphia, Pennsylvania 19104-6069*

SHOBANA S. MANI (4), *Section of Genetics and Developmental Biology, Cornell University, Ithaca, New York 14853-2703*

NATHAN L. MATA (25), *Department of Psychiatry, University of Texas Southwestern Medical Center, Dallas, Texas 75235*

TAKAHIKO MATSUDA (31), *Department of Biophysics and Biochemistry, Graduate School of Science, University of Tokyo, Bunkyo-ku, Tokyo 113-0033, Japan*

HIROYUKI MATSUMOTO (33), *Department of Biochemistry and Molecular Biology, and NSF EPSCoR Oklahoma Biotechnology Network Laser Mass Spectrometry Facility, University of Oklahoma Health Sciences Center, Oklahoma City, Oklahoma 73190*

H. R. MATTHEWS (10, 14), *Physiological Laboratory, University of Cambridge, Cambridge CB2 3EG, United Kingdom*

ANA MENDEZ (11), *Departments of Ophthalmology and Cell and Neurobiology, Mary D. Allen Laboratory for Vision Research, Doheny Eye Institute, University of Southern California School of Medicine, Los Angeles, California 90033*

SHANNATH MERBS (42), *Department of Ophthalmology, Johns Hopkins University School of Medicine, Baltimore, Maryland 21205*

ROBERT S. MOLDAY (44), *Department of Biochemistry and Molecular Biology, University of British Columbia, Vancouver, British Columbia V6T 1Z3, Canada*

NICHOLAS MUZYCZKA (48), *Department of Molecular Genetics and Microbiology, Gene Therapy Center, University of Florida College of Medicine, Gainesville, Florida 32610*

KOJI NAKANISHI (28), *Department of Chemistry, Columbia University, New York, New York 10027*

ENRICO NASI (3), *Department of Physiology, Boston University School of Medicine, Boston, Massachusetts 02118*

JEREMY NATHANS (42), *Departments of Molecular Biology and Genetics, Neuroscience, and Ophthalmology, Howard Hughes Medical Institute, Johns Hopkins University School of Medicine, Baltimore, Maryland 21205*

HIROSHI OHGURO (32), *Department of Ophthalmology, Sapporo Medical University School of Medicine, Chuoku, Sapporo Hokkaido 060-8543, Japan*

TOSHIYUKI OKANO (17), *Department of Biophysics and Biochemistry, Graduate School of Science, University of Tokyo, Bunkyo-ku, Tokyo 113-0033, Japan*

ANNIE E. OTTO-BRUC (16), *Department of Ophthalmology, University of Washington School of Medicine, Seattle, Washington 98195-6485*

KRZYSZTOF PALCZEWSKI (21, 23, 24, 29), *Departments of Ophthalmology, Chemistry, and Pharmacology, University of Washington School of Medicine, Seattle, Washington 98195-6485*

CRAIG A. PARISH (30), *Department of Biological Chemistry and Molecular Pharmacology, Harvard Medical School, Boston, Massachusetts 02115*

DAVID R. PEPPERBERG (13), *Department of Ophthalmology and Visual Sciences, College of Medicine, University of Illinois, Chicago, Illinois 60612*

ARTHUR S. POLANS (7), *Department of Ophthalmology and Visual Sciences, Madison Medical School, University of Wisconsin, Madison, Wisconsin 53792*

IGNACIO PROVENCIO (18), *Department of Anatomy and Cell Biology, Uniformed Services University, Bethesda, Maryland 20815*

ROBERT R. RANDO (20, 30), *Department of Biological Chemistry and Molecular Pharmacology, Harvard Medical School, Boston, Massachusetts 02115*

JENNIFER A. RAY (25), *Division of Life Sciences, University of Texas, San Antonio, Texas 78249-0662*

PAMELA A. RAYMOND (39), *Department of Anatomy and Cell Biology, University of Michigan Medical School, Ann Arbor, Michigan 48109-0616*

T. MICHAEL REDMOND (46), *Laboratory of Retinal Cell and Molecular Biology, National Eye Institute, National Institutes of Health, Bethesda, Maryland 20892*

KEVIN D. RIDGE (6), *Center for Advanced Research in Biotechnology, National Institute of Standards and Technology and University of Maryland Biotechnology Institute, Rockville, Maryland 20850*

FRED RIEKE (12), *Department of Physiology and Biophysics, University of Washington, Seattle, Washington 98195-7290*

SUSAN W. ROBINSON (37), *Howard Hughes Medical Institute and Department of Pharmacology, University of Texas Southwestern Medical Center, Dallas, Texas 75235-9050*

RONALD ROEPMAN (45), *Department of Human Genetics, University Hospital Nijmegen, Nijmegen 6500 HB, The Netherlands*

MARK D. ROLLAG (18), *Department of Anatomy and Cell Biology, Uniformed Services University, Bethesda, Maryland 20815*

ANNA ROMERT (22), *Ludwig Institute for Cancer Research, S-17177 Stockholm, Sweden*

ALBERTO RUIZ (26), *Department of Neurobiology, University of California, Los Angeles, California 90095*

JOHN C. SAARI (19, 23, 36, 38), *Departments of Ophthalmology and Biochemistry, University of Washington School of Medicine, Seattle, Washington 98195-6485*

DAULET K. SATPAEV (2), *Department of Molecular and Cellular Pharmacology and Neuroscience Program, University of Miami School of Medicine, Miami, Florida 33136*

DIANA SCHICK (45), *Department of Pharmacology and Toxicology, Medical College of Wisconsin, Milwaukee, Wisconsin 53226*

DAVID SCHIELTZ (29), *Department of Molecular Biotechnology, University of Washington, Seattle, Washington 98195*

CHARISSE SCHLUETER (4), *Department of Biochemistry and Molecular Biology, State University of New York Health Science Center, Syracuse, New York 13210*

SUSAN L. SEMPLE-ROWLAND (35), *University of Florida Brain Institute, Gainesville, Florida 32610-0244*

LINDSAY T. SHARPE (42), *Forschungsstelle für Experimentelle Ophthalmologie, Universitäts-Augenklinik Abteilung II, D-72076 Tübingen, Germany*

LYNN C. SHAW (49), *Department of Molecular Genetics and Microbiology, University of Florida College of Medicine, Gainesville, Florida 32610*

ANDRAS SIMON (22), *Department of Biochemistry and Molecular Biology, University College London, London WC1E 6BT, United Kingdom*

VLADLEN Z. SLEPAK (2), *Department of Molecular and Cellular Pharmacology and Neuroscience Program, University of Miami School of Medicine, Miami, Florida 33136*

MARIA L. SOUTO (28), *Universidad de La Laguna, Santa Cruz, Mexico*

HARTMUT STECHER (21), *Department of Ophthalmology, University of Washington School of Medicine, Seattle, Washington 98195-6485*

ANDREW STOCKMAN (42), *Department of Psychology, University of California, San Diego, La Jolla, California 92093-0109*

LUBERT STRYER (8), *Department of Neurobiology, Stanford University School of Medicine, Stanford, California 94305-5125*

DAVID SUGDEN (18), *Endocrinology and Reproduction Research Group, Physiology Division, School of Biomedical Sciences, Kings College London, London SE1 1UL, United Kingdom*

RICHARD SWANSON (3), *Merck, Sharpe and Dohme Research Laboratories, West Point, Pennsylvania 19486*

MICHAEL R. TAYLOR (36), *Department of Biochemistry, University of Washington School of Medicine, Seattle, Washington 98195*

DEBRA A. THOMPSON (3), *Department of Ophthalmology, University of Michigan Medical School, Ann Arbor, Michigan 48109-0616*

KEN TRAN (47), *Jules Stein Eye Institute, University of California School of Medicine, Los Angeles, California 90095*

ANDREW T. C. TSIN (25), *Division of Life Sciences, University of Texas, San Antonio, Texas 78249-0662*

HEATHER A. VAN EPPS (36), *Department of Biochemistry, University of Washington School of Medicine, Seattle, Washington 98195*

PAUL R. VAN GINKEL (7), *Department of Ophthalmology and Visual Sciences, Madison Medical School, University of Wisconsin, Madison, Wisconsin 53792*

J. PRESTON VAN HOOSER (16, 38), *Department of Ophthalmology, University of Washington School of Medicine, Seattle, Washington 98195-6485*

ELIA T. VILLAZANA (25), *Division of Life Sciences, University of Texas, San Antonio, Texas 78249-0662*

T. MINH VUONG (15), *Aurora Biosciences, San Diego, California 92121*

TERESA M. WALKER (7), *Department of Ophthalmology and Visual Sciences, Madison Medical School, University of Wisconsin, Madison, Wisconsin 53792*

PATRICK O. WHALEN (49), *Department of Molecular Genetics and Microbiology, University of Florida College of Medicine, Gainesville, Florida 32610*

MARK E. WHITTEN (29), *Departments of Chemistry and Biochemistry, University of Washington, Seattle, Washington 98195*

JORIS WINDERICKX (43), *Departments of Medicine and Genetics, University of Washington, Seattle, Washington 98196*

ANETTE WINSTON (20), *Department of Biological Chemistry and Molecular Pharmacology, Harvard Medical School, Boston, Massachusetts 02115*

TOMOHIKO YAMAGUCHI (43), *Departments of Medicine and Genetics, University of Washington, Seattle, Washington 98196*

CLYDE K. YAMASHITA (47), *Jules Stein Eye Institute, University of California School of Medicine, Los Angeles, California 90095*

WEIMING YAN (49), *Department of Molecular Genetics and Microbiology, University of Florida College of Medicine, Gainesville, Florida 32610*

JOHN R. YATES III (29), *Department of Molecular Biotechnology, University of Washington, Seattle, Washington 98195*

KOHEI YOKOYAMA (29), *Departments of Chemistry and Biochemistry, University of Washington, Seattle, Washington 98195*

DONALD J. ZACK (40), *Departments of Ophthalmology, Neuroscience, and Molecular Biology and Genetics, Johns Hopkins University School of Medicine, Baltimore, Maryland 21287*

SERGEI ZOLOTUKHIN (48), *Department of Molecular Genetics and Microbiology, Gene Therapy Center, University of Florida College of Medicine, Gainesville, Florida 32610*

Preface

Molecular methods related to visual transduction and the visual cycle have grown in number and evolved considerably since they were covered in Volume 81 of *Methods in Enzymology* in 1982. During this time the fundamental principles elicited through the study of visual signal transduction have been extended far into the biology of other transduction systems initiated by numerous types of small chemical molecules.

Unprecedented progress has been made in vision research since 1982, and newly developed methods to study vertebrate phototransduction and the visual cycle are summarized in this Volume 316 (Part B) of *Methods in Enzymology* and its companion Volume 315 (Part A). The methods included are very broadly defined. In this volume the methods appear in sections on characterization of calcium-binding proteins and calcium measurements in photoreceptor cells; methods to identify posttranslational and chemical modifications; assays to study phototransduction *in vitro* and *in vivo;* characterization of enzymes of the visual cycle; analysis of animal models; and methods to study inherited retinal diseases: from the defective gene to its function and repair.

In Volume 315 they appear in sections on characterization of opsins; methods to study properties of proteins that interact with rhodopsin; characterization of photoreceptor protein phosphatases, phosphodiesterase, and guanylyl cyclase; cyclic nucleotide-gated channels; and Na^+/Ca^{2+}-K^+ exchanger and ABCR transporter.

It is my pleasure to thank the authors, the reviewers, Drs. M. I. Simon and J. N. Abelson (the editors-in-chief), and the publisher who made this project possible. In particular, I would like to thank members of my laboratory for their patience and support during the hectic period of editing these volumes.

KRZYSZTOF PALCZEWSKI

METHODS IN ENZYMOLOGY

VOLUME I. Preparation and Assay of Enzymes
Edited by SIDNEY P. COLOWICK AND NATHAN O. KAPLAN

VOLUME II. Preparation and Assay of Enzymes
Edited by SIDNEY P. COLOWICK AND NATHAN O. KAPLAN

VOLUME III. Preparation and Assay of Substrates
Edited by SIDNEY P. COLOWICK AND NATHAN O. KAPLAN

VOLUME IV. Special Techniques for the Enzymologist
Edited by SIDNEY P. COLOWICK AND NATHAN O. KAPLAN

VOLUME V. Preparation and Assay of Enzymes
Edited by SIDNEY P. COLOWICK AND NATHAN O. KAPLAN

VOLUME VI. Preparation and Assay of Enzymes (*Continued*)
Preparation and Assay of Substrates
Special Techniques
Edited by SIDNEY P. COLOWICK AND NATHAN O. KAPLAN

VOLUME VII. Cumulative Subject Index
Edited by SIDNEY P. COLOWICK AND NATHAN O. KAPLAN

VOLUME VIII. Complex Carbohydrates
Edited by ELIZABETH F. NEUFELD AND VICTOR GINSBURG

VOLUME IX. Carbohydrate Metabolism
Edited by WILLIS A. WOOD

VOLUME X. Oxidation and Phosphorylation
Edited by RONALD W. ESTABROOK AND MAYNARD E. PULLMAN

VOLUME XI. Enzyme Structure
Edited by C. H. W. HIRS

VOLUME XII. Nucleic Acids (Parts A and B)
Edited by LAWRENCE GROSSMAN AND KIVIE MOLDAVE

VOLUME XIII. Citric Acid Cycle
Edited by J. M. LOWENSTEIN

VOLUME XIV. Lipids
Edited by J. M. LOWENSTEIN

VOLUME XV. Steroids and Terpenoids
Edited by RAYMOND B. CLAYTON

VOLUME XVI. Fast Reactions
Edited by KENNETH KUSTIN

VOLUME XVII. Metabolism of Amino Acids and Amines (Parts A and B)
Edited by HERBERT TABOR AND CELIA WHITE TABOR

VOLUME XVIII. Vitamins and Coenzymes (Parts A, B, and C)
Edited by DONALD B. MCCORMICK AND LEMUEL D. WRIGHT

VOLUME XIX. Proteolytic Enzymes
Edited by GERTRUDE E. PERLMANN AND LASZLO LORAND

VOLUME XX. Nucleic Acids and Protein Synthesis (Part C)
Edited by KIVIE MOLDAVE AND LAWRENCE GROSSMAN

VOLUME XXI. Nucleic Acids (Part D)
Edited by LAWRENCE GROSSMAN AND KIVIE MOLDAVE

VOLUME XXII. Enzyme Purification and Related Techniques
Edited by WILLIAM B. JAKOBY

VOLUME XXIII. Photosynthesis (Part A)
Edited by ANTHONY SAN PIETRO

VOLUME XXIV. Photosynthesis and Nitrogen Fixation (Part B)
Edited by ANTHONY SAN PIETRO

VOLUME XXV. Enzyme Structure (Part B)
Edited by C. H. W. HIRS AND SERGE N. TIMASHEFF

VOLUME XXVI. Enzyme Structure (Part C)
Edited by C. H. W. HIRS AND SERGE N. TIMASHEFF

VOLUME XXVII. Enzyme Structure (Part D)
Edited by C. H. W. HIRS AND SERGE N. TIMASHEFF

VOLUME XXVIII. Complex Carbohydrates (Part B)
Edited by VICTOR GINSBURG

VOLUME XXIX. Nucleic Acids and Protein Synthesis (Part E)
Edited by LAWRENCE GROSSMAN AND KIVIE MOLDAVE

VOLUME XXX. Nucleic Acids and Protein Synthesis (Part F)
Edited by KIVIE MOLDAVE AND LAWRENCE GROSSMAN

VOLUME XXXI. Biomembranes (Part A)
Edited by SIDNEY FLEISCHER AND LESTER PACKER

VOLUME XXXII. Biomembranes (Part B)
Edited by SIDNEY FLEISCHER AND LESTER PACKER

VOLUME XXXIII. Cumulative Subject Index Volumes I–XXX
Edited by MARTHA G. DENNIS AND EDWARD A. DENNIS

VOLUME XXXIV. Affinity Techniques (Enzyme Purification: Part B)
Edited by WILLIAM B. JAKOBY AND MEIR WILCHEK

VOLUME XXXV. Lipids (Part B)
Edited by JOHN M. LOWENSTEIN

VOLUME XXXVI. Hormone Action (Part A: Steroid Hormones)
Edited by BERT W. O'MALLEY AND JOEL G. HARDMAN

VOLUME XXXVII. Hormone Action (Part B: Peptide Hormones)
Edited by BERT W. O'MALLEY AND JOEL G. HARDMAN

VOLUME XXXVIII. Hormone Action (Part C: Cyclic Nucleotides)
Edited by JOEL G. HARDMAN AND BERT W. O'MALLEY

VOLUME XXXIX. Hormone Action (Part D: Isolated Cells, Tissues, and Organ Systems)
Edited by JOEL G. HARDMAN AND BERT W. O'MALLEY

VOLUME XL. Hormone Action (Part E: Nuclear Structure and Function)
Edited by BERT W. O'MALLEY AND JOEL G. HARDMAN

VOLUME XLI. Carbohydrate Metabolism (Part B)
Edited by W. A. WOOD

VOLUME XLII. Carbohydrate Metabolism (Part C)
Edited by W. A. WOOD

VOLUME XLIII. Antibiotics
Edited by JOHN H. HASH

VOLUME XLIV. Immobilized Enzymes
Edited by KLAUS MOSBACH

VOLUME XLV. Proteolytic Enzymes (Part B)
Edited by LASZLO LORAND

VOLUME XLVI. Affinity Labeling
Edited by WILLIAM B. JAKOBY AND MEIR WILCHEK

VOLUME XLVII. Enzyme Structure (Part E)
Edited by C. H. W. HIRS AND SERGE N. TIMASHEFF

VOLUME XLVIII. Enzyme Structure (Part F)
Edited by C. H. W. HIRS AND SERGE N. TIMASHEFF

VOLUME XLIX. Enzyme Structure (Part G)
Edited by C. H. W. HIRS AND SERGE N. TIMASHEFF

VOLUME L. Complex Carbohydrates (Part C)
Edited by VICTOR GINSBURG

VOLUME LI. Purine and Pyrimidine Nucleotide Metabolism
Edited by PATRICIA A. HOFFEE AND MARY ELLEN JONES

VOLUME LII. Biomembranes (Part C: Biological Oxidations)
Edited by SIDNEY FLEISCHER AND LESTER PACKER

VOLUME LIII. Biomembranes (Part D: Biological Oxidations)
Edited by SIDNEY FLEISCHER AND LESTER PACKER

VOLUME LIV. Biomembranes (Part E: Biological Oxidations)
Edited by SIDNEY FLEISCHER AND LESTER PACKER

VOLUME LV. Biomembranes (Part F: Bioenergetics)
Edited by SIDNEY FLEISCHER AND LESTER PACKER

VOLUME LVI. Biomembranes (Part G: Bioenergetics)
Edited by SIDNEY FLEISCHER AND LESTER PACKER

VOLUME LVII. Bioluminescence and Chemiluminescence
Edited by MARLENE A. DELUCA

VOLUME LVIII. Cell Culture
Edited by WILLIAM B. JAKOBY AND IRA PASTAN

VOLUME LIX. Nucleic Acids and Protein Synthesis (Part G)
Edited by KIVIE MOLDAVE AND LAWRENCE GROSSMAN

VOLUME LX. Nucleic Acids and Protein Synthesis (Part H)
Edited by KIVIE MOLDAVE AND LAWRENCE GROSSMAN

VOLUME 61. Enzyme Structure (Part H)
Edited by C. H. W. HIRS AND SERGE N. TIMASHEFF

VOLUME 62. Vitamins and Coenzymes (Part D)
Edited by DONALD B. MCCORMICK AND LEMUEL D. WRIGHT

VOLUME 63. Enzyme Kinetics and Mechanism (Part A: Initial Rate and Inhibitor Methods)
Edited by DANIEL L. PURICH

VOLUME 64. Enzyme Kinetics and Mechanism (Part B: Isotopic Probes and Complex Enzyme Systems)
Edited by DANIEL L. PURICH

VOLUME 65. Nucleic Acids (Part I)
Edited by LAWRENCE GROSSMAN AND KIVIE MOLDAVE

VOLUME 66. Vitamins and Coenzymes (Part E)
Edited by DONALD B. MCCORMICK AND LEMUEL D. WRIGHT

VOLUME 67. Vitamins and Coenzymes (Part F)
Edited by DONALD B. MCCORMICK AND LEMUEL D. WRIGHT

VOLUME 68. Recombinant DNA
Edited by RAY WU

VOLUME 69. Photosynthesis and Nitrogen Fixation (Part C)
Edited by ANTHONY SAN PIETRO

VOLUME 70. Immunochemical Techniques (Part A)
Edited by HELEN VAN VUNAKIS AND JOHN J. LANGONE

VOLUME 71. Lipids (Part C)
Edited by JOHN M. LOWENSTEIN

VOLUME 72. Lipids (Part D)
Edited by JOHN M. LOWENSTEIN

VOLUME 73. Immunochemical Techniques (Part B)
Edited by JOHN J. LANGONE AND HELEN VAN VUNAKIS

VOLUME 74. Immunochemical Techniques (Part C)
Edited by JOHN J. LANGONE AND HELEN VAN VUNAKIS

VOLUME 75. Cumulative Subject Index Volumes XXXI, XXXII, XXXIV–LX
Edited by EDWARD A. DENNIS AND MARTHA G. DENNIS

VOLUME 76. Hemoglobins
Edited by ERALDO ANTONINI, LUIGI ROSSI-BERNARDI, AND EMILIA CHIANCONE

VOLUME 77. Detoxication and Drug Metabolism
Edited by WILLIAM B. JAKOBY

VOLUME 78. Interferons (Part A)
Edited by SIDNEY PESTKA

VOLUME 79. Interferons (Part B)
Edited by SIDNEY PESTKA

VOLUME 80. Proteolytic Enzymes (Part C)
Edited by LASZLO LORAND

VOLUME 81. Biomembranes (Part H: Visual Pigments and Purple Membranes, I)
Edited by LESTER PACKER

VOLUME 82. Structural and Contractile Proteins (Part A: Extracellular Matrix)
Edited by LEON W. CUNNINGHAM AND DIXIE W. FREDERIKSEN

VOLUME 83. Complex Carbohydrates (Part D)
Edited by VICTOR GINSBURG

VOLUME 84. Immunochemical Techniques (Part D: Selected Immunoassays)
Edited by JOHN J. LANGONE AND HELEN VAN VUNAKIS

VOLUME 85. Structural and Contractile Proteins (Part B: The Contractile Apparatus and the Cytoskeleton)
Edited by DIXIE W. FREDERIKSEN AND LEON W. CUNNINGHAM

VOLUME 86. Prostaglandins and Arachidonate Metabolites
Edited by WILLIAM E. M. LANDS AND WILLIAM L. SMITH

VOLUME 87. Enzyme Kinetics and Mechanism (Part C: Intermediates, Stereochemistry, and Rate Studies)
Edited by DANIEL L. PURICH

VOLUME 88. Biomembranes (Part I: Visual Pigments and Purple Membranes, II)
Edited by LESTER PACKER

VOLUME 89. Carbohydrate Metabolism (Part D)
Edited by WILLIS A. WOOD

VOLUME 90. Carbohydrate Metabolism (Part E)
Edited by WILLIS A. WOOD

VOLUME 91. Enzyme Structure (Part I)
Edited by C. H. W. HIRS AND SERGE N. TIMASHEFF

VOLUME 92. Immunochemical Techniques (Part E: Monoclonal Antibodies and General Immunoassay Methods)
Edited by JOHN J. LANGONE AND HELEN VAN VUNAKIS

VOLUME 93. Immunochemical Techniques (Part F: Conventional Antibodies, Fc Receptors, and Cytotoxicity)
Edited by JOHN J. LANGONE AND HELEN VAN VUNAKIS

VOLUME 94. Polyamines
Edited by HERBERT TABOR AND CELIA WHITE TABOR

VOLUME 95. Cumulative Subject Index Volumes 61–74, 76–80
Edited by EDWARD A. DENNIS AND MARTHA G. DENNIS

VOLUME 96. Biomembranes [Part J: Membrane Biogenesis: Assembly and Targeting (General Methods; Eukaryotes)]
Edited by SIDNEY FLEISCHER AND BECCA FLEISCHER

VOLUME 97. Biomembranes [Part K: Membrane Biogenesis: Assembly and Targeting (Prokaryotes, Mitochondria, and Chloroplasts)]
Edited by SIDNEY FLEISCHER AND BECCA FLEISCHER

VOLUME 98. Biomembranes (Part L: Membrane Biogenesis: Processing and Recycling)
Edited by SIDNEY FLEISCHER AND BECCA FLEISCHER

VOLUME 99. Hormone Action (Part F: Protein Kinases)
Edited by JACKIE D. CORBIN AND JOEL G. HARDMAN

VOLUME 100. Recombinant DNA (Part B)
Edited by RAY WU, LAWRENCE GROSSMAN, AND KIVIE MOLDAVE

VOLUME 101. Recombinant DNA (Part C)
Edited by RAY WU, LAWRENCE GROSSMAN, AND KIVIE MOLDAVE

VOLUME 102. Hormone Action (Part G: Calmodulin and Calcium-Binding Proteins)
Edited by ANTHONY R. MEANS AND BERT W. O'MALLEY

VOLUME 103. Hormone Action (Part H: Neuroendocrine Peptides)
Edited by P. MICHAEL CONN

VOLUME 104. Enzyme Purification and Related Techniques (Part C)
Edited by WILLIAM B. JAKOBY

VOLUME 105. Oxygen Radicals in Biological Systems
Edited by LESTER PACKER

VOLUME 106. Posttranslational Modifications (Part A)
Edited by FINN WOLD AND KIVIE MOLDAVE

VOLUME 107. Posttranslational Modifications (Part B)
Edited by FINN WOLD AND KIVIE MOLDAVE

VOLUME 108. Immunochemical Techniques (Part G: Separation and Characterization of Lymphoid Cells)
Edited by GIOVANNI DI SABATO, JOHN J. LANGONE, AND HELEN VAN VUNAKIS

VOLUME 109. Hormone Action (Part I: Peptide Hormones)
Edited by LUTZ BIRNBAUMER AND BERT W. O'MALLEY

VOLUME 110. Steroids and Isoprenoids (Part A)
Edited by JOHN H. LAW AND HANS C. RILLING

VOLUME 111. Steroids and Isoprenoids (Part B)
Edited by JOHN H. LAW AND HANS C. RILLING

VOLUME 112. Drug and Enzyme Targeting (Part A)
Edited by KENNETH J. WIDDER AND RALPH GREEN

VOLUME 113. Glutamate, Glutamine, Glutathione, and Related Compounds
Edited by ALTON MEISTER

VOLUME 114. Diffraction Methods for Biological Macromolecules (Part A)
Edited by HAROLD W. WYCKOFF, C. H. W. HIRS, AND SERGE N. TIMASHEFF

VOLUME 115. Diffraction Methods for Biological Macromolecules (Part B)
Edited by HAROLD W. WYCKOFF, C. H. W. HIRS, AND SERGE N. TIMASHEFF

VOLUME 116. Immunochemical Techniques (Part H: Effectors and Mediators of Lymphoid Cell Functions)
Edited by GIOVANNI DI SABATO, JOHN J. LANGONE, AND HELEN VAN VUNAKIS

VOLUME 117. Enzyme Structure (Part J)
Edited by C. H. W. HIRS AND SERGE N. TIMASHEFF

VOLUME 118. Plant Molecular Biology
Edited by ARTHUR WEISSBACH AND HERBERT WEISSBACH

VOLUME 119. Interferons (Part C)
Edited by SIDNEY PESTKA

VOLUME 120. Cumulative Subject Index Volumes 81–94, 96–101

VOLUME 121. Immunochemical Techniques (Part I: Hybridoma Technology and Monoclonal Antibodies)
Edited by JOHN J. LANGONE AND HELEN VAN VUNAKIS

VOLUME 122. Vitamins and Coenzymes (Part G)
Edited by FRANK CHYTIL AND DONALD B. MCCORMICK

VOLUME 123. Vitamins and Coenzymes (Part H)
Edited by FRANK CHYTIL AND DONALD B. MCCORMICK

VOLUME 124. Hormone Action (Part J: Neuroendocrine Peptides)
Edited by P. MICHAEL CONN

VOLUME 125. Biomembranes (Part M: Transport in Bacteria, Mitochondria, and Chloroplasts: General Approaches and Transport Systems)
Edited by SIDNEY FLEISCHER AND BECCA FLEISCHER

VOLUME 126. Biomembranes (Part N: Transport in Bacteria, Mitochondria, and Chloroplasts: Protonmotive Force)
Edited by SIDNEY FLEISCHER AND BECCA FLEISCHER

VOLUME 127. Biomembranes (Part O: Protons and Water: Structure and Translocation)
Edited by LESTER PACKER

VOLUME 128. Plasma Lipoproteins (Part A: Preparation, Structure, and Molecular Biology)
Edited by JERE P. SEGREST AND JOHN J. ALBERS

VOLUME 129. Plasma Lipoproteins (Part B: Characterization, Cell Biology, and Metabolism)
Edited by JOHN J. ALBERS AND JERE P. SEGREST

VOLUME 130. Enzyme Structure (Part K)
Edited by C. H. W. HIRS AND SERGE N. TIMASHEFF

VOLUME 131. Enzyme Structure (Part L)
Edited by C. H. W. HIRS AND SERGE N. TIMASHEFF

VOLUME 132. Immunochemical Techniques (Part J: Phagocytosis and Cell-Mediated Cytotoxicity)
Edited by GIOVANNI DI SABATO AND JOHANNES EVERSE

VOLUME 133. Bioluminescence and Chemiluminescence (Part B)
Edited by MARLENE DELUCA AND WILLIAM D. MCELROY

VOLUME 134. Structural and Contractile Proteins (Part C: The Contractile Apparatus and the Cytoskeleton)
Edited by RICHARD B. VALLEE

VOLUME 135. Immobilized Enzymes and Cells (Part B)
Edited by KLAUS MOSBACH

VOLUME 136. Immobilized Enzymes and Cells (Part C)
Edited by KLAUS MOSBACH

VOLUME 137. Immobilized Enzymes and Cells (Part D)
Edited by KLAUS MOSBACH

VOLUME 138. Complex Carbohydrates (Part E)
Edited by VICTOR GINSBURG

VOLUME 139. Cellular Regulators (Part A: Calcium- and Calmodulin-Binding Proteins)
Edited by ANTHONY R. MEANS AND P. MICHAEL CONN

VOLUME 140. Cumulative Subject Index Volumes 102–119, 121–134

VOLUME 141. Cellular Regulators (Part B: Calcium and Lipids)
Edited by P. MICHAEL CONN AND ANTHONY R. MEANS

VOLUME 142. Metabolism of Aromatic Amino Acids and Amines
Edited by SEYMOUR KAUFMAN

VOLUME 143. Sulfur and Sulfur Amino Acids
Edited by WILLIAM B. JAKOBY AND OWEN GRIFFITH

VOLUME 144. Structural and Contractile Proteins (Part D: Extracellular Matrix)
Edited by LEON W. CUNNINGHAM

VOLUME 145. Structural and Contractile Proteins (Part E: Extracellular Matrix)
Edited by LEON W. CUNNINGHAM

VOLUME 146. Peptide Growth Factors (Part A)
Edited by DAVID BARNES AND DAVID A. SIRBASKU

VOLUME 147. Peptide Growth Factors (Part B)
Edited by DAVID BARNES AND DAVID A. SIRBASKU

VOLUME 148. Plant Cell Membranes
Edited by LESTER PACKER AND ROLAND DOUCE

VOLUME 149. Drug and Enzyme Targeting (Part B)
Edited by RALPH GREEN AND KENNETH J. WIDDER

VOLUME 150. Immunochemical Techniques (Part K: *In Vitro* Models of B and T Cell Functions and Lymphoid Cell Receptors)
Edited by GIOVANNI DI SABATO

VOLUME 151. Molecular Genetics of Mammalian Cells
Edited by MICHAEL M. GOTTESMAN

VOLUME 152. Guide to Molecular Cloning Techniques
Edited by SHELBY L. BERGER AND ALAN R. KIMMEL

VOLUME 153. Recombinant DNA (Part D)
Edited by RAY WU AND LAWRENCE GROSSMAN

VOLUME 154. Recombinant DNA (Part E)
Edited by RAY WU AND LAWRENCE GROSSMAN

VOLUME 155. Recombinant DNA (Part F)
Edited by RAY WU

VOLUME 156. Biomembranes (Part P: ATP-Driven Pumps and Related Transport: The Na,K-Pump)
Edited by SIDNEY FLEISCHER AND BECCA FLEISCHER

VOLUME 157. Biomembranes (Part Q: ATP-Driven Pumps and Related Transport: Calcium, Proton, and Potassium Pumps)
Edited by SIDNEY FLEISCHER AND BECCA FLEISCHER

VOLUME 158. Metalloproteins (Part A)
Edited by JAMES F. RIORDAN AND BERT L. VALLEE

VOLUME 159. Initiation and Termination of Cyclic Nucleotide Action
Edited by JACKIE D. CORBIN AND ROGER A. JOHNSON

VOLUME 160. Biomass (Part A: Cellulose and Hemicellulose)
Edited by WILLIS A. WOOD AND SCOTT T. KELLOGG

VOLUME 161. Biomass (Part B: Lignin, Pectin, and Chitin)
Edited by WILLIS A. WOOD AND SCOTT T. KELLOGG

VOLUME 162. Immunochemical Techniques (Part L: Chemotaxis and Inflammation)
Edited by GIOVANNI DI SABATO

VOLUME 163. Immunochemical Techniques (Part M: Chemotaxis and Inflammation)
Edited by GIOVANNI DI SABATO

VOLUME 164. Ribosomes
Edited by HARRY F. NOLLER, JR., AND KIVIE MOLDAVE

VOLUME 165. Microbial Toxins: Tools for Enzymology
Edited by SIDNEY HARSHMAN

VOLUME 166. Branched-Chain Amino Acids
Edited by ROBERT HARRIS AND JOHN R. SOKATCH

VOLUME 167. Cyanobacteria
Edited by LESTER PACKER AND ALEXANDER N. GLAZER

VOLUME 168. Hormone Action (Part K: Neuroendocrine Peptides)
Edited by P. MICHAEL CONN

VOLUME 169. Platelets: Receptors, Adhesion, Secretion (Part A)
Edited by JACEK HAWIGER

VOLUME 170. Nucleosomes
Edited by PAUL M. WASSARMAN AND ROGER D. KORNBERG

VOLUME 171. Biomembranes (Part R: Transport Theory: Cells and Model Membranes)
Edited by SIDNEY FLEISCHER AND BECCA FLEISCHER

VOLUME 172. Biomembranes (Part S: Transport: Membrane Isolation and Characterization)
Edited by SIDNEY FLEISCHER AND BECCA FLEISCHER

VOLUME 173. Biomembranes [Part T: Cellular and Subcellular Transport: Eukaryotic (Nonepithelial) Cells]
Edited by SIDNEY FLEISCHER AND BECCA FLEISCHER

VOLUME 174. Biomembranes [Part U: Cellular and Subcellular Transport: Eukaryotic (Nonepithelial) Cells]
Edited by SIDNEY FLEISCHER AND BECCA FLEISCHER

VOLUME 175. Cumulative Subject Index Volumes 135–139, 141–167

VOLUME 176. Nuclear Magnetic Resonance (Part A: Spectral Techniques and Dynamics)
Edited by NORMAN J. OPPENHEIMER AND THOMAS L. JAMES

VOLUME 177. Nuclear Magnetic Resonance (Part B: Structure and Mechanism)
Edited by NORMAN J. OPPENHEIMER AND THOMAS L. JAMES

VOLUME 178. Antibodies, Antigens, and Molecular Mimicry
Edited by JOHN J. LANGONE

VOLUME 179. Complex Carbohydrates (Part F)
Edited by VICTOR GINSBURG

VOLUME 180. RNA Processing (Part A: General Methods)
Edited by JAMES E. DAHLBERG AND JOHN N. ABELSON

VOLUME 181. RNA Processing (Part B: Specific Methods)
Edited by JAMES E. DAHLBERG AND JOHN N. ABELSON

VOLUME 182. Guide to Protein Purification
Edited by MURRAY P. DEUTSCHER

VOLUME 183. Molecular Evolution: Computer Analysis of Protein and Nucleic Acid Sequences
Edited by RUSSELL F. DOOLITTLE

VOLUME 184. Avidin–Biotin Technology
Edited by MEIR WILCHEK AND EDWARD A. BAYER

VOLUME 185. Gene Expression Technology
Edited by DAVID V. GOEDDEL

VOLUME 186. Oxygen Radicals in Biological Systems (Part B: Oxygen Radicals and Antioxidants)
Edited by LESTER PACKER AND ALEXANDER N. GLAZER

VOLUME 187. Arachidonate Related Lipid Mediators
Edited by ROBERT C. MURPHY AND FRANK A. FITZPATRICK

VOLUME 188. Hydrocarbons and Methylotrophy
Edited by MARY E. LIDSTROM

VOLUME 189. Retinoids (Part A: Molecular and Metabolic Aspects)
Edited by LESTER PACKER

VOLUME 190. Retinoids (Part B: Cell Differentiation and Clinical Applications)
Edited by LESTER PACKER

VOLUME 191. Biomembranes (Part V: Cellular and Subcellular Transport: Epithelial Cells)
Edited by SIDNEY FLEISCHER AND BECCA FLEISCHER

VOLUME 192. Biomembranes (Part W: Cellular and Subcellular Transport: Epithelial Cells)
Edited by SIDNEY FLEISCHER AND BECCA FLEISCHER

VOLUME 193. Mass Spectrometry
Edited by JAMES A. MCCLOSKEY

VOLUME 194. Guide to Yeast Genetics and Molecular Biology
Edited by CHRISTINE GUTHRIE AND GERALD R. FINK

VOLUME 195. Adenylyl Cyclase, G Proteins, and Guanylyl Cyclase
Edited by ROGER A. JOHNSON AND JACKIE D. CORBIN

VOLUME 196. Molecular Motors and the Cytoskeleton
Edited by RICHARD B. VALLEE

VOLUME 197. Phospholipases
Edited by EDWARD A. DENNIS

VOLUME 198. Peptide Growth Factors (Part C)
Edited by DAVID BARNES, J. P. MATHER, AND GORDON H. SATO

VOLUME 199. Cumulative Subject Index Volumes 168–174, 176–194

VOLUME 200. Protein Phosphorylation (Part A: Protein Kinases: Assays, Purification, Antibodies, Functional Analysis, Cloning, and Expression)
Edited by TONY HUNTER AND BARTHOLOMEW M. SEFTON

VOLUME 201. Protein Phosphorylation (Part B: Analysis of Protein Phosphorylation, Protein Kinase Inhibitors, and Protein Phosphatases)
Edited by TONY HUNTER AND BARTHOLOMEW M. SEFTON

VOLUME 202. Molecular Design and Modeling: Concepts and Applications (Part A: Proteins, Peptides, and Enzymes)
Edited by JOHN J. LANGONE

VOLUME 203. Molecular Design and Modeling: Concepts and Applications (Part B: Antibodies and Antigens, Nucleic Acids, Polysaccharides, and Drugs)
Edited by JOHN J. LANGONE

VOLUME 204. Bacterial Genetic Systems
Edited by JEFFREY H. MILLER

VOLUME 205. Metallobiochemistry (Part B: Metallothionein and Related Molecules)
Edited by JAMES F. RIORDAN AND BERT L. VALLEE

VOLUME 206. Cytochrome P450
Edited by MICHAEL R. WATERMAN AND ERIC F. JOHNSON

VOLUME 207. Ion Channels
Edited by BERNARDO RUDY AND LINDA E. IVERSON

VOLUME 208. Protein–DNA Interactions
Edited by ROBERT T. SAUER

VOLUME 209. Phospholipid Biosynthesis
Edited by EDWARD A. DENNIS AND DENNIS E. VANCE

VOLUME 210. Numerical Computer Methods
Edited by LUDWIG BRAND AND MICHAEL L. JOHNSON

VOLUME 211. DNA Structures (Part A: Synthesis and Physical Analysis of DNA)
Edited by DAVID M. J. LILLEY AND JAMES E. DAHLBERG

VOLUME 212. DNA Structures (Part B: Chemical and Electrophoretic Analysis of DNA)
Edited by DAVID M. J. LILLEY AND JAMES E. DAHLBERG

VOLUME 213. Carotenoids (Part A: Chemistry, Separation, Quantitation, and Antioxidation)
Edited by LESTER PACKER

VOLUME 214. Carotenoids (Part B: Metabolism, Genetics, and Biosynthesis)
Edited by LESTER PACKER

VOLUME 215. Platelets: Receptors, Adhesion, Secretion (Part B)
Edited by JACEK J. HAWIGER

VOLUME 216. Recombinant DNA (Part G)
Edited by RAY WU

VOLUME 217. Recombinant DNA (Part H)
Edited by RAY WU

VOLUME 218. Recombinant DNA (Part I)
Edited by RAY WU

VOLUME 219. Reconstitution of Intracellular Transport
Edited by JAMES E. ROTHMAN

VOLUME 220. Membrane Fusion Techniques (Part A)
Edited by NEJAT DÜZGÜNEŞ

VOLUME 221. Membrane Fusion Techniques (Part B)
Edited by NEJAT DÜZGÜNEŞ

VOLUME 222. Proteolytic Enzymes in Coagulation, Fibrinolysis, and Complement Activation (Part A: Mammalian Blood Coagulation Factors and Inhibitors)
Edited by LASZLO LORAND AND KENNETH G. MANN

VOLUME 223. Proteolytic Enzymes in Coagulation, Fibrinolysis, and Complement Activation (Part B: Complement Activation, Fibrinolysis, and Nonmammalian Blood Coagulation Factors)
Edited by LASZLO LORAND AND KENNETH G. MANN

VOLUME 224. Molecular Evolution: Producing the Biochemical Data
Edited by ELIZABETH ANNE ZIMMER, THOMAS J. WHITE, REBECCA L. CANN, AND ALLAN C. WILSON

VOLUME 225. Guide to Techniques in Mouse Development
Edited by PAUL M. WASSARMAN AND MELVIN L. DEPAMPHILIS

VOLUME 226. Metallobiochemistry (Part C: Spectroscopic and Physical Methods for Probing Metal Ion Environments in Metalloenzymes and Metalloproteins)
Edited by JAMES F. RIORDAN AND BERT L. VALLEE

VOLUME 227. Metallobiochemistry (Part D: Physical and Spectroscopic Methods for Probing Metal Ion Environments in Metalloproteins)
Edited by JAMES F. RIORDAN AND BERT L. VALLEE

VOLUME 228. Aqueous Two-Phase Systems
Edited by HARRY WALTER AND GÖTE JOHANSSON

VOLUME 229. Cumulative Subject Index Volumes 195–198, 200–227

VOLUME 230. Guide to Techniques in Glycobiology
Edited by WILLIAM J. LENNARZ AND GERALD W. HART

VOLUME 231. Hemoglobins (Part B: Biochemical and Analytical Methods)
Edited by JOHANNES EVERSE, KIM D. VANDEGRIFF, AND ROBERT M. WINSLOW

VOLUME 232. Hemoglobins (Part C: Biophysical Methods)
Edited by JOHANNES EVERSE, KIM D. VANDEGRIFF, AND ROBERT M. WINSLOW

VOLUME 233. Oxygen Radicals in Biological Systems (Part C)
Edited by LESTER PACKER

VOLUME 234. Oxygen Radicals in Biological Systems (Part D)
Edited by LESTER PACKER

VOLUME 235. Bacterial Pathogenesis (Part A: Identification and Regulation of Virulence Factors)
Edited by VIRGINIA L. CLARK AND PATRIK M. BAVOIL

VOLUME 236. Bacterial Pathogenesis (Part B: Integration of Pathogenic Bacteria with Host Cells)
Edited by VIRGINIA L. CLARK AND PATRIK M. BAVOIL

VOLUME 237. Heterotrimeric G Proteins
Edited by RAVI IYENGAR

VOLUME 238. Heterotrimeric G-Protein Effectors
Edited by RAVI IYENGAR

VOLUME 239. Nuclear Magnetic Resonance (Part C)
Edited by THOMAS L. JAMES AND NORMAN J. OPPENHEIMER

VOLUME 240. Numerical Computer Methods (Part B)
Edited by MICHAEL L. JOHNSON AND LUDWIG BRAND

VOLUME 241. Retroviral Proteases
Edited by LAWRENCE C. KUO AND JULES A. SHAFER

VOLUME 242. Neoglycoconjugates (Part A)
Edited by Y. C. LEE AND REIKO T. LEE

VOLUME 243. Inorganic Microbial Sulfur Metabolism
Edited by HARRY D. PECK, JR., AND JEAN LEGALL

VOLUME 244. Proteolytic Enzymes: Serine and Cysteine Peptidases
Edited by ALAN J. BARRETT

VOLUME 245. Extracellular Matrix Components
Edited by E. RUOSLAHTI AND E. ENGVALL

VOLUME 246. Biochemical Spectroscopy
Edited by KENNETH SAUER

VOLUME 247. Neoglycoconjugates (Part B: Biomedical Applications)
Edited by Y. C. LEE AND REIKO T. LEE

VOLUME 248. Proteolytic Enzymes: Aspartic and Metallo Peptidases
Edited by ALAN J. BARRETT

VOLUME 249. Enzyme Kinetics and Mechanism (Part D: Developments in Enzyme Dynamics)
Edited by DANIEL L. PURICH

VOLUME 250. Lipid Modifications of Proteins
Edited by PATRICK J. CASEY AND JANICE E. BUSS

VOLUME 251. Biothiols (Part A: Monothiols and Dithiols, Protein Thiols, and Thiyl Radicals)
Edited by LESTER PACKER

VOLUME 252. Biothiols (Part B: Glutathione and Thioredoxin; Thiols in Signal Transduction and Gene Regulation)
Edited by LESTER PACKER

VOLUME 253. Adhesion of Microbial Pathogens
Edited by RON J. DOYLE AND ITZHAK OFEK

VOLUME 254. Oncogene Techniques
Edited by PETER K. VOGT AND INDER M. VERMA

VOLUME 255. Small GTPases and Their Regulators (Part A: Ras Family)
Edited by W. E. BALCH, CHANNING J. DER, AND ALAN HALL

VOLUME 256. Small GTPases and Their Regulators (Part B: Rho Family)
Edited by W. E. BALCH, CHANNING J. DER, AND ALAN HALL

VOLUME 257. Small GTPases and Their Regulators (Part C: Proteins Involved in Transport)
Edited by W. E. BALCH, CHANNING J. DER, AND ALAN HALL

VOLUME 258. Redox-Active Amino Acids in Biology
Edited by JUDITH P. KLINMAN

VOLUME 259. Energetics of Biological Macromolecules
Edited by MICHAEL L. JOHNSON AND GARY K. ACKERS

VOLUME 260. Mitochondrial Biogenesis and Genetics (Part A)
Edited by GIUSEPPE M. ATTARDI AND ANNE CHOMYN

VOLUME 261. Nuclear Magnetic Resonance and Nucleic Acids
Edited by THOMAS L. JAMES

VOLUME 262. DNA Replication
Edited by JUDITH L. CAMPBELL

VOLUME 263. Plasma Lipoproteins (Part C: Quantitation)
Edited by WILLIAM A. BRADLEY, SANDRA H. GIANTURCO, AND JERE P. SEGREST

VOLUME 264. Mitochondrial Biogenesis and Genetics (Part B)
Edited by GIUSEPPE M. ATTARDI AND ANNE CHOMYN

VOLUME 265. Cumulative Subject Index Volumes 228, 230–262

VOLUME 266. Computer Methods for Macromolecular Sequence Analysis
Edited by RUSSELL F. DOOLITTLE

VOLUME 267. Combinatorial Chemistry
Edited by JOHN N. ABELSON

VOLUME 268. Nitric Oxide (Part A: Sources and Detection of NO; NO Synthase)
Edited by LESTER PACKER

VOLUME 269. Nitric Oxide (Part B: Physiological and Pathological Processes)
Edited by LESTER PACKER

VOLUME 270. High Resolution Separation and Analysis of Biological Macromolecules (Part A: Fundamentals)
Edited by BARRY L. KARGER AND WILLIAM S. HANCOCK

VOLUME 271. High Resolution Separation and Analysis of Biological Macromolecules (Part B: Applications)
Edited by BARRY L. KARGER AND WILLIAM S. HANCOCK

VOLUME 272. Cytochrome P450 (Part B)
Edited by ERIC F. JOHNSON AND MICHAEL R. WATERMAN

VOLUME 273. RNA Polymerase and Associated Factors (Part A)
Edited by SANKAR ADHYA

VOLUME 274. RNA Polymerase and Associated Factors (Part B)
Edited by SANKAR ADHYA

VOLUME 275. Viral Polymerases and Related Proteins
Edited by LAWRENCE C. KUO, DAVID B. OLSEN, AND STEVEN S. CARROLL

VOLUME 276. Macromolecular Crystallography (Part A)
Edited by CHARLES W. CARTER, JR., AND ROBERT M. SWEET

VOLUME 277. Macromolecular Crystallography (Part B)
Edited by CHARLES W. CARTER, JR., AND ROBERT M. SWEET

VOLUME 278. Fluorescence Spectroscopy
Edited by LUDWIG BRAND AND MICHAEL L. JOHNSON

VOLUME 279. Vitamins and Coenzymes (Part I)
Edited by DONALD B. MCCORMICK, JOHN W. SUTTIE, AND CONRAD WAGNER

VOLUME 280. Vitamins and Coenzymes (Part J)
Edited by DONALD B. MCCORMICK, JOHN W. SUTTIE, AND CONRAD WAGNER

VOLUME 281. Vitamins and Coenzymes (Part K)
Edited by DONALD B. MCCORMICK, JOHN W. SUTTIE, AND CONRAD WAGNER

VOLUME 282. Vitamins and Coenzymes (Part L)
Edited by DONALD B. MCCORMICK, JOHN W. SUTTIE, AND CONRAD WAGNER

VOLUME 283. Cell Cycle Control
Edited by WILLIAM G. DUNPHY

VOLUME 284. Lipases (Part A: Biotechnology)
Edited by BYRON RUBIN AND EDWARD A. DENNIS

VOLUME 285. Cumulative Subject Index Volumes 263, 264, 266–284, 286–289

VOLUME 286. Lipases (Part B: Enzyme Characterization and Utilization)
Edited by BYRON RUBIN AND EDWARD A. DENNIS

VOLUME 287. Chemokines
Edited by RICHARD HORUK

VOLUME 288. Chemokine Receptors
Edited by RICHARD HORUK

VOLUME 289. Solid Phase Peptide Synthesis
Edited by GREGG B. FIELDS

VOLUME 290. Molecular Chaperones
Edited by GEORGE H. LORIMER AND THOMAS BALDWIN

VOLUME 291. Caged Compounds
Edited by GERARD MARRIOTT

VOLUME 292. ABC Transporters: Biochemical, Cellular, and Molecular Aspects
Edited by SURESH V. AMBUDKAR AND MICHAEL M. GOTTESMAN

VOLUME 293. Ion Channels (Part B)
Edited by P. MICHAEL CONN

VOLUME 294. Ion Channels (Part C)
Edited by P. MICHAEL CONN

VOLUME 295. Energetics of Biological Macromolecules (Part B)
Edited by GARY K. ACKERS AND MICHAEL L. JOHNSON

VOLUME 296. Neurotransmitter Transporters
Edited by SUSAN G. AMARA

VOLUME 297. Photosynthesis: Molecular Biology of Energy Capture
Edited by LEE MCINTOSH

VOLUME 298. Molecular Motors and the Cytoskeleton (Part B)
Edited by RICHARD B. VALLEE

VOLUME 299. Oxidants and Antioxidants (Part A)
Edited by LESTER PACKER

VOLUME 300. Oxidants and Antioxidants (Part B)
Edited by LESTER PACKER

VOLUME 301. Nitric Oxide: Biological and Antioxidant Activities (Part C)
Edited by LESTER PACKER

VOLUME 302. Green Fluorescent Protein
Edited by P. MICHAEL CONN

VOLUME 303. cDNA Preparation and Display
Edited by SHERMAN M. WEISSMAN

VOLUME 304. Chromatin
Edited by PAUL M. WASSARMAN AND ALAN P. WOLFFE

VOLUME 305. Bioluminescence and Chemiluminescence (Part C)
Edited by THOMAS O. BALDWIN AND MIRIAM M. ZIEGLER

VOLUME 306. Expression of Recombinant Genes in Eukaryotic Systems
Edited by JOSEPH C. GLORIOSO AND MARTIN C. SCHMIDT

VOLUME 307. Confocal Microscopy
Edited by P. MICHAEL CONN

VOLUME 308. Enzyme Kinetics and Mechanism (Part E: Energetics of Enzyme Catalysis)
Edited by VERN L. SCHRAMM AND DANIEL L. PURICH

VOLUME 309. Amyloids, Prions, and Other Protein Aggregates
Edited by RONALD WETZEL

VOLUME 310. Biofilms
Edited by RON J. DOYLE

VOLUME 311. Sphingolipid Metabolism and Cell Signaling (Part A)
Edited by ALFRED H. MERRILL, JR., AND YUSUF A. HANNUN

VOLUME 312. Sphingolipid Metabolism and Cell Signaling (Part B)
Edited by ALFRED H. MERRILL, JR., AND YUSUF A. HANNUN

VOLUME 313. Antisense Technology (Part A: General Methods, Methods of Delivery, and RNA Studies)
Edited by M. IAN PHILLIPS

VOLUME 314. Antisense Technology (Part B: Applications)
Edited by M. IAN PHILLIPS

VOLUME 315. Vertebrate Phototransduction and the Visual Cycle (Part A)
Edited by KRZYSZTOF PALCZEWSKI

VOLUME 316. Vertebrate Phototransduction and the Visual Cycle (Part B)
Edited by KRZYSZTOF PALCZEWSKI

VOLUME 317. RNA–Ligand Interactions (Part A: Structural Biology Methods) (in preparation)
Edited by DANIEL W. CELANDER AND JOHN N. ABELSON

VOLUME 318. RNA-Ligand Interactions (Part B: Molecular Biology Methods) (in preparation)
Edited by DANIEL W. CELANDER AND JOHN N. ABELSON

VOLUME 319. Singlet Oxygen, UV-A, and Ozone (in preparation)
Edited by LESTER PACKER AND HELMUT SIES

VOLUME 320. Cumulative Subject Index Volumes 290–319 (in preparation)

VOLUME 321. Numerical Computer Methods (Part C) (in preparation)
Edited by MICHAEL L. JOHNSON AND LUDWIG BRAND

VOLUME 322. Apoptosis (in preparation)
Edited by JOHN C. REED

VOLUME 323. Energetics of Biological Macromolecules (Part C) (in preparation)
Edited by MICHAEL L. JOHNSON AND GARY K. ACKERS

VOLUME 324. Branched Chain Amino Acids (Part B) (in preparation)
Edited by ROBERT A. HARRIS AND JOHN R. SOKATCH

Section I

Photoreceptor Proteins

[1] Inhibition of Rhodopsin Phosphorylation by S-Modulins: Purification, Reconstitution, and Assays

By SATORU KAWAMURA

Introduction

S-Modulin is a calcium-binding protein found in frog retinal rods.[1] S-modulin activity was first detected while performing an electrophysiological measurement of cGMP phosphodiesterase (PDE) activity, using a truncated preparation of frog rod outer segment (tROS) that can be internally perfused with a bathing solution. This measurement suggested the presence of a factor that regulates PDE activation in a Ca^{2+}-dependent manner. It was also suggested that the factor binds to the disk membrane at high Ca^{2+} concentrations. This characteristic of the factor was used to purify S-modulin.

In a later study, it was shown that S-modulin regulates PDE activation by inhibiting rhodopsin phosphorylation at high Ca^{2+} concentrations.[2] Rhodopsin phosphorylation is a reaction that deactivates light-activated rhodopsin.[3] Because the cytoplasmic Ca^{2+} concentration decreases under light-adapted conditions,[4-8] the Ca^{2+}-dependent regulation of rhodopsin phosphorylation by S-modulin is thought to be one of the mechanisms regulating the light sensitivity of rods.[2]

At almost the same time that S-modulin was reported, another Ca^{2+}-binding protein was reported. This protein, named recoverin, was found in bovine retina.[9] Later study showed that recoverin is the bovine homolog of S-modulin.[10-12]

---

[1] S. Kawamura and M. Murakami, *Nature* (*London*) **349**, 420 (1991).
[2] S. Kawamura, *Nature* (*London*) **362**, 855 (1993).
[3] J. Chen, C. L. Makino, N. S. Peachey, D. A. Baylor, and M. I. Simon, *Science* **267**, 374 (1995).
[4] K.-W. Yau and K. Nakatani, *Nature* (*London*) **313**, 579 (1985).
[5] S. T. McCarthy, J. P. Younger, and W. G. Owen, *Biophys. J.* **67**, 2076 (1994).
[6] M. P. Gray-Keller and P. B. Detwiler, *Neuron* **13**, 849 (1994).
[7] J. I. Korenrot and D. L. Miller, *Vision Res.* **29**, 939 (1989).
[8] A. P. Sampath, H. R. Matthews, M. C. Cornwall, and G. L. Fain, *J. Gen. Physiol.* **111**, 53 (1998).
[9] A. M. Dizhoor, S. Ray, S. Kumar, G. Niemi, M. Spencer, D. Brolley, K. A. Walsh, P. P. Philipov, J. B. Hurley, and L. Stryer, *Science* **251**, 915 (1991).
[10] S. Kawamura, O. Hisatomi, S. Kayada, F. Tokunaga, and C.-H. Kuo, *J. Biol. Chem.* **268**, 14579 (1993).

This chapter describes (1) an electrophysiological method by which to detect S-modulin activity, (2) purification of S-modulin and its cone homolog s26 (frog visinin[13]) from frog retina and purification of recoverin from bovine retina, (3) purification of S-modulin, s26, and recoverin that are expressed in *Escherichia coli*, and (4) reconstitution and assay of S-modulin activity.

Electrophysiological Detection of S-Modulin Activity in Truncated Rod Outer Segment

Isolation of Rod Outer Segment Still Attached to Inner Segment

A bullfrog (*Rana catesbeiana*) is dark adapted in a light-tight container for at least 2 hr. The animal is then decapitated and pithed. The eyeball is enucleated and cut at the equator. The anterior half is removed with a pair of forceps. Often the retina is left in the posterior half. The retina is carefully removed with the forceps from the pigment epithelium. Because the retina is firmly connected to the eyeball at the site where the retinal axons come out of the retina, this portion should be cut with a pair of small scissors. Sometimes the retina comes off the eyeball with the anterior half; in this case, the isolation of the retina is easier.

The isolated retina is placed in a petri dish filled with a few milliliters of Ringer solution [115 mM NaCl, 2.5 mM KCl, 1 mM CaCl$_2$, 2 mM MgCl$_2$, 10 mM N-2-hydroxyethylpiperazine-N'-2-ethanesulfonic acid (HEPES), pH 7.5]. The retina is torn gently into small pieces with two pairs of forceps. A frog rod outer segment (ROS) still attached to the inner segment (ROS-IS) can be isolated by this procedure. The yield of ROS-IS is rather low and most of the cells are ROSs that lack the inner segment. In any event, 1 ROS-IS in every 100–200 ROSs can be easily found under a microscope. Some ROS-ISs do not survive during the course of the preparation of tROS. If the outer segment of an ROS-IS looks bent, that cell does not give a good result. A satisfactory ROS-IS has a straight-looking outer segment.

Preparation of Suction Electrode

The outer segment portion of each ROS-IS is sucked into a fire-polished glass electrode for the measurement of the membrane current. The elec-

---

[11] E. N. Gorodovikova, A. A. Gimelbrant, I. I. Senin, and P. P. Philippov, *FEBS Lett.* **349**, 187 (1994).
[12] V. A. Klenchin, P. D. Calvart, and M. D. Bownds, *J. Biol. Chem.* **270**, 16147 (1995).
[13] K. Yamagata, K. Goto, C.-H. Kuo, H. Kondo, and N. Miki, *Neuron* **2**, 469 (1990).

trode is prepared in the following way, using a Narishige (Tokyo, Japan) puller. First, the middle portion of a glass tube (1.2 × 90 mm; Narishige) is heated and allowed to lengthen by ~7 mm under the force of gravity. The thin middle portion is then reheated and quickly lengthened, again by gravity, by an additional ~3 cm; this produces a pair of glass tubes, each with a tip diameter of less than 1 μm at the pulled end. The tip portion is cut away with a file at the point at which the diameter is close to 20–30 μm. The cut end is fire polished, using a Narishige microforge, so that the inner diameter of the glass is ~6–7 μm, which is close to the diameter of a frog ROS. The shape of the orifice cannot be controlled and therefore it may not fit well to the shape of an ROS. For this and other reasons, the creation of an electrode suitable for tROS experiments involves some trial and error.

Preparation of Truncated Rod Outer Segment

The suction electrode is filled with a choline solution (115 mM choline chloride, 2.5 mM KCl, 2 mM MgCl$_2$, 10 mM HEPES, pH 7.5).[14] The Ca^{2+} concentration is usually set at 1 μM, using an EGTA–Ca^{2+} buffering system (0.195 mM CaCl$_2$ plus 0.2 mM EGTA). A perfusion chamber (Fig. 1)[14] is placed on the stage of an inverted microscope. The suspension of isolated ROSs and ROS-ISs is introduced to the perfusion chamber. After the ROSs and ROS-ISs have been allowed to sediment, the Ringer solution of the suspension is replaced with choline solution by perfusion of the chamber. After the replacement has been accomplished, the electrode is introduced into the choline solution in the chamber. Under visual control with the aid of an infrared television (IR-TV) monitor, the outer segment portion of an ROS-IS is sucked into the suction electrode (Fig. 2A)[15] by means of negative pressure through an electrode holder of the head amplifier (List, Darmstadt, Germany). The length of the ROS under suction should be determined: the longer the length, the larger the current. In this case, the signal is large, but the internal perfusion is less effective. In this laboratory, typically 80% of the length of an ROS is sucked into the electrode.

The choline solution in the chamber is then replaced with a potassium gluconate buffer (K-gluc buffer: 115 mM potassium gluconate, 2.5 mM KCl, 2 mM MgCl$_2$, 0.2 mM EGTA, 0.1 mM CaCl$_2$, pH 7.5). The K-gluc buffer is perfused continuously from one side of the chamber to the other, throughout the experiment, to ensure that test solutions flow in one direction (see Fig. 1).

[14] S. Kawamura and M. Murakami, *J. Gen. Physiol.* **94**, 649 (1989).
[15] S. Kawamura and M. Murakami, *Neurosci. Res. Suppl.* **10**, S15 (1989).

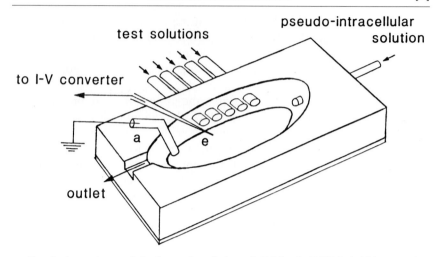

FIG. 1. Apparatus used for internal perfusion of tROSs. A tROS is held in a suction electrode (e) and is placed in the stream of a test solution. The bottom of the chamber is a thin glass plate through which ROSs are viewed with an IR-TV camera installed in an inverted microscope. The inside of the chamber is grounded through an agarose bridge (a). [Reproduced from the *Journal of General Physiology* **94,** 649–668, 1989. Reprinted by permission of the Rockefeller University Press.]

After substitution of the solution, the exposed part of the ROS-IS, which consists of the basal part of an ROS plus an inner segment, is truncated with a thin glass rod under visual control (Fig. 2B). When the pressure on the ROS-IS is appropriate, the ROS-IS does not move in the electrode during truncation. However, after truncation the tROS is even less stable than the ROS-IS, and it sometimes moves. Thus, pressure control is important in creating stable tROS preparations. In addition, the size and shape of the orifice of the electrode seems to be important in performing stable measurements.

Internal Perfusion of Truncated Rod Outer Segment

As shown in Fig. 1, the tip of the electrode is placed in the stream of a test solution. Internal perfusion of the tROS is attained by diffusion of the bathing solution into the inside of tROS through the truncated open end. Five test solutions can be examined by moving the position of the chamber relative to the electrode.

Because a tROS retains rhodopsin, transducin, cGMP phosphodiesterase (PDE), and cGMP-gated channels, it elicits an electrical response

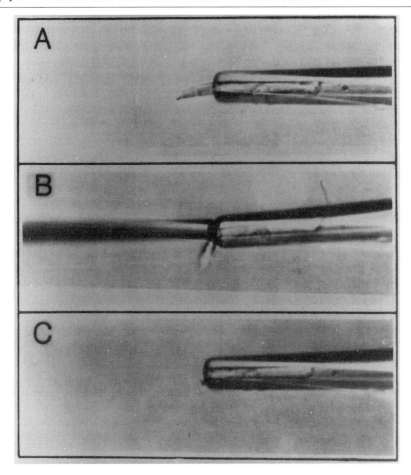

FIG. 2. Preparation of tROSs. The outer segment of an ROS-IS is sucked into the electrode (A). The exposed part is truncated with a thin glass rod under visual control (B). The resultant preparation (tROS) (C) has an open end through which the inside of the cell is internally perfused with a bathing solution. [Reprinted from *Neuroscience Research Supplement* **10,** S15–S22, 1989. Reprinted with permission from Elsevier Science.]

when cGMP and GTP are supplied.[16] In Fig. 3 truncation is made, as indicated by the arrow, at a high (1 μM) Ca^{2+} concentration. On application of 1 mM cGMP, an outward current starts to flow, which is due to the opening of the cGMP-gated channel. The membrane voltage is clamped at 0 mV and therefore the current carried by K^+ flows

[16] K.-W. Yau and K. Nakatani, *Nature* (*London*) **317,** 252 (1985).

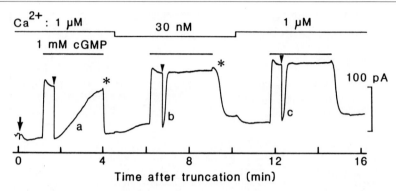

FIG. 3. Membrane current record in a tROS preparation. Truncation occurred at time 0, indicated by the arrow. On application of cGMP, an outward current (upward) started to flow. A weak light flash (arrowhead) blocked this current. Measurements were done by varying the Ca^{2+} concentration. During the course of the measurement, the baseline shifted slightly. Asterisks indicate that at the time of the removal of cGMP, PDE activity is higher at high Ca^{2+} concentrations. [Reprinted with permission from *Nature (London)* **349**(6308), 420–423, 1991. Copyright © Macmillan Magazines, Ltd.]

through the channel from the inside of the tROS to the outside. When a weak light flash is given (arrowhead), the current decreases (Fig. 3a). This decrease is due to hydrolysis of cGMP. After the light flash, the current recovers gradually. The diffusion of cGMP from a test solution to the inside of a tROS is rather fast, which can be seen as the rising phase of the outward current. The concentration of GTP used is 100 μM, which is high enough to support the activation of transducin but too low to see the effect of the synthesis of cGMP by guanylate cyclase.[14] Therefore, the gradual recovery after the light flash is due to a gradual decrease in PDE activity. In other words, by measuring the membrane current in a tROS, the time course of PDE activation and inactivation can be measured.

When a similar experiment is repeated at a low Ca^{2+} concentration (30 nM; Fig. 3b), the amplitude of the flash response (arrowhead) is slightly smaller and the recovery is much faster than that measured at 1 μM Ca^{2+} concentration. This result indicates that the PDE activation is smaller and its inactivation time course is faster at low Ca^{2+} concentrations.

The preceding Ca^{2+}-dependent regulation of PDE activity is an irreversible reaction in tROSs: even when the Ca^{2+} concentration is restored to 1 μM, the PDE activation is small and the recovery is fast (Fig. 3c). One possibility that accounts for this irreversibility is the loss of a factor that regulates PDE activation in a Ca^{2+}-dependent manner. It may be postulated that the factor binds to the disk membrane at high Ca^{2+} concentrations but becomes soluble at low Ca^{2+} concentrations. Once the factor becomes soluble, it is washed out from the tROS during internal perfusion and the Ca^{2+} sensitivity of PDE activation is lost. On the basis of this idea—that the binding of the factor to disk membranes is Ca^{2+} dependent—S-modulin was purified for the first time.[1]

Purification

Preparation of Frog Retinal Soluble Proteins

Bullfrogs (*R. catesbeiana*) are dark adapted in light-tight containers for at least 2 hr. Owing to the migration of the screening pigment,[17] melanin granules often attach to the retina during its isolation when dark adaptation is not complete. Contamination by melanin granules introduces yellowish products during preparation of S-modulin and, when excessive, S-modulin preparation becomes difficult. When a retina contaminated with melanin granules is freeze–thawed, the amount of yellowish products increases. It is thus recommended that the soluble proteins be isolated immediately after isolation of the retina. The pigment migration shows a daily cycle and becomes minimum at night, and thus the contamination is minimum when the retina is prepared at night or in the evening. It is recommended that the retina be isolated under dim red light to minimize the contamination. The procedure for the isolation of a retina is the same as described in the section, Isolation of Rod Outer Segment Still Attached to Inner Segment (above).

The isolated retina is placed, photoreceptor side up, in a petri dish (~3 cm in diameter) containing K-gluc buffer (~3 ml). To ensure that the Ca^{2+} concentration is maintained at low levels even when many retinas are placed in the dish, a concentrated EGTA solution is added, with the final EGTA concentration being 1 mM. ROSs are brushed off the retina, and this procedure is repeated for ~10 retinas. The buffer solution containing ROSs is collected in a centrifuge tube (~50 ml) and fresh buffer is supplied to the dish. The isolation of ROSs is repeated.

ROSs thus obtained have probably been broken during the process of isolation, and thus soluble proteins are already present in the buffer solution. However, to ensure that this is actually the case, the collected

---

[17] A. Fein and E. Z. Szuts, "Photoreceptors; Their Role in Vision." Cambridge University Press, New York, 1982.

buffer solution is passed through a thin needle (28 gauge) ~10 times to disrupt the structure of the ROSs. After this manipulation, the buffer solution is centrifuged for 30 min at 100,000g at 4°. The supernatant, containing S-modulin and s26, is used for the subsequent purification. The less the contamination by melanin granules, the fewer yellowish products there well be. Some of the yellowish products are at the surface of the buffer solution after the centrifugation, and are removed easily by decantation; some of the yellowish products still remain in the supernatant and they are not completely eliminated even after centrifugation at 250,000g for many hours. For practical reasons, therefore, the supernatant after the 100,000g centrifugation is used in the subsequent purification.

Preparation of Bovine Retinal Soluble Proteins

Bovine eyeballs are purchased from a local slaughterhouse. The eyeball is cut at the equator with a razor blade. The anterior half, together with the aqueous humor, is easily removed. The remaining retina is isolated more easily from bovine than frog retinas, even though the bovine optic nerve should be cut in the same manner as the frog nerve. The isolated retinas (about 50) are shaken vigorously in a 50-ml centrifuge tube containing ~40 ml of K-gluc buffer supplemented with 1 mM EGTA. The contents are filtered through a nylon cloth and the filtrate containing ROSs and soluble proteins is centrifuged (10,000g, 30 min). The supernatant is used for the purification of recoverin.

Purification of S-Modulin, Its Cone Homolog s26 (Frog Visinin), and Recoverin

When S-modulin was purified for the first time, its characteristic of Ca^{2+}-dependent binding to disk membranes was used as the basis for purification.[1] However, later studies showed that S-modulin binds to a phenyl-Sepharose column in a Ca^{2+}-dependent manner.[18] Because column chromatography is simpler than the disk membrane method, a phenyl-Sepharose column is now used. After phenyl-Sepharose column chromatography, an anion-exchange column is employed. With this two-step procedure, S-modulin, s26, and recoverin can be purified satisfactorily.

---

[18] S. Kawamura, K. Takamatsu, and K. Kitamura, *Biochem. Biophys. Res. Commun.* **186,** 411 (1992).

Because S-modulin is sticky, it is recommended that tubes, etc., be coated with ovalbumin to avoid a loss of S-modulin during purification.[1] Bovine serum albumin is not used as the coating material because it affects some enzyme reactions, for example, PDE activation in rods.[19]

Phenyl-Sepharose Column Chromatography

The column (diameter, 1 cm; length, 15 cm) is equilibrated with K-gluc buffer supplemented with 1 mM $CaCl_2$.[18] The soluble protein fraction, either frog (S-modulin and s26) or bovine (recoverin), is brought to 2 mM $CaCl_2$ in the presence of 1 mM EGTA by adding stock 1 M $CaCl_2$ solution just before application to the column. After application, the column is washed extensively with K-gluc buffer supplemented with 1 mM $CaCl_2$.

S-Modulin, s26, and recoverin bind to the column.[18,20–22] They are eluted by applying an elution buffer solution containing a low concentration of Ca^{2+}. To prepare for anion-exchange column chromatography, the salt concentration of the elution buffer is made low (40 mM Tris or potassium gluconate, 2 mM $MgCl_2$, 3 mM EGTA, 1 mM dithiothreitol, pH 7.5). S-modulin and s26 in frog preparations, and recoverin in bovine preparations, are eluted in a small volume (e.g., see Fig. 4A for S-modulin and s26).

Anion-Exchange Column Chromatography

A DEAE-Sepharose column is used.[18,22] The column is equilibrated with the elution buffer mentioned in the previous section. The salt concentration required for the elution of S-modulin from the DEAE column is much lower than that required for s26 or recoverin. In the case of S-modulin, the required salt concentration for the elution is about 20 mM. For this reason, S-modulin does not bind to the column and is recovered in a pass-through fraction when the above-described elution buffer is used (Fig. 4Ba). The purity of S-modulin thus obtained, however, is usually more than 90%. When more highly purified S-modulin is needed, the salt concentration of the elution buffer should be lowered to 5 mM and an NaCl gradient elution should be applied. In the cases of s26 and recoverin, the salt concentrations required for elution from the DEAE column are close to 100 mM and an NaCl gradient elution is necessary.

---

[19] B. E. Buzdygon and P. A. Liebman, *J. Biol. Chem.* **259,** 14567 (1984).
[20] S. Kawamura, *Photochem. Photobiol.* **56,** 1173 (1992).
[21] A. S. Polans, J. Buczytko, J. Crabb, and K. Palczewski, *J. Cell Biol.* **112,** 981 (1991).
[22] S. Kawamura, O. Kuwata, M. Yamada, S. Matsuda, O. Hisatomi, and F. Tokunaga, *J. Biol. Chem.* **271,** 21359 (1996).

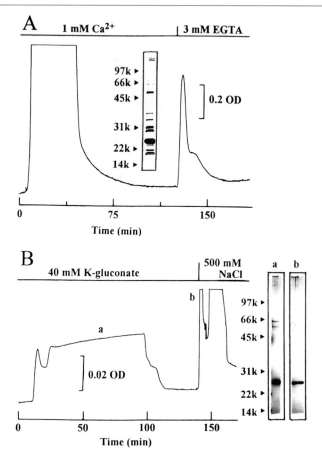

Fig. 4. Purification of S-modulin and s26. (A) Frog retina soluble proteins were applied to a phenyl-Sepharose column. The proteins bound to the column at 1 mM Ca^{2+} were eluted with a 40 mM K-gluc buffer containing 3 mM EGTA. *Inset:* An SDS–PAGE pattern of the eluate with the EGTA solution. (B) The eluate with the EGTA solution in (A) was applied to a DEAE-Sepharose column. S-modulin was obtained in the pass-through fraction (a) and s26 was obtained in a fraction of increased salt concentration (b). *Insets:* SDS–PAGE pattern of the pass-through fraction (a) and of the fraction of increased salt concentration (b). [Reprinted from the *Journal of Biological Chemistry* **271**, 21359–21364, 1996, with permission of the publisher.]

Approximately 5 μg of S-modulin can be obtained per frog.[18] The relative molar abundance of rhodopsin, S-modulin, and s26 is roughly 100:7:5 in frog retina,[22] and the relative molar abundance of rhodopsin and recoverin is 100:9 in bovine retina.[10] According to these ratios, the cytoplasmic S-modulin concentration is estimated to be 140 μM.[22]

Purification of S-Modulin, s26, and Recoverin Expressed in Escherichia coli

An expression system is now available for S-modulin,[23] s26,[22] and recoverin[24] in *E. coli*. Because the expressed proteins are the major proteins in these systems, purification of these proteins is straightforward. The expressed proteins, however, are present in inclusion bodies, and a protein-refolding step is required. Satisfactory purification is attained by phenyl-Sepharose column chromatography, and further purification is attained by DEAE-Sepharose column chromatography.

Molecular Properties of S-Modulin, s26, and Recoverin and Their Localization in Retina

The apparent molecular masses of S-modulin,[1] s26,[20] and recoverin[9] are ~26 kDa on sodium dodecyl sulfate–polyacrylamide gel electrophoresis (SDS–PAGE). According to the amino acid sequence analysis of their cDNAs, S-modulin[10] and recoverin[9] consist of 201 amino acid residues and s26 consists of 195 amino acids.[22] They have four potential Ca^{2+}-binding sites known as EF-hand structures, but only EF2 and EF3 can bind Ca^{2+}.[25] The reported half-effective Ca^{2+} concentration varies from several hundred nanomolar to several micromolar units.[12,26,27] The isoelectric point of S-modulin is 5.8,[1] and those of s26 and recoverin are lower (i.e., more acidic).

At high Ca^{2+} concentrations, S-modulin tends to aggregate when stored even at $-80°$. As can be inferred from the chromatographic behavior on a phenyl-Sepharose column, the Ca^{2+}-bound form of S-modulin seems to have a hydrophobic region on the surface of the molecule. It is recommended that these proteins be stored at low Ca^{2+} concentrations to prevent aggregation.

The three-dimensional structure of recoverin has been determined both in the Ca^{2+}-free (inactive) form and in the Ca^{2+}-bound (active) form.[28–30]

---

[23] O. Hisatomi, T. Ishino, S. Matsuda, K. Yamaguchi, Y. Kobayashi, S. Kawamura, and F. Tokunaga, *Biochem. Biophys. Res. Commun.* **234**, 173 (1997).

[24] S. Ray, S. Zozulya, G. A. Niemi, K. M. Flaherty, D. Brolley, A. M. Dizhoor, D. B. McKay, J. Hurley, and L. Stryer, *Proc. Natl. Acad. Sci. U.S.A.* **89**, 5705 (1992).

[25] J. B. Ames, T. Porumb, T. Tanaka, M. Ikura, and L. Stryer, *J. Biol. Chem.* **270**, 4526 (1995).

[26] C.-K. Chen, J. Inglese, R. J. Lefkowitz, and J. B. Hurley, *J. Biol. Chem.* **270**, 18060 (1995).

[27] N. Sato and S. Kawamura, *J. Biochem.* **122**, 1139 (1997).

[28] K. M. Flaherty, S. Zozulya, L. Stryer, and D. B. McKay, *Cell* **75**, 709 (1993).

[29] T. Tanaka, J. B. Ames, T. S. Harvey, L. Stryer, and M. Ikura, *Nature (London)* **376**, 444 (1995).

[30] J. B. Ames, R. Ishima, T. Tanaka, J. I. Gordon, L. Stryer, and M. Ikura, *Nature (London)* **389**, 198 (1997).

An immunohistochemical study has shown that S-modulin is present in frog rods and that s26 (frog visinin) is present in frog cones.[22] S-modulin and s26 distribute uniformly throughout the rods and cones, respectively. Immunoreactivity to S-modulin and s26 does not occur in other parts of the frog retina. In the case of recoverin, immunoreactivity against recoverin is observed in both rods and cones[9,21] and its cone homolog has not been found. In addition, a certain cone bipolar cell is immunopositive.[31]

Reconstitution

Regulation of Phosphodiesterase Activation by S-Modulin

The effect of purified S-modulin on PDE activation has been examined[2] by the pH assay method.[32,33] The method is based on the fact that the hydrolysis of cGMP accompanies the release of one H^+ (pK_a 6.3), which reduces the pH of the solution. The extent of the pH decrease depends on the pH-buffering capacity of the solution. This laboratory always uses a 10 mM HEPES buffer solution (pH 7.5–7.8), in which hydrolysis of 1 mM cGMP decreases the pH of the solution by approximately 0.15.[34]

Frog ROSs are brushed off retinas in K-gluc buffer in complete darkness with the aid of an infrared image converter. They are passed through a thin needle (28 gauge) three times and are washed with buffer three times (10,000g, 3 min) to eliminate endogenous S-modulin. ROS membranes thus obtained are stored overnight at 4° in the presence of 0.1 mM ATP in a light-tight box.

The pH measurement is carried out in complete darkness with the aid of an infrared image converter. The ROS membrane suspension (180 μl) is mixed with 5 μl of 20 mM GTP, 5 μl of 20 mM ATP and 10 μl of 80 mM cGMP in a V-bottom vial and the mixture is stirred with a small stirring bar. The rhodopsin concentration is adjusted to ~10 μM. Approximately 1 min after the addition of the chemicals, a weak light flash is given (Fig. 5B, arrowheads). Owing to light-induced hydrolysis of cGMP, the pH of the solution decreases (upward deflection of the trace). In the absence of exogenous S-modulin, the profile of the pH decrease is the same at a high (1 μM) and a low (30 nM) Ca^{2+} concentration (Fig. 5B; 0 S-modulin). When purified S-modulin (Fig. 5A) is supplied, PDE activation becomes Ca^{2+} dependent (Fig. 5B; 5, 10, 20 μM S-modulin).

[31] A. H. Milam, D. M. Dacey, and A. M. Dizhoor, *Vis. Neurosci.* **10,** 1 (1993).
[32] P. A. Liebman and T. Evanczuk, *Methods Enzymol.* **81,** 532 (1982).
[33] M. W. Kaplan and K. Palczewski, in "Methods in Neuroscience" (P. A. Hargrave, ed.), p. 205. Academic Press, San Diego, California, 1993.
[34] S. Kawamura and M. D. Bownds, *J. Gen. Physiol.* **77,** 571 (1981).

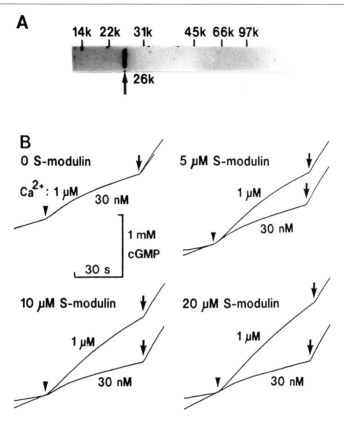

FIG. 5. Effect of purified S-modulin on the activation of PDE. (A) An SDS–PAGE pattern of purified S-modulin. Stained with Coomassie Brilliant Blue. (B) PDE activation was monitored by the pH assay method at various S-modulin concentrations and high and low Ca^{2+} concentrations. Arrowheads indicate the times at which a weak light flash was delivered, and arrows indicate the time of delivery of a saturating light flash. [Reprinted with permission from *Nature* (*London*) **362**(6423), 855–857, 1993. Copyright © Macmillan Magazines, Ltd.]

Inhibition of Rhodopsin Phosphorylation by S-Modulin

Because PDE activation is attained through the phototransduction cascade, the Ca^{2+}-dependent regulation of PDE activation described above must occur at some point during this cascade. It has been determined that the site of S-modulin action is the reaction of rhodopsin phosphorylation.[2] The following manipulations are performed in the dark with the aid of an infrared image converter to examine the effect of S-modulin on rhodopsin phosphorylation.

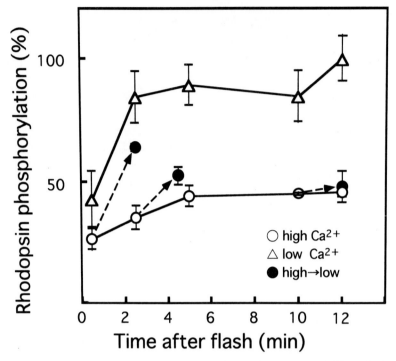

FIG. 6. Time course of rhodopsin phosphorylation in the presence of ATP. Rhodopsin phosphorylation was measured at various time intervals after a light flash. The measurement was done at a high Ca^{2+} concentration (open circles) and at a low Ca^{2+} concentration (open triangles). In some experiments, the Ca^{2+} concentration was reduced during incubation (dashed arrows and filled circles). [Reproduced from the *Journal of Biochemistry* **122**, 1139–1145, 1997. Reprinted with permission.]

Preparation of Frog Rod Outer Segment Membranes

ROS membranes are prepared as described above and stored at 4° overnight in the presence of 0.1 mM ATP. ATP added during storage is eliminated by centrifugation just before the phosphorylation assay, and the membranes are suspended in K-gluc buffer so that the rhodopsin concentration is 20–30 μM. An ROS membrane suspension (15 μl) containing GTP and various Ca^{2+} buffers is mixed with 10 μl of a [γ-^{32}P]ATP solution (20–30 MBq μmol^{-1} of ATP) in the dark in a glass tube. The use of a glass tube is to minimize the Cerenkov radiation that activates rhodopsin in the absence of external light stimulation. The resultant reaction mixture (25 μl) contains rhodopsin (10 μM), ATP (100 μM), and GTP (500 μM). The Ca^{2+} concentration is buffered by the EGTA–Ca^{2+} buffer system[14] and is

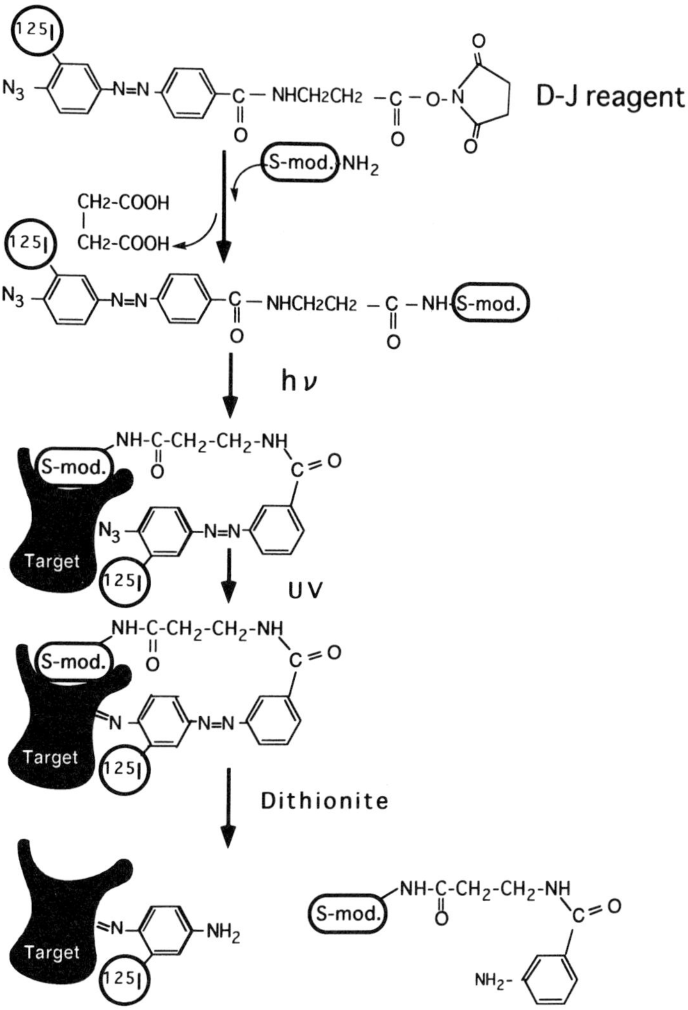

FIG. 7. Method used for identification of the target molecule of S-modulin. S-Modulin was complexed with D–J reagent and the complex was mixed with ROS membranes. Because the D–J reagent is labeled with ^{125}I, the target molecule is also labeled with this isotope (for details, see text). [Reproduced from the *Journal of Biochemistry* **122,** 1139–1145, 1997. Reprinted with permission.]

calibrated with Ca^{2+} buffer solutions (Calbuf-2; World Precision Instruments, Sarasota, FL). (Calcium ions form a water-insoluble $CaCO_3$ precipitate sometimes found even in a newly purchased bottle. The formation of this precipitate introduces an overestimate of the Ca^{2+} concentration

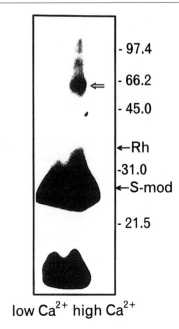

FIG. 8. Autoradiograph of candidate(s) for S-modulin target molecule. The protein(s) binding to S-modulin was identified using D–J reagent (see text). A protein band at ~60 kDa was specifically labeled at a high Ca^{2+} concentration. [Reproduced from the *Journal of Biochemistry* **122,** 1139–1145, 1997. Reprinted with permission.]

in the calibration and therefore these standards should be used with caution.)

Rhodopsin Phosphorylation

The reaction mixture, prepared as described above, is kept in the dark for approximately 30 sec, and then a light flash bleaching 3.5×10^6 rhodopsins per ROS (0.115% bleach) is given. The reaction is quenched by addition of 200 μl of 10% trichloroacetic acid (ice chilled). The subsequent manipulations are carried out in the light. The sample is centrifuged (10,000g, 10 min, 4°) and the precipitate is washed with K-gluc buffer twice by centrifugation. The precipitate is subjected to SDS–PAGE. Because the heat treatment of ROS membranes introduces the aggregation of rhodopsins, the precipitate is applied to the gel immediately after dissolving it in SDS sample buffer. The gel is stained with Coomassie Brilliant Blue and the radioactivity in the rhodopsin band is counted.

The time course of rhodopsin phosphorylation in the presence of 10

μM S-modulin is shown in Fig. 6.[27] At a low Ca^{2+} concentration (47 nM; Fig. 6, open triangles), rhodopsin phosphorylation reaches a steady state level ~2 min after a light flash. At a high Ca^{2+} concentration (10 μM; Fig. 6, open circles), rhodopsin phosphorylation takes place with a similar time course but the amount of phosphorylation is about half that observed at the low Ca^{2+} concentration. The result shows that rhodopsin phosphorylation is inhibited by the Ca^{2+}-bound form of S-modulin. The inhibition by S-modulin is reversible in terms of the Ca^{2+} concentration: when the Ca^{2+} concentration is reduced during incubation, the phosphorylation increases (Fig. 6, filled circles). The reversibility, however, is a function of time after a light flash. At the time when the phosphorylation reaches a steady state, reduction of the Ca^{2+} concentration no longer increases the phosphorylation. The result suggests that autophosphorylation of rhodopsin kinase, the target molecule of S-modulin/recoverin (see below), is responsible for the loss of the kinase activity.[27,35,36]

Identification of Target Molecule of S-Modulin

Proteins participating in the reaction of rhodopsin phosphorylation are light-activated rhodopsin and rhodopsin kinase, and therefore one of them is the target molecule of S-modulin. To identify the target molecule, Denney–Jaffe reagent (D–J reagent) is used.[27] One end of this reagent forms a covalent bond with an amino group of S-modulin (Fig. 7). The S-modulin/D–J reagent complex is mixed with ROS membranes at high and low Ca^{2+} concentrations. It is to be expected that only at a high Ca^{2+} concentration will S-modulin bind to the target molecule. At the other end of the reagent there is an azide group that is activated by UV light. Illumination with UV light will produce the formation of a complex of S-modulin and the target molecule linked through the D–J reagent. At a position close to the azide group ^{125}I is introduced in this reagent and, therefore, the target molecule is radiolabeled. After extraction of membrane proteins, the target molecule is identified by SDS–PAGE. The result shows that a ~60-kDa protein is specifically labeled only at high Ca^{2+} concentrations (Fig. 8). Although it was not possible at the time to identify the ~60-kDa protein as rhodopsin kinase, the result was consistent with the idea that S-modulin binds to rhodopsin kinase (~60 kDa) but not light-activated rhodopsin (~40 kDa). From studies done in other laboratories[26,36–38] and another study done in

[35] J. Buczytko, C. Gutmann, and K. Palczewski, *Proc. Natl. Acad. Sci. U.S.A.* **88**, 2568 (1991).
[36] D. K. Satpaev, C.-K. Chen, A. Scotti, M. I. Simon, J. B. Hurley, and V. Z. Slepak, *Biochemistry* **37**, 10256 (1998).
[37] E. N. Gorodovikova and P. P. Philippov, *FEBS Lett.* **335**, 277 (1993).
[38] K. Sanada, F. Shimizu, K. Kameyama, K. Haga, T. Haga, and Y. Fukada, *FEBS Lett.* **384**, 227 (1996).

this laboratory,[27] it is now evident that the target molecule of S-modulin is rhodopsin kinase.

Recoverin, visinin, and other members of the S-modulin protein family so far tested all inhibit rhodopsin phosphorylation at high Ca^{2+} concentrations.[39] The result therefore suggests that S-modulin family proteins are involved in the Ca^{2+}-dependent regulation of a certain type of G protein-coupled receptor kinase.

[39] S. Kawamura, *Neurosci. Res.* **20,** 293 (1994).

[2] Analysis of Protein–Protein Interactions in Phototransduction Cascade Using Surface Plasmon Resonance

By DAULET K. SATPAEV and VLADLEN Z. SLEPAK

Introduction

The excitation of rhodopsin by light initiates a cascade of protein–protein interactions regulating the intracellular level of cGMP and ultimately resulting in hyperpolarization of the photoreceptor membrane. Our knowledge about the dynamics of this process has come mostly from recordings of the ion currents through the cGMP-gated cation channels as well as from studies of enzyme kinetics, i.e., rates of hydrolysis and synthesis of cGMP and GTPase activity of transducin. However, the protein–protein interactions that modulate these activities are usually characterized by their affinities, i.e., equilibrium binding constants (K_D). The affinity alone cannot describe an interaction because *in vivo* binding might never reach equilibrium. The K_D is determined by the association and dissociation rates: $K_D = k_d/k_a$, where the k_a and k_d are the on- and off-rate constants. Therefore, understanding the dynamics of photoresponse at the molecular level requires evaluation of protein–protein binding kinetics. One of the few methods that can directly measure the rates of protein–protein interactions is surface plasmon resonance (SPR). Since the first commercial SPR detector (BIACORE, Uppsala, Sweden) was introduced in 1991,[1,2] the method

[1] U. Jönsson, L. Fägerstam, B. Ivarsson, B. Johnsson, R. Karlsson, K. Lundh, S. Löfås, B. Persson, H. Roos, I. Rönnberg, S. Sjölander, E. Stenberg, R. Ståhlberg, C. Urbaniczky, H. Östlin, and M. Malmqvist, *BioTechniques* **11**(5), 620 (1991).
[2] M. Malmqvist, *Nature (London)* **361,** 186 (1993).

has been applied in many areas of biochemistry (reviewed in Refs. 3–5). In a typical experiment, one of the interacting molecules (the ligand) is immobilized on the surface of the BIACORE sensor chip, and the other molecule (analyte) is delivered across in solution. Binding of an analyte to the ligand elevates its local concentration, gradually increasing the refractive index of the medium close to the surface, and this affects the properties of free electrons (plasmons) in the gold layer of the chip. These changes are proportional to the amount of the molecular complex at the sensor surface, and the rate of ligand–analyte binding is recorded in real time. Here, we describe the methods used to study the kinetics and regulation of some protein–protein interactions involved in the phototransduction cascade.

Ligand Immobilization

The first step in an experiment using a BIACORE instrument is to immobilize the ligand on the sensor surface. Coupling is based on the same principles developed for affinity chromatography, and a variety of sensor chips from BIACORE provide several alternatives. The ligand can be immobilized via a covalent bond between the molecule and an activated chip surface, or captured through an affinity interaction, such as avidin–biotin, antibody–antigen, and Ni^{2+}-His_6.

Direct Coupling through Amine Groups

The gold surface of the basic sensor chip (sensor chip CM5; BIACORE) is covered with linear carboxymethylated dextran, which produces a hydrophilic environment and net negative charge that reduces nonspecific adsorption. The carboxymethyl ($-CH_2-COOH$) groups must be activated to form a covalent bond with the free amine groups of the ligand. In a standard protocol (BIACORE manual), the surface of the CM5 chip is activated by an injection of 35 μl of a fresh 1:1 mixture of 0.2 N-(3-dimethylaminopropyl)-N-ethylcarbodiimide (EDC) and 0.4 M N-hydroxysuccinimide (NHS) at 5 μl/min. This results in formation of N-hydroxysuccinimide esters that readily react with primary amines such as ε-amino groups of lysine residues or α-amino group of the N terminus. Prior to immobilization, the ligand should be transferred into a solution free of Tris or other compounds containing primary amines. Since most proteins are negatively charged

---

[3] J. H. Lakey and E. M. Raggett, *Curr. Opin. Struct. Biol.* **8**(1), 119 (1998).
[4] D. H. Margulies, D. Plaksin, S. N. Khilko and M. T. Jelonek, *Curr. Opin. Immunol.* **8**(2), 262 (1996).
[5] P. Schuck, *Annu. Rev. Biophys. Biomol. Struct.* **26**, 541 (1997).

and experience electrostatic repulsion from the carboxymethylated dextran matrix, coupling is drastically facilitated if the protein is attracted (preconcentrated) to the surface. This is achieved by lowering the pH below the isoelectric point of the protein and reducing the ionic strength of the buffer. For example, 30 μl of 0.5 μM recoverin in 10 mM sodium acetate (pH 4.5) is injected at a 5-μl/min flow rate immediately after the EDC/NHS activation. During the coupling process, the level of immobilization is monitored as the increase in the SPR signal and can be adjusted as required for a particular experiment. The amount of attached protein (surface density) is one of the key parameters important for obtaining correct kinetic measurements. Similar to the amount of bound analyte, it is measured in resonance units, RU (1000 RU = 1 ng/mm^2). The immobilization levels shown in this chapter are appropriate for the studies of regulation of protein–protein binding (e.g., inhibition by low Ca^{2+} concentration) and are too high for precise kinetic analysis. After the desired surface density is reached, unreacted succinimide ester groups at the surface are blocked by a 5-min injection of 100 mM ethanolamine, pH 8.5, or 0.5 M Tris-HCl, pH 8.5. Noncovalently bound protein is removed by a 1-min pulse of 6 M guanidine hydrochloride. A similar protocol was used to immobilize the photoreceptor phosphodiesterase γ subunit (PDEγ) for studies of its interaction with transducin α subunit and PDE$\alpha\beta$.[6]

Direct Coupling through Sulfhydryl Groups

Cysteine residues are usually less abundant in proteins than lysines, and therefore coupling through thiol groups is likely to result in a less heterogeneous or even uniform orientation of the coupled protein on the surface.[7] For example, recoverin has only 1 cysteine residue, at position 39, and 25 lysines. For thiol coupling, buffers should be free of reducing agents such as 2-mercaptoethanol and dithiothreitol (DTT). Figure 1 illustrates a sensogram recorded during thiol immobilization of recoverin. First, the surface of a CM5 chip is activated by an injection of the EDC/NHS mixture as described above. The hydroxysuccinimide esters are then converted into thiol-reactive groups by means of a bifunctional reagent, such as 2-(2-pyridinyldithio)ethanamine hydrochloride (PDEA; BIACORE). This is achieved by injecting 80 mM PDEA in 0.1 M borate buffer, pH 8.5, for 5 min at a 5-μl/min flow rate immediately after the EDC/NHS solution. Recoverin at concentrations of 0.5–5 μM in 10 mM sodium acetate, pH

---

[6] V. Z. Slepak, N. Artemiev, Y. Zhu, C. L. Dumke, J. Sondek, L. Sobacan, H. E. Hamm, M. D. Bownds, and V. Y. Arshavsky, *J. Biol. Chem.* **270**, 4037 (1995).

[7] S. C. Schuster, R. V. Swanson, L. A. Alex, R. B. Bourret, and M. I. Simon, *Nature (London)* **365**, 343 (1993).

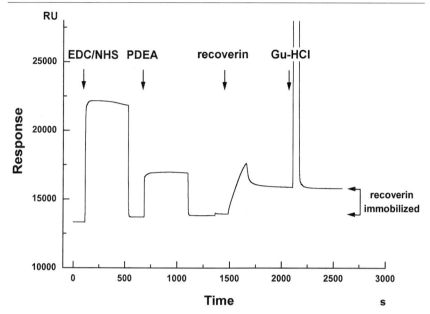

FIG. 1. Immobilization of recoverin on the CM chip. Injection times of EDC/NHS, PDEA, recoverin, and guanidine hydrochloride are indicated with vertical arrows. Sharp shifts in SPR signal during the injections of EDC/NHS, PDEA, and guanidine are due to the difference in bulk refractive indices of these solutions compared with the running buffer. On injection of recoverin (20 μg/ml) in 10 mM sodium acetate, pH 4.5, the SPR signal gradually rises, reflecting the protein coupling. The amount of bound recoverin is determined as the difference between SPR readings before its injection and after the guanidine hydrochloride wash (shown with horizontal arrows).

4.5, is injected after PDEA and is covalently coupled via the thiol exchange reaction. Typically, a 2.5-min injection of recoverin at a concentration of 1 μM yields approximately 500 RU of coupled protein. A fraction of recoverin (less than 5%) is usually adsorbed at the surface noncovalently and can be removed by a 1-min pulse of 6 M guanidine hydrochloride. To block residual activated thiol exchange groups, the surface is treated by an injection of 20 mM cysteine for 5 min.

Recoverin immobilization via cysteine was used by us to study the interaction of recoverin with G protein-coupled receptor kinase 1 (GRK1; rhodopsin kinase)[8,9] and by Lange and Koch to study the interaction of

[8] D. K. Satpaev, C.-K. Chen, A. Scotti, M. I. Simon, J. B. Hurley, and V. Z. Slepak, *Biochemistry* **37,** 10256 (1998).

[9] K. Levay, D. K. Satpaev, A. N. Pronin, J. L. Benovic, and V. Z. Slepak, *Biochemistry* **37,** 13650 (1998).

recoverin with liposomes.[10] Coupling of recoverin via NH_2– or SH– is equally effective with respect to the total amount of protein immobilized. However, the NH_2-immobilized ligand could bind 5- to 10-fold less GRK1 than the same amount of SH-coupled recoverin. This was apparently due to random immobilization and masking of the GRK1-binding site because the kinetics of GRK1 interaction with NH_2- and SH-coupled recoverin, as well as regulation by calcium, were almost identical (V. Z. Slepak, unpublished observation, 1996).

Coupling of Biotinylated Proteins to Immobilized Streptavidin

Some proteins cannot be immobilized on the chip directly because they cannot be attracted to the surface (preconcentrated), or they irreversibly denature under the coupling conditions. In such cases, biotinylation is a good alternative because it can be performed in solution and under conditions preserving protein activity. Biotinylated protein,[9,11–13] oligonucleotide,[14] lipid,[10,15] or peptide[16] is readily and stably immobilized at the surface of streptavidin-coated chips (SA chip; BIACORE). Several commercially available biotinylating agents can be used to modify the ligand via amine or sulfhydryl groups, and by choosing a particular reagent, one can achieve alternative orientation of the ligand on the surface. We have used biotinylation for immobilization of calmodulin and transducin $\beta\gamma$ subunits.

Calmodulin cannot be coupled directly to the surface of a CM5 chip owing to its acidic nature (pI ~4), which apparently causes its electrostatic repulsion from the carboxymethylated dextran. Preconcentration of calmodulin on the chip surface would require a pH <3.5, at which the coupling reaction is strongly inhibited. For biotinylation of calmodulin at its cysteine residue, we use spinach calmodulin (Sigma, St. Louis, MO), which contains a unique cysteine at position 26. Iodoacetyl-LC-biotin (Pierce, Rockford, IL) is prepared as a 1 mM solution in dimethyl sulfoxide (DMSO) and added to 3 nmol of calmodulin in 80 μl of 20 mM HEPES–50 mM NaCl,

[10] C. Lange and K. W. Koch, *Biochemistry* **36**, 12019 (1997).
[11] D. R. Kim and C. S. McHenry, *J. Biol. Chem.* **271**, 20690 (1996).
[12] J. Xu, D. Wu, V. Z. Slepak, and M. I. Simon, *Proc. Natl. Acad. Sci. U.S.A.* **92**, 2086 (1995).
[13] C. M. Craft, J. Xu, V. Z. Slepak, X. Zhan-Poe, X. Zhu, R. Brown, and R. N. Lolley, *Biochemistry* **37**(45), 15758 (1998).
[14] B. Persson, K. Stenhag, P. Nilsson, A. Larsson, M. Uhlen, and P. Nygren, *Anal. Biochem.* **264**(1), 34 (1997).
[15] L. Masson, A. Mazza, and R. Brousseau, *Anal. Biochem.* **218**(2), 405 (1994).
[16] B. E. Snow, R. A. Hall, A. M. Krumins, G. M. Brothers, D. Bouchard, C. A. Brothers, S. Chung, J. Mangion, A. G. Gilman, R. J. Lefkowitz, and D. P. Siderovski, *J. Biol. Chem.* **273**, 17749 (1998).

pH 7.4 at a 1:5 molar ratio. The reaction mixture is kept in the dark at ambient temperature for 1 hr and cysteine is then added to a final concentration of 10 mM to inactivate the unreacted biotinylating agent. The mixture (120 μl) is then desalted on a 1-ml Sephadex G-15 column and is ready for immobilization. Calmodulin can also be biotinylated at NH_2– groups. For this reaction, 3 nmol of calmodulin in 80 μl of 20 mM HEPES–50 mM NaCl, pH 7.4 is mixed with a 1:1 molar amount of NHS-LC-biotin (Pierce), which is prepared immediately prior to use as a 1 mM solution in water. The reaction is allowed to proceed for 1 hr on ice and then quenched by the addition of 1 M Tris, pH 7.4, to a final concentration of 20 mM, and the sample is desalted. Biotinylated calmodulin can be stored at $-80°$ without significant loss of activity. We do not recommend the use of commercially available preparations of biotinylated calmodulin because at least some of them are contaminated by other biotinylated proteins.

The incorporation of biotin into calmodulin is tested by an overlay assay, using peroxidase-conjugated streptavidin as a probe. Biotinylated protein ($\sim$0.2 μg) is resolved by sodium dodecyl sulfate–polyacrylamide gel electrophoresis (SDS–PAGE) and transferred onto a nitrocellulose membrane, which is then blocked with 5% nonfat milk in Tris-buffered saline (TBS)–0.1% Tween 20 (TBS–Tween) for 1 hr and probed with a 1:8000 dilution of streptavidin–peroxidase conjugate (Amersham, Arlington Heights, IL) in TBS–Tween for 1 hr. After a wash with TBS–Tween, the band is visualized by the ECL (enhanced chemiluminescence) system (Amersham). Calmodulin biotinylation at NH_2– or SH– groups was similar with respect to labeling efficiency (Fig. 2A) as well as the interaction with G protein-coupled receptor kinases (GRKs).[9]

Transducin $\beta\gamma$ subunit complex (G$\beta\gamma$) cannot be coupled to the chip covalently because of its irreversible inactivation at the low pH needed for preconcentration. Moreover, direct biotinylation of purified G$\beta\gamma$ also leads to the loss of binding to Gα_t or phosducin, apparently owing to modification of important sites. On the basis of the idea that such sites can be protected from excessive modification by protein–protein interactions,[17,18] we developed a procedure for labeling holotransducin while it is associated with rod outer segment (ROS) membranes.[12,13] Purified bleached bovine ROS (200 μl of membrane pellet) is resuspended in 10 volumes of 20 mM HEPES (pH 7.4)–100 mM NaCl–2 mM $MgCl_2$ and 0.5 mM 2-mercaptoethanol and NHS-LC-biotin (Pierce) are added to a final concentration of 1 mM. The reaction is allowed to proceed for 1 hr at 4° with mixing of the suspension

---

[17] R. E. Kohnken and J. D. Hildebrandt, *J. Biol. Chem.* **264**, 20688 (1989).
[18] J. Dingus, M. D. Wilcox, R. Kohnken, and J. D. Hildebrandt, *Methods Enzymol.* **237**, 457 (1994).

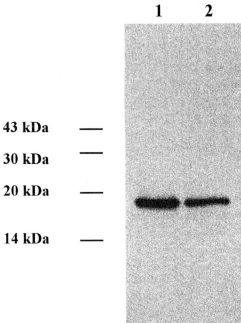
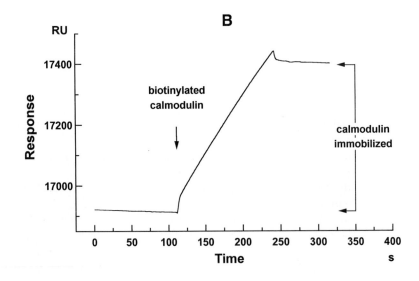

and then quenched by the addition of Tris-HCl, pH 7.4, to a final concentration of 50 mM. The membranes are then washed twice in isotonic buffer [10 mM HEPES (pH 7.4), 100 mM NaCl, 2 mM MgCl$_2$, and 1 mM 2-mercaptoethanol] and twice in hypotonic buffer [5 mM HEPES (pH 7.4), 1 mM EDTA, and 1 mM 2-mercaptoethanol]. The differential elution of transducin α and $\beta\gamma$ subunits is carried out as described.[19,20] To elute Gα_t, the membranes are resuspended in 5 ml of isotonic buffer containing 0.2 mM GTP and immediately subjected to centrifugation in an SW50 rotor (Beckman, Fullerton, CA) at 35,000 rpm for 20 min. Membranes are collected, and for a more complete extraction of Gα_t, this step is repeated. To elute G$\beta\gamma$, the pellet is then resuspended in 5 ml of hypotonic buffer and centrifuged at 45,000 rpm for 40 min. The supernatant containing biotinylated G$\beta\gamma$ is then loaded on a 200-μl hexyl agarose (ICN, Costa Mesa, CA) column equilibrated with hypotonic buffer. The beads are washed by hypotonic buffer to remove traces of low molecular weight biotin and G$\beta\gamma$ is eluted with 300 mM NaCl in isotonic buffer. G$\beta\gamma$ can also be biotinylated on thiol groups, using iodoacetyl-LC-biotin (2-mercaptoethanol is excluded from all the buffers), and purified in the same way. From 2 mg of ROS the yield is approximately 100 μg of G$\beta\gamma$; that is nearly 90% pure on the basis of Coomassie staining. Incorporation of biotin is verified by an overlay assay with horseradish peroxidase (HRP)–streptavidin conjugate as described above for calmodulin. The best results are obtained if at least 50% of the biotin is incorporated into Gβ. Some preparations of ROS yield G$\beta\gamma$ heavily contaminated by other biotinylated proteins, which are not detected by Coomassie staining, but, surprisingly, can constitute up to 90% of the biotinylated proteins revealed by the streptavidin overlay assay. The purity of ligand or analyte is not necessary for the detection of specific binding, and given that appropriate controls are provided these preparations can be used for the study of G$\beta\gamma$ protein–protein interactions. However, for quantitative kinetic analysis, we recommend further purification of the

[19] J. Bigay and M. Chabre, *Methods Enzymol.* **237**, 139 (1994).
[20] J. Bigay and M. Chabre, *Methods Enzymol.* **237**, 449 (1994).

FIG. 2. Immobilization of biotinylated protein. (A) Analysis of calmodulin biotinylation by overlay assay. After SDS–PAGE, the protein was transferred onto nitrocellulose membrane. The membrane was blocked with 5% nonfat milk and probed with HRP–streptavidin. The bands were visualized by chemiluminescence, using ECL as described in text. Lanes 1 and 2, calmodulin biotinylated at NH$_2$– and SH– groups, respectively. Molecular weight standards are shown to the left. (B) Immobilization of biotinylated calmodulin. Biotinylated calmodulin (20 μg/ml) in 20 mM Tris (pH 7.4)–100 mM NaCl–0.5 mM Ca^{2+} is injected across the SA chip surface.

biotinylated G$\beta\gamma$, e.g., by chromatography on immobilized Gα or through the use of a cleaner ROS preparation for the isolation.

Biotinylated protein is immobilized by injection across an SA chip, following an initial conditioning of the surface by three 1-min pulses of 1 M NaCl–50 mM NaOH solution, as recommended by Biacore, or as we found, even without such treatment. A sensogram recorded during the immobilization should be a straight ascending line. This reflects the fast on-rate and high affinity (K_D 10^{-15} M) of streptavidin–biotin binding, which is mostly limited by the diffusion of the biotinylated ligand to the surface. The optimal concentration of the biotinylated protein in the sample is determined empirically; it is usually rather low, e.g., 1–10 μg/ml. The surface density of the immobilized ligand is monitored and adjusted to the desired level by additional injections. The maximal capacity of the surface for biotinylated ligand is determined by the amount of streptavidin on the surface (the commercial SA chip has 4000 RU of streptavidin coupled). For example, immobilization of calmodulin, which is a smaller protein than streptavidin, can reach 1000 RU and we could couple more than 2000 RU of biotinylated G$\beta\gamma$. If the sensogram recorded on the coupling of biotinylated ligand forms a plateau significantly lower than expected on the basis of the maximal capacity, the sample is likely to be contaminated by low molecular weight biotin or nonspecific proteins that have multiple biotinylated sites. After immobilization, calmodulin remains active for several days and because both calmodulin and streptavidin are stable molecules, the surface can withstand multiple regeneration cycles by HCl or guanidine hydrochloride. In contrast, the lifetime of G$\beta\gamma$ on the chip at room temperature is about 12 hr. For repetitive experiments, a fresh portion of G$\beta\gamma$ should be immobilized on a new channel of the chip.

Other Immobilization Techniques

Recombinant proteins expressed as glutathione *S*-transferase (GST) or oligo-His-tag fusions in *Escherichia coli* can be immobilized through anti-GST and anti-His-tag antibodies covalently attached to the CM chip. Oligo-His-tagged proteins can also be bound to Ni^{2+}-NTA chips.[21–23] Some sensor chips utilize hydrophobic interactions, for example, HPA chips (BIACORE)

---

[21] P. D. Gershon and S. Khilko *J. Immunol. Methods* **183**(1), 65 (1995).

[22] L. Nieba, S. E. Nieba-Axmann, A. Persson, M. Hamalainen, F. Edebratt, A. Hansson, J. Lidholm, K. Magnusson, A. F. Karlsson, and A. Pluckthun, *Anal. Biochem.* **252**(2), 217 (1997).

[23] P. Lindner, K. Bauer, A. Krebber, L. Nieba, E. Kremmer, C. Krebber, A. Honegger, B. Klinger, R. Mocikat, and A. Pluckthun, *Biotechniques* **22**, 140 (1997).

can be used to create stable alkanethiol/phospholipid hybrid monolayers.[24] Promising advances have been made with noncommercial SPR detectors in the study of the rhodopsin–transducin interaction in a lipid bilayer.[25,26]

Activity of Immobilized Ligand

The key issue in ligand immobilization concerns the preservation of the biological activity of the coupled molecule. Since the amount of ligand expressed in resonance units is known, one can ascertain its functional activity based on the stoichiometry of interaction and molecular weights of the interactants. If a saturating amount of analyte does not match the amount of ligand immobilized, a portion of the coupled protein might be inactive or inaccessible. Alternative experimental designs (which molecule is coupled and which is soluble) as well as coupling techniques should be compared to evaluate the effect of immobilization on the interaction.

Analysis of Protein–Protein Binding on BIACORE

BIACORE records analyte–ligand binding as response (RU)-versus-time curves, called *sensograms*. A typical sensogram consists of an ascending phase when the complex is being formed, a plateau when binding reaches equilibrium, and a descending phase, which corresponds to ligand–analyte complex decay during buffer washout. To monitor nonspecific binding, a signal from a reference channel with an irrelevant molecule attached should be recorded. Prior to the analysis of the sensogram, the control response can be subtracted from the data recorded from the channel with the coupled ligand. Nonspecific binding, which is generally low in the BIACORE, can be suppressed by 100 mM NaCl combined with a mild detergent (0.01–0.1%), such as P20 (BIACORE), and bovine serum albumin (BSA) as a carrier protein.

Under the correct experimental conditions, the ligand–analyte interaction is limited by the on and off rates characteristic of a given analyte–ligand pair. Fitting of the curves provides data on binding kinetics. Analysis of the effects of low molecular weight compounds, protein phosphorylation, etc., can shed light on the regulation of the protein–protein interaction. SPR is also useful for structure–function analysis since mutants that lack

---

[24] A. L. Plant, M. Brigham-Burke, E. C. Petrella, and D. J. O'Shannessy, *Anal. Biochem.* **226**, 342 (1995).

[25] S. Heyse, O. P. Ernst, Z. Dienes, K. P. Hofmann, and H. Vogel, *Biochemistry* **37**, 507 (1998).

[26] Z. Salamon, Y. Wang, J. L. Soulages, M. F. Brown, and G. Tollin, *Biophys. J.* **71**, 283 (1996).

measurable functional activity other than specific protein–protein binding can be studied. Below we provide the examples of such applications of SPR for protein interactions in the phototransduction cascade.

Kinetics of Protein–Protein Interactions

Figure 3 illustrates the different binding kinetics of the two ligands, recoverin and calmodulin, with the same analyte, GRK1 (Fig. 3A). The GRK1–recoverin interaction is characterized by rapid association (k_a of 10^5 M^{-1} sec^{-1}) and rapid dissociation (k_d of 10^{-1} sec^{-1}) rates. In contrast, the GRK1–calmodulin interaction is characterized by much slower association (k_a of 10^4 M^{-1} sec^{-1}) and dissociation (k_d of 10^{-3} sec^{-1}). Owing to the slower dissociation rate, the affinity of GRK1 for calmodulin is approximately 10 times higher than for recoverin (K_D of 100 nM versus 1 μM). Figure 3B shows that two different analytes, $G\alpha_t$ and phosducin, also bind to a common ligand, $G\beta\gamma$, with different kinetics. The binding of $G\alpha_t$–GDP to $G\beta\gamma$ features a rather rapid association (k_a of 1.5×10^5 M^{-1} sec^{-1}) and slow dissociation (k_d of 4.2×10^{-4} sec^{-1}); K_D ~10 nM (Fig. 3B). While the dissociation rate of phosducin is approximately the same as that of $G\alpha_t$, its association is unusually slow (k_a of 2×10^3 M^{-1} sec^{-1}).[12,13] The coupling of $G\beta\gamma$ via thiol groups (biotinylation via cysteine residues) resulted in the same kinetic pattern as with the NH_2-modified $G\beta\gamma$. Therefore, it is not likely that the difference in binding kinetics between $G\alpha_t$ and phosducin is caused by $G\beta\gamma$ tethering to the surface.

Knowledge of the kinetic constants of interaction between different partners of the same cascade could suggest that they have a distinct role in signal transduction. For example, the relatively slow kinetics of the $G\beta\gamma$–phosducin interaction suggest that it might be less important for quenching the photoexcitation than for a slower process such as light adaptation.

Approximate values of the kinetic constants k_d and k_a as well as K_D can in theory be obtained from a single sensogram. First, the dissociation rate (k_d) is determined from the descending phase of the sensogram. Next, if the concentration of the analyte (C) is known, for the pseudo-first order interaction k_a can be determined by fitting data to the equations

$$R = R_{eq}[1 - e^{-(k_a C + k_d)(t-t_0)}]$$

$$R = \frac{k_a C R_{max}}{k_a C + k_d}[1 - e^{-(k_a C + k_d)(t-t_0)}]$$

where ($t - t_0$) is the time since the beginning of the injection or dissociation), R_{eq} is the steady state response level and R_{max} is the maximum response level (saturation of the ligand), which are also calculated. Once k_d and k_a are determined, the equilibrium constant K_D can be calculated from their ratio.

However, in actual experiments, the association can be limited by the rate of delivery of analyte (diffusion) to the immobile ligand, and with the decay of the complex, slow diffusion of the dissociating analyte from the chip can lead to its rebinding. This would result in measurements of a lower association rate as well as dissociation rate, which is slower than the natural decay of the complex. These mass transport artifacts can be particularly significant if the k_a of the complex is above $10^5 \, M^{-1} \, \text{sec}^{-1}$ and when the surface density is high. An abnormal shape of the sensogram, such as a straight ascending phase or biphasic dissociation, as well as poor fits of the experimental data, might indicate that the binding process is mass transport limited. Another common occurrence that may alter binding kinetics is heterogeneity of either analyte or immobilized ligand. Mass transport limitations are more reliably detected and distinguished from a complicated binding process (two binding sites, dimerization, etc.) if the data are recorded over a range of analyte concentrations and several surface densities. Determination of the rate of ligand–analyte complex formation (dR/dt) under various conditions allows the correct values of k_a and k_d to be extracted by excluding the regions of sensograms affected by mass transport. Researchers have developed methods that analyze the data from an entire sensogram and simultaneously fit several binding curves. These sophisticated algorithms ("global analysis") can account for diffusion, bulk refractive index change, and other factors contributing to the experimentally acquired raw data. Particular algorithms, kinetic models, and software are not discussed here because several articles and reviews devoted to kinetic analysis of SPR data in general have been published.[27–33] From the point of experimental design, complying with the following recommendations should allow mass transport effects to be minimized: (1) increase the flow rate to the maximum permitted by sample availability; this will enhance diffusion. Under mass transport-unlimited conditions the values of kinetic constants are independent of flow rate; (2) reduce the ligand immobilization level to the minimum necessary for reliable detection (10–100 RU); this

---

[27] D. J. O'Shannessy and D. J. Winzor, *Anal. Biochem.* **236,** 275 (1996).
[28] T. A. Morton, D. G. Myszaka, and I. M. Chaiken, *Anal. Biochem.* **227,** 176 (1995).
[29] D. G. Myszka, P. R. Arulanantham, T. Sana, Z. Wu, T. A. Morton, and T. L. Ciardelli, *Protein Sci.* **5**(12), 2468 (1996).
[30] L. D. Roden and D. G. Myszka, *Biochem Biophys. Res. Commun.* **225,** 1073 (1996).
[31] M. Fivash, E. M. Towler, and R. J. Fisher, *Curr. Opin. Biotechnol.* **9,** 97 (1998).
[32] P. M. A. Schuck, *Anal. Biochem.* **240**(2), 262 (1996).
[33] D. G. Myszka, T. A. Morton, M. L. Doyle, and I. M. Chaiken, *Biophys. Chem.* **64,** 127 (1997).

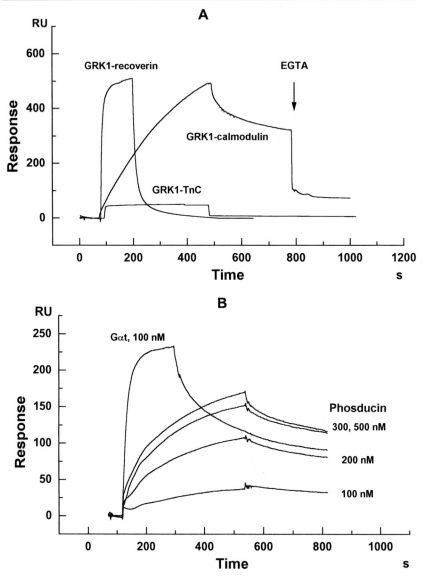

FIG. 3. Protein–protein interactions are characterized by different kinetics. (A) GRK1 (1 μM) is injected across surfaces with immobilized recoverin, calmodulin, or troponin C (negative control, not shown) at 5 μl/min. The running buffer contained 20 mM Tris-HCl (pH 7.5), 100 mM NaCl, 0.5 mM CaCl$_2$, 1 mM MgCl$_2$, 0.005% surfactant P20 (Biacore), and BSA (0.01 mg/ml). Troponin C, a muscle EF-hand Ca^{2+}-binding protein structurally related to calmodulin, displayed no binding. The calmodulin surface is regenerated by a short pulse of

will lessen rebinding and crowding effects; (3) verify sample homogeneity; and (4) collect binding data for a series of consecutive injections of the analyte at several concentrations, typically ranging from 0.1 to 10 K_D of the interaction.

Multiple consecutive injections of analyte at different concentrations can be performed with the same sensor surface. Some interactions, for example, GRK1–recoverin, have fast dissociation rates and the surface is ready for the next injection in several minutes. In contrast, the dissociation rates of GRK1–calmodulin, G$\beta\gamma$–phosducin, and other interactions are rather slow, and the recovery of the surface may take a long time. Prior to the next injection, the bound analyte is removed from the surface by an injection of a solution that dissolves the complex. EGTA can remove the GRK1 from calmodulin and recoverin; these proteins can also withstand numerous 1-min pulses of 6 M guanidine-hydrochloride without loss of the ability to interact with GRKs in a Ca^{2+}-dependent manner. Surfaces with covalently coupled antibodies can usually be regenerated by low pH, for example, 1-min pulses of 10 mM HCl. Other regeneration procedures may be designed on the basis of the properties of a particular ligand. Many proteins, for instance G$\beta\gamma$, cannot refold after denaturation and thus regeneration of the surface with chaotropic agents or pH is not possible. For accurate kinetic analysis, it is best if the duration of each cycle allows the binding to reach equilibrium, i.e., the plateau of the sensogram. The range of analyte concentrations used should approach the saturation of the ligand on the surface, i.e., define the R_{max} experimentally. The analyte concentration at which binding at equilibrium equals 1/2 R_{max} is the K_D. The K_D can also be independently determined as the k_d/k_a ratio, using the rate constants provided by the BIAevaluation (BIACORE) or other software. Ideally, the K_D values obtained from the kinetics and at equilibrium should be in agreement.

Regulation of Protein–Protein Interactions

Because low molecular weight compounds contribute little to the specific SPR signal, the BIACORE is an ideal tool with which to examine the effects of small molecules on protein–protein interactions.

---

25 mM EGTA to remove residual bound kinase at the end of the cycle. (B) Biotinylated transducin G$\beta\gamma$ subunits are immobilized on an SA sensor chip as described in text. Gα_t and phosducin at the concentrations shown are injected across the surface in running buffer containing 20 mM HEPES, 100 mM NaCl, 1 mM MgCl$_2$, 10 μM GDP, BSA (0.01 mg/ml), and 0.005% P20.

Figure 4 demonstrates that the binding of recoverin to GRK1 is directly regulated by low molecular weight ligands of these proteins, Ca^{2+} (Fig. 4), and adenine nucleotides (Fig. 5). The affinity of isolated myristoylated recoverin for calcium is 17 μM.[34,35] At the same time, the EC_{50} for Ca^{2+} for the recoverin-mediated inhibition of rhodopsin phosphorylation by GRK1 *in vitro* is severalfold less, 1.5–3.0 μM.[36,37] It has been suggested that this difference could be due to the presence of the membranes, i.e., that binding of recoverin to the lipid might result in an increase in its local concentration and perhaps affinity for calcium.[35,38,39] Indeed, initial studies of recoverin properties showed that its association with ROS membranes occurred with a Ca^{2+} EC_{50} of 2.1 μM.[40] More recently, Lange and Koch examined the Ca^{2+} dependence of the recoverin interaction with lipid vesicles, using SPR, and found that the EC_{50} for Ca^{2+} was in the low micromolar range, 4 or 7 μM, depending on the configuration of the assay, i.e., whether recoverin or the liposomes were immobilized on the chip.[10] Our direct recoverin–GRK1 binding assay on the BIACORE revealed that the EC_{50} for Ca^{2+} was in the submicromolar range, 450 nM for myristoylated recoverin and 150 nM for nonacylated recoverin (Fig. 4C).[8] In this assay, the samples containing 0.25 μM GRK1 were prepared in buffers with defined free Ca^{2+} concentrations and injected across the surfaces with immobilized myristoylated or nonmyristoylated recoverin (Fig. 4A and B). Under the same conditions the interaction of GRK1 and other GRKs with calmodulin has a higher EC_{50} for Ca^{2+}, 1–3 μM.[9] The relatively low EC_{50} for Ca^{2+} in the SPR experiments can be explained by the increase in the affinity of recoverin for Ca^{2+} when recoverin is bound to GRK1. This is further supported by structure–function studies demonstrating that the binding of full-length GRK1 requires less Ca^{2+} than do fragments containing the recoverin-binding site (see below).[9] Thus, not only the membranes, but also the target of recoverin, GRK1, might play a role in shifting the affinity of recoverin for Ca^{2+} into the physiological range of Ca^{2+} concentrations.

To test whether the GRK1–recoverin interaction is regulated by the ATP or ADP-bound state of the enzyme, GRK1 (0.25 μM) was incubated for 15 min at 20° with various concentrations of ATP, ADP(NH)P, ADP, or

---

[34] J. B. Ames, T. Porumb, T. Tanaka, M. Ikura, and L. Stryer, *J. Biol. Chem.* **270**, 4526 (1995).
[35] A. N. Baldwin and J. B. Ames, *Biochemistry* **37**, 17408 (1998).
[36] C.-K. Chen, J. Inglese, R. J. Lefkowitz, and J. B. Hurley, *J. Biol. Chem.* **270**, 18060 (1995).
[37] V. Klenchin, P. D. Calvert, and M. D. Bownds, *J. Biol. Chem.* **270**, 16147 (1995).
[38] J. Ames, R. Ishima, T. Tanaka, J. I. Gordon, L. Stryer, and M. Ikura, *Nature (London)* **389**, 198 (1997).
[39] P. D. Calvert, V. A. Klenchin, and M. D. Bownds, *J. Biol. Chem.* **270**, 24127 (1995).
[40] S. A. Zozulya and L. Stryer, *Proc. Natl. Acad. Sci. U.S.A.* **89**, 11569 (1992).

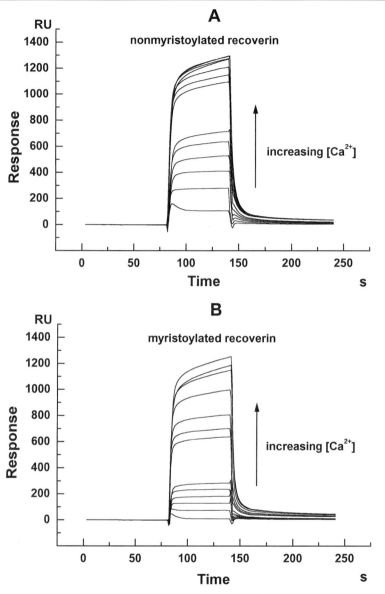

FIG. 4. Regulation of GRK1–recoverin interaction by Ca^{2+}. Overlay plot of sensograms recorded at increasing concentrations of Ca^{2+} on binding of GRK1 to nonacylated (A) and myristoylated (B) recoverin. GRK1 at 0.25 μM is prepared in 13 calcium buffers as described in text; free Ca^{2+} concentrations were 0.00, 0.017, 0.038, 0.065, 0.100, 0.150, 0.225, 0.350, 0.602, 1.38, 3.0, 5.0, and 39.8 μM. (C) The response at equilibrium (R_{eq}) values at each concentration were normalized against those obtained at 1 mM calcium (100%) and plotted as a function of log[Ca^{2+}]. Open squares, nonmyristoylated recoverin; filled squares, myristoylated recoverin.

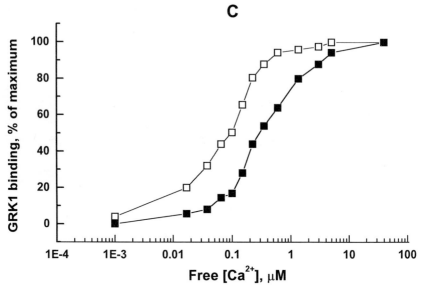

Fig. 4. (*continued*)

GTP and XTP as controls and injected across the surface with immobilized recoverin in the running buffer (which also contained 0.5 mM CaCl$_2$). Figure 5A shows that ATP strongly inhibits the GRK1–recoverin interaction, while ADP enhances it. Since the nonhydrolyzable analog of ATP did not influence recoverin–GRK1 binding[8] and protein phosphatase can completely reverse the ATP effect, we concluded that ATP-mediated inhibition is due to autophosphorylation of GRK1. To monitor autophosphorylation, 0.25 μCi of [γ-^{32}P]ATP (New England Nuclear, Boston, MA) is added as the tracer to the reaction. For dephosphorylation, the ^{32}P-labeled GRK1 is separated from ATP by desalting on Sephadex G-15, and then incubated with 0.2 units of protein phosphatase PP2A (Upstate Biotechnology, Lake Placid, NY) for 5 min. Dephosphorylation occurs, however, even in the presence of ATP, indicating that the rate of PP2A-catalyzed GRK1 dephosphorylation might be even faster than that of GRK1 autophosphorylation. The autophosphorylation and dephosphorylation reactions are monitored by SDS–PAGE followed by autoradiography, and aliquots of the same reaction mixtures are analyzed on the BIACORE (Fig. 5B). It is important to point out here that the profound effect of GRK1 autophosphorylation on GRK1–recoverin binding cannot be easily detected in conventional tests of rhodopsin phosphorylation because these assays contain ATP.

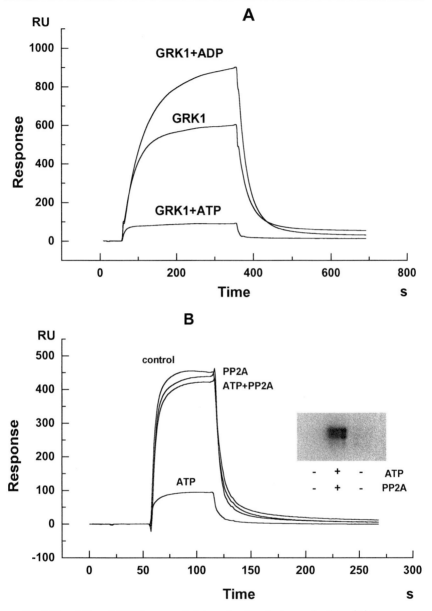

FIG. 5. Regulation of GRK1–recoverin interaction by adenine nucleotides. (A) Sensograms recorded on binding of GRK1 to recoverin in the presence of 100 μM ATP and ADP or in the absence of nucleotide. (B) GRK1 incubated with or without [^{32}P]ATP and treated with protein phosphatase PP2A as described in text. Aliquots of the samples were injected across the recoverin surface or analyzed by SDS–PAGE autoradiography (*inset*). Dephosphorylation of GRK1 by PP2A restored its binding to recoverin.

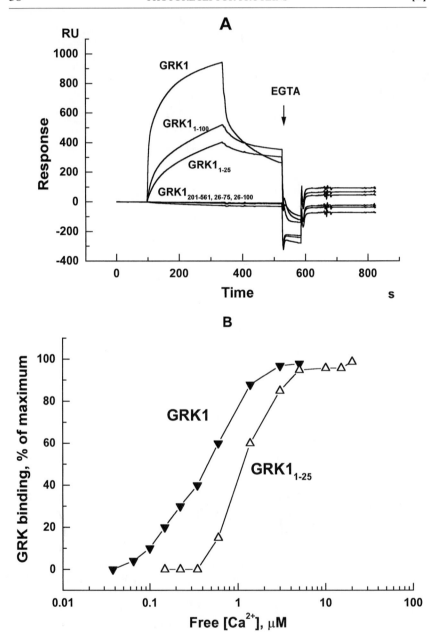

These experiments demonstrate how SPR can help to dissect a mechanism consisting of several interrelated molecular events that can be difficult to unravel by traditional biochemical methods.

Structure–Function Analysis

A common problem in protein structure–function research using mutagenesis is to set apart mutations that indeed target the specific activity of the protein and those that cause a nonspecific change in protein folding. The BIACORE can rapidly and quantitatively characterize a relatively large number of analytes, and so it is useful for screening and comparison of mutants. Furthermore, when a functional (e.g., enzymatic) activity of a mutant is eliminated and cannot be assayed, SPR can verify that the molecule is still capable of a characteristic protein–protein interaction and thus is not denatured. We used this strategy for the mutational analysis of the GTPase-activating protein (GAP) activity of the γ subunit of cGMP phosphodiesterase (PDEγ), when in addition to the GAP assays the ability of mutants to bind to PDE$\alpha\beta$ and Gα_t was tested. Substitution of the Trp-70 residue in PDEγ resulted in complete loss of its GAP activity, but the mutant could still interact with Gα_t and PDE$\alpha\beta$.[6]

SPR allows the study of protein fragments that do not have any functional activity but specific protein–protein binding, and the characteristics of such mutants can be quantitatively compared with those of the wild-type protein and other mutants with respect to binding kinetics and its regulation. Therefore, in addition to bona fide mapping of functionally important residues and domains of a protein molecule, new insight can be gained into the mechanism of its function. Figure 6 shows a comparison of recoverin binding to GRK1 and two of its N-terminal fragments (expressed as GST fusions) containing the recoverin-binding site. Surprisingly, these GRK1 fragments display even stronger affinity for recoverin than does the full-length kinase. The association of fragments is hindered, but binding is tighter owing to much slower decay of the formed complex. This suggests that although the severed C terminus of GRK1 does not bind to recoverin, it might be important for rapid recoverin–GRK1 binding kinetics and may

---

FIG. 6. Recoverin interaction with GRK1 fragments. (A) Binding of GRK1 fragments (GST fusions) to recoverin is compared with binding of the full-length GRK1. GST–GRK1(1–25) and GST–GRK1(1–100) specifically bind to recoverin in the presence of Ca^{2+} and can be removed by excess EGTA; the GST fusions lacking the 25 N-terminal amino acids of GRK1 do not bind to recoverin. (B) Ca^{2+} dependence of the interaction of the full-length GRK1 and GRK1(1–25) with recoverin (the data are plotted as in Fig. 4C). Binding of the full-length GRK1 (filled triangles) to recoverin has a lower EC_{50} for Ca^{2+} as compared with the GRK1(1–25) fragment (open triangles).

influence the N terminal-binding site allosterically. It should be noted here that the higher binding affinity of separated domains compared with the full-length protein is rather unusual. For example, the GST fusion proteins of phosducin fragments bound to G$\beta\gamma$ with a lower affinity than did the full-length phosducin[12,13]; some fragments of PDEα bound to PDEγ with approximately 1000-fold lower affinity (approximately 10^{-6} M) than did the native PDE$\alpha\beta$ (10^{-9} M) (V. Z. Slepak and N. Artemiev, unpublished, 1995).

Figure 6B compares the Ca^{2+} dependence of recoverin binding to GRK1 with that of GRK1(1–25). The experiments were carried out as described above, with binding recorded at various Ca^{2+} concentrations. The sample and running buffers contained 20 mM HEPES (pH 7.4), 100 mM NaCl, 1 mM MgCl$_2$, 0.005% surfactant P20, BSA (0.01 mg/ml), and 0.1 mM CaCl$_2$. The analytes, GRK1(1–25)–GST fusion and GRK1, were diluted >40-fold from the Ca^{2+}- and EGTA-free stock solutions to a concentration of 0.25 μM in the Ca^{2+} buffer solutions and were injected over the surface with myristoylated recoverin for 2 min at 10 μl/min. The half-saturating concentration of Ca^{2+} required for binding of the fragment was about 3.5 μM, which is almost 10-fold higher than that necessary for the full-length kinase. This result can be interpreted as follows. The EC_{50} for Ca^{2+} reflects the Ca^{2+} level necessary for the conformational change of recoverin that favors its binding to the target. Although the apparent affinity of recoverin for Ca^{2+} is likely to be altered (enhanced) by the GRK1(1–25) fragment, it can serve as a reference point reflecting the analyte-independent (intrinsic) affinity of immobilized recoverin for Ca^{2+}. Since the binding of the full-length kinase requires less Ca^{2+}, it is likely that in the GRK1-bound state the affinity of recoverin for Ca^{2+} is higher than in the GRK1-free or fragment-bound state. Similar differences in EC_{50} for Ca^{2+} between the full-length kinase and its fragments have also been found for the interactions of GRK1, GRK2, and GRK5 with calmodulin.[9] Although many calmodulin targets (as well as troponin C and other Ca^{2+}-binding proteins) have been demonstrated to enhance calmodulin–Ca^{2+} binding, this phenomenon has not been previously demonstrated for recoverin or GRKs.

Acknowledgments

D.K.S. is supported by an American Heart Association Florida Affiliate, Inc., Postdoctoral Fellowship (9703012). V.Z.S. is an Initial Investigator Award (9603008) from the American Heart Association, Florida, Affiliate, Inc., and a recipient of a Pharmaceutical Research and Manufacturers of America Foundation Faculty Development Award in Basic Pharmacology and a Grant-in-Aid (GA98029) from the Fight for Sight Research to Prevent Blindness America Foundation.

[3] Expression of Phototransduction Proteins in *Xenopus* Oocytes

By BARRY E. KNOX, ROBERT B. BARLOW, DEBRA A. THOMPSON, RICHARD SWANSON, and ENRICO NASI

Principle

Xenopus oocytes provide a unique system for the expression of phototransduction proteins.[1] These large cells can be microinjected with RNA or DNA, and express a wide variety of proteins in a functional state. They are amenable to micropipette impalement for electrophysiological recording and intracellular injections. Moreover, oocytes contain a G protein-controlled pathway that regulates Cl^- channels in the plasma membrane through intracellular calcium concentration. The oocyte G protein is able to interact with a wide variety of receptors, and thus has made the oocyte a convenient system for expression of seven transmembrane receptors. It is possible to inject either RNA derived from a cloned cDNA or cellular RNA, containing a complex mixture of proteins, some involved in signal transduction and biosynthesis. Thus, proteins that require a specific protein for proper folding or signaling can be identified or coexpressed.[2] We have utilized this system to express both vertebrate[3,4] and invertebrate light-sensitive pigments.[5] In addition, we have characterized the expression of a transducin subunit and show the membrane association of the protein.

Materials and Solutions

$OR2\text{-}Ca^{2+}$: Combine
  NaCl, 82.5 mM
  KCl, 2.5 mM
  $MgCl_2$, 1 mM
  Sodium HEPES, 10 mM
  and adjust to pH 7.6 with NaOH

---

[1] N. Dascal, *CRC Crit. Rev. Biochem.* **22**, 317 (1987).
[2] L. M. McLatchie, N. J. Fraser, M. J. Main, A. Wise, J. Brown, N. Thompson, R. Solari, M. G. Lee, and S. M. Foord, *Nature (London)* **393**, 333 (1998).
[3] H. G. Khorana, B. E. Knox, E. Nasi, R. Swanson, and D. Thompson, *Proc. Natl. Acad. Sci. U.S.A.* **85**, 7917 (1988).
[4] B. E. Knox, H. G. Khorana, and E. Nasi, *J. Physiol.* **466**, 157 (1993).
[5] E. J. Mole, J. Schaefer, K. Mathiesz, V. E. Dionne, B. E. Knox, and R. B. Barlow, *Biol. Bull.* **191**, 264 (1996).

modified Barth saline (MBS): Combine
  NaCl, 88 mM
  KCl, 1 mM
  NaHCO$_3$, 2.4 mM
  Sodium HEPES (pH 7.6), 15 mM
  Ca(NO$_3$)$_2$, 0.3 mM
  CaCl$_2$, 0.41 mM
  MgSO$_4$, 0.82
  Sodium penicillin, 10 μg/ml
  Streptomycin sulfate, 10 μg/ml

Dialyzed fetal bovine serum: Heat inactivate fetal bovine serum (FBS) at 56° for 45 min and dialyze the serum, using 30,000-Da cutoff dialysis tubing extensively against 1× MBS. Sterilize by filtration and store at $-20°$

FBS–MBS (5%): Combine
  Dialyzed fetal bovine serum, 10 ml
  MBS, 90 ml

MBS–sodium pyruvate: supplement MBS with 5 mM sodium pyruvate (Sigma, St. Louis, MO)

KBS: Combine
  KCl, 0.2 M
  KP$_i$ (pH 7), 20 mM

MMR (10×): For 1 liter, combine
  NaCl, 1000 mM   58.44 g
  KCl, 20 mM   1.49 g
  CaCl$_2$, 20 mM   2.94 g
  MgCl$_2$, 10 mM   2.04 g
  HEPES, 50 mM   11.92 g
and adjust to pH 7.4 with 10 N NaOH

Tricaine: For a stock solution, dissolve 5 g of tricaine (3-amino benzoic acid ethyl ester methane sulfonate; Sigma) in 50 ml of H$_2$O and store at $-2°$. Dilute to 0.1% in 0.1× MMR to anesthetize the frog prior to surgery

Collagenase (Sigma type 1A): Prepare a 2-mg/ml solution immediately prior to use in OR2-Ca^{2+}

11-*cis*-Retinal: Prepare a 5–50 mM stock solution in ethanol; store under argon at $-70°$

Lysis solution: Combine
  Dodecyl maltoside (Anatrace, Maumenee, OH), 1% (w/v)

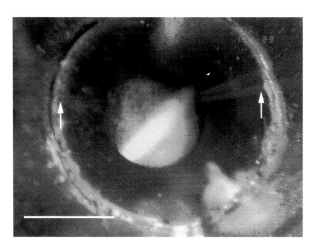

FIG. 1. Stage VI oocyte impaled by two electrodes (*arrows*) in the recording chamber. The animal (dark) and vegetal (light) hemispheres are separated by a lighter stripe at the equator, indicative of a final stage oocyte. Bar: 1 mm.

are elicited by exposing the oocyte to light delivered through a fiber optic light guide with suitable filters to control the wavelength of illumination.

Preparation of Oocyte Membranes

Wash injected oocytes (20–50) extensively with MBS and resuspend the oocytes in KBS (1 ml per 50 cells) containing aprotinin, benzamidine, pepstatin, and leupeptin (50 μg/ml each). Homogenize the oocytes in a microcentrifuge tube, using a pestle fitted for these tubes (Kontes, Vineland, NJ). Centrifuge the homogenate in a microcentrifuge for 10 min at 15,000 rpm at 4°. Discard the supernatant, resuspend the pellet with 1 ml of KBS containing protease inhibitors, and recentrifuge. Resuspend the pellet in 0.2 ml of KBS containing protease inhibitors. Layer the membranes on a sucrose step gradient [1 ml of 50% (w/v) and 1 ml of 20% (w/v) sucrose in KBS] and centrifuge in an SW/TL55 rotor at 30,000 rpm for 30 min at 4°. Collect the membranes at the 20%/50% interface, dilute with 1 ml of KBS, and collect membranes by centrifuging at 10,000 rpm for 20 min at 4° in a microcentrifuge. Resuspend the membranes in 100–200 μl of lysis solution and proceed to immunoprecipitation.

Immunoprecipitation of Labeled Protein

After RNA injection, oocytes (5–10) are placed in 0.2 ml of MBS containing 5% FBS and 0.5 mCi of Tran[35]S-Label (mixture of ^{35}S-labeled amino acids from *Escherichia coli*; ICN, Costa Mesa, CA) per milliliter in 96-well bacteriological plates. The medium should be replaced every day. After 3–4 days, collect the oocytes and wash three times with 1 ml of MBS in a microcentrifuge tube. Remove as much of the liquid as possible and add 50–100 μl of lysis solution per oocyte. Homogenize in a microcentrifuge tube, using small Teflon pestle (Kontes), and incubate on ice for 30 min. Centrifuge the homogenate for 20 min at maximum speed in the microcentrifuge at 4°. Carefully remove the clear yellow supernatant and add antibody. For precipitation of mammalian rhodopsin (Fig. 2), 10 μg of ID4 coupled to agarose[3] is added and gently mixed overnight at 4°. The ID4 IgG–agarose should be washed once with an excess of nonradioactive oocyte extract prior to addition to the sample. To precipitate other proteins (e.g., transducin subunit; Fig. 3), a polyclonal antibody (IgG gives the lowest background) can be used at 1–10 μg of IgG per milliliter. After overnight incubation, protein A–agarose (enough to bind all of the IgG, usually 5–10 μl of agarose is sufficient) is added and gently mixed at room temperature for 1–2 hr. The protein A–agarose should be incubated overnight at 4°

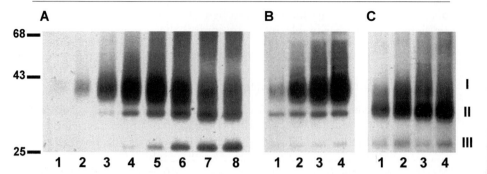

FIG. 2. Immunoprecipitation of bovine opsin from *Xenopus* oocytes. (A) Dependence of opsin synthesis on the concentration of RNA injected. Opsin was immunoprecipitated with anti-opsin IgG from pools of 15 oocytes injected with 50 nl of various concentrations of *in vitro*-synthesized RNA (lanes 1–8, 0.75, 2.3, 6.8, 20, 61, 183, 550, and 1650 μg/ml). The total amount of opsin synthesized was maximal at 100 μg/ml and the fraction of protein in band I decreased from 100% at the lowest concentrations to 55% at the highest. Molecular mass markers are indicated (in kDa) and the various forms of opsin are designated I, II, and III. (B and C) Time course of expression. Oocytes were injected with RNA (B, 5 ng) or DNA (C, 10 ng) and incubated in ^{35}S-labeled amino acids for the indicated number of days prior to immunoprecipitation. The amount of opsin increased throughout the incubation for both DNA and RNA; however, the relative proportion of the various opsin forms did not significantly change.

with a 10-fold excess volume of oocyte block solution and then washed extensively in lysis buffer. The agarose beads are collected by low-speed centrifugation and washed five times with lysis buffer, changing the microcentrifuge one time. Bound protein can be eluted with 1% sodium dodecyl sulfate (SDS) or by using competing peptide.[3]

Application

When oocytes are injected with bovine opsin RNA, they produce several polypeptide forms that differ in N-glycosylation[3] (Fig. 2). Band I contains complex carbohydrates and represents the mature form of the protein. Band II contains high-mannose carbohydrate and is a partially processed form of the protein, while band III is unglycosylated. The relative proportion of the various opsin forms depends on the amount of RNA injected (Fig. 2A), suggesting that there is a limiting component in the proper processing of opsin. Synthesis of opsin was relatively constant in both RNA-injected (Fig. 2B) and DNA-injected (Fig. 2C) oocytes, with substantial (1–5 ng/cell, using a radioimmunoassay[3]) synthesis after 4 days. The relative distribution of opsin isoforms was different in RNA- compared with DNA-

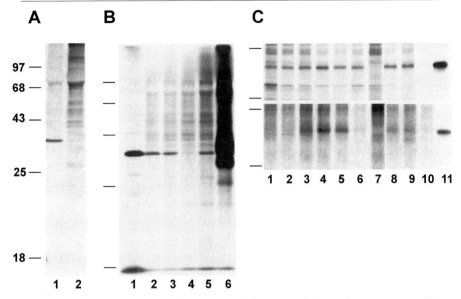

FIG. 3. Immunoprecipitation of bovine transducin α subunit from *Xenopus* oocytes. (A) Oocytes (43) were injected with RNA encoding bovine transducin α subunit (lane 1) or were uninjected (lane 2), cultured for 4 days with ^{35}S-labeled amino acids, and immunoprecipitated with an anti-transducin α subunit antibody. The transducin-specific band migrates at 39 kDa. (B) Expression of transducin α subunit in oocytes injected with DNA (lanes 2 and 3), oocytes left uninjected (lane 4), or oocytes injected with RNA (lane 5). Similar amounts of transducin α subunit are produced with either RNA or DNA, and the product comigrates with the peptide produced in rabbit reticulocyte lysate (lane 1). To compare expression levels, oocytes were injected with opsin RNA (lane 6). The opsin accumulates to much higher levels (>100-fold) than transducin. Molecular mass standards are in kilodaltons. (C) Association of transducin α subunit with oocyte membranes. Various concentrations of transducin α-subunit RNA (lanes 1–7: 12.5, 25, 50, 100, 200, 400, and 0 μg/ml) and DNA (lanes 8 and 9: 1 and 2 mg/ml) were injected into oocytes. Lanes 7 and 10 represent uninjected control oocytes. The membrane (*top*) and soluble (*bottom*) fractions were isolated and immunoprecipitated. At low concentrations of RNA (<50 μg/ml), the transducin α subunit is associated with the membrane. At higher concentrations, the amount of subunit in the membrane is constant and the additional subunit is found in the soluble fraction. As control, the peptide produced in rabbit reticulocyte lysate is shown (lane 10).

injected oocytes. The DNA-injected oocytes always exhibited less of the mature opsin than RNA-injected cells. Thus, it is important to determine the optimum conditions necessary to obtain proper processing. The opsin produced in oocytes was able to bind retinal and exhibited the functional properties expected for bovine opsin.[3]

Oocytes injected with transducin α-subunit RNA produced a protein with a molecular mass of 39 kDa (Fig. 3A). The oocytes were more sensitive

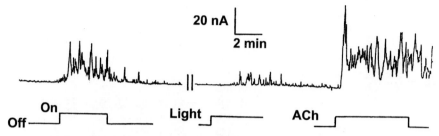

FIG. 4. Light response in an oocyte that had been injected 4 days previously with opsin RNA. Rhodopsin pigment was generated by incubation with 20 μM 11-*cis*-retinal for approximately 30 min. The perfusion solution was MBS containing high Ca^{2+} and the oocyte was held at 0 mV during pigment generation and stimulation. A saturating, white light stimulus produced a fluctuating, outward current that initiated with a delay of 30–60 sec and eventually subsides. After an additional period of incubation with 11-*cis*-retinal in the dark (represented by the vertical bars), the oocyte was stimulated again, with a significantly reduced response. The cell remained responsive to other stimuli, as indicated by the treatment with muscarinic agonist, acetylcholine (ACh). This demonstrates the desensitization of light response mediated by rhodopsin.

to overexpression of transducin than opsin, often dying on the second day. The expression could be observed using either DNA or RNA (Fig. 3B), but protein accumulated to much lower extent than opsin. The transducin α subunit was found predominantly in the membrane fraction at low expression levels, but was found in the soluble fraction in oocytes injected with <1.25 ng of RNA (Fig. 3C). Oocytes injected with DNA exhibited trans-

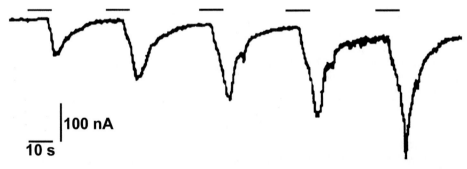

FIG. 5. Light-induced current responses from an oocyte that had been injected with *Limulus* retinal poly(A)$^+$ RNA 4 days previously. After incubation with 11-*cis*-retinal, the oocyte, held at -40 mV in MBS (normal calcium), was repeatedly stimulated with white light. The oocyte produced robust inward currents that exhibited no tendency to desensitize or necessity for additional incubation with 11-*cis*-retinal. Bars over the current recording indicate the duration of the light stimulus.

ducin α subunit in both fractions, indicating that localization of the subunit is limited by some component of the oocyte biosynthetic apparatus.

Oocytes injected with bovine opsin RNA or DNA and incubated with 11-cis-retinal responded to light, exhibiting transient fluctuations typical of the endogenous Cl⁻ conductance[4] (Fig. 4). The currents in Fig. 4 are outward, since the holding potential was more positive than the chloride resting potential. Elevated holding potential and high extracellular calcium helps enhance the responses.[4] One striking feature of the response was its absence until 4 days post-RNA injection, even though there was significant opsin synthesis by the second day. It is not clear what limits the activity. The light response appears to desensitize, even after long periods of recovery and pigment regeneration. Since the cell maintains the ability to respond to acetylcholine, it appears that the desensitization involves changes in rhodopsin after the initial exposure to light.

In contrast to the experiments with bovine opsin, oocytes injected with RNA encoding an invertebrate (*Limulus*) opsin did not exhibit any responses, even though the polypeptide was synthesized efficiently. This result suggests that there is some component missing from the oocytes that is necessary for correct biosynthesis. When *Limulus* retinal poly(A) RNA was injected into oocytes, robust light responses were obtained[5] (Fig. 5). These responses are inward because the holding potential was −40 mV and did not require elevated external calcium. Moreover, there was no apparent loss of responsiveness with stimulation. The light responses were obtained even without regeneration periods. These results are consistent with the *in vivo* data that *Limulus* opsin is bistable.[10] Moreover, the retinal RNA contains the components necessary to produce a functional invertebrate visual pigment.

Xenopus oocytes efficiently translate a number of retinal proteins and provide a suitable system for studying the characteristics of light responses. With this system, it is possible to recombine several components in the same, physiologically accessible cell to study functional interactions. Finally, it is possible to use oocytes as part of a functional assay to identify novel cDNAs that participate in the biosynthesis or functional assembly of light transduction systems.

Acknowledgment

Supported by grants from the National Eye Institute and Research to Prevent Blindness.

---

[10] J. E. Lisman and Y. Sheline, *J. Gen. Physiol.* **68**, 487 (1976).

[4] *Xenopus* Rod Photoreceptor: Model for Expression of Retinal Genes

By SUCHITRA BATNI, SHOBANA S. MANI, CHARISSE SCHLUETER, MING JI, and BARRY E. KNOX

Xenopus offers a number of specific advantages for studying the molecular mechanisms that regulate photoreceptor metabolism and gene expression. Embryos develop *in vitro,* thus allowing precise manipulation at any time after fertilization.[1-3] Retinal development proceeds quickly: a functional retina forms in 3-4 days.[4,5] Moreover, the retina continues to grow after lamination in proliferative zones called *ciliary margins.*[6-8] The photoreceptors are large, and ideally suited for electrophysiology and cell biology studies. The retina contains a large proportion of cones, 45%[9,10] compared with 3-7% in mammals. Rapid methods for expression of genes in the developing retina are available: microinjection of early blastomeres,[11] transfection,[12,13] and transgenesis.[14,15] Photoreceptor metabolism can be studied *in vitro* for up to 1 week in eye cups, isolated retinas, and isolated photoreceptor layers.[16-21] A circadian clock in the photoreceptor layer[17] modulates

---

[1] P. D. Niewkoop and J. Faber, "Normal Table of *Xenopus laevis* (Daudin)," 2nd Ed. North-Holland Publishing, Amsterdam, 1967.
[2] V. Hamburger, "A Manual of Experimental Embryology," revised. University of Chicago Press, Chicago, 1960.
[3] R. Rugh, "Experimental Embryology," 3rd Ed. Burgess, Minneapolis, Minnesota, 1962.
[4] C. E. Holt, T. W. Bertsch, H. M. Ellis, and W. A. Harris, *Neuron* **1**, 15 (1988).
[5] P. Wikovsky, E. Gallen, J. Hollyfield, H. Ripps, and C. D. B. Bridges, *J. Neurophys.* **39**, 1272 (1976).
[6] T. A. Reh, *J. Neurosci.* **7**, 3317 (1987).
[7] J. Hollyfield, *Dev. Biol.* **24**, 264 (1971).
[8] K. Straznicky and R. M. Gaze, *J. Embryol. Exp. Morphol.* **26**, 67 (1971).
[9] L. Saxen, *Ann. Acad. Scientiarum Fennicae.* **1**, (1954).
[10] J. Zhang, J. Kleinschmidt, P. Sun, and P. Witkovsky, *Vis. Neurosci.* **11**, 1185 (1994).
[11] D. Melton, *Proc. Natl. Acad. Sci. U.S.A.* **82**, 144 (1985).
[12] C. E. Holt, N. Garlick, and E. Cornel, *Neuron* **4**, 203 (1990).
[13] S. Batni, L. Scalzetti, S. A. Moody, and B. E. Knox, *J. Biol. Chem.* **271**, 3179 (1996).
[14] K. Kroll and E. Amaya, *Development* **122**, 3173 (1996).
[15] B. E. Knox, C. Schlueter, B. Sanger, C. Green, and J. Besharse, *FEBS Lett.* **243**, 117 (1998).
[16] J. Besharse, P. M. Iuvone, and M. E. Pierce, *in* "Progress in Retinal Research" (N. Osborne and G. Chader, eds.), Vol 7, p. 21. Pergamon Press, New York, 1983.
[17] G. Cahill and J. Besharse, *Neuron* **10**, 573 (1993).
[18] J. Besharse and D. Dunis, *Exp. Eye Res.* **36**, 567 (1983).
[19] G. Cahill and J. Besharse, *J. Neurosci.* **11**, 2959 (1991).

several aspects of *Xenopus* photoreceptor metabolism.[16] In fact, a great deal of what is known about rod outer segment disk formation has been learned in *Xenopus* or other frogs.[22] Genes or cDNAs for many phototransduction proteins are highly conserved. Given the similar structure and developmental program between *Xenopus* and mammalian retina, the frog retina is an important model for studies of human disease. This chapter describes techniques for introducing genes into *Xenopus* retina by transient transfection and transgenesis, permitting analysis of promoters, overexpression of mutant genes (e.g., rhodopsin point mutants), or chimeric proteins for the study of protein sorting and processing.

Transient Transfection of *Xenopus* Embryos

Principle

This assay is based on transient transfections of *Xenopus* embryos, allowing the introduction of plasmid DNA in embryonic retinal tissue (Fig. 1). The most important factor for efficient transfection is consistent access of DNA–lipid complexes to mitotic retinal precursor cells. A number of barriers exist in *Xenopus* embryos[23]: a double-layered epidermis covering the developing eye vesicles and separating the developing retina from the outer environment; a multicellular presumptive lens derived from the inner epidermis; and mesodermal and endodermal tissues that separate the presumptive retina from the ventricular spaces of the brain. Previous work has shown that, on treatment with trypsin, DNA could be introduced into retinal precursor cells of dissected embryo heads.[5,13] The combination of EDTA[24] and trypsin[7] treatments prior to the addition of DNA–lipid mixture permits access to retinal cells in stage 26–28 embryos. Both agents disrupt cell–cell interactions and would thus expose the cells for transfection. An alternative approach for introducing DNA into retinal precursor cells utilizes microinjection of DNA–lipid mixtures into primitive eyebud vesicles.[25]

A second complication encountered in transfections of a complex mixture of embryonic cells is the choice of a promoter for quantitation of

---

[20] M. Stiemke, R. A. Landers, M. R. Al-Ubaidi, M. E. Rayborn, and J. G. Hollyfield, *Dev. Biol.* **162**, 169 (1994).
[21] D. Lahiri, R. Landers, and J. Hollyfield, *J. Morphol.* **223**, 325 (1995).
[22] D. Deretic, *Eye* **12**, 526 (1998).
[23] P. Hausen and M. Riebesell, "The Early Development of *Xenopus laevis*." Springer-Verlag, New York, 1991.
[24] W. A. Harris and L. Messersmith, *Neuron* **9**, 357 (1992).
[25] R. Dorsky, D. Rapaport, and W. Harris, *Neuron* **14**, 487 (1995).

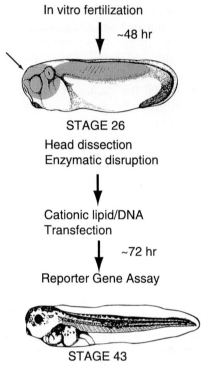

Fig. 1. Embryo transfection scheme.

transfection efficiency of retinal precursors. The use of general promoters, such as those from Rous sarcoma virus (RSV) or cytomegalovirus (CMV), are unsuitable because the superficial epithelial cells of the head were the predominant cells transfected. However, the *Xenopus* rhodopsin upstream region or subfragments can be used without a normalization control in embryo transfections, if sufficient number of trials are performed.[26] Cell-specific expression can then be monitored by transfection of other embryonic tissues such as mesodermal and endodermal trunk tissues.

Materials and Solutions

Pregnant mare serum gonadotropin (PMSG): Dissolve 2000 units of pregnant mare serum gonadotropin (PMSG; Sigma, St. Louis, MO) directly in a bottle with 5 ml of sterile H_2O to make a 400-U/ml stock. For injections, use 0.25 ml of PMSG and 0.25 ml of sterile H_2O per frog. Stable for several weeks at 4°

---

[26] S. Batni, S. Moody, J. Besharse, and B. E. Knox, submitted for publication.

Human chorionic gonadotropin (hCG): Dissolve 15,000 units of human chorionic gonadotropin (hCG; Sigma) directly in a bottle with 3 ml of sterile H_2O to make a 5000-U/ml stock. This stock solution is stable for several months at 4°. For injections, dilute the hCG to 1500 U/ml: use 0.14 ml of hCG stock and 0.36 ml of sterile H_2O per frog at 4°

FBS–MMR (10%): Combine
  Fetal bovine serum (heat inactivated at 56° for 45 min), 10 ml
  MMR (10×), 10 ml
  Sterile H_2O, 79.5 ml
  Penicillin–streptomycin solution [GIBCO (Grand Island, NY) stock solution: penicillin (10,000 units/ml) and streptomycin (10 mg/ml)], 0.5 ml

Cysteine-hydrochloride (2%): Dissolve 2 g of cysteine-hydrochloride (Sigma) in 100 ml of H_2O; adjust to pH 7.8 with 10 N NaOH

MMR (10×): For 1 liter, combine

NaCl, 1000 mM	58.44 g
KCl, 20 mM	1.49 g
$CaCl_2$, 20 mM	2.94 g
$MgCl_2$, 10 mM	2.04 g
HEPES, 50 mM	11.92 g

Adjust to pH 7.4 with 10 N NaOH

MMR (10×), divalent free: Prepare as described above, without $CaCl_2$ and $MgCl_2$

MMR (10×), divalents plus EDTA: Prepare 10× MMR-divalent free containing 10 mM EDTA

Transfection medium: For 100 ml, combine

L15-medium (Life Technologies, Gaithersburg, MD), 65%	65 ml of medium plus 35 ml of sterile H_2O
L-Glutamine (Life Technologies), 1 mM	1 ml of 100 mM stock solution
Gentamicin sulfate (Sigma), 2.5 µg/ml	0.5 ml of a 0.5-mg/ml stock solution

Embryo lysis buffer: Combine
  Tris–PO_4, 25 mM
  1,2-diaminocyclohexane-N,N,N′,N′-tetra acetic acid, 2 mM

Glycerol, 10% (w/v)
Triton X-100, 1%
Dithiothreitol (DTT), 2 mM
and adjust to pH 7.8 with H_3PO_4

Preparation of Embryos

To prepare embryos,[27,28] we have found the following points to be useful.

Adults. Three to 7 days prior to fertilization, inject the female in the dorsal lymph sac with pregnant mare serum gonadotropin to induce maturation of the oocytes. Use a tuberculin syringe to draw up the solution containing 100 U in 0.5 ml of sterile H_2O. Hold the female in damp cheesecloth with the rear slightly elevated, and inject the solution with a 27-gauge/1.25-in. needle into the dorsal lymph sac. Make sure the solution does not leak out. Return the female to an isolation section of the tank and do not feed. There is no need to treat the males with hormones before removing the testes.

Induction of Egg-Laying. Place the female in a small tank half-filled with aged tap H_2O at 21–23°. Eight to 12 hr prior to egg collection, inject the female with 700 U of hCG in the dorsal lymph sac.

Testes. Place the male in an ice–water bath and allow the frog to chill for about 15 min. When the frog is insensitive, use the guillotine to decapitate the animal, pithing both cranial and spinal cord nerves. Quickly open the abdominal cavity, remove both testes, rinse with ice-cold 10% fetal calf serum (FCS)–1× MMR, and store in a sterile 50-ml Falcon tube in an ice bucket, in about 20 ml of 10% FBS–1× MMR containing antibiotics. The testes can be stored for several days if kept on ice.

Egg Collection. Allow the female to lay a group of eggs (about 100 or so) before beginning the harvest. Wrap the female in damp cheesecloth, and grip with the forefingers, spreading and immobilizing the rear legs. Gently squeeze the abdomen to release the eggs. Allow the eggs to drop into a glass petri dish containing 20 ml of 1× MMR. Be careful not to allow any skin secretions to contaminate the eggs collected in the dish. Collect about 100 eggs per 10-cm dish. When the eggs cease (usually after 200–300), return the frog to the tank, and allow the animal to rest for 30–60 min (or until more eggs are laid) before trying collection again. *Note: Xenopus* eggs and sperm have a limited period of time to be competent for fertilization in low ionic strength buffers (e.g., 0.1× MMR); they exhibit a pronounced decrease in fertilization rate after 10

---

[27] D. Wolf and J. Hedrick, *Dev. Biol.* **25,** 348 (1971).
[28] J. Newport and M. Kirschner, *Cell* **30,** 675 (1982).

min. Sperm suspensions are most stable in 1× MMR at 0°. The eggs are also sensitive to ionic strength, but can be stored at room temperature for 1 or 2 hr in 1× MMR prior to fertilization with little loss in efficiency of fertilization.

Fertilization. Quickly rinse the eggs three times with 10 ml of 1× MMR. After the last wash, remove as much of the liquid as possible. Remove a piece of a testis, crush slightly with forceps, and then rub it over the eggs to release sperm. Allow the sperm to bind for 1 min, and then add 5–10 ml of H_2O to initiate fertilization. Incubate the eggs for 20–30 min. Remove the water and add 15 ml of 1× MMR, and place the eggs in the 22° incubator. It is important to remove obviously damaged or dying eggs as soon as possible from the dish, and to continue clearing malformed embryos as soon as possible.

Removal of Jelly Coat. Fertilization can be scored by observing the animal pole contraction and egg rotation. After the majority of the eggs exhibit signs of fertilization, the jelly coat can be removed by treating the fertilized eggs with 2% cysteine hydrochloride, pH 7.8. The jelly coat can also be removed at later times, and this is advised for albino embryos. Remove the 1× MMR solution, and add enough cysteine to cover the eggs completely (about 10 ml). Gently swirl the dish to mix and continue to agitate them occasionally for the next 5 min, until the eggs lose the jelly coat and begin closer contact. Remove the cysteine solution, rinse the eggs three times with 1× MMR, and finally add 10 ml of 1× MMR. For the embryos to undergo normal gastrulation, the ionic strength of the medium must be low. After 4–6 hr of development, remove the 1× MMR and add 10 ml of 0.1–0.2× MMR. Remove all dead and malformed embryos from the dish and incubate at 16–23°.

Embryo Transfections

Embryos at stages 24–28 are selected and the vitelline membrane should be manually removed. Embryonic heads are prepared by severing at the ear placode and removing the cement gland in 1× MMR. Twelve to 20 embryo heads are prepared in 5–10 min and processed subsequently as a group. All subsequent treatments are performed in 24-well plates. The group of heads is washed once with 1 ml of divalent-free 1× MMR, incubated in 1 ml of divalent-free 1× MMR containing 1 mM EDTA (pretreatment), and then with 1 ml of divalent free 1× MMR containing trypsin (0.5 mg/ml) (Life Technologies) and 1 mM EDTA. The time period for pretreatment and trypsinization varies from 2 to 8 min, depending on the stage, and is unique for different batches of embryos. Proceed to perform the pretreatment in trypsin–EDTA solution until the head mesenchyme is disrupted, severed margins spread, and ventricular spaces become partially exposed to solution. This can be easily monitored by observing the heads with a dissecting microscope during pretreatment. After these treatments, the heads are washed twice with 1 ml of divalent-free 1× MMR and once with 1 ml of 1× MMR. Immediately after the final wash, the heads are incubated with a fresh mixture of DNA and lipid transfection reagent in a total volume of 0.5 ml of culture medium [65% L-15 medium (Life Technologies) containing gentamicin (25 μg/ml) and 2 mM L-glutamine]. Lipofectin (Life Technologies) or 1,2-dioleoyl-3-trimethylammoniumpropane (DOTAP; (Boehringer Mannheim, Indianapolis, IN) can be used as the lipid for transfections, although DOTAP has given more consistent results. The amount of DNA used ranges from 6 to 12 μg per transfected well, with the DNA : lipid ratio being 1 : 3. DNA used for transfections can be purified either by CsCl or by the Qiagen (Chatsworth, CA) protocol. Typically, the treated heads are divided into two groups (6–10 heads each) and exposed to DNA in separate wells (referred to as a *matched pair*). Heads are incubated at 21–23° for 20 hr, the DNA-containing medium was replaced by culture medium containing penicillin (10 U/ml), streptomycin (10 μg/ml), and 10% heat-inactivated fetal bovine serum, and incubation at 21–23° continued to 78–80 hr postfertilization. The heads in each transfection well are washed three times with 1 ml each of 1× MMR, frozen in dry ice for 15 min, and stored at −70°. At the time of harvest, heads usually lack the epithelial layer (skin) and exhibit three types of morphology (Fig. 2, see color insert): (1) head with everted eyes, (2) mounds with everted eyes, or (3) mounds with pigment, but no obvious everted eyes. Transfected heads usually exhibit a variable amount of retinal pigmented epithelium and remnants from lens structures lenses. Typically, >75% of the trypsinized heads survive the treatment and transfection procedure and can be used for assaying the reporter.

Luciferase Assay. Heads are homogenized (10 μl/head) in lysis buffer and centrifuged for 15 min at 14,000 rpm, and duplicate aliquots are used to measure luciferase activity according to standard protocols (Promega, Madison, WI). The luciferase assay gives 50,000–100,000 relative light units per picogram of purified luciferase (Sigma) per 30 sec, using the single-channel luminometer (Berthold, Wildbad, Germany). Protein estimation is done with the Bradford reagent (Bio-Rad, Hercules, CA) and bovine serum albumin (BSA) standards, and ranged from 10 to 20 μg/head. Transfection experiments using two or more preparations of an individual plasmid are routinely performed with embryos from two or three female donors. Luciferase activity is reported as the mean (including all trials, given in figure legend) ± SEM. To determine relative promoter efficiency, activities from different plasmids, generated in individual transfection trials, are

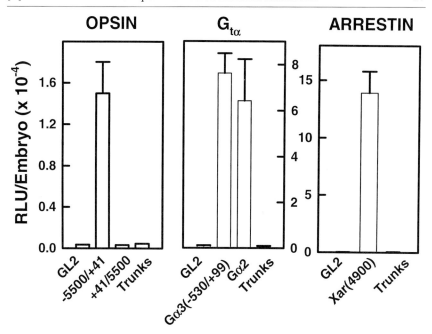

FIG. 3. Luciferase activity from rod gene promoters. Luciferase activity (RLU/embryo) from transfections in heads or trunks, using upstream sequences from the *Xenopus* rod opsin, G_t α subunit and arrestin genes, or GL2 (promoterless control). The activity measured from GL2 was not significantly different from controls without luciferase. Measurements were taken from different transfections, and represent four to eight individual experiments per construct.

compared using single-factor ANOVA ($\alpha = 0.05$; Excel software, Microsoft, Redmond, WA).

Application: Analysis of Cis-Acting Promoter Elements

With this assay, it is possible to compare activity from plasmids that contain different promoter constructs. For example, luciferase activity can be measured from transfections using 5' upstream sequences of *Xenopus* rhodopsin, rod transducin α subunit, and rod arrestin genes to drive reporter gene expression (Fig. 3). This approach permits rapid comparisons of a number of different constructs at once, and thus is quite suited for detailed promoter analysis of retinal genes. The high sensitivity of luciferase permits the assay of promoters that drive expression to varying levels. Coupled with the relatively preserved development of retina structures, this approach is an attractive alternative for the rapid characterization of *cis*-acting elements of retinal gene promoters

Transgenic *Xenopus*

Principle

The production of transgenic *Xenopus* involves the integration of transgene DNA into the sperm nuclei prior to transplantation into the egg.[14,15] The integration is carried out *in vitro* with a restriction enzyme that facilitates incorporation of linear DNA into chromosomes. To increase the efficiency of the integration, sperm nuclei are incubated with an interphase egg extract that partially decondenses the chromatin, permitting better access of the restriction enzyme and the plasmid DNA to the chromosomes. In addition, the egg extract also contains an activity that ligates the transgene into the chromatin and joins the linear transgene DNA molecules as random concatemers, which is apparently the major form of integrated DNA. Sperm nuclei treated in this manner are transplanted into mature eggs by microinjection, using a continuous flow injection apparatus. Eggs that receive single, undamaged nuclei will undergo normal cleavage pattern and can be selected by inspection. Eggs that receive damaged, multiple, or no nuclei will develop abnormally and can be easily identified. Normally developing embryos and tadpoles can then be screened for the presence of the transgene by several methods: expression of nonmosaic fluorescent reporter gene (e.g., green fluorescent protein) in living tadpoles, by polymerase chain reaction (PCR) of genomic DNA from tail segments or after eye collection by reverse transcriptase (RT)-PCR.

The major variable in the production of transgenic *Xenopus* is the quality of the eggs used in the transplantation reaction. Healthy females kept on a regular breeding schedule (approximately every 3 months) appear to produce healthy eggs consistently. However, some females kept under optimum conditions are not quality donors, and the colony should be culled to remove consistently poor producers. Feeding and care of tadpole and adult *Xenopus* have been previously described,[29] and there appears to be a wide latitude of rearing conditions. This laboratory has found that tadpoles are best kept separately after about 2 weeks at a temperature of 18–20° and reach metamorphosis in about 3 months. We have consistently obtained sexually mature transgenic animals 11–12 months posttransplantation.

There are three parts to the protocol. First is the preparation of interphase egg extract, which is stable at $-70°$ and thus can be prepared in advance of the transplantation. The second part is the preparation of the sperm nuclei, which should be done just prior to the transplantation, usually

---

[29] M. Wu and J. Gerhart, *Methods Cell Biol.* **36**, 3 (1991).

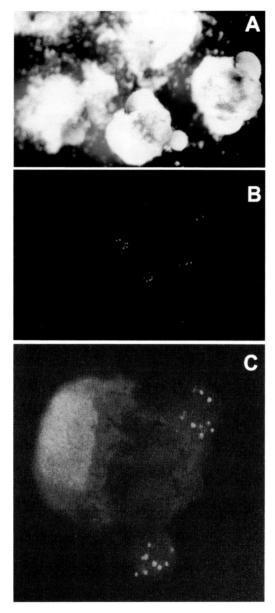

FIG. 2. Transfected embryo heads. Embryos were transfected with a plasmid containing 4.9 kb of the 5′ portion of the *Xenopus* rod arrestin gene (including intron 1) driving expression of GFP. Transfected heads were cultured for 2 days and then photographed under bright-field illumination (A), blue light with GFP emission filter set (B), or with both bright field and fluorescence (C). In (A), two intact heads with everted eyes can be seen in focus, and several mounds in the background. Green fluorescent protein (GFP) expression is observed only in the everted eye structures, where the rods are located.

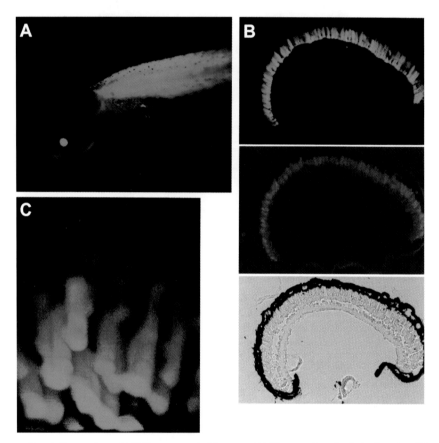

FIG. 4. Expression of GFP in transgenic *Xenopus*. (A) Transgenic tadpole (~6 weeks old) that was created with two plasmids, one containing the *Xenopus* opsin promoter[15] and the other the cardiac actin promoter,[7] both driving expression of GFP. Fluorescence can be seen in the eye and muscles of this animal. (B) Expression of GFP in retina. Bright-field image (*bottom*) of a radial cross-section through the central retina of a stage 50 transgenic tadpole generated using pXOP(−508/+41)gfp. The retinal pigment epithelium, photoreceptors, and other cells of the retina are visible. Rods and cones can be distinguished in the photoreceptor layer by the oil droplets. Fluorescence images of the same section showing GFP expression (*top panel*) limited to the rod photoreceptors, coincident with antirhodopsin fluorescence (*middle panel*). The GFP expression is found throughout the rod cell, since it is not confined to the outer segment as opsin. (C) Confocal image of rod cells, synaptic region foreground, inner and outer segments into the page. Created from 75 optical sections (5 micron thick) obtained using a living tadpole. Rod dimensions (~7 × ~20 microns).

the day before the injections. The final part is transplantation, which takes place throughout the day, as eggs are collected from the adult females.

Materials and Solutions

All stock solutions are stored at −20°. Methods are taken from Kroll and Amaya,[14] with minor modifications as described previously.[15]

Extract buffer salts (XB salts) (×20): Combine
KCl, 2 M
$MgCl_2$, 20 mM
$CaCl_2$, 2 mM

Extract buffer: Combine
XB salts, 1×
Sucrose, 50 mM
Potassium HEPES, 10 mM
and adjust to pH 7.7 with KOH

Cysteine (Sigma): 2% (w/v) in 1× XB and titrated to pH 7.8

CSF–XB: Combine
XB salts, 1×
$MgCl_2$, 1 mM
Potassium HEPES (pH 7.7), 10 mM
Sucrose, 50 mM
EGTA, 5 mM

Protease inhibitor mix (PIC): Combine all of the following at 10 mg/ml:
Leupeptin
Chymostatin
Pepstatin

Energy mix: Combine
Creatine phosphate (Boehringer Mannheim), 150 mM
ATP, 20 mM
$MgCl_2$, 20 mM

Nuclear preparation buffer (NPB), 2×:
Sucrose, 250 mM
HEPES, 15 mM (adjust to pH 7.7)
EDTA, 1 mM
Spermidine trihydrochloride (Sigma), 0.5 mM
Spermine tetrahydrochloride (Sigma), 0.2 mM
Dithiothreitol (Sigma), 1 mM

Sperm dilution buffer (SDB): Combine
  Sucrose, 250 mM
  KCl, 75 mM
  Spermidine trihydrochloride, 0.5 mM
  Spermine tetrahydrochloride, 0.2 mM
and adjust to pH 7.5 with NaOH. Store at $-20°$ in 0.5-ml aliquots

Sperm storage buffer: Combine
  NPB, 1×
  BSA, 0.3%
  Glycerol, 30% (w/v)

$CaCl_2$, 1 M
Leupeptin [10-mg/ml stock in dimethyl sulfoxide (DMSO)]
Phenylmethylsulfonyl fluoride (PMSF, 0.3 M stock in ethanol)
L-α-Lysophosphatidylcholine (10-mg/ml stock in H_2O; Sigma type I)
Bovine serum albumin [10% (w/v) stock in H_2O, fraction V; Sigma]
Hoechst No. 33342 (10-mg/ml stock in H_2O; Sigma)
$MgCl_2$, 100 mM
MMR (0.4×) plus Ficoll [6% (w/v), type 400; Sigma]
MMR (0.1×) plus 6% Ficoll

Interphase Egg Extract

This preparation is a cytoplasmic fraction that has been allowed to progress into interphase, and promotes sperm nuclei swelling and some chromatin decondensation without DNA replication. Eggs should be collected from 10 adult females as described above, gently obtained by squeezing the primed females. Eggs from individual frogs should be inspected and discarded if they contain a significant proportion of damaged or nonviable eggs, or if the eggs are obtained as long strings or clumps. The remaining, high-quality eggs are combined and washed in 1× MMR several times. The total volume of eggs prior to dejellying should be at least 100 ml per preparation. Typically it is possible to obtain 200–300 ml of eggs from a group of healthy females in 1 day. Dejelly the eggs in 2% cysteine in 1× XB salts in 50-ml centrifuge tubes, replacing the cysteine at least once during the treatment. When the eggs settle on each other, wash extensively in extract buffer to remove the cysteine. Wash the eggs twice in CSF–XB containing a 1:1000 dilution of the protease inhibitor mix (PIC). Carefully transfer the eggs to round-bottom 14-ml centrifuge tubes (Lab Products, Raleigh, NC), using a wide-bore pipette. Remove as much liquid as possible and add 1 ml of G.E. Versilube F-50 (Andpak-EMA, San Jose, CA). Pack the eggs at 4° by centrifugation at 150g for 1 min and then at 600g for an additional 30 sec [use, e.g., a Beckman (Fullerton, CA) JS13.1 swinging bucket rotor]. Remove as much liquid as possible from the eggs. Crush the eggs by centrifugation for 10 min at 10,000g in the cold, which will result in three layers: floating lipid, central cytoplasm, and lower yolk granules. Remove the cytoplasmic layer, add a 1:1000 volume of protease mix (PIC), and recentrifuge. Collect the cytoplasmic fraction, add a 1:20 volume of the energy mix, and transfer to SW50.1 (Beckman) centrifuge tubes. Add $CaCl_2$ to a final concentration of 0.4 mM to induce the cytoplasmic fraction to enter interphase and incubate for 15 min at room temperature. Centrifuge the tubes at 50,000 rpm for 2.5 hr at 4°. The fraction will separate into four layers: floating lipid, cytosol, membranes, and ribosomes at the bottom. Collect the cytosol and recentrifuge at 100,000g for 20 min at 4°. Aliquot (25–50 μl) the cytosolic extract in microcentrifuge tubes, quick freeze in dry ice, and store at −70°.

Sperm Nuclei Preparation

To prepare sperm nuclei,[14,30] remove the testes as described above and place them in a 35-mm tissue culture dish containing cold 1× MMR. Carefully dissect the testes to remove fat tissue and blood vessels, using No. 5 forceps. Rinse the testes in 1× MMR and several times in cold 1× NPB. In a clean petri dish, mince the testes thoroughly, until a fine paste is obtained. Resuspend the minced testes in 2 ml of 1× NPB by gently pipetting, using a wide-bore Pasteur pipette. Filter the sperm suspension through cheesecloth (eight layers) placed into a funnel and collect the solution into a 15-ml tube (round-bottom polypropylene tubes; Fisher, Pittsburgh, PA). Collect the remaining sperm in 1× NPB and wash the cheesecloth into the 15-ml tube, squeezing any residual liquid into the 15-ml tube. Collect the sperm by centrifugation at 3000g for 15 min at 4°. Wash the pelleted sperm with 8 ml of 1× NPB and recentrifuge. Resuspend the pellet in 1 ml of 1× NPB with a cut plastic pipette tip, warm the suspension to room temperature for 5 min, and add 50 μl of lysolecithin (10 mg/ml; Sigma). Mix gently and incubate for 5 min at room temperature. Dilute the sperm with 10 ml of cold 1× NPB–3% BSA containing leupeptin (10 μg/ml) and 0.3 mM PMSF and centrifuge at 3000g for 10 min at 4°. Carefully wash the pellet with cold 1× NPB–0.3% BSA and recentrifuge. Resuspend the sperm nuclei in 300–600 μl of 1× NPB–0.3% BSA–30% (w/v) glycerol (sperm storage buffer) and transfer the suspension to a fresh microcentrifuge tube. Make a 1:20–1:100 dilution of the sperm nuclei and add Hoechst dye to 0.1 μg/ml to determine the nuclei density. The density

---

[30] A. Murray, *Methods Cell Biol.* **36**, 581 (1991).

of the sperm nuclei is typically $5-20 \times 10^4/\mu l$ and can be concentrated if necessary by centrifugation. Sperm nuclei can be stored at 4° and used for 2 days.

Restriction Enzyme-Mediated Integration Reaction

A standard restriction enzyme-mediated integration (REMI) reaction[14] is set up as follows. Four microliters of sperm nuclei ($1-5 \times 10^5$ nuclei) and 5 μl of linearized plasmid DNA (100–300 ng/μl) are incubated in a 0.5-ml microcentrifuge tube at room temperature for 5 min. All tubes and tips must be coated to prevent adherence of the nuclei (Sigmacoat; Sigma). One microliter of restriction enzyme (diluted into 1× restriction enzyme reaction buffer), 2 μl of 100 mM $MgCl_2$, and 25 μl of egg extract are then added and incubated at room temperature for 3 min. With cut pipette tips, dilute the reaction to 100 nuclei/μl in sperm dilution buffer for microinjection.

There are a number of issues that must be considered for effective transgenesis. First is the choice of restriction enzyme. The plasmid containing the transgene to be integrated into the sperm nuclei must be linearized with a restriction enzyme that cuts outside of the transcription unit, including poly(A) addition sites and promoter elements. There does not appear to be a significant amount of nuclease digestion during the REMI procedure, and therefore loss of nucleotides at the ends of the plasmid is not a major concern. In considering potential enzymes, it is important to note that the restriction enzyme will be included in the reaction of the DNA and sperm nuclei with the egg extract. Therefore, the enzyme should be generally tolerant to salt concentrations above 100 mM. It also appears that enzymes that generate blunt ends are not as efficient as those that generate cohesive ends. The concentration of enzyme to be used in the reaction should be determined empirically, since high enzyme concentrations will lead to significant damage of the chromosomes and failure to produce viable embryos. Typically, 0.1–0.5 unit of restriction enzyme per reaction is a good starting range for testing. Second, reaction conditions may need to be adjusted depending on the preparation of egg extract. Although the activity of egg extracts can be verified by observing the decondensation of sperm nuclei, using a fluorescence microscope, quantifying the activity of a preparation is difficult. Therefore, the amount of extract and length of reaction time should be tested with each batch to optimize the production of embryos.

Nuclear Transplantation

Nuclei are transplanted into eggs with a continuous flow syringe pump (Harvard Apparatus, Hollison, MA) and 80- to 100-μm-wide glass needles (siliconized prior to pulling) connected through tygon tubing filled with mineral oil (embryo tested; Sigma), held in place by a micromanipulator. Needles are loaded by means of a cut micropipette tip fitted with a small length of flexible tubing, to which the glass needle can be attached. It is most convenient to collect eggs from primed females (see above) just prior to initiating the REMI reaction. Typically, ~500 eggs are collected and treated with 2% cysteine (in 1× MMR) to remove the jelly coat. After the cysteine treatment (which takes about 5–10 min), the eggs are washed extensively with 1× MMR and damaged eggs are removed. Batches of eggs that contain many lysed, mottled, or deformed cells are discarded. Healthy eggs are transferred to petri dishes containing 1-mm-square nylon mesh in 0.4× MMR containing 6% (w/v) Ficoll (type 400; Sigma). Usually, the eggs are permitted to stand in the Ficoll for 5–10 min prior to injection. Nuclei are transplanted into the eggs by brief (1 sec) injection into the animal pole. The flow rate can be adjusted to 10 nl/sec with an electronic controller. It is usually easiest to move the dish of eggs slowly, placing them in rows in line with the glass needle. After all of the eggs are injected, they are transferred to several petri dishes and maintained in 0.4× MMR containing 6% Ficoll at room temperature.

Selection of Embryos

After 1–2 hr, the injected eggs should be inspected. Usually a portion of the injected eggs will begin first cleavage by this time. Cleaving transplanted embryos should be gently transferred by wide-bore pipettes to a separate dish of 0.4× MMR containing 6% Ficoll. Eggs that receive damaged nuclei or more than one nucleus will exhibit abnormal cleavage patterns (multiple cleavage planes, incomplete or shallow furrows, pseudocleavages, and multiple spindle formation). Because of the large size of the injection needle and the resulting damage to the vitelline membrane, the embryos often develop blebs of tissue at the site of injection. In normally developing embryos, these will eventually fall off or can be removed manually with forceps at later stages of development. It is important to remove abnormal or damaged embryos from the selected embryos, as healthy embryos can be adversely affected. At stage 10 (or >6 hr after transplantation), the embryos should be transferred to 0.1× MMR containing 6% Ficoll. After gastrulation, the embryos can be transferred to 0.1× MMR containing gentamicin (50 μg/ml). Feeding should begin 1 week after transplantation, and consists of powdered frog brittle. Tanks should be kept free of large pieces of food and cleaned once per week.

If the transgene encodes a fluorescent vital marker such as green fluorescent protein (GFP), transgenic tadpoles can be detected by observation under a dissecting microscope fitted with fluorescence module. Usually, expression from extrachromosomal DNA will be mosaic and short-lived; thus likely transgenic animals can be selected by nonmosiac expression several weeks after transplantation. With the *Xenopus* rhodopsin promoter, GFP expression can be observed in the eye 4 days after transplantation.[15] Fluorescence from a general promoter such as CMV can be observed shortly after the midblastula transition. The characteristics of the specific promoter will determine the level and timing of expression. The transgene number and site(s) of integration can be determined by Southern analysis. Typically, the DNA integrates as concatemers (two to six copies) at several locations in any one transgenic animal, although this can be partially controlled by the *in vitro* REMI conditions (e.g., DNA concentration or 5'-end phosphorylation). The transgene expression level can also be determined after sacrifice of the animal, extraction of RNA, and RT-PCR with specific primers to detect the transgene. Thus, it is possible to compare phenotypic effects produced by the introduction of various transgenes that express at different levels by studying multiple transgenic animals, which is equivalent to studying different transgenic lines in other species.

Application

By this approach, *Xenopus* can be readily produced that express transgenes from one or more plasmids (Fig. 4A, see color insert). The efficiency of obtaining transgenic embryos ranges from 25 to 50% of the successfully transplanted nuclei. In fact, the major variable appears to be the quality of eggs used for transplantation. Typically, 1–10% of the injected eggs will develop to gastrula and ~30% will continue to develop normally. However, there is a wide variation in the yield of embryos, depending on the female donor. Transgenic *Xenopus* are well suited for the study of promoters, especially cell-specific expression. For example, a fragment of the 5' flanking region of the *Xenopus* rhodopsin gene is capable of driving high-level expression of GFP in the rods (Fig. 4B, see color insert). Moreover, the optical properties of the eye make it amenable for the observation of individual cells in the living animal by confocal microscopy (Fig. 4C, see color insert). Finally, the expression of dominant negative proteins (e.g., rhodopsin mutants) provides the opportunity to manipulate retinal function in a tractable organism.

Acknowledgment

Supported by grants from the National Eye Institute.

[5] Fusion between Retinal Rod Outer Segment Membranes and Model Membranes: Functional Assays and Role for Peripherin/rds

By KATHLEEN BOESZE-BATTAGLIA

Introduction

Membrane fusion is defined as the consolidation of two membrane bilayers and a subsequent mixing of the two previously separated aqueous compartments. Membrane fusion processes are mediated and regulated by a growing family of soluble and integral membrane proteins termed *fusion proteins*.[1,2] Fusion proteins share structural and functional properties and aid in the thermodynamically unfavorable fusion event by promoting hydrophobic interactions that favor fusion.[1-3]

Within retinal photoreceptor rod cells membrane fusion is a component step of at least three essential cellular processes. Fusion is necessary for the delivery of proteins and lipids in vesicles from the rod inner segment (site of synthesis) to the rod outer segment (ROS). Two additional fusion processes preserve the unique architecture of the outer segment by maintaining the outer segment at a constant length. The coordinated processes of disk morphogenesis and compensatory disk shedding[4] require the fusion of two opposing membranes: the fusion of two outgrowing inside-out disk rims for morphogenesis,[5] and disk–plasma membrane fusion for disk packet formation.[6] Fusion during disk packet formation is documented in microscopy studies in which an analysis of dye penetration into distinct regions of the ROS found that large molecules do not enter the narrow bands of the dye-stained region of the ROS, suggesting a fusion of the plasma membrane with the disk membranes.[7,8] This fusion is mediated by a fusion protein unique to photoreceptors: peripherin/rds. In this chapter we describe the protocols used in photoreceptor cell-free fusion assays

---

[1] N. Düzgüneş, *in* "Subcellular Biochemistry" (D. Roodyn, ed.), p. 195. Plenum Press, London, 1985.
[2] E. Pecheur, J. SainteMarie, A. Bienvenue, and D. Hoekstra, *J. Membr. Biol.* **167,** 1 (1999).
[3] J. M. White, *Science* **258,** 917 (1992).
[4] R. W. Young, *Invest. Ophthalmol. Vis. Sci.* **15,** 700 (1976).
[5] R. H. Steinberg, S. K. Fisher, and D. H. Anderson, *J. Comp. Neurol.* **190,** 501 (1980).
[6] K. Boesze-Battaglia, A. D. Albert, and P. L. Yeagle, *Biochemistry* **31,** 3733 (1992).
[7] B. Matsumoto and J. C. Besharse, *Invest. Ophthalmol. Vis. Sci.* **26,** 628 (1985).
[8] J. C. Besharse and G. Hageman, *Invest. Ophthalmol. Vis. Sci.* **31,** 57 (1990).

and the characterization of peripherin/rds as a rod cell-specific fusion protein.[6,9–13]

General Principles Involved in Design of Fusion Assays

Membrane fusion must satisfy two criteria: (1) the merger of two membrane bilayers and (2) the subsequent mixing of the aqueous contents.[14] To characterize the molecular mechanism of fusion, it is necessary to evaluate the two phenomena separately by distinct albeit complementary techniques. In biological systems, the measurement of aqueous contents mixing is often difficult or impossible to achieve and thus is often inferred from the detection of lipid mixing. Unfortunately, the formation of a fusion pore independent of aqueous contents mixing (i.e., hemifusion), can be misinterpreted as fusion.[15] The careful design and rigorous interpretation of control experiments avoid such misinterpretations. Aqueous contents mixing in some biological systems is now detectable with the development of molecular genetics techniques that rely on the cytoplasmic activation of reporter genes on fusion of two distinct cell populations.[16] Fusion processes are measured *in vitro* by a variety of techniques described elsewhere.[17,18]

The most sensitive and reliable measurements of the kinetics of a fusion process require biophysical techniques,[19] the most well characterized of which is fluorescence spectroscopy. Fluorescence-based biological fusion assays are broadly divided into two categories: probe dilution[20] and resonance energy transfer,[21] both of which measure the mixing of the lipids within the membranes. Both of these techniques have been utilized in the design of photoreceptor specific cell-free fusion assays.

[9] K. Boesze-Battaglia, F. Kong, O. P. Lamba, F. P. Stefano, and D. S. Williams, *Biochemistry* **22,** 6835 (1997).
[10] K. Boesze-Battaglia, O. P. Lamba, A. Napoli, S. Sinha, and Y. Guo, *Biochemistry* **37,** 9477 (1998).
[11] K. Boesze-Battaglia, *Invest. Ophthalmol. Vis. Sci.* **38,** 487 (1997).
[12] K. Boesze-Battaglia and P. L. Yeagle, *Invest. Ophthalmol. Vis. Sci.* **33,** 484 (1992).
[13] K. Boesze-Battaglia, F. P. Stefano, M. Fenner, and A. A. Napoli, Jr., *Biochim. Biophys. Acta,* in press (2000).
[14] E. A. Liberman and V. A. Nenashev, *Biofizika* **13,** 193 (1968).
[15] J. R. Monck and J. M. Fernandez, *Curr. Opin. Cell Biol.* **8,** 524 (1996).
[16] O. Nussbaum, C. C. Broder, and E. A. Berger, *J. Virol.* **68,** 5411 (1994).
[17] N. Düzgüneş, *Methods Enzymol.* **220,** (1993).
[18] N. Düzgüneş, *Methods Enzymol.* **221,** (1993).
[19] N. Düzgüneş, J. Wilschut, and D. Papahadjopoulos, *in* "Physical Methods in Biological Membranes and Their Model Systems" (F. Conti, W. E. Blumberg, J. DeGier, and F. Pocchiari, eds.), p. 193. Plenum, New York, 1985.
[20] P. M. Keller, S. Person, and W. Snipes, *J. Cell Sci.* **28,** 167 (1977).
[21] D. Hoekstra, T. D. Boer, K. Klappe, and J. Wilschut, *Biochemistry* **23,** 5675 (1984).

Rod Outer Segment Membrane Fusion

In practice, lipid mixing is measured by the incorporation of a fluorescent probe into the membrane bilayer and a change in fluorescence emission, accessed either through resonance energy transfer pairs, or via a release of self-quenching, on fusion of the labeled membrane with a suitable target membrane. In photoreceptors, the fusion assay used most often is based on a relief of octadecylrhodamine B chloride (R_{18}) self-quenching as two membrane bilayers mix. When the R_{18}-labeled membrane fuses with unlabeled target membrane, the lipid-like probe is effectively diluted by its subsequent lateral diffusion within the target membrane. Probe dilution is detected as a progressive linear increase in fluorescence intensity, which is proportional to the extent of fusion.[21] This technique allows kinetic and quantitative measurements of fusion between labeled membranes and both artificial and biological membranes.[19–21] The typical protocol for R_{18} labeling of bovine ROS plasma membrane and ensuing fusion follows.

Preparation and Labeling of Rod Outer Segment Plasma Membrane

Materials

Purified ROS plasma membrane vesicles
Octadecylrhodamine B chloride (R_{18}; Molecular Probes, Eugene, OR)
Sephadex G-75 (Pharmacia, Piscataway, NJ)
Column buffer: 100 mM NaCl, 10 mM glycine, 0.1 mM EDTA, pH 7.5
Calcium chelating buffer: 5 mM HEPES, 1 mM EDTA, pH 7.4

ROS disks and plasma membranes are isolated from either fresh or frozen bovine retinas.[22] The plasma membrane vesicles are purified free from ROS disk membranes by binding to ricin$_{120}$–agarose (Sigma, St. Louis, MO) and separated by continuous sucrose density gradient centrifugation.[22] The ROS plasma membrane bound to ricin–agarose is recovered as a pellet in the gradient. The plasma membrane is eluted from the ricin–agarose in a Pasteur pipette column with 1 M galactose in 0.1 M sodium borate, pH 8.0. The resulting plasma membrane vesicles are washed free of galactose (spin at 50,000 rpm for 40 min at 10°) and resuspended in calcium chelating buffer.[23] The phospholipid content of the membrane is determined[24,25] and the final concentration adjusted to 2.0 mM phosphate. ROS plasma membrane vesicles must be labeled and used within a 24-hr period; otherwise vesicles become leaky and give spurious results. Both the preparation

---

[22] K. Boesze-Battaglia and A. D. Albert, *Exp. Eye Res.* **49**, 699 (1989).
[23] S. Robertson and J. Potter, *J. Pharmacol.* **5**, 63 (1984).
[24] G. R. Bartlett, *J. Biol. Chem.* **234**, 466 (1959).
[25] B. J. Litman, *Biochemistry* **13**, 2545 (1973).

and labeling of the ROS plasma membrane are performed under dim red light.

R_{18} Labeling of Rod Outer Segment Plasma Membrane

The freshly isolated ROS plasma membrane vesicles are labeled with R_{18}.[6] A stock solution of R_{18} (10 mg/ml) is prepared in chloroform–methanol (1:1, v/v) and stored at $-20°$. An aliquot of this solution is removed, dried under a stream of nitrogen, and reconstituted in a minimal volume of ethanol. R_{18} is incorporated into the ROS plasma membrane at self-quenching concentrations, equivalent to approximately 5 mol% relative to the phospholipid content of the ROS plasma membrane. Typically, a 2-ml suspension of ROS plasma membrane (rhodopsin concentration, 1 mg/ml) is added to 10 μl of 10 nmol of R_{18}, in ethanol, under vigorous vortexing. Since the final incubation volume is 2 ml, the final ethanol concentration is less than 1% (v/v). The mixture is rocked gently at room temperature for 30–60 min in the dark.[6]

The unincorporated probe is removed by chromatography on a Sephadex G-75 column (total volume of the column, 3–4 ml). The unincorporated R_{18} adsorbs to the top of the column, and the R_{18}-labeled plasma membrane is recovered in the void volume and identified by absorbance at 214 nm. Free probe can also be removed with bovine serum albumin,[11] followed by sucrose density gradient centrifugation. In addition to being more time consuming, the bovine serum albumin (BSA) procedure may also remove membrane lipids and alter bilayer properties. For the remainder of this chapter R_{18}-labeled plasma membrane is abbreviated as R_{18}–PM. To quantitate the incorporation of R_{18} into the plasma membrane, the labeled vesicles are extracted with chloroform–methanol (2:1).[26] The fluorescence intensity of an aliquot of the extract is measured and compared with the intensity of known amounts of R_{18} in chloroform, which increase linearly up to concentrations of 1 nmol/ml.[27]

Preparation of Target Membranes

The accurate measurement of membrane fusion processes by relief of self-quenching requires that the concentration of target membrane be at least 30- to 100-fold greater than that of the R_{18}-labeled species.[28] In our experiments a surface area ratio between R_{18}–PM and target membranes of 1:50 or 1:100 yields satisfactory results, although the ratio of labeled

---

[26] J. Folch, M. Lees, and M. G. A. Sloane-Stanley, *J. Biol. Chem.* **226**, 497 (1957).
[27] D. Hoekstra and K. Klappe, *J. Virol.* **58**, 87 (1986).
[28] M. J. Clague, C. Schoch, L. Zech, and R. Blumenthal, *Biochemistry* **29**, 1303 (1990).

membrane to unlabeled membrane should be optimized for each system.[28] R_{18}–PM fuse with a variety of target membranes including ROS disk membranes, disk rim-specific vesicles, and disk lipid recombinant vesicles containing peripherin/rds. The preparation of these target vesicles is described below.

Disk Membranes. Disk membranes are routinely isolated during the preparation of plasma membrane from a continuous sucrose density gradient at 30% sucrose.[22] In our experience, 50 frozen bovine retinas routinely yield 20–25 mg of disk membrane rhodopsin. The disk membranes are washed free of sucrose and resuspended in calcium chelating buffer[23] to a rhodopsin concentration of 2.5 mg/ml for phosphate determination and diluted with chelating buffer to a final phosphate concentration of 1 mM for fusion assays.

Disk Rim-Specific Vesicles. Rim-specific vesicles have a protein content that closely mimicks the protein content of the disk rim region, i.e., enriched in peripherin/rds,[29,30] rom-1,[31] and the rim–ABC protein,[32] with negligible levels of rhodopsin. These vesicles are prepared from disk membranes isolated as described in the preceding section. The disk membranes are solubilized with octylglucoside (OG) and the solubilized mixture subjected to concanavalin A affinity chromatography.[33,34] The unbound fraction from the concanavalin A column[9] contains total disk lipids and the peripherin/rds and rom-1, which do not bind to the column. This unbound fraction is collected and concentrated to a final volume of 5–8 ml, using an Amicon (Danvers, MA) concentrator (YM10 filter). Disk rim-specific vesicles form spontaneously after the removal of OG by dialysis for 24–48 hr against 1 M NaCl, 10 mM HEPES, pH 7.4, with two changes of buffer. The rim-specific vesicles are not used for fusion unless the residual OG concentration (determined as described[35]) is less than 0.05 mol% relative to phospholipid. If the OG concentration is higher, the vesicles are dialyzed for an additional 24 hr in the presence of SM-2 BioBeads (Bio-Rad, Hercules, CA). After dialysis, the vesicles are subject to five freeze–thaw cycles [liquid nitrogen (freeze)/room temperature (thaw)]. The volume of vesicles is adjusted with calcium chelating buffer to a final phospholipid concentration of 1 mM.

Peripherin/rds Recombinants. Peripherin/rds is purified from bovine retinas by a combination of concanavalin A affinity chromatography and

---

[29] G. Connell and R. S. Molday, *Biochemistry* **29**, 4691 (1990).
[30] K. Arikawa, L. Molday, R. S. Molday, and D. S. Williams, *J. Cell. Biol.* **116**, 659 (1992).
[31] O. L. Mortiz and R. S. Molday, *Invest. Ophthalmol. Vis. Sci.* **37**, 352 (1996).
[32] M. Illing, L. Molday, and R. S. Molday, *J. Biol. Chem.* **272**, 10303 (1997).
[33] A. J. Adams, M. Tanaka, and H. Shichi, *Exp. Eye Res.* **27**, 595 (1978).
[34] B. J. Litman, *Methods Enzymol.* **81**, 150 (1982).
[35] M. A. Jermyn, *Anal. Biochem.* **68**, 332 (1975).

chromatofocusing techniques.[9] Nonphosphorylated peripherin/rds is isolated from the chromatofocusing column at its pI of 4.7, and phosphorylated peripherin/rds at a pI of 4.21. Peripherin/rds and phosphoperipherin/rds recombinants are prepared by detergent dialysis.[36] In the preparation of peripherin/rds recombinants, the purified protein is recombined with vesicles prepared from extracted disk membrane lipids.[12] Since retinal has been shown to induce lipid-mediated fusion in photoreceptors, we reduce the retinal Schiff-base linkage with $NaCNBH_3$,[37] thereby eliminating any retinal-induced effects on fusion. To prepare disk lipid small unilamellar vesicles (SUVs) for recombination, freshly prepared 2 M $NaCNBH_3$ in 1 M acetic acid is added to freshly isolated disk membranes in a 2:1 (v/v) ratio in the dark and allowed to sit at room temperature for 45 min. The treated disk membranes are washed and resuspended in 10 mM HEPES, pH 7.4, and extracted with 2:1 chloroform–methanol.[26] The extracted lipids are dried under a stream of nitrogen, lyophilized overnight, and resuspended in 10 mM HEPES, pH 7.4. SUVs are prepared from the liposomes.[38] If necessary, the amount of residual retinal(ol) can be quantitated by high-performance liquid chromatography (HPLC).[39]

The disk lipid SUVs are recombined with purified peripherin/rds, isolated by chromatofocusing. The peripherin/rds-containing fractions are combined and the pH adjusted to 7.0 with 0.1 M imidazole, pH 8.0. The sample is concentrated to one-tenth its original volume with a Centricon 30 filter device (30,000 MW cutoff filter). The ultrafiltration device is centrifuged at 7500g at 4°, time variable, using a fixed-angle rotor. The concentrated purified peripherin/rds in 20–30 mM OG is added to disk lipid SUVs with vigorous vortexing and the mixture placed in an ice bath for 1 hr. The resulting solution should be cloudy, indicating that the OG concentration in the recombinant is below the critical micelle concentration of the detergent. Routinely, a ratio of 1 mol of peripherin/rds to 100–300 mol of phospholipid is used. The recombinant is then dialyzed against 10 mM HEPES, 0.5 M NaCl, pH 7.4, for 24–48 hr, with three changes of buffer, with the final buffer change containing SM-2 BioBeads (Bio-Rad). Under these experimental conditions, less than 0.05% OG remains associated with the recombinants. The recombinant-containing peripherin/rds in the disk lipid vesicles is puri-

---

[36] M. Jackson and B. J. Litman, *Biochemistry* **21**, 5601 (1982).

[37] R. S. Fager, P. Sejnowski, and E. W. Abrahamson, *Biochem. Biophys. Res. Commun.* **47**, 1244 (1972).

[38] Y. Barenholtz, D. Gibbs, B. J. Litman, T. Thompson, and F. D. Carlson, *Biochemistry* **16**, 2806 (1977).

[39] K. Boesze-Battaglia, S. J. Fliesler, L. Jun, J. E. Young, and P. L. Yeagle, *Biochim. Biophys. Acta* **1111**, 256 (1992).

fied on a continuous 10–40% (w/w) sucrose density gradient, spun at 25,000 rpm for 8 hr at 4°. The recombinant is routinely isolated as a single band at approximately 20–30% sucrose, depending on the final phospholipid/protein ratio. Peripherin/rds in the recombinant is oriented asymmetrically, such that all of the peripherin/rds molecules are inserted with their extradiskal side out. This orientation is confirmed by immunofluorescence with anti-peripherin/rds monoclonal antibody 2B6 to the C terminus (generously provided by R. Molday[29]) or by trypsinolysis.[9,10] The peripherin/rds recombinants are subject to five freeze–thaw cycles: rapid freezing in liquid nitrogen followed by thawing at room temperature to form large unilamellar vesicles called peripherin/rds LUVs. Disk lipid vesicles containing no peripherin/rds or rhodopsin recombinants, which do not to fuse with R_{18}–PM, are used as controls.[9,12,36]

R_{18} Fusion Assay

Fusion between R_{18}–PM and target membranes is detected as an increase in fluorescence intensity at 586 nm on dilution of R_{18} throughout the target membrane. In designing these assays it is essential that the amount of lipid in the R_{18}–PM be 50-fold lower than the amount of phospholipid in the target membranes. Therefore, we routinely adjust the phospholipid concentration of the target membrane to 1 mM. The assays are performed in a total volume of either 3 or 1 ml. The calcium concentration of the target membrane is adjusted by the addition of $CaCl_2$ to the cuvette prior to or simultaneously with the initiation of fusion. Fluorescence and light scattering are monitored with λ_{ex} 560 nm and λ_{em} 586 nm on a Perkin-Elmer (Norwalk, CT) LS50 B spectrofluorimeter or equivalent.

The target membranes are equilibrated to the appropriate temperature while stirring in a thermostatted cuvette turret. Fusion is initiated by the addition of 50 μl of R_{18}–PM to this suspension of target membrane. The change in R_{18} fluorescence is monitored continuously and increases linearly as the probe dilutes into the target membrane (see Fig. 1). The increase in observed fluorescence intensity is proportional to the membrane fusion.[6,21] The fluorescence intensity of the target membrane without addition of plasma membrane is taken as baseline fluorescence and is equal to 0% fusion. Fluorescence at 100%, equal to 100% fusion, is determined by the addition of 100 μl of 10% Triton X-100 to the membrane mixture. Triton X-100 does not interfere with rhodamine fluorescence and the only correction factor required is that due to dilution. The initial rate of fusion (IRF) is determined as the percent increase in fluorescence intensity as a function of time[21] as shown in Eq. (1).

$$\text{IRF} = [\text{slope } R_{18}/(F_0 - F_f)] \times 100 \tag{1}$$

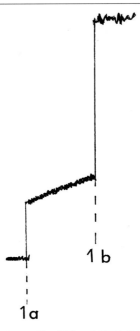

FIG. 1. Membrane fusion between R_{18}–PM and ROS disk membranes. A representative tracing of the change in fluorescence observed with the R_{18} lipid mixing assay, at 37°, is shown. (a) Fusion is initiated with the simultaneous addition of R_{18}–PM and calcium ($[Ca^{2+}]_{free}$ = 45 mM) to the disk membranes. (b) The addition of Triton X-100. [Reprinted with permission from K. Boesze-Battaglia, A. D. Albert, and P. L. Yeagle, *Biochemistry* **31**, 3733 (1992); American Chemical Society.]

where slope R_{18} is equal to the linear change in R_{18} fluorescence over time, F_0 is the fluorescence intensity at baseline (i.e., in the absence of labeled R_{18}–PM), and F_f is the fluorescence intensity on the addition of Triton X-100. If a lag time is observed the slope R_{18} is determined only after the lag phase.

Analysis of Fluorescence Data

The parameters obtained from the fluorescence data used to characterize the fusion process are as follows: (1) the initial rate of fusion, determined from the slope of the fluorescence increase (Figs. 1 and 2), (2) the extent of fusion, determined on an overnight incubation between R_{18}–PM and target membranes, and (3), in some cases, the lag time prior to the initiation of fusion (Fig. 2).

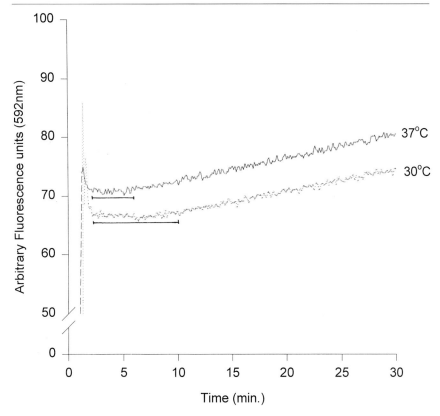

FIG. 2. Membrane fusion between R_{18}–PM and peripherin/rds recombinants. A representative tracing of the increase in fluorescence intensity of R_{18} at 592 nm over time (minutes) on the addition of R_{18}–PM (at 15 sec) to peripherin/rds recombinants is shown at 30 and 37°. The horizontal solid lines indicate the lag times. The initial rate of fusion is calculated from the slope of the change in fluorescence intensity after the lag time indicated. [Reprinted with permission from K. Boesze-Battaglia, O. P. Lamba, A. Napoli, S. Sinha, and Y. Guo, *Biochemistry* **37,** 9477 (1998); American Chemical Society.]

Characteristics of Fusion between R_{18}-Labeled Plasma Membrane and Target Membranes

Fusion between R_{18}–PM and the target membranes share common characteristics: fusion increases linearly over time, requires nanomolar levels of calcium, and is completely inhibited by the addition of 100 µl of 1 *M* EGTA. Fusion is temperature dependent, with the highest initial rates at 37° and almost no detectable fusion above 45° or below 30°.

TABLE I
COMPARISON OF R_{18}-LABELED PLASMA MEMBRANE FUSION WITH VARIOUS TARGET MEMBRANES[a]

Target membrane	Addition	Lag time (sec)	IRF (%) = ΔF/time
Disks	10 μM Ca^{2+}	<20	2.9 ± 0.04
Rim-specific vesicles	10 μM Ca^{2+}	20–50	0.26 ± 0.02
Peripherin/rds LUVs	10 μM Ca^{2+}	120	0.30 ± 0.03

[a] Fusion of R_{18}–PMs with disk membranes,[12] isolated rim-specific vesicles, and peripherin/rds recombinant LUVs[23] at 37°. The initial rate of fusion is given as the percent change in fluorescence intensity per unit time (minutes with disk membranes and seconds with rim-specific vesicles and peripherin/rds LUVs).

Disk Membranes. A representative tracing of the increase in fluorescence intensity observed upon the fusion of R_{18}–PM with isolated disk membranes[6] is shown in Fig. 1. Fusion is initiated by the simultaneous addition (Fig. 1, 1a) of R_{18}–PM and calcium to a suspension of disk membranes at 37°. Maximal fusion is determined by the addition of Triton X-100 (Fig. 1, 1b). The slope of the fluorescence increase is used to calculate the initial rate of fusion. The fusion between disk membranes and R_{18}–PM can be inhibited by modifications of endogenous peripherin/rds, such as trypsinolysis of the disk membranes,[9,10] the addition of anti-peripherin monoclonal antibody 2B6,[9] or the addition of low concentrations of the peptide PP-5,[10] corresponding to amino acids 311 to 325 of bovine peripherin/rds.

Disk Rim-Specific Vesicles or Peripherin/rds Recombinants. When R_{18}–PM fuse with either rim-specific vesicles or peripherin/rds recombinants, a qualitatively different fluorescence pattern is observed as shown in Fig 2. When fusion of R_{18}–PM with peripherin/rds recombinants is measured at 37° a lag phase lasting 120 sec occurs, during which time little or no fusion is observed.[10] This lag time increases to almost 300 sec when the temperature is lowered to 30°. Similar results are obtained with rim-specific vesicles, with the exception that the lag phase is shorter in duration; for example, at 37°, the lag phase is 60–100 sec. A comparison of these parameters is given in Table I.

To ensure that the observed lag phase results from fusion-related events, both the temperature dependence and the effect of proteolysis or phospholipase treatment on fusion should be analyzed. A lag in the onset of fusion should increase when the temperature is lowered. If the fusion is protein mediated, the lag phase would also change in duration on proteolysis of the proposed fusion protein. Conversely, if the fusion is lipid mediated then any alteration in lipid properties after phospholipase treatment would affect a true lag phase.

Critical Notes Regarding R_{18} Dequenching Assay

The use of endogenous probes for the measurement of membrane fusion requires that a number of critical factors be considered and controls designed to account for specious results. Under some experimental conditions, a spontaneous transfer of R_{18} between membranes can occur in the absence of membrane merging and fusion.[40] Such transfer has not been seen to any appreciable extent with R_{18}–PM, when these membranes are incubated with phosphatidylcholine LUVs.[6] In our experience, spontaneous transfer of R_{18} from plasma membrane vesicles to phosphatidylcholine SUVs or LUVs is indicative of permeable vesicles due to prolonged storage (longer than 24 hr).

Fusion should also be confirmed by alternative techniques, i.e., sucrose density gradient centrifugation (allows the separation of fused from unfused species[11,12]) or microscopy.[17,18] Alternatively, the addition of high concentrations of unlabeled plasma membrane should inhibit R_{18} dequenching, indicating that the increase in fluorescence was due to the merger of two membranes. Conversely, binding of R_{18}-labeled membranes to the target membrane should result in no dequenching or in a nonlinear change in R_{18} fluorescence. Investigators are strongly encouraged, when possible, to compare different fusion assays in the same system and to characterize known fusion properties of the membranes (i.e., specific inhibitors, temperature dependence, cation dependence, membrane specificity).

Characterization of Molecular Basis of Peripherin/rds-Mediated Fusion

Peripherin/rds is the first rod cell-specific fusion protein to be identified. Fusion is mediated through at least one region of peripherin/rds, from residues 311 to 325. A peptide analog to this region, called PP-5, promotes membrane adhesion and membrane destabilization,[9,10,13] two prerequisite steps for fusion.[41]

An initial strategy in the identification of a candidate fusion protein and subsequent characterization of a fusogenic region within such a protein relies on proteolysis studies. Trypsinolysis of peripherin/rds led to the

---

[40] A. Loyter, V. Citovsky, and R. Blumenthal, *Methods Biochem. Anal.* **33**, 128 (1988).
[41] J. Bentz and H. Ellens, *Colloids Surf.* **30**, 65 (1988).

identification of a fusion-promoting region.[9,10] The target membranes are incubated with trypsin (final concentration, 0.2 μg/ml) for 30 min in the dark at 37°. The reaction is stopped by the addition of trypsin inhibitor (final concentration, 0.4 mg/ml), and the samples are washed in 10 mM HEPES, pH 7.2, and centrifuged at 60,000 rpm for 40 min. The pellet is resuspended in 10 mM HEPES, pH 7.2, for the fusion assay; the supernatant containing the tryptic fragments is reserved on ice. Trypsinolysis of disk membranes and peripherin/rds recombinants under these conditions results in the cleavage of a 12.5-kDa band immunoreactive with anti-peripherin/rds antibody 2B6, which was found to correspond to the C terminus of peripherin/rds. Cleavage of this peptide fragment inhibited fusion of disks and peripherin/rds recombinants with R_{18}–PM.[10] Fusion activity is restored with the subsequent addition of the tryptic fragments at a ratio equal to that in the native membrane.[10] The size of the tryptic fragments is determined by gradient gel electrophoresis[42] and/or HPLC.[43] Such experiments provide evidence that the tryptic fragments contain a fusion-promoting region of peripherin/rds that retained biological activity, as determined on the basis of the restoration of fusion activity, and led to the synthesis of overlapping C-terminal peptides (Table II).

Peptide-Induced Aqueous Contents Mixing as Measure of Fusion

The fusion-promoting activity of tryptic fragments and synthetic peptides is evaluated directly in an assay measuring the mixing of the aqueous contents of model membranes. This approach allows the investigator to determine if a peptide satisfies both criteria for fusion, i.e., membrane lipid and aqueous contents mixing. A fluorescence-based contents mixing assay using 8-aminonaphthalene-1,3,6-trisulfonic acid and *p*-xylene bispyridinium bromide (ANTS and DPX, respectively) has been chosen since these membranes are not biologically active. This assay is based on the principle that the fluorescence of one probe, ANTS, is quenched by a second probe, DPX; such probes are called *quenching pairs*.[44] The two fluorophores are encapsulated separately into two different populations of vesicles. On fusion and contents mixing a collisional quenching of the ANTS occurs and is detected as a decrease in fluorescence.[44,45]

---

[42] H. Schagger and G. von Jagow, *Anal. Biochem.* **166,** 368 (1987).
[43] J. H. McDowell, J. P. Nawrocki, and P. A. Hargrave, *Biochemistry* **32,** 4968 (1993).
[44] H. Ellens, J. Bentz, and F. C. Szoka, *Biochemistry* **23,** 1532 (1984).
[45] F. Szoka, F. Olson, T. Heath, W. Vail, E. Mayhew, and D. Papahadjopoulos, *Biochim. Biophys. Acta* **601,** 559 (1980).

TABLE II
ALIGNED AMINO ACID SEQUENCE OF SYNTHETIC C-TERMINAL PERIPHERIN/rds PPETIDES[a]

Peptide	Amino acid sequence
PP-3	V-E-A-E-G-E-D-A-G-Q-A-P-A-A-G
PP-4	S-V-K-K-L-G-K-G-N-Q-V-E-A
PP-5	V-P-E-T-W-K-A-F-L-E-S-V-K-K-L
PP-6	L-K-S-V-P-E-T-W-K-A-F-L
PP-7	W-K-A-F-L
Residue No.	H$_2$N-308------311-------315--------321------------331--------------345-COOH

[a] Left to right.

Materials

ANTS: 8-Aminonaphthalene-1,3,6-trisulfonic acid, disodium salt (Molecular Probes); store in dark

DPX: *p*-Xylene bispyridium bromide (Molecular Probes); store in dark

PS: Phosphatidylserine (bovine brain; Avanti Polar Lipids, Alabaster, AL)

N-Methyl-DOPE: N-Monomethyldioleoylphosphatidylethanolamine (Avanti Polar Lipids)

ANTS buffer: 10 mM Glycine, 45 mM NaCl, 25 mM ANTS, pH 9.5

DPX buffer: 10 mM Glycine, 90 mM DPX, pH 9.5

Preparation of ANTS- or DPX-Containing Vesicles

Large unilamellar vesicles (LUVs) encapsulating either ANTS or DPX are prepared from 95 mol% N-methyl-DOPE and 5 mol% PS (bovine brain).[45,46] The dried lipids are resuspended in buffer containing either ANTS or DPX, subjected to five freeze–thaw cycles in liquid nitrogen, and extruded through a polycarbonate filter (Poretics, Pleasanton, CA; 467 mm, pore size 0.1 μm), using a Nuclepore (Pleasanton, CA) 47-mm in-line filter holder.[45] Encapsulated fluorescent probe is separated from unincorporated probe on a Sephadex G-50 column eluted with 100 mM NaCl, 10 mM glycine, and 0.1 mM EDTA, pH 9.5. The vesicles containing the encapsulated probe elute with the void volume. Both the ANTS-containing and the DPX-containing vesicles are resuspended in 10 mM glycine, pH 9.5, to a final 1 mM phosphate concentration. The pH of the solutions must be accurate since N-monomethyl-DOPE will form nonbilayer structures below pH 8.0, resulting in vesicle aggregation.

ANTS–DPX Contents Mixing Assay

Fluorescence intensity is monitored at λ_{ex} 380 nm and λ_{em} 510 nm. In a 3-ml fusion reaction mixture, the thermostatted cuvette contains 30 μl of ANTS-containing vesicles, 270 μl of DPX-containing vesicles, and 2.7 ml of column buffer (100 mM NaCl, 10 mM glycine, and 0.1 mM EDTA, pH 9.5). Fusion is initiated by lowering the pH of the fusion mixture from 9.5 to 4.5 (Fig. 3a) by the addition of 2 M sodium acetate–acetic acid (25–50 μl). Fusion is measured as a decrease in fluorescence intensity as the aqueous contents are mixed and an ANTS–DPX complex is formed. Baseline fluorescence is the intensity obtained with the shutters closed. The 100% fluorescence level is the initial fluorescence intensity before

[46] H. Ellens, J. Bentz, and F. C. Szoka, *Biochemistry* **24**, 3099 (1985).

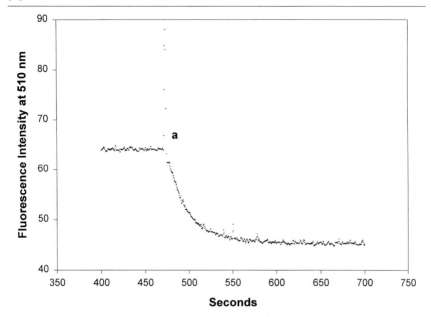

FIG. 3. Membrane fusion detected by aqueous contents mixing of ANTS- and DPX-containing vesicles. Shown is a representative tracing of the decrease in fluorescence at 510 nm over time (seconds) as the pH of the fusion mixture (ANTS vesicles and DPX vesicles) is lowered with 2 M sodium acetate (a) at 37°.

lowering the pH. The IRF between ANTS- and DPX-containing vesicles is calculated as shown in Eq. (2)[47]:

$$\text{IRF} = [\text{slope}_{\text{ANTS/DPX}}/(F_f - F_0)] \times 100 \qquad (2)$$

where $\text{slope}_{\text{ANTS/DPX}}$ is equal to the linear change in ANTS fluorescence over time, F_0 is the fluorescence intensity prior to the initiation of fusion with acetic acid and F_f is the fluorescence intensity with the shutters closed.

Effect of Peripherin/rds Peptides on ANTS–DPX Fusion

To determine if the various synthetic peptides to the C terminus of peripherin/rds are fusogenic, the ANTS–DPX fusion mixture is preincubated with the peptide at the desired temperature prior to the initiation of fusion. Routinely, the effect of a single peptide concentration on fusion is studied within a temperature range of 28 to 42°. The peptide concentrations used are in the millimolar to nanomolar range. Such studies, while necessary

[47] D. Kelsey, T. Flanagan, J. E. Young, and P. L. Yeagle, *Virology* **182,** 690 (1991).

to describe a mechanistic pathway, should be performed to complement the fusion assays described above.

Pitfalls

ANTS–DPX Fusion. Transient perturbations, analogous to hemifusion, can be distinguished from fusion by ANTS–DPX leakage assays.[48] The ANTS–DPX assay cannot be used to measure aqueous contents mixing of SUVs because ANTS binds excessively to such vesicles.[49]

Synthetic Peptides. To confirm the specificity of a peptide-induced fusion, a scrambled peptide (same amino acids in a different sequence), a fusogenic peptide from a different system, e.g., a viral fusion peptide, or a nonspecific peptide at high concentration should be tested. The fusogenic peptide should produce an effect at physiologically relevant peptide concentrations. Peptide-induced fusion in a model membrane system should affect fusion in the biological cell-free assay system. In photoreceptors, PP-5 inhibits fusion between disk and R_{18}–PM when added in a 1 : 1 ratio with peripherin/rds.[10] The secondary structure of the peptide should be evaluated in the fusion mixture and an effort made to denature this structure and show that the denatured species is unable to induce fusion.

Peripherin/rds Mutagenesis

Fusion-promoting region(s) of peripherin/rds can also be identified by mutagenesis. Peripherin/rds mutagenesis studies require a viable cell expression system in which the fusion competence of the cells or of the proteins harvested from the cells can be determined. The protocol for the determination of this fusion competency in Madin–Darby canine kidney (MDCK) cells is given below. A similar strategy can potentially be employed to characterize other cell expression systems.

Establishment of Fusion Competency of MDCK Cells Expressing Human Peripherin/rds

MDCK strain II cells stably transfected with human peripherin/rds with an Xpress N-terminal tag[50] have been generously provided by R. Kim. The peripherin/rds is localized to the basolateral MDCK plasma membrane, and oriented with the N-terminal epitope tag and the C terminus on the

---

[48] J. Zhao, S. Kimura, and Y. Imanishi, *Biochim. Biphys. Acta* **1283**, 37 (1996).
[49] N. Düzgüneş and J. Bentz, in "Spectroscopic Membrane Probes" (L. M. Loew, ed.), Vol. 1, p.117. CRC Press, Boca Raton, Florida, 1988.
[50] R. Y. Kim, *Invest. Ophthalmol. Vis. Sci.* **39**, S677 (1998).

extracellular surface.[50] MDCK cells expressing human peripherin/rds are grown to confluence in Dulbecco's modified Eagle's medium (DMEM), supplemented with 10% fetal bovine calf serum, and isolated by scraping. The cells cannot be harvested with trypsin, since trypsinolysis hydrolyzes the fusogenic domain of peripherin/rds. The cells are resuspended in 145 mM NaCl, 5 mM KCl, 10 mM HEPES, pH 7.4. Fusion between whole MDCK cells and R_{18}–PM is measured as described above. On the addition of ROS R_{18}–PM to peripherin/rds-expressing MDCK cells at 37° an increase in R_{18} fluorescence was observed. The initial rate of fusion under these conditions is 0.35 ± 0.040 ΔF(fluorescence)/sec. Fusion between R_{18}–PM and peripherin/rds-expressing cells is inhibited by the addition of 1.0 mM EGTA and by trypsinolysis of the MDCK cell plasma membranes. Fusion is not detected with R_{18}–PM and MDCK cells containing no peripherin/rds. The R_{18} fluorescence profile seen in these assays is qualitatively similar to that shown in Fig. 2, with a lag time of 60–90 sec.

Resonance Energy Transfer-Based Lipid Mixing Assay

The R_{18} dequenching assay described above requires large quantities of MDCK cells and may not be usable if fewer cells are available. An alternative technique requiring less material and relying on resonance energy transfer (RET) may be used. Detection of lipid mixing by RET requires that two membrane preparations be labeled with a fluorescent probe at non-self-quenching concentrations. In the RET assay, pairs of fluorophores are used in which one fluorophore, designated the energy donor, has an emission band that overlaps with the excitation band of the second fluorophore, called the energy acceptor.[51,52] When these two fluorophores are in close proximity, as would occur in membrane bilayers on fusion and lipid mixing, there is a transfer of the excited state energy from a donor to an acceptor.[53] Thus the acceptor fluorophore fluoresces as though it has been excited directly. A large number of lipid probes satisfies the criteria for efficient energy transfer, and they are called energy transfer couples.[18] RET between two fluorophores has a number of advantages[51–53] when compared with the R_{18} lipid mixing assay for use in the MDCK cell system: (1) The RET method is extremely sensitive, requiring only small quantities of target membrane since the ratio of the two membranes undergoing fusion is equal; and (2) RET assays are adapted to be performed in

---

[51] L. Stryer, *Annu. Rev. Biochem.* **47**, 819 (1978).
[52] T. Z. Forster, *Naturforsch. A* **4A**, 321 (1949).
[53] D. Hoekstra, *Biochemistry* **21**, 1055 (1982).

96-well plates, using a fluorescence plate reader in a total fusion reaction volume of 200 μl.[54]

Two approaches have been described in using RET techniques. In the first, a single membrane species is labeled with two fluorophores[55]; in the second, the fluorophores are incorporated into two separate membrane species and fusion is detected as an increase in fluorescence intensity of the acceptor fluorophore.[54,56] In the MDCK cell system we have employed the latter approach, using R_{18}–PM and F_{18}-labeled MDCK cell plasma membranes, with a fluorescence plate reader at room temperature.[54,56]

Labeling of MDCK Cells and Rod Outer Segment Plasma Membrane with Fluorescent Membrane Probes

Purified ROS plasma membrane vesicles are labeled with R_{18} exactly as described above, with the exception that the concentration of R_{18} incorporated is equal to 0.8 to 1.0% relative to the membrane phospholipid. Unincorporated R_{18} is removed by chromatography on a Sephadex G-75 column (total volume of the column, 1–2 ml).

The stably transfected MDCK cells expressing human peripherin/rds are grown to confluence and isolated by scraping. The cells are resuspended in 145 mM NaCl, 5 mM KCl, 10 mM HEPES, pH 7.4, and labeled with 5-(N-octadecanoyl)aminofluorescein (F_{18}), following the same protocol as described above for R_{18}-labeled plasma membrane. The F_{18} is added at 0.80–1.0 mol% relative to phospholipid in the MDCK cells. Routinely, phospholipid phosphate[24,25] is determined for each experiment and correlated to cell number. The F_{18} is removed by chromatography on a Sephadex G-75 column (total volume of the column, 1–2 ml).

R_{18}–F_{18} Fusion Assay

Fusion between R_{18}-labeled plasma membrane and F_{18}-labeled MDCK cells is measured on a Perkin-Elmer LS 50B spectrofluorimeter equipped with a fluorescence plate reader (model L225 0137). Fusion is initiated by the addition of R_{18}–PM to F_{18}-labeled MDCK cells present in the 96-well

---

[54] M. A. Partearroyo, E. Cabezon, J. L. Nieva, A. Alonso, and R. M. Foni, *Biochim. Biophys. Acta* **1189**, 175 (1994).

[55] A. J. M. Driessen, D. Hoekstra, G. Scherphof, R. D. Kalicharan, and J. Wilschut, *J. Biol. Chem.* **260**, 10880 (1985).

[56] V. Litwin, K. Nagashima, A. M. Ryder, C. Chun-Huey, J. M. Carver, W. C. Olson, M. Alizon, K. Hasel, P. J. Maddon, and G. P. Alloway, *J. Virol.* **70**, 6437 (1996).

plates. The change in fluorescence intensity is measured continuously for 2–10 min with λ_{ex} 470 nm (F_{18} excitation) and at λ_{em} 524 nm (emission wavelength F_{18}) and λ_{em} 592 nm (emission wavelength R_{18}). The fluorescence intensity obtained with double-labeled plasma membrane is taken as maximal fusion. The extent of fusion (R) is calculated as shown in Eq. (3)[54]:

$$R = [(I_{592}/I_{524}) + I_{592}]_t - [(I_{592}/I_{524}) + I_{592}]_c \qquad (3)$$

where I_{524} and I_{592} are the fluorescence intensities at 524 and 592 nm, respectively. The subscripts c and t represent the initial time point and any time thereafter, respectively. On fusion, fluorescence intensity increases at 592 nm and decreases at 524 nm, with the λ_{ex} of 470 nm. To confirm that resonance energy transfer is occurring the emission scan of R_{18} from 500 to 680 nm is recorded. This assay may be performed on a larger scale in a thermostatted cuvette or on coverslips.[56]

Advantages of MDCK Cell Line

Peripherin/rds expressed in COS-1 cells is localized to the intracellular membranes and is abnormally glycosylated.[57] For studying fusion processes the MDCK cell line has some important advantages over COS-1 cells: (1) In MDCK cells, peripherin/rds is also expressed on the plasma membrane surface, thus allowing whole cells to be used in fusion assays and easier manipulation of the membranes; (2) MDCK cells are stably transfected with human peripherin/rds, allowing a direct correlation between the fusion-promoting activity of peripherin/rds and human peripherin/rds disease-linked mutations; and (3) the human peripherin/rds expressed in MDCK cells is glycosylated and forms dimers as occurs in native photoreceptors.[50]

Pitfalls of Expression Systems

COS-1 and MDCK cells allow production of large quantities of peripherin/rds and peripherin/rds mutants for recombination and subsequent fusion studies. In addition to the commonly encountered shortcomings of cell expression systems, localization of peripherin/rds to the intracellular membranes of COS-1 cells may result in artifactual characterization of peripherin/rds-mediated fusion processes owing to the presence of endogenous proteins involved in other fusion processes. For example, in COS-1 cells peripherin/rds is localized to the intracellular membranes,

---

[57] A. F. X. Goldberg, L. Moritz, and R. S. Molday, *Biochemistry* **34**, 14213 (1995).

which also contain a family of intracellular fusion proteins[58,59] that may inhibit or enhance peripherin/rds-mediated fusion. In lieu of another expression system, such shortcomings can be overcome by characterizing in detail the specificity of the fusion pathway. Although the overexpression of the fusion protein may be essential for the isolation of large quantities of protein, membrane fusion processes dependent on this protein may actually be inhibited by excessive amounts of expressed protein. This possibility becomes more likely if a specific oligomeric form of the protein is required for fusion competency.

Pitfalls of Resonance Energy Transfer Assays

To identify and exclude extensive membrane aggregation as a contributor to energy transfer, EGTA or EDTA, known to decrease peripherin/rds-promoted vesicle aggregation, would selectively decrease RET efficiency due to aggregation. In pure liposome preparations, changes in the behavior of the membrane lipids may alter the quantum efficiency of various fluorophores, with changes in fluorescence observed due to factors other than fusion.[60]

Analysis of Individual Steps Contributing to Membrane Fusion

Once a region of a candidate fusion protein has been shown to promote fusion, the ability of this region to mediate the individual steps in fusion can be evaluated, taking into consideration the four steps of membrane fusion[41]: (1) aggregation of the fusing membranes, (2) close approach of the lipid bilayers, (3) destabilization of the membrane bilayer at the point of fusion (two bilayers that are closely opposed may not necessarily fuse), and (4) mixing of the components of the lipid bilayer and of the aqueous contents, to form a new structure. The protocols below describe how two mechanistic criteria necessary for fusion, membrane aggregation and destabilization,[61] were met for PP-5.

Membrane Aggregation–Adhesion

The Ca^{2+}-dependent aggregation of SUVs or LUVs can be monitored continuously as a change in absorbance at 380 nm in a Perkin-Elmer 2.0 UV/Vis spectrometer or equivalent equipped with a thermostatted cu-

---

[58] M. Liniai, *J. Neurochem.* **69**, 1781 (1997).
[59] T. Sollner, S. W. Whiteheart, M. Brunner, H. Erdjument-Bromage, S. Geromanos, P. Tempst, and J. E. Rothman, *Nature (London)* **362**, 318 (1993).
[60] D. Hoekstra and N. Düzgüneş, *Methods Enzymol.* **220**, 15 (1993).
[61] Y. Kliger, A. Aharon, D. Rapaport, P. Jones, R. Blumenthal, and Y. Shai, *J. Biol. Chem.* **272**, 13496 (1997).

vette holder. As the vesicles aggregate and become larger an increase in scattered light is detected as a change in absorbance, recorded every 0.1 min for 10 min.[23] The SUVs are composed of phosphatidylcholine–phosphatidylserine–cholesterol (PC:PS:Chol, 4:4:1; final phosphate concentration, 1.5 mM).[10,13] Aggregation is initiated with the addition of CaCl$_2$ (final concentration, 16 μM Ca^{2+}). Reversible aggregation is distinguished from fusion by the addition of 1 M EDTA to a final concentration of 33 mM and the absorbance is monitored for an additional 3 min. To determine the effect of individual peptides on aggregation, the vesicles are preincubated with peptide for 10 min at the indicated temperatures prior to the addition of calcium. The kinetic parameters of membrane adhesion can be directly calculated from the fluorescence data obtained in the R$_{18}$ lipid mixing experiments, assuming a mass action kinetic model.[62]

Membrane Bilayer Destabilization

Membrane bilayer destabilization can be inferred fluorimetrically with the fluorescent dye 3,3′-diethylthiodicarbocyanine iodide (diS-C$_2$-5; Molecular Probes) as a collapse in a valinomycin-induced diffusion potential in model membranes.[61,63] SUVs composed of PC:PS:Chol in a 4:4:1 molar ratio are prepared by probe sonication in the presence of K$^+$-containing buffer (50 mM K$_2$SO$_4$, 10 mM HEPES–SO$_4$, pH 6.8).[38] An aliquot of the SUVs (phospholipid concentration, 36 nmol) is added to 10 ml of isotonic buffer (50 mM Na$_2$SO$_4$, 10 mM HEPES–SO$_4$, pH 6.8) containing 10 μl of the fluorescent probe diS-C$_2$-5 (stock, 1 mM) and incubated at 37° until a stable baseline fluorescence is established. The addition of valinomycin to a final concentration of 10^{-7} M selectively permeabilizes the vesicles to K$^+$, creating a negative diffusion potential inside the vesicles and resulting in the quenching of the fluorescence of the dye. Peptides from a 1-mg/ml stock in either distilled H$_2$O or K$^+$-free buffer are added in 25-, 75-, or 100-μl aliquots. Fluorescence is recorded at λ_{ex} 620 nm and λ_{em} 670 nm for 90 min after the addition of the various peptides. An increase in fluorescence intensity on the addition of peptides is indicative of a dissipation of the diffusion potential due to peptide-induced destabilization. To quantitate the total fluorescence recovered, fluorescence is monitored before the addition of valinomycin, after the addition of valinomycin, and after the addition of the desired peptide. The percent fluorescence recovery is calculated[61] as shown in Eq. (4):

---

[62] S. Nir, K. Klappe, and D. Hoekstra, *Biochemistry* **25**, 8261 (1986).
[63] L. M. Loew, I. Rosenberg, M. Bridge, and C. Gitler, *Biochemistry* **22**, 837 (1983).

$$F_t = [(I_t - I_0)/(I_f - I_0)] \times 100 \tag{4}$$

I_t is the fluorescence after the addition of the peptide, at time t, I_0 is fluorescence after the addition of valinomycin, and I_f is the fluorescence intensity prior to the addition of valinomycin. Mellitin (final concentration, 9.0 μg/ml) is added to confirm that the increase in fluorescence observed is due to a collapse in the diffusion potential. Mellitin should dissipate the valinomycin-induced diffusion potential even after the addition of those peptides that have no destabilizing effect.

Membrane destabilization is inferred from the fluorescence experiments described above. Changes in bilayer structure can be measured directly using ^{31}P NMR[64-66] and freeze–fracture electron microscopy.[67]

Conclusions

This chapter has summarized the technical advantages of fluorescence-based fusion assays as applied to retinal rod cells and shown how these assays led to the identification of peripherin/rds as a photoreceptor-specific fusion protein. It is important to note an added benefit of this approach, namely that the data obtained can be readily analyzed in the framework of a mass action kinetic model for fusion.[68,69] Such a model depicts the overall fusion reaction as consisting of two distinct steps: a second-order aggregation reaction and the actual fusion event, which is first order. Spontaneous disk–plasma membrane fusion in the ROS[11] has been shown to conform to such a model. Therefore, by using a variety of experimental conditions the rate constants of the aggregation step and fusion can be calculated, thereby providing greater understanding of the molecular mechanism of fusion in the ROS.

Acknowledgments

This work was supported by National Institute of Health Grant EY10420. The author thanks Dr. Robert Molday for generously providing anti-peripherin/rds 2B6 antibody. The author is grateful to Dr. R. Schimmel for continuous support, thoughtful discussions, and critical reading of the manuscript.

[64] A. D. Albert, A. Sen, and P. L. Yeagle, *Biochim. Biophys. Acta* **771**, 28 (1984).
[65] L. Mollevanger and W. J. DeGrip, *FEBS Lett.* **169**, 256 (1984).
[66] W. J. DeGrip, E. H. S. Drenthe, C. J. A. Van Echteld, B. De Kruijf, and A. J. Verkleij, *Biochim. Biophys. Acta* **558**, 330 (1979).
[67] S. W. Hui, T. P. Stewart, L. T. Boni, and P. L. Yeagle, *Science* **212**, 921 (1981).
[68] S. Nir, T. Stegmann, and J. Wilschut, *Biochemistry* **25**, 257 (1986).
[69] J. Bentz, S. Nir, and D. G. Covell, *Biophys. J.* **54**, 449 (1988).

[6] Bovine Retinal Nucleoside Diphosphate Kinase: Biochemistry and Molecular Cloning

By Najmoutin G. Abdulaev, Dmitri L. Kakuev, and Kevin D. Ridge

Introduction

Signal transduction in vertebrate photoreceptor cells begins with absorption of light by the visual pigment rhodopsin and culminates in the closure of ion channels on the plasma membrane. Consecutive binding and hydrolysis of several guanine nucleotides support the flow of information between these two events. The high levels of GTP required for G-protein activation and cGMP synthesis in the retina are likely to be supported by nucleoside diphosphate kinase (NDP kinase). The activity of the enzyme has been measured in retinal layers and outer segments,[1,2] but the enzyme itself has not been biochemically or structurally characterized. NDP kinase presumably constitutes an integral part of the cGMP cycle (cGMP → 5'-GMP → GDP → GTP → cGMP) and catalyzes the phosphorylation of nucleoside diphosphates to nucleoside triphosphates by a ping–pong mechanism involving a high-energy phosphorylated enzyme intermediate. The high-energy phosphate is usually supplied by ATP.[3]

Although the key role of guanine nucleotides in visual transduction is firmly established, what remains unclear is how the phototransduction cascade blends with the biochemical pathways of nucleotide metabolism. The purpose of this chapter is to highlight the role of NDP kinase in visual transduction and to provide systematic approaches for the study of its biochemical properties.[4] The focus is on the purification of the enzyme from bovine retina, its functional characterization, the cloning of two distinct isoforms from a bovine retinal library, and their heterologous expression in *Escherichia coli*.

---

[1] S. J. Berger, G. W. DeVries, J. G. Carter, D. W. Schulz, P. N. Passonneau, O. H. Lowry, and J. A. Ferrendelli, *J. Biol. Chem.* **255,** 3128 (1980).
[2] S. W. Hall and H. Kuhn, *Eur. J. Biochem.* **161,** 551 (1986).
[3] R. P. Agarwal, B. Robison, and R. E. Parks, Jr., *Methods Enzymol.* **51,** 376 (1978).
[4] N. G. Abdulaev, G. N. Karaschuk, J. E. Ladner, D. L. Kakuev, A. V. Yakhyaev, M. Tordova, I. O. Gaidarov, V. I. Popov, J. H. Fujiwara, D. Chinchilla, E. Eisenstein, G. L. Gilliland, and K. D. Ridge, *Biochemistry* **37,** 13958 (1998).

Purification and Characterization of Bovine Retinal Nucleoside Diphosphate Kinase

Purification of Nucleoside Diphosphate Kinase from Bovine Retina

The first stage of the NDP kinase purification procedure represents a modified guanylate kinase isolation procedure from Hall and Kuhn.[2] Two hundred frozen bovine retina (W. L. Lawson, Lincoln, NE) are thawed and suspended in 170 ml of 10 mM (Na,K,H)PO$_4$, pH 7.6, containing 0.2 mM MgCl$_2$, 0.2 mM EGTA, 0.2 mM phenylmethylsulfonyl fluoride (PMSF), and 0.02% (w/v) NaN$_3$. After stirring for 30 min at 4° in the light, the concentration of NaCl and MgCl$_2$ is adjusted to 150 and 4 mM, respectively. The suspension is stirred for another 30 min at 4° and the insoluble material is removed by centrifugation at 30,000g for 1 hr at 4°. The supernatant is centrifuged again at 100,000g for 30 min at 4° to remove trace amounts of membranes. An equal volume of (NH$_4$)$_2$SO$_4$ solution saturated at 60° is added to the supernatant and stirred for 2 hr at 4°. The precipitate is removed by centrifugation at 40,000g for 40 min at 4° and the supernatant is brought to 75% saturation with (NH$_4$)$_2$SO$_4$. After stirring overnight at 4°, the solution is centrifuged and the pellet, which contains NDP kinase, is resuspended in 10 mM Tris-HCl, pH 7.3, containing 2 mM MgCl$_2$, 0.1 mM EDTA, 1 mM dithiothreitol (DTT), and 300 mM NaCl (TMED buffer). Any insoluble material is removed by centrifugation at 100,000g for 30 min at 4°. The clear supernatant is applied to a Blue Sepharose CL-6B (Pharmacia Biotech, Piscataway, NJ) column (15 × 1.5 cm) equilibrated with TMED buffer. The column is extensively washed with TMED and the NDP kinase is eluted with TMED containing 2 mM GTP.

The yield of purified enzyme from this procedure is typically ~1 mg and the specific activity is ~763 units/mg (Table I). No appreciable difference is observed in the yield of NDP kinase or in its functional properties after purification is performed from light- or dark-adapted retina. Importantly, purified NDP kinase is free of both guanylate kinase and adenylate kinase activities. The purified enzyme exhibits a doublet on sodium dodecyl sulfate–polyacrylamide gel electrophoresis (SDS–PAGE), with the molecular masses of the two polypeptides ~17.5 and 18.5 kDa (Fig. 1A, lane 3). Both of these polypeptides are barely visible in the crude retinal extract or in the pellet obtained after 75% saturation with (NH$_4$)$_2$SO$_4$ (Fig. 1A, lanes 1 and 2, respectively), further emphasizing their relatively low abundance compared with other soluble retinal proteins. Immunoblot analysis of the purified enzyme using anti-NDP kinase polyclonal antibodies also shows the two polypeptides (Fig. 1B). The isoelectric focusing pattern of NDP

TABLE I
PURIFICATION OF NDP KINASE FROM BOVINE RETINA

Fraction	Total activity[a] (units)	Protein (mg)	Specific activity (units/mg)	Purification (-fold)	Recover[b] (%)
Crude extract	6909.6	3840	1.80	1	100
Supernatant[c]	6828.4	1015	6.73	3.7	98.8
Pellet[d]	5316	595	8.93	5	76.9
Blue Sepharose	839.5	1.1	763.18	424	12.1

[a] One unit of activity is defined as the amount of enzyme required to convert 1 μmol of CDP and ATP to CTP and ADP per minute at 25° in a coupled pyruvate kinase/lactate dehydrogenase assay.
[b] Recovery relates the total activity and extent of purification to the specific activity.
[c] The values shown correspond to the supernatant obtained after 50% $(NH_4)_2SO_4$ precipitation.
[d] The values shown correspond to the pellet obtained at 75% $(NH_4)_2SO_4$ precipitation.

kinase shows seven or eight protein bands with a p*I* range from 7.4 to 8.2 (Fig. 1C). Such a plurality of protein bands can be explained by the existence of NDP kinase oligomers composed of different proportions of the constitutive subunits and/or by varying degrees of phosphorylation within these oligomers.

Quaternary Structure Analysis

The molecular mass of NDP kinase in solution can be determined by comparing its migration with known protein standards by analytical gel-filtration chromatography (Fig. 2). Gel filtration of native (or recombinant) NDP kinase is performed on a Superose-12 HR (Pharmacia Biotech) column (30 × 1 cm) attached to a BioCAD Sprint perfusion chromatography workstation (PerSeptive Biosystems, Framingham, MA) or a fast protein liquid chromatography (FPLC) system (Pharmacia Biotech). The column is equilibrated with TMED buffer containing bovine serum albumin (1 mg/ml) at a flow rate of 0.2 ml/min to block nonspecific binding. The column is subsequently washed with TMED buffer and calibrated several times with 1-mg/ml solutions of the following molecular mass protein standards (Sigma, St. Louis, MO): rabbit muscle aldolase (158 kDa; Fig. 2a), rabbit muscle lactate dehydrogenase (140 kDa; Fig. 2b), bovine serum albumin (66 kDa; Fig. 2c), ovalbumin (45 kDa; Fig. 2d), chymotrypsinogen A (25 kDa; Fig. 2e), and cytochrome *c* (12.5 kDa; Fig. 2f). A standard curve is constructed by plotting K_{av} versus log M_r. K_{av} is calculated from the elution volumes V_e, from the total column volume V_t, and from the void volume V_0 using Eq. (1):

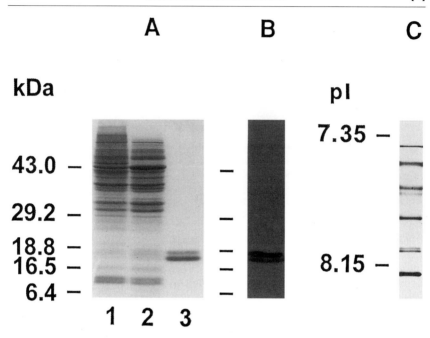

FIG. 1. Electrophoretic analysis of NDP kinase. (A) Coomassie blue stain of proteins in crude retinal extract (lane 1) and after 75% saturation with $(NH_4)_2SO_4$ (lane 2), and purified NDP kinase (lane 3) separated by SDS–PAGE. (B) Immunoblot analysis of purified NDP kinase separated by SDS–PAGE. The polypeptides were detected with the polyclonal anti-NDP kinase antibody, and visualized by chemiluminescence. (C) Coomassie blue stain of purified NDP kinase separated by isoelectric focusing. Positions of molecular size standards for (A) and (B) and pI standards for (C) are shown on the left.

$$K_{av} = V_e - V_0/V_t - V_0 \qquad (1)$$

V_0 is determined by measuring the eluted volume of blue dextran, and V_t is determined by measuring the eluted volume of tryptophan. This analysis yields a molecular mass of 96 ± 2 kDa for NDP kinase (Fig. 2), which closely approaches that of a hexamer. Notably, this subunit stoichiometry is characteristic of virtually all eukaryotic NDP kinases.[5]

Carbohydrate Analysis

The rather small difference in apparent molecular mass between the two NDP kinase polypeptides (~1 kDa) raises the possibility that they differ in some posttranslational modification(s). In fact, both NDP kinase

---

[5] A.-M. Gilles, E. Presecan, A. Vonica, and I. Lascu, *J. Biol. Chem.* **266,** 8784 (1991).

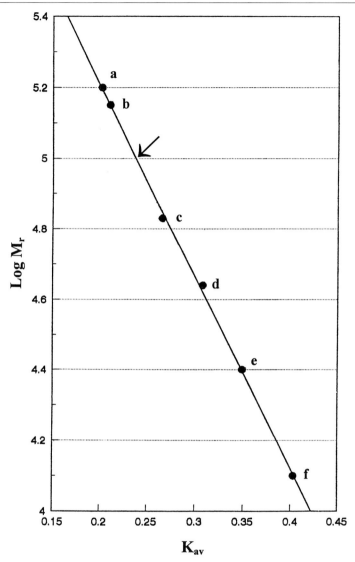

FIG. 2. Size-exclusion chromatography of retinal NDP kinase. The arrow indicates the position of NDP kinase relative to the molecular weight protein standards (details in text).

polypeptides appear to contain equivalent amounts of Gal, Man, GlcNAc, Fuc, and GalNAc saccharides, which are characteristic of O-linked glycosylation. The carbohydrate content is analyzed by separating the purified NDP kinase subunits by SDS–PAGE, electroblotting them onto poly(vinyl difluoride) membranes (Millipore, Bedford, MA), staining the proteins with Ponceau S, and excising the individual polypeptides with a razor blade. The membrane strips are destained with 1% acetic acid, washed extensively with deionized water, and dried. Acid hydrolysis of oligosaccharides in the separated NDP kinase polypeptides is performed as described.[6] The oligosaccharides are coupled with 7-amino-4-methylcoumarin,[7] and the monosaccharide fluorescent derivatives are separated on an Ultrasphere ODS (Beckman-Coulter, Fullerton, CA) column (22 × 0.21 cm) with a mobile phase consisting of 6.8% (v/v) 2-propanol, 3.5% (v/v) acetonitrile, and 0.01% (v/v) trifluoroacetic acid at a flow rate of 0.2 ml/min. The results of this analysis also show that the total content of carbohydrate in each polypeptide accounts for 2–3% of protein weight. Consequently, retinal NDP kinase can be referred to as a glycoprotein with a low content of oligosaccharides. Although it is possible that some properties of this enzyme could be explained by the presence of carbohydrates, it is evident that the difference in molecular mass of the NDP kinase subunits may not be attributed to a distinction in oligosaccharide content.

Functional Assay of Nucleoside Diphosphate Kinase

NDP kinase is assayed spectrophotometrically using a coupled pyruvate kinase–lactate dehydrogenase enzyme system.[2,8] The assays are done in a 1-ml reaction mixture containing 50 mM Tris-HCl (pH 7.6), 5 mM MgCl$_2$, 0.05 mM KCl, 0.1 mM phosphoenolpyruvate, 0.5 mM ATP, 0.1 mM TDP, NADH (0.1 mg/ml), 2 units of pyruvate kinase, and 2.5 units of lacate dehydrogenase. The reaction is initiated by addition of 0.5–5 μg of NDP kinase at 25°. Monitoring the decrease in absorbance at 334 nm follows the oxidation of NADH, which reflects ADP formation by NDP kinase. The specific activity of 1 unit of enzyme is defined as the turnover of 1 μmol of substrate in 1 min per milligram of protein. To measure the apparent Michaelis constant (K_m) and maximum velocity (V_{max}) for the diphosphate nucleotides CDP and TDP, a constant ATP concentration equal to 2 mM is used. For determination of the K_m and V_{max} for the

---

[6] H. Takemoto, S. Hase, and T. Ikenaka, *Anal. Biochem.* **145**, 245 (1985).

[7] A. Ya. Khorlin, S. D. Shiyan, V. A. Markin, V. V. Nasonov, and M. N. Mirzayanova, *Bioorg. Khim.* **12**, 1203 (1986).

[8] R. E. Parks and R. P. Agarwal, *in* "The Enzymes" (P. D. Boyer, ed.), Vol. 8, p. 307. Academic Press, New York, 1973.

triphosphate nucleotides GTP and ATP, 1.4 mM CDP is used as the fixed substrate. The K_m and V_{max} values are estimated from double-reciprocal plots by the method of Florini and Vestling.[9] This method allows the determination of kinetic constants for two-substrate enzyme systems. Like other eukaryotic NDP kinases, retinal NDP kinase also shows broad specificity for the nucleotide substrate. The K_m of the enzyme for ATP (104 μM) is the lowest among the nucleotides tested, while that for GTP is only about three times higher. Unfortunately, because of this relatively small difference in the K_m values, it is not possible to obtain accurate and reproducible kinetic measurements using ATP and GDP as the donor and acceptor substrates, respectively.

Molecular Cloning and Functional Expression of Bovine Retinal Nucleoside Diphosphate Kinase

Cloning of Nucleoside Diphosphate Kinase Isoforms from Bovine Retinal Library

Direct sequence analysis of NDP kinase does not show cleavage of any amino acids, suggesting that the NH_2-terminal residue in each of the polypeptides is blocked. The nature of this blocking group remains to be determined. However, digestion of NDP kinase with trypsin after heat denaturation at elevated pH allows exhaustive cleavage of the protein and provides facile peptide purification and sequencing. Using this sequence information and taking into account the codon usage of several known retinal proteins, oligonucleotides are synthesized and used to screen a bovine retinal cDNA library. The tryptic peptides are prepared by heat denaturing purified NDP kinase (1 mg) in 100 mM $NaHCO_3$, pH 9.0, at 95° for 10 min and then subjecting them to trypsin digestion (1:50, trypsin:NDP kinase) at 37° for 6 hr. The digest is applied to a reversed-phase high-performance liquid chromatography (HPLC) Zorbax C_{18} column (25 × 0.5 cm; Waters, Milford, MA) and eluted with a linear gradient of 0–70% (v/v) acetonitrile containing 0.1% (v/v) trifluoroacetic acid. Several peak fractions are collected and further purified on the same column. The peptides are sequenced by Edman degradation on a liquid-phase sequencer (Applied Biosystems, Foster City, CA). Two oligonucleotides based on the sequences of the tryptic peptides, as well as two additional probes corresponding to the most conserved sequences of NDP kinases

---

[9] J. R. Florini and C. S. Vestling, *Biochim. Biophys. Acta* **25**, 575 (1957).

from different sources, are synthesized. The oligonucleotide probes are 5'-end labeled with ^{32}P using T4 polynucleotide kinase.[10]

A bovine retinal cDNA library in bacteriophage λ-ZAP (provided by M. Applebury, Massachusetts Eye and Ear Infirmary, Boston, MA) is used for the screening. To screen the library, 4×10^6 bacteriophage are plated, and filter replicas are prepared using nitrocellulose filters (Millipore). The hybridization protocol is adapted from Sambrook et al.[10] Briefly, the procedure is carried out in the minimal volume of the hybridization buffer with addition of purified radioactively labeled oligonucleotide probe for 1 hr at 67° and then incubated overnight with slow cooling and shaking. The filters are washed in 4× SSC solution (1× SSC is 0.15 M NaCl plus 0.015 sodium citrate) containing 0.1% SDS. Autoradiography is performed using Hyperfilm-βmax film (Kodak, Rochester, NY) with an intensifying screen for 48 hr at −70°. Isolation of phage DNA and subcloning of cDNA inserts in the pBluescript M13 vector (Stratagene, La Jolla, CA) are done as described.[11] The plasmids are amplified in E. coli XL-1 Blue, the DNA is purified, and the inserts are sequenced. The results of this screening yield two NDP kinase clones, termed NBR-A and NBR-B (they are designated as A and B in analogy with NDP kinases from other sources).[12] The clone containing NBR-A has a 696-bp-long insert while that containing NBR-B has an 863-bp insert (Fig. 3). Sequence analysis of NBR-A and NBR-B shows considerable base differences in both the 5' and 3' noncoding regions, suggesting that they are indeed isoforms of the enzyme. Further, NBR-A and NBR-B differ from each other in four codons. Two nucleotide substitutions are responsible for the amino acid diversity of the NDP kinase isoforms: Ile-21 and His-135 in NBR-A are replaced by methionine and arginine residues, respectively, in NBR-B (Fig. 3). Two additional codon changes, at Pro-72 and Gly-106, do not result in amino acid substitutions. The deduced amino acid sequences of the NDP kinase isoforms each consist of 152 residues and the calculated molecular masses are 17,262 kDa for NBR-A and 17,299 kDa for NBR-B.

Expression and Purification of NBR-A and NBR-B Nucleoside Diphosphate Kinases

The cDNAs are ligated into the pALTER-*Ex*2 vector (Promega, Madison, WI) at the unique *Eco*RI restriction site in the multiple cloning region.

---

[10] J. Sambrook, E. F. Fritsch, and T. Maniatis, "Molecular Cloning: A Laboratory Manual," 2nd Ed. Cold Spring Harbor Laboratory Press, Cold Spring Harbor, New York, 1989.

[11] J. M. Short, J. A. Sorge, and W. D. Huse, *Nucleic Acids Res.* **16,** 7583 (1988).

[12] The nucleotide sequence data have been deposited with the EMBL Sequence Data Bank and are available under accession number X92956 for NBR-A and X92957 for NBR-B.

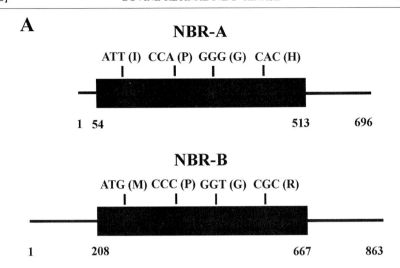
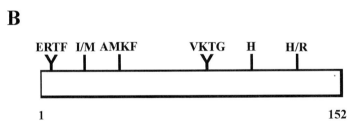

FIG. 3. Schematic representation of retinal NDP kinase nucleotide and amino acid sequences. (A) The codons and corresponding amino acids (single-letter code) in NBR-A that differ from those in NBR-B are highlighted. (B) The positions of amino acid diversity between NBR-A and NBR-B, the two potential O-glycosylation sites, the proposed ATP-binding region of the cyclic nucleotide-dependent protein kinases, and the active site histidine residue are highlighted.

Proper insertion and orientation of NBR-A and NBR-B in the vector are confirmed by restriction enzyme analysis. *Escherichia coli* JM109 (DE3) is transformed with the vector pALTER-Ex2 harboring the NBR-A and NBR-B cDNAs. The transformation mixtures are spread on LB–agar plates containing tetracycline and after 16–18 hr at 37°, colonies are isolated and cultured overnight at 37° in LB broth containing tetracycline. The overnight cultures are used to inoculate 500 ml of LB broth containing tetracycline in a 1 liter-capacity shake flask. After further growth at 37° for 16 hr, the cells are harvested by centrifugation at 1500g for 5 min, and disrupted in a French pressure cell (800–900 lb/in^2). The supernatant is used for purification of the recombinant enzymes.

The purification protocol for retinal NDP kinase can be easily adapted for preparation of NBR-A and NBR-B expressed in *E. coli*. Typically, the yields of purified NBR-A and NBR-B are 18–20 mg from a 500-ml shake flask culture, and the specific activities are in the range of ~857 and ~1063 units/mg, respectively. Similar to retinal NDP kinase, the K_m of the expressed enzymes for ATP is the lowest among the nucleotides tested, with GTP showing a threefold higher difference.

NBR-A and NBR-B each show a single polypeptide chain of ~17 kDa on SDS–PAGE followed by immunoblot analysis using anti-NDP kinase polyclonal antibodies raised against the retinal enzyme (Fig. 4, lanes 2 and 3). Similar to retinal NDP kinase, the isoelectric focusing patterns of NBR-A and NBR-B show six or seven major protein bands. However, the p*I* range for NBR-A is 8.0–8.5 while that for NBR-B is 6.5–7.5. Analytical gel filtration of the NBR-A and NBR-B isoforms shows molecular masses of 97 and 95 kDa, respectively. Like retinal NDP kinase, this is suggestive of a hexameric arrangement for the expressed enzymes.

Comments

To understand the biochemical events controlling the synthesis of guanine nucleotides involved in visual transduction, the isolation and characterization of retinal enzymes involved in nucleotide metabolism are desirable. Bovine retinal NDP kinase, like those from other sources,[13–15] shows two distinct protein bands with apparent molecular masses of 17.5 and 18.5 kDa when analyzed by SDS–PAGE. In contrast, both NBR-A and NBR-B show a single polypeptide chain of ~17 kDa. It is interesting that the correspondence between separate polypeptides and the products of gene isoforms has been shown only for human erythrocyte NDP kinase.[5] Both the SDS–PAGE and cDNA cloning results indicate that there exists at least two NDP kinase isoforms in bovine retina, as is the case for other higher eukaryotes.[16,17] However, NBR-A and NBR-B

[13] E. Presecan, A. Vonica, and I. Lascu, *FEBS Lett.* **250,** 629 (1989).
[14] N. Kimura and N. Shimada, *J. Biol. Chem.* **263,** 4647 (1988).
[15] J. A. Nickerson and W. W. Well, *J. Biol. Chem.* **259,** 11297 (1984).
[16] N. Shimada, N. Ishikawa, Y. Mukanata, T. Toda, K. Watanabe, and N. Kimura, *J. Biol. Chem.* **268,** 2583 (1993).
[17] T. Urano, K. Takamiya, K. Furukawa, and H. Shiku, *FEBS Lett.* **309,** 358 (1992).

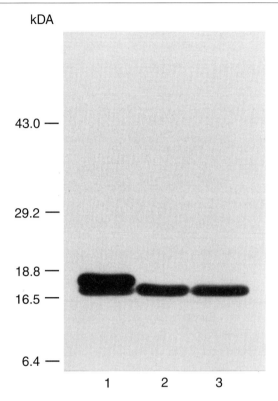

FIG. 4. Immunoblot analysis of purified bovine retinal NDP kinase (lane 1), purified NBR-A (lane 2), and purified NBR-B (lane 3). Positions of molecular size standards are shown on the left.

are remarkably similar in their coding nucleotide and amino acid sequences, with identities of 99.1 and 98.8%, respectively. Therefore, the ~1-kDa difference in apparent molecular weight between the two retinal NDP kinase polypeptides suggests that they are either subject to different, as yet unknown, posttranslational modifications or bovine retina contains still another NDP kinase isoform(s). Analytical gel filtration of the retinal and expressed NDP kinases suggests that the enzymes exist as a hexamer. This is further substantiated by the three-dimensional structures of NBR-A and NBR-B.[4,18]

[18] J. E. Ladner, N. G. Abdulaev, D. L. Kakuev, M. Tordova, K. D. Ridge, and G. L. Gilliland, *Acta Crystallogr. D* **55,** 1127 (1999).

TABLE II
AMINO ACID SEQUENCE IDENTITY BETWEEN NDP KINASES FROM VARIOUS SOURCES[a]

Source	1	2	3	4	5	6	7	8	9
1. awd, Drosophila melanogaster									
2. NDP kinase, Dictyostelium discoideum	60.0								
3. NDP kinase, Myxococcus xanthus	48.8	43.1							
4. NBR-A, bovine retina	74.4	58.1	47.2						
5. NBR-B, bovine retina	47.4	58.1	47.2	98.8					
6. NDP kinase α, rat liver	47.4	55.6	44.1	85.8	85.8				
7. NDP kinase β, rat liver	73.8	58.1	44.1	88.9	88.9	90.1			
8. pl8, rat mucosal mast cells	76.9	58.1	47.2	87.0	87.0	91.9	84.5		
9. nm23-H1, human	78.1	61.9	46.0	90.1	89.5	86.4	92.0	87.0	
10. nm23-H2, human	78.1	59.4	46.0	87.0	87.0	93.2	86.4	95.0	89.9

[a] Amino acid sequences within the coding region are compared and the sequence pair distances determined by the CLUSTAL method.

Both subunits of the enzyme from bovine retina contain ~2–3% (w/w) carbohydrate that consists of Gal, Man, GlcNAc, Fuc, and GalNAc saccharides. All eukaryotic NDP kinases contain a potential O-glycosylation site (ERTF, amino acid residues 5–8) and another consensus O-glycosylation motif, VKTG (amino acid residues 84–87), is preserved in proteins from mammals and *Drosophila melanogaster*. These potential glycosylation sites are also on the outside of the molecule and readily accessible for modification. In fact, these sites are adjacent to one another; the distance between the C_α of Glu-5 and Val-84 is 7 Å.[4,18] Differences in the extent of amino acid residue phosphorylation are implied from the isoelectric focusing experiments on both the retinal and expressed enzymes. The His-118 residue, a phosphorylation site on NDP kinase, is present in all known NDP kinases. NBR-A and NBR-B both contain an AMKF sequence (amino acid residues 37–40) that is nearly conserved as the ATP-binding region in cyclic nucleotide-dependent protein kinases. Lys-39 in this region is invariant in nearly all NDP kinases and is essential for catalytic activity.[19] The results of comparative sequence analysis (Table II) show that both the NBR-A and NBR-B sequences can be aligned rather well with those of human nm23-H1 protein (90.1 and 89.5% homology, respectively) and the rat β

---

[19] N. Kimura, N. Shimada, K. Nomura, and K. Watanabe, *J. Biol. Chem.* **265**, 15744 (1990).

isoform of NDP kinase (88.9% homology for both isoforms). Almost all sequences are homologous to the sequence of NDP kinase in question, without any internal gaps. The exceptions are the enzymes from *Dictyostelium discoideum* and *Myxococcus xanthus.*

Kinetic studies of retinal and expressed NDP kinases showed that the enzymes display low substrate specificity, as illustrated for other NDP kinases. However, NDP kinase can function in a low ATP-containing environment (K_m of retinal NDP kinase for ATP, ~104 μM) with high velocity, supplying a sufficient level of nucleoside triphosphates in visual cells. The specific activity of retinal NDP kinase, ~763 units/mg, is considerably high when compared with other guanine nucleotide pathway enzymes. For example, the specific activity of retinal guanylate kinase is ~340 units/mg.[2] The turnover number of NDP kinase is 229 mol of CTP produced per mole of enzyme per second or 1374 mol of CTP produced per mole of enzyme per second for a hexamer. This is considerably greater than the turnover number of guanylate kinase (~130 mol of GDP produced per mole of enzyme per second). These findings suggest that even relatively small amounts of guanylate kinase and NDP kinase in the rod outer segment (ROS) are greater in terms of absolute activity than guanylate cyclase.[20–22] Therefore, these two enzymes may support local cGMP requirements at the expense of ATP, which is produced abundantly in the ROS as a result of anaerobic glycolysis,[23] without the need for shuttling GMP from the ROS to the rod inner segment for conversion to GTP.

The biochemical and structural characterization of NDP kinase from bovine retina will allow us to investigate further its role in the regulation of the GTP supply to G proteins and other components involved in visual phototransduction. The availability of the cDNA sequences for retinal NDP kinase should allow for the overexpression of mutant forms of the enzyme in sufficient quantities for three-dimensional structure determination.

Acknowledgments

We recognize the contributions of John Fujiwara and Tony Ngo in some aspects of these studies. This work was supported by the National Institute of Standards and Technology. Certain commercial materials, instruments, and equipment are identified in this manuscript

[20] I. O. Gaidarov, O. N. Suslov, and N. G. Abdulaev, *FEBS Lett.* **335,** 81 (1993).
[21] I. O. Gaidarov, O. N. Suslov, T. V. Ovchinnikova, and N. G. Abdulaev, *Biorg. Khim.* **20,** 367 (1994).
[22] A. Ames, T. F. Walseth, R. A. Heyman, M. Barad, R. M. Graeff, and N. D. Goldberg, *J. Biol. Chem.* **261,** 13034 (1986).
[23] S.-C. Hsu and R. S. Molday, *J. Biol. Chem.* **266,** 21745 (1991).

in order to specify the experimental procedure as completely as possible. In no case does such identification imply a recommendation or endorsement by the National Institute of Standards and Technology, nor does it imply that the materials, instruments, or equipment identified is necessarily the best available for the purpose.

Section II

Calcium-Binding Proteins and Calcium Measurements in Photoreceptor Cells

[7] Calcium-Binding Proteins and Their Assessment in Ocular Diseases

By ARTHUR S. POLANS, RICARDO L. GEE, TERESA M. WALKER, and PAUL R. VAN GINKEL

Introduction

Calcium ions (Ca^{2+}) have evolved as the principal nondegradable second messengers in nature; they participate in virtually every aspect of cellular life including the regulation of cell division on fertilization, determination of the differentiation of specialized cells, and the final demise of those cells through apoptotic or necrotic pathways. Interpretation of the vast array of calcium signals impinging on the cell depends on spatiotemporal information about the sources of the signals as well as the localization of specific target molecules.[1,2]

Often these cellular targets are calcium-binding proteins that alter their configuration on the chelation of calcium and in turn bind to and alter the activity of other biologically relevant molecules such as enzymes or ion channels.[3–6] In vertebrate photoreceptor cells, for example, guanylate cyclase-activating protein 1 (GCAP1), a member of the EF-hand family of calcium-binding proteins, regulates the activity of the enzyme guanylate cyclase and thereby the concentration of the internal transmitter, cyclic GMP.[7,8] As the concentration of intracellular calcium diminishes on illumination, GCAP1 stimulates guanylate cyclase to synthesize cyclic GMP, thus opening cation channels in the plasma membrane of the photoreceptor cell outer segment and reestablishing the dark resting potential of the cell. Another calcium-binding protein, perhaps calmodulin, appears to modulate the affinity of the cation channel for cyclic GMP, thereby effecting the time

---

[1] D. E. Clapham, *Cell* **80,** 259 (1995).
[2] D. D. Ginty, *Neuron* **18,** 183 (1997).
[3] C. W. Heizmann and W. Hunziker, *Trends Biochem. Sci.* **16,** 98 (1991).
[4] C. W. Heizmann and K. Braun, *Trends Neurochem. Sci.* **15,** 259 (1992).
[5] R. J. P. Williams, *Cell Calcium* **20,** 87 (1996).
[6] I. Niki, H. Yokokura, T. Sudo, M. Kato, and H. Hidaka, *J. Biochem.* **120,** 685 (1996).
[7] W. A. Gorczyca, M. P. Gray-Keller, P. B. Detwiler, and K. Palczewski, *Proc. Natl. Acad. Sci. U.S.A.* **91,** 4014 (1994).
[8] K. Palczewski, I. Subbaraya, W. A. Gorczyca, B. S. Helekar, C. C. Ruiz, H. Ohguro, J. Huang, X. Zhao, J. W. Crabb, R. S. Johnson, K. A. Walsh, M. P. Gray-Keller, P. B. Detwiler, and W. Baehr, *Neuron* **13,** 395 (1994).

course for the recovery of the dark current and the reestablishment of the intracellular concentration of calcium[9] (for a review see Polans et al.[10]).

Aside from studies of phototransduction, these and other calcium-binding proteins are involved in ocular pathologies arising in humans. Recoverin, for example, is an autoantigen in a degenerative disease of the retina known as cancer-associated retinopathy.[11] The aberrant expression of recoverin in some tumors arising outside of the eye leads to an autoimmune response that inadvertently causes the destruction of photoreceptor cells, the normal site of recoverin expression.[12] In addition to this "remote effect" of cancer, different calcium-binding proteins have been associated with different primary tumors, and their levels of expression can correlate with the malignant and metastatic phenotypes.[13,14] S100A6, also referred to as calcyclin, is associated with more malignant melanocytic tumors of the skin and eye but not, for example, cancer of the prostate.[13]

Calcium-binding proteins, therefore, play critical roles in the normal function and assorted pathologies of many cells including those of ocular origin. Their study may reveal rational sites for intervention in the progression of a disease by drug treatment or gene therapy.

Methods and Results

The study of calcium-binding proteins first requires their purification and identification, and methods related to these objectives are described in the next section of this chapter. Calcium-binding parameters such as the number of binding sites and their affinities and the subsequent conformational changes induced by calcium then can be assessed, and methods related to these characterizations are also considered in this chapter. Finally, cellular targets need to be identified and their interactions with their regulatory calcium-binding protein delineated. This final endeavor is the most challenging, and the final portion of this chapter describes some traditional methods related to this objective along with newer, alternative approaches.

[9] Y.-T. Hsu and R. S. Molday, Nature (London) **361**, 76 (1993).
[10] A. Polans, W. Baehr, and K. Palczewski, Trends Neurosci. **19**, 547 (1996).
[11] A. S. Polans, J. Buczylko, J. Crabb, and K. Palczewski, J. Cell Biol. **112**, 981 (1991).
[12] A. Polans, D. Witkowska, T. Haley, D. Amundson, L. Baizer, and G. Adamus, Proc. Natl. Acad. Sci. U.S.A. **92**, 9176 (1995).
[13] E. C. Ilg, B. W. Schäfer, and C. W. Heizmann, Int. J. Cancer **68**, 325 (1996).
[14] G. M. Maelandsmo, V. A. Florenes, T. Mellingsaeter, E. Hovig, R. S. Kerbel, and O. Fodstad, Int. J. Cancer **74**, 464 (1997).

Purification of Calcium-Binding Proteins

Several forms of affinity chromatography have been devised for the purification of calcium-binding proteins that depend on interactions with venoms[15] or synthetic antagonists.[16,17] Alternatively, a number of calcium-binding proteins alter their conformation and become significantly more hydrophobic on binding calcium; this provides an easy method for simultaneously isolating several different calcium-binding proteins from cells or complex tissues by taking advantage of their calcium-dependent binding to a hydrophobic matrix such as phenyl-Sepharose.[18] As an example, the isolation of several calcium-binding proteins from ocular tumor cells is described here. The separation of calcium-binding proteins from retinal and other ocular tissues has been described elsewhere.[11,12,18]

Preparation of Cellular Extracts

Uveal melanoma tumor tissue (3–16 mm^3) obtained from enucleated eyes or cell lines (10^7–10^8 cells) derived from biopsies of human tumors are disrupted in a buffer consisting of 50 mM HEPES, 1 mM EDTA, and 100 mM NaCl, pH 7.5, using a Teflon homogenizer. Aprotinin and leupeptin are added to limit proteolysis. (In some cases, the salt concentration is lowered to 10 mM in order to assist with cellular disruption and the release of additional protein.) After centrifugation at 39,000g for 30 min at 4° in a JA-17 rotor (Beckman Instruments, Fullerton, CA), the volume of the supernatant is determined and calcium chloride added to a final concentration of 2 mM.

Phenyl-Sepharose Chromatography

A phenyl-Sepharose (Pharmacia Fine Chemicals, Piscataway, NJ) column is prepared at 4°. Columns vary in size from 1 × 1 cm to 1 × 8 cm, depending on the number of cells or the amount of tissue extract to be applied to the column. Typically, extracts from 10^7–10^8 cells are applied to a 1 × 4 cm column. The column is prepared and equilibrated for a minimum of 3 hr with buffer (50 mM HEPES, 2 mM calcium, and 100 mM NaCl, pH 7.5) prior to use. The extract is applied to the column at a flow rate of 15 ml/hr and washed with the equilibration buffer until the A_{280}

---

[15] R. L. Kincaid, *Methods Enzymol.* **139,** 3 (1987).
[16] M. Asano, Y. Suzuki, and H. Hidaka, *J. Pharmacol. Exp. Ther.* **220,** 191 (1982).
[17] Y. Watanabe, R. Kobayashi, T. Ishikawa, and H. Hidaka, *Arch. Biochem. Biophys.* **292,** 563 (1992).
[18] A. S. Polans, J. Crabb, and K. Palczewski, *Methods Neurosci.* **15,** 248 (1993).

reaches baseline. In some instances the period of washing is shortened, because some calcium-binding proteins elute from the column during extensive washing. Bound proteins are then eluted from the column by washing with the same HEPES buffer containing 10 mM EDTA. Fractions (1 ml) are collected and aliquots subjected to fractionation on a 15% sodium dodecyl sulfate (SDS)–polyacrylamide gel. The higher percentage of acrylamide is recommended initially, because a number of the smaller S-100 calcium-binding proteins will remain unresolved and migrate with the tracking dye in lower percentage gels. Figure 1A illustrates the phenyl-Sepharose eluate derived from a spindle cell ocular melanoma cell extract. SDS–polyacrylamide gel electrophoresis (PAGE) reveals six major protein bands with molecular masses of approximately 67, 39, 35, 18, 12, and 9.5 kDa. Each of these proteins is identified as a calcium-binding protein (described below).

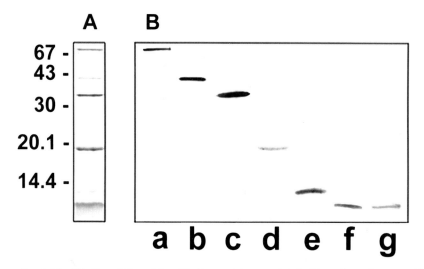

FIG. 1. Identification of the calcium-binding proteins associated with ocular melanoma cell lines. Homogenates of cell lines derived from a spindle cell ocular melanoma (OCM-1) were extracted with EDTA. A soluble fraction was adjusted with calcium, applied to a phenyl-Sepharose column, and material that bound to the column was eluted with EDTA as described in Methods and Results. Aliquots of the final eluates were subjected to SDS–polyacrylamide gel electrophoresis and stained with Coomassie Brilliant Blue. The eluate from OCM-1 is depicted in (A). Molecular mass standards were as follows: phosphorylase b (94 kDa), bovine serum albumin (67 kDa), ovalbumin (43 kDa), carbonic anhydrase (30 kDa), soybean trypsin inhibitor (20.1 kDa), and α-lactalbumin (14.4 kDa). (B) An aliquot from the phenyl-Sepharose eluate of OCM-1 depicted in (A) was transferred to Immobilon and immunostained as described in Methods and Results with antibodies specific for annexin VI (lane a), cap g (b), annexin V (c), calmodulin (d), S100A11 (e), S100B (f), and S100A6 (g). [P. R. van Ginkel, R. L. Gee, T. M. Walker, C. Hu, C. W. Heizmann, and A. S. Polans, *Biochim. Biophys. Acta* **1448,** 290 (1998).]

Reversed-Phase High-Performance Liquid Chromatography

Some calcium-binding proteins isolated by phenyl-Sepharose chromatography can be purified to homogeneity by subsequent anion-exchange chromatography with either DEAE-Sepharose or Mono Q (Pharmacia Fine Chemicals). The photoreceptor-specific calcium-binding protein recoverin is purified by this strategy.[11] Alternatively, some calcium-binding proteins differ significantly in the number of free sulfhydryl groups and can be separated, after phenyl-Sepharose chromatography, by an organomercurial column and dithiothreitol (DTT) for elution.[12,19] However, the majority of calcium-binding proteins enriched by phenyl-Sepharose chromatography typically require further purification by reversed-phase high-performance liquid chromatography (RP-HPLC).

The eluate from a phenyl-Sepharose column is applied directly to a C_4 column (4.6 × 150 mm, W-Porex 5; Phenomenex, Torrance, CA). After washing with buffer A [0.1% (v/v) trifluoroacetic acid (TFA) in water] at 0.6 ml/min, a gradient is developed with buffers A and B [80% CH_3CN, TFA (0.08%, v/v)]. Depending on the number of proteins to be resolved and their hydrophobic characteristics, the gradient can be linear over a relatively short period of time (40 min), or consist of a more extensive program (0–47% buffer B, 5 min; 47–57% buffer B, 85 min; 57–100% buffer B, 100 min). Figure 2 shows the RP-HPLC profile obtained for the phenyl-Sepharose eluate illustrated in Fig. 1A.

Fractions obtained from each peak are neutralized with 0.1 M Tris base and aliquots are subjected to SDS–PAGE. The 39-kDa protein obtained in the phenyl-Sepharose eluate is separated as two peaks (1a and 1b), while peaks 2a and 2b contain the 18-kDa protein. The remainder of the proteins depicted in the phenyl-Sepharose eluate are isolated as single peaks by RP-HPLC. Peak 3 consists of the 12-kDa protein, peak 4 a 9.5-kDa protein, peak 5 an additional 9.5-kDa protein, peak 6 the 67-kDa protein, and peak 7 the 35-kDa protein.

Sequence Analysis

Purified proteins from the C_4 column are reduced, alkylated with 4-vinylpyridine, and then succinylated. These procedures and the specific cleavages at either methionyl, lysyl, glutamyl, or aspartyl residues are described elsewhere.[20–22] Sequence analysis by Edman degradation is per-

---

[19] A. S. Polans, K. Palczewski, M. A. Asson-Batres, H. Ohguro, D. Witkowska, T. L. Haley, L. Baizer, and J. W. Crabb, *J. Biol. Chem.* **269,** 6233 (1994).

[20] J. W. Crabb, L. F. Armes, S. A. Carr, C. M. Johnson, G. D. Roberts, and R. S. Bordoli, *Biochemistry* **25,** 4988 (1986).

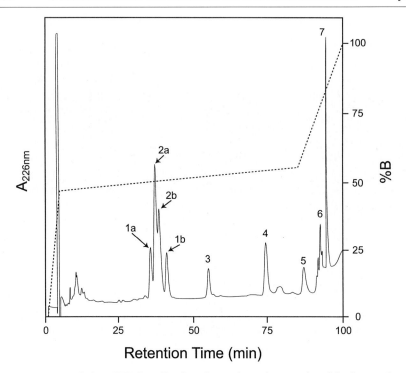

FIG. 2. Reversed-phase HPLC purification of an ocular melanoma phenyl-Sepharose eluate. Protein eluted from the phenyl-Sepharose column by the addition of EDTA (Fig. 1A) was separated on a C_4 column, as described in text. Seven major fractions were identified, and the constituent proteins were characterized by SDS–PAGE. The proteins from the phenyl-Sepharose eluate were contained in the following peaks: peaks 1a and 1b contained a 39-kDa protein, peaks 2a and 2b an 18-kDa protein, peak 3 a 12-kDa protein, peak 4 a 9.5-kDa protein, peak 5 a 9.5-kDa protein, peak 6 a 67-kDa protein, and peak 7 a 35-kDa protein. The gradient conditions indicated by the dashed line are provided in text.

formed with an Applied Biosystems (Foster City, CA) gas-phase sequencer.[21] Digested peptides purified by RP-HPLC also are dissolved in 50% methanol in 0.1% formic acid and infused into the ion source of a triple quadrupole mass spectrometer (Sciex API III; PE/Sciex, Thornhill, Ontario, Canada) fitted with a nebulization-assisted electrospray ionization source. Electrospray mass spectrometry (ES/MS) and tandem mass spectra

---

[21] J. W. Crabb, C. M. Johnson, S. A. Carr, L. G. Armes, and J. C. Saari, *J. Biol. Chem.* **263**, 18678 (1988).
[22] K. Palczewski, J. Buczylko, M. W. Kaplan, A. S. Polans, and J. W. Crabb, *J. Biol. Chem.* **266**, 12949 (1991).

(MS/MS) were compiled with the data obtained by Edman analysis in order to definitively identify each of the calcium-binding proteins purified by phenyl-Sepharose chromatography and RP-HPLC.

The following proteins are identified in this manner: annexin VI (67 kDa), cap g (39 kDa), annexin V (35 kDa), calmodulin (18 kDa in the presence of EGTA), S100A11 (12 kDa), and two comigrating proteins at 9.5 kDa (S100B from peak 4 of Fig. 2 and S100A6 from peak 5). The reasons for the separation of cap g and calmodulin each into two peaks have not been determined. It may reflect differences in posttranslational modifications of the same polypeptide chain or more likely the extent of bound calcium and concomitant hydrophobicity of the proteins.

It should be noted that sequence analysis has confirmed that all of the uveal melanoma proteins initially isolated by phenyl-Sepharose chromatography are calcium-binding proteins. They consist of proteins of either the EF-hand class of calcium-binding protein (cap g, calmodulin, S100A11, S100B, and S100A6) or the annexin family (annexin V and annexin VI). Proteins isolated by phenyl-Sepharose chromatography using retinal homogenates likewise have been demonstrated to be calcium-binding proteins.[18]

Characterization of Calcium-Binding Proteins

Antibody Production

Synthetic peptides corresponding to the nine and eight carboxy-terminal amino acids from the human sequences of S100A11 and S100A6, respectively, are conjugated to keyhole limpet hemocyanin (ultrapure immunojet; Pierce, Rockford, IL) using glutaraldehyde.[23] An additional C-terminal lysine residue in each peptide optimizes conjugation. In addition, proteins purified by RP-HPLC are dialyzed against phosphate-buffered saline (PBS) and used as immunogens. New Zealand White rabbits are immunized with 100 μg of each antigen mixed with Freund's complete adjuvant. Animals are boosted at 10- to 14-day intervals with 10–50 μg of antigen mixed with incomplete adjuvant. Antibodies are affinity purified with the corresponding peptide or protein coupled to CNBr-activated Sepharose.[24] Additional antibodies are obtained elsewhere.[25]

---

[23] H. M. Geysen, S. J. Barteling, and R. H. Meloen, *Proc. Natl. Acad. Sci. U.S.A.* **82,** 178 (1985).
[24] K. Palczewski, J. Buczylko, L. Lebioda, J. W. Crabb, and A. S. Polans, *J. Biol. Chem.* **268,** 6004 (1993).
[25] P. R. van Ginkel, R. L. Gee, T. M. Walker, D. Hu, C. W. Heizmann, and A. S. Polans, *Biochim. Biophys. Acta* **1448,** 290 (1998).

To analyze the specificity of the antibodies, proteins from the phenyl-Sepharose eluate depicted in Fig. 1A are electrotransferred to Immobilon-P (Millipore, Bedford, MA) and incubated with primary antibody (1 μg/ml), a 1:5000 dilution of alkaline phosphatase-conjugated anti-immunoglobulin IgG (Sigma, St. Louis, MO), and a mixture of nitroblue tetrazolium salt and 5-bromo-4-chloro-3-indolylphosphate as substrate. Figure 1B illustrates a Western blot stained with antibodies generated against annexin VI (lane a), cap g (lane b), annexin V (lane c), calmodulin (lane d), S100A11 (lane e), S100B (lane f), and S100A6 (lane g). These antibodies are used in subsequent Western blot analyses to demonstrate the differential expression of this complement of calcium-binding proteins in normal uveal melanocytes, cutaneous melanoma, and ocular melanomas of both spindle and epithelioid cell morphology.[25]

Immunohistochemistry

Antibodies also are employed in localization studies of the various calcium-binding proteins. Cells are placed in Nunc (Naperville, IL) culture chamber slides and fixed with 4% (v/v) paraformaldehyde, 5% (w/v) sucrose in 0.13 M sodium phosphate buffer, pH 7.4, for 30 min. To enhance the penetration of antibodies, cells are permeabilized by incubation for 4 min with 0.5% (v/v) Triton X-100 in phosphate-buffered saline. Nonspecific sites are blocked by incubation with 3% (w/v) bovine serum albumin (BSA), 0.1% (w/v) glycine, 0.1% (v/v) normal goat serum in phosphate-buffered saline. Cells are incubated sequentially with primary antibody (1 μg/ml), biotinyl anti-immunoglobulin IgG (3 μg/ml; Vector Laboratories, Burlingame, CA), and streptavidin–Texas red conjugates (1 μg/ml; Amersham, Arlington Heights, IL).

Figure 3 shows the immunocytochemical labeling of spindle cell (Fig. 3A) and epithelioid cell (Fig. 3B) ocular melanoma cell lines with anti-S100A11 antibodies. Both cell lines are stained extensively in the perinuclear region as well as the remainder of their cytoplasm. Staining extends throughout the cell processes, including the elongated processes from spindle cell subtypes. The immunocytochemical profile is similar to that seen in previous studies of cap g and actin[25]; S100A11 is related to S100A4 and both may be regulators of the actin cytoskeleton during rapid cell growth and migration associated with malignancy and metastatic events.

Calcium-Binding Parameters

Flow dialysis procedures are adapted from Colowick and Womack[26] and modified according to Haiech et al.[27] Briefly, dialysis tubing is boiled

---

[26] S. P. Colowick and F. C. Womack, *J. Biol. Chem.* **244**, 774 (1969).
[27] J. Haiech, C. Klee, and J. G. Demaille, *Biochemistry* **20**, 3890 (1981).

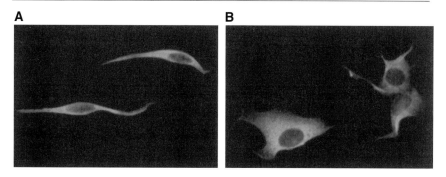

FIG. 3. Immunohistochemistry of S100A11 in spindle cell and epithelioid cell ocular melanomas. OCM-1 (A) and OCM-3 (B) cells were cultured in chamber slides for 1–2 days, fixed, and processed for immunostaining with antibodies specific for S100A11 as described in Methods and Results. Omission of the primary antibody resulted in no detectable staining.

in 5% (w/v) $NaHCO_3$ and then in deionized water. Contaminating calcium is removed from the buffer (10 mM HEPES, 100 mM KCl, pH 7.5) by Chelex chromatography. EDTA is removed from protein samples by multiple passages over Sephadex G-25 PD-10 columns. Protein samples dialyzed against buffer also are treated by Chelex chromatography just prior to use. (The concentration of calcium is determined in all samples and solutions by atomic absorption spectrophotometry.)

Protein (1–2 mg/ml in 2 ml) is placed into the upper chamber of a flow dialysis unit described by Feldman.[28] The upper and lower chambers (0.1–0.2 ml) are temperature controlled at 25°. The flow rate through the lower chamber is set at 1 ml/min, using a peristaltic pump. Nonisotopic calcium is added in steps to the upper chamber (also containing trace $^{45}Ca^{2+}$) to yield a final concentration of 0–700 μM. Aliquots are removed periodically from the lower chamber for scintillation counting. Data are analyzed using either the Adair–Klotz equation or the Hill equation.

Figure 4 shows a fractional saturation curve for the binding of calcium to the photoreceptor-specific calcium-binding protein recoverin. The curve is asymptotic at $n = 2$, indicating that recoverin has two calcium-binding sites. Initial studies of the recoverin sequence identified three potential EF-hand calcium-binding domains.[29] Further inspection, however, reveals that the first of these EF-hands contains the substitution of a highly invariant aspartate residue at the $+X$ position for calcium binding. Owing to this

[28] K. Feldman, *Anal. Biochem.* **88**, 225 (1978).
[29] A. Dizhoor, S. Ray, S. Kumar, G. Niemi, M. Spencer, D. Brolley, K. Walsh, P. Philipov, J. Hurley, and L. Stryer, *Science* **251,** 915 (1991).

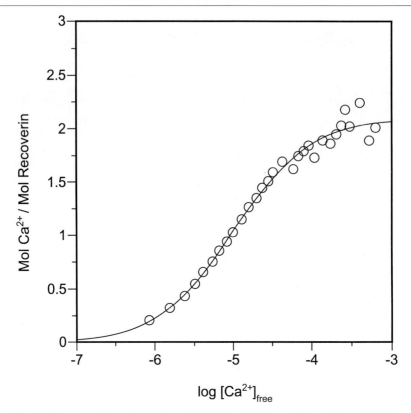

FIG. 4. Recoverin (27 μM) in 10 mM HEPES, pH 7.6, containing 100 mM NaCl was adjusted stepwise in calcium concentration from 0 to 700 μM in the upper chamber of a flow dialysis apparatus. Unbound ^{45}Ca^{2+} also present in the upper chamber passed through the dialysis membrane and was collected for scinitillation counting. The number of moles of calcium bound per mole of recoverin as a function of the free calcium concentration was calculated, and the experimental points are presented as open circles. The solid line represents the computer-derived fit from the Adair–Klotz equation. [J. F. Maune, C. B. Klee, and K. Beckingham, *J. Biol. Chem.* **267,** 5286 (1992).]

and other modifications, it is unlikely that the first EF-hand can act as a high-affinity site for calcium binding.

Conformational Changes as a Function of Calcium Binding

The structures of several calcium-binding proteins have been solved by nuclear magnetic resonance (NMR) and X-ray crystallography.[30-34] Other

---

[30] N. C. J. Strynadka and M. N. G. James, *Annu. Rev. Biochem.* **58,** 951 (1989).

structural approaches for the study of calcium-binding proteins measure such parameters as optical rotary dispersion and UV absorbance. Conformational changes also can be detected by spectrofluorometry, circular dichroism, or by measuring a shift in gel mobility.

The intrinsic fluorescence of a protein is due primarily to its constituent tryptophan and tyrosine residues, and fluorescent emission spectra can be measured easily while varying the calcium concentration. The concentration of free calcium can be determined as described above for flow dialysis, or calculated.[35] Changes in either the intensity of fluorescence or the λ_{max} of emission often are associated with the binding of calcium. In Fig. 5 the spectra of 1 μM recoverin are recorded with a Perkin-Elmer (Norwalk, CT) LS 50B spectrofluorophotometer with cells of 0.25- or 1-cm path length and using slit widths of 10 nm. Fortunately, both tyrosine and tryptophan residues lie in the vicinity of the two calcium-binding regions of recoverin, and changes in fluorescence can be used as a sensitive indicator of conformational changes. As shown in Fig. 5, both the intensity and λ_{max} of emission changed as a function of the calcium concentration. Interestingly, changes at the elevated concentrations of calcium could not be reversed by the addition of EGTA. This may reflect aggregation of the protein or other thermodynamic changes independent of the binding of calcium to the two EF-hand binding sites.

In addition to changes in the fluorescence of a protein, differences in the absorption of left and right circularly polarized light also can reflect changes in the structure of a protein that arise on the binding of a ligand such as calcium. Circular dichroism (CD) thus can be used to determine changes in the secondary structure of a calcium-binding protein during titrations with calcium. Figure 6 illustrates the CD spectra in the vacuum UV region for recoverin in 1 mM EGTA and two different concentrations of calcium. Analysis of these spectra shows that the maximum α-helical content of recoverin (67%) occurs in the presence of 0.1 mM calcium. The α-helical content decreases to 38% in the presence of 1 mM EGTA, while these changes are paralleled by alterations in β-sheet, β-turn, and other

---

[31] W. E. Meador, A. R. Means, and F. A. Quiocho, *Science* **262,** 1718 (1993).

[32] T. Tanaka, J. B. Ames, T. S. Harvey, L. Stryer, and M. Ikura, *Nature (London)* **376,** 444 (1995).

[33] A. C. Drohat, D. M. Baldisseri, R. R. Rustandi, and D. J. Weber, *Biochemistry* **37,** 2729 (1998).

[34] M. Sastry, R. R. Ketchem, O. Crescenzi, C. Weber, M. J. Lubienski, H. Hidaka, and W. J. Chazin, *Structure* **6,** 223 (1998).

[35] W. C. Johnson, K. Palczewski, W. A. Gorczyca, J. H. Riazance-Lawrence, D. Witkowska, and A. S. Polans, *Biochim. Biophys. Acta* **1342,** 164 (1997).

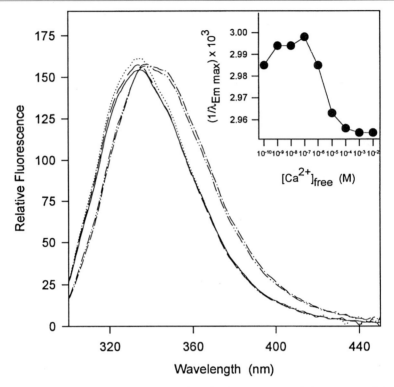

FIG. 5. Fluorescence spectra of recoverin. The fluorescence emission spectra of 1 μM recoverin were recorded with excitation at 291 nm as described in Methods and Results. The various curves represent the fluorescent emission spectra in 10^{-10} (· · ·), 10^{-7} (---), 10^{-6} (—), 10^{-5} (- · -), and 10^{-3} (- · · -) M free calcium. Inset: Semireciprocal plot of λ_{max} of the fluorescence emission versus the concentration of free calcium. [W. C. Johnson, K. Palczewski, W. A. Gorczyca, J. H. Riazance-Lawrence, D. Witkowska, and A. S. Polans, Biochim. Biophys. Acta **1342**, 164 (1997).]

structures.[35] Both CD and fluorescence measurements, therefore, verify that recoverin undergoes significant conformational changes on binding calcium.

Calcium-induced structural changes in recoverin (and the other calcium-binding proteins identified in these studies) alter the hydrophobicity of these proteins, as revealed initially by their enhanced binding to phenyl-Sepharose. New hydrophobic sites presumably underlie their interactions with specific target molecules. Unfortunately, the identity of such target molecules is known with some assurance in only a few instances. To understand the function of a calcium-binding protein and thus interpret the complex calcium signaling that occurs in a cell, it is essential that additional target molecules be identified.

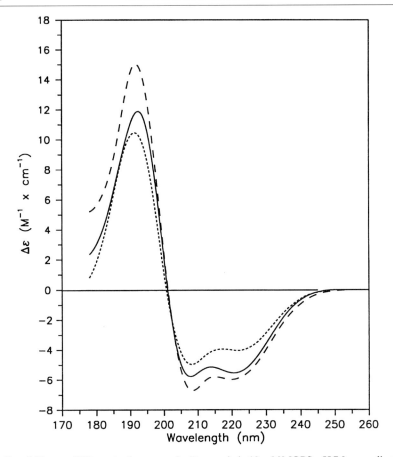

FIG. 6. Vacuum UV spectra for recoverin. Recoverin in 10 mM MOPS, pH 7.0, was adjusted to 1 mM Ca^{2+} (—), 0.1 mM Ca^{2+} (---), or 1 mM EGTA (· · ·) and the spectra measured as described previously. [W. C. Johnson, K. Palczewski, W. A. Gorczyca, J. H. Riazance-Lawrence, D. Witkowska, and A. S. Polans, *Biochim. Biophys. Acta* **1342,** 164 (1997).]

Identifying Target Molecules for Calcium-Binding Proteins

Gel Overlay

Briefly, polyacrylamide gels of protein extract are fixed in 40% (v/v) methanol, 10% (v/v) acetic acid for 30 min, rinsed briefly in distilled water, and incubated in 10% (v/v) ethanol for 2 hr to remove excess SDS.[36] Gels then are incubated at 37° in 0.1 M imidazole, pH 7.0, for 10 min, fol-

---

[36] J. R. J. Glenney and K. Weber, *J. Biol. Chem.* **255,** 10551 (1980).

lowed by solution A [20 mM imidazole, 0.2 M KCl, 0.5% (w/v) gelatin, 0.02% (w/v) azide, pH 7.0] containing either 1 mM CaCl$_2$ or 1 mM EGTA for 10 min. Gels then are incubated at the same temperature for 4–8 hr in solution A typically containing ^{125}I-labeled calcium-binding protein (Bolton–Hunter reagent; Pierce). A general purpose ^{35}S-labeling reagent (Amersham) can in some instances substitute for iodination. Unbound label is removed by washing in large volumes of solution A over a 1- to 2-day period. Gels are dried and exposed to X-ray film or a phosphor imaging screen. Proteins that bind radiolabeled sample in a calcium-dependent manner are purified for identification by Edman degradation and mass spectrometry, as described previously.[19,35]

Figure 7 illustrates the binding of radiolabeled S100A11 to a fraction of cellular lysate obtained from a spindle cell ocular melanoma cell line. In the presence of 1 mM calcium S100A11 binds primarily to a protein of approximately 35 kDa (Fig. 7, lane a). This interaction appears to depend on calcium, because in the presence of 1 mM EGTA the labeling is almost eliminated (Fig. 7, lane b). S100A11 has been shown in other studies to

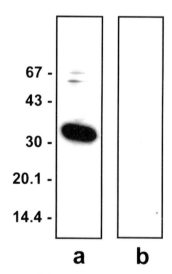

Fig. 7. S100A11 binding to a cellular fraction of ocular melanoma cell lysate detected by gel overlay. Proteins from a spindle cell ocular melanoma cell lysate were fractionated by SDS–PAGE and incubated with radiolabeled S100A11 as described in Methods and Results. Lanes a and b are autoradiograms of gels incubated with S100A11 in the presence of 1 mM CaCl$_2$ (a) or 1 mM EGTA (b).

bind to annexin I (38 kDa) in a calcium-dependent manner.[37] We have not yet determined whether annexin I is present in our preparation of ocular melanoma cell lysate, or whether annexin V (35 kDa) or an entirely different protein of approximately 35 kDa binds S100A11 in these experiments.

Cross-Linking and Coimmunoprecipitation

Cells or tissues are incubated for 10 min at room temperature with 1 mM DSP [dithiobis(succinimidyl propionate)], a cleavable cross-linking reagent that can penetrate cell membranes.[38] After washing with buffer, samples are lysed with 1% Triton X-100 in Tris-buffered saline containing protease inhibitors. Five microliters of antibody (specific for a calcium-binding protein) is added to the lysate and incubation is conducted for 1 hr at 4°. Protein G–agarose is then added to the solution and the incubation continued. The beads are washed sequentially with lysis buffer, Tris-buffered saline, and 50 mM Tris-HCl, pH 7.4, in order to reduce nonspecific binding. The sample then is heated at 100° for 5 min in the presence of SDS–gel solubilization buffer (without 2-mercaptoethanol) in order to elute protein from the beads. After centrifugation, the supernatant is subjected to analysis by SDS–PAGE either in the presence or absence of reducing agent.

Protein bands corresponding to the IgGs used during immunoprecipitation can be recognized by their well-defined masses that change in a predictable manner on reduction (Fig. 8, lanes a and b); they also are labeled by anti-immunoglobulin antibodies on Western blots. Additional bands, either silver stained on the gel or dye stained on polyvinylidene difluoride (PVDF) membrane, (other than the calcium-binding protein) represent potential target molecules.

In the absence of reducing agent, putative target molecules will remain cross-linked to the calcium-binding protein (Fig. 8, lane a). These cross-linked species will bind antibodies to the calcium-binding protein on Western blots but at higher molecular masses (owing to the cross-linked target protein). In the presence of reducing agent, however, the cross-linker will be cleaved and the putative target protein will dissociate from the calcium-binding protein (Fig. 8, lane b). In the absence of calcium (Fig. 8, lanes c and d), no interaction may occur between a calcium-binding protein and its cellular target, and therefore no cross-linked product will be observed.

Protein bands, other than IgG and the calcium-binding protein, corresponding to target proteins can be cut from the gel or from the PVDF

[37] M. Naka, Z. X. Qing, T. Sasaki, H. Kise, I. Tawara, S. Hamaguchi, and T. Tanaka, *Biochim. Biophys. Acta* **1223,** 348 (1994).
[38] C. Gamby, M. C. Waage, R. G. Allen, and L. Baizer, *J. Biol. Chem.* **271,** 26698 (1996).

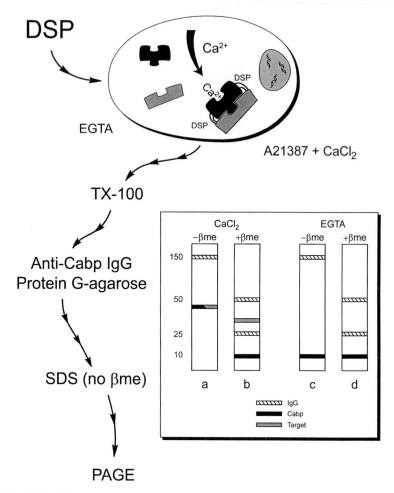

FIG. 8. Identifying cellular targets of a calcium-binding protein by cross-linking and coimmunoprecipitation. Depicted here is the interaction of a calcium-binding protein (Cabp) with its cellular target on the chelation of calcium. The complex is stabilized by the addition of a cleavable cross-linking reagent, DSP, as described in Methods and Results. After cell lysis, the complex is immunoprecipitated with an antibody specific for the calcium-binding protein and then eluted from the beads by treatment with SDS. Aliquots are fractionated by SDS–PAGE in the presence or absence of reducing agent (2-mercaptoethanol, βme). IgG used for immunoprecipitation migrates as a 150-kDa band in the absence of βme (lane a) but on reduction migrates as two bands (lane b; 50 kDa, heavy chain; 25 kDa, light chain). The calcium-binding protein (9 kDa in this example) complexes with its target (35 kDa) to produce a 45-kDa band in the absence of reducing agent (lane a). The individual components are separated on cleavage of the cross-linking reagent with βme (lane b). In some cases the calcium-binding protein may not be in sufficient proximity to its cellular target in the absence of calcium and, therefore, it will not be stabilized as a complex by the cross-linking reagent. Under these conditions only the IgG and calcium-binding protein will be detected by SDS–PAGE (lanes c and d).

membrane and prepared for sequence analysis.[39] Briefly, after dehydration in acetonitrile, gel pieces are dried and a small volume of 100 mM NH$_4$HCO$_3$ containing 10 mM DTT is added. Incubation is conducted at 56° for 1 hr. Iodoacetamide, 55 mM in NH$_4$CO$_3$ buffer, is then exchanged for the original solution and incubation is conducted in the dark for 45 min. The gel pieces are sequentially dehydrated with acetonitrile and rehydrated with NH$_4$CO$_3$ two times. Dried gel pieces are rehydrated in digestion buffer consisting of 50 mM NH$_4$CO$_3$, 5 mM CaCl$_2$, and trypsin (12.5 ng/μl). The solution is removed after 1 hr and the same buffer without enzyme is added and incubation continued overnight at 37°. Peptides are recovered with NH$_4$CO$_3$ buffer and three changes of 5% (v/v) formic acid in 50% (v/v) acetonitrile and then dried. Peptides are resuspended in 5% (v/v) formic acid, purified by RP-HPLC, and sequenced by Edman analysis and/or mass spectrometry.[19,35]

Once identified, a purified form of the target protein can be obtained, and its specific binding verified by coimmunoprecipitation using specific antibodies and protein G–agarose.[40] Briefly, putative target molecules are iodinated (Bolton-Hunter reagent; Pierce) or radiolabeled by alternative methods. The calcium-binding protein and the radiolabeled target protein then are incubated in phosphate-buffered saline. The incubation solution contains either 1 mM EGTA or varying concentrations of CaCl$_2$. Incubation is conducted for 1 hr at room temperature. Affinity-purified antibodies to the calcium-binding protein and protein G-Sepharose are added sequentially, each for a 30- to 60-min incubation. After centrifugation, the beads are washed five times with the same buffers containing 0.1% Triton X-100. The pelleted beads are processed for scintillation counting to determine the amount of putative target molecule that binds in a calcium-dependent manner. If antibodies are available to the putative target molecule (or can be generated), coimmunoprecipitation studies can be performed with radiolabeled calcium-binding protein, target protein, and antibodies to the putative target molecule; this approach is useful when the antigenic sites on the calcium-binding protein are masked by the target molecule.

Alternative Methods

If the traditional methods of gel overlay, cross-linking, and coimmunoprecipitation are unsuccessful, then alternative approaches for identifying target proteins can be pursued. The yeast two-hybrid system was developed to detect interacting proteins, and to provide the cloned genes for identifi-

[39] A. Shevchenko, M. Wilm, O. Vorm, and M. Mann, *Anal. Chem.* **68,** 850 (1996).
[40] Y. Watanabe, N. Usuda, S. Tsugane, R. Kobayashi, and H. Hidaka, *J. Biol. Chem.* **267,** 17136 (1992).

cation.[41] This approach has been used successfully to identify target molecules for calcium-binding proteins.[42]

Alternatively, protein complexes containing a histidine tag have been isolated and identified by mass spectrometry.[43,44] Cells transfected with a plasmid encoding a hexahistidine-tagged calcium-binding protein can be lysed and complexes then isolated by nickel nitrilotriacetic acid affinity chromatography. (Immunoaffinity chromatography can follow to purify the complex further.) Once the complex is dissociated and resolved by SDS–PAGE, individual protein bands are reduced in gel and alkylated, followed by digestion with trypsin as previously described. Peptides are extracted, desalted, and identified by electrospray mass spectrometry either by using a sequence database or by tandem mass spectrometry.[44] These experiments can be performed initially in the presence of either EGTA or calcium ionophore (as in Fig. 8) to alter the intracellular concentration of calcium as a further indicator of the specificity of the protein interactions in the complex.

Concluding Remarks

Organisms from the four kingdoms express calcium-binding proteins with similar structural features, indicating an evolutionary drive to harness calcium for a multitude of signal transduction pathways. Unfortunately, in most instances large portions of these pathways remain unknown and the molecular details describing the binding of calcium and the subsequent interactions with cellular targets are bereft of elucidation. However, just as studies of phototransduction have formed the framework for investigations of more complex and inaccessible tissues, the various functions of calcium and calcium-binding proteins deciphered in the retina and other ocular tissues may lead to a clearer understanding of general principles applicable to a wide variety of cells and tissues. Studies of GCAP1, for example, may be pertinent to other cellular processes involving calcium for which the molecular mechanisms of target recognition and regulation are difficult to evaluate. Other calcium-binding proteins expressed in ocular tissues, including recoverin, calbindin,[45] and a host of S100-like proteins,

[41] C. Chien, P. L. Bartel, R. Sternglanz, and S. Fields, *Proc. Natl. Acad. Sci. U.S.A.* **88**, 9578 (1991).
[42] T. Wu, C. W. Angus, X. L. Yao, C. Logan, and J. H. Shelhamer, *J. Biol. Chem.* **272**, 17145 (1997).
[43] D. A. Fancy, K. Melcher, S. A. Johnston, and T. Kodadek, *Chem. Biol.* **3**, 551 (1996).
[44] G. Neubauer, A. Gottschalk, P. Fabrizio, B. Seraphin, R. Luhrmann, and M. Mann, *Proc. Natl. Acad. Sci. U.S.A.* **94**, 385 (1997).
[45] T. L. Haley, R. Pochet, L. Baizer, M. D. Burton, J. W. Crabb, M. Parmentier, and A. S. Polans, *Visual Neurosci.* **12**, 301 (1995).

likewise offer unique opportunities to contribute to our understanding of calcium signaling during normal cellular processes and during the progression of a disease.

Several strategies have been provided in this chapter for the purification and characterization of calcium-binding proteins. These strategies comprise only the first step toward unraveling the assortment of biological processes involving calcium. As in other areas of biochemistry, however, new methods must be devised for the study of protein–protein interactions, because currently the most significant obstacle to our understanding of calcium signaling is the minimal number of methods available for identifying bona fide cellular targets.

Acknowledgments

We thank Erik Nielsen for preparation of the figures. This research was supported by grants from Research to Prevent Blindness, Inc. (RPB), to the Department of Ophthalmology and Visual Sciences, University of Wisconsin. A.S.P. is the recipient of a Jules and Doris Stein Professorship from RPB.

[8] Molecular Structure of Membrane-Targeting Calcium Sensors in Vision: Recoverin and Guanylate Cyclase-Activating Protein 2

By JAMES B. AMES, MITSUHIKO IKURA, and LUBERT STRYER

Introduction

An important class of calcium-binding proteins present in retinal photoreceptor cells includes recoverin from mammalian rods,[1] guanylate cyclase-activating proteins (GCAP-1, GCAP-2, and GCAP-3) from mammalian rods and cones,[2–4] and guanylate cyclase inhibitory protein (GCIP) from

---

[1] A. M. Dizhoor, S. Ray, S. Kumar, G. Niemi, M. Spencer, D. Brolley, K. A. Walsh, P. P. Philipov, J. B. Hurley, and L. Stryer, *Science,* **251,** 915 (1991).

[2] K. Palczewski, I. Subbaraya, W. A. Gorczyca, B. S. Helekar, C. C. Ruiz, H. Ohguro, J. Huang, X. Zhao, J. W. Crabb, R. S. Johnson, K. A. Walsh, M. P. Gray-Keller, P. B. Detwiler, and W. Baehr, *Neuron* **13,** 395 (1994).

[3] A. M. Dizhoor, D. G. Lowe, E. V. Olshevskaya, R. P. Laura, and J. B. Hurley, *Neuron* **12,** 1345 (1994).

[4] F. Haeseleer, I. Sokal, N. Li, M. Pettenati, R. Nagesh, D. Bronson, R. Wechter, W. Baehr, and K. Palczewski, *J. Biol. Chem.* **274,** 6526 (1999).

frog rods.[5] Their amino acid sequences indicate that they are members of the EF-hand superfamily[6] that includes calmodulin and troponin C (Fig. 1). The recoverin branch of the EF-hand superfamily includes neuronal Ca^{2+} sensors such as neurocalcin, frequenin, and hippocalcin (reviewed in Ref. 7). Indeed, there is a homolog in yeast,[8] indicating that these calcium sensors arose early in the evolution of eukaryotes. The common features of these proteins are (1) an amino-terminal myristoylation consensus sequence, (2) an approximately 200-residue chain containing four EF-hand motifs, and (3) the sequence CPXG in the first EF-hand that markedly impairs its capacity to bind Ca^{2+}.

Mass spectrometric analysis of retinal recoverin and the GCAP proteins revealed that they are myristoylated at the amino terminus.[2,9,10] Recoverin contains an N-terminal myristoyl (14:0) or related fatty acyl group (12:0, 14:1, 14:2). Retinal recoverin and myristoylated recombinant recoverin, but not unmyristoylated recoverin, bind to membranes in a Ca^{2+}-dependent manner.[11,12] Likewise, bovine neurocalcin and hippocalcin contain an N-terminal myristoyl group and both exhibit Ca^{2+}-induced membrane binding.[13,14] These findings led to the proposal that recoverin and the neural homologs possess a Ca^{2+}-myristoyl switch (Fig. 2). The covalently attached fatty acid is highly sequestered in recoverin in the calcium-free state.[15] The binding of calcium to recoverin leads to the extrusion of the fatty acid,[16,17] making it available to interact with lipid bilayer membranes or other hydrophobic sites.

GCAP-2 exhibits different membrane-targeting properties[10] from those

---

[5] N. Li, R. N. Fariss, K. Zhang, A. Otto-Bruc, F. Haeseleer, D. Bronson, N. Qin, A. Yamazaki, I. Subbaraya, A. H. Milam, K. Palczewski, and W. Baehr, *Eur. J. Biochem.* **15,** 591 (1998).
[6] N. D. Moncrief, R. H. Kretsinger, and M. Goodman, *J. Mol. Evol.* **30,** 522 (1990).
[7] J. B. Ames, T. Tanaka, L. Stryer, and M. Ikura, *Curr. Opin. Struct. Biol.* **6,** 432 (1996).
[8] K. Devlin, C. M. Churcher, B. G. Marrell, M. A. Rajandream, and S. V. Walsh, unpublished (1995); locus NCS1SCHPO, Swiss-Prot accession no. Q09711.
[9] A. M. Dizhoor, L. H. Ericcson, R. S. Johnson, S. Kumar, E. Olshevskaya, S. Zozulya, T. A. Neubert, L. Stryer, J. B. Hurley, and K. A. Walsh, *J. Biol. Chem.* **267,** 16033 (1992).
[10] E. V. Olshevskaya, R. E. Hughes, J. B. Hurley, and A. M. Dizhoor, *J. Biol. Chem.* **272,** 14327 (1997).
[11] S. Zozulya, and L. Stryer, *Proc. Natl. Acad. Sci. U.S.A.* **89,** 11569 (1992).
[12] A. M. Dizhoor, C. K. Chen, E. Olshevskaya, V. V. Sinelnikova, P. Phillipov, and J. B. Hurley, *Science* **259,** 829 (1993).
[13] D. Ladant, *J. Biol. Chem.* **270,** 3179 (1995).
[14] K. Kobayashi, S. Takamatso, S. Saitoh, and T. Noguchi, *J. Biol. Chem.* **268,** 18898 (1993).
[15] T. Tanaka, J. B. Ames, T. S. Harvey, L. Stryer, and M. Ikura, *Nature* (*London*) **376,** 444 (1995).
[16] J. B. Ames, T. Tanaka, M. Ikura, and L. Stryer, *J. Biol. Chem.* **270,** 30909 (1995).
[17] J. B. Ames, R. Ishima, T. Tanaka, J. I. Gordon, L. Stryer, and M. Ikura, *Nature* (*London*) **389,** 198 (1997).

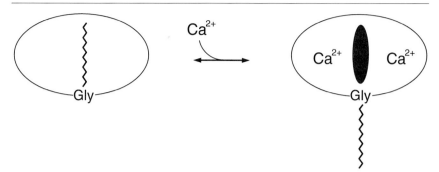

FIG. 2. Schematic diagram of calcium–myristoyl switch in recoverin. The binding of two Ca^{2+} ions promotes the extrusion of the myristoyl group and exposure of other hydrophobic residues (marked by the blackened oval).

of recoverin. The myristoylated and unmyristoylated forms of GCAP-2 bind to membranes at low Ca^{2+} concentrations (<100 nM). The GCAP-2 membrane interaction is highly sensitive to ionic strength and is much stronger at higher salt concentrations (>50 mM NaCl). Interestingly, the myristoylated form of GCAP-2 appears cytosolic at high Ca^{2+} and low ionic strength (<50 mM NaCl), although the Ca^{2+}-bound GCAP-2 does bind to membranes at physiological ionic strength. Recoverin and the GCAPs regulate different proteins. The Ca^{2+}-bound form of recoverin blocks the phosphorylation of photoexcited rhodopsin *in vitro*,[18,19] thereby prolonging the photoresponse.[20,21] Myristoylated recoverin is much more effective than unmyristoylated protein in inhibiting the deactivation of photoexcited rhodopsin.[20] In contrast, the GCAP proteins activate photoreceptor guanylate cyclases at low Ca^{2+} (<100 nM).[2,3] GCAP-2 in addition inhibits cylase at high Ca^{2+}.[10] Furthermore, the unmyristoylated form of GCAP-2 is nearly as effective as the myristoylated protein in activating the cyclase *in vitro*.[10]

In this chapter we present a detailed structural analysis of recoverin and GCAP-2. The Ca^{2+}-induced structural changes in these proteins are important for elucidating their membrane-targeting mechanisms and for understanding the molecular mechanism of Ca^{2+}-sensitive regulation of phototransduction.

---

[18] C. K. Chen, J. Inglese, R. J. Lefkowitz, and J. B. Hurley, *J. Biol. Chem.* **270**, 18060 (1995).
[19] V. A. Klenchin, P. D. Calvert, and M. D. Bownds, *J. Biol. Chem.* **270**, 16147 (1995).
[20] M. A. Erickson, L. Lagnado, S. Zozulya, T. A. Neubert, L. Stryer, and D. A. Baylor, *Proc. Natl. Acad. Sci. U.S.A.* **95**, 6474 (1997).
[21] M. P. Gray-Keller, A. S. Polans, K. Palczewski, and P. B. Detwiler, *Neuron* **10**, 523 (1993).

Methods

Preparation of Isotopically Labeled Recoverin and Guanylate Cyclase-Activating Protein 2

Myristoylated recombinant recoverin protein uniformly labeled with nitrogen-15 and carbon-13 is expressed in *Escherichia coli* DH5α (containing plasmids pTREC2, for recoverin expression, and pBB131, for expression of yeast *N*-myristoyltransferase) grown in M9 minimal medium (containing $^{15}NH_4Cl$ and [6-^{13}C] glucose) with ampicillin and kanamycin (100 mg/liter each) at 30°. At an OD_{600} of 0.7, cultures are supplied with [14-^{13}C] myristic acid (2 mg/liter; Isotec, Miamisburg, OH) and 0.5 mM isopropyl-β-D-thiogalactoside (IPTG) to induce expression of *N*-myristoyltransferase. After 45 min, the cultures are heat induced by rapidly shifting the temperature to 42° to induce expression of recombinant recoverin. After 2 hr of heat induction, the cells are harvested by centrifugation (8000g, 10 min, 4°), resuspended in 10 ml of buffer A [100 mM KCl, 1 mM dithiothreitol (DTT), 1 mM $MgCl_2$, 50 mM HEPES, pH 7.5] per liter of bacterial culture, and stored at −70°.

Approximately 10 mg of isotopically labeled, myristoylated recoverin is purified per liter of bacterial culture. One volume of buffer A supplemented with 2.0 mM EGTA, 0.1 mM phenylmethylsulfonyl fluoride (PMSF), and egg lysozyme (0.2 mg/ml) is added to the thawed cell slurry, and cells are disrupted by sonication for 5 min (sonifier 450, preparative size probe, 80% power, 50% duty cycle; Branson, Danbury, CT). Streptomycin sulfate (5% stock solution) is added over a 10-min period, with stirring, to a final concentration of 0.1% (w/v). Cell debris is removed by centrifugation (30,000g, 20 min, 4°). The cleared lysate is supplemented with 1 mM $CaCl_2$ and applied to a 50-ml phenyl-Sepharose 4B (Pharmacia, Piscataway, NJ) column at a flow rate of 2 ml/min. The column is washed with 500 ml of buffer A containing 1 mM $CaCl_2$ (or until the UV absorbance at 280 nm of the flow-through returns to baseline). Recoverin is eluted as a narrow peak in a volume of 50–100 ml, using buffer A containing 2 mM EGTA. The eluate is diluted threefold with ice-cold water and applied directly to an 80-ml Q-Sepharose Fast Flow (Pharmacia) column equilibrated with buffer B (1 mM DTT, 1 mM EGTA, 20 mM Tris-HCl, pH 8.0). Recoverin is eluted at approximately 100 mM KCl, using a gradient of 0–200 mM KCl in buffer B at a flow rate of 2 ml/min. Protein fractions are analyzed by sodium dodecyl sulfate–polyacrylamide gel electrophoresis (SDS–PAGE), pooled, and concentrated up to 20 mg/ml by centrifugal concentrators (Centriprep-10; Amicon, Danvers, MA). Recoverin can be stored at 4° for several weeks or kept at −70° for long-term storage for up to 1 year.

Unmyristoylated recombinant GCAP-2 protein uniformly labeled with nitrogen-15 and/or carbon-13 is expressed in *E. coli* BLR(DE3)pLysS (containing pET11d expression vector) grown in M9 minimal medium as described above with ampicillin (100 mg/liter) at 37°. At an OD_{600} of 0.5, cultures are supplied with 0.7 mM IPTG to induce expression of recombinant GCAP-2. After 3 hr of induction, the cells are harvested by centrifugation and resuspended in buffer A as described above. The cell slurry is sonicated as described above and the recombinant GCAP-2 protein (insoluble inclusion bodies) sediments with the cell debris by centrifugation (30,000g, 20 min, 4°). GCAP-2 is extracted from the pellet fraction by suspending the pellet in 10 ml of 8 M urea containing 1 mM EDTA and 50 mM DTT at 4° for 30 min. The urea-solubilized protein solution is dialyzed against 2 liters of buffer A overnight at 4°. Precipitated material is removed by centrifugation (30,000g for 20 min at 4°). Recombinant GCAP-2 protein is then purified by gel filtration on a Sephacryl S100 column in buffer A at a flow rate of 1 ml/min. Fractions containing GCAP-2 are diluted threefold with ice-cold water and are applied to a 50-ml DEAE-Sepharose column equilibrated in buffer C (1 mM EDTA, 1 mM DTT, 50 mM imidazole, pH 6.1) at 25°. GCAP-2 is eluted from the DEAE-Sepharose column at approximately 0.4 M KCl at pH 6.1, using a gradient of 0 to 0.5 M KCl in buffer C at a flow rate of 2 ml/min. The GCAP-2 protein fractions are combined and concentrated using Amicon YM10 membranes under nitrogen pressure to 10 mg/ml and stored at $-70°$. The typical yield of purified protein is 5–10 mg/liter of bacterial culture.

Protein samples for the nuclear magnetic resonance (NMR) experiments are prepared by placing 10 mg of isotopically labeled protein in 0.5 ml of a 95% H_2O–5% ^2H_2O solution containing 50 mM KCl, 10 mM [10-^{2}H] dithiothreitol, and 10 mM $CaCl_2$ (Ca^{2+}-bound state) or 1 mM [28-^{2}H] EDTA (Ca^{2+}-free state) at pH 7.0. The buffer in samples of recoverin and GCAP-2 can be rapidly exchanged with disposable desalting columns (Econo-Pak 10DG; Bio-Rad, Hercules, CA) equilibrated with the desired buffer.

Nuclear Magnetic Resonance Spectroscopy

All NMR experiments are performed with a Varian (Santa Clara, CA) Unity-Plus 500 or Unity-600 spectrometer equipped with a four-channel interface and a triple-resonance probe with an actively shielded z gradient, together with a pulse-field gradient accessory. The NMR data are processed by the software package NMRPipe and NMRDraw (F. Delaglio, National Institutes of Health, Bethesda, MD), and spectral analysis is performed with the software Capp and Pipp.[22]

---

[22] D. S. Garrett, R. Powers, A. M. Gronenborn, and G. M. Clore, *J. Magn. Reson.* **95**, 214 (1991).

The structures of recombinant recoverin and GCAP-2 uniformly labeled with carbon-13 and nitrogen-15 are determined by heteronuclear NMR spectroscopy. To elucidate the structure, resonances in the NMR spectra are first assigned to specific amino acid residues. Triple-resonance NMR experiments correlating ^{15}N, ^{13}C, and ^{1}H resonances (including HNCO, HNCACB, CBCACONNH, HBHACONNH, and CBCACOCAHA) are performed and analyzed to facilitate making the assignment of backbone resonances.[23] The backbone assignments serve as the basis for assigning the side-chain resonances by analyzing ^{15}N- and ^{13}C-edited total correlation spectroscopy (TOCSY) and nuclear Overhauser effect spectroscopy (NOESY) experiments.[24] The complete sequence-specific assignments then serve as the basis for analyzing ^{15}N- and ^{13}C-edited NOESY experiments to establish more than 3000 proton–proton distance relationships throughout the protein. In addition, dihedral angle information is deduced from the analysis of j couplings ($^3J_{NH\alpha}$) and chemical shift data.[23] Finally, the NMR-derived distances and dihedral angle information are used to calculate the three-dimensional structure, using distance geometry and restrained molecular dynamics. The structure calculations are performed according to the YASAP protocol[25] within the program X-PLOR,[26] as described previously.[24]

Three-Dimensional Structures of Recoverin

Atomic resolution structures of the various forms of recoverin were determined to elucidate the detailed molecular mechanism of the Ca^{2+}-myristoyl switch. The X-ray crystal structure of recombinant unmyristoylated recoverin first showed it to contain a compact array of EF-hand motifs,[27] in contrast to the dumbbell shape of calmodulin[28] and troponin C.[29] The linker between the two domains, one containing EF-1 and EF-2, the other containing EF-3 and EF-4, is U-shaped rather than α helical. Ca^{2+} is bound to EF-3 and Sm^{3+} (used to derive phases) is bound to EF-2. The other two EF hands possess novel features that prevented ion binding. EF-1 is disabled by a Cys–Pro sequence in the binding loop. EF-4 contains

---

[23] J. B. Ames, T. Tanaka, L. Stryer, and M. Ikura, *Biochemistry* **33**, 10743 (1994).
[24] T. Tanaka, J. B. Ames, M. Kainosho, L. Stryer, and M. Ikura, *J. Biomol. Nucl. Magn. Reson.* **11**, 135 (1998).
[25] M. Nilges, A. M. Gronenborn, A. T. Brunger, and G. M. Clore, *Protein Eng.* **2**, 27 (1988).
[26] A. T. Brunger, "X-PLOR: A System for X-Ray Crystallography and NMR," version 3.1. Yale University Press, New Haven, Connecticut, 1992.
[27] K. M. Flaherty, S. Zozulya, L. Stryer, and D. B. McKay, *Cell* **75**, 709 (1993).
[28] Y. S. Babu, C. E. Bugg, and W. J. Cook, *J. Mol. Biol.* **204**, 191 (1988).
[29] O. Herzberg, and M. N. James, *Nature* (*London*) **313**, 653 (1985).

```
              10         20         30         40         50         60         70
GCAP2  GQQFSWEEA EEN----GAVGAAD AAQLQEWYKK FLEECPSGTL FMHEFKRFFK VPDNEE-ATQY VEAMFRAFDT
GCAP1  GNIMDGKSV EE---------LS STECHQWYKK FMTECPSGQL TLYEFRQFFG LKNLSPWASQY VEQMFETFDF
GCAP3  GNGKSIAGD QKA---------VP TQETHVWYRT FMMECPSGLQ TLHEFKTLLG LQGLNQKANKH IDQVYNTFDT
GCIP   GQVASMPHR CG---------TY VLELHEWYRK FVEECPSGLI TLHEFRQFFS DVTVGENSSEY AEQIFRALDN
REC    GNSKSGALS KEILEELQLNTKFT EEELSSWYQS FLKECPSGRI TRQEFQTIYS KFFPEADPKAY AQHVFRSFDA

              80         90        100        110        120        130        140
GCAP2  NGDNTIDFLE YVAALNLVLR GTLEHKLKWT FKIYDKDRNG CIDRQELLDI VESIYKLKKA CSVEVEAEQQ
GCAP1  NKDGYIDFME YVAALSLVLK GKVEQKLRWY FKLYDVDGNG CIDRDELLTI IRAIRAIN-- --------CS
GCAP3  NKDGFIDFLE FIAAVNLIMQ EKMEQKLKWY FKLYDKDGNG SIDKNELLDM FMAVQALNG- ----------
GCIP   NGDGIVDFRE YVTAISMLAH GTPEDKLKWS FKLYDKDGDG AITRSEMLEI MRAVYKMSVV ASL-----TK
REC    NSDGTLDFKE YVIALHMTSA GKTNQKLEWA FSLYDVDGNG TISKNEVLEI VTAIFKMI-- -SPEDTKHLP

              150        160        170        180        190        200
GCAP2  GKLLTPEEVV DRIFLLVDEN GDGQLSLNEF VEGARRDKWV MKMLQMDLNP SSWISQQRRK SAMF    204
GCAP1  DSTMTAEEFT DTVFSKIDVN GDGELSLEEF MEGVQKDQML LDTLTRSLDL TRIVRRLQNG EQDEEGASGR
GCAP3  QQTLSPEEFI NLVFHKIDIN NDGELTLEEF INGMAKDQDL LEIVYKSFDF SNVLRVICNG KQPDMETDSS
GCIP   VNPMTAEECT NRIFVRLDKD QNAIISLQEF VDGSLGDEWV RQMLECDLST VEIQKMTKHS HLPARSSRER
REC    EDENTPEKRA EKIWGFFGKK DDDKLTEKEF IEGTLANKEI LRLIQFE--P QKVKEKLKEK KL      202

GCAP1  ETEAAEADG   205
GCAP3  KSPDKAGLGKVKMK 209
GCIP   LFHANT      202
```

FIG. 1. Amino acid sequence alignment of bovine GCAP-2 (accession no. U32856) with bovine GCAP-1 (accession no. P46065), human GCAP-3 (accession no. AF110002), frog GCIP (accession no. AF047884), and bovine recoverin (accession no. P21457). The 29-residue EF-hand motifs are highlighted in color: EF-1 (green), EF-2 (red), EF-3 (cyan), EF-4 (yellow). Regions of regular secondary structure (α helices and β strands) are indicated schematically.

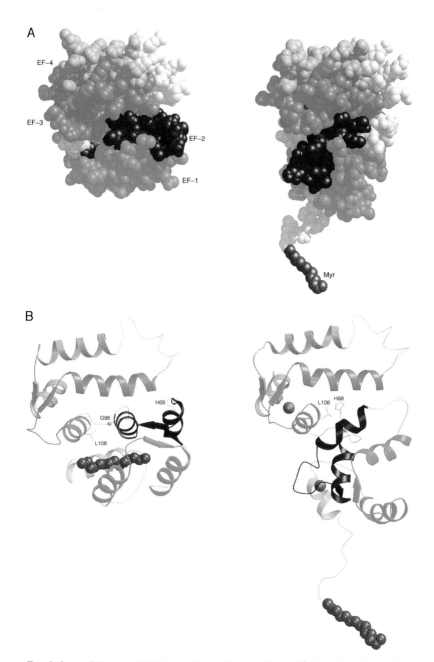

FIG. 3. Space-filling model (A) and ribbon diagram (B) of Ca^{2+}-free (left, 1iku.pdb) and Ca^{2+}-bound (right, 1jsa.pdb) myristoylated recoverin. The C-terminal domains of the two forms are aligned to show the Ca^{2+}-induced 45° rotation of the N-terminal domain. This and the other color illustrations were generated with Midas,[29a] Molscript,[29b] and Raster3d.[29c]

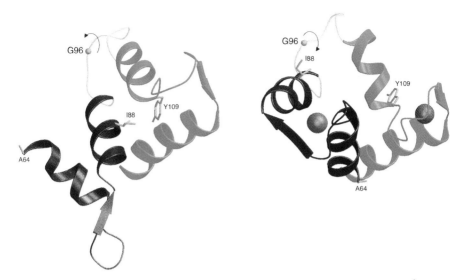

Fig. 4. Schematic ribbon representation of the structure of EF-2 and EF-3 of Ca^{2+}-free (left) and Ca^{2+}-bound (right) myristoylated recoverin. Rotation about Gly-96 dramatically alters the interaction between EF-2 and EF-3, as highlighted by the displacement of Ile-88 and Tyr-109.

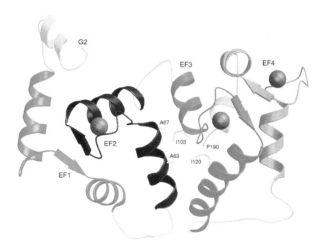

Fig. 7. Schematic ribbon diagram of the structure of unmyristoylated GCAP-2 with three bound Ca^{2+}. The side chain atoms of residues at the domain interface (Ala-63, Ala-67, Ile-103, and Ile-120) are indicated.

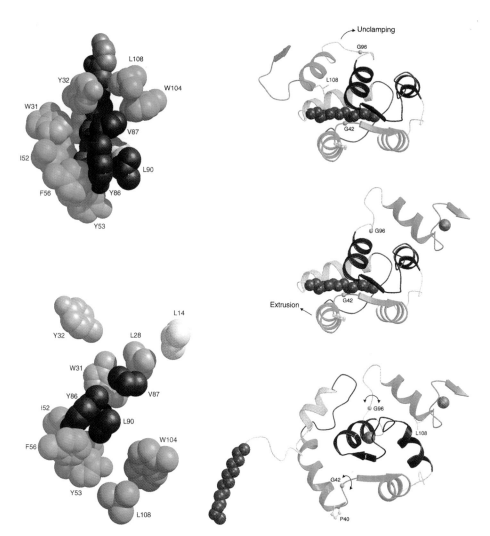

FIG. 5. Space-filling representation of hydrophobic residues that clamp the myristoyl group in Ca^{2+}-free recoverin (top). The clamp is released in Ca^{2+}-bound recoverin (bottom).

FIG. 6. A possible pathway for the calcium-induced extrusion of the myristoyl group of recoverin. The upper (no bound Ca^+) and lower (two bound Ca^{2+}) structures were determined by NMR. The middle structure (one bound Ca^{2+}) is hypothetical. We propose that rotation about Gly-96, which leads to unclamping, is followed by rotation about Gly-42, which leads to extrusion of the myristoyl group.

an internal salt bridge in the binding loop that competes with Ca^{2+} binding. Myristoylated recoverin, the physiologically active form, has thus far eluded crystallization.

The structures of myristoylated recoverin in the Ca^{2+}-free and Ca^{2+}-bound states were determined by NMR spectroscopy (Fig. 3[29a-c]). In the Ca^{2+}-free state, the myristoyl group is sequestered in a deep hydrophobic cavity in the N-terminal domain.[15] The cavity is formed from five α helices. The two helices of EF-1 (residues 26–36 and 46–56), the exiting helix of EF-2 (residues 83–93), and entering helix of EF-3 (residues 100–109) lie perpendicular to the fatty acyl chain and form a box-like arrangement that surrounds the myristoyl group laterally. A long, amphipathic α helix near the N terminus (residues 4–16) packs closely against and runs antiparallel to the fatty acyl group, and serves as a lid on top of the four-helix box. The N-terminal residues Gly-2 and Asn-3 form a tight hairpin turn that connects the myristoyl group to the N-terminal helix. This turn positions the myristoyl group inside the hydrophobic cavity and gives the impression of a cocked trigger. The bond angle strain stored in the tight hairpin turn may help eject the myristoyl group from the pocket once Ca^{2+} binds to the protein.

The structure of myristoylated recoverin with two Ca^{2+} bound shows the amino-terminal myristoyl group to be extruded (Fig. 3).[17] The N-terminal eight residues are solvent exposed and highly flexible and thus serve as a mobile arm to position the myristoyl group outside the protein when Ca^{2+} is bound. The flexible arm is followed by a short α helix (residues 9–17) that precedes the four EF-hand motifs, arranged in a tandem array as was seen in the X-ray structure. Calcium ions are bound to EF-2 and EF-3. EF-3 has the classic "open conformation" similar to the Ca^{2+}-occupied EF hands in calmodulin[28] and troponin C.[29] EF-2 is somewhat unusual. The root mean squared (RMS) deviation of the main chain atoms of EF-2 is 3.7 Å in comparing recoverin with calmodulin. Interestingly, the helix packing angle of Ca^{2+}-bound EF-2 (120°) in recoverin more closely resembles that of the Ca^{2+}-free EF hands (in the "closed conformation") found in calmodulin and troponin C. The overall topology of Ca^{2+}-bound myristoylated recoverin is similar to the X-ray structure of unmyristoylated recoverin with one Ca^{2+} bound.[27] The RMS deviation of the main chain atoms in the EF-hand motifs is 1.5 Å in comparing Ca^{2+}-bound myristoylated recoverin with unmyristoylated recoverin.

[29a] T. Ferrin, C. Huang, L. Jarvis, and R. Langridge, *J. Mol. Graphics* **6,** 13 (1988).
[29b] P. J. Kraulis, *J. Appl. Crystallogr.* **24,** 946 (1991).
[29c] D. J. Bacon and W. F. Anderson, *J. Mol. Graphics* **6,** 219 (1988).

Calcium-Induced Conformational Changes in Recoverin

The structures of Ca^{2+}-free and Ca^{2+}-bound recoverin are compared in Fig. 3. A striking feature of these structures is the large rotation of the two domains. The C-terminal domains of the two forms are similar, apart from minor changes in the Ca^{2+}-binding loop and entering helix of EF-3. The N-terminal domain, in contrast, undergoes a striking rearrangement that leads to the extrusion of the myristoyl group.

Extrusion of the myristoyl group requires the binding of Ca^{2+} to EF-2 and EF-3.[11] The binding of Ca^{2+} to EF-3 decreases its interhelical angle by 15°, similar to the Ca^{2+}-induced "opening" of EF hands seen in calmodulin and troponin C. Calcium binding to EF-2 does not change its interhelical angle much, but instead Ca^{2+} binding causes the exiting helix to twist 30° clockwise about its central helical axis.[30] This Ca^{2+}-induced helical twisting in EF-2 is novel and has not been observed previously in other members of the superfamily. The Ca^{2+}-induced conformational changes in EF-3 and EF-2 alter the interaction of these EF hands at the domain interface and promote a conformational change near Gly-96 in the interdomain linker. The interface between the two domains is rearranged completely by rotation at Gly-96, leading to a 45° rotation of one domain with respect to the other (Fig. 3).

The structural consequences of Ca^{2+}-induced rotation at Gly-96 are highlighted in Fig. 4. In the Ca^{2+}-free state, the exiting helix of EF-2 contains a series of hydrophobic residues (Ile-88, Ala-89, His-91, and Met-92) that interact with hydrophobic residues in both the entering and exiting helices of EF-3 (Leu-108, Tyr-109, Ile-125, and Ala-128). The entering helix of EF-2 is exposed to solvent in the Ca^{2+}-free protein, and does not interact with EF-3. Conversely, in the Ca^{2+}-bound protein, the entering helix of EF-2 (marked by Ala-64 in Fig. 4) interacts intimately with EF-3, whereas the exiting helix of EF-2 (marked by Ile-88 in Fig. 4) does not. In particular, Ala-64, Tyr-65, and His-68 of the EF-2 entering helix interact with the hydrophobic residues Leu-108, Ile-125, Ala-128, and Met-132 of EF-3. Furthermore, the calcium-binding loops of EF-2 and EF-3 are 20 Å closer when Ca^{2+} is bound. The switching of helices accompanying the domain rotation accounts for the cooperativity of Ca^{2+} binding to EF-2 and EF-3.

The Ca^{2+}-induced rotation at Gly-96 also leads to a large displacement of residues necessary for the unclamping of the myristoyl group (Fig. 5). Many aromatic and other hydrophobic residues (notably Leu-28, Trp-31, Tyr-32, Ile-52, Phe-49, Tyr-53, Phe-56, Tyr-86, Leu-90, Trp-104, and Leu-108) fit snugly around the myristoyl group in the Ca^{2+}-free state. Swivel-

---

[30] K. Yap, J. B. Ames, M. Swindells, and M. Ikura, *Trends Biochem. Sci.*, in press (1999).

ing about Gly-96 causes a large relative displacement of the C-terminal residues (Trp-104 and Leu-108) away from the amino-terminal myristoyl group (Fig. 5 and 6). The Ca^{2+}-induced displacement of several clamping residues (Leu-14, Leu-28, Tyr-32, Trp-104, and Leu-108) results in the unclamping of the myristoyl group.

The extrusion of the myristoyl group is accomplished by a second important swivel at Gly-42 in the loop between the helices of EF-1 (Fig. 6). Rotation here moves the entering helix of EF-1 outward, causing it to pull on residues Gly-2 to Glu-26 on the N-terminal side of the helix. This displacement allows the myristoyl group to swing out of its binding pocket. Ejection of the myristoyl group is also made possible by the melting of part of the N-terminal helix. In the Ca^{2+}-free form, the N-terminal helix (Lys-5 to Glu-16) fits tightly against and is stabilized by the sequestered myristoyl group. In the Ca^{2+}-bound state, residues Gly-2 to Ala-8 have become part of the flexible arm that places the myristoyl group outside and gives it freedom to insert into a bilayer membrane or other hydrophobic site.

Mechanism of Ca^{2+}–Myristoyl Switch

We propose a plausible kinetic pathway of the Ca^{2+}-induced conformational changes described above to illustrate the atomic-level mechanism of the Ca^{2+}–myristoyl switch (Fig. 6). Initially, the protein is in the Ca^{2+}-free state with the myristoyl group sequestered. The first step is the binding of Ca^{2+} to EF-3, forming an intermediate with one Ca^{2+} bound. The X-ray structure of unmyristoylated recoverin shows Ca^{2+} is bound to EF-3 and not bound to EF-2.[27] Also, our Ca^{2+}-binding studies reveal that EF-3 binds Ca^{2+} with much higher intrinsic affinity than EF-2.[31] We propose that the initial binding of Ca^{2+} to EF-3 induces conformational changes in this EF hand that lead, in a small proportion of molecules, to the swiveling about Gly-96, which, in turn, causes unclamping of the sequestered myristoyl group (Figs. 5 and 6). The swiveling about Gly-96 also alters the interaction between EF-2 and EF-3 at the domain interface (Fig. 4), which may induce a structural change in EF-2 necessary to increase its Ca^{2+}-binding affinity. The binding of a second Ca^{2+} ion to EF-2 induces conformational changes in this EF hand that result in the swiveling about Gly-42 in EF-1. The swiveling about Gly-42 directly causes the myristoyl group to be lifted out of the binding pocket and places the fatty acid toward the exterior, where it can interact with membrane targets.

---

[31] J. B. Ames, T. Porumb, T. Tanaka, M. Ikura, and L. Stryer, *J. Biol. Chem.* **270,** 4526 (1995).

Three-Dimensional Structure of Guanylate Cyclase-Activating Protein 2

The structures of recoverin whetted our appetite for atomic resolution views of the GCAP proteins. None of these proteins has yet been crystallized. NMR spectroscopic studies of GCAP-2 were performed instead. Ideally, one would like to solve the structure of the Ca^{2+}-free and Ca^{2+}-bound forms of the myristoylated protein, the physiologic species, but this is not yet feasible because of their low solubility. We decided to begin the structural analysis of this important group of retinal calcium sensors by focusing on Ca^{2+}-bound unmyristoylated GCAP-2, which is soluble and gives clearly resolved NMR spectra. Moreover, the structure of the unmyristoylated form of GCAP-2 is likely to be biologically pertinent. Unmyristoylated GCAP-2 is nearly as effective as myristoylated GCAP-2 in activating guanylyl cyclase at low Ca^{2+} and inhibiting it at high Ca^{2+}.[10] Hence, structural studies of unmyristoylated GCAP-2 should reveal the Ca^{2+}-induced conformational changes underlying its regulation of cyclase.

The structure of Ca^{2+}-bound, unmyristoylated GCAP-2 is presented in Fig. 7. The entire polypeptide chain has been traced except for the disordered region at the carboxyl terminus (residues 191–204). The structure near the amino terminus (residues 2–18) and the region between EF-3 and EF-4 (residues 132–144) are rather poorly defined owing to dynamic disordering. GCAP-2 is a compact protein (radius of gyration 17 Å) made of two domains separated by a flexible linker. Each domain contains a pair of EF hands defined from the amino terminus: EF-1 (Ala^{22}–Val^{51}), EF-2 (Thr^{58}–Leu^{87}), EF-3 (Leu^{96}–Lys^{126}), and EF-4 (Glu^{147}–Arg^{176}). EF-1 and EF-2 interact intimately to form the N-terminal domain, and EF-3 and EF-4 interact to form the C-terminal domain. The entering helix of EF-2 (residues 58–68) packs against the helices of EF-3 (residues 96–104 and 114–127) at the interface between the two domains, forming a cleft. The linker between the domains is U-shaped, which positions the four EF hands in a compact tandem array like that found in Ca^{2+}-bound recoverin. Indeed, the overall folding of Ca^{2+}-bound unmyristoylated GCAP-2 closely resembles that of Ca^{2+}-bound unmyristoylated recoverin[27] and Ca^{2+}-bound myristoylated recoverin.[17] The RMS deviation of the main chain atoms (in the EF-hand motifs) is 2.2 Å in comparing GCAP-2 with recoverin.

Three Ca^{2+} ions are bound to GCAP-2, as anticipated on the basis of its amino acid sequence and site-directed mutagenesis.[32] Ca^{2+} is not bound to EF-1 because the binding loop is distorted from a favorable Ca^{2+}-binding geometry by Pro-36 at the fourth position of the 12-residue loop. Also, the

---

[32] A. M. Dizhoor and J. B. Hurley, *J. Biol. Chem.* **271,** 19346 (1996).

third residue in the loop (Cys-35) is not suitable for ligating Ca^{2+}. EF-2 binds Ca^{2+} and adopts a favorable structure despite the tight turn centered at Asn-74 (position 6 of the loop). The structure of EF-3 is strikingly similar to that of EF-3 in Ca^{2+}-bound recoverin and calmodulin. The RMS deviations of the 116 main chain atoms of EF-3 are 0.66 Å in comparing GCAP-2 with recoverin, and 0.80 Å in comparing GCAP-2 with calmodulin. The structure of EF-4 is quite different from that of recoverin. In recoverin, the second residue in the loop (Lys-161) forms a salt bridge with residue 12 (Glu-171) that disables Ca^{2+} binding. In GCAP-2, the second residue of the EF-4 loop (Glu-159) is negatively charged and cannot form a salt bridge that would impede Ca^{2+} binding. Furthermore, residues 1 and 3 of the EF-4 loop (AspD-158 and Asn-160) contain oxygen atoms in their side chains that can ligate Ca^{2+}, in contrast with the corresponding residues of recoverin (Gly-160 and Lys-162). Thus, Ca^{2+} binds to EF-4 much as it does to EF-2 and EF-3.

GCAP-2 has a solvent-exposed, hydrophobic surface formed by residues from EF-1 and EF-2. The exposed patch of hydrophobic residues is formed by the clustering of several aromatic side chains (Trp-27, Phe-31, Phe-45, Phe-48, Phe-49, and Tyr-81), and several aliphatic residues (Leu-24, Leu-40, Ile-76, Val-82, Leu-85, and Leu-89). These exposed hydrophobic residues are highly conserved in members of the family[7] and form a similar nonpolar patch in Ca^{2+}-bound recoverin.[17,27] The exposed hydrophobic patch of GCAP-2 may serve a role in regulating guanylyl cyclase. Site-directed mutagenesis studies reveal that many of these exposed residues are important in the cyclase interaction.[33] In particular, replacement of residues 78–110 (the exiting helix of EF-2) with corresponding residues of neurocalcin results in a chimeric protein that fails to inhibit guanylyl cyclase, but activates it at high Ca^{2+} levels. Also, the replacement of residues in EF-1 (residues 24–49) with the corresponding residues of neurocalcin renders the chimera completely inactive.

The structure of GCAP-2 near the amino terminus (residues 2–18) appears different from that of recoverin. There is virtually no sequence similarity between recoverin and the GCAPs in this region. The orientation of the amino-terminal helix (shown in pink in Fig. 7) is different in recoverin and GCAP-2. This helix in recoverin extends close to the interdomain linker, whereas it interacts primarily with the entering helix of EF-1 in GCAP-2. We note, however, that these apparent structural differences in the amino-terminal region between recoverin and GCAP-2 may result from the low precision of our structure in this region because of dynamic disorder-

[33] E. V. Olshevskaya, S. Boikov, J. B. Hurley, and A. M. Dizhoor, *J. Biol. Chem.* **274**, 10823 (1999).

ing. Deletion of the amino-terminal region by mutagenesis has no effect on the function of GCAP-2,[33] consistent with our finding that this region is structurally disordered.

The GCAP-2 structure is likely to be similar to that of GCAP-1 (40% sequence identity), GCAP-3 (35% identity), and GCIP (37% identity), because the overall main chain structure appears so similar to recoverin (2.2 Å RMS deviation; 30% identity). Most of the hydrophobic residues in the hydrophobic core and in the exposed patch are highly conserved. Also conserved are the residues that ligate Ca^{2+} in the EF-hand loops. Interestingly, important residues in the entering helix of EF-2 at the domain interface (Ala-57, Ala-63, and Ala-67) are not conserved. Other structurally important and nonconserved residues include Asn-74, Leu-79, Thr-93, His-95 and Thr-100. Considerable differences are also found in the amino-terminal (residues 2–18) and carboxyl-terminal (residues 191–204) regions. These differences suggest that the interaction and/or orientation between the N-terminal and C-terminal domains might be different in GCAP-1, GCAP-3, and GCIP. Indeed, a point mutation at the domain interface causes different phenotypes in GCAP-1 and GCAP-2. The mutation (Y99C) causes GCAP-1 to be constitutively active,[34,35] resulting in autosomal dominant cone dystrophy.[36] In contrast, the corresponding mutation in GCAP-2 (Y104C) does not alter its Ca^{2+} sensitivity and is not linked with cone degeneration.[34,35]

In summary, the overall main chain structure of GCAP-2 is similar to that of Ca^{2+}-bound recoverin except for structural differences near the amino terminus (residues 2–18) and the binding of Ca^{2+} to EF-4. We see an exposed hydrophobic patch of residues belonging to EF-1 and EF-2 that may play an important role in regulating guanylyl cyclase. Our next goal is to solve the structure of Ca^{2+}-free GCAP-2, a formidable challenge because of its lower stability and solubility, to elucidate fully the Ca^{2+}-induced structural changes that enable GCAP-2 to activate guanylyl cyclases in the absence of Ca^{2+}.

Acknowledgments

This work was supported by grants to J.B.A. and L.S. from the NIH, and to M.I. from the Medical Research Council of Canada. M.I. is a Howard Hughes Medical Institute international research scholar.

[34] A. M. Dizhoor, S. G. Boikov, and E. V. Olshevskaya, *J. Biol. Chem.* **273,** 17311 (1998).
[35] I. Sokal, N. Li, I. Surgucheva, M. J. Warren, A. M. Payne, S. S. Bhattacharya, W. Baehr, and K. Palczewski, *Mol. Cell* **2,** 129 (1999).
[36] A. M. Payne, S. M. Downes, D. A. R. Bessant, R. Taylor, G. E. Holder, M. J. Warren, A. C. Bird, and S. S. Bhattacharya, *Hum. Mol. Genet.* **7,** 273 (1998).

[9] Measurement of Light-Evoked Changes in Cytoplasmic Calcium in Functionally Intact Isolated Rod Outer Segments

By Peter B. Detwiler and Mark P. Gray-Keller

Introduction

It is well known in vertebrate photoreceptors that light-evoked changes in two intracellular second messengers, cGMP and Ca^{2+}, play essential roles in the generation, recovery, and adaptation of the electrical light response. The enzyme cascade activated by light leads to a decrease in both cGMP and Ca^{2+}, giving rise to intracellular second-messenger signals that have the opposite sign of the cyclic nucleotide and Ca^{2+} signals used in most other cell types.

In darkness a standing Ca^{2+} current circulates through the photoreceptor outer segment, a cellular compartment that houses all the transduction apparatus and is the only light-sensitive part of the receptor. Calcium flows into the rod outer segment (ROS) through cyclic nucleotide-gated (CNG) channels that are opened by cGMP and is pumped out again by an $Na^+:Ca^{2+},K^+$ exchanger. The photoisomerization of rhodopsin activates an amplified G-protein-coupled signal transduction cascade that increases cGMP hydrolysis by stimulating cGMP-specific phosphodiesterase (PDE). The resulting decrease in cGMP causes CNG channels to close. This reduces Ca^{2+} influx without affecting its efflux by continued $Na^+:Ca^{2+},K^+$ exchange and internal free Ca^{2+} (Ca_i^{2+}) falls. The light-evoked decrease in Ca_i^{2+} is a feedback signal that is thought to regulate a number of steps in the transduction cascade. The best understood action of the feedback signal is on guanylyl cyclase, which is stimulated by the decrease in Ca_i^{2+} and leads to the resynthesis of cGMP that is necessary for the timely recovery of resting dark current after light exposure. Calcium is clearly important in photoreceptor function and there have been a number of studies that have used Ca^{2+}-sensitive indicators to measure its steady state level in dark and light.[1–9] Here we describe the methods we have used to measure steady

---

[1] P. A. McNaughton, L. Cervetto, and B. J. Nunn, *Nature* (*London*) **322**, 261 (1986).
[2] G. M. Ratto, R. Payne, W. G. Owen, and R. Y. Tsien, *J. Neurosci.* **8**, 3240 (1988).
[3] J. I. Korenbrot and D. L. Miller, *Vision Res.* **29**, 939 (1989).
[4] L. Lagnado, L. Cervetto, and P. A. McNaughton, *J. Physiol.* **455**, 111 (1992).
[5] M. P. Gray-Keller and P. B. Detwiler, *Neuron* **13**, 849 (1994).
[6] S. T. McCarthy, J. P. Younger, and W. G. Owen, *Biophys. J.* **67**, 2076 (1994).

state Ca_i^{2+} as well as the transient changes in Ca_i^{2+} evoked by subsaturating flashes and steps of light.[5,7]

Protocol

The approach we have used consists of two parts: the first is to introduce Ca^{2+}-sensitive fluorescent dye into functionally intact isolated ROS, using internal dialysis via a whole-cell voltage clamp.[10–12] The second part is to use standard techniques to measure and calibrate Ca^{2+}-sensitive changes in dye fluorescence.

Whole-Cell Recording and Internal Dialysis

Because the transduction process and the Ca^{2+} changes that regulate it are confined to the outer segment, this part of the cell is separated from the rest of the photoreceptor by breaking the short, thin ciliary neck that connects the rod outer and inner segments. The broken end of the connecting cilium reseals rapidly and the isolated ROS is fully enclosed by an intact surface membrane that ions cross only by either permeating cGMP-gated channels or being pumped by $Na^+:Ca^{2+},K^+$ exchange. Our studies were done on isolated ROS from the adult nocturnal lizard (*Gecko gecko*) but we have used the same methods to record from isolated salamander and frog ROS.

Animals are dark adapted overnight and all manipulations are done in darkness with infrared illumination (>890 nm) and night vision goggles (NVEC-800HP; Night Vision Equipment) to maintain the adaptational state of the retina. After abrupt decapitation and rapid pithing both eyes are removed, hemisected, and the back half cut into quarters that are stored on ice in lizard Ringer solution (Table I) bubbled with 100% O_2. Rod outer segments are dissociated in ~0.5 ml of oxygenated Ringer solution by gently stretching a small piece of retina isolated from one of the quarter pieces of the hemisected eyecups. About 30 μl of dissociated cells is transferred to a drop (~250 μl) of Ringer solution held by surface tension between two parallel glass coverslips (No. 1 thickness) that are about

---

[7] M. P. Gray-Keller and P. B. Detwiler, *Neuron* **17**, 323 (1996).
[8] S. T. McCarthy, J. P. Younger, and W. G. Owen, *J. Neurophysiol.* **76**, 1991 (1996).
[9] A. P. Sampath, H. R. Matthews, M. C. Cornwall, and G. L. Fain, *J. Gen. Physiol.* **111**, 53 (1998).
[10] W. A. Sather and P. B. Detwiler, *Proc. Natl. Acad. Sci. U.S.A.* **84**, 9290 (1987).
[11] W. A. Sather, "Phototransduction in Detached Rod Outer Segments." Ph.D. Thesis, University of Washington, Seattle, 1988.
[12] G. Rispoli, W. A. Sather, and P. B. Detwiler, *J. Physiol.* **465**, 513 (1993).

TABLE I
Composition[a] of Internal Dialysis and External Bath Solutions[b]

Internal Solutions

	KAsp	KCl	MgCl$_2$	HEPES	ATP	GTP	Indo	CaCl$_2$	EGTA	pCa$_{free}$
Standard solution	110	10	6.05	5	5	1	0.1	0	0	8.41
R_{min}	15	10	6.05	5	5	1	0.1	0	50	11.23
R_{max}	120	0	6.05	5	5	1	0.1	20	0	1.8
$R_{mid\#1}$	30	10	6.3	5	5	1	0.1	17.5	25	6.69
$R_{mid\#2}$	40	10	6.25	5	5	1	0.1	20	22.7	6.2

External Solutions

	NaCl	KCl	MgSO$_4$	HEPES	CaCl$_2$	EGTA	LiCl
Lizard Ringer	160	3.3	1.7	2.8	1	0	0
Li$^+$ Ringer	0	3.3	1.7	2.8	1	0	159
Li$^+$/OCa^{2+} Ringer	0	3.3	1.7	2.8	0	1.57	159

[a] Composition in millimolar units.
[b] All solutions are pH 7.4 and osmolality is adjusted to 320 mOs with sucrose.

3 mm apart. The two coverslips (5 × 20 mm) are supported at one of their narrow ends and form the floor and ceiling of a recording chamber that is open on three sides. The chamber is connected to a Ringer solution reservoir to compensate for evaporative fluid loss and mounted on an inverted microscope [Zeiss (Thornwood, NY) Axiovert] in a light-tight refrigerated box (~16 ± 1°). Isolated ROS are viewed with an infrared-sensitive television camera (Ultricon; RCA, Lancaster, PA), using long-wavelength (>890 nm) illumination, and only crisp outer segments with sharply demarcated surface membranes are selected for recordings. Observations from the bottom of the chamber are performed with a ×40 oil immersion objective (Achrostigmate, 1.3 NA, ∞; Zeiss) and light stimuli are delivered from above, using a standard ×10 microscope objective.

Patch pipettes are made from 100-μl micropipette glass (VWR Scientific, San Francisco, CA) that is drawn to a small tip (1–2 μm in internal diameter) on a pipette puller (List L/M-3P-A; Medical Systems, Greenvale, NY) and fire-polished to 8–12 MΩ. Whole-cell pipettes are filled with standard internal solution (Table I) that has been passed through a 0.2-μm pore size Supor Acrodisc filter (Gelman Sciences, Ann Arbor, MI). Under visual control a pipette is pressed against the surface membrane of a suitable ROS to form a cell-attached gigaohm seal (typically 10 to 40 GΩ). After sealing, the pipette and isolated ROS at the tip are raised above the floor of the chamber (Fig. 1) and gentle pulses of suction are applied to the back of the pipette to rupture the membrane at the tip of the electrode. The

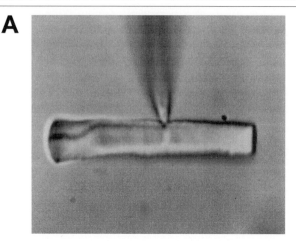

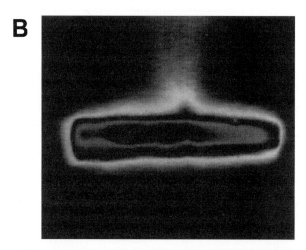

FIG. 1. Intracellular dialysis of an isolated rod outer segment. Photomicrographs show an isolated lizard rod outer segment at the end of a whole-cell pipette filled with a standard internal solution containing 100 μM fluorescein conjugated to 10-kDa dextran. After ~10 min of internal dialysis the outer segment was illuminated with infrared (>890 nm) in (A) and UV (365 nm) in (B). The intracellular incorporation of the fluorescent dye is apparent in the fluorescence image taken with UV.

formation of an electrical connection between the pipette and the inside of the cell was signaled by an instantaneous 30- to 40-pF increase in the capacitance of the recording system. The electrical events that accompany the formation of a gigaohm seal and breakthrough to whole-cell recording are shown in Fig. 2A (inset). A patch-clamp amplifier (List EPC-7; Medical

Systems) is used in the voltage-clamp mode to hold the membrane potential at a fixed value (about -30 mV) and record current flow across the outer segment surface membrane.

Breakthrough from cell-attached to whole-cell recording also provides a pathway for compounds to diffuse into ROS cytoplasm from the pipette filling solution. This is illustrated in Fig. 1, which shows photomicrographs taken with infrared (A) and ultraviolet (B) illumination of an isolated ROS that has been dialyzed with standard internal solution containing 100 μM Indo conjugated to 10-kDa dextran (Molecular Probes, Engene, OR). We have not examined the molecular weight dependence of dye coupling in any detail, but have observed rapid staining with dye conjugated to 70-kDa dextran.

When the pipette filling solution is supplemented with the nucleoside triphosphates that are normally supplied by the inner segment, i.e., ATP and GTP, breakthrough is followed by the development of a standing inward dark current (Fig. 2). Cyclic nucleotide-gated channels are opened and current, carried in part by Ca^{2+} ($\sim 15\%$), flows into the ROS as cGMP is synthesized by guanylyl cyclase using GTP supplied by the dialysis solution. As internal Ca^{2+} increases, guanylyl cyclase activity declines and ultimately the circulating current settles at a steady state level that represents a balance between the basal dark rates of cGMP synthesis by guanylyl cyclase and hydrolysis by PDE. The dialysis solution also contains ATP, which plays an important role in quenching light-activated rhodopsin and is necessary for normal recovery after light exposure.[10–12]

One of the advantages of using isolated detached ROS for these recordings is that there are only two sources of membrane current. One is the current entering the cell through CNG channels, which is directly proportional to cytoplasmic free [cGMP]. Thus changes in this current track changes in free [cGMP] and do so with higher sensitivity and faster time resolution than is possible by any other method for measuring intracellular changes in cyclic nucleotides. With a dose–response curve that relates current and cGMP, it is possible to express changes in current as changes in free cGMP. In dialyzed ROS the cGMP affinity of CNG channels does not appear to be modulated[5,13] (but also see Refs. 14 and 15). The relationship between circulating current (I_{cc}) and [cGMP] is given by

$$[\text{cGMP}_{\text{free}}] = \left[\frac{I_{cc} \times (K_{1/2})^3}{I_{cc}^{\max} - I_{cc}} \right]^{1/3} \qquad (1)$$

[13] M. P. Gray-Keller and P. B. Detwiler, *Behav. Brain Sci.* **18,** 475 (1993).
[14] S. E. Gordon, D. L. Brautigan, and A. L. Zimmerman, *Neuron* **9,** 739 (1992).
[15] Y. T. Hsu and R. S. Molday, *Nature (London)* **361,** 76 (1993).

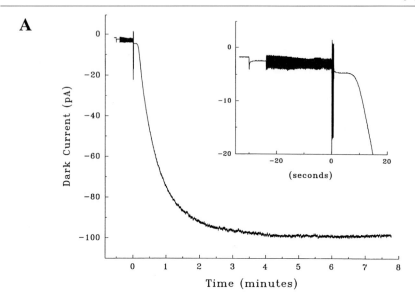
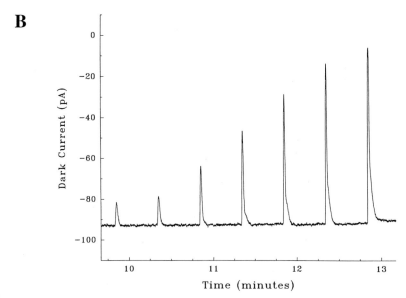

FIG. 2. Establishment of whole-cell recording and development of a standing light-sensitive inward dark current in an isolated ROS. The recording in (A) shows breakthrough from cell-attached to whole-cell recording and the development of inward circulating dark current. *Inset:* Electrical events associated with the rupture of the membrane at the tip of the pipette and electrical contact with the inside of the ROS. After forming a cell-attached seal a negative

where $K_{1/2}$ is 26 μM, maximum current (I_{cc}^{max}) is -2.5 nA, and the Hill coefficient is 3.[12] With these parameter values we estimate that in the dark [cGMP] is about 8.5 μM for $I_{cc} = -85$ pA and that it is reduced by $\sim$30 nM at the peak of a single photon response.

The second source of membrane current, which accounts for about 8% of total resting circulating dark current, is generated by $Na^+ : Ca^{2+}, K^+$ exchange. The stoichiometry of the exchanger ($4Na^+ : 1Ca^{2+}, 1K^+$) is electrogenic,[16] i.e., it transports one net inward ($+$) charge per extruded Ca^{2+} ion. Because the influx and efflux of Ca^{2+} are equal at steady state, the exchange current represents a measure of the rate of Ca_i^{2+} turnover. In darkness, Ca^{2+} circulates through a lizard ROS, with a 1-pl cytoplasmic volume and -85 pA dark current, at the rapid rate of 70 μM/sec, which corresponds to the resting Ca_i^{2+} level ($\sim$550 nM) being turned over about 140 times per second.

The inward resting dark current is reduced by light and generates light responses that resemble those recorded from intact rods in lower vertebrates.[11,12] In dialyzed ROS single-photon responses suppress about 1% of the dark current and reach a peak in about 1 sec. Both parameters are reduced during background light adaptation in a manner consistent with a large number of previous studies using a variety of rod preparations and recording methods. Figure 2B shows a family of light responses evoked by brief flashes of increasing intensity.

Calcium Measurements

Ca^{2+}-dependent fluorescence measurements are made in isolated ROS dialyzed with standard internal solution supplemented with 100 μM Indo

---

[16] L. Cervetto, L. Lagnado, R. J. Perry, D. W. Robinson, and P. A. McNaughton, *Nature (London)* **337**, 740, (1989).

---

holding potential was increased from the liquid junction potential of the pipette filling solution (about -10 mV) to -30 mV. After a capacitative current transient a -20 mV voltage step evoked $\sim$0.7 pA shift in steady state current which corresponds to a $\sim$30 gigaohm seal resistance. A short time later negative voltage pulses (10 mV, 50 msec) were delivered while gentle suction was applied to the back of the pipette to break the membrane patch and form a whole-cell recording, shown by the abrupt increase in the amplitude of the capacitative current transient. This was followed by the development of inward current as the cell used GTP supplied by the internal dialysis solution (Table I) to synthesize cGMP and open CNG channels. The whole-cell recording is continued in (B), which shows light responses evoked by increasingly bright flashes (20 msec, 514 nm) that bleached from $\sim$3.4 to $\sim$330 rhodopsins per flash in roughly equal increments. The dialysis solution used in this recording did not contain Indo-dextran, but was otherwise the same as the standard solution listed in Table I.

conjugated to 10-kDa dextran (Molecular Probes) (see Table I). This is a membrane-impermeant ratiometric fluorescent Ca^{2+} indicator that, when excited by short-wavelength light (365 nm), fluoresces with an emission maximum that shifts from ~475 nm (Ca^{2+} free) to ~400 nm (Ca^{2+} bound). By monitoring the ratio of fluorescence intensity at two different wavelengths, quantitative measurements of $[Ca_i^{2+}]$ are possible that are insensitive to variations in cell thickness, photobleaching, and dye loading or leakage.

Fluorescence signals from single dialyzed detached ROS are measured with a modified Photoscan ratio fluorescence system from Photon Technology International (Monmouth Junction, NJ). Light from a 75-W xenon arc lamp is collected, reflected by a cold mirror to remove the longest wavelength components, and passed through a blue filter (300 to 400 nm) followed by a 365 (±15)-nm bandpass interference filter. The optical setup is shown in Fig. 3. UV light is delivered by a quartz fiber optic coupler to an epifluorescence filter cube and directed by a 390-nm longpass dichroic

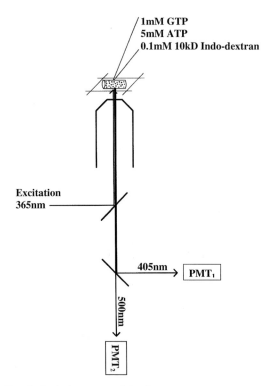

FIG. 3. Optical setup used for fluorescence measurements.

mirror through a ×40 oil immersion objective (see above) to illuminate a dialyzed ROS at the tip of a whole-cell pipette. Fluorescence emission is collected with the same objective and relayed by intermediate optics to adjustable rectangular slits that are positioned using a charge-couple device (CCD) camera with infrared illumination (>890 nm) to accept light only from the ROS by masking out the pipette and the rest of the visual field. A dichroic mirror (440-nm longpass) and bandpass emission filters at 405 (±40)- and 495 (±20)-nm separate light that comes through the slit into nominal 405- and 500-nm photons that are counted by separate photomultiplier (PMT) tubes. All optical filters are purchased from Omega Optical (Brattleboro, VT). The signals from the PMTs are analyzed with a 486 PC using software (Oscar, Photo Technology International) designed for the photoscan system. Counts at 405 and 500 nm are determined at 2-msec intervals and the 405:500-nm ratio is computed. When Ca^{2+} estimates are made during steady background illumination, the background light is turned off 20 msec before making the fluorescence measurements; during this time there is no change in circulating current.

In Vivo Calibration of Indo-Dextran Fluorescence

The 405:500-nm counts ratio is converted to free Ca^{2+} by the relation[17]

$$Ca^{2+}_{free} = K* \left(\frac{R - R_{min}}{R_{max} - R} \right) \qquad (2)$$

where R is the measured ratio of 405:500-nm fluorescence and $K*$ is the K_d of the dye multiplied by the ratio ($S_f:S_b$) of 500-nm fluorescence when all the dye is Ca^{2+} free (S_f) or Ca^{2+} bound (S_b). R_{min} and R_{max} are the 405:500-nm ratios when all the dye is Ca^{2+} free and Ca^{2+} bound, respectively.

R_{min}, R_{max}, and $K*$ are determined *in vivo* by the following procedures.[5] R_{min} is measured in ROS dialyzed with standard internal solution (Table I) to which 50 mM EGTA has been added (~6 pM free Ca^{2+}). ROS are bathed in normal Ringer solution and ~5 min before making the fluorescence measurement Ca^{2+} influx is shut off by using a saturating step of 520-nm light to permanently close all the cGMP-gated channels without affecting Ca^{2+} efflux by the exchanger. R_{min} determined in this way has an average value of 0.29 ± 0.04 ($n = 5$). To measure R_{max}, ROS are dialyzed with standard control solution containing 20 mM added Ca^{2+} (Table I) and bathed in Ringer solution containing Li^+ in place of Na^+ (Li^+ Ringer solution). Lithium blocks $Na^+:Ca^{2+},K^+$ exchange and thus prevents Ca^{2+}

---

[17] G. Grynkiewicz, M. Poenie, and R. Tsien, *J. Biol. Chem.* **260**, 3440 (1985).

from being pumped out of the cell by the exchanger. Under these conditions ROS are unable to support a light-sensitive current, consistent with elevated internal Ca^{2+}.[12] The average value of R_{max} is 4.3 ± 0.2 ($n = 4$).

To estimate K^*, ROS are dialyzed with internal solutions in which free Ca^{2+} is highly buffered (Table I) to either 204 nM ($R_{mid\#1}$) or 630 nM ($R_{mid\#2}$). ROS are bathed in Li^+ Ringer solution with zero added Ca^{2+} ($Li^+/0Ca^{2+}$ Ringer solution) to prevent Ca^{2+} influx and efflux. The large inward current that is caused by zero external Ca^{2+} is prevented by briefly (~30 sec) exposing 5 μl of dissociated ROS in normal Ringer solution to saturating 520-nm light. This permanently closes all the cGMP-gated channels and allows ROS to be transferred to the recording chamber filled with $Li^+/0Ca^{2+}$ Ringer solution without affecting circulating current.

Under these conditions we assume that free Ca_i^{2+} in the dialyzed ROS is the same as in the buffered R_{mid} solutions. These values of Ca_i^{2+} along with the measurements of R, R_{min}, and R_{max} are used in Eq. (2) to calculate K^*, which is 2669 ± 159 nM ($n = 4$) and 2939 ± 200 nM ($n = 3$) for $R_{mid\#1}$ and $R_{mid\#2}$, respectively. The mean of these, 2784 ± 110 nM ($n = 7$), corresponding to a K_d of ~928 nM, is used to convert 405:500-nm fluorescence ratios to Ca^{2+} concentrations. The calibration constants do not change significantly over the course of a series of experiments. The values redetermined at the end of a 1-year study are as follows: $R_{min} = 0.272 ± 0.005$ ($n = 5$); $R_{max} = 4.1 ± 0.6$ ($n = 6$); and $K^* = 3168 ± 315$ nM ($n = 7$).

Average background fluorescence in the absence of dye is measured in several ROS for each different dialysis solution and experimental or calibration protocol. Background fluorescence, which is typically ~20% of the fluorescece observed in the presence of Indo-dextran, is subtracted from the corresponding fluorescence measurements made with solutions containing dye. The relationship between the ratio of 405:500-nm counts and free Ca_i^{2+} is shown in Fig. 4, which plots the *in vivo* calibration curve for Indo-dextran.

Ca^{2+} Snapshots

Fluorescence measurements in rods are complicated by the fact that they are photoreceptors and the light used to excite dye fluorescence also excites the phototransduction cascade. This closes CNG channels and leads to a fall in Ca^{2+}, the variable being measured. To circumvent this problem we took advantage of the fact that photomultiplier tubes are faster than rods. This is shown in the top part of Fig. 5, which compares the onset of the light response evoked by a step of 365-nm light in a dialyzed ROS (filled circles) with the responses of two photomultiplier tubes to the fluorescence excited by the UV step. The photomultiplier tubes detected the fluorescence

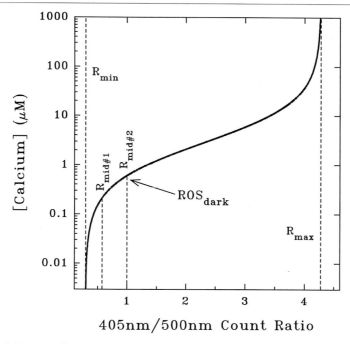

FIG. 4. *In vivo* calibration curve for Indo-dextran. The solid curve shows the relationship between free Ca^{2+} and the 405 nm-to-500 nm (405:500-nm) fluorescence ratio, using Eq. (2) and the calibration procedure described in text. R_{min} and R_{max} are 405:500-nm fluorescence intensity ratios when all the dye is Ca^{2+} free and Ca^{2+} bound, respectively. The fluorescence intensity ratios measured when ROSs were dialyzed with internal solutions in which free Ca^{2+} was highly buffered to either 204 or 630 n*M* are indicated by $R_{mid\#1}$ and $R_{mid\#2}$, respectively.

signals instantaneously, while a 20- to 30-msec delay preceded the onset of the rod photoresponse. The excitation light was bright enough to close all the CNG channels in the outer segment and evoked a saturated electrical response that did not recover from saturation for the duration of the experiment. Because the 405:500-nm florescence ratio was determined every 2 msec, the first 10–15 data points gave an estimate of internal Ca^{2+} during the 20- to 30-msec delay before the beginning of the electrical light response. These initial readings allow a snapshot of Ca^{2+}-sensitive fluorescence to be taken before the rod has had time to respond to the short-wavelength light used to excite the indicator dye. This is illustrated in Fig. 5 (inset), which shows that the initial Ca^{2+} estimates were noisy but relatively stable during the delay period preceding the onset of the electrical response and then decay monotonically (Fig. 5, lower trace) as Ca^{2+} was pumped out of the cell by continued $Na^+:Ca^{2+},K^+$ exchange. This was not the case, however,

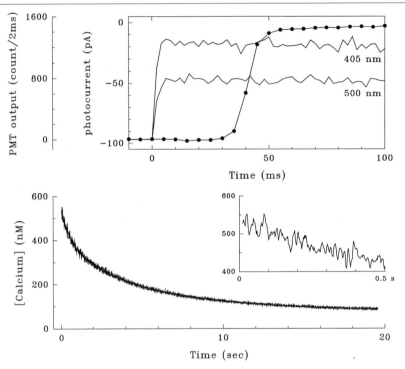

FIG. 5. Photomultiplier tubes are faster than rods. *Top:* Electrical response of an isolated ROS dialyzed with standard internal solution with 100 μM Indo-dextran (closed circles) to a step of UV light in comparison with the excitation of fluorescence at 405 and 500 nm as detected by two photomultiplier tubes (PMTs). *Bottom:* Time course of the decline in free Ca^{2+} that follows the development of the saturating electrical response evoked by the UV step. Ca^{2+} is relatively constant during 20- to 30-msec delay period that precedes the onset of the electrical response (*inset*) and then decays monotonically.

when Indo was not conjugated to dextran.[5] While free Indo and Indo-dextran reported similar initial resting Ca^{2+} levels (611 ± 68 and 543 ± 25 nM, respectively), free Indo showed a large, rapid rise in Ca^{2+} (~600 nM in ~7 msec) followed by a much slower decline than was reported by Indo-dextran. The reason for the difference in the two forms of the dye is not known, but because the measurements made with free Indo appeared to be contaminated by an artifactual rise in Ca^{2+} we used Indo-dextran for all our measurements.

There are a number of reasons for considering that the Ca^{2+} measurements made with Indo-dextran and the methods used to calibrate it are valid. First, the mean resting Ca^{2+} level in darkness was estimated to be 554 ± 25 nM (n = 28). This agrees with both earlier and subsequent estimates of ~500 and 670 nM, respectively.[4,6,9] Second, as expected, the time course of the decline in measured Ca^{2+} was similar to the decline in Na$^+$:Ca^{2+},K$^+$ exchange current in saturating light.[5] Both were described by the sum of two exponential processes that contributed equally and had time constants that differed by about a factor of 10, i.e., ~0.5 and ~5 sec. Third, the Ca^{2+} dependence of the rate of Na$^+$:Ca^{2+},K$^+$ exchange (I_{exh}) is given by

$$I_{exh} = I_{exh}^{sat} \times \frac{Ca^{2+}}{K_m + Ca^{2+}} \quad (3)$$

where K_m is the Ca^{2+} affinity of the exchanger and the saturated rate of exchange (I_{exh}^{sat}) is 17.8 pA. Because I_{exh} is a fixed percentage (~8%) of the circulating current, the relationship between Ca^{2+} and steady-state I_{cc} can be used to estimate K_m, where

$$K_m = \left[\frac{I_{exh}^{sat} - 0.08 \times I_{cc}}{0.08 \times I_{cc}}\right] Ca_i^{2+} \quad (4)$$

The mean value of K_m estimated in this way was 902 ± 34 nM ($n = 52$), which agrees with previous estimates ranging from 0.9 to 1.6 μM,[4,18] and thus provides further confidence in the accuracy of the Ca^{2+} measurements on which it is based.

Light-Evoked Changes in Ca^{2+}

The methods described above can be used to estimate steady state Ca$_i^{2+}$ in darkness or when circulating current is reduced by different intensities of steady background light. The measurements are based on taking a snapshot of Ca$_i^{2+}$ during the delay that precedes the onset of the saturating photoresponse that is produce by the fluorescence excitation light. Because circulating current does not recovery after exposure to the short-wavelength excitation light, it is possible to make only one Ca^{2+} measurement per experiment. To monitor the transient change in Ca$_i^{2+}$ that occurs during a subsaturating flash response or during the peak-to-plateau transition of a subsaturating step response, initial Ca^{2+} levels were measured in different ROS at different fixed times after delivering a conditioning flash or step of a set intensity.[5] Several measurements are made at each of the fixed time points, allowing one to obtain an average estimate of Ca$_i^{2+}$ at that time during the electrical response produced by the conditioning stimulus. We have done this for a single intensity flash and step response. In both cases changes Ca$_i^{2+}$ did not faithfully track the light-induced changes in

[18] P. P. M. Schnetkamp, *J. Gen. Physiol.* **98**, 555 (1991).

current. For example, during a flash response that produced a 70% peak reduction in circulating current and recovered fully in ~5 sec, Ca_i^{2+} was reduced by about 40% at the peak and recovered to a sustained plateau during which Ca^{2+} was reduced by ~20%. The plateau of reduced Ca^{2+} persisted long after (~15 sec) the full recovery of the electrical response. That there is a maintained decrease in Ca_i^{2+} when circulating current has fully recovered is unexpected and the mechanism responsible is not understood. It is important, however, for a full understanding of Ca^{2+} dynamics and the molecular elements involved in controlling circulating current that these observations be explored further.

[10] Laser Spot Confocal Technique to Measure Cytoplasmic Calcium Concentration in Photoreceptors

By Hugh R. MATTHEWS and Gordon L. FAIN

Introduction

This chapter describes the application of spot confocal fluorescence microscopy to the measurement of cytoplasmic calcium concentration (Ca_i^{2+}) from isolated vertebrate photoreceptors. Calcium is well established to play a crucial role in controlling the sensitivity of the light response. In darkness, calcium ions enter the outer segment as a component of the dark current[1,2] and are extruded by an exchange extrusion mechanism that utilizes the sodium and potassium gradients to transport Ca^{2+} out across the outer segment membrane.[3-5] During the light response the circulating current is suppressed in a graded manner, leading to a reduction in Ca^{2+} influx and a decrease in the outer segment calcium concentration.[6]

This decrease in Ca_i^{2+}, was first measured directly by McNaughton *et al.*, using the luminescent photoprotein aequorin in isolated salamander rods.[7] The rationale for using aequorin as the Ca^{2+} indicator was to minimize stimulation of the rod photoreceptor by the light emitted during the course

---

[1] A. L. Hodgkin, P. A. McNaughton, and B. J. Nunn, *J. Physiol.* **358**, 447 (1985).
[2] K.-W. Yau and K. Nakatani, *Nature* (*London*) **309**, 352 (1984).
[3] K.-W. Yau and K. Nakatani, *Nature* (*London*) **311**, 661 (1984).
[4] L. Cervetto, L. Lagnado, R. J. Perry, D. W. Robinson, and P. A. McNaughton, *Nature* (*London*) **337**, 740 (1989).
[5] P. P. Schnetkamp, D. K. Basu, and R. T. Szerencsei, *Am. J. Physiol.* **257**, C153 (1989).
[6] K.-W. Yau and K. Nakatani, *Nature* (*London*) **313**, 579 (1985).
[7] P. A. McNaughton, L. Cervetto, and B. J. Nunn, *Nature* (*London*) **322**, 261 (1986).

of the measurement. However, the relative insensitivity of aequorin at physiological Ca^{2+} concentrations and the nonlinearity of its response made accurate determination of the dark level of Ca_i^{2+} difficult, and only permitted measurement of its changes in Ca_i^{2+} when Ca_i^{2+} was first pharmacologically elevated.[8] Subsequently Ca_i^{2+} was measured with Ca^{2+}-sensitive fluorescent dyes in populations of rod photoreceptors from the isolated retinae of toad[9] and bullfrog.[10–12] By collecting fluorescence from a large number of photoreceptors, it was possible to minimize the stimulation of each individual rod by the light used to excite the fluorescent dye. However, the use of a population of photoreceptors meant that the fluorescence signal could only indirectly be attributed to the photoreceptor outer segment. Nevertheless, these measurements indicated that in amphibian rods Ca_i^{2+} is on the order of several hundred nanomolar in darkness, and declines to nearly zero over a time course of seconds to tens of seconds.

The first measurements of Ca_i^{2+} from single photoreceptors were made by Gray-Keller and Detwiler, who used the ratiometric fluorescent Ca^{2+} dye indo-1 loaded with a patch pipette into the detached outer segments of *Gecko* rods.[13] Their results demonstrated that in this preparation Ca_i^{2+} falls from about 550 nM in darkness to around 50 nM in bright light, declining with two exponential components of approximately equal amplitude and time constants of around 0.5 and 5 sec, respectively. An inevitable feature of this, and other, measurements of Ca_i^{2+} from single photoreceptors using the currently available fluorescent indicators is that the fluorescence-exciting beam was sufficiently intense also to bleach a substantial fraction of the photopigment. Consequently, only a single measurement of Ca_i^{2+} could be made from each detached outer segment. Furthermore, in these experiments on detached outer segments it was necessary to infer the original dark level of Ca_i^{2+} by setting the dark current to a value near the physiological level.

To extend such measurements of Ca_i^{2+} to the intact amphibian rod, and to make equivalent measurements from the much smaller outer segments of intact cone photoreceptors or mammalian rods and cones, a spatially precise technique of high sensitivity is required. Furthermore, because the kinetics of the light-induced change in Ca_i^{2+} would be expected to be much faster in amphibian cone or mammalian rod photoreceptors than in amphibian or reptile rods, a wide recording bandwidth is also necessary. We have

[8] L. Lagnado, L. Cervetto, and P. A. McNaughton, *J. Physiol. (London)* **455,** 111 (1992).
[9] J. I. Korenbrot and D. L. Miller, *Vis. Res.* **29,** 939 (1989).
[10] S. T. McCarthy, J. P. Younger, and W. G. Owen, *Biophys. J.* **67,** 2076 (1994).
[11] G. M. Ratto, R. Payne, W. G. Owen, and R. Y. Tsien, *J. Neurosci.* **8,** 3240 (1988).
[12] J. P. Younger, S. T. McCarthy, and W. G. Owen, *J. Neurophysiol.* **75,** 354 (1996).
[13] M. P. Gray-Keller and P. B. Detwiler, *Neuron* **13,** 849 (1994).

addressed these problems by adapting a laser spot confocal technique, originally developed by J. Vergara and colleagues to study Ca^{2+} release from the sarcoplasmic reticulum of skeletal muscle,[14] for the measurement of Ca_i^{2+} from vertebrate photoreceptors.[15,16]

Overview of Spot Confocal Technique

Spot confocal microscopy employs a laser to illuminate a restricted region of interest within an isolated cell that has been loaded with a single-wavelength fluorescent Ca^{2+} indicator such as fluo-3. The emitted fluorescence is then imaged onto a small-area detector that is optically conjugate with the exciting spot. This approach allows the dye to be excited and fluorescence collected from a small volume-element that is delimited in the plane of the preparation by the extent of the laser spot, and defined in depth by the confocality of the detector.

This technique offers several advantages over other methods for measuring Ca_i^{2+}. First, the use of a single-wavelength Ca^{2+} indicator allows high time resolution, because it is not necessary to change between excitation or emission wavelengths as is the case for some ratiometric dyes (such as fura-2),[17,18] or to acquire an entire video frame as would be the case in scanning confocal microscopy. Consequently measurements of Ca_i^{2+} can be made over a bandwidth limited primarily by the dissociation rate constant of the dye. Furthermore, the accuracy of the measurement is not compromised at values of Ca_i^{2+} much lower than the dissociation constant of the dye, as would be the case for a ratiometric Ca^{2+} indicator. Fluo-3, originally developed in the laboratory of R. Y. Tsien from the Ca^{2+} buffer BAPTA 1,2-bis(*o*-aminophenoxy)ethane-*N,N,N',N'*-tetraacetic acid,[18] offers the advantage over other single-wavelength Ca^{2+} indicators of an especially high ratio of maximum to minimum fluorescence intensity, thereby minimizing the Ca^{2+}-independent component of the fluorescence signal. A second advantage is that fluorescence can be collected simultaneously from the entire area defined by the illuminating laser spot, allowing modest and rapid changes in Ca_i^{2+} to be detected not only from amphibian rods[15] but also

---

[14] A. L. Escobar, J. R. Monck, J. M. Fernandez, and J. L. Vergara, *Nature (London)* **367**, 739 (1994).

[15] A. P. Sampath, H. R. Matthews, M. C. Cornwall, and G. L. Fain, *J. Gen. Physiol.* **111**, 53 (1998).

[16] A. P. Sampath, H. R. Matthews, M. C. Cornwall, J. Bandarchi, and G. L. Fain, *J. Gen. Physiol.* **113**, 267 (1999).

[17] G. Grynkiewicz, M. Poenie, and R. Y. Tsien, *J. Biol. Chem.* **260**, 3440 (1985).

[18] A. Minta, J. P. Kao, and R. Y. Tsien, *J. Biol. Chem.* **264**, 8171 (1989).

from the small outer segments of amphibian cones[16] and mammalian rods.[19] Furthermore, the spatial precision of the spot confocal technique allows the signal from strongly fluorescing subcellular structures such as the ellipsoid region of the photoreceptor inner segment to be readily excluded from the measurement.

The physical implementation of the spot confocal technique is illustrated schematically in Fig. 1. The beam from an argon ion laser [model 98 (Lexel Laser, Fremont, CA) or model 60 (American Laser, Salt Lake City, UT)], tuned to 514 nm, illuminates the 50-μm pinhole of a spatial filter assembly (Newport, Irvine, CA). To allow the maximum laser flux to pass through the pinhole, a converging lens is placed before the spatial filter to reduce the beam diameter from its original value of about 1 mm, while still ensuring uniform illumination of the pinhole aperture. After it passes through the pinhole, the beam is recollimated by a ×4 microscope objective (Nikon, Melville, NY) and enters the inverted microscope [Zeiss IM (Carl Zeiss, Oberkochen, Germany) or Nikon Eclipse 300 (Nikon, Kingston, UK)] through the epifluorescence port via a dichroic mirror (525DRLP; Omega Optical, Brattleboro, VT). A ×40 oil immersion objective lens [1.3 NA; either 160-mm tube length (Nikon, Melville, NY) or 1.3 NA, CFI 60 infinity corrected (Nikon, Kingston, UK)] forms a reduced image of the pinhole to illuminate a spot approximately 8 μm in diameter on the outer segment of an isolated rod or cone photoreceptor, whose inner segment is drawn into a suction pipette so that the photocurrent can be recorded.[20] To ensure optical stability, laser, microscope, and detector are all mounted on the same air-supported vibration isolation table (Newport).

Fluorescence is evoked by a series of brief laser exposures controlled by an electromagnetic shutter (Vincent Associates, Rochester, NY), driven by the data acquisition system. The fluo-3 fluorescence passes through a 530-nm emission filter (530EFLP; Omega Optical) and is imaged onto the limiting aperture of the detector, as discussed in more detail below. The analog signal from the detector is lowpass filtered at up to 250 Hz with an active eight-pole Bessel filter (Frequency Devices, Haverhill, MA; or Kemo, Beckenham, UK) and digitized at up to 1000 Hz [PClamp 6 (Axon Instruments, Foster City, CA); or Cambridge Research Systems, Rochester, UK].

Argon Ion Laser

To make a faithful measurement of Ca_i^{2+} the intensity of the laser beam should be as stable as possible. However, many argon ion lasers produced for the reprographics or entertainment industries are not suitable for this

[19] A. P. Sampath, H. R. Matthews, and G. L. Fain, *Invest. Ophthalmol. Vis. Sci.* **39,** S983 (1998).
[20] D. A. Baylor, T. D. Lamb, and K.-W. Yau, *J. Physiol.* **288,** 589 (1979).

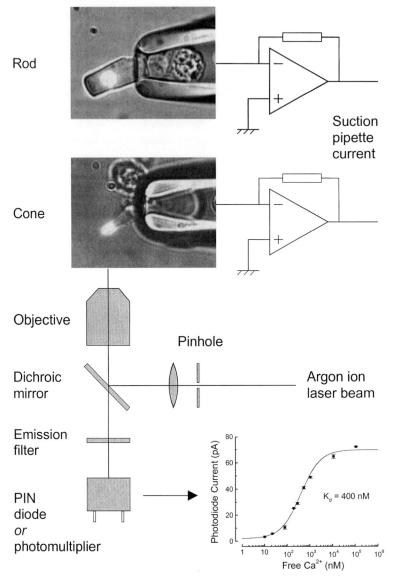

FIG. 1. Schematic diagram illustrating the measurement of Ca_i^{2+} from isolated salamander rods and cones, using the spot confocal technique. The beam from an argon ion laser (514-nm line) illuminates the pinhole of the spatial filter, is recollimated, and enters the inverted microscope through the epifluorescence port. A dichroic mirror (525 nm) redirects the beam through the oil immersion objective lens (×40, 1.3 NA) to illuminate a spot (diameter, 7.7 μm) on the outer segment of a rod or cone loaded with fluo-3 AM whose inner segment is held in a suction pipette. The fluo-3 fluorescence evoked by the laser spot passes through

purpose, because their power supplies are insufficiently stable to deliver the low optical noise that is required to make these measurements. In particular, many lasers designed for "light-show" use provide the 20-A low-voltage supply required to heat the plasma tube filament directly from a transformer connected to the ac line. This results in a fluctuation of more than 25% in light output at twice the line frequency. It is therefore necessary to choose a scientific-grade laser with a dc-driven filament, which also uses a stable source for the high voltage (100–300 V) and high current (4–40 A) that drives the plasma tube discharge.

We have employed two different lasers for these experiments. The first is a water-cooled "small frame" argon ion laser with a nominal maximum power output of 4 W (model 98; Lexel Laser). While its power output is considerably higher than is required for these measurements, the linear pass-bank of the Lexel power supply ensures stability of the plasma tube discharge current, providing comparatively low optical noise. While the relatively long cavity of the Lexel provides excellent beam pointing stability, its large physical size demands a substantial air-supported vibration isolation table. Moreover, its demands for electrical power and cooling are prodigious, because it requires a dedicated three-phase supply at 40 A and 1 gal of cooling water per minute.

The second laser that we have used is an air-cooled argon ion laser with a maximum power output of 25 mW at 514 nm (model 60; American Laser). This laser employs a dc filament supply, and uses a switch-mode buck converter to stabilize the plasma tube discharge current. Furthermore, it includes a light feedback loop to stabilize the laser beam intensity. While the electrical and cooling requirements of this laser are considerably more modest, beam pointing stability is inferior to the Lexel owing to the shorter laser cavity. Consequently, to couple the beam into the pinhole aperture it is necessary to use a lens with a focal length of 5 cm or less, in order to avoid fluctuations in spot intensity owing to rapid wandering of the beam.

Because of the high currents and voltages used by these lasers, it is necessary to be meticulous about grounding and shielding to prevent the pickup of interference into the suction electrode or other electrical recording from the laser power supply. This has proved to be especially true of the air-cooled laser, which requires separate "star" grounding of all

---

the emission filter (530-nm longpass) and is collected confocally by either a PIN photodiode or a photomultiplier. *Inset:* Calibration curve measured *in situ* with fluo-3 free acid solutions of known Ca^{2+} concentration in a hemocytometer. Solid curve is the Michaelis–Menten relation [Eq. (1)] fitted by a Marquardt–Levenberg least-squares algorithm (K_d 400 nM). [Modified from Ref. 15.]

components of the laser to a single point, and the shielding of all its connecting cables with nickel-copper alloy Monel screening mesh. The transverse node structure of the laser is set up for $TEM_{0,0}$ operation so as to yield a Gaussian beam profile. For both lasers an intracavity wavelength tuning prism allows a single argon line to be selected. In the majority of experiments the laser is tuned to the 514-nm line; however, some measurements have also been made using the brighter 488-nm argon line. The flexibility afforded by wavelength tuning makes such an "external mirror" laser more suitable for these experiments than "sealed mirror" designs in which wavelength-selective mirrors are integral with the plasma tube. In addition, the tunability of the laser allows other Ca^{2+} indicators to be used, such as those with an Oregon Green chromophore (Molecular Probes, Eugene, OR).

Argon ion lasers are unfortunately relatively short-lived, offering a plasma tube life of only a few thousand hours, which depends strongly on the magnitude of the discharge current. It is therefore desirable to run the laser at the lowest possible discharge current in order to maximize the operating life of the tube. However, the relative level of optical noise increases as the discharge current is decreased. We found the best compromise between low noise and long life to be at a discharge current of about 6–7 A for the American Laser Corporation laser and 25–30 A for the Lexel laser. Between measurements it is desirable to reduce the operating current to an "idle" level, which extends tube life to tens of thousands of hours. Once a plasma tube has failed, it can normally be reconditioned by a specialist at significantly lower cost than the purchase of a replacement. Often this involves little more than "regassing" with fresh argon, but more extreme measures can be required after the migration of metal into the tube bore, contamination of the Brewster-angle windows, or distortion of the filament. Some modes of failure are essentially terminal: for example, catastrophic failure of the cooling water jacket leading to water entering the plasma tube bore!

The unattenuated intensity of the laser is normally considerably higher than can be used to make a Ca^{2+} measurement without significantly bleaching the Ca^{2+} indicator fluo-3. Therefore the laser beam is typically attenuated by 2–4 $\log_{10}$ units with a reflective neutral density filter (Newport; or New Focus, Santa Clara, CA) to yield a flux of less than 2×10^{11} photons μm^{-2} sec^{-1} at the plane of the pinhole image. It is best to place this filter before the shutter that controls the laser exposure, so that it is continuously illuminated by the beam. This avoids the possibility of small changes in the density of the filter owing to local heating after the onset of laser illumination. For a high-power laser such as the Lexel, it is important to remember that the reflected beam from the filter must be directed at a noncombustible beam stop such as a fire brick. It is also important to shield against stray beams from specular or diffuse reflections. Amber Plexiglas absorbs both the 488-nm and 514-nm argon lines and so can be used to enclose the beam path but still permit filter settings and beam steering optics to be seen.

To relay the beam to the spatial filter assembly behind the microscope, it is necessary first to raise the beam to the height of the epifluorescence port with a periscope constructed from kinematic mirror mounts (New Focus), and then to reflect the beam through a right angle so that it enters the microscope through the epifluorescence port. Laser beam-steering mirrors, the converging lens, the pinhole, and the recollimating lens are then adjusted iteratively until they are all aligned with the optical axis of the microscope. The alignment of the pinhole with the laser beam is then checked on a regular basis (often daily) during each series of experiments. When setting up the laser it is essential to use protective goggles specified for an argon laser of the appropriate power. It is often convenient to use "alignment goggles," which allow a small (and eye-safe) fraction of the beam to pass so as to allow the beam to be visualized when aligning it with the beam steering optics and the pinhole.

Because the wavelength of the laser is close to the wavelength of peak absorption of rhodopsin, exposure of a rod photoreceptor to the laser will bleach the vast majority of the photopigment within the area of the laser spot. The percentage of photopigment bleached in salamander rods by the 514-nm light from the argon ion laser was estimated from the photosensitivity at 520 nm for vitamin A_2-based pigments in free solution,[21] corrected for the difference in dichroism in free solution and in disk membranes (6.2×10^{-9} μm^2).[22] In salamander red-sensitive cones the photosensitivity was corrected by the ratio of absorbance at 514 and 610 nm predicted from the pigment nomogram.[21] In both cases even a brief laser exposure can be calculated to bleach >99% of the photopigment within the area of the laser spot. In the case of a salamander rod, the laser spot covered only a relatively small fraction of the outer segment, although less extensive bleaching is also likely to have taken place in the remainder of the outer segment owing to scattered laser light. In the case of salamander cones, the laser spot will have bleached the majority of the much smaller outer segment. Within the area illuminated by the laser spot, the fractional bleach may have been reduced somewhat by photoregeneration from bleaching intermediates.[23,24]

---

[21] H. J. A. Dartnall, in "Handbook of Sensory Physiology" (H. J. A. Dartnall, ed.), pp. 122–145. Springer-Verlag, Berlin, 1972.
[22] G. J. Jones, A. Fein, E. F. J. MacNichol, and M. C. Cornwall, *J. Gen. Physiol.* **102,** 483 (1993).
[23] T. P. Williams, *J. Gen. Physiol.* **47,** 679 (1964).
[24] E. N. Pugh, *J. Physiol.* **248,** 393 (1975).

However, this seems likely to have been a relatively modest effect, since even the shortest 20-msec laser pulses were considerably longer than the lifetimes that have been reported for intermediates from which substantial photoregeneration can take place in vertebrate photoreceptors.

Detectors for Emitted Fluorescence

We have employed two different strategies to measure the emitted fluorescence. In the first, a 200 × 200 μm PIN photodiode (model HR008; United Detector Technologies, Hawthorne, CA) is aligned with the pinhole image formed by the camera port of the inverted microscope, and its photocurrent recorded at zero bias voltage (photovoltaic mode) by the headstage of a capacitative feedback patch-clamp amplifier (Axopatch 200A; Axon Instruments). The small diameter of the photodiode detector serves to exclude defocused light from above or below the plane of the pinhole image.

In the second strategy, a 200- or 400-μm-diameter silica multimode optical fiber is used to relay the emitted fluorescence from the image plane to a remote detector. In this case the limiting aperture is provided by the diameter of the fiber core. This approach allows a physically more bulky detector to be employed without the requirement that it be located adjacent to the microscope. We have used two detectors for this purpose: either a commercially available PIN diode femtowatt photoreceiver (type 2151; New Focus) or a photomultiplier with a small-area S20 photocathode (model 9130/100A; Thorn EMI, Ruislip, UK) in conjunction with a low-noise current-to-voltage converter (model PDA700; Terahertz Technology, Oriskany, NY). Commercial fiberoptic patch cords provide insufficient screening against stray light, so it is necessary to run the fiber within opaque black polyvinyl chloride (PVC) cable sleeving to prevent saturation of such a sensitive detector by ambient room lighting. The photodiode or optical fiber is mounted on a three axis translator (Newport) coupled to the camera port of the microscope, so that it can be aligned precisely with the fluorescence image of the pinhole.

Photodiodes and photomultipliers have specific advantages and disadvantages that depend on the intensity of the emitted fluorescence. Silicon photodiodes have a relatively high quantum efficiency of about 60% at 530 nm. Although the intrinsic noise equivalent power of a small-area photodiode can be as low as 10 fW/Hz$^{1/2}$ at 530 nm, the overall measurement is normally dominated by the noise in the current-to-voltage converter. Furthermore, as the area of the photodiode increases, so does its capacitance, allowing progressively more current noise to be driven by the voltage noise of the amplifier. The combination of the 200 × 200 μm UDT photodiode and the Axopatch 200A amplifier yields a peak-to-peak noise level in darkness of little more than 0.1 pA over a 250-Hz bandwidth. In contrast, the noise level of the New Focus femtowatt photoreceiver over a 125-Hz bandwidth is about 0.3 pA, peak-to-peak, in darkness (Fig. 2A), reflecting both a somewhat noisier amplifier and the larger photodiode required to couple to the optical fiber. In contrast, the photomultiplier is virtually silent in darkness (Fig. 2B), reflecting its low dark count of 35 counts per second (cps), which results in a low dark current. The low rate of single-photon

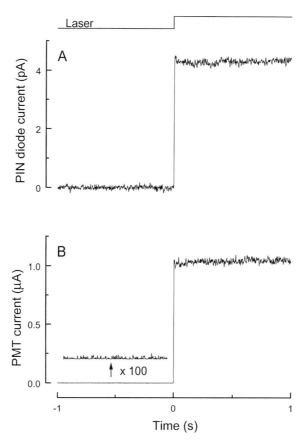

FIG. 2. Comparison of noise level in (A), a PIN diode femtowatt photoreceiver (model 2151; New Focus, Santa Clara, CA), and (B), a low dark count photomultiplier (model 9130/100A; Thorn EMI, Ruislip, UK). Fluorescence was evoked from a dried film of the fluorescent dye FM1-43 on a coverslip located on the stage of the spot confocal microscope with the same laser spot intensity in both cases. Data have been filtered dc–125 Hz with a Bessel filter ("Pulse" response dc–250 Hz, VBF8 Mk4; Kemo, Beckenham, UK). Inset in (B) shows the noise in darkness magnified by a factor of 100. Upper trace: laser monitor.

events in darkness largely results from the use of magnetic focusing within the photomultiplier tube to restrict the effective photocathode diameter to only 2.5 mm.

Once the laser shutter opens, the noise levels in the signals from the two detectors become more comparable for a physiologically realistic fluorescence signal, the noise in the photomultiplier signal reflecting in large part random fluctuations in photon catch. For a given fluorescence intensity quantal fluctuations will be greater for the photomultiplier by a factor of about two. The reason for this is that the photomultiplier, despite the extended long-wavelength sensitivity of its S20 photocathode, has a quantum efficiency of only about 15% at 530 nm, as compared with a quantum efficiency of about 60% for a silicon photodiode. Nevertheless, as a consequence of its low noise in darkness the photomultiplier is capable of detecting levels of fluorescence that would be quite unmeasurable with a photodiode detector. At intermediate intensities it may be possible to obtain a further improvement in the signal-to-noise ratio by using an optimized avalanche photodiode, which has a quantum efficiency similar to that of a PIN diode, but that provides a gain of several hundred within the detector itself, thereby reducing the relative contribution from the noise of the current-to-voltage converter.

Preparation and Dye Loading of Isolated Photoreceptors

Photoreceptors are isolated by mechanical dissociation from the dark-adapted retina. The procedure described below has been developed for isolating photoreceptors from the salamander retina but can also be applied with minor modifications to the isolation of individual receptors or small clumps of receptors from other species, such as mouse, guinea pig, and chicken.

Aquatic tiger salamanders (*Ambystoma tigrinum*) are maintained between 7 and 10° on a 12-hr light–dark cycle. Animals are dark adapted overnight, killed under dim red illumination by decapitation, followed by rostral and caudal pithing, and the eyes enucleated. All subsequent dissection is carried out under infrared illumination. Eyes are hemisected and the eyecup cut into two pieces across the optic disk. Each piece of eyecup is then placed in a silicone rubber-coated dish (Sylgard 184; Dow Corning, Midland, MI) containing amphibian Ringer solution [111 mM NaCl, 2.5 mM KCl, 1 mM CaCl$_2$, 1.6 mM MgCl$_2$, 3 mM HEPES–NaOH, 10 μM EDTA–NaOH, 10 mM glucose, bovine serum albumin (BSA, 40 mg/ml), BSA, pH 7.7] and stored under refrigeration until required. At the time of the experiment the retina is peeled from the eyecup, placed photoreceptor side up on the Sylgard coating, and chopped lightly with a small piece of razor blade. A fixed volume of the resulting cell suspension is collected with a transfer pipette and moved to a microcentrifuge tube containing the membrane-permeant acetoxymethyl ester of the Ca^{2+} indicator fluo-3 dissolved in amphibian Ringer solution (without glucose or BSA), to achieve a final concentration of fluo-3 AM of 10 μM after mixing. The dye-containing cell suspension is then transferred to the recording chamber, allowed to settle on the coverslip that forms the chamber base, and incubated for 30 min at room temperature (23°) prior to recording. After this period, excess dye is flushed from the recording chamber via the bath perfusion system.

Measurement of Light-Induced Decline in Calcium

Figure 3 shows examples of Ca^{2+} signals recorded from isolated salamander photoreceptors. Each panel shows the fluorescence evoked by two successive laser exposures presented to a dark-adapted rod (Fig. 3A), blue-sensitive cone (Fig. 3B), or red-sensitive cone (Fig. 3C) which had previously been loaded with fluo-3 by incubation with the membrane-permeable AM ester of the dye. In each case the onset of laser illumination evoked a high level of fluo-3 fluorescence from the previously dark-adapted photoreceptor, which then progressively declined to a much lower level. The intense laser light also bleached the vast majority of the photopigment within the region illuminated by the laser spot. For a rod photoreceptor this substantial bleach was sufficient to hold the circulating current recorded by the suction pipette in saturation for 10 or more minutes thereafter (not shown), whereas in a cone partial recovery of the circulating current took place within a few tens of seconds. The decline in fluorescence evoked by the first exposure of a dark-adapted photoreceptor to the laser spot was much more rapid in cones than in rods. In each case the decay in fluo-3 fluorescence could be fitted by the sum of two exponential components, with time constants which averaged 260 and 2200 msec in a rod, 140 and 1400 msec in a blue-sensitive cone, and 43 and 640 msec in a red-sensitive cone.[15,16] The amplitudes of the two exponential components were approximately equal in rods and blue-sensitive cones, whereas in red-sensitive cones the fast component was 1.7-fold larger than the slow component. In contrast to the decline in fluorescence evoked by the first laser exposure, the second elicited only a low and maintained fluorescence signal. The slight initial relaxation that was seen in the cone fluorescence signals is likely to reflect the partial recovery of circulating current that had taken place by the time of the second laser exposure. The relative constancy of the fluorescence evoked by the second and subsequent laser exposures indicates that the decay in fluo-3 fluorescence that is seen when a dark-adapted photoreceptor is first

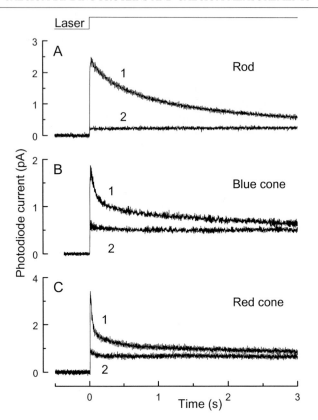

FIG. 3. Measurement of the decline in fluo-3 fluorescence evoked by laser light from (A) a salamander rod, (B) a blue-sensitive salamander cone, and (C) a red-sensitive salamander cone. Fluorescence signal measured by a small-area PIN photodiode coupled to an Axopatch 200A patch clamp amplifier. Trace 1, first laser exposure in the dark-adapted photoreceptor; trace 2, second laser exposure presented 10–60 sec later, while the circulating current remained largely suppressed. The declining fluorescence signals in traces labeled 1 were fitted with two exponentials with time constants of (A) 380 and 2000 msec; (B) 90 and 2300 ms; and (C) 30 and 670 msec, using a least-squares algorithm. Upper trace, laser light monitor. [Data replotted from Refs. 15 and 16.]

exposed to the laser spot represents the light-induced decline in Ca_i^{2+} that follows suppression of the dark current.[15]

A potential drawback of such single-wavelength measurements is that if significant fluo-3 bleaching were to take place during the laser exposure, the corresponding change in fluorescence would be indistinguishable from that caused by a real decrease in Ca_i^{2+}. However, several pieces of evidence indicate that significant dye bleaching is unlikely to have taken place during these measurements. First, the kinetics of the decline of the fluo-3 signal differ for each type of photoreceptor, even though the intensity of the laser spot was similar in each case. Such systematic differences would not be expected if fluo-3 were being destroyed by bleaching, a process that should be governed solely by laser intensity. Second, if the total laser exposure is substantially reduced, either by attenuating the laser beam intensity or by presenting repeated brief laser pulses instead of a continuous exposure, then the time course of the subsequent decline in fluorescence is little affected for a given type of photoreceptor. These approaches will each have reduced the laser exposure by at least a further order of magnitude. Third, in a rod, if after the initial laser exposure the laser spot is displaced to a region of the outer segment that has not previously been excited directly, then fluo-3 fluorescence is depressed there also, even though the dye in this region has not previously been exposed to the laser spot. Fourth, repeated laser exposures show that fluo-3 fluorescence remains depressed after the first exposure of a dark-adapted photoreceptor to the laser, consistent with the notion that Ca_i^{2+} has already reached its minimum level once the circulating current has been completely suppressed. Application of exogenous 11-cis-retinal, however, restores the fluorescence signal along with circulating current and sensitivity. Taken together, these observations indicate that the decline in fluo-3 fluorescence during saturating illumination of a dark-adapted photoreceptor is a real physiological phenomenon and not an artifact of fluo-3 bleaching.

In Situ Calibration of Dye Fluorescence

To quantify how the intensity of the fluorescence excited by the laser spot from the single-wavelength dye fluo-3 changes as a function of Ca_i^{2+}, we require both the dissociation constant (K_d) of fluo-3 and the maximum and minimum fluorescence that can be elicited *in situ* in the presence of high and low Ca^{2+} (F_{max} and F_{min}).[18] The K_d of the Ca^{2+} indicator was determined in the spot confocal microscope with a hemocytometer containing 100 μM fluo-3 free acid in pseudointracellular solutions of buffered Ca^{2+} (World Precision Instruments, Sarasota, FL). The sigmoidal variation of the fluorescence signal (F) with free Ca^{2+} concentration (Fig. 1, inset) was fitted by the Michaelis–Menten equation for Ca^{2+} binding:

$$F = F_{min} + \frac{F_{max} - F_{min}}{1 + (K_d/[Ca^{2+}])} \tag{1}$$

to yield a value for K_d of 400 nM. However, it remains possible that a slightly different value for K_d may apply in the intracellular environment of the photoreceptor itself.[25]

[25] P. P. Schnetkamp, X. B. Li, D. K. Basu, and R. T. Szerencsei, *J. Biol. Chem.* **266,** 22975 (1991).

The fluo-3 fluorescence signal evoked by saturating laser light from isolated rod and cone photoreceptors was calibrated by artificially manipulating Ca_i^{2+} to determine the maximum and minimum fluorescence that could be elicited from the dye in the region of the laser spot. A typical calibration experiment for a salamander rod is shown in Fig. 4. In each case fluorescence was excited by a sequence of brief 20-msec laser pulses instead of a continuous exposure, in order to ensure that dye bleaching was insignificant during the multiple laser exposures used for these measurements. Figure 4A and 4B illustrate the responses to the first two sequences of laser pulses. When the rod was first stimulated by the exciting beam, fluo-3 fluorescence declined to a reduced level that was then maintained

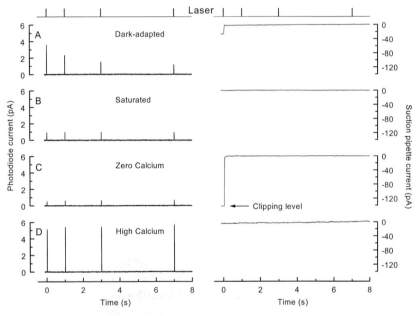

FIG. 4. *In situ* calibration of Ca^{2+}-dependent fluorescence from a rod loaded with fluo-3 AM. *Upper trace:* laser monitor; laser illumination delivered as a sequence of four 20-msec pulses. *Left:* fluo-3 fluorescence. *Right:* Suction pipette current. (A) First laser exposure of a dark-adapted rod. (B) Second laser exposure presented shortly thereafter while the circulating current remained completely suppressed. (C) Outer segment exposed to 0 Ca^{2+} solution containing 25 μM ionomycin to determine minimum fluo-3 fluorescence (F_{min}). Arrow indicates the clipping level of the suction pipette amplifier. (D) Outer segment exposed to isotonic Ca^{2+} solution containing 25 μM ionomycin to determine maximum fluo-3 fluorescence (F_{max}). These parameters allow the initial and final fluorescence levels from this experiment to be translated into values for Ca_i^{2+} of 474 nM in darkness and 35 nM when the circulating current was completely suppressed. [Reproduced from Ref. 15 with permission.]

for subsequent laser pulses. During this period, the circulating current remained suppressed by the substantial bleach induced by the laser. After these measurements, the recording chamber was perfused with solutions designed to greatly lower or greatly elevate Ca_i^{2+} to levels that would result in minimum (F_{min}) or maximum (F_{max}) values of fluo-3 fluorescence. In each case these solutions contained 25 μM ionomycin to shunt Ca^{2+} fluxes across the outer segment membrane. In Fig. 4C the outer segment was exposed to a solution containing nominally 0 Ca^{2+}, which included Na^+ to support extrusion of Ca^{2+} via $Na^+/Ca^{2+}-K^+$ exchange (111 mM NaCl, 2.5 mM KCl, 2.05 mM MgCl$_2$, 3 mM HEPES, and 2 mM EGTA, pH 7.7). This manipulation led to a dramatic elevation of the circulating current, which had previously been completely suppressed in Ringer solution by the preceding laser exposures. This observation suggests that Ca_i^{2+} had indeed been lowered further from its already-reduced level during response saturation, leading to further relief of the inhibition of guanylyl cyclase by Ca^{2+}.[26] After this solution change, the fluorescence evoked by each laser pulse decreased to a level slightly below that seen during saturating illumination. In Fig. 4D, the outer segment was exposed to a solution containing isotonic Ca^{2+} and 25 μM ionomycin, from which Na^+ had been omitted to prevent extrusion of Ca^{2+} via $Na^+/Ca^{2+}-K^+$ exchange (76.6 mM CaCl$_2$, 2.5 mM KCl, and 3 mM HEPES, pH 7.7). This manipulation increased the fluorescence evoked by each laser pulse to a level well above the reduced levels seen while the circulating current was suppressed in Ringer solution, and also somewhat greater than that evoked from the dark-adapted rod at the onset of laser illumination.

Fluorescence values such as those in Fig. 4C and 4D were taken as approximating to F_{min} and F_{max}, respectively, and were used in conjunction with the Fluo-3 dissociation constant to convert the intensities of fluo-3 fluorescence into levels of Ca_i^{2+} both in the dark and after complete suppression of the circulating current by the laser. This procedure gave values for Ca_i^{2+} of 670 ± 70 nM in a dark-adapted salamander rod and 30 ± 3 nM after saturating laser illumination (mean ± SEM).[15] In red-sensitive salamander cones, Ca_i^{2+} fell from 410 ± 37 nM in darkness to 5.5 ± 2.4 nM during response saturation.[16] It should be noted that the rapid recovery of cone photoreceptors after photopigment bleaching makes it difficult to hold the response in saturation for an extended period, despite the high intensity of the laser spot. Despite the lower values for Ca_i^{2+} in cones, examination of their ratio reveals that the dynamic range over which Ca_i^{2+} varies is more than three times larger in the cone than in the rod.[16]

In these calibration measurements for both rods and cones the

[26] K.-W. Koch and L. Stryer, *Nature (London)* **334,** 64 (1988).

F_{max}/F_{min} ratio determined *in situ* was consistently less than that obtained in the cuvette. Because more aggressive disruption of the outer segment membrane with saponin (40 mg/ml) did not result in a further increase in fluorescence, it seems unlikely that F_{max} was consistently underestimated by the calibration procedure. We therefore interpret the reduced F_{max}/F_{min} ratio as representing a residual Ca^{2+}- and light-insensitive "pedestal" of fluorescence, which may represent fluo-3 bound or compartmentalized within the rod outer segment. However, a similar residual level of fluorescence was observed from rods when whole-cell patch-clamp recording was used to incorporate the membrane-impermeant pentapotassium salt of fluo-3, which would be expected to be less readily compartmentalized within the disks, while the cone outer segment contains no comparable compartment that might accumulate the dye. We therefore believe that the "pedestal" of fluorescence most probably originated from nonspecific binding of dye within the outer segment cytoplasm.

Changes in Calcium during Photopigment Cycle

As already described, exposure of the photoreceptor outer segment to the laser spot bleaches the majority of the photopigment within the area of the laser spot. While this might at first sight appear to be a major drawback, it is possible to make a virtue of this necessity and to use the spot confocal technique to study the changes in Ca_i^{2+} that take place in rod and cone photoreceptors during the photopigment cycle. For rods, this can be achieved by measuring fluo-3 fluorescence from a single photoreceptor when dark adapted, after exposure to laser light sufficiently intense to bleach a substantial fraction of the photopigment, and after regeneration with 11-*cis*-retinal. These measurements demonstrate that when the dark current and sensitivity are depressed after bleaching, the fluo-3 fluorescence signal is depressed also, whereas regeneration of the photopigment with 11-*cis*-retinal restores all of these parameters to near their original dark-adapted values.[15]

We have also obtained similar results for changes in Ca_i^{2+} during the photopigment cycle in red-sensitive salamander cones.[16] The main difference between the two types of photoreceptor is that in a cone both the stabilization of the photoresponse after bleaching and the subsequent regeneration of the photopigment by 11-*cis*-retinal are extremely rapid, taking place within a few minutes instead of the tens of minutes required in a rod. The rapid and reproducible pigment regeneration that occurs in the continuous presence of an ethanolic solution of 11-*cis*-retinal allows repeated measurements of Ca_i^{2+} to be made from the same cone. We have used this approach to investigate the relationship between the circulating current and outer segment Ca^{2+} concentration in red-sensitive cones.[16] When the circulating current was progressively suppressed by background light of increasing intensity, fluo-3 fluorescence decreased linearly, ultimately falling to a pedestal level once the response to the background saturated. This progressive decline in fluorescence is consistent with a proportional variation between the outer segment Ca^{2+} concentration and the magnitude of the circulating current during background adaptation.[12,16]

This approach does not, however, allow measurement of the dynamic changes in Ca_i^{2+} that take place during the response to a flash. To achieve this in future we will require a method that allows us to measure Ca_i^{2+} without bleaching an appreciable proportion of the photopigment. It may be possible to achieve this goal by judicious choice of the photoreceptor, the fluorescent dye, and the wavelength of the laser, so as to minimize photoreceptor bleaching during the measurement.

Acknowledgments

This work was supported by Grant EY-01844 from the National Eye Institute of the National Institutes of Health, and by a Project Grant from the Wellcome Trust. We thank J. L. Vergara, A. P. Sampath, and M. C. Cornwall for their participation in many of our early efforts to implement spot confocal Ca^{2+} measurements, and A. P. Sampath and M. C. Cornwall for collaboration in experiments on changes in Ca_i^{2+} during the photopigment cycle. We also thank A. P. Sampath for comments on an earlier draft of the manuscript.

Section III

Phototransduction

[11] Functional Study of Rhodopsin Phosphorylation in Vivo

By ANA MENDEZ, NATALIJA V. KRASNOPEROVA, JANIS LEM, and JEANNIE CHEN

Introduction

Phosphorylation of rhodopsin is a key reaction in the termination of the light response. It is the first step in rhodopsin inactivation, beginning 100 msec after the light stimulus.[1,2] A subsequent step, the binding of arrestin to phosphorylated rhodopsin, is necessary to complete inactivation.[3] Both steps limit the lifetime of photoactivated rhodopsin and have a major influence on the gain of the transduction cascade.

Extensive work has been done to address basic aspects of rhodopsin phosphorylation (see Ref. 4 for a review). Its importance in rhodopsin shutoff was first identified by *in vitro* reconstitution studies.[5] It was reported that up to nine phosphates could be incorporated per rhodopsin molecule *in vitro*,[6] and that heavily phosphorylated rhodopsin activated transducin with a reduced efficiency.[7–11] The kinase responsible for this activity, rhodopsin kinase (RK), was identified and extensively characterized.[12–14] The sites of phosphorylation on rhodopsin were localized at several serine and threonine residues at the cytoplasmic carboxyl terminus of rhodopsin,[15,16]

---

[1] J. Chen, C. L. Makino, N. S. Peachey, D. A. Baylor, and M. I. Simon, *Science* **267,** 374 (1995).
[2] C. K. Chen, M. E. Burns, M. Spencer, G. A. Niemi, J. Chen, J. B. Hurley, D. A. Baylor, and M. I. Simon, *Proc. Natl. Acad. Sci. U.S.A.,* **96,** 3718 (1999).
[3] J. Xu, R. L. Dodd, C. L. Makino, M. I. Simon, D. A. Baylor, and J. Chen, *Nature (London)* **389,** 505 (1997).
[4] J. B. Hurley, M. Spencer, and G. A. Niemi, *Vis. Res.* **38,** 1341 (1998).
[5] P. A. Liebman and E. N. Pugh, Jr., *Nature (London)* **287,** 734 (1980).
[6] U. Wilden and H. Kuhn, *Biochemistry* **21,** 3014 (1982).
[7] J. L. Miller and E. A. Dratz, *Vis. Res.* **24,** 1509 (1984).
[8] U. Wilden, S. W. Hall, and H. Kuhn, *Proc. Natl. Acad. Sci. U.S.A.* **83,** 1174 (1986).
[9] J. L. Miller, D. A. Fox, and B. J. Litman, *Biochemistry* **25,** 4983 (1986).
[10] V. Yu. Arshavsky, M. P. Antoch, and P. P. Philippov, *FEBS Lett.* **224,** 19 (1987).
[11] N. Bennett and A. Sitaramayya, *Biochemistry* **27,** 1710 (1988).
[12] K. Palczewski, J. H. McDowell, and P. A. Hargrave, *J. Biol. Chem.* **263,** 14067 (1988).
[13] R. T. Premont, J. Inglese, and R. J. Lefkowitz, *FASEB J.* **9,** 175 (1995).
[14] K. Palczewski, *Eur. J. Biochem.* **248,** 261 (1997).
[15] P. Thompson and J. B. Findlay, *Biochem. J.* **220,** 773 (1984).
[16] K. Palczewski, J. Buczylko, M. W. Kaplan, A. S. Polans, and J. W. Crabb, *J. Biol. Chem.* **266,** 12949 (1991).

and the most favored sites *in vitro* were reported as Ser-338 and Ser-343 by three independent studies.[17–19]

A study by Ohguro and colleagues reported different preferred sites of rhodopsin phosphorylation from live animals exposed to light. The carboxyl terminus of rhodopsin was isolated from photoreceptor disk membranes and analyzed by chromatographic separation and mass spectrometry.[20] Only monophosphorylated species of rhodopsin were detected in the study, at either Ser-338 or Ser-334 depending on whether the mice had been exposed to a flash or to constant light. However, an alternative method developed by J. Hurley *et al.*, based on liquid chromatography-electrospray mass spectrometry (LC-MS) of mouse retinas bleached *ex vivo*, allowed detection of diphosphorylated forms of rhodopsin at bleaching levels of 17% or higher.[4] Consistent with the latter result is the report that sections from light-adapted retina are immunoreactive with an antibody that recognizes only multiply phosphorylated rhodopsin.[21]

One caveat of reconstituted systems, or the use of biochemical methods for the analysis of rhodopsin phosphorylation, is the difficulty of following real-time reaction kinetics. As mentioned above, phosphorylation of rhodopsin takes place 0.1 sec after the flash,[1] and dephosphorylation of certain residues could happen equally fast. Therefore, there is a possibility that the phosphorylated residues detected by biochemical methods may not be the actual ones that trigger rhodopsin shutoff. Furthermore, it is not clear whether rhodopsin shutoff is regulated by the number of phosphates that is added to the photolyzed rhodopsin, and/or the timing of phosphate incorporation. Certainly even millisecond differences in the kinetics of phosphate incorporation could affect the gain of phototransduction.

An alternative approach to the study of rhodopsin phosphorylation consists of analyzing the effect of mutations at the rhodopsin carboxyl-terminus phosphorylation sites in the photoreceptors of transgenic mice by single-cell recordings. This combination of transgenic mouse technology with electrophysiological measurements allows real-time reaction kinetics to be examined in the living cell.[1,2] Single-photon responses, elicited by individual rhodopsin molecules, can be recorded from mice expressing

---

[17] J. H. McDowell, J. P. Nawrocki, and P. A. Hargrave, *Biochemistry* **32,** 4968 (1993).
[18] D. I. Papac, J. E. Oatis, Jr., R. K. Crouch, and D. R. Knapp, *Biochemistry* **32,** 5930 (1993).
[19] H. Ohguro, K. Palczewski, L. H. Ericsson, K. A. Walsh, and R. S. Johnson, *Biochemistry* **32,** 5718 (1993).
[20] H. Ohguro, J. P. Van-Hooser, A. H. Milam, and K. Palczewski, *J. Biol. Chem.* **270,** 14259 (1995).
[21] G. Adamus, Z. S. Zam, J. H. McDowell, G. P. Shaw, and P. A. Hargrave, *Hybridoma.* **7,** 237 (1988).

both endogenous and mutant rhodopsin. The comparison of normal and abnormal responses allows one to establish the mutant phenotype.

In this chapter we describe how to generate transgenic mice expressing rhodopsin mutants at the phosphorylation sites and how to characterize the lines in terms of transgene expression and retinal morphology. In addition, we describe the criteria used in the selection of mice for electrophysiological recordings so that light responses elicited by mutant rhodopsins can be unequivocally assigned.

Methods

Transgene Construct

A collection of opsin mutants has been designed in which the phosphorylation sites at positions 334, 338, and 343 are mutated one by one or in combination, or in which all of the phosphorylation sites at the carboxyl terminus are mutated or removed.

The desired mutations are introduced in an 11-kb *Bam*HI mouse genomic clone that encompasses the entire transcriptional unit of the opsin gene with 5 kb of upstream and 1.5 kb of downstream flanking sequences.[1] This fragment includes the sequences that have been shown to direct expression of the β-galactosidase reporter gene to rod photoreceptors in transgenic mice.[22] Indeed, the spatial and temporal expression pattern of this transgene corresponds to that of the endogenous opsin.[23]

An example is shown in Fig. 1 for three of the constructs: Ser334ter, in which Ser-334 is substituted by a termination codon, thereby eliminating all of the phosphorylation sites; S343A, a single mutant at one of the serines; and the completely substituted mutant (CSM), in which all the phosphorylation sites are substituted to alanine.

In the design of each construct, several considerations need to be taken into account. First, the transgene must be distinguished from the endogenous gene in the genotyping process. This can be done either by using polymerase chain reaction (PCR) primers whose 3' ends match the specific mutations (therefore amplifying only the transgene) or by making use of restriction sites engineered in the transgene that allow its identification in PCR products derived from genomic DNA. For that purpose, silent mutations are introduced in each transgene to generate new restriction sites, as shown in Fig. 1. Second, the expression level of the mutant opsin in each line must be determined. As discussed below, this can be done by using

[22] J. Lem, M. L. Applebury, J. D. Falk, J. G. Flannery, and M. I. Simon, *Neuron* **6,** 201 (1991).
[23] B. J. Woodford, J. Chen, and M. I. Simon, *Exp. Eye Res.* **58,** 631 (1994).

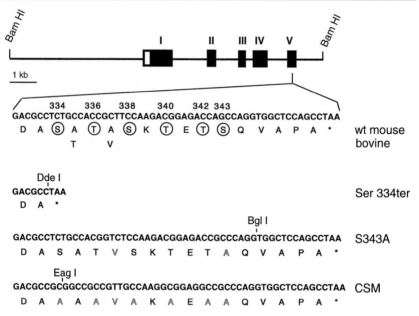

FIG. 1. Examples of three different rhodopsin phosphorylation site mutant constructs. *Top:* The 11-kb *Bam*HI fragment of the mouse opsin gene. Solid boxes indicate the protein-coding region. Open box is the 5' untranslated sequence. *Middle:* Enlarged map of exon 5 sequence encoding the last 17 amino acids of mouse opsin. Sites that can be phosphorylated are circled. Differences in amino acid at codons 335 and 337 between mouse and bovine opsins are indicated. *Bottom:* Same region from three transgene constructs, showing the engineered DNA sequence changes and the resulting coding differences (outlined letters) and introduced restriction sites. The A337V substitution generates a recognition epitope for MAb 3A6 in the S343A mutant, and is maintained in the completely substituted mutant (CSM) for consistency.

reverse transcription (RT)-PCR to determine the mutant-to-total transcript ratio. Third, distinction of heterologous and endogenous opsins at the protein level would allow the study of the expression pattern of the mutant opsins. If specific antibodies for the mutated region are not available, the introduction of a tag for antibody recognition in the heterologous opsin should be considered at this point. A conservative amino acid substitution, A337V, has been introduced in our constructs to confer to the mouse opsin a linear antigenic determinant for 3A6 monoclonal antibody (MAb) present in bovine and human opsins (residues 337–341, VSKTE).[24] This linear

[24] R. S. Hodges, R. J. Heaton, J. M. R. Parker, L. Molday, and R. S. Molday, *J. Biol. Chem.* **263**, 11768 (1988).

epitope therefore distinguishes bovine and human opsins (VSKTE) from mouse opsin (ASKTE). The A337V change has been introduced in all constructs (Fig. 1) for consistency, although mutations at either Ser-338 or Thr-340 are expected to decrease significantly the antibody-binding affinity.[24]

Preparation of DNA for Microinjection

The plasmid containing the transgene is prepared by lysozyme–alkaline lysis of bacterial cultures followed by purification in CsCl gradients.[25] The 11-kb insert is then obtained by *Bam*HI restriction digestion and isolated by preparative agarose gel electrophoresis.

DNA for microinjection must be free of all contaminants that may be toxic to the eggs and of all particulate matter that could clog the injection pipette. A QiaexII kit (Qiagen, Chatsworth, CA) is used for isolation of DNA from agarose slices. The DNA is further purified by passage through an ion-exchange chromatography column (Elutip-D; Schleicher & Schuell, Keene, NH). Eluted DNA is then precipitated with ice-cold prefiltered ethanol, pelleted, washed with 70% (v/v) ethanol, air dried, and resuspended in 100 μl of a prefiltered microinjection buffer [10 mM Tris-HCl (pH 7.4), 0.1 mM EDTA in high-quality distilled water]. A DNA solution of approximately 100 ng/μl, which is 100 times more concentrated than needed for microinjection, is normally obtained with 20 μg of plasmid starting material.

The DNA concentration is determined in an agarose gel by comparative ethidium bromide staining. Different amounts of the stock DNA solution are loaded onto different lanes. The intensity of ethidium bromide-stained bands is compared with a series of dilutions of a standard DNA of known concentration (Φ λ DNA digested with *Hin*dIII) in several other lanes. By loading 250 ng of Φ λ/HIII DNA, the mass of each band is predicted to be 23.1 kb (119.25 ng), 9.4 kb (48.5 ng), 6.5 kb (33.75 ng), 4.3 kb (22.5 ng), 2.3 kb (12 ng), 2.0 kb (10.5 ng), and 0.56 kb (3 ng). The intensity of the 11-kb band from the transgene is compared with the intensity of the band closest in size (9.4 kb). A microinjection stock of DNA is prepared at a concentration of 1 ng/μl, divided into aliquots, and stored at $-20°$ until use. Immediately before microinjection, DNA is filtered through a 0.22-μm pore size filter column to remove particulates.

---

[25] J. Sambrook, E. F. Fritsch, and T. Maniatis, "Molecular Cloning: A Laboratory Manual." Cold Spring Harbor Laboratory Press, Cold Spring Harbor, New York, 1989.

Generation of Transgenic Mice

Transgenic mice are generated by established procedures.[26,27] When selecting the mouse strain for the study, the fact that ambient light levels are sufficient to damage photoreceptors in albino rodents should be taken into account. Furthermore, albinism exacerbates photoreceptor degeneration induced by the expression of a mutant opsin in transgenic mice.[28] Pigmented mice were therefore chosen for our studies.

Briefly, eggs for microinjection are obtained from donor females (C57BL/6J × DBA/2J F_1 hybrid), previously superovulated and mated to stud males (C57BL/6J × DBA/2J F_1 hybrid). Microinjection is thus performed on F_2 hybrid zygotes from C57BL/6J and DBA/2J F_1 strains. About 12 hr postcoitum (p c) the oviducts of the donor females are removed and the fertilized one-cell eggs are collected and placed into culture. The purified transgene fragment, diluted to 1 ng/μl in microinjection buffer [10 mM Tris-HCl (pH 7.4), 0.1 mM EDTA] is then microinjected into one of the pronuclei, typically the larger male pronucleus. Eggs that survive injection are implanted into a pseudopregnant recipient surrogate mother (CD-1 female mated to vasectomized male and used 0.5 day pc). Implants are done at the one-cell stage, immediately after microinjection, or at the two-cell stage after overnight culture.

For all the opsin mutant constructs the efficiency of transgenic mouse production is the same as described by most investigators.[27] About one-third of the microinjected eggs that are implanted survive to term and are born 3 weeks later. Typically, 20–35% of the pups that are born have integrated the transgene into their genomes. These transgenic mice are bred to C57BL/6 mice to maintain the pigmented background. The resulting transgenic lines are therefore genetically heterogeneous.

The mechanism of transgene integration into the host chromosomes is unknown. From an early study of the organization of inserted DNAs in transgenic mice it can be inferred that integration occurs prior to the first round of DNA replication in 70% of the transgenic mice, but it can also occur after the first round of DNA replication (in the remaining 30%).[29]

The copy number of the integrated transgene can vary from one to

---

[26] B. Hogan, R. Beddington, F. Costantini, and E. Lacy, "Manipulating the Mouse Embryo: A Laboratory Manual." Cold Spring Harbor Laboratory Press, Cold Spring Harbor, New York, 1994.

[27] S. J. Waller, M. Y. Ho, and D. Murphy, in "DNA Cloning 4: Mammalian Systems" (D. M. Glover and B. D. Hames, eds.), p. 185. The Practical Approach Series. Oxford University Press, New York, 1995.

[28] M. I. Naash, H. Ripps, S. Li, Y. Goto, and N. Peachey, J. Neurosci. **16**, 7853 (1996).

[29] R. D. Palmiter and R. L. Brinster, Annu. Rev. Genet. **20**, 465 (1986).

several hundred. For a particular founder, there is usually only one integration site where multiple copies of the transgene are arranged in a head-to-tail tandem array. However, the transgene can also be integrated at two or more loci in the founder genomic DNA.

In our experience, multiple integration of the transgene occurs in about 25% of founder animals. In these cases the transgene segregates in subsequent rounds of breeding, yielding a percentage of transgenic mice typically higher than 50%. Given the high frequency at which multiple integrations are observed, we consider it worthwhile to perform routine Southern blot analyses of F_1 generations. The Southern strategy applied to segregate different lines from the same founder is outlined in one of the following sections.

Founder Screening

Mice born from microinjected eggs are screened by PCR for integration of the transgene into their genomes. Transgene-positive mice derived from the injected eggs are called "founders" because they give rise to the subsequent colonies of transgenic mice. A founder with a single integration site produces one independent line with its characteristic expression level that is typically stable throughout propagation of the line. Founders with multiple integration sites produce multiple independent lines. Because different lines express the transgene at different levels, it is critical that these lines be sorted for proper interpretation of the results.

The PCR screening strategy is different for each construct. For the screening of the Ser334ter construct, for instance, primers are designed to amplify specifically the transgene: Rho2, 5′ TGGGAGATGACGACGCC-TAA 3′ (forward); and Rho3, 5′ TGAGGGAGGGGTACAGATCC 3′ (reverse). The underlined sequence in Rho2 matches the specific mutations introduced in the transgene.

For the analysis of the completely substituted mutant (CSM), two primers are designed to recognize and amplify endogenous and transgenic sequences with the same efficiency: Rho5940, 5′ GTGAGGGGACATGC-TGGAGGTGAGGC 3′ (forward); and mRh5, 5′ GGAGCCTGCAT-GACCTCATCC 3′ (reverse). The PCR amplification product is then digested with *Eag*I enzyme, which specifically recognizes the transgene. A comparison of the intensity of digested to nondigested DNAs allows detection of different lines with obvious differences in the copy number of the integrated transgene.

The PCR protocol for genotyping is as follows. Briefly, a small piece of tail from each mouse is digested at 55° overnight in 200 μl of proteinase K buffer [50 mM Tris (pH 8.0), 100 mM EDTA, 0.5% (w/v) sodium dodecyl sulfate (SDS), proteinase K (0.5 mg/ml)]. Hair and particulates are pelleted by centrifugation of the samples at 13,000g for 2 min. Supernatants are transferred to new tubes and DNA is precipitated by addition of a 0.5 volume of 8 M ammonium acetate and 2 volumes of 96% (v/v) ethanol. DNA is allowed to air dry and resuspended in 100 μl of buffer [10 mM Tris-HCl (pH 7.4), 0.1 mM EDTA]. One microliter of DNA is amplified by PCR. PCRs are conducted in a final volume of 20 μl [1× PCR buffer (Life Technologies, Gaithersburg, MD), 0.2 mM dNTP mix, 1.5 mM MgCl$_2$, a 0.25 μM concentration of each primer, and 0.6 U of *Taq* polymerase (Life Technologies)]. Samples are first heated at 95° for 5 min, and then cycled 30 times (94° for 1 min, 63° for 45 sec, and 72° for 1 min). If restriction digestion is required (CSM), digestion of half of the PCR product follows with the corresponding restriction enzyme (*Eag*I).

Identifying Founders with Multiple Integration Sites of the Transgene

When the percentage of mice positive for the transgene is higher than 50% in one line, there is a high probability that the transgene is integrated at two or more loci in the genome. In this case the transgene segregates during subsequent rounds of breeding, giving rise to independent lines with different expression levels.

A Southern blot analysis of the integrated DNA is necessary to distinguish the various emerging lines. Both the choice of restriction enzymes for digestion of genomic DNA and the choice of the hybridization probe are important in the experiment. The 11-kb *Bam*HI transgene typically integrates in a head-to-tail array. From the restriction map, an enzyme is chosen that cleaves the transgene array, yielding fragments of predicted sizes. The cleavage will also yield "junction fragments" of novel length from the two ends of the array where the transgene is flanked by genomic DNA. For detection of these flanking fragments, a probe must be designed that maps close to either the 5' end or the 3' end of the transgene. An example is shown in Fig. 2 for one line that showed three different integration sites. *Eco*RI enzyme is used to digest the genomic DNA. It cuts twice in the transgene (Fig. 2A), yielding the predicted bands of 5 and 4.7 kb from the head-to-tail array. The probe used, a 1.2-kb *Nco*I fragment close to the 5' end of the transgene, recognizes the junction fragment at the 5' end but not the 5-kb internal fragment or the 3'-junction fragment (Fig. 2A). Three "junction fragments" of 2.2, 3, and 4.5 kb are observed in the founder DNA lane, indicating that the transgene integrated at three different sites in the genome. One of the lines, identified by the 2.2-kb junction fragment, segregates in the F$_1$ progeny. The 4.5 and 3-kb bands cosegregate in the F$_1$, suggesting that these integration sites may be closely linked.

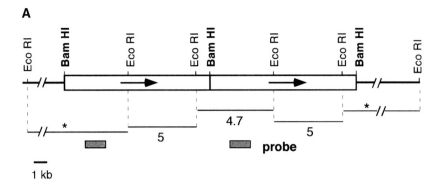
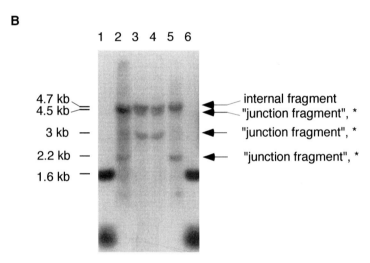

FIG. 2. Southern blot strategy to identify different independent lines emerging from one founder in the F_1 generation due to integration of the transgene at different loci. (A) Putative representation of the integrated transgene. Boxes represent two copies of the transgene (11-kb *Bam*HI genomic fragment of mouse opsin gene) integrated into the genome in a head-to-tail tandem array. *Eco*RI enzyme is used to digest genomic DNA. A 1.2-kb *Nco*I DNA fragment that maps near the 5' end of the transgene is used as a probe. Only the flanking DNA fragments will be characteristic of different lines, reflecting different integration sites in the genome ("junction fragment," *broken lines in the diagram). The others will typically be observed in all lines, their intensity correlating with integration copy number. (B) Southern blot result of *Eco*RI-digested genomic DNAs from one founder (lane 2) and its F_1 progeny (lanes 3, 4, and 5). The 4.7-kb band is observed in all lanes. The 4.5-, 3-, and 2.2-kb bands are all observed in the founder DNA lane. The 2.2-kb band segregated in the F_1. The 4.5- and 3-kb bands cosegregated in the F_1. Lanes 1 and 6 are 1-kb ladder molecular weight markers (GIBCO-BRL, Gaithersburg, MD).

Quantitation of Transgene Expression

Once several independent lines for each construct are established, it is important to determine the transgene expression level in each line.

The expression level of mutant rhodopsin in each line is analyzed by determining the mutant-to-total opsin transcript ratio by RT-PCR. An engineered restriction site is used to distinguish the mutant transgene mRNA from the wild-type mRNA. Figure 3 shows how the percentage of mutant transcript is determined for one of the transgenic lines expressing an opsin with all the serine and threonine residues at the carboxyl terminus substituted to alanine (CSM, completely substituted mutant). Briefly, RNA from individual mouse retinas is extracted from transgene-positive mice and littermate controls with Triazol (Life Technologies). After isopropanol precipitation, the RNA is dissolved in water, and the whole volume is used in a cDNA synthesis reaction. For a 20-μl reaction, RNA (preheated to 65°, 10 min) is incubated with oligo(dT)$_{12-18}$ (10 μg/ml) and 200 units of Moloney murine leukemia virus (Mo-MuLV) RT (Life Technologies) in 50 mM Tris-HCl (pH 8.3), 75 mM KCl, 10 mM dithiothreitol (DTT), 3 mM MgCl$_2$, and a 0.5 mM concentration of each dNTP for 60 min at 37°. After the RT reaction is completed, RNAs are hydrolyzed by addition of NaOH to a final concentration of 0.1 M and incubated at 68° for 15 min. After neutralization by addition of Tris-HCl, pH 7.6, to 0.1M, cDNAs are precipitated with ethanol in the presence of 10 μg of glycogen as carrier. Twenty percent of the cDNA product from one retina is used as a template for PCR amplification with primers mRh4 (5' GCTCTTCCATCTA-TAACCCGG 3', forward) and mRh5 (5' GGAGCCTGCATGACCT-CATCC 3', reverse), as shown in Fig. 3B. These primers flank intron 4. The resulting PCR product, 263 bp in size, represents a mixture of endogenous and transgene sequences. Half of the sample is digested with *Eag*I, and the other half is incubated without enzyme. To assess the completion of enzyme digestion, a 1236-bp fragment, obtained by PCR amplification of the corresponding transgene plasmid clone with the same primers, is used as a control. Digested and undigested samples are size fractionated in a 3% NuSieve agarose gel (FMC, Philadelphia, PA). In the digested sample, the product derived from the transgene is cleaved at the engineered restriction site to yield fragments of 110 and 153 bp, whereas the product derived from the endogenous rhodopsin gene lacks the restriction site and remains as a fragment of 263 bp. By quantitating the intensity of the 263-bp band in digested vs undigested samples (lanes 3 and 4, Fig. 3C) in the digitized image (Eagle Eye software; Stratagene, La Jolla, CA), the endogenous-to-total transcript ratio is calculated.

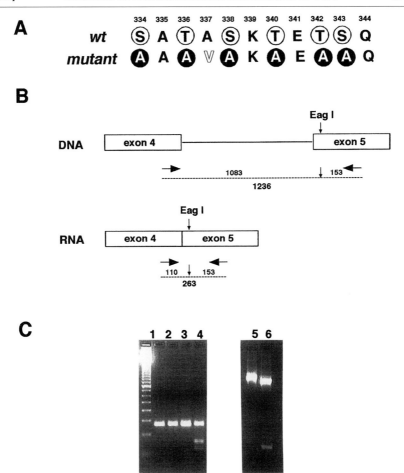

FIG. 3. Method for quantitation of transgene expression in the CSM line. (A) Amino acid sequence of endogenous vs mutant rhodopsin at the COOH terminus. Potential phosphorylation sites are circled. (B) Diagram showing the position of the primers used in the RT-PCR, and the predicted sizes (in base pairs) for DNA and cDNA amplification products. An *Eag*I restriction site was engineered in the transgene that allowed identification of the PCR product derived from the mutant transcript. (C) Examples of reaction products from a control mouse retina (lanes 1 and 2) and a littermate transgenic mouse retina (lanes 3 and 4), incubated either with (lanes 2 and 4) or without (lanes 1 and 3) *Eag*I enzyme and resolved in a 3% NuSieve agarose gel. Lanes 5 and 6, PCR product from the corresponding mutant opsin genomic fragment, incubated without or with *Eag*I enzyme, respectively.

TABLE I
CORRELATION BETWEEN TRANSGENE EXPRESSION AND TIME COURSE OF RETINAL DEGENERATION IN TRANSGENIC LINES EXPRESSING VARIOUS RHODOSPIN MUTANTS[a]

Construct	Line	Mutant transcript (% of total)	Time (weeks) for loss of 50% of photoreceptors
S338A	A	2	>12
	B	75	<3
	C	9	>12
S343A	D	15	>12
	E	5	>12
	F	15	>12
S338A/S343A	G	25	ND
S334A/S338A/S343A	H	6	>12
	I	15	>12
	J	50	<3
	K	20	>6
CSM	L	ND	<3
	M	ND	<3
	N	40	<4
	O	35	<4
	P	ND	<2
CSM/A338S	Q	60	<3
	R	ND	<3
	S	ND	<3

[a] The percentage of mutant transcript was determined by RT-PCR as described in text. The degeneration rate is indicated as the number of weeks in which the length of outer nuclear layer is halved. ND, Not determined.

Opsin Overexpression and Retinal Degeneration

One commonly encountered problem that emerges when establishing opsin transgenic lines is that opsin overexpression leads to retinal degeneration.[1,28,30,31] It is therefore necessary to examine the retinal morphology from independent transgenic lines for each construct and to select those in which the photoreceptors are preserved for the electrophysiological measurements.

Table I shows a correlation between mutant opsin expression levels and time course of retinal degeneration for several independent lines established for a collection of opsin mutants at the carboxyl-terminus phosphorylation sites. Regardless of the mutations introduced in the transgene, all lines show

[30] T. Li, W. K. Snyder, J. E. Olsson, and T. P. Dryja, *Proc. Natl. Acad. Sci. U.S.A.* **93,** 14176 (1996).
[31] C. H. Sung, C. Makino, D. Baylor, and J. Nathans, *J. Neurosci.* **14,** 5818 (1994).

a progressive degeneration of the photoreceptor layer. Among different transgenic lines, the time course of degeneration varies from several weeks to several months, and correlates more tightly to the transgene expression level than to the specific mutation.

Three groups can be roughly distinguished: (1) lines in which the expression of mutant opsin is up to 15% of the total opsin. Retinas from these lines show progressive shortening and disorganization of the rod outer segment layer over the course of months; (2) lines in which the mutant opsin ranges from 15 to 40–50% of the total. In this group of intermediate expression levels, the number of photoreceptor nuclei is halved in the course of weeks; and (3) lines with high expression levels of the transgene, in which the outer nuclear layer is absent by postnatal days 15–18.

Lines from the first group are useful for single-cell recordings because photoreceptors are largely intact for several months. However, the single-photon responses elicited from the mutant opsins are a minor component of the total, making it necessary to collect more data from the same cell.

Lines from the second group show an intermediate rate of degeneration. Photoreceptor cell death begins at the second or third postnatal week. Suction electrode recording from these lines is possible within a certain time window. This time window must be determined by monitoring the retinal morphology at various postnatal ages. An example of one line expressing 40% of mutant opsin (CSM) is shown in Fig. 4. At 16 days the number of nuclei is approximately halved, and rod outer segments (ROS) are significantly shortened and disorganized. Recording from these mice is possible at 16 to 20 postnatal days.

Lines from the third group degenerate too fast to be useful for electrophysiological recordings. At 15 postnatal days these lines show only one layer of photoreceptor nuclei at the outer nuclear layer and no rod outer segments. It is therefore difficult to determine the expression level of the transgene in these lines, because at 15 postnatal days there are almost no photoreceptor cells for RNA extraction. These lines are obtained with a high frequency (Table I).

Use of Opsin Knockout Background to Slow Retinal Degeneration

There are at least two variables that can affect the rate of retinal degeneration in a transgenic mouse model carrying an opsin mutation: one is the deleterious effect of the mutation itself, and the other is opsin gene dosage. Photoreceptors are apparently sensitive to increased opsin expression, as even overexpression of wild-type opsin in transgenic mice can lead to retinal

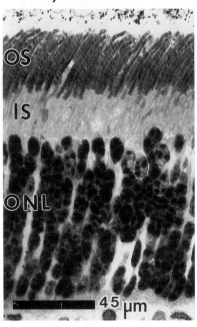 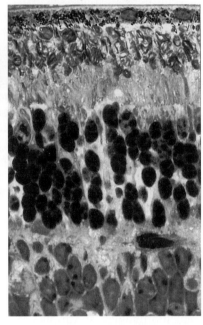

FIG. 4. Light micrograph of a retinal section of line N. Expression level of the transgene (CSM) in this line was 40%. One-micrometer Epon sections of central retina from a transgenic mouse and a littermate control (wt) at postnatal day 16 are compared. OS, Outer segment; IS, inner segment; ONL, outer nuclear layer.

degeneration.[31,32] The rate of degeneration has been shown to correlate with expression levels. Retinas from a transgenic line expressing the wild-type opsin transgene at a ratio of 5.2:1 to the endogenous opsin mRNA degenerate in less than 2 weeks.[31] Another line with an expression of 1:1 shows normal retinal morphology up to 15 months of age. However, when the transgene is bred to homozygosity to increase the expression level to 2:1, a slow course of photoreceptor degeneration is observed.[32]

Thus, in the lines with an intermediate level of expression (second group), where the ratio of transgene to endogenous opsin mRNA is less than 1:1, the retinal degeneration can result from a combination of a specific pathogenic effect of the mutation and a deleterious effect due to overexpression of opsin. To separate the effect of the opsin mutations in

[32] J. E. Olsson, J. W. Gordon, B. S. Pawlyk, D. Roof, A. Hayes, R. S. Molday, S. Mukai, G. S. Cowley, E. L. Berson, and T. P. Dryja, *Neuron* **9**, 815 (1992).

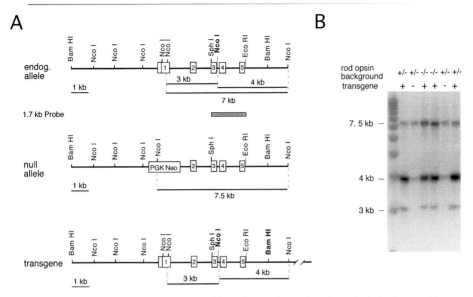

FIG. 5. Identification of transgene-positive mice on the opsin +/− and −/− background by Southern blot. (A) Partial restriction maps of the opsin endogenous locus, null allele, and integrated transgene, showing the DNA fragments yielded by NcoI digestion that were recognized by the 1.7-kb SphI–EcoRI probe (gray bar). Boxes represent the coding sequences. The NcoI site in boldface is polymorphic in mice: it is absent in the null allele of opsin but present in the transgene, and may be present or absent in the endogenous locus of mice from an outbred colony. PGK Neo, neomycin resistance selection marker. (B) Southern blot showing the bands of the expected size from mice that are +/− or −/− at the opsin locus, and positive or negative for the transgene.

the CSM (lacking all the phosphorylation sites) from the effect of gene dosage on the retinal architecture, several independent CSM transgenic lines have been crossed to opsin −/− mice.

The opsin knockout mice used in the study have been generated and characterized by J. Lem and colleagues.[33] Mice expressing the transgene in the +/− opsin background are obtained in the first round of breeding. These mice are crossed to opsin −/− mice to obtain transgene expression on the +/− or −/− backgrounds. The strategy for the genotyping of the resulting F_2 progeny is outlined in Fig. 5. Mouse genomic DNA is digested with NcoI, resolved in an agarose gel, transferred to a nylon membrane, and probed with the SphI–EcoRI 1.7-kb genomic DNA fragment of the opsin gene (Fig. 5). The NcoI restriction pattern from the endogenous opsin

[33] J. Lem, N. V. Krasnoperova, P. D. Calvert, B. Kosaras, D. A. Cameron, M. Nicolo, C. L. Makino, and R. L. Sidman, *Proc. Natl. Acad. Sci. U.S.A.* **96,** 736 (1999).

gene or the transgene is different from that of the null allele of opsin –/– mice because the knockout allele carries the phosphoglycerate kinase-neomycin (PGK Neo) selection casette (Fig. 5A). The *Nco*I site between exons 3 and 4 of the opsin gene is polymorphic: it could be present or absent in the endogenous gene in our outbred mouse population. It is present in the transgene derived from the 11-kb *Bam*HI opsin genomic clone, and absent in the null allele of opsin knockout mice. In this experiment the endogenous gene yields two bands of 3 and 4 kb that are recognized by the probe (Fig. 5A, top; Fig. 5B, lanes +/–); the transgene yields two bands of 3 and 4 kb that are more intense owing to the multiple copies of the transgene integrated in a tandem array (Fig. 5A, bottom; Fig. 5B, lanes + for the transgene), and the opsin null allele yields a band of 7.5-kb (Fig. 5A, middle; Fig. 5B, lanes +/– and –/–). Whether one mouse has one or two copies of the null allele (+/– or –/– genotype) is established by comparison of the 7.5-kb band intensity among different lanes. The same amount of genomic DNA (10 μg) is loaded in each lane.

Three CSM lines (N, O, and P from Table I) were bred to opsin –/– mice, and their retinal morphology compared at a given age in the +/+, +/–, and –/– backgrounds (Fig. 6). Similar to published results,[33,34] rod outer segments of opsin +/– mice are shorter than wild type at 1 month of age, whereas no outer segment is formed in the absence of opsin (Fig. 6 A–C). For lines N (40% expression) and O (35% expression), retinal morphology is progressively improved as the level of endogenous opsin is reduced (see Fig. 6D–F and G–I). In these lines, the number of photoreceptor cells in the opsin –/– background is close to normal at 1 month of age, and rod outer segments are longer and less disorganized (Fig. 6F and I). Line P, a high expressing line in which the outer nuclear layer is absent at 15 postnatal days in the wild-type background (Fig. 6J), does not show any improved morphology at the same age in either the +/– or –/– background (Fig. 6 K and L).

The percentage of mutant transcript increases as the endogenous opsin expression is reduced. Percentages of mutant transcript in the +/– background are 58% in line N, 54% in line O, and 70% in line P, as determined by RT-PCR. As expected, they represent 100% of the transcript in the –/– background in the three cases.

We conclude from these results that the mutations introduced in the CSM opsin are not responsible for the retinal degeneration. Rather, the

---

[34] M. M. Humphries, D. Rancourt, G. J. Farrar, P. Kenna, M. Hazel, R. A. Bush, P. A. Sieving, D. M. Sheils, N. McNally, P. Creighton, A. Erven, A. Boros, K. Gulya, M. R. Capecchi, and P. Humphries, *Nature Genet.* **15,** 216 (1997).

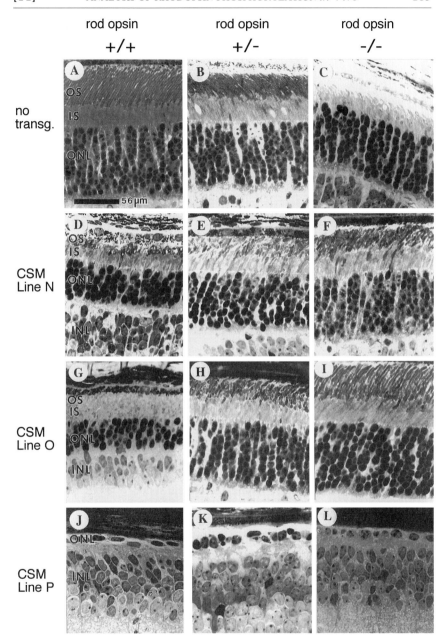

FIG. 6. Effect of opsin dosage on retinal structure. Sections of central retina from mice from three independent lines expressing the CSM over the +/+, +/−, and −/− opsin background. Line P expressed the transgene at the highest level (>50%). (A–I), Retinas from 1-month-old mice; (J–L) retinas from postnatal day 15 mice.

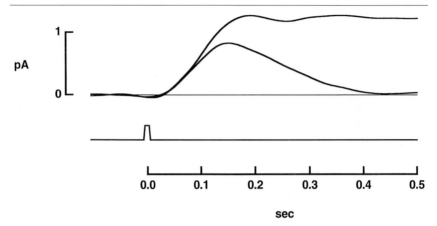

FIG. 7. Kinetics of prolonged responses in the Ser334ter mutant. The upper trace is the average of 49 prolonged responses elicited by the truncated rhodopsin at Ser-334 that was superimposed on the normal single-photon response by aligning their initial rising phases. Ser334ter responses have an amplitude nearly twice that of the normal single-photon response, and a mean duration of 5.2 sec. [Reprinted with permission from J. Chen, C. L. Makino, N. S. Peachey, D. A. Baylor, and M. I. Simon, 1995. Mechanism of Rhodopsin Inactivation *in Vivo* as Revealed by a COOH-Terminal Truncation Mutant. *Science* **267,** 374–377. Copyright (1995) American Association for the Advancement of Science.]

degeneration phenotype observed in different lines of the CSM in the opsin +/+ background is likely due to opsin overexpression.

The carboxyl-terminal domain of rhodopsin has been implicated in vectorial transport of post-Golgi vesicles containing rhodopsin to the rod outer segments of photoreceptor cells.[35,36] We found that mutating all of the phosphorylation sites at the carboxyl terminus did not affect rhodopsin trafficking, as the CSM mutant was capable of reconstituting a largely normal outer segment in the absence of endogenous rhodopsin. We conclude from this result that the phosphorylation sites are not required in the signal for rhodopsin transport.

Recording from the transgenic lines in the opsin +/− background presents some advantages vs recording in the wild-type background: (1) the higher percentage of mutant rhodopsin, (2) the rescue effect in the morphology that expands the time window for physiological analysis, and (3) the presence of endogenous rhodopsin that elicits normal responses, thereby providing a built-in control. On the other hand, having the transgene in the opsin −/− background allows unambiguous assignment of the mutant phenotype. Both approaches are complementary in the study of rhodopsin phosphorylation.

[35] D. Deretic, *Eye* **12,** 526 (1998).
[36] A. W. Tai, J. Z. Chuang, C. Bode, V. Wolfrum, and C. H. Sung, *Cell.* **97,** 877 (1999).

Single-Cell Recordings from Photoreceptors of Transgenic Mice

Because of the tremendous amplification gain of the photocascade, light capture by a single rhodopsin molecule can give rise to a measurable electrical signal. To detect single-photon responses the light source is adjusted so that it produces, on average, less than one photoisomerization per flash. Under these experimental conditions, most flashes will not cause any photoisomerization (resulting in "failures"), and some flashes will result in one or more photoisomerizations ("successes"). Elementary responses can be identified by plotting the amplitude histogram of responses and setting a criterion level between peaks, as the first peak at nonzero amplitude. The mean response amplitude and the variance of the elementary responses can be computed.[37]

In a transgenic mouse rod expressing both normal and mutant rhodopsin molecules, the probability of eliciting responses from either of these molecules will be determined by their relative expression levels. In the case of the Ser334ter mutation, the abnormal responses are prolonged (Fig. 7). Thus, the mutant phenotype is easy to assign.

To establish the mutant phenotype in transgenic lines in which the kinetics of inactivation are not drastically affected, the variance of the group of elementary responses recorded from transgenic mouse rods will be calculated and compared with that of wild-type littermate controls. The variance of the response is expected to be higher in transgenic rods if the mutation compromises the reproducibility of the single-photon response. In this case, it would increase as the ratio of mutant to endogenous rhodopsin increases in the +/− and −/− rod opsin backgrounds. By identifying the mutations that compromise the reliability of the single-photon response, it should be possible to identify the phosphorylation sites that are essential for rhodopsin shutoff.

Acknowledgments

Results shown from electrical recordings were obtained by Clint Makino in the laboratory of Dr. Baylor at Stanford University. Ser-334ter transgenic mice were obtained at Dr. Simon's laboratory at Caltech.

We thank Francis Conception, Renard Dubois, and Anthony Rodriguez for technical support. We are grateful to Robert and Laurie Molday for the generous gift of 3A6 antibody. We thank Nancy Wu and the USC Transgenic Core Facility for the generation of transgenic mice. This work was supported in part by grants from Research to Prevent Blindness, the Ruth and Milton Steinbach Fund, and NEI (EY 12155) (J.C.), and from NEI (EY 12008), the Foundation Fighting Blindness, Research to Prevent Blindness, and the Massachusetts Lions Eye Research Fund (J.L.). A.M. was the recipient of a postdoctoral fellowship from the Education and Science Ministry of Spain.

---

[37] D. A. Baylor, T. D. Lamb, and K. W. Yau, *J. Physiol.* **288**, 613 (1979).

[12] Mechanisms of Single-Photon Detection in Rod Photoreceptors

By FRED RIEKE

Introduction

At low light levels the visual system detects and counts photons with a reliability limited by statistical fluctuations in the number of absorbed photons and photoreceptor noise. This remarkable sensitivity is not merely an obscure laboratory phenomenon; reliable photon detection is crucial for normal rod vision, much of which occurs at light levels where individual rod photoreceptors receive photons rarely. The rod's ability to detect single photons has been appreciated for many years. In the 1940s, Hecht et al.[1] found that dark-adapted humans could see flashes producing fewer than 10 absorbed photons spread over an area of the retina containing about 500 rods. This sensitivity is possible only if the rods detect single photons. Subsequent behavioral experiments by Sakitt[2] suggest that the rods produce distinguishable responses to zero, one, and two absorbed photons, permitting the visual system to count photon absorptions. Sakitt's work also provides an estimate of the dark noise that limits absolute visual sensitivity.

Behavioral measurements of the sensitivity of rod vision guide studies of the biophysical and biochemical mechanisms of photon detection as they impose stringent constraints on how single photons are transduced by the rods and how the resulting signals are processed within the retina. Three important constraints on phototransduction are: (1) the single-photon response must be amplified to produce a macroscopic electrical signal from activation of a single rhodopsin molecule; (2) the dark noise in the rod's phototransduction cascade must not consume the single-photon response and render it undetectable; and (3) individual single-photon responses must have similar shapes so that one photon can be reliably distinguished from two. Although the ability of the visual system to detect and count photons has been appreciated for many years, only in the last 10–15 years have we begun to understand the mechanisms that permit the rod to meet these functional requirements. This chapter describes some of the experimental and theoretical methods that have made

---

[1] S. Hecht, S. Shlaer, and M. Pirenne, *J. Gen. Physiol.* **25,** 819 (1942).
[2] B. Sakitt, *J. Physiol.* **233,** 131 (1972).

this progress possible and points out some of the important unresolved issues.

Experimental Methods

Three experimental challenges in studying the rod's single-photon response are as follows: (1) keeping the cells fully dark adapted through all steps in the experiment, (2) developing recording techniques with the resolution to detect the single-photon response and the stability to collect enough responses to study the response statistics, and (3) interpreting the properties of the response in terms of the elements of the transduction cascade.

Tissue Preparation

Current techniques for measuring the rod's single-photon response require separating the retina from the pigment epithelium to provide direct access to the rod outer segment. Once the retina is removed from the pigment epithelium, rhodopsin can no longer be regenerated and the rods cannot fully dark adapt after bright light exposure. We maintain the cells in a dark-adapted state by performing all the necessary experimental procedures with infrared light and infrared/visible image converters. Rods are about a factor of 10^9 less sensitive to 880-nm light than to 500-nm light, but even this residual sensitivity to infrared light can cause adaptation. Toad rods begin to adapt at steady light intensities that produce 1–10 photoisomerizations (Rh*) per second[3,4]; when these relatively dim lights are extinguished the rods recover their dark sensitivity quickly and completely. Thus a conservative goal is to use infrared light that produces less than 1 Rh*/sec during all experimental procedures.

For experiments on toad or salamander rods we keep the animal in complete darkness for at least 12 hr prior to the experiment. The animal is quickly decapitated and pithed and the eyes are removed using night vision goggles with a built-in infrared illuminator (5001 Night Invader; ITT Night Vision, Roanoke, VA). All subsequent procedures are performed under a dissecting microscope (SMZ-2B; Nikon, Garden City, NY) equipped with infrared/visible converters (NiteMare 4100 Pocketscope; B. E. Meyers, Redmond, WA) and an illuminator made with an infrared LED (276-143C RadioShack, Ft. Worth, TX) whose output is collimated and

[3] D. A. Baylor, G. Matthews, and K.-W. Yau, *J. Physiol.* **309**, 591 (1980).
[4] K. Donner, D. R. Copenhagen, and T. Reuter, *J. Gen. Physiol.* **95**, 733 (1990).

filtered (RG-850 filter; Schott Glass, Duryea, PA) to eliminate any light reaching the retina with wavelength shorter than 850 nm. The intensity of the infrared light reaching the preparation corresponds to about 0.1 Rh*/sec. Each eye is hemisected under the dissecting microscope using a double-edged razor blade. The back half of each eye is then cut into several pieces and put in Ringer solution, where the retina is gently peeled from the pigment epithelium. Pieces of retina are stored in a light-tight container at 4° for up to 36 hr. Isolated rods are obtained by shredding a small piece of retina (roughly 1–2 mm^2) with fine needles in a 100-μl drop of Ringer solution. The drop is then transferred to a recording chamber mounted on the stage of an inverted microscope equipped with an infrared sensitive camera (e.g., Cohu 4815-2000; San Diego, CA). The cells are visualized using >850-nm light with an intensity in the image plane of the microscope corresponding to less than 0.5 Rh*/sec.

General Recording Issues

The basic operation of the rod is shown in Fig. 1. In darkness a circulating current carried primarily by Na$^+$ ions flows into the outer segment through cGMP-gated channels in the surface membrane. This circulating current sets the voltage across the cell membrane to a relatively depolarized level, about −40 mV, and causes continual release of neurotransmitter from the synaptic terminal. Incident light activates the photopigment rhodopsin, which triggers the series of biochemical reactions that make up the phototransduction cascade. The end result of activation of the transduction cascade is a reduction in the cGMP concentration, closure of channels in the surface membrane, and a decrease in the circulating current. This current decrease permits the rod to hyperpolarize and slows the rate of transmitter release from the synaptic terminal.

An important consequence of the rod's operation is that the light-induced changes in membrane voltage are a property of both the circulating current and voltage-activated conductances in the inner segment[5,6]; in contrast, the outer segment current is only weakly voltage dependent,[6] and thus provides an electrical signal controlled almost entirely by the transduction process. To measure the light response of rods the circulating current is rerouted through a current-measuring amplifier. These measurements are typically made using patch-clamp recordings from detached outer segments

[5] C. R. Bader, D. Bertrand, and E. A. Schwartz, *J. Physiol.* **331**, 253 (1982).
[6] D. A. Baylor and B. J. Nunn, *J. Physiol.* **371**, 115 (1986).

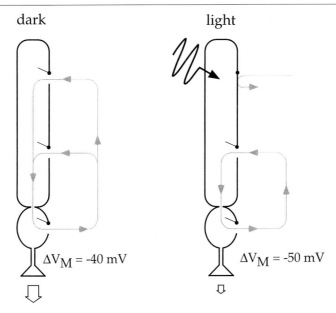

FIG. 1. Basic operation of rod photoreceptor. In darkness a circulating current flows into the outer segment and out the inner segment, depolarizing the rod and causing continual transmitter release from the synaptic terminal. This current is carried primarily by the movement of Na^+ ions, which flow down their electrochemical gradient into the outer segment and are pumped out the inner segment. Light activates the transduction cascade and suppresses the circulating current, permitting the cell to hyperpolarize and slowing transmitter release.

or suction electrode recordings from intact rods or truncated outer segments. Each recording technique faces several challenges:

The single-photon response is small: In a toad rod the single-photon response has a maximum amplitude of about 1 pA, comparable to the current flowing through a typical ion channel while it is open. Measuring this small signal is difficult, a problem that is exacerbated because it is desirable that the measured responses be free of distortions from instrumental noise and thus represent the true signals of the rod. The single-photon response can be measured accurately if the instrumental noise is considerably smaller than the cellular noise in the rod current. The cellular noise is dominated by occasional discrete photon-like events generated by spontaneous isomerization of rhodopsin and continuous current fluctuations caused by spontaneous phosphodiesterase (PDE) activation.[3,7] In a toad rod the continuous noise component has a root-mean-square amplitude

---

[7] F. Rieke and D. A. Baylor, *Biophys. J.* **71**, 2553 (1996).

of about 0.1 pA; this sets a practical goal for the maximum allowable instrumental noise. The limited bandwidth of the single-photon response is helpful in reaching this low noise level, as the measured currents can be low-pass filtered at 5–10 Hz without significantly affecting the light responses.

Long recordings are required to measure response statistics: To investigate how the rod generates a reproducible response to each absorbed photon it is necessary to characterize the trial-to-trial response variability and how this variability changes when the operation of the transduction cascade is altered. Measuring the variability of the single-photon response requires collecting a minimum of a few hundred responses to a fixed dim flash; the large number of responses is required in part because it is not possible to deliver a light stimulus that deterministically produces a single isomerized rhodopsin (see below). In an amphibian rod, we typically collect 200–400 responses over a period of 1–2 hr. In addition to the problems of instrumental noise already discussed, these long recordings require that the rod response does not change systematically during the course of the measurements, as such changes could be mistaken for response variability. To check stability during the recording the response to a moderate intensity flash is measured every 5–10 min; experiments are aborted if this response changes significantly. During data analysis, we reject experiments that show systematic changes in the single-photon response over time.

Interpretation of results in terms of events in transduction cascade: Electrophysiology provides a limited view of the transduction process: the single-photon current response is measured, and not the activity of the components of the transduction cascade that produce the response. Several methods are helpful in understanding how the transduction cascade shapes the measured currents; these include studies of phototransduction in transgenic mice, experiments on internally dialyzed outer segments, and modeling approaches that combine evidence from electrophysiology and biochemistry. An important characteristic of recording techniques, and one that differs between techniques, is the ability to manipulate elements of the transduction cascade.

Suction Electrode Recording

One approach to recording the rod's light-sensitive current is to draw the outer segment into a tight-fitting glass electrode (Fig. 2A). The electrode collects the current flowing into the outer segment and changes in this current in response to light can be directly monitored. Baylor *et al.*[8,9]

---

[8] D. A. Baylor, T. D. Lamb, and K.-W. Yau, *J. Physiol.* **288,** 589 (1979).
[9] D. A. Baylor, T. D. Lamb, and K.-W. Yau, *J. Physiol.* **288,** 613 (1979).

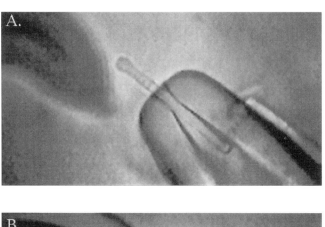
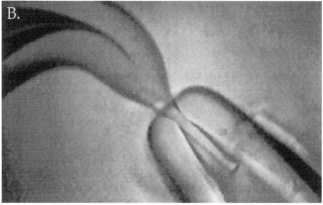
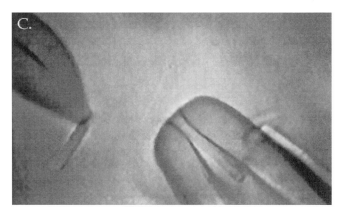

FIG. 2. Truncation of a toad rod. In truncated outer segment experiments the rod is first drawn into the suction electrode (A). The inner segment and a small piece of the outer segment are then cut off with a sharp glass probe (B), leaving the outer segment open and easily dialyzed with various solutions (C).

developed this technique and used it to make the first recordings of single-photon responses from vertebrate rods. Suction electrode recordings can be very stable, permitting a cell to be studied for several hours without a noticeable change in its responses. Suction electrode techniques also permit experiments on small cells, such as mammalian rods, that are difficult or impossible to record from using other methods.

The success of suction electrode recording depends greatly on the electrode itself. Three characteristics of a good suction electrode are a round (not oval) opening, so that the outer segment is not distorted as it is drawn in; walls that taper relatively quickly to the tip opening, so that the electrode resistance is minimized and the current is effectively collected; and a nonstick surface, so the outer segment is not damaged as it is drawn in. We make suction electrodes from borosilicate glass tubing pulled on a horizontal electrode puller. The electrodes are cut with a diamond knife to a diameter four to five times larger than the desired diameter of the final opening (e.g., for toad a 5 to 6-μm-diameter opening is used and the electrodes are cut initially to a diameter of 20–30 μm). The cut electrodes are fire polished until the opening reaches the desired size; electrodes that are more than 10% out of round are discarded. The electrodes are coated with silane to prevent the outer segment from sticking. For recording, the suction electrode is filled with Ringer solution and placed in an electrode holder with one port connected to a pressure/suction source made from glass syringes. Electrical connections to the bath ground and suction electrode are made by Ringer solution-filled agar bridges that contact calomel half-cells. A voltage-clamp amplifier (Axopatch 200B; Axon Instruments, Foster City, CA) is used to measure the circulating current while the bath and electrode are held at the same voltage.

An important property of suction electrode recordings is the electrical resistance of the seal formed between the cell membrane and the wall of the suction electrode. The seal resistance is typically about 5 MΩ, roughly five times the resistance of the electrode itself; this is in contrast to the seal resistances of >1 GΩ achieved in conventional patch-clamp recordings. The relatively low seal resistance causes two limitations of suction recording. First, some of the light-sensitive current flows across the seal resistance and thus goes undetected; if the seal is five times the electrode resistance, about 20% of the current continues to flow in its normal current loop and 80% flows through the current measuring amplifier. Second, a major component of the instrumental noise in suction electrode recordings comes from the thermal movement of ions across the seal resistance. This noise—Johnson noise—is inherent in any resistor at room temperature and its variance scales inversely with the resistance. The instrumental noise can be isolated from cellular noise by exposing the rod to a bright light that closes all the channels in the outer segment membrane (Fig. 3A). Instrumental noise measured in this way is usually close to the Johnson noise limit and considerably less than the cellular dark noise of the rod (Fig. 3B); however, noise sources such as outer segment channel noise can have an amplitude similar to or smaller than the seal noise and thus are difficult to resolve and study by suction electrode methods. Attempts to increase the seal resistance by making the opening in the suction electrode smaller or inducing the cell to stick to the electrode usually result in damage to the outer segment as it is drawn into the electrode.

The main advantages of suction electrode recording—stability and noninvasiveness—also present some disadvantages. Suction electrode recording does not permit the contents of the outer segment to be changed, making it difficult to alter the operation of particular elements of the transduction cascade. Studies of phototransduction in transgenic mice provide one way around this limitation.[10-12] Suction electrode recording also does not allow control of the intracellular voltage; instead, changes in the outer segment current lead to changes in voltage as they would under normal conditions (Fig. 1).

Truncated Outer Segment Recording

One of the main limitations of suction electrode recordings from intact cells is the difficulty in manipulating the transduction cascade. Yau and Nakatani[13] introduced a variation of suction electrode recording—the truncated outer segment preparation—that permits the solution inside the outer segment to be changed and the operation of individual elements of the transduction cascade to be altered. A rod is drawn into a suction electrode and the inner segment and a small piece of outer segment are cut off with a sharp probe (Fig. 2), providing diffusional access to the inside of the outer segment. After truncation the contents of the outer segment can be changed by flowing different solutions across the cut end and allowing them to exchange with the solution inside the outer segment by diffusion. These solution changes require only 5–10 sec to complete, permitting several conditions to be tested in the same outer segment and even allowing solution changes to be made during the flash response. We fill the suction electrode with a Ringer solution containing low Ca^{2+} (0.25 mM Ca^{2+} and 1 mM

---

[10] J. Chen, C. L. Makino, N. S. Peachey, D. A. Baylor, and M. I. Simon, *Science* **267,** 374 (1995).
[11] J. Xu, R. L. Dodd, C. L. Makino, M. I. Simon, D. A. Baylor, and J. Chen, *Nature (London)* **389,** 505 (1997).
[12] S. H. Tsang, M. E. Burns, P. D. Calvert, P. Gouras, D. A. Baylor, S. P. Goff, and V. Y. Arshavsky, *Science* **282,** 117 (1998).
[13] K.-W. Yau and K. Nakatani, *Nature (London)* **317,** 252 (1985).

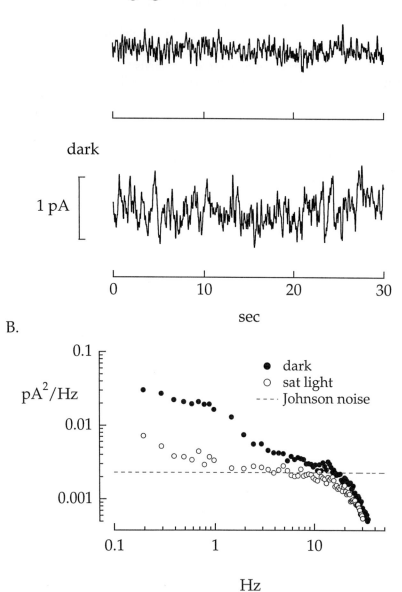

Fig. 3. Comparison of instrumental and cellular dark noise in a suction electrode recording from a salamander rod. (A) Sections of current record measured in saturating light and in darkness. The saturating light closed all the channels in the outer segment surface membrane,

EGTA) and dialyze the outer segment with a solution containing mostly arginine–glutamate. This choice of ionic compositions ensures a large driving force for Na^+ through the cGMP-gated channels in the outer segment.

Usually the first reaction to truncation experiments is surprise that they work at all and concern that important components of the transduction cascade are leaking from the outer segment. Long-term stable recordings are an important limitation of truncation experiments, but light responses can often be measured for 30 min before they begin to change systematically. It is likely that this eventual rundown is due to diffusional loss of components of the transduction cascade. During truncation experiments the bath and the suction electrode are clamped at the same voltage to isolate light-induced changes in current. This preparation does not, however, permit the voltage dependence of the light-activated currents to be studied, as changing the outer segment voltage with respect to the bath creates unmanageable instrumental noise due to the low resistance seal between the outer segment and the electrode. The processes shaping the flash response are also altered in truncation experiments, as diffusion of cGMP and Ca^{2+} from the solution bathing the cut end of the outer segment adds an additional component to the balance of cGMP creation and destruction and Ca^{2+} influx and efflux. Comparison of responses in truncated outer segments and intact cells requires modeling the spatial profiles of Ca^{2+} and cGMP.[7]

Dialyzed Outer Segment Recording

Patch-clamp recordings from isolated outer segments provide another means of characterizing how manipulations of the transduction cascade affect the rod's light response. This technique has been used extensively by Detwiler and colleagues[14–16] and is described in detail by Detwiler and Gray-Keller ([9] in this volume[16a]).

[14] K. Palczewski, G. Rispoli, and P. B. Detwiler, *Neuron* **8,** 117 (1992).
[15] G. Rispoli, W. A. Sather, and P. B. Detwiler, *J. Physiol.* **465,** 513 (1993).
[16] M. P. Gray-Keller and P. B. Detwiler, *Neuron* **13,** 849 (1994).
[16a] P. B. Detwiler and M. P. Gray-Keller, *Methods Enzymol.* **316,** 9, 1999 (this volume).

---

eliminating the circulating current (Fig. 1) and isolating instrumental noise. The record in darkness contains this instrumental noise as well as cellular dark noise. The records have been digitally low-pass filtered at 5 Hz. (B) Power spectral densities of currents recorded in saturating light (○) and in darkness (●) as in (A). The dashed line shows the Johnson noise level (0.0023 pA^2/Hz) expected for the seal resistance of 7 $M\Omega$ in this experiment. Currents have been filtered at 20 Hz (eight-pole Bessel low-pass).

Patch-clamp experiments, like truncation experiments, permit the solution inside the outer segment to be changed. In patch-clamp experiments the desired internal solution is used to fill the electrode; when the membrane occluding the electrode tip is ruptured the solution diffuses into the outer segment. It is difficult to compare light responses in the same outer segment with different internal solutions, as this requires perfusing the patch electrode to change its contents during recording. The resistance between the electrode and outer segment is usually >1 GΩ in patch-clamp experiments. This high seal resistance permits the transmembrane voltage of the rod to be controlled without a large leak current and produces low instrumental noise, permitting, for example, noise due to stochastic channel gating to be measured.[17,18] As for experiments on truncated outer segments, light responses can usually be measured for about 30 min before changing significantly in shape.

Identification of Single-Photon Responses

One of the primary difficulties in studying single-photon responses is that it is not possible, on command, to cause the isomerization of one and only one rhodopsin molecule. Several stochastic processes contribute to trial-to-trial fluctuations in the number of isomerized rhodopsin molecules produced by a dim flash of nominally fixed intensity; these include the generation of photons by a typical incoherent light source, the attenuation of the light by neutral density filters, and the absorption of photons by rhodopsin. Thus, while a flash that produces *on average* a single photoisomerization can be delivered, some flashes will produce no photoisomerizations, some one, and so on. This raises two issues: determining how many photoisomerizations on average the flash produces, and isolating the single-photon responses of the rod from responses to multiple photons.

Estimating Average Number of Effective Photon Absorptions

Two methods are used to estimate the average number of effective photon absorptions (i.e., number of photoisomerized rhodopsin molecules) produced by a dim flash. The first relies on measuring the photon flux (in photons μm^{-2} sec^{-1}) reaching the preparation and estimating the rod collecting area (in μm^2). We measure and control the photon flux with a light power meter (268R; Graseby Optronics, Orlando, FL) and calibrated neutral density filters. The rate of photoisomerizations for a

[17] R. D. Bodoia and P. B. Detwiler, *J. Physiol.* **367,** 183 (1984).
[18] P. Gray and D. Attwell, *Proc. R. Soc. Lond. B* **223,** 379 (1985).

given photon flux is determined by the rod collecting area. Several factors influence the collecting area: the absorption cross-section of an individual rhodopsin molecule, the rhodopsin concentration in the outer segment, the outer segment dimensions, and the probability that rhodopsin isomerizes on photon absorption and triggers an electrical response.[19] The absorption properties of the rod can also be measured directly by measuring the fraction of light absorbed by the rod before and after bleaching rhodopsin. In toad and salamander rods the estimated collecting areas are 15–20 μm^2.

A second method to determine the average number of photoisomerizations produced by a flash makes use of the Poisson statistics that govern photon absorption and the reproducibility of the rod single-photon response (see below). A defining characteristic of a Poisson process is that the variance in the event count is equal to the mean count. In the case of repeated trials of a dim flash, the variance in the number of photoisomerizations is equal to the mean number. Thus if each photoisomerization produces a response $\hat{r}(t)$ and a flash producing an average of $\bar{n}$ photoisomerizations is delivered, the mean response will be $\bar{r}(t) = \bar{n}\hat{r}(t)$ and the time-dependent variance of the ensemble of responses will be $\sigma_r^2(t) = \bar{n}\hat{r}^2(t)$. By measuring the mean flash response $\bar{r}(t)$ and the time-dependent ensemble variance of the flash response $\sigma_r^2(t)$ the mean number of photoisomerizations per flash can be estimated as $\bar{n} = \bar{r}^2(t)/\sigma_r^2(t)$ and the single-photon response as $\hat{r}(t) = \bar{r}(t)/\bar{n}$. This procedure is illustrated in Fig. 4. Figure 4A shows the mean response and Fig. 4B the ensemble variance of the flash responses and the variance measured in darkness; the difference (light − dark) is the ensemble variance of the flash response itself. Figure 4C compares the variance of the flash response with the square of the mean response; the ratio of the scales of the left and right axes provides an estimate of the mean number of photoisomerizations per flash of $\bar{n} = 0.6$. This is comparable with the estimate of 0.8 obtained from calibration of the light intensity assuming a collecting area of 15 μm^2.

Isolating Single-Photon Responses

Trial-to-trial fluctuations in the single-photon response place important constraints on the operation of the phototransduction cascade.[20,21] To separate the variability in the single-photon response itself from variability due to fluctuations in the number of photons absorbed, we first record several hundred responses to repetitions of a dim flash. Next, we separate single-

---

[19] G. J. Jones, M. C. Cornwall, and G. L. Fain, *J. Gen. Physiol.* **108,** 333 (1996).
[20] L. Lagnado and D. A. Baylor, *Neuron* **8,** 995 (1992).
[21] F. Rieke and D. A. Baylor, *Biophys. J.* **75,** 1836 (1998).

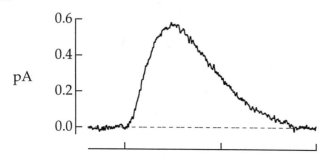

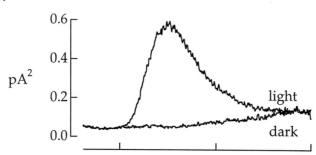

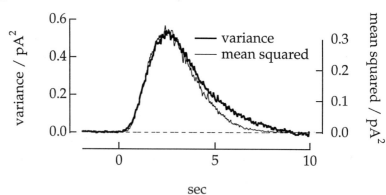

FIG. 4. Estimating the average number of photoisomerizations produced by a dim flash. (A) Average response to a dim flash measured in an intact toad rod, using a suction electrode. The flash was 10 msec in duration, delivered at time 0. The flash strength was 0.053 photon μm^{-2}; for a collecting area of 15 μm^2 this corresponds to 0.8 Rh*. (B) Time-dependent

photon responses from responses to multiple photons and to zero photons by constructing a histogram of the response amplitudes, such as that shown in Fig. 5A. The peaks in the histogram correspond respectively to zero, one, and two photoisomerizations. As these peaks are reasonably distinct, responses to single photons ("singles") can be separated from responses to zero or multiple photons. For example, in Fig. 5A responses with amplitudes between 0.43 and 1.63 pA were taken as singles. This procedure isolates from the original several hundred responses a smaller group of responses, 79 in this case, that are predominantly responses to single photons; 50 of these responses are superimposed in Fig. 5B. Differences between individual single-photon responses represent variability in the response of the transduction cascade to a single isomerized rhodopsin molecule. This variability in the single-photon response is surprisingly small given our intuition from other signals originating from single particles,[9,21] e.g., the time to decay of a radioactive particle, which exhibits large trial-to-trial fluctuations. Understanding the origin of this reproducibility is an important unresolved question in phototransduction.

Interpretation of Results in Terms of Transduction Cascade

Models of the transduction cascade provide a useful tool for interpreting the results of physiological measurements of the cellular current response in terms of the underlying biochemistry. The transduction cascade is usually described in qualitative terms: e.g., photoisomerized rhodopsin activates transducin, which in turn activates phosphodiesterase, etc. But the function of the cascade can be described by a set of coupled differential equations (see Fig. 6) with many of the important rate constants fixed or constrained by quantitative physiological or biochemical data.[7,22,23] These differential equations can be solved either numerically or approximated and solved analytically.[21,22] Because these models have few or no free parameters, this approach provides testable predictions for the transduction process based on the underlying biochemistry.

[22] E. N. Pugh and T. D. Lamb, *Biochem. Biophys. Acta* **1141,** 111 (1993).
[23] Y. Koutalos, K. Nakatani, and K.-W. Yau, *J. Gen. Physiol.* **106,** 891 (1995).

---

ensemble variance of the dim flash responses ("light") and responses measured in darkness ("dark"). (C) Variance of the flash response [light − dark from (B)] and square of the mean response from (A). The scale factor between the left and right axes (i.e., the mean response squared divided by the variance) provides an estimate of the mean number of photoisomerizations produced by the flash (see text), 0.6 in this experiment.

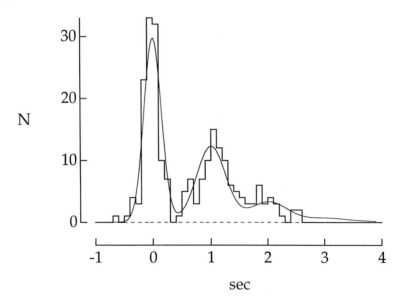

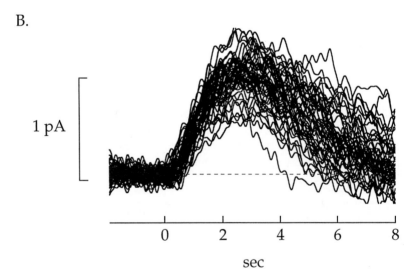

Fig. 5. Separation of single-photon responses from failures and responses to multiple photons. (A) Histogram of peak response amplitudes from 225 responses to a fixed dim flash. The smooth curve fit to the measured histogram was calculated by assuming that the noise in darkness and noise in the elementary response amplitude are independent and additive

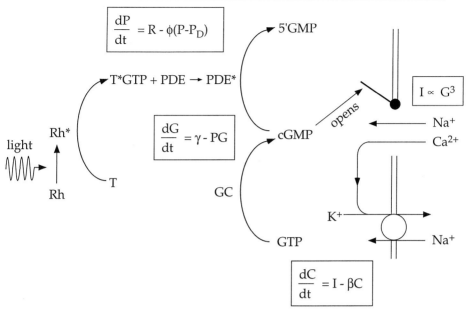

FIG. 6. Differential equations modeling the transduction cascade. Each of the boxed equations describes part of the transduction cascade. Symbols: R, rhodopsin catalytic activity; P, PDE activity; P_D, basal PDE activity; ϕ, decay rate constant of PDE; G, cGMP concentration; γ, rate of cGMP creation by guanylate cyclase; I, membrane current; C, Ca^{2+} concentration; and β, rate constant for Ca^{2+} extrusion. These coupled differential equations can be solved either numerically or approximated and solved analytically; see Pugh and Lamb[22] or Rieke and Baylor[21] for details.

Models of the transduction process such as the one outlined in Fig. 6 serve several roles in studying the single-photon response. First, studies of different classes of models can help pose experimental questions. D. Baylor and I have used this approach to study the reproducibility of the rod's single-photon response.[21] Second, models can provide mechanistic explanations for properties of the light response based on measurements from more reduced preparations—biochemical measurements or experiments on truncated outer segments. Pugh and Lamb used this approach to

---

and that the number of photoisomerizations per flash obeys Poisson statistics.[9,21] The peaks in the histogram correspond to responses to zero, one, and two photoisomerizations. (B) Fifty superimposed responses with amplitudes between 0.43 and 1.63 pA. From the amplitude histogram in (A) these are primarily single-photon responses. Each response has been shifted vertically to correct for baseline drift and digitally low-pass filtered at 5 Hz.

account for the amplification of the single-photon response,[22] and Koutalos et al. combined modeling and experiment to study the relative contributions of different mechanisms mediating light adaptation in rods.[23]

Summary

Rod photoreceptors detect and encode incident photons exceptionally well. They collect sparse photons with high efficiency, maintain a low dark noise, and generate reproducible responses to each absorbed photon. The mechanisms involved in single-photon detection—control of the effective lifetime of a single active receptor molecule, amplification of the activity of this single molecule by a second-messenger cascade, and reliable transmission of small synaptic signals—recur throughout the nervous system. Indeed, several other sensory systems reach or approach limits set by quantization of their input signals. For example, olfactory receptors can detect single odorant molecules.[24]

Although our understanding of visual transduction and signal processing has advanced rapidly during the past 10–15 years, fundamental questions still remain: What mechanisms are responsible for the reproducibility of the rod's elementary response? What are the tradeoffs of speed and sensitivity in the transduction cascade? How are the rod single-photon responses reliably transmitted to the rest of the visual system? Future technical innovations, particularly better methods to monitor the activity of intermediate steps in transduction, will play an important role in providing answers.

[24] A. Menini, C. Picco, and S. Firestein, *Nature* (*London*) **373,** 435 (1995).

[13] Electroretinographic Determination of Human Rod Flash Response in Vivo

By DAVID R. PEPPERBERG, DAVID G. BIRCH, and DONALD C. HOOD

Introduction

The electroretinogram (ERG), a multicomponent electrical signal that can be recorded at the cornea of the vertebrate eye, originates from the responses of retinal neurons to a test flash.[1] The first component of the ERG elicited by a brief flash of moderate or high intensity is a cornea-

[1] R. Granit, *J. Physiol.* **77,** 207 (1933).

negative response, termed the *a*-wave, that in mammals reaches its (negative) peak at ~5–20 msec after flash presentation. Abundant evidence indicates that the leading edge (rising phase) of the rod-mediated *a*-wave closely monitors the massed photocurrent response of the rod photoreceptors, i.e., the flash-induced reduction of rod circulating current.[2–11] However, the onset of postreceptor ERG components including the cornea-positive *b*-wave shapes the *a*-wave peak and subsequently obscures the rod photoreceptor response (see, e.g., Hood and Birch,[12] Robson and Frishman,[13] and Wachtmeister[14]).

The noninvasive nature of ERG recording allows the study of photoreceptor activity in human subjects through analysis of the *a*-wave response. ERG studies have provided information on the sensitivity (e.g., biochemical amplification) of early activating stages in the rod phototransduction process, and on the maximal excursion of the rod response, under a variety of illumination conditions in both normal subjects and in patients with retinal disease[8–10,15–17] (reviewed by Hood and Birch[18]). However, conventional ERG investigation of the rod photoreceptor response is severely constrained by the shortness of the postflash period that precedes *b*-wave intrusion. That is, *in vitro* photocurrent data show that in human rods, as in the rods of other mammalian species, the time scale of the response is several hundred milliseconds with weak flashes and increases with flash strength.[19–21] The period of development of the *a*-wave leading edge, ~20

---

[2] R. D. Penn and W. A. Hagins, *Nature (London)* **223,** 201 (1969).
[3] H. Heynen and D. van Norren, *Vision Res.* **25,** 697 (1985).
[4] H. Heynen and D. van Norren, *Vision Res.* **25,** 709 (1985).
[5] D. C. Hood and D. G. Birch, *Visual Neurosci.* **5,** 379 (1990).
[6] D. C. Hood and D. G. Birch, *Invest. Ophthalmol. Vis. Sci.* **31,** 2070 (1990).
[7] M. E. Breton and D. P. Montzka, *Doc. Ophthalmol.* **79,** 337 (1992).
[8] A. V. Cideciyan and S. G. Jacobson, *Invest. Ophthalmol. Vis. Sci.* **34,** 3253 (1993).
[9] D. C. Hood and D. G. Birch, *Vision Res.* **33,** 1605 (1993).
[10] M. E. Breton, A. W. Schueller, T. D. Lamb, and E. N. Pugh, Jr., *Invest. Ophthalmol. Vis. Sci.* **35,** 295 (1994).
[11] J. G. Robson and L. J. Frishman, *Visual Neurosci.* **12,** 837 (1995).
[12] D. C. Hood and D. G. Birch, *Visual Neurosci.* **8,** 107 (1992).
[13] J. G. Robson and L. J. Frishman, *J. Opt. Soc. Am. A* **13,** 613 (1996).
[14] L. Wachtmeister, *Prog. Ret. Eye Res.* **17,** 485 (1998).
[15] D. C. Hood and D. G. Birch, *Invest. Ophthalmol. Vis. Sci.* **35,** 2948 (1994).
[16] A. V. Cideciyan and S. G. Jacobson, *Vision Res.* **36,** 2609 (1996).
[17] D. G. Birch and D. C. Hood, in "Degenerative Diseases of the Retina" (R. Anderson, J. Hollyfield, and M. LaVail, eds.), p. 359. Plenum, New York, 1995.
[18] D. C. Hood and D. G. Birch, *Doc. Ophthalmol.* **92,** 253 (1997).
[19] D. A. Baylor, B. J. Nunn, and J. L. Schnapf, *J. Physiol.* **357,** 575 (1984).
[20] K. Nakatani, T. Tamura, and K.-W. Yau, *J. Gen. Physiol.* **97,** 413 (1991).
[21] T. W. Kraft, D. M. Schneeweis, and J. L. Schnapf, *J. Physiol.* **464,** 747 (1993).

msec or less, is short in comparison with this overall time scale of the rod response.

The present chapter describes a "paired-flash" ERG method that circumvents the constraint just noted, and in human subjects yields approximate determination of the full time course of the massed rod response to a test flash of arbitrary intensity.[22–24] A similar paired-flash method has also been used in studies of *in vivo* rod responses in experimental animals.[25–29]

Principle of Paired-Flash Method

In vitro data from rod photoreceptors show that the burst of cGMP hydrolysis produced by a bright flash leads to a rapid, complete suppression of the circulating current and a resulting maximal, i.e., saturating, photocurrent response (reviewed by Yau[30]). The paired-flash ERG method described here involves the presentation of a bright probe flash at a defined time after a test flash, and determination of the prevailing response to the *test* stimulus by analysis of the *probe* flash response. The central notion underlying the method is that the bright probe flash rapidly drives the rods to saturation and that the amplitude of the probe-generated ERG *a*-wave titrates the prevailing circulating current. Figure 1A diagrammatically illustrates the concept. For clarity, the cone photoreceptor contribution to the *a*-wave response is not considered in Fig. 1. The curve in the upper part of Fig. 1A is the hypothetical ERG response to a brief test flash delivered at time zero. Labels identify the *a*-wave of this response and the subsequent development of the *b*-wave and oscillatory potentials. Figure 1 further illustrates the later presentation of a brief, high-intensity flash. This second flash, termed the *probe,* elicits a second ERG response that includes a rapidly developing *a*-wave. The probe-generated *a*-wave is presumed to reach its peak within a brief interval that precedes substantial intrusion by

---

[22] D. G. Birch, D. C. Hood, S. Nusinowitz, and D. R. Pepperberg, *Invest. Ophthalmol. Vis. Sci.* **36**, 1603 (1995).

[23] D. R. Pepperberg, D. G. Birch, K. P. Hofmann, and D. C. Hood, *J. Opt. Soc. Am. A* **13**, 586 (1996).

[24] D. R. Pepperberg, D. G. Birch, and D. C. Hood, *Visual Neurosci.* **14**, 73 (1997).

[25] D. G. Birch, W. Kedzierski, S. Nusinowitz, J. L. Anderson, and G. H. Travis, *Invest. Ophthalmol. Vis. Sci.* **36**, S641 (1995).

[26] A. L. Lyubarsky and E. N. Pugh, Jr., *J. Neurosci.* **16**, 563 (1996).

[27] Y. Goto, N. S. Peachey, N. E. Ziroli, W. H. Seiple, C. Gryczan, D. R. Pepperberg, and M. I. Naash, *J. Opt. Soc. Am. A* **13**, 577 (1996).

[28] J. G. Robson, L. J. Frishman, and S. Viswanathan, *Invest. Ophthalmol. Vis. Sci.* **38**, S886 (1997).

[29] J. R. Hetling and D. R. Pepperberg, *J. Physiol.* **516**, 593 (1999).

[30] K.-W. Yau, *Invest. Ophthalmol. Vis. Sci.* **35**, 9 (1994).

the probe-generated ERG b-wave, and the peak of this a-wave response is presumed to correspond with saturation of the rod photocurrent response.

The lower part of Fig. 1A shows hypothetical probe flash responses (hooklike curves) obtained with variation of t, the time of probe flash presentation, in a series of paired-flash trials. Shown at the left is the hypothetical response to the probe flash delivered in the absence of a recent test flash ("probe-alone" response). These probe responses are positioned with their peaks aligned at the putative constant state representing photocurrent saturation. Filled circles in the diagram represent the baselines from which the probe responses depart. If the prevailing rate of change of the test-flash-induced response at time t is relatively small, little if any change in this test flash response will occur in the brief interval between time t and attainment of the a-wave peak of the probe response. Under these conditions, the peak amplitude of the probe response as referred to the pre-probe baseline will approximate the circulating current at time t. From this determination of the probe response amplitude, and from the maximal amplitude exhibited by the probe-alone response, one obtains by subtraction an amplitude A that represents the rod response to the test flash at time t. That is,

$$A(t) = A_{mo} - A_m(t) \qquad (1)$$

where $A_m(t)$ is the probe response amplitude determined in the paired-flash trial, A_{mo} is the amplitude of the probe-alone response, and $A(t)$ is the amplitude at time t of the derived response to the test flash. The family of determinations of $A(t)$ obtained with variation of the interflash interval t (Fig. 1B) yields, in turn, the complete "derived" rod response to the test flash.

The a-wave response to the bright probe flash requires a brief period to reach its peak amplitude. With probe flash strengths of 10^4 scotopic troland-seconds (sc td-sec) or higher, as in the experiments described below, this period is typically about 8 msec. The procedure just outlined ignores the nonsimultaneity of the probe flash presentation time t and the determination time of the probe a-wave response. That is, determination of the derived response at time t $[A(t)]$ is based on measurement of the probe response amplitude at time $(t + \approx 8$ msec). The error introduced by effectively equating t with $(t + \approx 8$ msec) is negligible when the interflash interval t greatly exceeds 8 msec, i.e., under most of the conditions investigated below. However, with short interflash intervals, consideration of this ≈ 8-msec period becomes important for accurate representation of the derived response.[29] Unless otherwise stated (Fig. 5, below), post-test-flash times quoted for derived response amplitudes presented here are the interflash intervals t.[22-24]

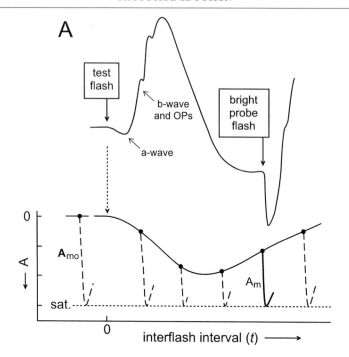

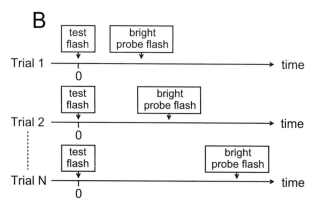

Fig. 1. Diagrammatic illustration of the paired-flash method. (A) Hypothetical ERG response to a test flash presented at time 0 and to a bright probe flash presented at a later time t. Labels identify the negative-going a-wave and the subsequent b-wave and oscillatory potentials (OPs) of the test-flash-generated response. The secondary ERG produced by the bright probe flash includes a rapidly developing a-wave. The solid hooklike curve in the lower part of the panel reproduces the probe-generated a-wave response of the illustrated ERG. Dashed hooklike curves positioned at $t > 0$ symbolize probe-generated a-waves obtained by altering the time of probe flash presentation, i.e., by altering the interflash interval t as

Experimental Procedures

General Description

Most of the procedures for obtaining paired-flash ERGs are similar to those for corneal recording of the conventional, i.e., single-flash, ERG. All experiments are conducted in accordance with the principles embodied in the Declaration of Helsinki, and with institutional regulations and guidelines including those for informed consent. Prior to the ERG testing session, the pupil of the eye to be tested is dilated by the application of 1% (w/v) tropicamide and 2.5% (w/v) phenylephrine hydrochloride. An opaque patch is then placed over the eye for 45 min to allow complete dark adaptation. Immediately before the experiment, the patch is removed in the ERG darkroom laboratory, and 0.5% (w/v) proparacaine hydrochloride (topical anesthetic) is applied to the cornea. A bipolar recording electrode (Gold-Lens; Diagnosys LLC, Littleton, MA), the corneal contact surface of which has been lubricated with 2.5% (w/v) methylcellulose solution, is then placed on the cornea, and a cup-style ground electrode is attached to the forehead.

The subject is seated at a ganzfeld dome that delivers full-field stimuli from two flashguns. Light from one of these two flashguns, representing the test flash, is ordinarily of short wavelength ("blue" flash; Wratten 47B filter; λ_{max} = 449 nm; Eastman Kodak, Rochester, NY) and preferentially stimulates rod photoreceptors. The second flashgun provides the "probe" flash. As needed (see below), the probe stimulus consists of either a short-wavelength flash ("blue" probe; spectrum identical to that of the test flash) or a relatively long-wavelength flash ("red" probe; Wratten 26; λ_{cut-on} = 605 nm) that with neutral density attenuation is photopically matched to the blue probe, i.e., is matched to the blue probe with respect to its effectiveness in stimulating cone photoreceptors.[9] Light from the probe flashgun also passes through a heat-absorbing filter. Strengths of the flashes associated with a given set of filters are calibrated by the use of an integrating photometer (model 40X; United Detector Technology, Hawthorne, CA). Instrumen-

---

schematically shown in (B). In (A) the dashed hooklike curve to the left of $t = 0$ symbolizes the a-wave response to the probe flash presented alone. Peaks of the probe-generated a-waves are aligned at a fixed ordinate value presumed to represent rod photocurrent saturation (sat.). Filled circles represent the baselines from which the probe-generated a-waves depart. The amplitude A_m of each probe-generated a-wave is taken as a measure of the rod circulating current at time t; A_{mo}, the maximal amplitude, is determined by the probe-alone response. The measured values of $A_m(t)$ and of A_{mo} yield $A(t)$, the derived rod response to the test flash [filled circles and connecting curve; see Eq. (1)]. This conceptual diagram ignores both the contribution of cones to the a-wave responses and the determination of probe-generated a-wave amplitudes at times slightly preceding the response peak. See text for further details.

tation and procedures relevant to recording ERGs from infants are generally similar to those just described. Here the corneal recording electrode is of reduced size (infant GoldLens; Diagnosys LLC), the pupil is dilated with pediatric eye drops, and the subject, in a supine position, views a ganzfeld dome positioned above the head.[31]

Determination of the time course of the derived response to a fixed test flash, or of the amplitude–intensity relation of the derived response at a fixed interflash interval, requires a series of paired-flash trials. It is important within the series to separate successive trials by a dark-adaptation period to permit the subject to recover completely from the flashed stimuli of the preceding trial. For normal subjects, and for the strengths of the probe flashes used in the experiments described here, we have found a dark-adaptation period of 1 min or longer (depending on the test flash strength) necessary to ensure full recovery. Paired-flash determinations of the incremental rod response to a test flash can also be carried out in the presence of steady background light.[24] Background illumination in the ganzfeld dome of our system is provided by two 12-V incandescent bulbs operated at a color temperature of about 2650 K. After the experiment, the rod-mediated component of each probe flash response (see below) is analyzed for amplitude at a fixed time near or somewhat preceding the a-wave peak (typically, 6.4–8.4 msec), and amplitudes of the derived rod response to the test flash are determined from Eq. (1).[24] Determinations of stimulus strength in scotopic troland-seconds (flashes) and scotopic trolands (backgrounds) are based on measurement of the diameter of the fully dilated pupil during the experiment.

Additional Technical Considerations

Any of a number of high-intensity photographic units can be used to generate the bright probe flash required in the paired-flash experiments. The probe flash unit of our system consists of a Novatron model 2150 flash head and model 1600 power supply (Novatron, Dallas, TX). The test flash is provided by a second, identical Novatron unit or (for relatively weak test flashes) a Grass photostimulator (model PS-22; Astro-Med/Grass, Quincy, MA). Each of the two flash heads is mounted in a holder that spans a port in the ganzfeld dome. Light from each unit is spectrally shaped (see above) and attenuated (see below) by filters positioned in the holder. The dimensions of the Novatron and Grass flash heads are similar, and the two heads can be readily exchanged. The shot-to-shot variation in flash strength of the Novatron unit is less than 4%. At the maximum power setting, the duration (half-height) of the Novatron flash is about 1.3 msec, i.e., well

---

[31] S. Nusinowitz, D. G. Birch, and E. E. Birch, *Vision Res.* **38,** 627 (1998).

within the photoreceptor integration time, and the unit is fully recharged within 5 sec. However, the length of this recharge period necessitates the use of two separate flash units, as the interval between the test and probe flashes chosen for a given trial is often far less than 5 sec. The use of separate flash units is made necessary also by the need for independent control of the intensity and wavelength of the two flashes. The neutral density filters used in our early studies were Wratten (gelatin) filters. However, even with the use of an infrared-absorbing filter, gelatin filters are gradually degraded by the heat of the flash unit. Our solution was to prepare a set of "aperture filters" from rectangular sections of Bakelite (thickness, 3 mm), an extremely durable, heat-resistant phenolic. By varying the area of the aperture it was possible to produce a set of filters that, as calibrated by direct measurement of the ganzfeld dome luminance, spanned a convenient range of effective attenuation.

The timing of the flashed stimuli is controlled by a computer equipped with a timing board (PC-TIO-10; National Instruments, Austin, TX). Responses from the ERG electrodes are amplified (model AM-502 differential amplifier; Tektronix, Beaverton, OR) with a bandpass (3 dB down) typically of 2 Hz to 10 kHz. The amplified data are digitized (routinely used sampling rate, 5 kHz) and stored in a computer. The probe flash waveform obtained in a paired-flash trial is acquired over an interval that typically extends to 50 msec after flash presentation. The experimental data are also routinely recorded on tape (bandpass of dc to 9 kHz; model 420 instrumentation recorder; Vetter Instruments, Rebersburg, PA) for later analysis of long-duration signals such as the full ERG response. Near-real-time visualization of the family of probe responses obtained during the experiment is advantageous. For example, variably among subjects and among trials, the test flash elicits a blink reflex at ~70–90 msec after flash presentation that introduces noise into the ERG response. The blink reflex can present a problem for a paired-flash determination when it occurs within the period of recording of the response to the probe flash; visualization of the response immediately after its collection allows evaluation of the need for repetition of the paired-flash trial. With practice, most subjects can learn to minimize this reflex, even to intense flashes. Analysis of the data employs Igor Pro (Wavemetrics, Lake Oswego, OR), a software package compatible with both Windows and Macintosh platforms.

Analysis of Probe Flash Response

Subtraction of Cone Contribution

Paired-flash derivation of the rod response to the test flash depends on accurate determination of the rod-mediated response to the bright probe flash, as the amplitude of this rod-mediated probe response [A_m in Eq. (1)] measures the remaining rod circulating current. In the human eye, cone photoreceptors contribute to the *a*-wave generated by relatively bright flashes; under dark-adapted conditions, this cone contribution represents up to ~20% of the peak amplitude of the overall *a*-wave response.[5,6] Obtaining the derived rod response to the probe flash thus necessitates determining and then subtracting the cone contribution.

A later section (Time Course of Derived Response) describes a "cone subtraction" procedure workable in the case of substantial variation of the cone contribution within a series of paired-flash trials. However, under conditions relevant to two key types of paired-flash experiments, the relatively low photic sensitivity of the cones and their relatively fast recovery kinetics combine to simplify this subtraction procedure. The first of these is determination of the time course of the derived response to a weak test flash, i.e., one for which the peak amplitude of the derived response is well below photocurrent saturation. Here one can expect relatively little cone stimulation by the test flash, and the cone contribution to the bright probe flash of a given paired-flash trial will be essentially constant, i.e., independent of the interflash interval *t*. The second type of experiment is the determination of recovery kinetics after a strong, i.e., rod-saturating, test flash. In this case the cones are fully recovered by the time of departure of the rods from saturation, and the cone contribution to the probe flash response during the period of rod recovery will also be essentially constant (independent of *t*). In the two types of experiments just described, each of the series of paired-flash trials routinely employs a short-wavelength (blue) flash of fixed high intensity as the probe stimulus. Also recorded during the experiment is a single waveform obtained in response to a long-wavelength (red) probe flash, the strength of which has been set to achieve a photopic match (equal cone-stimulating activity) with the blue probe flash.[5,6,9] The response to the red probe is presumed to represent the (constant) cone contribution to the group of blue probe responses obtained in the paired-flash trials, and is computationally subtracted from each of the blue probe responses to yield the presumed rod-mediated component.

The strength of the blue probe flash used in our experiments is typically set within the range of 10^4–2.5×10^4 sc td-sec. Figure 2 shows the response to a representative blue probe flash (1.6×10^4 sc td-sec) and illustrates a test of the photopic match being used. Waveform 1 in Fig. 2, which shows the *a*-wave and subsequent upswing of the response produced by the blue probe flash presented alone (i.e., in the absence of any other recent stimulus), is expected to contain contributions from both the rod and the cone photoreceptors. Waveform 2 is the response produced by a red probe flash of putatively equal photopic strength. These data are compared with

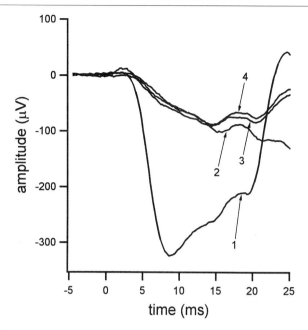

FIG. 2. Responses to photopically matched short-wavelength (blue) and long-wavelength (red) flashes of high intensity presented at time zero. Trace 1: Response to the blue flash. The flash strength (1.6×10^4 sc td-sec) is representative of that of the blue probe flash used in paired-flash experiments. Trace 2: Response to the photopically matched red flash. Trace 3: Response to the blue flash presented 3 sec after a brilliant conditioning flash (3.4×10^5 sc td-sec). Trace 4: Response to the photopically matched red flash presented 3 sec after the same conditioning flash. Data replotted from Fig. 2 of Pepperberg et al.[24], with permission of Cambridge University Press.

waveforms 3 and 4 obtained, respectively, on presentation of the blue or the red probe flash 3 sec after a brilliant conditioning flash that produced a prolonged (about 7-sec) saturation of the rods. The similarity of waveforms 2, 3, and 4 during the ~10-msec period immediately after the probe flash, i.e., throughout the rising phase of waveform 1, indicates that the relatively small responses 2, 3, and 4 are cone-mediated and supports the notion that the red and blue probe flashes are photopically matched. Importantly, the rod-mediated component of the response to the red probe flash is negligible under the present experimental conditions, as the scotopic strength of this red stimulus is only ~1/300 of the scotopic strength of the blue probe flash. That is, for example, the red flash photopically matched to a 1.6×10^4 sc td-sec blue flash has a scotopic strength of ~50 sc td-sec, and, even under dark-adapted conditions, the rod a-wave produced by a flash of this strength is minute during the ~10-msec period relevant to the present analysis of probe flash responses.

b-Wave and Other Postreceptor Components

Onset of the *b*-wave, together with oscillatory potentials and other postreceptor potentials, shapes the peak of the *a*-wave response in a manner dependent on both the test flash strength and the state of adaptation. The effects of these components on paired-flash determinations can be illustrated by considering how they might influence the kinetics of probe flash responses obtained with differing test flash strengths and a fixed interflash interval. Here, increasing the test flash strength can be expected to alter and ultimately saturate the *b*-wave of the test-flash-generated ERG (see, e.g., Hood and Birch[12,32]). This must have some nonzero effect on the response to the bright probe flash and, thus, on the derived amplitude of the test flash response. Experimental evidence indicates, however, that this effect is small for the case of weak test flashes. For example, rod-mediated probe responses obtained with a fixed interflash interval of 170 msec (i.e., obtained at a post-test-flash time near the peak of the derived rod response) exhibit generally similar normalized kinetics over their rising and near-peak phases. That is, these probe responses are roughly scaled versions of one another (Fig. 6 of Pepperberg *et al.*[24]). An approximate kinetic similarity is evident also among rod-mediated probe responses obtained with a fixed, relatively weak test flash and differing interflash intervals (Fig. 2C of Pepperberg *et al.*[24]). These findings argue against a large effect of altered *b*-wave kinetics on determination of the probe response amplitude and thus on the derived response to the test flash.

In addition to the cornea-positive *b*-wave, components of the same (cornea-negative) polarity as the photoreceptor response may contribute to the rod ERG generated by the probe flash.[13,33–35] A test for the contribution of postreceptor, negative ERG components is to examine how background light affects the response to a flash of fixed high intensity, i.e., a flash comparable to the probe flash used in paired-flash experiments. It is known, for example, that the scotopic threshold response (STR) is adapted out, i.e., the flash-generated STR is eliminated, by backgrounds much weaker than those that significantly reduce the rod circulating current.[35,36] Thus, if a postreceptor response contributed substantially to the *a*-wave produced by the bright flash under dark-adapted conditions (i.e., in the absence of background light), a decrease in the *a*-wave response from the maximum, fully dark-adapted level should be evident with relatively weak

---

[32] D. C. Hood and D. G. Birch, *J. Opt. Soc. Am. A* **13,** 623 (1996).
[33] A. J. Sillman, H. Ito, and T. Tomita, *Vision Res.* **9,** 1435 (1969).
[34] R. H. Steinberg, L. J. Frishman, and P. A. Sieving, *Prog. Ret. Res.* **10,** 121 (1991).
[35] L. J. Frishman, M. G. Reddy, and J. G. Robson, *J. Opt. Soc. Am. A* **13,** 601 (1996).
[36] L. J. Frishman and P. A. Sieving, *Vision Res.* **35,** 435 (1995).

backgrounds. Figure 3 shows results obtained when the response to a fixed-intensity flash (2.5×10^4 sc td-sec) was measured in the presence of differing backgrounds (0–680 sc td).[18] A substantial decrease from the dark-adapted amplitude (decrease from the dashed horizontal line in Fig. 3B) is apparent

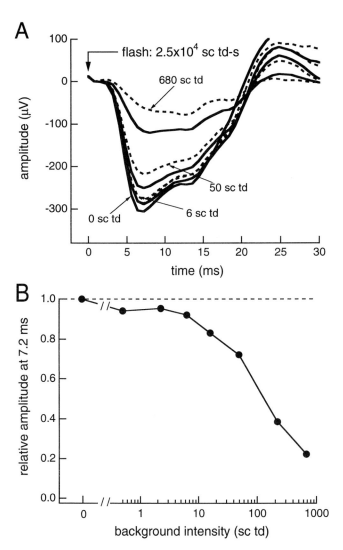

FIG. 3. Effect of background light (0–680 sc td) on the response to a flash of fixed high intensity (2.5×10^4 sc td-sec). (A) Flash responses obtained with differing backgrounds. (B) Relative amplitudes of the flash response determined 7.2 msec after flash presentation.

with backgrounds of ~50 sc td and higher. On the basis of *in vitro* photocurrent data from rods[19–21] and estimates of the *in vivo* photoisomerizing strength of a given stimulus,[10,37] this decrease is attributable to a reduction of the photocurrent excursion, i.e., to a decrease in the rod response itself. However, backgrounds below ~6 sc td produce only a small reduction from the dark-adapted amplitude. The data of Fig. 3 are consistent with those obtained from monkey in a similar type of experiment[28] and in experiments involving pharmacological isolation of the photoreceptor response.[38] Together, the results suggest that postreceptor negative components do not substantially skew determinations of the probe flash response in the present method, where the *a*-wave amplitude is measured shortly (<10 msec) after the probe flash.

Desensitization of Rod-Mediated Probe Response

A further possible source of error in the paired-flash method relates to the assumption that the probe-driven saturation of the rods proceeds in kinetically invariant fashion. That is, the normalized rate at which the probe flash blocks the rod circulating current is presumed to be essentially constant. This assumption would no longer be valid if, for example, the rods were desensitized by steady background light or by the test flash being used in the experiment. Such desensitization would lead to a skewed determination of the derived response to the test flash. We have tested this possibility for the case of recovery from a near-saturating test flash, by investigating the dependence of the derived response amplitude at a fixed interflash interval ($t = 500$ msec) on the strength of the probe flash (Fig. 5B of Pepperberg *et al.*[24]). Varying the probe flash strength over an approximately 10-fold range (2.5×10^3–2.4×10^4 sc td-sec) was found to have little effect on the probe response amplitude and, thus, on the amplitude of the derived response to the test flash. A hint of desensitization is apparent shortly after extremely intense test flashes, in the somewhat reduced rate of rise of the probe flash response (e.g., Figs. 2 and 5 of Birch *et al.*,[22] and present Fig. 3). However, such a change in kinetics is not conclusive evidence of rod desensitization, as the *b*-wave contribution to the probe flash response can also be substantially altered by a bright test flash.[29]

Time Course of Derived Response

Figure 4 illustrates derived responses obtained by the use of the simple cone subtraction procedure described above, i.e., that in which the cone

[37] W. L. Makous, *J. Opt. Soc. Am. A* **14,** 2323 (1997).
[38] J. A. Jamison, R. A. Bush, and P. A. Sieving, *Invest. Ophthalmol. Vis. Sci.* **37,** S136 (1996).

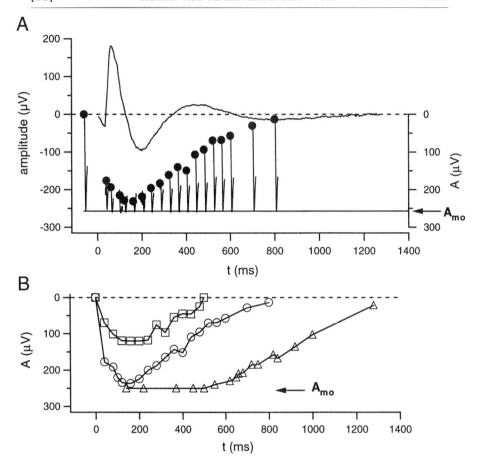

FIG. 4. Derived rod response to a fixed-intensity flash. (A) Test flash of 44 sc td-sec. Hooklike traces are rod-mediated probe responses obtained in paired-flash trials with variation of the interflash interval t. Trace to the left of $t = 0$ is the probe-alone response. The response to a photopically matched red probe flash (not shown) was subtracted from the raw probe responses to yield the illustrated rod-mediated components. Filled circles represent baselines from which the probe responses depart. Presentation of the test flash alone yielded the ERG response illustrated at the top. (B) Derived rod responses to test flashes of 11 sc td-sec (squares), 44 sc td-sec [circles; same experiment as that of (A)], and 320 sc td-sec (triangles). Data obtained from a single subject. Results shown by squares and circles are replotted from Fig. 5 of Pepperberg et al.[24], with permission of Cambridge University Press.

component of the probe flash response is taken as constant. The experiment of Fig. 4A involved presentation of a 44-sc td-sec test flash and a 1.2×10^4 sc td-sec blue probe flash in each of a series of paired-flash trials, and computational subtraction, from each of the raw responses to the blue probe, of the cone-mediated response to the photopically matched red probe. The rod-mediated probe responses resulting from this procedure, including that obtained for the probe-alone response (trace at the left), are positioned with their peaks at the assumed fixed level associated with photocurrent saturation (cf. Fig. 1). The pre-probe-flash baseline values determined from these responses (filled circles), as referred to the value obtained from the probe-alone response, represent the derived rod response to the test flash. The waveform in the upper part of Fig. 4A is the ERG response to the test flash delivered alone and is shown for comparison with the paired-flash results.

In Fig. 4B the derived response to the 44-sc td-sec test flash (circles) is shown together with the results of similar experiments that employed test flashes of 11 and 320 sc td-sec (squares and triangles, respectively). The derived response increases in peak amplitude and becomes longer in duration with increasing test flash strength, and the kinetics of the response are generally similar to those of photocurrent flash responses recorded from human rods *in vitro*.[21,24] Analyses of the apparent saturation period and the subsequent recovery phase that characterize the derived response to a relatively bright test flash (e.g., triangles in Fig. 4B) have been conducted in both human subjects and in mice, with test flashes that extend into the range of significant rhodopsin bleaching.[22,23,26,27,39]

As noted above, the Fig. 4 experiments employed subtraction of a single response, one representing a cone contribution assumed constant under the investigated conditions, to isolate the rod-mediated component of the probe flash responses. A technically more complex but conceptually similar procedure can be used to determine the rod-mediated probe response under conditions associated with a rapidly changing cone component, e.g., those prevailing at early times (short interflash interval) after a moderate to bright test flash. Two related factors must be considered under conditions of this type. First, the cone photoreceptor response itself is expected to vary significantly with small changes in t, the interflash interval. Second, the *b*-wave and other postreceptor components of the cone-driven response may produce time-varying (i.e., t-dependent) changes in *b*-wave intrusion and, thus, variation in the shape of the probe response.

Figure 5 describes the use of this modified procedure to obtain derived amplitudes of the rod response at a fixed early time after a test flash.

[39] M. M. Thomas and T. D. Lamb, *J. Physiol.* **518**, 479 (1999).

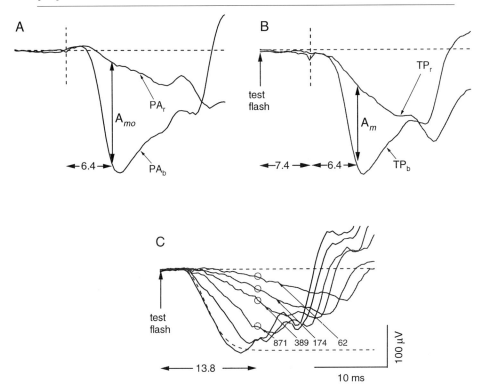

FIG. 5. Paired-flash determination of the rod response amplitude with short interflash interval and a relatively strong test flash. Data obtained from a single subject. (A) Responses to the blue probe alone (PA_b) and the red probe alone (PA_r). Here and in (B), the dashed vertical line is the time of probe flash presentation. The difference between the two responses 6.4 msec after flash presentation (vertical arrow) is taken as A_{mo}, the maximal amplitude of the rod-mediated probe response. (B) Responses obtained in two paired-flash trials, each of which involved presentation of a test flash (174 sc td-sec) and, 7.4 msec later (vertical dashed line), a bright probe flash. The probe used in trial TP_b was a bright blue flash; that used in trial TP_r was a photopically matched red flash. The difference between the two responses 6.4 msec after probe flash presentation (vertical arrow) was taken as the amplitude A_m of the rod-mediated probe response. (C) Rod-mediated a-wave responses to test flashes that ranged from 62 to 3.5×10^3 sc td-sec. The dashed curve represents the fit of a computational a-wave model to the response at 3.5×10^3 sc td-sec. Circles indicate paired-flash determinations of A/A_{mo} at the post-test-flash time of 13.8 msec, obtained for the 174-sc td-sec test flash as described in (B), and similarly obtained for test flashes of 62, 389, and 871 sc td-sec. For illustration with the a-wave data, the value of A_{mo} was equated with the maximal excursion predicted by the a-wave model (asymptote of dashed curve). See text for further details.

Responses PA_b and PA_r in Fig. 5A are probe-alone responses obtained, respectively, with the blue and photopically matched red probe flashes used in the illustrated experiment. As in the experiments described above, the difference between these two probe-alone responses (here, evaluated 6.4 msec after the probe flash) was taken as A_{mo}, the maximal excursion of the rod circulating current (vertical double-headed arrow in Fig. 5A). Also as in the Fig. 4 experiments, that of Fig. 5 involved paired-flash trials conducted with a blue test flash and, at a fixed later time (here 7.4 msec), a blue probe flash. Here, however, at each of the test flash strengths investigated, we also conducted a paired-flash trial that employed the identical test flash (blue) and interflash interval (7.4 msec) but used a photopically matched red probe flash rather than the blue probe flash. Records TP_b (test flash and blue probe flash) and TP_r (test flash and red probe flash) in Fig. 5B show the paired-flash ERG responses obtained, respectively, with the blue and red probe flashes, and with the blue test flash strength in both trials set at 174 sc td-sec. By analogy with the cone subtraction procedure used in the Fig. 4 experiments, the difference between responses TP_b and TP_r determined 6.4 msec after probe flash presentation (vertical double-headed arrow in Fig. 5B) is taken as the rod-mediated probe response amplitude A_m at the post-test-flash time of 13.8 msec (7.4–msec interflash interval plus 6.4 msec). With Eq. (1), the measured values of A_{mo} and A_m (Fig. 5A and B) yield determination of the normalized derived response A/A_{mo} 13.8 msec after the 174-sc td-sec test flash.

Figure 5C illustrates an evaluation of the cone subtraction procedure just described. Waveforms in Fig. 5C represent rod-mediated a-wave responses to a series of test flashes, i.e., those obtained on correction for the cone contribution to the test-alone response. The dashed curve indicates the predicted response to a test flash of 3.5×10^3 sc td-sec; this curve was obtained by fitting (ensemble fit) a computational model of the a-wave leading edge to the family of waveforms.[9,40] Open circles positioned at the post-test-flash time of 13.8 msec indicate paired-flash determinations of A/A_{mo}, the normalized derived rod response, at four test flash strengths (those for which the a-wave peak occurred later than 13.8 msec) including the 174-sc td-sec case described in Fig. 5B. These A/A_{mo} data have been scaled so as to equate the maximal excursion determined from the a-wave data (asymptote of the dashed curve) with A_{mo}, the maximal excursion determined 6.4 msec after the probe flash (Fig. 5A). The correspondence of the paired-flash results (circles) with the a-wave responses provides evidence of the validity of the procedure used here to derive the rod-

---

[40] T. D. Lamb and E. N. Pugh, Jr., *J. Physiol.* **449,** 719 (1992).

mediated response to a relatively strong test flash at short post-test-flash times.

The derived rod response to a test flash within the range of $\sim 10^2$–10^6 sc td-sec is characterized by a period (T_{sat}) of near-complete saturation (i.e., of near-complete suppression of the rod circulating current) and by a subsequent, approximately exponential recovery phase.[22,23] As shown by the Fig. 6A results obtained from an adult subject, fitting an exponential function to the paired-flash data yields determinations of the saturation period T_{sat} and of τ, an exponential time constant describing recovery (see caption to Fig. 6). Because the paired-flash method involves rapid titration of the circulating current, i.e., development of the probe-flash-generated a-wave response within about 8 msec, the derived response obtained with the method is relatively insensitive to low-frequency noise and baseline drift. It is thus possible, for example, to study the kinetics of rod recovery even in infant subjects. Figure 6B shows results obtained from an infant of age 6 weeks. Despite amplitudes that are considerably smaller than those obtained from the adult subject under similar stimulus conditions (Fig. 6A), the time courses of recovery are comparable. In contrast to the maturation evident for the activation stages of rod phototransduction,[31,41] the data of Fig. 6 suggest that processes underlying recovery of the rod response may be essentially fully developed at or soon after birth.

Amplitude–Intensity Relation

To determine the amplitude–intensity relation (response function) of the derived rod response, the interflash interval is held constant, and the derived amplitude is examined in relation to the strength of the test flash. Figure 7A shows results of determinations in normal subjects at an interflash interval of 170 msec, i.e., near the peak of the derived response. The smooth curve is an exponential saturation function[19,42] with half-saturation at a flash strength ($I_{0.5}$) of 7.6 sc td-sec (see caption to Fig. 7). The generally good fit of this curve to the data is consistent with the correspondence of this type of function with *in vitro* photocurrent data. Determining the photoisomerizing strength of the test flash at a given point on the amplitude–intensity relation, e.g., at half-saturation of the response, requires knowledge of the number of rhodopsins per rod isomerized (activated) by a flash of unit strength, i.e., by a flash strength of 1 scotopic troland-second (sc td-sec). This conversion factor is based on a number of estimates such as those for preretinal absorption, the size and packing of the rods, and

[41] A. B. Fulton and R. M. Hansen, *Curr. Eye Res.* **11,** 1193 (1992).
[42] T. D. Lamb, P. A. McNaughton, and K.-W. Yau, *J. Physiol.* **319,** 463 (1981).

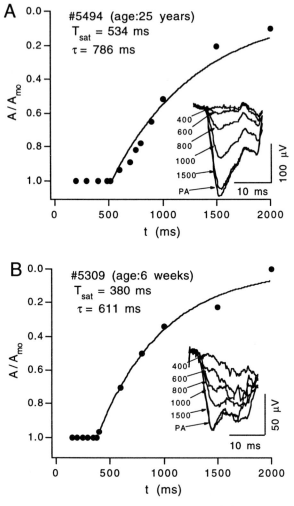

FIG. 6. Rod recovery time courses obtained from an adult (A) and from an infant of age 6 weeks (B). Data within each panel plot the relative amplitude A/A_{mo} of the derived rod response to a fixed-intensity test flash in relation to the interflash interval t. Test flash strengths in the two experiments were comparable (200–300 sc td-sec). The smooth curve fitted to each set of data plots the relation, $A/A_{mo} = \exp[-(t - T_{sat})/\tau]$, where T_{sat} is the period of apparent rod saturation that precedes recovery and τ is a recovery time constant. The inset of each panel shows representative rod-mediated probe responses; labels identify the interflash interval, in milliseconds. Response PA is the probe-alone response.

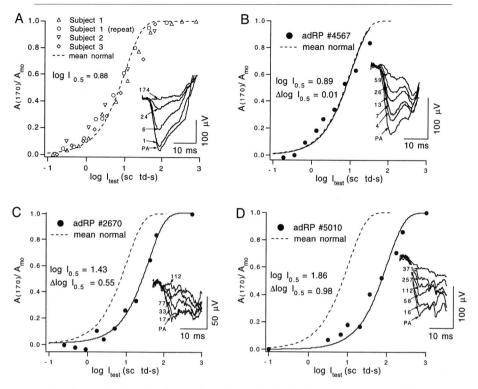

FIG. 7. Amplitude–intensity functions for the derived rod response obtained with a fixed interflash interval (170 msec). (A) Data collected from three normal subjects (see legend). The dashed curve plots the saturating exponential relation, $A(170)/A_{mo} = 1 - \exp[-(\ln 2) I_{test}/I_{0.5}]$, where I_{test} is the test flash strength and $I_{0.5}$ is the test flash strength at half-saturation. For the curve in (A), $I_{0.5} = 7.6$ sc td-sec (i.e., log $I_{0.5} = 0.88$). *Inset:* Rod-mediated probe responses obtained with test flashes of the indicated strength, in scotopic troland-seconds. Response PA is the probe-alone response. Data replotted from Fig. 6 of Pepperberg et al.[24] with permission of Cambridge University Press. (B–D) Results obtained from three patients with autosomal dominant retinitis pigmentosa. The solid curves in (B–D) describe the fit of the preceding exponential relation to the data obtained from the patients. The dashed curves replot the curve of (A) (data from normal subjects). Values of log $I_{0.5}$ and of $\Delta \log I_{0.5}$, the logarithmic shift of the exponential relation relative to that of (A), are shown. Insets show representative rod-mediated probe responses. See text for further details.

the quantum efficiency of the rod. Estimates for the human eye indicate that 1 sc td-sec is equivalent, on average, to between 4.3 and 8.6 photoisomerizations per rod.[10,37] As the half-saturation flash strength $I_{0.5}$ is about 7.6 sc td-sec in normal subjects (Fig. 7A), use of these estimates yields the range of about 33 to 66 photoactivated rhodopsins per rod, on average, at half-saturation of the near-peak (170 msec) amplitude of the human rod flash response.

Clinical Applications

Properties of the response derived from paired-flash measurements provide insight into photoreceptor abnormalities in diseases such as retinitis pigmentosa, and extend the analysis of phenotype beyond that possible by conventional ERG characterization. Figure 7B–D shows amplitude–intensity data obtained from three patients with autosomal dominant retinitis pigmentosa (adRP). As in Fig. 7A (results from normal subjects), an exponential saturation function (see caption to Fig. 7) has been fitted to each set of data to yield values for the half-saturation parameter $I_{0.5}$ and, thus, for the relative position of the function along the axis representing the logarithm of the test flash strength (log I_{test}). Functions from these three patients reflect the wide variation in sensitivity loss (Δlog $I_{0.5}$ in Fig. 7B–D) evident even among patients of similar age. Patient 4567 (Fig. 7B) has a sector form of adRP with field loss limited to the superior far periphery. She does not report night blindness, and final dark-adapted rod threshold determined psychophysically is normal at 7° eccentricity. Patient 2670 (Fig. 7C) has a form of adRP caused by a point mutation in codon 185 of the rhodopsin gene (Cys185Arg). His visual field is constricted to 25° in diameter, but within the central field he retains considerable rod function with an elevation of only 0.5 log unit in dark-adapted rod threshold. Patient 5010 (Fig. 7D) has a Pro23His rhodopsin mutation. Although his field is larger than that of patient 2670 and his rod-mediated probe flash responses are larger (compare responses shown in the insets of Fig. 7C and D), his amplitude–intensity relation exhibits a desensitizing shift considerably greater than that of patient 2670 (compare values of Δlog $I_{0.5}$ in Fig. 7C and D).

Summary and Areas for Future Investigation

Studies to date investigating the paired-flash technique in human subjects indicate that this method yields, to good approximation, the full time course of the massed rod photoreceptor response to a test flash.[22–24] Support for this conclusion comes primarily from the correspondence of kinetic, sensitivity, and light adaptation properties of the paired-flash-derived response with photocurrent response properties of human and other mammalian rods *in vitro*.[19–21] Results obtained with the paired-flash method motivate further work in both human subjects and experimental animals to refine this *in vivo* approach for studying rod phototransduction. The present chapter has identified a number of considerations likely to be important for future development of the technique, including the refinement of procedures to subtract the cone contribution at short interflash intervals, and to better correct for postreceptor contributions and desensitization of the probe flash response. Improvements and extensions of the paired-flash technique should prove valuable for both fundamental and clinically oriented studies of the rod photoreceptor response.

A related and particularly interesting avenue for future work is paired-flash determination of the *in vivo* response properties of *cone* photoreceptors. Some information of this type obtained from human subjects has been reported.[43,44] In these experiments, contributions of the rods to the measured responses have been suppressed by the use of intense, i.e., rod-saturating, backgrounds. As in the case of the paired-flash-derived response of the rods, that determined for cones[43] shows quantitative agreement with single-cell data.[45] Thus far, however, relatively few conditions of background and test flash strength have been investigated, and much remains to be done with regard to determining the contributions of different cone types to the derived response.

Acknowledgments

We thank Dr. J. R. Hetling for helpful comments on the manuscript. Research in our laboratories is supported by Grants EY-05494, EY-05235, EY-09076, and HD-22380 from the National Institutes of Health; by Research to Prevent Blindness, Inc. (New York, NY); and by The Foundation Fighting Blindness (Baltimore, MD).

[43] D. C. Hood, D. G. Birch, and D. R. Pepperberg, *in* "Vision Science and Its Applications," 1996 Optical Soc. Amer. Technical Digest Series, Vol. 1, p. 64. (Optical Society of America, Washington, D.C.).
[44] A. V. Cideciyan, X. Zhao, L. Nielsen, S. C. Khani, S. G. Jacobson, and K. Palczewski, *Proc. Natl. Acad. Sci. U.S.A.* **95,** 328 (1998).
[45] J. L. Schnapf, B. J. Nunn, M. Meister, and D. A. Baylor, *J. Physiol.* **427,** 681 (1990).

[14] Electrophysiological Methods for Measurement of Activation of Phototransduction by Bleached Visual Pigment in Salamander Photoreceptors

By M. Carter Cornwall, G. J. Jones, V. J. Kefalov, G. L. Fain, and H. R. Matthews

Introduction

This chapter describes single-cell recording methods for the study of bleaching adaptation and the visual cycle in isolated amphibian rods and cones, and illustrates the physiological effects that bleached visual pigment and photopigment regeneration have on the state of receptor adaptation. Bleaching adaptation is defined as the physiological state of the retina (or more specifically, of retinal rods and cones) that is observed in darkness after exposure to bright light of sufficient intensity to photoactivate or bleach a substantial fraction of the visual pigment. The visual cycle is defined as the series of reactions occurring in the eye through which visual pigment proceeds from photoactivation to regeneration. In the intact eye, these two processes are related in the following way. Photon absorption by the visual pigment not only leads to the activation of the transduction cascade, but also to the eventual dissociation of the chromophore from the protein moiety of the visual pigment to form all-*trans*-retinal and free opsin. The visual pigment is then slowly regenerated through a series of reactions requiring a few minutes in cones, and tens of minutes in rods.

Pigment regeneration in the vertebrate retina is a complex process that involves both photoreceptors and the pigment epithelium. Reviews summarizing work on the chemistry of visual pigments as well as the physiological effects that bleached pigment has on salamander rods have appeared.[1–3] Reduction of all-*trans*-retinal to all-*trans*-retinol must occur in the bleached outer segments of photoreceptors before the opsin-binding site becomes accessible for regeneration. all-*trans*-Retinol is then removed from the outer segment and is translocated to the pigment epithelium. There it is converted back to 11-*cis*-retinal and then returned to the photoreceptor outer segment, where it combines with opsin to form visual pigment.

The advantage of studying the relationship of bleaching adaptation to

---

[1] R. R. Rando, *Photochem. Photobiol.* **56,** 1145 (1992).
[2] R. K. Crouch, G. J. Chader, B. Wiggert, and D. R. Pepperberg, *Photochem. Photobiol.* **64,** 613 (1996).
[3] G. L. Fain, H. R. Matthews, and M. C. Cornwall, *Trends Neurosci.* **19,** 502 (1996).

the visual cycle in isolated retinal photoreceptors is that the physiology of the rods and cones after exposure to bright bleaching light can be studied while they are maintained physically separate from the pigment epithelium where many of the principal reactions of the visual cycle reside. This allows the effects of bleached visual pigment on the transduction cascade to be studied separately from the reactions that regulate photopigment regeneration and the subsequent recovery of sensitivity. There are other advantages to the study of isolated cells as well. First, the physiology of rods and cones can be studied separately. This is particularly important since, although it is known that the sensitivity is much lower and recovery after bright light faster in cones than in rods, little is known of the exact mechanisms by which these phenomena take place, and how they differ in these two cell types. Second, as is illustrated here, cellular parameters such as sensitivity, dark current, photoresponse kinetics, and the rates of cyclic GMP phosphodiesterase and guanylyl cyclase can be measured at various times during the desensitization that follows bright light, as well as during the recovery of sensitivity that occurs when the cells are exposed to solutions containing 11-*cis*-retinal or its analogs. Descriptions of the methods for the study of bleaching and regeneration in isolated cells are presented below.

Preparation

The techniques we describe here were developed for the study of rods and cones of the larval tiger salamander, *Ambystoma tigrinum*. However, they are applicable to the study of receptors of many other species, provided that the cells to be studied are of sufficient size and are sufficiently robust to withstand these manipulations. Before the experiment, animals are dark-adapted in a light-tight, oxygenated container for 12-hr. All subsequent manipulations are performed in dim red light or under infrared illumination with the aid of infrared image converters or image intensifiers.

At the beginning of the experiment, the animal is decapitated and pithed, both rostrally and caudally. Eyes are removed and hemisected, and the retinas are dissected from the eyecup and placed in a saline solution containing 110 mM NaCl, 2.4 mM KCl, 1.6 mM CaCl$_2$, 1.2 mM MgCl$_2$, 0.5 mM NaHCO$_3$, 15 mM dextrose, 0.01% bovine serum albumin (BSA), and 5 mM HEPES, pH 7.8. When cells are to be held for more than 12 hr, the medium should be supplemented with amino acids [1% solution, MEM amino acids (50×); Sigma, St. Louis, MO] and vitamins (0.5%, MEM vitamins (100×); Sigma).

After removal from the pigment epithelium, the retinal tissue is chopped into small pieces with fine scissors or a razor blade and triturated by repeated passage in and out of a Pasteur pipette, the tip of which has been heat polished to a diameter of about 0.5 mm. A few drops of the resulting suspension of retinal tissue are placed between two glass coverslips: one coverslip forms the bottom of a recording chamber located on the stage of an inverted microscope; the other forms the bottom surface of a protective cap, fitted to a ×10 microscope objective on the condenser assembly of the microscope. The fluid meniscus between these two glass surfaces confines the superfusion medium to a small volume around the cells.

A few minutes after the introduction of the suspension to the chamber, individual cells settle and gently adhere to the bottom of the chamber. These cells can then be continuously viewed during subsequent experimental manipulations with an infrared-sensitive television monitoring system fitted to the microscope. Electrophysiological recordings are then made by methods established previously.[4] A rod or a cone, identified by its distinct morphology, is sucked inner segment first into the tip of a recording pipette. The pipette has been heat polished to an inner diameter slightly smaller than the maximum diameter of the cell and filled with saline solution. It is connected via a silver–silver chloride junction to the input of a recording amplifier (e.g., List Electronic, Darmstadt, Germany). The current recorded by this amplifier is converted to a voltage and amplified and then filtered with an eight-pole active Bessel filter at 20 Hz (Frequency Devices, Haverhill, MA). The data are then digitized, stored, and subsequently analyzed on a PC-type computer using (for example) pClamp data acquisition and analysis software (Axon Instruments, Foster City, CA) and Origin graphics and data analysis software (Microcal Software, Northampton, MA).

Discrete light flashes as well as bleaching and background lights are provided by a dual-beam optical stimulator. Each beam is illuminated by separate 120-W tungsten–halogen filaments and is composed of a condenser lens, electromagnetic shutter, neutral density filters, interference filters, and a field lens. Neutral density and interference filters are calibrated over a range of 300–700 nm with a spectrophotometer. The separate light beams are united by a combining cube placed in the light path and projected onto the back of the objective positioned in the condenser mount of the inverted microscope.

Estimation of Visual Pigment Bleaching

The rate at which pigment is bleached during a step of light is equal to the concentration of pigment times the product of stimulus intensity and the photosensitivity of the visual pigment. The photosensitivity is given by the product of the absorption cross-section and the quantum efficiency for

[4] D. A. Baylor, T. D. Lamb, and K. W. Yau, *J. Physiol.* (*London*) **288,** 589 (1979).

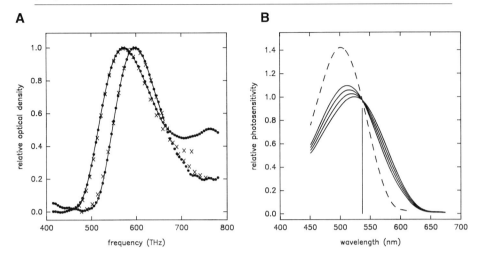

FIG. 1. (A) Rhodopsin and porphyropsin absorption spectra. The rhodopsin spectrum is from microspectrophotometry of outer segments of *Bufo marinus* red rods. Data points are averaged from 15 individual spectra. The curve is an eighth-order polynomial fitted by sum-of-squares minimization. The peak absorption is at frequency 597 THz; wavelength, 502 nm. The crosses are from the Dartnall nomogram for vitamin A_1-based visual pigments. The porphyropsin spectrum is from microspectrophotometry of outer segments of *Necturus maculosa* red rods. Data points are an average spectrum after smoothing and normalization. The curve is an eighth-order polynomial fitted by sum-of-squares minimization. The peak absorption is at frequency 573 THz; wavelength, 523 nm. The crosses are from the nomogram by Bridges for vitamin A_2-based visual pigments. [Original experimental data kindly provided by E. F. MacNichol, Jr.] (B) Composite relative photosensitivity spectra for mixtures of rhodopsin and porphyropsin. Dashed curve is the rhodopsin absorption spectrum from (A), after normalization and scaling by a factor of 1.42 to account for the higher peak photosensitivity of rhodospin relative to that of porphyropsin. Solid curves are the normalized porphyropsin absorption spectrum of (A) combined with the rhodopsin spectrum for porphyropsin-to-rhodopsin ratios corresponding to 100, 90, 80, and 70% porphyropsin. The vertical line indicates the isosbestic point at wavelength 537 nm.

bleaching.[5] After a light step of duration t, and intensity I_B, the fraction of pigment bleached is therefore

$$F = 1 - \exp(-I_B P t) \tag{1}$$

where P is the photosensitivity. Methods for estimation of the photosensitivity are discussed below.

Larval tiger salamander rods usually contain a mixture of rhodopsin and porphyropsin visual pigments, with the latter predominating. Figure 1A shows absorption spectra of amphibian rhodopsin and porphyropsin

---

[5] H. J. A. Dartnall, Photosensitivity. *In* "Handbook of Sensory Physiology," VII/1. Springer-Verlag, Berlin, 1972.

obtained by microspectrophotometry (points, data from E. F. MacNichol, Jr.), which correspond well with the results from the corresponding visual pigment nomograms of Dartnall[6] and Bridges[7] (Fig. 1A, crosses). The curves fitted to the data points are used in Fig. 1B to indicate the absorption spectra of salamander rods containing porphyropsin alone and zero or a small fraction of rhodopsin (solid lines). For comparison, the absorption spectrum of rhodopsin is also shown (dashed line). The isosbestic wavelength for mixtures of the two pigments is 537 nm (vertical line).

We have measured the absorption cross section of intact salamander rods directly from the responses to dim flashes in dark-adapted cells. The amplitude of these responses is expected to follow a Poisson distribution, for which the expected value of the variance is equal to the mean. This means that the number of unit events (single-photon events) can be estimated from the ratio of the variance of these events to their mean value squared. Figure 2A shows, superimposed, the average response of a salamander rod to 99 presentations of a dim flash, and the average of 99 interleaved blanks, i.e., periods with no stimulation. To obtain a good signal-to-noise ratio, the responses were somewhat above the region where the responses are linear with flash intensity. However, the amplitudes were corrected for this nonlinearity with the inverse Michaelis relation,[8] because peak response amplitude in this range closely follows a Michaelis relation when plotted against flash intensity. Figure 2B shows the point-to-point variance of these linearized responses and blanks. The lower value of the variance at the beginning and end of these traces is a consequence of the algorithm used to remove drift from the recordings. Figure 2C shows the excess variance in the flash responses with the square of the averaged response scaled for a best fit. The scaling factor indicates 8.8 average single-photon events per flash, and therefore a measured collecting area of 21.4 μm^2.

The mean collecting area for 19 rods was 18.2 ± 4.0 μm^2 (Fig. 3). This, however, assumed that all the variance in dim flash responses is due to variance in the number of single-photon events produced by dim flashes. Baylor et al. showed that about 5% of the variance should be attributed to variance in the size of unit events.[9] This means that the measured collecting area should be revised upward by this amount, to 19 ± 4 μm^2.

Measurements were made of the diameter and length of the outer segments from photographs of video images of the cells taken in visible

---

[6] H. J. A. Dartnall, *Br. Med. Bull.* **9,** 24 (1953).
[7] C. D. B. Bridges, *Vision Res.* **7,** 349 (1967).
[8] D. A. Baylor, A. L. Hodgkin, and T. D. Lamb, *J. Physiol.* (*London*) **242,** 685 (1974).
[9] D. A. Baylor, T. D. Lamb, and K. W. Yau, *J. Physiol.* (*London*) **288,** 613 (1979).

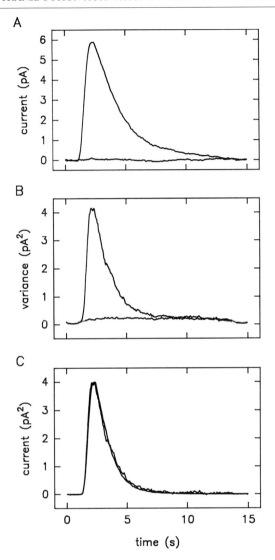

FIG. 2. Measurement of salamander rod outer segment collecting area. (A) Superimposed average of 99 responses to a dim flash (10-msec duration; wavelength, 537 nm; 0.41 photons μm^{-2} at time 1 sec) and average of 99 interleaved 15-sec periods without stimulation. The responses were corrected for nonlinear summation as described in text. (B) Superimposed point-to-point variance of the average response and of the average baseline. (C) The square of the average response fitted to the variance increase associated with the flashes, using the Marquardt–Levenberg algorithm to minimize the sum of differences squared. The scaling factor indicated 8.8 single-photon events per flash on average, and therefore a collecting area of 21.4 μm^2.

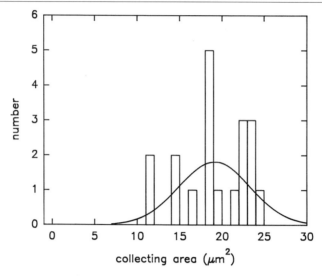

Fig. 3. Histogram of measurements of the collecting area of salamander rod outer segments. Measurements were made at 537 nm. The curve is a Gaussian with mean and standard deviation equal to the mean and standard deviation of the measurements (18.2 and 4.0 μm^2, respectively).

light at the end of experiments. From these, we calculated the outer segment volume, assuming a cylindrical shape (Fig. 4). These measurements were used to estimate the specific optical density for salamander rods.

The specific optical density for rod outer segments is defined as the optical density per unit distance for plane polarized light traveling transversely across the outer segment, with the plane of polarization in the same

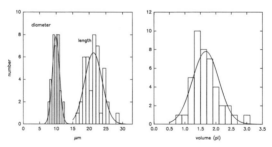

Fig. 4. Histograms of the diameter and length of salamander rod outer segments, and the volume calculated from these, assuming a cylindrical shape. Measurements were made from photographs of video images, using a photograph of a stage micrometer for calibration. The Gaussian curves indicate the means and standard deviations for 42 cells: diameter, 9.9 ± 1.1 μm; length, 21.3 ± 2.6 μm; volume: 1.67 ± 0.43 pl.

plane as that of the outer segment disks. This definition has utility in that the specific optical density will be the same for light traveling along the axis of the outer segment, i.e., in the natural direction. The effective collecting area for a cylindrical rod outer segment of diameter d, and length l, can then calculated from the following relation:

$$A_c = l(d/2)\gamma \int_0^\pi (1 - 10^{-\varepsilon d \sin\theta}) \sin\theta\, d\theta \qquad (2)$$

where γ is the quantum efficiency for bleaching (0.67; Dartnall[5]), and ε is the specific optical density. As written, Eq. (2) applies to polarized light. For nonpolarized light, two integrals are needed on the right-hand side of Eq. (2), with appropriate weighting for the fractions of the incident light polarized in the plane of the disks and polarized at right angles to the disks, and with different values for the specific optical density according to the dichroic radio for absorption in the outer segments.

The integral of Eq. (2) does not appear to have a closed solution. However, the first term in a series expansion,

$$A_c \approx (\ln 10) \frac{\pi d^2 l}{4} \varepsilon\gamma \qquad (3)$$

is useful. This approximation corresponds to the case of either the diameter or the specific optical density being small, i.e., to the case of negligible self-screening.

In the present context, in which measurements have been made of the collecting area, the diameter, and the radius, the unknowns in Eq. (2) all appear within the integral. In other words, it is the specific optical density and dichroic ratio that need to be evaluated. Equation (3) was solved numerically with the average values for outer segment dimensions, over a range of values for the specific optical density. The dichroic ratio was taken to be 3.5. The results are shown as the solid line in Fig. 5. The dashed line shows the results for negligible self-screening. A collecting area of 19 μm^2 corresponds to a specific optical density of 0.0118 μm^{-1} (at 537 nm). The dashed line of Fig. 5 shows solution of the approximation of Eq. (3), and indicates that inclusion of self-screening reduces estimates of the amount of light absorbed by ~10%. This estimate of the specific optical density of the salamander rod outer segment disks corresponding to the measured collecting area is remarkably insensitive to the dichroic ratio assumed. Repeating the calculations with a dichroic ratio varying over the range of 2.8 to 4.0 changes the estimate of specific optical density by less than 4%.

The specific optical density of salamander rod outer segment disks has been measured by microspectrophotometry to be 0.0128 μm^{-1} at 516 nm, corresponding to 0.0119 μm^{-1} at 537 nm. This is satisfactorily close to the

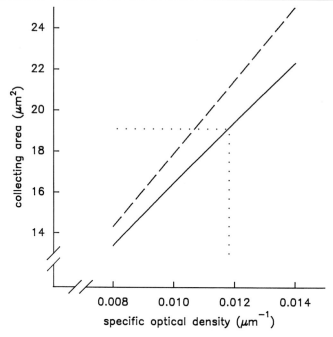

FIG. 5. Calculated collecting area of salamander rod outer segments, with dimensions given by the mean values of Fig. 3, according to Eq. (2), modified to take account of the polarization of the light beams, and a deviation from horizontal alignment of 15°. The dichroic ratio was taken to be 3.5, and the polarization ratio for the light stimulus was 0.181. A collecting area of 19 μm^2 corresponds to a specific optical density of 0.0118 μm^{-1}, when self-screening is taken into account (solid line). The dashed line indicates calculation of collecting area without self-screening.

present estimate. The experimental value of 0.0128 μm^{-1} is from Hárosi,[10] and assumes 100% porphyropsin. It would be slightly higher if the bleaching correction used by Hárosi accounted for the rods containing a fraction of visual pigment as rhodopsin. It does not appear to have been adjusted for depolarization by the condenser optics of the microspectrophotometer. The correction to 537 nm assumes a visual pigment mixture of 15% rhodopsin (see Fig. 1).

It should be noted that the present estimate of the specific optical density of the salamander rod outer segment is independent of the proportion of visual pigment present as rhodopsin or porphyropsin, because the measurement was made at the isosbestic point for absorption by these two pigments.

Knowledge of the specific density of the rod outer segment makes

[10] F. I. Hárosi, *J. Gen. Physiol.* **66**, 357 (1975).

possible an estimate of the concentration of visual pigment within the outer segment. This calculation requires use of the dichroic ratio for absorbance by the outer segment, but is not overly sensitive to the value used, or to the value of the molar extinct coefficient of either rhodopsin or porphyropsin. Assuming the dichroic ratio is 4.0, the total visual pigment extinction at 537 nm is $0.0118 + 0.0118 + 0.0118/4.0 = 0.02655$ μm^{-1}. Dividing by 3 to obtain the equivalent solution extinction, and adjusting units, gives 88.5 cm^{-1}. The average peak photosensitivities of rhodopsin and porphyropsin are 27,400 and 19,300 ℓ cm^{-1}.[5] Using the templates to adjust to wavelength 537 nm, and assuming a quantum efficiency of 0.67 [but noting from Eq. (2) that this now effectively cancels out] gives a molar extinction of 27,400 at 537 nm, and a concentration therefore of 3.2 mM. Changing the dichroic ratio to 3 increases this estimate by only $\sim$3%.

The estimate of outer segment visual pigment concentration of Hárosi is only slightly higher. It was based on a direct measure of specific optical density, and a similar estimate of molar extinction, but it was not independent of the quantum efficiency for bleaching.[10] A concentration of 3.2 mM, combined with an average volume of 1.7 pl (Fig. 4), implies that an average rod contains 3.3×10^9 opsin molecules, and has an *in situ* photosensitivity of 5.9×10^{-9} μm^2 (537 nm).

Salamander rods contain predominantly a vitamin A_2-based visual pigment,[10] for which the solution photosensitivity is 7.4×10^{-9} μm^2.[5] *In situ*, the photosensitivity will be smaller, predominently because of self-screening and the partial alignment of the visual pigment chromophore in the plane of the disk membrane. It can be shown that the *in situ* photosensitivity for transverse illumination of salamander rods with a peak optical density of 0.143 and a dichroic ratio of 4.0 (Hárosi[10]) will be reduced by $\sim$15% because of the self screening and by $\sim$10% because of the chromophore alignment, but this reduction will be partially compensated if the outer segments contain a small fraction of vitamin A_1-based visual pigment. A crude estimate for the photosensitivity is therefore 6×10^{-9}, at a wavelength of 520 nm. This corresponds reasonably well with the estimate at 537 nm described above. A reasonable correspondence between the solution photosensitivity and *in situ* values has been previously reported for both rods and cones: from microspectrophotometry in cones[11,12] and from the early receptor current in rods and cones.[13] That the photosensitivity of the visual pigment in isolated photoreceptors does not differ appreciably from that in solution

---

[11] B. D. Gupta and T. P. Williams, *J. Physiol.* (*London*) **430**, 483 (1990).
[12] G. J. Jones, A. Fein, E. F. MacNichol, Jr., and M. C. Cornwall, *J. Gen. Physiol.* **102**, 483 (1993).
[13] C. L. Makino, W. R. Taylor, and D. A. Baylor, *J. Physiol.* (*London*) **442**, 761 (1991).

implies there is little interference from photoproduct absorption during bleaching, photoreversal of bleaching, or regeneration of visual pigment.

Preparation of Retinoid Solutions

After bleaching in the intact eye, pigment regeneration occurs as 11-*cis*-retinal, supplied by the pigment epithelium, is transported to the retina bound to interphotoreceptor retinoid binding protein (IRBP). There, the chromophore combines with opsin in the outer segment membranes of rods and cones to form new visual pigment and restore sensitivity to light. This cannot occur in photoreceptor cells isolated from the retina. In an isolated rod or cone in the absence of the retinal pigment epithelium, full recovery of sensitivity can be achieved only by exposure to a solution containing exogenous 11-*cis*-retinal. Because retinoids are highly lipid soluble, we have used a number of different methods to supply these hydrophobic substances to the aqueous medium in which the photoreceptor cells are held. These are (1) suspension in lipid vesicles, (2) dissolution in Ringer solution containing ethanol, and (3) binding to bovine IRBP. Crystalline 11-*cis*-retinal is stored under argon at $-80°$ in a light-tight container. A stock solution is prepared by weighing approximately 10 mg of crystalline 11-*cis*-retinal into a small glass vial. Sufficient ethanol is then added to bring the solution volume up to about 1 ml, to make a solution of 10 mg/ml. Aliquots (30 μl) of the solution are then placed in individual small vials, dried under a gentle stream of nitrogen, capped under nitrogen, and stored in darkness at $-80°$ until use. Each of these vials contains about 300 μg (1.05 μmol).

11-cis-Retinal in phospholipid vesicles: The method we have used for preparation of lipid vesicles is similar to those described previously.[14,15] Vesicles are prepared by placing 25 mg of L-α-phosphatidylcholine dissolved in a chloroform–methanol solution (type V-E; Sigma) in a glass scintillation vial and evaporating the contents to dryness under a stream of dry nitrogen. Any residual organic solvent is then removed by lyophilizing the sample in darkness in the open vial for an additional 15 min. Fifteen milliliters of Ringer solution is then added to the vial, and the contents are subjected to intermittent sonication for 20 min in an ice bath (1 sec on/1 sec off) at 45 W using a 1.0-cm probe (Vibracell high-intensity ultrasonic processor; Sonics and Materials, Danbury, CT). The vesicle solution is then stored for up to 24 hr in a refrigerator. Under dim red light about 1 hr before the start of an experiment, 1.5 ml of the vesicle solution is placed in one of the small glass vials containing the lyophilized 11-*cis*-retinal. The contents

---

[14] S. Yoshikami and G. N. Noll, *Methods Enzymol.* **81,** 447 (1982).
[15] J. I. Perlman, B. R. Nodes, and D. R. Pepperberg, *J. Gen. Physiol.* **80,** 885 (1982).

are then sonicated for about 60 sec at low power, using a tapered microtip. If all of the retinoid that had been dried to the walls of the tube were incorporated into the solution, the resulting concentration would be as high as 700 µM. The exact concentration is determined spectrophotometrically by making a 10-fold dilution in ethanol. We assume a molar extinction coefficient for 11-*cis*-retinal of 25,000 M^{-1} cm^{-1} at 380 nm.[16] The vesicle stock solution is then diluted with Ringer solution before each experiment.

11-cis-Retinal in ethanol solution: Another method for treating bleached photoreceptors with retinal is to make a simple 0.1% dilution of the 11-*cis*-retinal–ethanol stock solution in Ringer solution. This gives a final concentration of about 35 µM 11-*cis*-retinal. Ethanol itself at 0.1% dilution causes a slight reduction in dark current and sensitivity of receptors (see below), but this effect is transient.

11-cis-Retinal in IRBP solution: Bleached photoreceptors can also be regenerated with 11-*cis*-retinal by exposing them to chromophore bound to the carrier molecule, IRBP, purified from bovine eyes. IRBP is a prominent protein found in the interphotoreceptor matrix and is thought to transport retinoids (11-*cis* as well as all-*trans* isomers) between the retina and the pigment epithelium during the visual cycle.[17] Procedures for extraction of IRBP have been described.[18] IRBP is then concentrated by dialysis against physiological saline solution (glucose omitted) and stored at −80° until use. We estimate the concentration of free IRBP and IRBP with bound retinoid spectrophotometrically using extinction coefficients for IRBP (120,000 M^{-1} cm^{-1} at 280 nm) and 11-*cis*-retinal (see above), assuming that retinoid binding to the protein does not significantly alter the extinction coefficient of either species.[19]

Each of these methods for the exogenous application of retinoids to isolated cells has its inherent advantages and disadvantages. Using phospholipid vesicles has the advantage of allowing treatment with high concentrations of retinal. We have used concentrations of 11-*cis*-retinal and other retinal analogs as high as 300 µM (measured in aqueous solution). Two disadvantages of this method are the possibility of introducing unwanted lipids into the plasma membrane of the cell, and of altering the concentrations of substances (e.g., ions) in the cytosol. Control experiments, however, have indicated that these are probably not significant concerns. The use of ethanol as a vehicle avoids the problems of vesicles; however, as noted

[16] R. Hubbard, P. K. Brown, and D. Bounds, *Methods Enzymol.* **18,** 615 (1971).
[17] G. J. Chader, *Invest. Ophthalmol. Vis. Sci.* **30,** 7 (1989).
[18] T. M. Redmond, B. Wiggert, F. A. Robey, N. Y. Nguyen, M. S. Lewis, L. Lee, and G. J. Chader, *Biochemistry* **24,** 787 (1985).
[19] A. J. Adler and C. D. Evans, *Prog. Clin. Biol. Res.* **190,** 65 (1985).

above, treatment of rods and cones with low concentrations of ethanol results in a transient reduction of the dark current. Also, because of their limited solubility in ethanolic–aqueous solution, retinoid concentration is limited to about 30 μM. The most physiological method for exogenous application of retinoids employs IRBP as the vehicle. Here, adverse physiological effects are minimized, but the use of high concentrations of retinoids is not possible, and it is difficult to obtain large quantities of this protein.

Illustration of Bleach-Adapted Rod and Cone

Utilizing the methods described above, we have investigated the photoresponse kinetics and sensitivity of isolated salamander photoreceptors under various conditions of adaptation. The experiment illustrated in Fig. 6, reprinted from the work of Jones et al., is typical.[20] It demonstrates the effect of bright bleaching light on the sensitivity and response properties of an isolated salamander rod. It also shows the effect of exogenous application of 11-*cis*-retinal in lipid vesicles in promoting recovery of sensitivity. Figure 6 A–C plots membrane current responses to flashes recorded before and after intense light, estimated by the methods given previously in this chapter to have bleached between 50 and 75% of the visual pigment, and after recovery following treatment with 11-*cis*-retinal. Immediately after pigment bleaching, no response could be elicited. Fifty minutes later, however, sensitivity and responsiveness had recovered to a new steady state from which no further recovery occurred. In this new condition, the dark current was reduced by about 40% and the time course of flash responses was somewhat accelerated. This is also shown in Fig. 6D, which demonstrates that in addition to the loss of dark current, the cell is also desensitized by about 2.5 log units, as evidenced by the shift to the right of the response–intensity relation from that measured prior to bleaching (Fig. 6D, open triangles) to that measured after bleaching (Fig. 6D, open squares). Figure 6E shows that the flash intensity just sufficient to elicit a detectable response (the response threshold) was elevated after bleaching by more than 2.5 log units. After treatment with 11-*cis*-retinal, the dark current, sensitivity, and threshold slowly recovered to their prebleach levels.

The effects that bright bleaching lights have on isolated salamander cones are similar to those observed in rods, but there are some differences. There are illustrated in Fig. 7, reproduced from Jin et al.[21] The insets in

[20] G. J. Jones, R. K. Crouch, B. Wiggert, M. C. Cornwall, and G. J. Chader, *Proc. Natl. Acad. Sci. U.S.A.* **86,** 9606 (1989).
[21] J. Jin, R. K. Crouch, D. W. Corson, B. M. Katz, E. F. MacNichol, and M. C. Cornwall, *Neuron* **11,** 513 (1993).

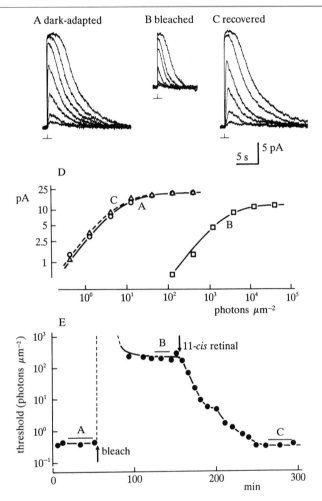

Fig. 6. Bleaching and recovery in an isolated salamander rod photoreceptor. (A–C) Responses to flashes recorded before and after bleaching and after recovery with 11-*cis*-retinal (flashes were 10 msec, 540 nm). Each trace is an average of five responses. Bleaching light was a 2-sec light step, 580 nm, 2.3×10^8 photons μm^{-2} sec^{-1}. (D) Response–intensity plots from (A–C). Flash sensitivity was measured as the flash intensity that produced a 1-pA peak response; these are plotted as thresholds in (E). (E) Complete time course of sensitivity changes in this cell. Horizontal bars indicate when responses in (A–C) were recorded. Bath perfusion was halted from just before addition of 5 μM 11-*cis*-retinal to just before the bar marked C. 11-*cis*-Retinal was applied using liposomes. [Reproduced from Jones et al.[20]]

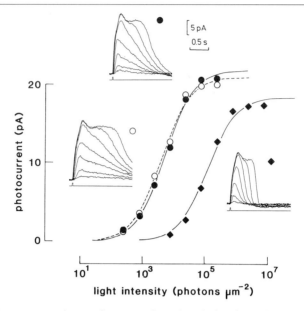

FIG. 7. Response waveform and response–intensity relations in a salamander cone before and after bleaching. *Insets:* Superimposed flash responses to 20-msec 600-nm light flashes. Closed circles: responses elicited in darkness 1 min before bleaching. Closed diamonds: responses elicited 30 min after a 600-nm light step that bleached 80% of the pigment. Open circles: responses 35 min after 11-*cis*-retinal was added to the bath (75 min after bleaching). Flash increased in ~0.5-log unit increments. Responses elicited after bleaching exhibited slightly decreased maximum amplitude and accelerated time courses. After the treatment with 11-*cis*-retinal, response amplitude recovered completely, and the kinetics of the responses were slowed. [Reproduced from Jin et al.[21]]

Fig. 7 illustrate ensembles of membrane current responses elicited by flashes in darkness before bleaching (upper center, Fig. 7), 30 min after a bleaching light calculated to have bleached 80% of the visual pigment (lower right, Fig. 7), and then 35 min after 11-*cis*-retinal was added to the bath in lipid vesicle solution (lower left, Fig. 7). Response–intensity relations are also shown in Fig. 7, plotted as response amplitude versus log flash intensity. As is the case for rods, bleaching of a substantial fraction of the visual pigment in this cone can be seen to have resulted in a persistent decrease in sensitivity also amounting to about 2.5 log units. This is apparent from the shift of the of the response–intensity relation after bleaching (filled diamonds, Fig. 7) to higher light intensities relative to that measured prior to bleaching (filled circles, Fig. 7). It can be seen from response–intensity relations before bleaching and after treatment with 11-*cis*-retinal that photopigment regeneration resulted in complete recovery of sensitivity and dark current.

Comparison of the data in Figs. 6 and 7 also reveals a number of significant differences between rods and cones. First, in darkness, flash responses are considerably faster and the absolute sensitivity is about 100-fold lower in cones than in rods.[12] Second, the decrease in the dark current after bleaching is much smaller in cones than in rods for a comparable bleach. Third, recovery of sensitivity after a substantial bleach (80–90%) (but before treatment with 11-*cis*-retinal) occurs much faster in cones, with the steady state value reached in a few seconds as compared with more than 30 min in rods. Finally, although not shown here, the recovery of sensitivity after treatment with 11-*cis*-retinal is much faster in cones than in rods for a comparable bleach. Complete recovery of sensitivity for a cone in which 90% of the pigment was bleached and subsequently regenerated with 100 μM 11-*cis*-retinal occurred within about 3 min. A rod under similar conditions required in excess of 30 min for complete recovery of sensitivity. The reasons for these differences in dark current suppression by bleaching light and in the time course of recovery of sensitivity before and after exposure to 11-*cis*-retinal are presently unknown.

Methods for Rapid Solution Changes

In an effort to understand the mechanisms that underlie changes in dark current, sensitivity, and flash response kinetics in bleached photoreceptors, we have made use of techniques, first developed by Hodgkin and Nunn, to measure the enzymatic rates of cyclic GMP phosphodiesterase and guanylyl cyclase in intact photoreceptors.[22] The method involves rapidly exposing the receptor outer segment to solutions of different composition, while recording extracellularly the change in membrane current. The theoretical basis for the use of these solutions for measuring the phosphodiesterase and cyclase rates is discussed in the next section. In this section we outline the way in which the superfusion system is constructed and illustrate its function. Three solutions are used for this analysis. The first is normal Ringer solution; the other two are test solutions, which we refer to as Li^+ solution and IBMX solution. Li^+ solution is made by substituting Li^+ for Na^+ on an equimolar basis. IBMX solution is made by adding 500 μM isobutylmethylxanthine (IBMX; Sigma) to normal Ringer solution.

The recording chamber in which the cells are placed is supplied with Ringer solution from a gravity-fed reservoir. Ringer solution is led out of the chamber via a narrow trough in which a wick of tissue paper is placed.

[22] A. L. Hodgkin and B. J. Nunn, *J. Physiol.* (*London*) **403,** 439 (1988).

We expose the outer segments of the photoreceptors to Li$^+$ and IBMX solutions by a rapid superfusion system that is constructed in the following way. A double-barreled micropipette is fabricated so that the two orifices at the tip are attached side by side and each is approximately 90 μm in diameter. These pipette tips are positioned in the perfusion chamber opposite to, and about 200 μm distant from, another pipette tip of similar dimension. This assembly is placed about 50–100 μm away from the pipette tip in which a photoreceptor has been drawn for current recording, as shown in the photograph in Fig. 8. The double-barrel pipette shown at the bottom and the single-barrel pipette shown at the top of Fig. 8 are mounted on separate micromanipulators. Solutions entering the chamber through the tips of the double-barrel pipettes are propelled into the chamber by gravity. They are siphoned out of the chamber through the outflow pipette. The way in which this arrangement confines the three different solutions to discrete regions in the bath is shown in Fig. 8, where, by the addition of phenol red to the solutions contained in the two inlet streams, their

FIG. 8. Photograph of superfusion apparatus located within the recording chamber for making rapid solution changes. Recording micropipette with a cone sucked into the tip, inner segment first, is shown entering from the right. Two inlet micropipette tips are shown entering from the bottom and a single outlet pipette tip is shown at the top. Solution containing phenol red allows visualization of separate plumes entering the bath from the inlet pipettes; these two solutions exit the bath through the single pipette at the top. Flow from the inlet pipettes is gravity fed from separate reservoirs. Flow in the egress pipette occurs via siphon.

outline is clearly delineated. Flow of both the inlet and outlet pipettes can be interrupted by separate solenoid-operated valves.

The separate streams of solution can be seen to exhibit laminar flow and provide discrete interfaces across which the recorded cell can be moved. Phenol red-containing solutions such as these are used at the beginning of each experiment to position the streams relative to the recording pipette and to be sure that the outflow siphon is operating correctly. Although early experiments were performed with phenol red in the superfusate solutions as shown in Fig. 8, this practice has been discontinued because we have observed that its presence slightly affects the measured cyclase and phosphodiesterase (PDE) rates. Little or no electrical artifact is observed when a cell is rapidly changed between streams having identical ionic composition.

The micromanipulators supporting the inlet and outlet pipettes are mounted on a micrometer-driven platform (model 16221m; Oriel, Stratford, CT) whose horizontal motion is regulated by a stepping motor connected to a microprocessor (model AX; Compumotor Division, Parker Hannifin, Rohnert Park, CA). The microprocessor is controlled by a program resident in an IBM PC computer. Rotations of the stepping motor produce horizontal motion of the two micromanipulators and cause the solution interfaces to move laterally in the bath. With proper positioning of the recording pipette in which the photoreceptor is held, rapid solution changes around the cell occur when these interfaces move past the static position at which the photoreceptor is held. The two principal advantages of this arrangement are the rapid time course of the solution change (<30 msec) and the minimal mechanical insult to the cell. The speed of the solution change can be determined by recording the change in the junction current measured from the recording pipette as it is being jumped from control solution into a test solution (e.g., Li^+ solution). Normally there is a significant delay before the actual change occurs, which is accounted for by the time necessary for the processing command following the trigger of the stepping motor, as well as the time necessary for the solution interface to approach the position at which the cell is held. This delay can be as much as 75 msec. However, once the solution change begins, the transition is normally 90% complete within about 20 msec.

Physiological Measurement of Phosphodiesterase and Cyclase Velocities

The method for estimating the rate constants of cyclase and phosphodiesterase in photoreceptors was first described by Hodgkin and Nunn.[22] We have used modifications of this method to determine the catalytic activity of bleached pigment in rods[23] and in cones.[24] Also, the effect of the noncovalent occupancy of the chromophore pocket of opsin has been studied with this technique in both rods[25] and cones.[24]

The basis for the method is as follows. The economy of cGMP in the cell is regulated by the relative rate constants of guanylyl cyclase, responsible for cGMP synthesis; and of cGMP phosphodiesterase, responsible for its hydrolysis:

$$\text{GTP} \xrightarrow{\alpha} \text{cGMP} \xrightarrow{\beta} \text{GMP} \tag{4}$$

where α and β are the rate constants of cyclase and phosphodiesterase (referred to as rate of cyclase and rate of phosphodiesterase). The light-sensitive current in photoreceptors, i, is controlled by the concentration of cyclic GMP according to the relation[22,26]

$$\frac{i}{i_m} = \frac{[\text{cGMP}]^N}{[\text{cGMP}]^N + K_{1/2}^N} \tag{5}$$

where i_m is the maximum membrane current, $K_{1/2}$ is the Michaelis constant for binding of cGMP to the light-sensitive channels, and N is the Hill coefficient, reflecting the cooperative binding of cGMP to the channels. From Eq. (4), the change in free [cGMP] in the cell is given by

$$\frac{d}{dt}[\text{cGMP}] = \alpha[\text{GTP}] - \beta[\text{cGMP}] \tag{6}$$

where it has been assumed that cGMP-binding sites are of high affinity and are saturated so that the buffering power of the cytoplasm for cGMP is close to unity.[22,23] Combining Eqs. (5) and (6) yields a relation between the current and the rates of cyclase and phosphodiesterase:

$$\frac{d}{dt}\left[K_{1/2}\left(\frac{i}{i_m - i}\right)^{1/N}\right] = \alpha[\text{GTP}] - \beta K_{1/2}\left(\frac{i}{i_m - i}\right)^{1/N} \tag{7}$$

One can estimate the rate of cyclase or phosphodiesterase in the intact photoreceptor in different conditions of adaptation or in the presence of drugs by suddenly blocking one of these two enzymes. The measured circulating current changes when the balance between cGMP synthesis and degradation is disrupted. The rate of current increase or decrease immedi-

---

[23] M. C. Cornwall and G. L. Fain, *J. Physiol. (London)* **480**, 261 (1994).
[24] M. C. Cornwall, H. R. Matthews, R. K. Crouch, and G. L. Fain, *J. Gen. Physiol.* **106**, 543 (1995).
[25] V. J. Kefalov, M. C. Cornwall, and R. K. Crouch, *J. Gen. Physiol.* **113**, 491 (1999).
[26] E. N. Pugh, Jr. and T. D. Lamb, *Vision Res.* **30**, 1923 (1990).

ately after the block provides an estimate of the rate at which cGMP is being synthesized or hydrolyzed by the unaffected enzyme.

The rate of phosphodiesterase in the dark-adapted state, during background light, after a bleach, or after pigment regeneration can be estimated by rapidly exposing the cell to Li^+ solution. An example of such an experiment is shown in Fig. 9, taken from Cornwall and Fain.[23] The top left panel of Fig. 9 illustrates the change in the photocurrent in a dark-adapted salamander rod after subtraction of the junction current recorded in bright saturating light. Li^+ can enter the cell through the light-sensitive channels with a permeability greater than Na^+.[27,28] This causes an initial transient increase in inward current. Subsequently, the substitution of Li^+ for Na^+ blocks the $Na^+/Ca^{2+}-K^+$ exchange mechanism,[27,28] and as a result $[Ca^{2+}]_i$ increases rapidly and inhibits the cyclase. The continuous hydrolysis of cGMP by phosphodiesterase leads to the observed exponential decline of the current.

The rate of the decrease in current provides an estimate of the rate of hydrolysis of cGMP by phosphodiesterase. It can be seen from the left panels of Fig. 9 that the decay of the current is accelerated as a result of the bleach and then returns to its dark-adapted rate after regeneration of the pigment with exogenous 11-*cis*-retinal. We can use this technique to estimate the rate of phosphodiesterase provided that the following assumptions are made: first, stepping the cell into Li^+ solution causes an increase in $[Ca^{2+}]_i$ that is large enough to produce a rapid and nearly complete block of cyclase ($\alpha \cong 0$); second, the cGMP channel characteristics ($K_{1/2}$ and N) as well as i_m do not change during and immediately after the solution change; and finally, $i_m \gg i$ (typical values are $i_m > 1000$ pA,[26] $i \sim 40$ pA[29]). These assumptions allows us to simplify Eq. (7) to an expression linking the current and the rate of phosphodiesterase:

$$\frac{d}{dt} i^{1/N} = -\beta i^{1/N} \qquad (8)$$

Equation (8) describes the exponential decline in the current observed on exposure of the cell to Li^+ solution. The right-hand panels of Fig. 9 show semilogarithmic plots of the normalized current as a function of time. Here, the slope of the linear decay is proportional to the rate of the phosphodiesterase. Bleaching accelerates phosphodiesterase, as evidenced by the steeper current decline. Regeneration of the pigment after application of 11-*cis*-retinal slows down phosphodiesterase to its dark-adapted rate (compare

---

[27] K. W. Yau and K. Nakatani, *Nature* (*London*) **309**, 352 (1984).
[28] A. L. Hodgkin, P. A. McNaughton, and B. J. Nunn, *J. Physiol.* (*London*) **358**, 447 (1985).
[29] D. A. Baylor, G. Matthews, and B. J. Nunn, *J. Physiol.* (*London*) **354**, 203 (1984).

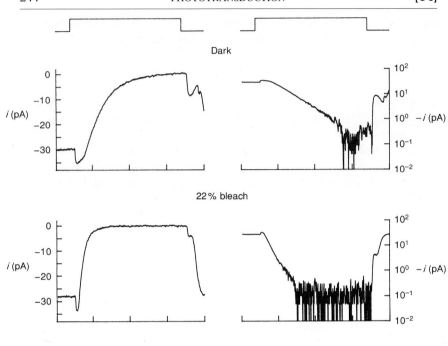

slope in Fig. 9, top right, with that in Fig. 9, bottom right). Solving Eq. (8) for the rate of phosphodiesterase and dividing it by the same expression for the dark-adapted cell gives the relative change of rate of phosphodiesterase:

$$\ln(J)/\ln(J_D) = \beta/\beta_D \qquad (9)$$

where J and J_D are the normalized currents measured in the manipulated and in the dark-adapted state. β is estimated from the slope of the straight line fitted to the normalized current trace of the right panel for each solution change (Fig. 9).

Alternatively, the rate of cyclase in photoreceptors can be estimated by using Ringer solution containing the phosphodiesterase inhibitor, IBMX.[22–24] Sudden exposure of the cell to saline solution containing 500 μM IBMX substantially inhibits phosphodiesterase, and the continued production of cGMP by cyclase results in an increase of the current. The rate of increase of current provides an estimate of the rate of cGMP synthesis by cyclase. Figure 10 (from Cornwall and Fain[23]) shows the effect of bleaching and pigment regeneration on the rate of cyclase in a rod. The increase of the current on exposing the cell to IBMX accelerated as a result of the bleach (left-hand panels, Fig. 10) and then returned to its dark-adapted rate on regeneration of the pigment with exogenous 11-*cis*-retinal. This can also be seen from the right-hand panels of Fig. 10, which plot $d(J^{1/3})/dt$ as a function of time ($J = i/i_D$). Because the derivative of $J^{1/3}$ is proportional to the rate of change in cGMP concentration [see Eq. (5)], its maximum represents the rate of synthesis of cGMP by guanylyl cyclase. We will once again make the assumptions that the cGMP channel characteristics ($K_{1/2}$ and N) and i_m do not change during the solution jump and that $i_m \gg i$. Equation (7) can then be simplified by setting $\beta = 0$ (blocked phosphodiesterase) and dividing by Eq. (5), to yield

$$\frac{K_{1/2}\dfrac{d}{dt}[(i)^{1/N}]}{K_{1/2}^D[(i_D)^{1/N}]} = \frac{\alpha[\text{GTP}]}{[\text{cGMP}]_D} = \alpha' \qquad (10)$$

where D indicates the parameters of the cell in its dark-adapted state.

FIG. 9. Estimation of rod photoreceptor phosphodiesterase rates before and after bleaching. *Left:* Current responses to steps from control Ringer solution into Li$^+$ solution under the following conditions (top to bottom): initial darkness; 42–44 min after a 520-nm 2-sec light exposure of intensity 2.0×10^7 photons μm^{-2} sec^{-1}, estimated to have bleached 22% of the visual pigment; 49–51 min after a second exposure to 520-nm light at the same intensity but for 4 sec, which together with the 2-sec exposure produced a total bleach of 52%; and 98–100 min after exposure to liposomes containing 11-*cis*-retinal. *Right:* The decline in current on a log scale. Current responses have been corrected for junction currents. [Reproduced from Cornwall and Fain.[23]]

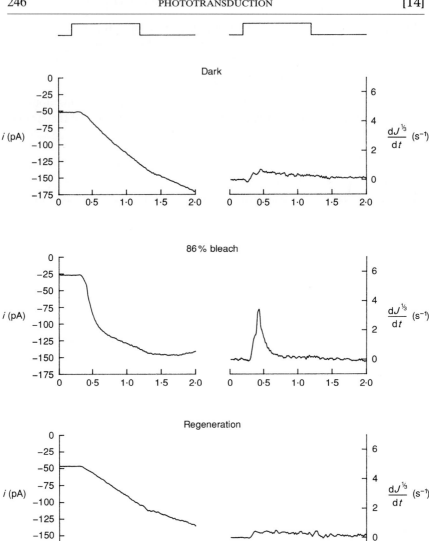

FIG. 10. Estimation of cyclase activity before and after bleaching in a rod. *Left:* Current responses to steps into IBMX solution in darkness, at steady state after a strong bleach, and after exposure to liposomes containing 11-*cis*-retinal. *Right:* The derivative of the one-third power of the normalized current for these responses. The bleaching exposure was produced by 16 sec of 520-nm light of intensity 2.0×10^7 photons μm^{-2} sec^{-1} (calculated to have bleached 86% of the visual pigment). Data in the middle panel were recorded 47 min after exposure to the bleaching light, and data in the lower panel, 50 min after addition of liposomes. [Reproduced from Cornwall and Fain.[23]]

We followed Hodgkin and Nunn,[22] and used α' as a measure of the change in the rate of cyclase (see also Cornwall and Fain[23]). The expression for the relative velocity of cyclase derived from Eq. (10) is

$$\frac{K_{1/2}\frac{d}{dt}[i^{1/3}]}{K_{1/2}^D\frac{d}{dt}[(i_D)^{1/3}]} = \frac{\alpha'}{\alpha'_D} = \frac{\alpha}{\alpha_D} \qquad (11)$$

where it is assumed that $N = 3$.[30] As can be seen from Eq. (11), in contrast to the Li$^+$ method, calculation of the relative rate of cyclase requires direct comparison of $K_{1/2}$ in the dark-adapted state of the cell and after pigment bleaching or other manipulations. However, $K_{1/2}$ of the cGMP-gated channel is not constant, but rather varies as a function of the concentration of calcium.[31–33] The decrease in [Ca^{2+}]$_i$ produced by the bleach in rods[34] and in cones[35] can cause a significant decrease of $K_{1/2}$. To improve the accuracy of the calculation of the rate of cyclase, an estimate of the change in $K_{1/2}$ caused by the bleaching of pigment can be incorporated in the analysis. In the case of rods, the change in $K_{1/2}$ after the bleach can be calculated from the corresponding drop in the dark current, using the relation between current and [Ca^{2+}]$_i$ in Nakatani et al.[33] However, we estimate that even substantial bleaches, accounting for the change in $K_{1/2}$, would only result in a change in the relative acceleration of cyclase by less than 10%. Using Eq. (10), α' is obtained in each experimental condition and the ratio α/α_D (equal to α'/α'_D) is calculated.

The data in Fig. 11 illustrate the use of step changes in IBMX solution to estimate the changes in cyclase rate that occur when a cone is exposed to bleaching light and is subsequently treated with 11-*cis*-retinal. These data are plotted in the same way as the data in Fig. 10. The left column of Fig. 11 illustrates the time course of the increase in current that occurs when the outer segment is rapidly exposed to IBMX solution; in the right column the derivative of the one-third power of the normalized current is shown. As in rods, when the cone was jumped into IBMX solution the current rapidly increased. This rate of increase was greater after the cell

[30] K. W. Yau and D. A. Baylor, *Annu. Rev. Neurosci.* **12**, 289 (1989).
[31] Y. T. Hsu and R. S. Molday, *Nature (London)* **361**, 76 (1993).
[32] S. E. Gordon, J. Downing-Park, and A. L. Zimmerman, *J. Physiol. (London)* **486**, 533 (1995).
[33] K. Nakatani, Y. Koutalos, and K. W. Yau, *J. Physiol. (London)* **484**, 69 (1995).
[34] A. P. Sampath, H. R. Matthews, M. C. Cornwall, and G. L. Fain, *J. Gen. Physiol.* **111**, 53 (1998).
[35] A. P. Sampath, H. R. Matthews, M. C. Cornwall, J. Bandarchi, and G. L. Fain, *J. Gen. Physiol.* **113**, 267 (1999).

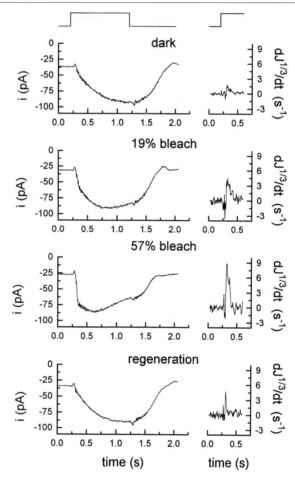

FIG. 11. Estimation of cyclase activity in a cone cell outer segment. *Left:* Steps into solution containing 500 μM IBMX in darkness, in steady state after exposure to bleaching light, and after regeneration with exogenous 11-*cis*-retinal. The traces represent current responses to steps from control solution into IBMX under the following conditions (top to bottom): in darkness; 2–3 min after a total 1-sec exposure to 600-nm light of intensity 3.4×10^7 photons μm^{-2} sec^{-1}, estimated to have bleached 19% of the visual pigment; 2–3 min after a total 4-sec exposure to the same 600-nm light, estimated to have increased the percentage of pigment bleached to 57%; and 10–11 min after exposure to liposomes containing 11-*cis*-retinal. *Right:* Derivative of the one-third power of the normalized current for the first 0.6 sec of the responses on the left. [Reproduced from Cornwall *et al.*[24]]

was exposed to lights that bleached a significant fraction of the visual pigment, and returned to that observed before the bleach following exposure to 11-*cis*-retinal. Thus, the changes observed in the cyclase rate under these conditions are qualitatively similar to those seen in rods.

Comparison of the data in Figs. 10 and 11 shows that cyclase rate in cones is more than twofold faster than in the rods both in darkness and after exposure to bleaching light. Two other differences between rods and cones can be observed with this technique, but are not shown in Fig. 11. First, the steady state accelerated cyclase rate observed after bleaching is achieved within seconds of exposure to the bleaching light in cones, whereas between 40 and 60 min is required for this to occur in bleached rods. Second, full recovery to the dark-adapted level of the cyclase rate in a bleached cell exposed to exogenous 11-*cis*-retinal occurs over a fourfold faster time course in cones than in comparably treated rods.[23,24]

Measurements of the PDE rate in cones have also been made using Li^+ jumps as discussed for rods, but both the experiments and their interpretation are significantly more difficult.[24,36] Because the PDE rate in cones is much faster than in rods, two assumptions that underlie the methods may no longer be valid. First, measurement of the PDE rate is possible only provided that the transient increase in the current produced by the greater permeability of the cGMP channel to Li^+ is complete before the current begins to decline, owing to blockage of cyclase. This is only marginally the case in dark-adapted cone photoreceptors. In cones exposed to either background or bleaching light the temporal overlap of the transient current produced by Li^+ substitution and the decline in current produced by blocking cyclase makes measurements of the rate of PDE impossible. Second, the block of guanylyl cyclase may not be sufficiently rapid and therefore may not allow an uncontaminated measurement of the PDE rate.

Involvement of $[Ca^{2+}]_i$ in Bleaching Adaptation

Intimately associated with the changes that take place in the cGMP economy of the photoreceptor after bleaching are the accompanying changes in cytoplasmic calcium concentration ($[Ca^{2+}]_i$). In the dark-adapted photoreceptor $[Ca^{2+}]_i$ is determined by the dynamic balance between the influx of Ca^{2+} through the outer segment conductance and its extrusion by Na^+/Ca^{2+}–K^+ exchange. The suppression of the circulating current during the response to light reduces Ca^{2+} influx, but its extrusion by Na^+/Ca^{2+}–K^+ exchange continues, leading to a light-induced fall in $[Ca^{2+}]_i$.[37] The persis-

---

[36] R. J. Perry and P. A. McNaughton, *J. Physiol.* (*London*) **433**, 561 (1991).
[37] K. W. Yau and K. Nakatani, *Nature* (*London*) **313**, 579 (1985).

tent excitation of the transduction cascade that follows bleaching would therefore be expected to lead to a sustained decrease in $[Ca^{2+}]_i$ through the persistent suppression of the circulating current that accompanies it. The way in which this bleach-induced fall in $[Ca^{2+}]_i$ can be measured directly is described in Chap. 10 in this volume.[37a] Here we deal instead with the techniques that can be used to determine its functional importance.

The role played by $[Ca^{2+}]_i$ in phototransduction has been established in large part by means of external solution changes designed to oppose the normal light-induced decline in $[Ca^{2+}]_i$.[38,39] The rapid solution changes required for this procedure can be carried out in the following way. The recording chamber is constructed from two parallel glass slides, separated by a Plexiglas spacer that encloses it on three sides, leaving the front open for access by the suction pipette.[9] An isolated photoreceptor is drawn inner segment first into the suction pipette, leaving the outer segment exposed. The tip of the suction pipette is placed in a flowing stream of Ringer solution emerging from one of up to four parallel tubes built into the back of the recording chamber. By moving the entire recording chamber sideways, using a computer-controlled stepper motor coupled with a lead screw to the microscope stage, the laminar interface between two neighboring streams of solution is rapidly stepped across the tip of the suction pipette, thereby exchanging the solution bathing the protruding outer segment of the photoreceptor within a few tens of milliseconds.

The influx and efflux of Ca^{2+} can be minimized simultaneously by stepping the photoreceptor outer segment into a solution in which Ca^{2+} had been reduced to submicromolar levels, and Na^+ replaced with another cation such as guanidinium, which does not support $Na^+/Ca^{2+}-K^+$ exchange. Superfusion with such a low-Ca^{2+}/0-Na^+ solution is believed to hold $[Ca^{2+}]_i$ close to its initial value before the solution change for a period of some 10–15 sec.[40,41] While this approach has proved fruitful in studying the role of $[Ca^{2+}]_i$ in light adaptation, it cannot be applied directly to bleaching adaptation for two reasons. First, in an amphibian rod the response stabilizes after bleaching over a protracted period of some tens of minutes, which is much longer than the period during which changes in $[Ca^{2+}]_i$ can be reliably opposed. Second, even after response stabilization the phosphodiesterase rate remains sufficiently elevated to suppress completely the circulating

---

[37a] H. R. Matthews and G. L. Fain, *Methods Enzymol.* **316**, 146 (2000).
[38] H. R. Matthews, R. L. Murphy, G. L. Fain, and T. D. Lamb, *Nature (London)* **334**, 67 (1988).
[39] K. Nakatani and K. W. Yau, *Nature (London)* **334**, 69 (1988).
[40] G. L. Fain, T. D. Lamb, H. R. Matthews, and R. L. Murphy, *J. Physiol. (London)* **416**, 215 (1989).
[41] A. Lyubarsky, S. Nikonov and E. N. Pugh, Jr., *J. Gen. Physiol.* **107**, 19 (1996).

current when $[Ca^{2+}]_i$ remains at its dark level, making it impossible to record a light response in this solution. The first of these problems can be addressed by using amphibian cones, in which the response stabilizes after bleaching within a few tens of seconds. The second can be resolved by

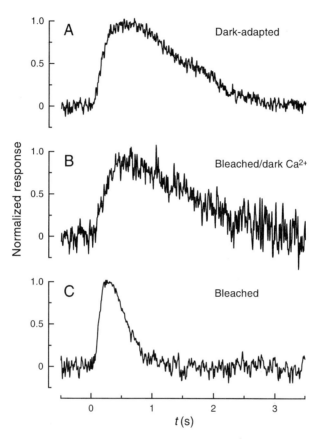

FIG. 12. Comparison of dim flash response kinetics in low-Ca^{2+}/0-Na^+ solution before and after bleaching. Responses to subsaturating flashes delivered at time zero with $[Ca^{2+}]_i$ held near the dark-adapted level by superfusion with low-Ca^{2+}/0-Na^+ solution (A), in darkness or (B), in the presence of 500 μM IBMX after a 44% bleach. (C) Response of the same cone to a subsaturating flash in low-Ca^{2+}/0-Na^+ solution after the bleach once $[Ca^{2+}]_i$ had been allowed to fall to the appropriate bleach-adapted level. Trace in (B) is a single response, while traces in (A) and (C) are the average of three and four responses, respectively. Each trace has been normalized in amplitude according to the response peak, after the subtraction of linear drift in the baseline. Bleaching light delivered 9.45×10^7 equivalent photons μm^{-2} at 620 nm. Subsaturating flashes, which typically suppressed one-third to one-half of the circulating current, delivered 1.90×10^2 (A), 1.26×10^4 (B), 2.48×10^3 (C) equivalent photons μm^{-2} at 620 nm. [Reproduced from Matthews et al.[42]]

including the phosphodiesterase inhibitor IBMX in the low-Ca^{2+}/0-Na^+ solution. The concentration of IBMX must be carefully matched to the intensity of the bleaching light so as to return the circulating current to its initial value in Ringer solution before the bleach, which corresponds to the restoration of the phosphodiesterase activity to its original dark-adapted rate.

An example of such an experiment is illustrated in Fig. 12, which compares the responses of a salamander cone to flashes of light in low-Ca^{2+}/0-Na^+ solution before and after bleaching,[3] modified from Matthews et al.[42] When $[Ca^{2+}]_i$ was held near its dark-adapted level by bleaching the cone in low-Ca^{2+}/0-Na^+ solution, the response measured in the presence of IBMX while the outer segment remained in this solution (Fig. 12B) was similar to that recorded in low-Ca^{2+}/0-Na^+ solution before the bleach (Fig. 12A). However, once the outer segment was returned to normal Ringer solution, thereby allowing $[Ca^{2+}]_i$ to fall after the bleach, the response kinetics in low-Ca^{2+}/0-Na^+ solution were speeded considerably (Fig. 12C). This result indicates that the light-induced reduction in $[Ca^{2+}]_i$ is necessary for the characteristic response acceleration that accompanies bleaching adaptation. It is likely to do so by relieving the inhibition of guanylyl cyclase by Ca^{2+}, thereby allowing the establishment of a new steady state between phosphodiesterase and cyclase rates after the bleach and an accelerated turnover of cyclic GMP, much as is believed to take place during light adaptation.

Conclusions

We have used the methods described here to gain important insights into the function of isolated rod and cone photoreceptors under controlled conditions of pigment bleaching and regeneration. However, it must be understood that in the intact retina, these are dynamic processes that interact continuously. Bleach-desensitized photoreceptors, as we have described them here, do not normally exist in the retina, except possibly in pathological conditions. A complete understanding of the physiology of the visual cycle will require correlation of physiological, biochemical, and psychophysical techniques.

---

[42] H. R. Matthews, G. L. Fain, and M. C. Cornwall, *J. Physiol.* (*London*) **490**, 293 (1996).

[15] Exploring Kinetics of Visual Transduction with Time-Resolved Microcalorimetry

By T. MINH VUONG

Introduction

In one conspicuous way, the cascade of visual transduction is different from other G-protein systems: the major proteins of vision are extremely abundant. One requirement for efficient photon capture is a high optical density that evolution has achieved by packing photoreceptor cells with large amounts of rhodopsin (R). To translate efficient photon capture into high sensitivity, the signaling species downstream from rhodopsin must be abundant too and, indeed, there are high levels of transducin (T) and cGMP phosphodiesterase (PDE) in rod outer segments (ROS). This bounty is often considered a gift from nature as it has greatly facilitated our dissection of visual biochemistry. Yet this has also been a mixed blessing. Under physiological conditions, little of the protein population of the rod cell is activated. The electrical response of a dark-adapted mammalian rod half-saturates when as few as 30 of its 10^7 rhodopsin molecules are photoexcited.[1] Because the transduction machinery is so bountiful, one does not need the high sensitivity of electrophysiological instrumentation to obtain some sort of *in vitro* response from an ROS suspension: one simply triggers the cascade with bright flashes. In many biochemical experiments, the extent of photoexcitation is immaterial as it does not appreciably influence the interpretation. To study certain important aspects of phototransduction such as its deactivation kinetics, however, one is required to do so in the physiological regime of dim light.

Aside from the electrical response of the rod cell, which has been amply studied by electrophysiological methods, we would like to time-resolve the biochemistry of the R*–T–PDE cascade at light levels comparable to those *in vivo*. Biochemically, the end result of this cascade is the massive hydrolysis of cGMP, which gives rise to three products: 5'-GMP, protons, and heat. All three can be detected biophysically. Potentially, one could exploit the slight difference in the absorption spectra of cGMP and 5'-GMP; but inevitably, the light source required for such a spectroscopic measurement would itself trigger the cascade. The pH electrode has long been a favorite tool to track cGMP hydrolysis from

---

[1] D. A. Baylor, B. J. Nunn, and J. L. Schnapp, *J. Physiol.* **357,** 575 (1984).

an ROS suspension,[2] but its lack of sensitivity prevents reliable measurements at light levels less than 5000 R*/rod. Homogeneous illumination is a more subtle issue. Bovine ROS prepared with standard protocols are always leaky. This allows exogenously added nucleotides and/or proteins to enter the cytoplasm but also lets endogenous proteins leak out into the bulk solution. One partial recourse is to use high ROS concentrations[3] of several milligrams of rhodopsin per milliliter. Cuvettes designed to accommodate the smallest commercially available pH electrodes are typically several millimeters thick.[4] This means that at 5 mg of rhodopsin per milliliter, a significant portion of the sample would receive essentially no light from the actinic flash.

The enthalpy of cGMP hydrolysis is about -12 kcal/mol.[5] With the proper instrumentation, this heat of reaction can be measured with enough time resolution and sensitivity to permit a fruitful kinetic exploration of the visual cascade at light levels resembling those *in vivo*. In this chapter we describe a time-resolved microcalorimeter that is capable of such measurements.

General Description

The sample chamber of the microcalorimeter is shown schematically in Fig. 1. The sensing element is a 3×3 cm (40-μm thickness) film of poled, pyroelectric polyvinylidene difluoride (PVDF) gold-metallized on both sides. Its back and front are epoxy-glued to a slab of thermally insulating balsa (8- or 10-mm thickness) and to a Mylar film (4-μm thickness), respectively. A minimal amount of glue is used in this operation. The front window of the sample chamber is made out of the same 4-μm Mylar, which is epoxy-glued to a square Plexiglas support and made taut by heat-shrinking. The chamber thickness is determined by a gasket made from three layers of laboratory Parafilm; the total thickness is about 0.4 mm. The window is bolted to a Teflon base with eight M2 hex sockets that traverse the balsa slab. ROS samples are introduced into this chamber via an inlet at the bottom of the window's Plexiglas support while air is allowed to escape through an outlet at the top. The vertical, rhombus orientation of the entire chamber prevents trapping of air bubbles during introduction and removal of samples. Typically, the volume of the chamber is about 300 μl. Pyroelec-

---

[2] P. A. Liebman and A. T. Evanczuk, *Methods Enzymol.* **81,** 532 (1982).
[3] P. Catty, C. Pfister, F. Bruckert, and P. Deterre, *J. Biol. Chem.* **267,** 17489 (1992).
[4] A. Otto-Bruc, B. Antonny, and T. M. Vuong, *Biochemistry* **33,** 15215 (1994).
[5] W. A. Hagins, P. D. Ross, R. L. Tate, and S. Yoshikami, *Proc. Natl. Acad. Sci. U.S.A.* **86,** 1224 (1989).

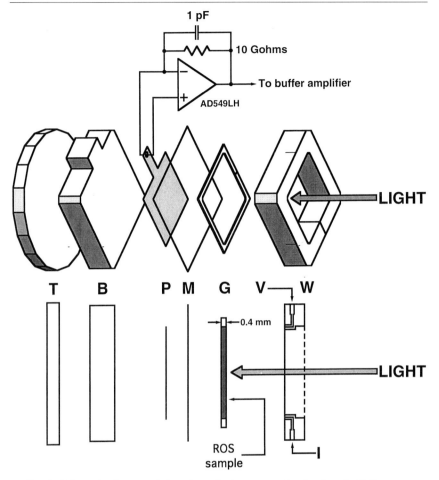

FIG. 1. Schematic diagram of the microcalorimeter sample chamber. T, Teflon support; B, thermally insulating balsa slab; P, pyroelectric PVDF film; M, 4-μm Mylar film that protects the front electrode of the PVDF from the ROS suspension; G, Parafilm gasket; W, front window whose inlet (I) is used to introduce the sample into the chamber; air moves in and out of the chamber via the vent (V).

tric PVDF film is purchased from Piezotech SA (Saint-Louis, France). Balsa wood can be bought from any hobbyist store as slabs 8 or 10 mm thick. The surface of such balsa is flat and smooth, which greatly facilitates the gluing of PVDF film to it. Thin, heat-shrinkable Mylar film is purchased from Goodfellow (Cambridge, England). One important idea behind this design is to spread the ROS sample into a layer as thin as feasible in order to achieve two goals. First, because the sensing element is a film, a large area maximizes sensitivity. Second, a minimal thickness allows for a uniform illumination of concentrated ROS suspensions. With this essentially two-dimensional geometry, we have sidestepped the inherent problem associated with cylindrical pH electrode cuvettes as discussed above: at 0.4-mm thickness, a 5-mg rhodopsin/ml sample has an optical density (OD) of only 0.2 at 500 nm. Two constraints determine the chamber surface area and thickness: (1) the area should not be so large that the thin Mylar film of the front window sags under the weight of the sample; and (2) the gasket should not be so thin that excessive capillarity prevents complete sample removal.

This sample chamber is placed inside a cavity between two brass blocks ($20 \times 20 \times 6$ cm; Fig. 2). The front brass block has a double-pane window through which passes the actinic light flash. ROS samples are pumped into and removed from the sample chamber with a Hamilton syringe through an injection port and associated plumbing located in the rear brass block. During measurements, the cavity is made air-tight (see below). The electronic circuit needed to convert current from the heat-sensing PVDF film

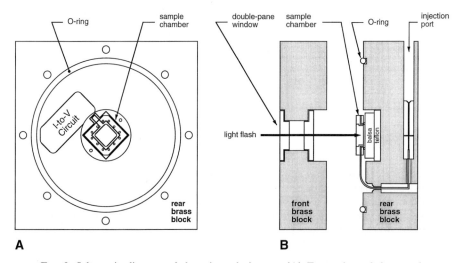

FIG. 2. Schematic diagram of the microcalorimeter. (A) Front view of the rear brass block. The sample chamber is bolted to this block inside a recessed cavity [see (B)]. The transimpedance circuit is located next to the PVDF film. (B) Side view of the front and rear brass blocks. The actinic light flash passes through the double-pane window in the front block. The injection port and plumbing are located in the rear block. When the two blocks are bolted together, a cavity is formed. Its air-tightness is ensured in part by the O ring. The sample chamber is electrically decoupled from the grounded brass block by being mounted on a Teflon support (see text).

into voltage is also mounted inside this cavity and as close to the sample chamber as possible. This brass block/sample chamber ensemble is placed inside a polystyrene box, which is enclosed by a brass Faraday cage. Actinic light comes from a photographic flash fitted with a 500-nm interference filter (10-nm bandwith) and a set of switchable neutral density filters. This photographic flash is controlled by the digitization computer, which triggers it at a precise sampling point during each acquisition. The instrument is installed on a vibration-isolated optical table (see below).

Poled Polyvinylidene Difluoride: Principle of Operation

The empirical formula for PVDF is

$$-CH_2-CF_2(CH_2-CF_2)_n-CH_2-CF_2-$$

As seen in Fig. 3, the β form of this structure has an electric dipole. PVDF is approximately 50% crystalline but even though the microcrystals are dipolar, they are randomly oriented so that the film itself bears no net dipole moment. To line up these microcrystals, two metallic electrodes are vapor deposited on both faces of the film. Then, while the PVDF film is

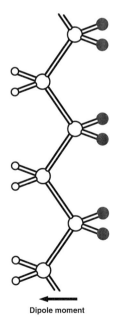

FIG. 3. The β form of polyvinylidene difluoride is dipolar. The gray spheres represent fluorine atoms.

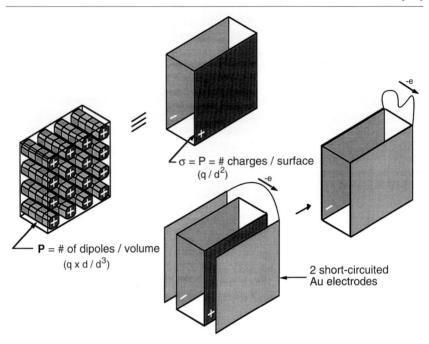

FIG. 4. The electrostatics of polar dielectrics. The actual dipoles inside the PVDF film can be represented equivalently by surface charges. When two gold electrodes are deposited onto the film faces, electrons flow from the negative side of the film to the positive side.

heated, it is subjected to a high-strength electric field via these two electrodes. On cooling, the microcrystals are locked in place and the electric field can be removed. This is known as "poling."

One way to characterize a polar dielectric is to specify the number of dipoles it carries per unit volume, i.e., its polarization **P**, which is a vector quantity in units of (dipoles/volume) or (charge × distance/distance^3). Instead of viewing a polar dielectric as harboring dipoles inside its bulk, one can equivalently view it as bearing fixed, immobile charges on its surfaces[6] (Fig. 4). As such it can be characterized by a surface charge density σ in units of (charges/area) or (charges/distance^2). σ is equal to $\pm P$ (note that their units are dimensionally identical). What happens when the two faces of a polar dielectric are coated with metallic electrodes? On the positive side of the dielectric, free electrons in the metal tend to gather at the metal/dielectric interface while on the negative side, they tend to be

[6] P. Lorrain and Dale Corson, in "Electromagnetic Fields and Waves," 2nd Ed., pp. 109–114. W. H. Freeman, San Francisco, 1970.

pushed away from the interface. If the electrodes are short-circuited, charges will flow so that on the positive side of the dielectric, there will be an excess of electrons at the expense of the negative side. The fixed surface charges on the dielectric have thus induced a charge separation by causing the transfer of free electrons from one electrode to the other (Fig. 4). We note that no voltage develops across these electrodes because the excess mobile charges in the metal are exactly balanced by the fixed charges in the dielectric.

From this discussion, we see that the extent of charge separation in the metal electrodes depends on the surface charge density σ of the dielectric or equivalently on its volume polarization **P**. Clearly, any changes in the dielectric volume will result in changes in P and equivalently in σ. For example, an increase in volume will lead to a decrease in P and σ, and consequently a decrease in the charge separation of the electrodes. To attain this new, lesser charge separation, free electrons will have to flow from the electrode on the positive side of the dielectric to the electrode on the negative side (Fig. 5). In essence then, a change in volume is converted to a flow of current. Because the dielectric volume can change via heating or cooling, a poled, metallized PVDF film is exquisitely sensitive to changes in temperature. We must recognize that current flows only between the two electrodes during changes in the dielectric volume, or indirectly its temperature. As soon as this volume reaches a new value and stops changing, current flow also ceases. In other words, by monitoring current flow,

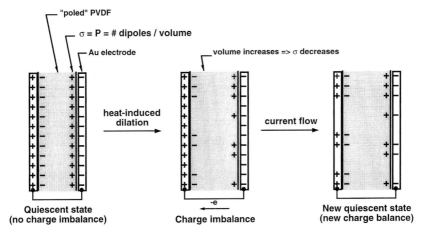

FIG. 5. Current flows whenever the volume of the PVDF film is modified. If the film dilates, this increase in volume will result in a drop in its surface density. The charge separation in the gold electrodes must decrease accordingly and in order to achieve this, a current will flow from the negative side of the PVDF to its positive side.

we measure the time derivative of temperature but not its actual value. We envision that light-induced cGMP hydrolysis by our ROS sample will cause a rise in temperature and hence, a measurable current flow between the two electrodes of the PVDF film. Clearly, for such heat to cause a significant rise in the film temperature, it must not be lost into the surroundings, or more realistically, its loss must be slow compared with its creation (see below). Given the two-dimensional geometry of the ROS sample, heat loss occurs mainly via the front and back surfaces. On the back side, such loss is minimized by the thermally nonconductive balsa slab. Most of the heat is then lost via the transparent front face, which is in thermal contact with the surrounding air. This loss could become a real problem should we make the front window out of a thermally conductive material such as glass. This is why we use a thin Mylar film of negligible thermal mass.

Obviously, the volume of a PVDF film can change owing to factors other than temperature. Changes in air pressure impinge directly on the film, producing unwanted currents many orders of magnitude larger than heat signals from an ROS sample: the sample chamber can act like a microphone. This is why the sample chamber is placed in a cavity that is made air-tight during measurements. Another noise source is mechanical vibrations of the building, which can be transmitted to the film via the liquid sample. This is why the instrument is placed on a vibration-isolated optical bench. Finally, temperature fluctuations in the ambient air are always much larger than any temperature changes induced in our ROS samples. By surrounding the sample chamber with the large thermal mass of 20 kg of brass and by making the cavity air-tight, these fluctuations can be avoided.

Electronics: Design, Noise Sources, Grounding, and Digitization

As a current source, a poled PVDF film is similar to a patch pipette. Our transimpedance circuit is essentially the same as one found in a typical electrode "head" of electrophysiology. The printed circuit board is designed to minimize leakage current. The operational amplifier is an AD549-LH from Analog Devices (Norwood, MA). A metal oxide 10-GΩ resistor is used for feedback. Electrical leads from the film electrodes to the circuit are not longer than 1 cm. On the side of the film, they are soldered to two small gold pads and the PVDF film is sandwiched between these with a clamping mechanism. Gold must be used for the PVDF electrodes, as with a different metal such as aluminum, the contacts tend to degrade with time, causing the transimpedance circuit to oscillate. Moreover, even with gold electrodes, the front surface of the film, which is in contact with the ROS suspension, must be protected by the 4-μm Mylar film. In our hands, a change in sample pH owing to cGMP hydrolysis causes large instabilities in the electronics, an effect that is probably electrochemical in nature. These gold electrodes are not deposited directly onto the PVDF but onto a first layer of tin or chromium. There are pinholes in the gold layer and it is likely from these small defects that the pH change wreaks its havoc. Moreover, whenever the chamber is dismantled for cleaning, this protective Mylar film prevents scratching of the thin gold electrode.

A 3×3 cm, high-impedance PVDF film easily picks up electromagnetic noise, 50-Hz hum in particular, from the electrical grid. The microcalorimeter is therefore enclosed in a brass box, i.e., a Faraday cage. The window on the front brass block is shielded with a fine copper mesh. The front electrode of the PVDF film is grounded (Figs. 1 and 5) to provide maximum shielding to the back electrode, which is connected to the input of the AD549-LH amplifier. The brass blocks, Faraday cage, and steel optical table are grounded via wide, flat copper braids to the building ground. The transimpedance circuit ground is kept separated from the building ground. To prevent electrical leakage from the PVDF film to the grounded brass blocks, the sample chamber is mounted on a Teflon support (Figs. 1 and 2).

The voltage output from the transimpedance circuit is buffered and amplified 10-fold by a low-noise operational amplifier (OP-27EN; Analog Devices) wired in a noninverting configuration. This amplifier is placed on the optical table, next to the microcalorimeter. The buffer amplifier output is RC-filtered (50- or 220-msec time constant) and fed into an instrumentation amplifier (AD624-CD; Analog Devices) with adjustable gain and offset. As the distance between buffer and instrumentation amplifiers is 10 m, the two are connected via a double-shielded coaxial cable. The cable shield links the grounds of the two amplifier circuits. The cable itself runs inside a tubular copper braid that connects the optical table to the case of the intrumentation amplifier. This grounding scheme is summarized in Fig. 6. Finally, to avoid all noise coupling from the electrical grid, the transimpedance, buffer, and instrumentation amplifiers are battery powered. Digitization is performed to a 12-bit precision with $10\times$ oversampling: The raw sampling rate is 1000 points/sec but during acquisition, every 10 points are averaged on the fly, giving an effective sampling rate of 100 points/sec and a significant reduction in high frequency noise.

Impulse Response, Time Resolution, and Calibration

The impulse response of the instrument ought to come from an ideal experiment in which a short pulse of heat is deposited onto a dye solution residing in the sample chamber. This heat pulse must heat only the dye and nothing else. Moreover, the OD of the dye at 500 nm must be close

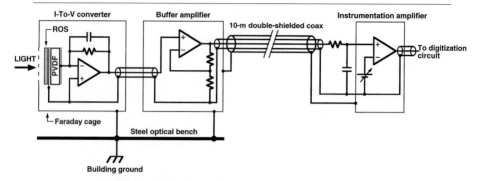

FIG. 6. Grounding scheme for the entire system. The goal is to separate the sensitive analog ground of the system from the noisy ground of the building. The Faraday enclosure for the microcalorimeter as well as the cases of all electronic instruments are grounded to the building but the circuits themselves float inside these shields. The 10-m coaxial cable that connects the two main parts of the system is itself shielded by a tubular copper braid, which acts as a Faraday cage.

to that of a 5-mg/ml ROS suspension. Such an ideal experiment can be closely approximated by performing two independent experiments. First, a bright flash is used to quickly heat up a Red Ponceau dye sample of OD 0.3, which is close to the 0.2 OD of a 5-mg/ml ROS suspension. At this OD, a significant portion of the light passes through the dye unscathed and directly heats up the PVDF film. However, this direct heating is equivalent to using the same light flash but attenuated by 0.3 OD to heat a sample of water. Thus the true impulse response can be obtained by subtracting the water signal from the total dye signal. In Fig. 7A, the upper trace is obtained with a 0.3-OD Red Ponceau dye while the lower trace comes from a sample of water that is illuminated with a 0.3-OD attenuated light flash. The dye-minus-water difference is shown in Fig. 7B and C.

How should we interpret these signals? We first note that each trace displays a positive as well as a negative lobe, the latter being most pronounced for the water trace. Because the instrument measures the time derivative of heat, the positive lobe corresponds to a heat gain or a rise in temperature, while the negative lobe denotes a heat loss or a drop in temperature. Mechanically speaking, the film expands during the positive lobe, the rate of expansion being highest at the peak of the lobe. This rate decreases on the downward portion and at the zero-crossing, the film stops expanding altogether: heat gain and loss exactly balance each other out at this point in time. During the negative lobe, the film slowly contracts. The trace comes back to baseline when the film has finally regained its initial thickness prior to the light flash. With just water in the sample chamber,

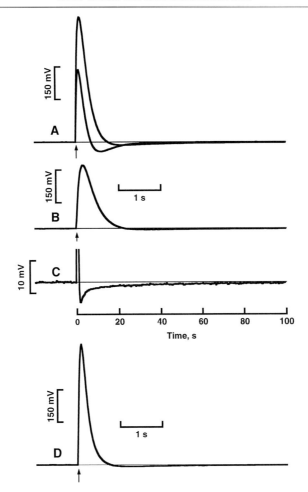

FIG. 7. (A) The upper trace comes from a Red Ponceau dye sample (OD 0.3 at 500 nm) that is quickly heated with a pulse of green light (arrow). The lower trace comes from a sample of water that is heated with a second pulse of light that is attenuated with a neutral density filter of OD 0.3 at 500 nm. The RC time constant for both experiments is 220 msec. (B) The lower trace in (A) is subtracted from the upper trace to obtain this difference trace, which is a good approximation of the ideal case, in which the light flash would only heat the dye and not the PVDF film itself. (C) The trace in (B) is extended 15-fold in the vertical axis and compressed 20-fold in the horizontal axis to show the time course of heat loss from the sample chamber, which is on the order of 100 sec. (D) The same experiments as in (A) are repeated albeit with an RC time constant of only 50 msec. The difference trace is shown here.

only the PVDF film is heated by the light flash while the nonabsorbing water stays cool. Hence this water is a heat sink that quickly cools the hot PVDF. This is why the negative portion of the water trace is so pronounced. Conversely, in the ideal situation, in which only the dye is heated, there is no drastic heat sink to draw away the deposited heat, the Mylar of the front window being of negligible thermal mass and the surrounding air not being an efficient heat sink such as water. In this case we expect heat loss to be much reduced and consequently a negative lobe that is more spread out in time. Effectively, the difference trace, which approximates the ideal experiment, does display a barely noticeable negative lobe (Fig. 7B). In fact this difference trace barely comes back to baseline after 100 sec (Fig. 7C), which tells us that this is about how long it takes for the heat deposited by the light flash to dissipate into the surrounding. In contrast, the full width at half-maximum (FWHM) of the positive lobe is less than 1 sec, which means that heat gain is much faster than heat loss for this instrument. How fast the instrument responds to the heat pulse is determined mainly by either one of the two RC filters used here. In Fig. 7B the RC time constant is 220 msec and the FWHM of the impulse response is 0.4 sec, but in Fig. 7D this FWHM decreases to 0.2 sec as the RC time constant is reduced to 50 msec. What kind of kinetics can we reliably measure with this instrument? As usual the signal must not be too fast compared with the subsecond FWHM of the positive lobe. In addition, because heat created in the sample chamber leaks away in about 100 sec, any signal component in this time domain will tend to be distorted. For example, let us consider a sample that produces heat at a constant rate, which in a perfectly insulated chamber should give rise to a plateau. In our more realistic case, where the chamber can hold its heat for only about 100 sec, such a plateau will droop around this characteristic time and in fact will disappear altogether after a time much longer than 100 sec. For signals that contain such slow components, some sort of deconvolution will be required.[7] Because these traces are derivatives of heat, their areas must represent heat itself. Then by conservation of energy, the areas of the positive and negative lobes must be roughly equal. Indeed, they are +62 and −52 mV · sec for the water trace. For the dye trace they are +170 and −150 mV · sec. For the difference trace, they are +130 and −120 mV · sec. In these measurements, the negative lobe is taken to be between the zero-crossing and 120 sec.

Calibration of the instrument means converting these area units of millivolt-seconds into a heat unit such as microcalories (μcal). One approximate way to do this is to heat an optically dense (e.g., 1–2 OD) sample of

---

[7] G. Langlois, C.-K. Chen, K. Palczewski, J. B. Hurley, and T. M. Vuong, *Proc. Natl. Acad. Sci. U.S.A.* **93**, 4677 (1996).

Red Ponceau dye with a light flash whose energy is measured with a calibrated photodiode. With this information, one can convert the area under the positive lobe from millivolt-seconds to microcalories and equivalently, the vertical axis from millivolts to microcalories per second. This calibrated unit of microcalories per second for the vertical axis brings out the fact that the instrument does not measure heat but its derivative. This absolute calibration is not without problems. For example, even though the photographic flash is fitted with a narrow-band interference filter, the optics being nontelecentric, the light that actually passes through this filter may have additional wavelengths other than the nominal 500 nm. This would make using the responsivity curve of the photodiode particularly tricky. A much better calibration method would be to install a heating element whose resistance is precisely known inside the water-filled sample chamber and to excite it with a current pulse of known energy. Because all of our work does not really call for a knowledge of the absolute energies involved, we have chosen to calibrate the instrument optically although we are well aware that the thus-obtained calibration factor can easily be incorrect by an order of magnitude.

Results

How should this microcalorimeter be used to follow the biochemistry of the visual cascade in real time? As cGMP is hydrolyzed by active PDE (PDE*) after a light flash, a heat $Q(t)$ is concomitantly produced. In other words, $Q(t) \propto [5'\text{GMP}](t)$. The microcalorimeter senses the rate of change of heat: it measures $d[5'\text{-GMP}]/dt$. The K_m for PDE* is about 100 μM and if the concentration of cGMP is much greater than this value, PDE* is saturated and its concentration at any moment is proportional to $d[5'\text{-GMP}]/dt$. In summary then, if

$$[\text{cGMP}] \gg K_m (70\text{–}150 \,\mu M) \quad \text{then} \quad dQ(t)/dt \propto d[5\text{-GMP}]/dt \propto [\text{PDE*}](t)$$

Therefore by measuring dQ/dt with the microcalorimeter, we can tell how much active PDE there is in the sample at any point in time. Consequently, the surface area under the microcalorimetric trace must be proportional to the total amount of PDE that becomes activated.

Bovine ROS are prepared according to method of Kühn.[8] The pellets are kept at $-70°$. They are resuspended in 100 mM KCl, 1 mM MgCl$_2$, 100 μM GTP, 2 mM NADPH, 1 mM dithiothreitol (DTT), 50 mM HEPES, pH 7.5 and disrupted through a 26-gauge Hamilton syringe needle. This

[8] H. Kühn, *Curr. Top. Membr. Transp.* **15**, 171 (1981).

stock suspension (10–15 mg of rhodopsin per milliliter) is kept in the dark on ice to allow any contaminating R* to decay away. When GTP and ATP are used, additional $MgCl_2$ is added to the sample so as to maintain a free Mg^{2+} concentration of 1 mM. When needed, the level of free calcium is buffered by using the method of Tsien and Pozzan.[9] In the final sample, bovine serum albumin (BSA, 0.01 mg/ml) is added to reduce the surface tension. This ensures that the sample can be completely removed from the chamber once the experiment is over, especially when the chamber has been in contact with ROS suspensions many times before. All operations are carried out either under infrared (IR) illumination or in total darkness. The sample is introduced into the chamber as gently as possible to minimize all mechanical and thermal disturbances to the PVDF film. After sample loading, it takes about 4 min for the sample to equilibrate thermally with the chamber. Typically, 5 to 10 signals are elicited from each sample, although usually only the first is used as the signals are essentially identical. Given the low light level used in these experiments, many more than 10 signals can in fact be obtained from the same sample.

In the experiments of Fig. 8, five samples containing rhodopsin (4.5 mg/ml), 1 mM GTP, 3 mM ATP, 10 mM cGMP, 1 mM free Mg^{2+}, and 10 nM free Ca^{2+} are subjected to five flashes of increasing intensities that photoexcited 5.3, 12, 25, 47, and 66 R* per rod, producing five heat pulses. As the total amount of active PDE is proportional to the illumination level,[7] we expect the surface area under the microcalorimeter trace to be proportional to the number of R* per rod. This linearity is evident in the inset of Fig. 8. The microcalorimeter is thus exquisitely sensitive, as it reliably detects PDE activities at light levels as low as 5–10 R* per rod. Moreover, in this physiological range between 5 and 70 R* per rod, the response is entirely linear. From the shape of these traces we can see that active PDE is created right after the light flash, reaches a maximal level about 1 sec later, and then decays down to baseline in about 5 sec. The population of active PDE decays after only a few seconds because of the added ATP, which is used by rhodopsin kinase to phosphorylate R* and help turn off the cascade. Thus, through the time evolution of this population of active PDE, we have caught a glimpse of how the visual cascade proceeds after a dim light flash.

In Fig. 9 we explore the effect of various forms of GTP in the absence of ATP, i.e., under conditions in which R* is long-lived because its phosphorylation is hindered. All four samples contain rhodopsin at 1 mg/ml, 15 mM cGMP, 1 mM Mg^{2+}, and 5 $\mu$$M$ Ca^{2+}. In samples 1 and 2 the GTP levels are 0 and 1 mM, respectively. There is 1 mM GTPαS(S_p) in sample

---

[9] R. Y. Tsien and T. Pozzan, *Methods Enzymol.* **172**, 230 (1989).

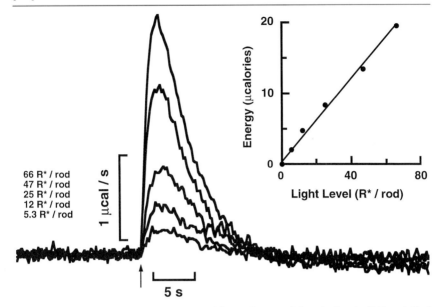

FIG. 8. Linearity and sensitivity. Five ROS samples containing rhodopsin (4.5 mg/ml), 1 mM GTP, 3 mM ATP, 1 mM free Mg^{2+}, and 10 nM free Ca^{2+} are photoexcited at 5.3, 12, 25, 47, and 66 R* per rod (arrow). The five traces are of the same shape and their surface areas are linear with the illumination level as shown in the inset.

3 and 0.2 mM GTPγS in sample 4. Clearly, no signal is elicited when no GTP is present. With 1 mM GTP, trace 1 attains a plateau after about 2 sec, which is to say that a steady state has been reached where the R*-catalyzed creation of active PDE is balanced by its destruction when transducin is inactivated by GTP hydrolysis. At this steady state, the amount of active PDE in the sample is a constant. The time taken to reach this steady state plateau is the time taken by transducin to hydrolyze its GTP.[10] If this is true, then a slower GTP hydrolysis should have two related effects: a higher plateau that takes longer to attain. Intrinsic hydrolysis of the analog GTPαS(S_p) by transducin is known to be about twofold slower than that of GTP.[11] When this analog is used in the experiment of trace 3, a plateau that is higher and takes longer to reach is indeed observed. To push this reasoning to its logical conclusion, if a nonhydrolyzable analog such as GTPγS is used, there should be no plateau but a sustained rise that represents a constant production of active PDE unhindered by any counteracting deactivation: more and more active PDE is created as time

[10] T. M. Vuong and M. Chabre, *Proc. Natl. Acad. Sci. U.S.A.* **88**, 9813 (1991).
[11] G. Yamanaka, F. Eckstein, and L. Stryer, *Biochemistry* **24**, 8094 (1985).

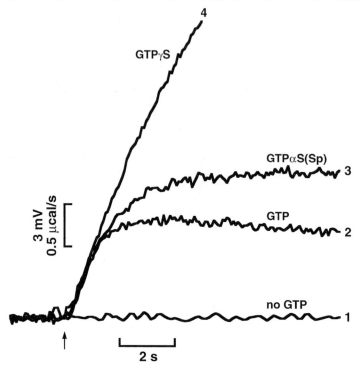

FIG. 9. The effect of GTP analogs. The four samples contain rhodopsin (1 mg/ml), 15 mM cGMP, 1 mM Mg^{2+}, and 5 μM Ca^{2+} and various amounts and types of GTP and GTP analogs. The photoexcitation level is 30 R* per rod. For samples 1, 2, 3, and 4 the guanine nucleotide levels are 0, 1 mM GTP, 1 mM GTPαS (S_p), and 0.2 mM GTPγS, respectively.

goes on. This is indeed the case, as is clearly displayed by trace 4. This experiment is also an extremely important control. That we can successfully detect these plateaus and especially the essentially constant rise in the presence of GTPγS confirms our previous finding that the sample chamber is truly well insulated in this time domain of 10–20 sec, because it substantially loses heat in 100 sec. Moreover, with this information in mind, we can say with confidence that the pulselike shape of the signals in Fig. 8 is truly due to the biochemistry of the samples and not to an excessive heat loss of the instrument.

Conclusion

The experiments shown here constitute an example of what is possible with this technique of time-resolved microcalorimetry. The sensitivity of the instrument allows us to obtain signals at physiological light levels of less than 100 R* per rod, which is important for two reasons: (1) unlike rhodopsin, transducin, and PDE, many important proteins of the cascade are soluble and not abundant. Consequently their concentrations inside the leaky ROS tend to be significantly diminished when compared with the *in vivo* situation. Two examples of such proteins are rhodopsin kinase and recoverin. To study their actions on R*, it is best to work at low light levels so that the stoichiometry of the situation is not too far from that in an intact rod.[7] In contrast, a typically unphysiological illumination that photoexcites 0.1% of the total rhodopsin would lead to a highly artificial situation in which the concentrations of R* and rhodopsin kinase should be about equal; (2) at fewer than 100 R* rod, so little of the pools of transducin and PDE is activated that one can make the valid approximation that these pools are essentially constant. This straightforward yet crucial approximation has greatly facilitated our efforts to devise and analyze kinetic models for the deactivation of the visual cascade.[7,10]

Acknowledgments

The author thanks Marc Chabre and the Centre National de la Recherche Scientifique (CNRS) for support during this work.

[16] Use of α-Toxin-Permeabilized Photoreceptors in *in Vitro* Phototransduction Studies

By ANNIE E. OTTO-BRUC, J. PRESTON VAN HOOSER, and ROBERT N. FARISS

Introduction

In vitro assays utilizing purified retinal proteins have been a cornerstone of biochemical studies of phototransduction. However, many studies have relied on purification and assay techniques that destroy photoreceptor integrity, and consequently, require the evaluation of dynamic enzymatic interactions under nonphysiological conditions. This increases the potential for experimental artifact arising from alterations in the relative concentration of critical components of the cascade or their inactivation during sample preparation. As a result, *in vivo* and *in vitro* studies of phototransduction have sometimes yielded substantially different results.[1,2] These discrepancies pose a critical challenge for those investigating regulatory mechanisms of phototransduction: how to design biochemical assays that preserve or replicate the complex intracellular milieu of living photoreceptors?

In vitro study of phototransduction, using intact photoreceptors, is an appealing alternative to conventional biochemical techniques. This strategy preserves conditions for enzymatic interactions that more closely resemble those found *in vivo*. However, this goal is technically challenging, because the intracellular environment of intact photoreceptors is not readily accessible to manipulation or analysis. To access the intracellular environment, the plasma membrane must be permeabilized, but in a way that does not adversely affect the normal biochemical interactions within photoreceptors. A variety of permeabilization strategies have been developed to overcome the physical barrier imposed by the plasma membrane. Among the most widely used techniques are electroporation, liposome-mediated transfer, detergent solubilization, microinjection, toxin-mediated poration, and biolistics.[3]

One of these techniques, toxin-mediated poration, has been used in a growing number of *in vitro* studies in both whole cells and membrane-bound organelles.[4–6] We have found this technique valuable in our own studies of rhodopsin phosphorylation in intact photoreceptors.[7] In our studies, the bacterial pore-forming toxin, α-toxin (staphlococcal α-hemolysin), was used to permeabilze the plasma membrane of photoreceptors. Although α-toxin (33.4 kDa), in its monomeric form, is soluble in an aqueous solution, these monomers self-assemble into heptameric pores that span the plasma membrane of the cell. With an internal diameter of 14 Å, these pores are size selective, allowing the diffusion of small molecules, ≤ 2 kDa (e.g., metal ions, nucleotides, simple sugars), across the plasma membrane, while preventing the diffusion of larger molecules (i.e., proteins) into or out of the cell (Fig. 1).[8]

---

[1] H. Ohguro, M. Rudnicka-Nawrot, J. Buczylko, X. Zhao, J. A. Taylor, K. A. Walsh, and K. Palczewski, *J. Biol. Chem.* **271,** 5215 (1996).

[2] H. Ohguro, J. P. Van Hooser, A. H. Milam, and K. Palczewski, *J. Biol. Chem.* **270,** 14259 (1995).

[3] I. Hapala, *Crit. Rev. Biotechnol.* **17,** 105 (1997).

[4] H. Bayley, *Sci. Am.* **277,** 62 (1997).

[5] S. Bhakdi, H. Bayley, A. Valeva, I. Walev, B. Walker, M. Kehoe, and M. Palmer, *Arch. Microbiol.* **165,** 73 (1996).

[6] S. Bhakdi and J. Tranum-Jensen, *Microbiol. Rev.* **55,** 733 (1991).

[7] A. E. Otto-Bruc, R. N. Fariss, J. P. Van Hooser, and K. Palczewski, *Proc. Natl. Acad. Sci. U.S.A.* **95,** 15014 (1998).

[8] L. Song, M. R. Hobaugh, C. Shustak, S. Cheley, H. Bayley, and J. E. Gouaux, *Science* **274,** 1859 (1996).

The utility of the α-toxin as a permeabilizing agent has been substantially enhanced through the engineering of gateable pores composed of mutant α-toxin proteins. A mutant α-toxin protein containing five consecutive histidine residues (residues 130–134), designated H5K8A, has been developed.[9] At micromolar concentrations, Zn^{2+} binds to these histidine residues, blocking movement of small molecules through the pores. Reduction of the free Zn^{2+} concentration, through addition of chelating agents such as EDTA, allows the pores to return to an unblocked state. The reversibility of these Zn^{2+}-sensitive pores can be exploited by loading cells with low molecular weight compounds (radiolabeled nucleotides, amino acids, aqueous dyes) while the pore is open, and then trapping these compounds inside the cells by closing the pores with Zn^{2+}.

Methods

Expression and Purification of α-Toxin

Two mutations are introduced in the α-toxin of *Staphylococcus aureus*: K8A and the replacement of amino acids 130–134 by five histidines.[9] The K8A mutation in α-toxin does not affect pore formation but increases the expression level of the protein in *Escherichia coli* by reducing proteolysis. The five histidines replacing amino acids 130–134, H5-K8A, allow a reversible closure of the pore by divalent zinc ions. *Escherichia coli* JM109 (DE3) transformed with pT7-Sf1A-H5 (H5-K8A mutant of α-toxin) are grown for 8 hr in 10 ml of LB/amp [Luria–Bertani medium containing ampicillin (100 μg/ml)] at 37°. This culture is then used to inoculate 2 liters of LB/amp. For expression, the cells are grown at 30° overnight to an OD_{570} of 1.3–1.4. The cells are centrifuged at 5000g for 20 min at 4°. The pellet is resuspended in 50 mM Tris, pH 8.0, containing 150 mM NaCl and 10 mM imidazole, and then sonicated and centrifuged at 125,000g for 60 min at 4°. The supernatant is further purified on Ni($^{2+}$)-NTA agarose (Qiagen, Chatsworth, CA) according to Walker and Bayley.[10] The protein is eluted with 100 mM imidazole and dialyzed against 10 mM HEPES, pH 7.5, containing 100 mM NaCl. Typically, 5–6 mg of α-toxin is obtained.

Assay for Rhodopsin Phosphorylation

Freshly enucleated bovine eyes are obtained from Schenk Packing Company (Stanwood, WA). Eyes are kept on ice in a light-tight container until

---

[9] B. Walker, J. Kasianowicz, M. Krishnasastry, and H. Bayley, *Protein Eng.* **7**, 655 (1994).
[10] B. Walker and H. Bayley, *Protein Eng.* **7**, 91 (1994).

dissection under dim red light. The anterior segment and vitreous are removed and the posterior portion of the eye, containing the neural retina, is gently inverted. A trephine punch (7.5 mm in diameter) is used to excise uniform pieces of retina. These retinal pieces are then transferred with fine-tipped forceps to petri dishes containing 0.5 ml of an isosmotic buffer [10 mM HEPES (pH 7.5), containing 120 mM NaCl, 3.5 mM KCl, 10 mM glucose, 0.2 mM CaCl$_2$, 0.2 mM MgCl$_2$, and 1 mM EDTA] containing 1 mM GTP and 1 mM NADPH. Typically, 50 to 100 μg of α-toxin (H5-K8A) per milliliter is added to the isosmotic buffer. Retina punches (three per condition) are incubated for 30 to 45 min at room temperature to allow pore formation. In control experiments, the α-toxin is omitted. Next, 1 mM [γ-^{32}P]ATP (40,000 cpm/nmol) is added (in some experiments 2–4 mM ZnSO$_4$ is added 5 min before or after addition of ATP). After 5 min, the punches are exposed to a single light flash (SunPak Thyristor 433D; SunPak Division, Hackensack, NJ) at half-power. The flash unit is positioned 15 cm from the punches and the light flash is 2.5 msec in duration. Under these conditions, approximately 10% of total rhodopsin is bleached. After an additional 5 to 10 min, the phosphorylation reaction is stopped by freezing the punches in dry ice–ethanol. For analysis, punches are homogenized in 1 ml of 10 mM HEPES, pH 7.5, containing 3 mM dodecyl-β-D-maltoside, 1 mM adenosine, 4 mM EDTA, 15 mM KF, and 10 mM ATP, and incubated at room temperature for 30 min, mixing gently. After centrifugation for 20 min at 15,400g and 4°, the supernatant is mixed with an anti-Rho monoclonal antibody (MAb) (C7)–Sepharose. Unbound proteins are removed and MAb (C7)–Sepharose is washed with 10 mM HEPES, pH 7.5, containing 0.4 mM dodecyl-β-D-maltoside. Bound proteins are eluted with 0.1 M glycine, pH 2.5, containing 0.4 mM dodecyl-β-D-maltoside. Eluted proteins are analyzed by sodium dodecyl sulfate–polyacrylamide gel electrophoresis (SDS–PAGE) and autoradiography (XOMAT-blue film; Eastman Kodak, Rochester, NY). In most experiments, the amount of rhodopsin loaded for SDS–PAGE is 200 pmol and the intensity of radiolabeling is about 5 pmol of phosphate, consistent with a bleaching of ~10% of rhodopsin (Rho).

Antibodies

Rabbit anti-α-toxin polyclonal antibodies (UW60) are raised in New Zealand White rabbits by subcutaneous immunization with ~50 μg of bacterially expressed α-toxin (in ~50 μl) mixed with an equal volume of Freund's complete adjuvant (Cocalico Biologicals, Reamstown, PA). Animals are given booster injections at 1- to 2-week intervals with 25 μg of α-toxin mixed with Freund's incomplete adjuvant. For Western blot

FIGURE 1

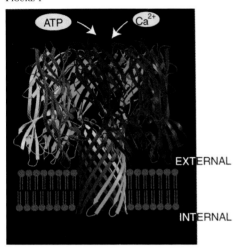

FIGURE 4

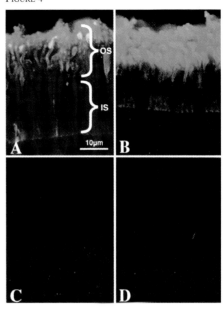

FIG. 1. Formation of a switchable pore in the plasma membrane of rod photoreceptors. α-Toxin of *Staphylococcus aureus* is a self-assembling, pore-forming protein. Heptameric α-toxin pores are inserted in the plasma membrane of rod cells, allowing nucleotide and ion exchange between intra- and extracellular compartments. [Reprinted with permission from L. Song, M. R. Hobaugh, C. Shustak, S. Cheley, H. Bayley, and J. E. Gouaux, *Science* **274,** 1859–1866. Copyright 1996 American Association for the Advancement of Science.]

FIG. 4. Permeability of rod plasma membranes containing the Zn^{2+}-sensitive form of the α-toxin pore (H5-K8A), as demonstrated by confocal fluorescence localization of a low molecular weight intracellular tracer. Retinal punches were incubated in the presence (A and B) or absence (C and D) of α-toxin. The addition of Zn^{2+} to the incubation media caused pore closure (B and D). The low molecular weight intracellular tracer N-(2-aminoethyl)biotinamide hydrochloride (neurobiotin) was added to all samples to assess permeability of α-toxin pores under various conditions. An α-toxin-specific PAb was used to immunolocalize this protein (green). α-Toxin was restricted to the plasma membrane surrounding rod inner segments (IS) and outer segments (OS) of treated retinas (A and B). Streptavidin–Cy3 was used to localize neurobiotin in these samples (red). Neurobiotin was present in the cytoplasm of rod OS and IS (A), confirming that α-toxin forms functional pores in these cells. Neurobiotin was prevented from diffusing through α-toxin pores closed by Zn^{2+} (B). [Reprinted with permission from A. E. Otto-Bruc, R. N. Fariss, J. P. Van-Hooser, and K. Palczewski, *Proc. Natl. Acad. Sci. U.S.A.* **95,** 15014–15019. Copyright 1998, National Academy of Sciences, U.S.A.]

analysis, monoclonal antibodies to rhodopsin kinase (G8), arrestin (S65-34; a gift from G. Adamus), and rhodopsin (4D2; a gift from R. S. Molday) and a polyclonal antibody to recoverin (asp; a gift from A. S. Polans) are diluted 1:5000 or 1:10,000.

Confocal Immunofluorescence Localization of α-Toxin and Neurobiotin on Sections of Bovine Retina

Trephine punches of neural retina collected from freshly enucleated bovine eyes are transferred to isosmotic buffer [10 mM HEPES (pH 7.5), containing 120 mM NaCl, 3.5 mM KCl, 10 mM glucose, 0.2 mM CaCl$_2$, 0.2 mM MgCl$_2$, and 1 mM EDTA] at room temperature. Purified α-toxin (H5-K8A) is added to selected samples (50 μg/ml final concentration) and mixed gently. After 20 min, ZnSO$_4$ (4 mM final concentration) is added to selected samples. After a 10-min incubation, the intracellular tracer N-(2-aminoethyl)biotinamide hydrochloride (neurobiotin; Vector Laboratories, Burlingame, CA) is added to all samples (4 mM final concentration). Samples are incubated in solution containing neurobiotin for 30 min. ZnSO$_4$ is then added to all samples (4 mM final concentration), which are then rinsed twice (20 min each) with gentle agitation in isosmotic buffer containing 4 mM Zn^{2+}. Retinal punches are rinsed briefly in 140 mM phosphate buffer, pH 7.2, and fixed in 4% paraformaldehyde in phosphate buffer overnight at 4°. For confocal immunofluorescence, punches are rinsed three times (15 min each) in 140 mM phosphate buffer, embedded in 5% low gelling temperature agarose (Sigma; St. Louis, MO), and sectioned in 100-μm slices on a VT1000E vibrating microtome (Leica GmbH, Bensheim, Germany). Sections are incubated in ICC buffer [phosphate-buffered saline (PBS, pH 7.2), 0.5% bovine serum albumin, 0.2% Triton X-100, 0.05% sodium azide] containing 5% normal goat serum for 20 min, to reduce nonspecific immunolabeling.[11] Sections are incubated overnight at 4° in ICC buffer containing an affinity-purified rabbit polyclonal antibody to α-toxin (UW60; 1:250 dilution). After repeated washing in ICC buffer, Cy2-conjugated goat anti-rabbit antibody (Jackson ImmunoResearch, West Grove, PA) (diluted 1:200 in ICC buffer) and Cy3-conjugated streptavidin (Jackson ImmunoResearch) (diluted 1:200 in ICC buffer) are added to sections of retina and incubated overnight at 4° in a foil-wrapped, humidified container. Sections are washed thoroughly in ICC buffer, mounted in 5% n-propyl gallate in glycerol (to retard photobleaching), and coverslipped. Sections are examined on a Bio-Rad (Hercules, CA) MRC-600 laser scanning confocal microscope configured for dual-channel labeling. Gain and black level

---

[11] I. L. Hale and B. Matsumoto, *Methods Cell Biol.* **38,** 289 (1993).

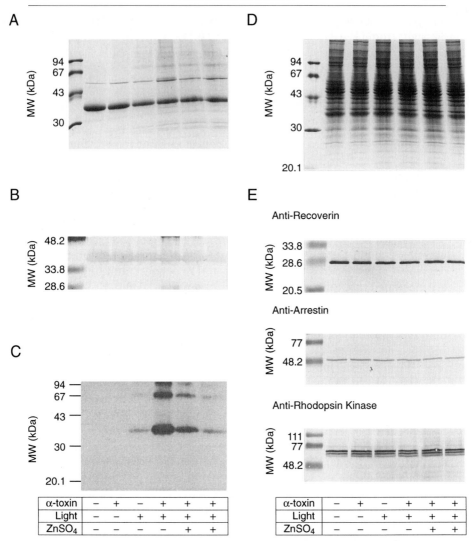

FIG. 2. Assay for Rho* phosphorylation in α-toxin-permeabilized retina. (A) SDS–PAGE. Retina punches (7.5 mm in diameter) were incubated under various experimental conditions. Punches were solubilized and Rho was purified by immunoaffinity, quantified by the Bradford method, and the same amounts of protein were subjected to 12% SDS–PAGE. The amounts of Rho were comparable under all conditions. (B) Western blot. A second MAb against Rho (4D2) identified the major protein in these samples as Rho. (C) Autoradiogram of the SDS–polyacrylamide gel. Samples kept in the dark showed no phosphorylation in the presence or absence of α-toxin (lanes 1 and 2). In samples exposed to light, an increase in ^{32}P incorporation in Rho* in retina permeabilized with α-toxin (100 μg/ml) is clearly visible (lanes 4 and 5) compared with nonpermeabilized retina (lane 3). When Zn^{2+} was added after [γ-^{32}P]ATP,

scan head controls are standardized before images are collected. Unprocessed digital images are converted to PICT format, using Photoshop 3.0.

Results and Discussion

An assay for rhodopsin phosphorylation was used to demonstrate α-toxin permeabilization of intact bovine photoreceptors. The α-toxin pores facilitate loading of the photoreceptors with [γ-^{32}P]ATP, allowing subsequent quantification of ^{32}P-labeled Rho after flash illumination. SDS–PAGE (Fig. 2D) and Western blots (Fig. 2E) of solubilized retina punches demonstrate that α-toxin-permeabilized photoreceptors retain soluble (arrestin) and membrane-associated proteins (recoverin, RK) at levels comparable to untreated retinal punches. Protein content within the photoreceptors is unaffected by permeabilization. These results are consistent with the restrictive size of the α-toxin pores. These pores are known to support transmembrane diffusion of small molecules while preventing diffusion of molecules greater than 2 kDa. After solubilization of the retinal punches, Rho was rapidly isolated by affinity chromatography in the presence of protein kinase and phosphatase inhibitors to prevent subsequent changes in phosphorylation levels. SDS–PAGE (Fig. 2A) and Western blots (Fig. 2B) demonstrate that affinity purification successfully isolated rhodopsin from these retinal punches. Autoradiography of ^{32}P-labeled Rho from a variety of incubation conditions confirms that Rho* phosphorylation is a light-dependent process. Furthermore, H5-K8A α-toxin pores in an open, Zn^{2+}-free state are necessary for ^{32}P incorporation in Rho. Preparations in which α-toxin was omitted, or in which the pore was blocked by Zn^{2+}, before addition of [γ-^{32}P]ATP, showed minimal levels of ^{32}P incorporation.

In our assays, the minimal Zn^{2+} concentration required to efficiently block the H5-K8A pores was 2 mM.

Incubation of retinal punches in α-toxin for 30 min was required to optimize pore-mediated diffusion of [γ-^{32}P]ATP into permeabilized photoreceptors (Fig. 3A). The efficiency of this process is influenced, in part, by the presence of other neuronal and glial cells in the punches, that, like photoreceptors, incorporate α-toxin in their plasma membranes. Dissocia-

---

an increase in ^{32}P incorporation in Rho was observed (lane 5). In contrast, the addition of Zn^{2+} before [γ-^{32}P]ATP diminished ^{32}P incorporation substantially (lane 6). (D and E) SDS–PAGE and Western blots of solubilized retinal punches. Before Rho purification, immunoblots of solubilized retina punches were labeled with antibodies against several retinal proteins (recoverin, arrestin, and rhodopsin kinase). α-Toxin permeabilization is required for the light-dependent phosphorylation, and permeabilized retina retains protein >20 kDa.

A

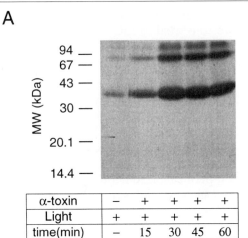

α-toxin	−	+	+	+	+
Light	+	+	+	+	+
time(min)	−	15	30	45	60

B

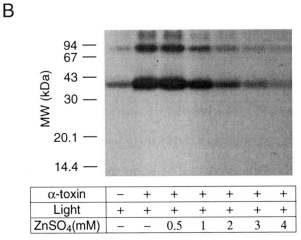

α-toxin	−	+	+	+	+	+	+
Light	+	+	+	+	+	+	+
ZnSO$_4$(mM)	−	−	0.5	1	2	3	4

FIG. 3. Optimal conditions for α-toxin incubation and Zn^{2+} closure of the pore. (A) Retina punches were incubated in the presence of α-toxin for various times. Phosphorylation of Rho* occurred as described in Methods. SDS–PAGE and autoradiography demonstrated optimal Rho* phosphorylation after a 30-min incubation with α-toxin. (B) Varying concentrations of ZnSO$_4$ were used to block the pore prior to the addition of [γ-^{32}P]ATP. ZnSO$_4$ concentrations ≥2 mM significantly inhibit the [γ-^{32}P]ATP diffusion through α-toxin pores and, therefore, the level of ^{32}P incorporation.

tion techniques could be used to isolate purer populations of photoreceptors, prior to α-toxin incubation. However, these techniques often damage the fragile structure of photoreceptors.

α-Toxin mediates bacterial virulence of *Staphylococcus aureus* and has been shown to induce necrotic and apoptotic cell death under *in vivo*[12] and *in vitro*[13–15] conditions. We have examined α-toxin-treated bovine retina by light and electron microscopy, looking for conventional histological indicators of these distinct forms of cell death. In photoreceptors exposed to α-toxin for 30 min, there is little evidence for increased necrotic (cellular swelling and lysis) or apoptotic cell death (chromatin condensation, nuclear fragmentation, cytoplasmic blebbing), compared with nontreated controls. The susceptibility of α-toxin-treated cells to severe damage or death should be influenced by a number of *in vitro* factors such as postmortem interval, α-toxin concentration and incubation time, buffer composition, and temperature. In physiological assays using α-toxin, these factors must be evaluated to optimize cell viability *in vitro*.

Immunofluorescent labeling and confocal microscopy were used to demonstrate the subcellular distribution and permeability of Zn^{2+}-sensitive (H5-K8A) α-toxin pores formed in bovine photoreceptors. Photoreceptors incubated in buffer containing α-toxin (Fig. 4A and B) were immunolabeled with an affinity-purified antibody for this pore-forming protein. α-Toxin immunolabeling (green) is heaviest in the plasma membrane surrounding outer segments (OS), with lighter labeling of the plasma membrane surrounding the rod inner segments (IS). No immunolabeling was observed in photoreceptors incubated in buffer from which the α-toxin was omitted (Fig. 4C and D). Neurobiotin, a low molecular weight compound used widely in intracellular labeling studies, was employed to evaluate the permeability of α-toxin pores in the presence or absence of Zn^{2+}ions. Neurobiotin (red; labeled with streptavidin–Cy3) is abundant throughout the cytoplasm of photoreceptor OS and IS that contain the pore in an open, Zn^{2+}-free state (Fig. 4A). Pretreatment of α-toxin-permeabilized photoreceptors with Zn^{2+}-containing buffer (4 mM) triggers the closure of these pores, thereby blocking diffusion of neurobiotin into the photoreceptor cytoplasm (Fig. 4B). Exclusion of neurobiotin from photoreceptors without α-toxin treatment (Fig. 4C and D) or retinas containing the α-toxin pore in a closed

---

[12] J. M. Moreau, G. D. Sloop, L. S. Engel, J. M. Hill, and R. J. O'Callaghan, *Curr. Eye Res.* **16,** 1221 (1997).

[13] D. Jonas, I. Walev, T. Berger, M. Liebetrau, M. Palmer, and S. Bhakdi, *Infect. Immun.* **62,** 1304 (1994).

[14] R. C. Duke, R. Z. Witter, P. B. Nash, J. D. Young, and D. M. Ojcius, *FASEB J.* **8,** 237 (1994).

[15] I. Walev, E. Martin, D. Jonas, M. Mohamadzadeh. W. Muller-Klieser, L. Kunz, and S. Bhakdi, *Infect. Immun.* **61,** 4972 (1993).

state (Fig. 4B) confirm that the permeability of this pore can be easily manipulated and that the treatment and incubation conditions alone do not adversely affect the integrity of the plasma membrane surrounding these cells.

Conclusion

α-Toxin-permeabilized photoreceptors are a valuable tool for studying phototransduction. The α-toxin-treated bovine retina provides a unique experimental approach to study phototransduction. α-Toxin, and other pore-forming cytotoxins, offer several distinct advantages over other permeabilization techniques. In contrast to electroporation and detergent solubilization, α-toxin permeabilization offers a far greater degree of control over the rate and magnitude of transmembrane diffusion.[16] In addition, α-toxin-mediated diffusion is bidirectional and gateable. This technique is rigidly size selective and allows introduction of small metabolites into the minimally compromised photoreceptors. Our studies have focused on phosphorylation of light-activated rhodopsin; however, this technique could be applied to the study of any step in the phototransduction cascade for which radioactive tracers or other analytical tools are available.

Acknowledgments

We thank Dr. H. Bayley for the α-toxin clones, Grace P. Yang for technical assistance, and Dr. Krzysztof Palczewski for guidance during these studies. Confocal microscopy was conducted with the assistance of Paulette Brunner and the Keck Center for Advanced Studies of Neural Signaling at the University of Washington. This research was supported by NIH Grants EY08061 and EY06935 (R.N.F.) and an award from Research to Prevent Blindness, Inc., to the Department of Ophthalmology at the University of Washington.

[16] J. C. Weaver, *J. Cell Biochem.* **51,** 426 (1993).

[17] Photoreceptors in Pineal Gland and Brain: Cloning, Localization, and Overexpression

By TOSHIYUKI OKANO and YOSHITAKA FUKADA

Introduction

Photoreceptors in extraretinal tissues such as pineal gland, deep brain, skin, and iris are involved in a variety of physiological functions other than vision: photoentrainment of the circadian rhythm, photoperiodicity, photoregulation of the body color, etc. Generally, the number of photoreceptor cells in these tissues is much smaller than that of rod and cone visual cells in the retina, and this has hampered studies on the extraretinal photoreception mechanism at a molecular level. Pinopsin, a chicken pineal photoreceptive molecule, was the first example of extraretinal opsins.[1] Several research groups including ours have identified other novel extraretinal opsins such as deep brain rhodopsin,[2] parapinopsin,[3] melanopsin,[4] and VA opsin.[5] In this chapter, we summarize strategies for cloning, localization, and overexpression of photoreceptive molecules present in the vertebrate extraretinal tissues.

cDNA or Genomic DNA Cloning

An important basis for the cloning of a novel extraretinal opsin gene is the amino acid identity among retinal and extraretinal opsins (Fig. 1). Vertebrate-type opsins form a large superfamily as reflected in the phylogenetic tree, and they are classified into seven groups to date; S, M1, M2 (including rhodopsins), L, P, PP, and VA. Amino acid identities among opsins clustered in the same group are higher than 65%, and the identities among opsins in different groups are also relatively high (40–50%). This enables identification and gene cloning of novel opsins. Importantly, the members of group S, M1, M2, or L have similar color sensitivities (absorption maxima) in the UV–violet, blue, green, and yellow–red regions, respectively. Melanopsin, which was cloned from the *Xenopus* melanophore,[4] is more closely related to invertebrate-type opsins than to vertebrate-type opsins (Fig. 1), suggesting that some vertebrates have invertebrate-type opsin gene(s) as well. In addition, a novel invertebrate opsin, SCOP2, which was cloned from scallop ocular tissue,[6] is equally related to vertebrate-type and invertebrate-type opsins (Fig. 1). Taking these observations into consideration, we can speculate that a novel opsin can be classified as one of the following cases:

A. A vertebrate-type opsin included in one of the seven groups (Fig. 1: boxed A)

---

[1] T. Okano, T. Yoshizawa, and Y. Fukada, *Nature (London)* **372**, 94 (1994).
[2] Y. Wada, T. Okano, A. Adachi, S. Ebihara, and Y. Fukada, *FEBS Lett.* **424**, 53 (1998).
[3] S. Blackshaw and S. H. Snyder *J. Neurosci.* **17**, 8083 (1997).
[4] I. Provencio, G. Jiang, W. J. DeGrip, W. P. Hayes, and M. D. Rollag, *Proc. Natl. Acad. Sci. U.S.A.* **95**, 340 (1998).
[5] B. G. Soni, A. R. Philp, R. G. Foster, and B. E. Knox, *Nature (London)* **394**, 27 (1998).
[6] D. Kojima, A. Terakita, T. Ishikawa, Y. Tsukahara, A. Maeda, and Y. Shichida, *J. Biol. Chem.* **272**, 22979 (1997).

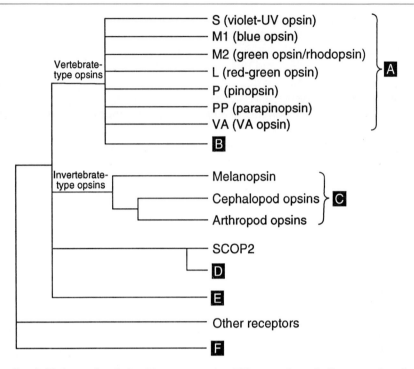

FIG. 1. Phylogenetic relationship among opsins. This tree schematically summarizes the phylogenetic relationship among opsins cloned from retinal and extraretinal tissues. For details on cases A–F, see text.

B. A vertebrate-type opsin forming a new group that is equally related to each of the seven groups (Fig. 1: boxed B)
C. An opsin closely related to melanopsin and invertebrate-type opsins (Fig. 1: boxed C)
D. An opsin closely related to SCOP2 (Fig. 1: boxed D)
E. An opsin equally related to vertebrate-type, invertebrate-type, and SCOP2 opsin (Fig. 1: boxed E)
F. A member of the G protein-coupled receptor family but not highly homologous to any of the known opsins (Fig. 1: boxed F)

When cloning a novel opsin gene, we select one of the hypotheses described above, based on the preceding immunohistochemical or physiological observations on the opsin of interest.

As an example of case A, we have cloned rhodopsin gene expressed in a small number of cerebrospinal fluid-contacting neurons in the lateral septum of the pigeon deep brain.[2] In this case, preceding studies suggested the presence of a green-sensitive opsin-like pigment in the pigeon deep brain. For polymerase chain reaction (PCR)-based cloning, we designed a pair of primers corresponding to sequences conserved among M2 opsins as follows:

```
Sense primer (M2F):      5'-ATG    AAC    GGI    ACT    GAA    GG-3'
                                   T                    G
                         N-Met¹    Asn    Gly    Thr    Glu    Gly⁶-C

Antisense primer (M2R):  5'-GT     IGC    AAA    AAA    ICC    CTC-3'
                                          G      G             T
                         C-Thr¹¹⁸  Ala    Phe    Phe    Gly    Glu¹¹³-N
```

PCR conditions: Thirty-five cycles of denaturation at 94° for 1 min, annealing at 50° for 1 min, and extension at 72° for 1 min, followed by an additional extension for 7 min at 72°. Polymerase used was Ampli-Taq or Ampli-Taq Gold (Perkin Elmer–Applied Biosystems, Foster City, CA). This reaction amplifies a 353-base pair (bp) stretch in the first exon of M2 opsin genes.

In case B, or when one is unable to predict the group to which an opsin gene of interest belongs (a variation of case A), the PCR primers are designed on the basis of the sequences highly conserved among vertebrate-type opsins. These are the loop region between the fifth and sixth transmembrane helices (TM-V and TM-VI) for the sense primer and the TM-VII region for the antisense primer[7] as follows.

Name	Sequence
	Sense primer

```
TK    5'-ACI    CAA    AAA    GCI    GAA    AAI    GAA    GTI    ACI    AAA    ATG    G-3'
         G      G              G      CG     G             T      GG
      N-Thr    Gln    Lys    Ala    Glu    Lys    Glu    Val    Thr    Lys    Met(Val)-C
                                          Arg                   Ser    Arg

TR    5'-ACI    CAA    AAA    GCI    GAA    AAI    GAA    GTI    ACI    CGI    ATG    G-3'
         G      G              G      CG     G             T
      N-Thr    Gln    Lys    Ala    Glu    Lys    Glu    Val    Thr    Arg    Met(Val)-C
                                          Arg                   Ser

SK    5'-ACI    CAA    AAA    GCI    GAA    AAI    GAA    GTI    AGT    AAA    ATG    G-3'
         G      G              G      CG     G             C             GG
      N-Thr    Gln    Lys    Ala    Glu    Lys    Glu    Val    Ser    Lys    Met(Val)-C
                                          Arg                          Arg

SR    5'-ACI    CAA    AAA    GCI    GAA    AAI    GAA    GTI    AGT    CGI    ATG    G-3'
         G      G              G      CG     G             C
      N-Thr    Gln    Lys    Ala    Glu    Lys    Glu    Val    Ser    Arg    Met(Val)-C
                                          Arg
```

[7] T. Yoshikawa, T. Okano, T. Oishi, and Y. Fukada, *FEBS Lett.* **424**, 69 (1998).

Antisense primer

R	5'-GAA	CTG	TTT CCG	GTT	CAT G	IAI	IAI	GTA A	IAT	IAT C	GGG-3' A
	C-Phe	Gln	Lys Arg Gln	Asn	Leu Met	Ile Phe Met Val Leu	Ile Phe Met Val Leu	Tyr	Ile	Ile Val	Pro-N

With the single antisense primer, each of the four sense primers (TK, TR, SK, and SR) must be tested, but one may use a mixture of the sense primers. The amplification reaction is performed under the same conditions as described above. These primers are designed to amplify a region in a single exon, so that they are applicable to both cDNA and genomic DNA templates. We have confirmed that these primers work well for genomic DNA templates from human, chicken, pigeon, quail, toad, and zebrafish. When using a cDNA template, one must pay attention to false-positive amplifications due to a small amount of genomic DNA contaminating the cDNA pool.

When one assumes case C or D, it is necessary to design new primers based on the updated sequences of opsins including melanopsin and/or SCOP2. In case E, a pair of primers used for the cloning of SCOP2 by Kojima et al.[6] is to be tested. They designed the primers on the basis of the two sequences widely conserved among vertebrate and invertebrate opsins as follows: Phe(Met/Ile/Leu/Phe)Ala(Trp/Tyr/Cys)ThrProTyr(Ala/Thr) in the TM-VI region and AsnPro(Met/Ile/Leu/Phe)(Met/Ile/Val)Tyr(Ala/Gly)(Met/Ile/Leu) in the TM-VII region.

Localization of Cloned Opsins in Extraretinal Tissues

To localize a novel opsin in extraretinal tissues, immunohistochemical analysis is preferable to *in situ* hybridization for several reasons. First, the immunohistochemical data give evidence of expression of the cloned gene. Second, the morphology of the extraretinal photoreceptive cell is usually so differentiated[2,7,8] that the functional opsin locates at specialized sites far from cell bodies. Third, compared with the protein levels, the mRNA levels

---

[8] R. Silver, P. Witkovsky, P. Horvath, V. Alones, C. J. Barnstable, and M. N. Lehmann, *Cell Tissue Res.* **253,** 189 (1988).

of rhodopsin,[9] chicken red,[10] and pinopsin[11] show diurnal changes owing to regulation by the circadian clock or the light conditions. Therefore, in sections prepared under fixed conditions, one may overlook the presence of photoreceptive cells by *in situ* hybridization.

Rod and cone visual pigments can be purified from chicken retina by several steps of column chromatography,[12] and they are used for production of specific antibodies. However, the protein levels of extraretinal opsins seem to be low. For example, a chick pineal gland contains only 2 ng of pinopsin,[11] and the opsin level in the deep brain region could be much lower. These amounts correspond to about one-millionth of the opsins in a retina. Thus recombinant fusion proteins are required for immunization or affinity purification of the resultant antibodies. For preparing fusion proteins, we have selected extreme carboxyl- or amino-terminal regions as the antigenic site because of their higher hydrophilicity and efficient expression of the recombinant fusion proteins.[2,7,13] Table I summarizes several antibodies recognizing extraretinal opsins. We usually prepare two kinds of fusion proteins for a single antigenic peptide. For example, maltose-binding protein (MBP) and glutathione S-transferase (GST) are independently fused with the same peptide, and they are used for immunization and affinity purification separately (Table I).

Overexpression

It is difficult to extract and purify a large quantity of extraretinal opsins from native tissues, so that characterization of novel opsins requires functional expression in the cultured cells. COS and 293S cells have been used for overexpression of rhodopsin and cone visual pigments so far,[14-16] and these original systems were slightly modified for functional expression of pinopsin.[1] We have used two combinations of cells and vectors. One combination is 293EBNA cells and the pREP vector, and the other is 293S cells and the pUSRα vector, both of which give similar yields of pinopsin

---

[9] J. I. Korenbrot and R. D. Fernald, *Nature* (*London*) **337,** 454 (1989).
[10] M. E. Pierce, H. Sheshberadaran, Z. Zhang, L. E. Fox, M. L. Applebury, and J. S. Takahashi, *Neuron* **10,** 579 (1993).
[11] Y. Takanaka, T. Okano, M. Iigo, and Y. Fukada, *J. Neurochem.* **70,** 908 (1998).
[12] T. Okano, Y. Fukada, I. D. Artamonov, and T. Yoshizawa, *Biochemistry* **28,** 8848 (1989).
[13] T. Okano, Y. Takanaka, A. Nakamura, K. Hirunagi, A. Adachi, S. Ebihara, and Y. Fukada, *Mol. Brain Res.* **50,** 190 (1997).
[14] J. Nathans, *Biochemistry* **29,** 937 (1990).
[15] S. L. Merbs and J. Nathans, *Nature* (*London*) **356,** 433 (1992).
[16] D. D. Oprian, *Methods Neurosci.* **15,** 301 (1993).

TABLE I
ANTIBODIES RECOGNIZING EXTRARETINAL OPSINS

Antibody	Immunized antigen	Affinity purification	Specificity
RhoN[a]	Chicken rhodopsin[b]	Purified by MBP–rhodopsin (N terminus)[a,c]	Rhodopsin (N terminus)
RhoC	Chicken rhodopsin[b]	Purified by MBP–rhodopsin (C terminus)[d]	Rhodopsin (C terminus)
P7[e]	MBP-p7[f]	Removal of anti-MBP antibodies[e,g]	Chick pinopsin (C terminus)
P9[e]	MBP-p9[h]	Removal of anti-MBP antibodies[e,g]	Chick pinopsin (C terminus)
tPinC[j]	MBP-PinC[j]	Purified by GST–PinC[i,k]	Toad pinopsin (C terminus)

[a] Y. Wada, T. Okano, A. Adachi, S. Ebihara, and Y. Fukada, *FEBS Lett.* **424**, 53 (1998).
[b] Purified from the chicken retina; T. Okano, Y. Fukada, I. D. Artamonov, and T. Yoshizawa, *Biochemistry* **28**, 8848 (1989).
[c] Affinity purified by a fusion protein composed of maltose-binding protein (MBP) and the N-terminal region (Met^1–Ala^{29}) of chicken rhodopsin.
[d] Affinity purified by a fusion protein composed of MBP and the C-terminal region (Ser^{298}–Ala^{351}) of chicken rhodopsin.
[e] T. Okano, Y. Takanaka, A. Nakamura, K. Hirunagi, A. Adachi, S. Ebihara, and Y. Fukada, *Mol. Brain Res.* **50**, 190 (1997).
[f] A fusion protein composed of MBP and the C-terminal region (Pro^{277}–Val^{362}) of chicken pinopsin.
[g] MBP–β-galactosidase fusion protein-immobilized column was used to remove anti-MBP antibodies.
[h] A fusion protein composed of MBP and the C-terminal region (Ile^{297}–Val^{362}) of chicken pinopsin.
[i] T. Yoshikawa, T. Okano, T. Oishi, and Y. Fukada, *FEBS Lett.* **424**, 69 (1998).
[j] A fusion protein composed of MBP and the N-terminal region (Gly^{314}–Ala^{346}) of toad pinopsin.
[k] A fusion protein composed of glutathione *S*-transferase (GST) and the N-terminal region (Gly^{314}–Ala^{346}) of toad pinopsin.

(approximately 2 μg of protein per 100-mm dish). We described below the construction of the expression plasmid and the method used to purify expressed pinopsin.

Construction of Expression Plasmid

An expression vector, pUSRα,[17] (kindly provided by F. Tokunaga of Osaka University), has been used for functional expression of rhodopsin

---

[17] S. Kayada, O. Hisatomi, and F. Tokunaga, *Comp. Biochem. Physiol.* **110B**, 599 (1995).

and cone visual pigments. This vector is modified from pUC-SRα[18] to have unique restriction sites of ClaI, HindIII, EcoRV, and EcoRI downstream of the SRα promoter.

The nucleotide sequence of pinopsin cDNA has been modified as shown in Fig. 2. Six histidine residues and a site for cleavage by factor Xa have been introduced at the amino-terminal region of pinopsin for affinity purification and for removal of the histidine tag. The factor Xa-site may be eliminated because several trials resulted in no efficient cleavage by factor Xa at the site. To the 5' end of the initiation ATG, an 11-bp stretch of chicken rhodopsin 5' untranslated sequence has been added for a possible enhancement of pinopsin expression. The pinopsin cDNA thus modified is inserted into HindIII–EcoRI sites of pUSRα to yield an expression plasmid termed pPNH (Fig. 3).

Growth of 293S Cells

The 293S cells are grown on 100-mm diameter plastic culture dishes in a CO_2 incubator at 37° in an atmosphere containing 5% CO_2. The medium is a mixture of 50% Dulbecco's modified Eagle's medium and 50% Ham's F12 (GIBCO-BRL, Rockville, MD), and it is supplemented with D-glucose (4.5 g/liter), streptomycin (100 mg/liter), penicillin (100,000 U/liter), and 10% (v/v) fetal calf serum (GIBCO-BRL). In a schedule for routine growth of the cells, each confluent plate is divided into four dishes with fresh medium. When replating, the cells in a dish are washed once with PBS (7.1 mM sodium phosphate, 136 mM NaCl, 4 mM KCl; pH 7.3) and treated with PBS (0.5 ml) containing 0.025% trypsin and 0.1% EDTA at 37° for 1–2 min. The cells are then suspended in 4 ml of the fresh culture medium, and each 1-ml suspension is plated onto a new dish supplemented with 9 ml of the medium.

Buffers

    Buffer P-10: 50 mM HEPES–NaOH (pH 6.6), 10 mM NaCl, leupeptin [50 kallikrein inhibitor units (KIU)/ml], aprotinin (4 μg/ml)
    Buffer P-130: 50 mM HEPES–NaOH (pH 6.6), 130 mM NaCl, leupeptin (50 KIU/ml), aprotinin (4 μg/ml)
    Buffer P-140: 50 mM HEPES–NaOH (pH 6.6), 140 mM NaCl, leupeptin (50 KIU/ml), aprotinin (4 μg/ml)

---

[18] Y. Takebe, M. Seiki, J. Fujisawa, P. Hoy, K. Yokota, K. Arai, M. Yoshida, and N. Arai, *Mol. Cell. Biol.* **8,** 466 (1988).

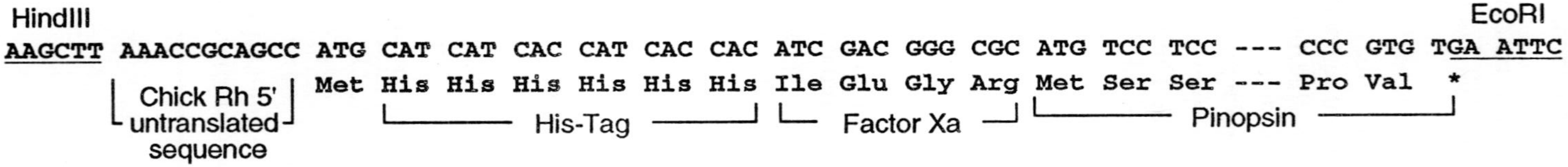

HindIII   EcoRI
AAGCTT AAACCGCAGCC ATG CAT CAT CAC CAT CAC ATC GAC GGG CGC ATG TCC TCC --- CCC GTG TGA ATTC
       └─Chick Rh 5'─┘ Met His His His His His His Ile Glu Gly Arg Met Ser Ser --- Pro Val *
         untranslated └────────His-Tag────────┘   └─Factor Xa─┘ └────────Pinopsin────────┘
           sequence

FIG. 2. Modification of pinopsin cDNA. A pinopsin cDNA was modified so as to encode pinopsin fused with a histidine tag and a factor Xa cleavage site at the N terminus.

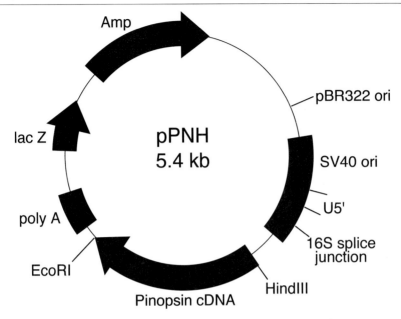

FIG. 3. pUSRα-derived plasmid pPNH for pinopsin expression.

Transfection Procedures

1. Twenty-four hours before transfection, plate cells at a density of 4×10^6 cells per 100-mm dish. A higher density of the cells results in a dramatic decrease in the yield of pinopsin.

2. For every three plates, prepare a DNA mixture containing 30 μg (10 μg per plate) of pPNH and 1.5 μg (0.5 μg per plate) of pRSV-Tag [an expression plasmid of simian virus 40 (SV40) large T antigen essential for replication of pPNH in 293 cells] in a glass tube, and adjust the volume to 1320 μl with 0.1× TE (10 mM Tris-HCl, 1 mM EDTA; pH 8.0). Add 30 μl of 70 mM Na$_2$HPO$_4$ and 1500 μl of 2× transfection cocktail (42 mM Hepes–NaOH, 290 mM NaCl; pH 7.1) to the DNA mixture.

3. Slowly add 150 μl of 2.5 M CaCl$_2$ aqueous solution to the DNA cocktail with gentle mixing. Allow the mixture to stand at room temperature for 30 min.

4. Add 1 ml of the DNA cocktail gently to the cultured cells (with 10 ml of medium) in a dish, and mix moderately.

5. Incubate the cells in the CO$_2$ incubator for 40 hr.

6. After incubation, suspend the cells in the medium by pipetting, and collect them by centrifugation at 2700g for 10 min at 4°.

7. Resuspend the pellet in 200 ml of buffer P-10 and centrifuge again. Store the pellet at −80° until use.

Extraction of Membrane Proteins

All the procedures are performed under dim red light (>660 nm).

1. Suspend several stocks of the cell pellet with 25 ml of buffer P-10 containing 25 μM 11-cis-retinal, and incubate the suspension at room temperature (15–25°) for 1 hr in order to reconstitute pinopsin with 11-cis-retinal. The concentration of 11-cis-retinal in a stock solution (in ethanol) should be higher than 2.5 mM so that the final concentration of ethanol in the working solution becomes lower than 1%. The retinal stock solution is stored at −80°.

2. Mix the suspension with 25 ml of buffer P-10 containing 2% dodecyl-β-D-maltoside (DM), and solubilize membrane proteins by pipetting. We have used a mixture of 0.75% 3-[(3-cholamidopropyl)dimethylammonio]-1-propane sulfonate (CHAPS) and egg yolk phosphatidylcholine (0.8 mg/ml) (CHAPS-PC) for solubilization of pinopsin[1]; pinopsin is stable in both detergent solutions, but we have not confirmed that the purification procedures described below are applicable to the CHAPS-solubilized pinopsin.

3. Centrifuge the suspension at 125,000g for 30 min at 4°, and collect the supernatant.

Subsequent chromatographic procedures are performed at room temperature under dim red light, and the buffers and samples are kept on ice unless otherwise stated.

DEAE-Sepharose FF Column Chromatography

On the basis of the amino acid sequence, pinopsin is expected to be positively charged at neutral pH.[1,19] Accordingly, pinopsin is effectively purified by passing the detergent extract through a DEAE-Sepharose anion-exchange column, to which the majority of acidic or neutral proteins are adsorbed.

1. Equilibrate a DEAE-Sepharose FF column (10 mm in diameter and 50 mm in height; bed volume, 4 ml) with 20 ml of buffer P-10 containing 1% DM.

2. Load the detergent extract of the cultured cells onto the column at a flow rate of 0.6 ml/min, and collect the pass-through fraction (~50 ml).

[19] T. Okano, D. Kojima, Y. Fukada, Y. Shichida, and T. Yoshizawa, *Proc. Natl. Acad. Sci. U.S.A.* **89**, 5932 (1992).

Nickel-Charged Agarose

For affinity purification of recombinant pinopsin with a histidine tag attached to the amino terminus, we use a metal-binding resin, Probond nickel-charged agarose (InVitrogen, San Diego, CA).

1. Equilibrate in a tube 2 ml of Probond nickel-charged agarose resin three times with 10 ml of buffer P-10 containing 1% DM.
2. Add the fraction passed through the DEAE-Sepharose FF to the equilibrated resin.
3. Incubate the mixture for 12 hr at 4° with gentle mixing.
4. Pour the resin into a column (7-mm diameter). About half of the recombinant pinopsin is not adsorbed to the resin, and is recovered in the pass-through fraction.
5. Wash the column with 150 ml of buffer P-140 containing 0.02% DM.
6. Wash the column with 50 ml of buffer P-140 containing 0.02% DM and 20 mM imidazole (pH 6.6).
7. Elute proteins bound to the column with 3 ml of buffer P-140 containing 0.02% DM and 200 mM imidazole (pH 6.6).

Dialysis and Centrifugation

To remove imidazole and to decrease the concentration of NaCl, the eluate from the nickel-charged agarose is dialyzed against 1 liter of buffer P-10 containing 0.02% DM at 4° with more than three changes of the dialysis buffer. Because the sample sometimes becomes turbid, it is centrifuged at 125,000g for 30 min at 4° to collect a clear supernatant. Confirm the absence of a measurable amount of pinopsin in the pellet.

SP-Sepharose FF

Pinopsin is a basic protein that adsorbs to a SP-Sepharose cation-exchange column at pH 6.6 in the presence of 10 mM NaCl. This column is effective in increasing the purity and concentration of pinopsin.

1. Equilibrate a SP-Sepharose FF column (7 mm in diameter and 13 mm in height; bed volume, 0.5 ml) with 3 ml of buffer P-10 containing 0.02% DM.
2. Load the dialyzed sample slowly onto the column.
3. Wash the column with 10 ml of buffer P-10 containing 0.02% DM.
4. Elute pinopsin with 3 ml of buffer P-130 containing 0.02% DM.

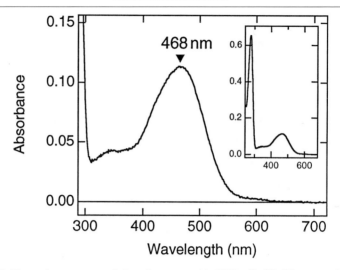

FIG. 4. Absorption spectrum of pinopsin expressed in 293S cells. Histidine-tagged pinopsin was expressed in 293S cells and purified by several steps of column chromatography. *Inset:* The absorption spectrum in the UV–Vis region.

Spectrophotometric Measurement

Concentration of pinopsin in the eluate from the SP-Sepharose FF column is estimated by measuring a difference absorption spectrum of the aliquot before and after complete photobleaching of pinopsin by orange light (>520 nm) in the presence of 50 mM hydroxylamine at 4°. Pinopsin is nearly stable in the presence of 50 mM hydroxylamine at 4°, while random Schiff bases are readily converted to retinal oximes.

Observation and Discussion

In a typical experiment, we can obtain about 0.1 OD_{468} · ml (a unit defined in ref. 12) of purified pinopsin from cells cultured in 200 plates. The absorbance spectrum of the eluate from the SP-Sepharose FF column shows that pinopsin has an absorption maximum at 468 nm (Fig. 4). This value is identical to that of the difference spectrum recorded before and after complete photobleaching of pinopsin.[1] We have confirmed that the recombinant pinopsin thus obtained functionally couples with rod transducin.[20] Biophysical studies of the photobleaching process of pinopsin by using low-temperature spectroscopy are also applicable to this purified sample.

[20] A. Nakamura, D. Kojima, H. Imai, A. Terakita, T. Okano, Y. Shichida, and Y. Fukada, *Biochemistry* **38,** 14738 (1999).

Acknowledgments

We thank Mr. Atsushi Nakamura and Dr. Daisuke Kojima (in our laboratory) for technical assistance in overexpression of pinopsin, Dr. Fumio Tokunaga (at Osaka University) for providing expression vector pUSRα, and Dr. Jeremy Nathans (at Johns Hopkins University) for the kind gifts of pRSV-Tag and 293S cells. This work was supported in part by Grants-in-Aid from the Japanese Ministry of Education, Science, Sports, and Culture.

[18] Cultured Amphibian Melanophores: A Model System to Study Melanopsin Photobiology

By MARK D. ROLLAG, IGNACIO PROVENCIO, DAVID SUGDEN, and CARLA B. GREEN

Introduction

Melanopsin is expressed by scattered perikarya among the amacrine and ganglion cells of the primate retina. This diffuse distribution complicates study of phototransduction mechanisms *in situ*. A model system that can be used to study regulation of melanopsin expression, regeneration of the retinaldehyde chromophore, and second-messenger coupling is needed. Candidate model systems can be found in lower vertebrates, where melanopsin is expressed in a greater variety of tissues than found in mammals. In amphibians, melanopsin is expressed in the retina, hypothalamus, iridial myocytes, and melanophores.[1] Amphibian melanophores and iridial myocytes, in particular, have attributes that make them useful for study of melanopsin expression and biochemistry. This chapter describes the *Xenopus* melanophore model system. We take advantage of the fact that the response parameter, melanosome movement, can be monitored by either video microscopy or spectroscopy techniques in real time without adversely affecting cellular activity. Observations can be made of melanophores maintained in defined salt solutions for a prolonged time (several days) at room temperature on an open laboratory bench. Transgenic techniques that modify melanopsin gene expression in amphibian melanophores are described.

The response of cultured *Xenopus* melanophores to light is shown in Fig. 1. The frame on the left shows the melanosome distribution in cells that have been maintained in dim red light. The frame on the right is the

---

[1] I. Provencio, G. Jiang, W. J. DeGrip, W. P. Hayes, and M. D. Rollag, *Proc. Natl. Acad. Sci. U.S.A.* **95**, 340 (1998).

dark   1 hour light

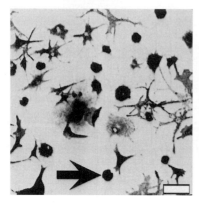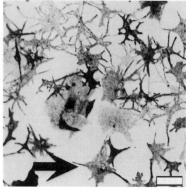

FIG. 1. Response of cultured *Xenopus* melanophores to light (approximately 1000 μW/cm^2). The melanophores were isolated and cultured as described in text. The response was obtained after incubating the cells with 100 nM all-*trans*-retinaldehyde in 60% Leibovitz L-15 medium (without serum) for 1 week. The two frames represent videomicroscopic images of the same 42 cells. The arrow in each frame identifies a cell that had a melanophore index value of 1.0 in the dark and a value of 5.0 in the light. Bar: 25 μm.

same set of cells after exposure to 1 hr of bright light from the microscope condenser (approximately 1000 μW/cm^2). In dim red light, the melanosomes tend to be clustered around the perinuclear region of the cell (Fig. 2). On exposure to light the melanosomes migrate to the cellular periphery. The action spectrum[2] for this response points to an opsin photopigment with peak absorption of about 460 nm. Photodispersion of melanosomes in cultured melanophores is dependent on vitamin A,[3] as would be expected if this response is mediated by an opsin-based photopigment.

There are two popular methods for quantifying the response of amphibian melanophores to stimulating agents. One, the melanophore index,[4] depends on manual scoring of the melanosome distribution in individual cells within a microscopic field. The other depends on a microtiter plate reader to measure the absorbance of light by a monolayer culture of melanophores.[5] The microtiter plate technique is particularly valuable for rapid, repeated screening of multiple samples, as is required in kinetic studies. The melanophore index, on the other hand, is useful for studies of small

---

[2] A. Daniolos, A. B. Lerner, and M. R. Lerner, *Pigment Cell Res.* **3,** 38 (1990).
[3] M. D. Rollag, *J. Exp. Zool.* **275,** 20 (1996).
[4] L. Hogben and D. Slome, *Proc. R. Soc. Lond. B* **108,** 1294 (1931).
[5] M. N. Potenza and M. R. Lerner, *Pigment Cell Res.* **5,** 372 (1992).

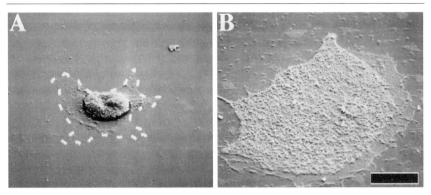

FIG. 2. Scanning electron micrographs of a *Xenopus* melanophore with aggregated melanosomes (A) and one with dispersed melanosomes (B). A dashed white line has been added to mark the peripheral edge of the cellular membrane when the melanosomes are aggregated. Bar: 25 μm.

numbers of cells, a constraint one experiences when testing photic responses in cells obtained from a single transgenic embryo. To obtain the melanophore index, a value of 1 is assigned to those cells with melanosomes aggregated in the perinuclear region of the cell. A value of 5 is assigned to those cells with melanosomes dispersed evenly throughout the cytoplasm. Values of 2, 3, and 4 reflect intermediate melanosome distributions; i.e., a melanophore index of 2 is assigned to cells with a dense core of melanosomes and some melanosomes in the major cytoplasmic extensions, a melanophore index of 3 is assigned to cells with a central core of melanosomes together with wide melanosome distribution throughout the cell, and a melanophore index of 4 is assigned to those cells with a variegated appearance, but no dense central core of melanosomes. In Fig. 1, the frame on the left has an average melanophore index of 2.5 ± 1.3 (SD) while that on the right has an average melanophore index of 4.6 ± 0.7 (SD). A more precise response parameter can be obtained by using image analysis software to calculate the area covered by melanosomes in a microscopic field.[6]

The biology of amphibian melanophores offers an opportunity to the photobiologist that may be unique. During development, melanophores in the tailfin of the tadpole switch their phenotype from photodispersion to photoaggregation of melanosomes.[7] The molecular basis for this phenotypic switch is not known. It might reflect differential expression of two photopigments within the same cell, differential coupling of a single photopigment

[6] D. Sugden, *Br. J. Pharmacol.* **104,** 922 (1991).
[7] R. Seldenrijk, D. R. W. Hup, P. N. E. de Graan, and F. C. G. van de Veerdonk, *Cell Tissue Res.* **198,** 397 (1979).

to two signal transduction pathways, or differential regulation of gene expression within neural crest cells derived from tailfin and body segments.

Data currently available support the hypothesis that the photoactivated melanophore photopigment is coupled to adenylate cyclase through a G_s intermediate for melanosome dispersion and a G_i intermediate for photoaggregation. Accordingly, during photodispersion of melanosomes, increased intracellular cAMP concentrations would activate cAMP-dependent protein kinase, which in turn would phosphorylate kinesin II[8] to cause melanosome dispersion. Decreased kinase activity in the face of constant protein phosphatase activity[9] would result in a return of the melanosomes to an aggregated state. This hypothesis is supported by several observations. When photosensitive melanophores are exposed to light there is an increase in intracellular cAMP concentrations correlated with photodispersion of melanosomes[2]; cGMP does not mimic the effects of light.[10] Inhibitors of cAMP-dependent protein kinase block the effect of light, but inhibitors of protein kinase C do not.[11] In the tailfin melanophores, the photoaggregation response is inhibited by pertussis toxin treatment,[12] consistent with coupling to G_i. Tailfin melanophores do not photoaggregate their melanosomes when intracellular cAMP is artificially clamped at high concentrations, but photoaggregate melanosomes normally when cGMP concentrations are pharmacologically manipulated.[12]

Preparation of Transgenic Embryos

A significant advantage of the amphibian melanophore model system is that transgenic techniques can be used to test hypotheses concerning molecular mechanisms for gene expression and signal transduction. Procedures to produce large numbers of transgenic *Xenopus* embryos were developed by Kroll and Amaya[13] and are summarized in Fig. 3. Accordingly, linearized plasmid DNA, containing the cDNA construct of interest, is integrated into the genome of decondensed sperm nuclei by restriction enzyme-mediated integration (REMI). An individual sperm nucleus with its incorporated transgene is microinjected into a frog oocyte, which develops into a transgenic embryo. Because the transgene is integrated into the sperm genome before fertilization, mosaic expression is avoided and the resultant first-generation progeny can be used for experimentation. In the

[8] M. C. Tuma, A. Zill, N. LeBot, I. Vernos, and V. Gelfand, *J. Cell Biol.* **143**, 1547 (1998).
[9] B. Cozzi and M. D. Rollag, *Pigment Cell Res.* **5**, 148 (1992).
[10] I. I. Geschwind, J. M. Horowitz, G. M. Mikuckis, and R. D. Dewey, *J. Cell Biol.* **74**, 928 (1977).
[11] T. S. McClintock, J. P. Rising, and M. R. Lerner, *J. Cell. Physiol.* **167**, 1 (1996).
[12] M. D. Rollag, *Photochem. Photobiol.* **57**, 862 (1993).
[13] K. L. Kroll and E. Amaya, *Development* **122**, 3173 (1996).

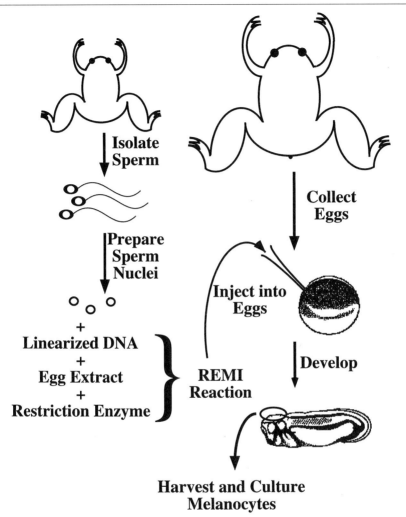

FIG. 3. A summary of the procedures used to prepare transgenic *Xenopus* melanophores.

example described in this chapter, the expression plasmid used for transgenesis consists of *Xenopus* melanopsin cDNA cloned in frame with a 6× Myc epitope tag. The complex is driven by the simian cytomegalovirus (CMV) promoter to cause ubiquitous and constitutively high-level transgene expression. After the transgenic embryos neurulate, the cells are dissociated and the melanophores isolated. The transgenic procedure is described in detail in Chapter 4 in this volume[14] and in other publications.[15,16] We highlight modifications we have made for the preparation of transgenic melanophores.

Materials

> Breeding pairs of adult male and female *Xenopus laevis:* Purchase from commercial suppliers (e.g., Nasco, Fort Atkinson, WI)
>
> High-speed egg extract: Eggs are collected from female *Xenopus* and used to purify a high-speed cytoplasmic fraction (as described in Chapter 4 in this volume[14]) that contains chromatin decondensing activity. The extract is aliquoted into 25-μl aliquots and stored at $-80°$
>
> Marc modified Ringer solution (MMR, 10×): 1 M NaCl, 20 mM KCl, 10 mM MgCl$_2$, 20 mM CaCl$_2$, 50 mM HEPES. Adjust with NaOH to pH 7.5. Sterilize by autoclaving. Store at room temperature
>
> Sperm dilution buffer (SDB): 250 mM sucrose, 75 mM KCl, 0.5 mM spermidine trihydrochloride, 0.2 mM spermine tetrahydrochloride. Adjust to pH 7.3–7.5 with NaOH. Store 1-ml aliquots at $-20°$
>
> Sperm nuclei preparation: Pith an adult male frog that has been anesthetized in 0.1% aqueous tricaine (ethyl 3-aminobenzoate) for 20 min. Isolate the testes, which are found at the bases of the large fat bodies in the abdominal cavity. Place one testis in 1× MMR containing 10% fetal calf serum and store at 4° for *in vitro* fertilization controls. Prepare the sperm nuclei from the remaining testis as described in Chapter 4 in this volume.[14] Store the sperm nuclei at 4° for up to 48 hr
>
> Hoechst stain No. 33342 (Sigma, St. Louis, MO)
>
> Hemocytometer
>
> Dissecting microscope with micromanipulator
>
> Injection buffer: 10 mM KPO$_4$ (pH 7.2), 125 mM potassium gluconate, 5 mM NaCl, 0.5 mM MgCl$_2$, 250 mM sucrose, 0.25 mM spermidine trihydrochloride, 0.125 mM spermine tetrahydrochloride. Store 1-ml aliquots at $-20°$
>
> Dejelly solution: 3 mM Dithiothreitol (DTT) in 1× MMR. Adjust to pH 8.9 with 1 N NaOH. Make fresh daily
>
> Injection medium: 6% Ficoll in 0.4× MMR, pH 7.5. Store at 4°, equilibrate to 16° before use

---

[14] S. Batni, S. S. Mani, C. Schlueter, and B. E. Knox, *Methods Enzymol.* **316**, [4], 1999.

[15] H. L. Sive, R. M. Grainger, and R. M. Harland, "Early Development of *Xenopus laevis:* A Course Manual." Cold Spring Harbor Laboratory Press, New York, 1999.

[16] B. E. Knox, C. Schlueter, B. M. Sanger, C. B. Green, and J. C. Besharse, *FEBS Lett.* **423**, 117 (1998).

NAM stock solution (2×): 220 mM NaCl, 4.0 mM KCl, 2.0 mM Ca(NO$_3$)$_2$, 2.0 mM MgSO$_4$, 0.2 mM EDTA, 4.0 mM NaH$_2$PO$_4$ (pH 7.5), 2.0 mM NaHCO$_3$

Linearized plasmid: The parental plasmid we use (pCS2+MT) is a gift of L. Snider (Fred Hutchinson Cancer Research Center, Seattle, WA). This vector contains a simian CMV promoter, 6× Myc epitope tags, a polylinker site, and simian virus 40 (SV40) polyadenylation signal. The melanopsin open reading frame is amplified by polymerase chain reaction (PCR) from *Xenopus* retina cDNA using Vent polymerase (New England BioLabs, Beverly, MA). Restriction sites are appended to the 5' end of the primers to facilitate directional cloning in-frame with the N-terminus 6× Myc epitope (forward primer with *Eco*RI restriction site in boldface: 5'-ATC GAT **GAA TTC** CCA GAA GAA TAA GAT TAC TCT GAC-3'; reverse primer with the *Xba*I restriction site in boldface: 5'-ATC GAT **CTA GAG** CAA ACA TCT GTT TAA AGT GGC-3'). PCR products are restriction cloned between the *Eco*RI and *Xba*I sites within the polylinker. The plasmid is sequenced and linearized at the *Sal*I site located immediately 5' of the promoter

Microinjection apparatus: Harvard (Hollison, MA) 11-in. syringe pump with 50-μl Hamilton gas-tight syringe

Tygon tubing (1/32-in. i.d., 3/32-in. o.d.)

Injection needles: Pull 1.0-mm o.d. borosilicate glass with 0.75-mm i.d. The ideal needle should have a gradual slope and should be broken so that a beveled end is created with an outer diameter at the tip of 80 μm. We use a model 700C vertical needle puller (David Kopf Instruments, Tujunga, CA) with a heater setting between 62.5 and 65

Mineral oil

Human chorionic gonadotropin (HCG)

Incubator maintained at 16°

First Day: Preparation

Prepare sperm nuclei and quantitate as described in Chapter 4 in this volume.[14] Attach Tygon tubing to a 50-μl gas-tight Hamilton syringe. Fill the tubing with mineral oil and position the syringe into the microinjection apparatus. Prepare injection dishes by pouring 1% agarose (in 0.4× MMR) into 35-mm petri dishes. Before the agarose solidifies, float a small square of plastic in the center of each dish to form a shallow well (we use a 1-cm^2 piece of a pipette tip box lid). In the evening, 12–16 hr before the nuclear transplantations are to begin, inject 800 units of human chorionic gonadotropin into the dorsal lymph sacs of two to four adult female frogs. Maintain the frogs overnight at 16°.

Second and Third Days: Restriction Enzyme-Mediated Integration and Nuclear Transplantation

It is best to have two people working together, one person collects and dejellies the eggs while the other performs the REMI reactions. When the REMI reaction is complete, both persons inject eggs. After 1 hr, more eggs are collected and another REMI reaction performed. This process is repeated for as long as the females keep laying good-quality eggs (usually four or five rounds per day).

Massage the abdomen of the previously injected female frogs to expel eggs into a glass petri dish containing 30–40 ml of dejelly solution. Gently swirl the eggs immediately after they are added to the dejelly solution, taking care not to let the eggs come into contact with air. When the jelly coats are dissolved (3–5 min), remove most of the dejelly solution and add excess 1× MMR. Swirl the eggs gently, and then exchange as much of the overlaying 1× MMR as can be removed without exposing the eggs to air with fresh 1× MMR. Repeat the 1× MMR wash three or four times to ensure that the dejelly solution is completely removed and the eggs are equilibrated to pH 7.5. Transfer the washed eggs into 6% Ficoll in 0.4× MMR equilibrated to 16° in the injection dish prepared on day 1. The eggs should be densely packed as a monolayer in the shallow well of the injection dish. Let the eggs incubate in the Ficoll solution at 16° for at least 6 min to stiffen the eggs. Prepare nontransgenic control embryos by expelling a small number of eggs from each female into a 60-mm petri dish containing a small amount of 1× MMR, pH 7.5. Macerate a small piece of testis with a forceps and lightly rub it over the surface of the eggs. Add water to dilute the MMR to 0.1× and place the *in vitro*-fertilized eggs into a 16° incubator.

For the REMI reaction, mix 4 μl of sperm stock (diluted to 100,000 nuclei/μl) with 1 μl of linearized plasmid (150–250 ng/μl) by gentle pipetting up and down through a wide-bore pipette tip. Incubate for 5 min at room temperature. During this 5-min incubation, prepare a mixture of 9 μl of SDB, 1 μl of 100 mM MgCl$_2$, 3 μl of high-speed egg extract, and 0.5 μl of the restriction enzyme used to linearize the plasmid diluted 1:50 in SDB. When the 5-min incubation is complete, add the preceding mixture to the sperm nuclei and plasmid DNA and mix gently with a wide-bore pipette tip. Incubate for 10 min at room temperature; longer incubations result in sperm nuclei with damaged DNA. At the end of the 10 min, dilute 9 μl of the REMI reaction into 332 μl of injection buffer. With a wide-bore pipette tip, gently pipette up and down to mix thoroughly without introducing air bubbles. This dilution gives a final concentration of 3 nuclei/ 5 nl.

Backload the injection needle with the REMI reaction and position the syringe in the microinjection apparatus. Adjust the flow rate to about 5 nl/ sec and inject an egg every second. This flow rate would theoretically introduce three nuclei per egg at a 1-sec cadence. Nevertheless, in our experience it gives the highest rate of normal cleavage (i.e., embryos with one nucleus per egg). We suspect that the nuclei aggregate or are otherwise not uniformly suspended. When the entire dish of eggs has been injected, place it in a 16° incubator and leave it undisturbed. After 3 hr at 16°, verify the success of the microinjections by watching for normal symmetrical cleavage. If the cleavage rate is low (i.e., less than 20%) the flow rate may have been too slow, with many eggs not receiving sperm nuclei. If the cleavage pattern is asymmetrical and extra cells have accumulated in the embryo, then too many nuclei may have been injected per egg. Using a loop of hair to gently manipulate the embryos in the dish, remove lysing and deformed embryos; transfer those cleaving normally to a new dish with 0.1× NAM containing gentamicin (see following description for isolation of melanophores). Keep the embryos in a 16° incubator overnight. We can inject about 1500 eggs/day with 30–50% of the embryos showing appropriate early cleavage and about 20% advancing to the neurula stage.

Isolation of Melanophore Cultures from *Xenopus* Embryos

Materials

Normal amphibian medium (NAM, 0.1×): 11 mM NaCl, 0.2 mM KCl, 0.1 mM Ca(NO$_3$)$_2$, 0.1 mM MgSO$_4$, 0.01 mM EDTA, 0.2 mM NaH$_2$PO$_4$, 0.1 mM NaHCO$_3$, gentamicin sulfate (25 mg/liter). Adjust to pH 7.5; filter sterilize

LP: 60% Leibovitz L-15 medium with penicillin G (100 U/ml) and streptomycin (100 μg/ml). Adjust to pH 7.5 and filter sterilize

Broad-spectrum antibiotic–antimycotic solution (CAG, 100×): Penicillin G (10,000 U/ml), streptomycin (10 mg/ml), amphotericin (25 μg/ml), ceftazidime (20 μg/ml), and gentamicin solution (1 mg/ml). The amphotericin may be suspended on beads; if so, it should not be filter sterilized

LP/CAG: Add 200 μl of 100× CAG to 20 ml of LP

Standard culture medium (LFMP): 50% Leibovitz L-15 medium, 10% fetal calf serum, insulin (10 μg/ml; approximately 0.0017 mM) α-MSH (0.1 μg/ml; approximately 60 nM), 380 mM L-asparagine, 100 mM hypoxanthine, 0.016 mM uridine, 160 mM thymidine, penicillin

G (100 U/ml), and streptomycin (100 μg/ml); adjust to pH 7.5 and filter sterilize

LFMP/CAG: Add 200 μl of 100× CAG to 20 ml of LFMP

Calcium, magnesium-free phosphate buffer (CMFB): 60% calcium, magnesium-free Dulbecco's phosphate buffer

25% Ficoll solution: Add 3.5 g of Ficoll 400DL and 1.4 ml of fetal bovine serum to 12 ml of tissue culture water

5% Ficoll solution: Add 1 ml of 25% Ficoll solution to 4 ml of CMFB

Trypsin solution (1×): Trypsin (5 mg/ml; ~5000 BAEE units/ml), 5 mM EDTA in CMFB

TrypIn-1K: Type I-S soybean trypsin inhibitor (1 mg/ml)

Low-speed swinging bucket centrifuge

Soft plastic pipettes

Flat-bottom multicluster plate (96-well)

Tissue culture flasks (25 cm^2)

Polypropylene tubes (sterile), 17 × 100 mm

Dissociation of Tissue and Primary Culture

To obtain cultured melanophores, let the fertilized *Xenopus* eggs develop in 0.1× NAM containing gentamicin until the vitelline membrane withdraws from the embryo (about 25 hr after fertilization). Wash the embryos three to five times with sterile 0.1× NAM. Remove the vitelline membrane and transfer the embryo to 200 μl of LP/CAG in a 96-well multicluster plate, using a wide-bore soft plastic transfer pipette. After transfer to a second 200 μl of LP/CAG solution, disrupt the embryonic tissue by gentle trituration through a 1-ml plastic pipette tip. The solution becomes milky owing to the large number of yolk platelets that are suspended. Place this milky solution as a drop in the center of a 25-cm^2 tissue culture flask. After 1 hr, gently flood the dish with 5 ml of LP/CAG, avoiding disturbance of the cells. It is our experience that cell survival is greatest when the cells are not widely dispersed and isolated. On the second day, replace the LP/CAG with fresh LP/CAG and on the third day exchange the LP/CAG with LFMP/CAG. Two days later (day 5), exchange the medium with standard culture medium (i.e., LFMP). Thereafter, feed the cells one to two times a week with a 50% exchange of medium. Before experimentation, the cells should be maintained at least 1 week in culture medium without antibiotics to ensure that the cultures are free of bacteria.

Ficoll Purification

When nonpigmented cells threaten overgrowth of the melanophore culture (usually after about 1 month), enrich the melanophores by Ficoll purification. To accomplish this enrichment, exchange the medium overlying the cultured cells in the 25-cm² flask with 4 ml of CMFB to remove residual serum proteins. Replace the CMFB with 3 ml of fresh CMFB. While the cells are incubating in CMFB, prepare a Ficoll step gradient by overlaying 0.5 ml of 5% Ficoll solution on top of 2 ml of 20% Ficoll solution in a 17 × 100 polypropylene round-bottomed tube. Twenty minutes after the addition of CMFB, add 1 ml of 1.0× trypsin solution (0.25× final dilution). Remove 3 ml of the diluted trypsin, leaving 1 ml behind. When the cells start to lift from the plastic (in about 5 min), add 1 ml of TrypIn-1K and layer the suspension onto the top of the Ficoll step gradient. Wash the culture flask with an additional 2 ml of CMFB to suspend all of the cells and add this wash to the top of the step gradient. Centrifuge the Ficoll gradient at 420 g for 5 min to isolate the melanophores as a loose pellet at the bottom of the tube. Remove the supernatant by gentle suction and resuspend the pellet in 100–200 μl of LP. Spot this suspension as a drop in the center of a new 25-cm² culture flask. Leave the flask undisturbed for about 30 min to allow the cells to attach, then gently flood the culture with 4 ml of culture medium. After two or three rounds of cell growth followed by Ficoll purification, one can generally obtain a pure culture of melanophores.

Maintenance of Cultures

Melanophores are subcultured by the trypsinization procedure described in the preceding section (Ficoll Purification). However, instead of layering the cells onto a Ficoll step gradient, the cells are layered onto a 200-μl cushion of 25% Ficoll solution in the bottom of a 5-ml polypropylene tube. The cells are pelleted at 170 g for 5 min. The supernatant overlying the 25% Ficoll is gently removed by suction and the cells are resuspended in 4 ml of culture medium before seeding onto a new 25-cm² flask. The Ficoll cushion minimizes cellular loss. In general, amphibian cells require lower ionic strength medium than is required for mammals (about 60–70% that of mammalian medium formulations). They grow well at cool room temperatures (18–22°), but can tolerate 4° for prolonged times. They die when the temperature exceeds 28–30°. Several medium formulations are available for amphibian cell culture. We prefer Leibovitz L-15 as a base for the culture medium because it is buffered by amino acids and does not require incubation in a CO_2-enriched atmosphere. At atmospheric CO_2 concentrations, however, amphibian cells do not efficiently produce purines[17]; hence, our addition of exogenous uridine. Insulin and melanocyte-

---

[17] G. D. Chinchar and J. H. Sinclair, *J. Cell. Physiol.* **96**, 333 (1978).

stimulating hormone (MSH) have been added as mitogens. Even so, the melanophores grow slowly, dividing once every 2–4 weeks in LFMP, and it can take up to 1 year to obtain sufficient numbers of cells to support standard biochemical techniques. Others have added conditioned medium[2] to increase the growth rate. Depending on one's needs, our procedure could be further optimized by a more careful dissection of the neural crest and reformulation of the culture medium.[18]

Evaluation of Transgene Incorporation

Transgene copy number and the number of insertion sites can be estimated from Southern blots of genomic DNA (see Kroll and Amaya[13]). We expect that most transgenes are randomly inserted into several sites in each genome and that negative position effects are compensated by the other inserted copies. Unusual effects resulting from specific insertion sites will likely be identified as outliers when a large number of independently generated cell lines are examined. Indeed, a major advantage of amphibian transgenesis is that a large number of transgenic embryos can be prepared in 1 day. Adding a Myc epitope to the transgene allows one to use immunohistochemical techniques to evaluate protein production. Alternatively, one can apply PCR to genomic DNA to ensure that cell lines contain the transgene. For melanopsin, a gene normally found in frogs, we use primers that span a 1.5-kb intron to identify cell lines with transgene incorporation (forward primer, 5′-ATG TTT CTG GCA ATT CGG AG-3′; reverse primer, 5′-TGT GAC GCA AGC ATA AGG AG-3′). Transgenic cells will show this intron-containing PCR product and an additional 170-bp product derived from the transgene. Noninvasive methods for transgene analysis are being developed. For example, one can use a green fluorescent protein reporter gene to serve as visual marker of transgene incorporation.[16]

Assaying Biological Response to Light by Video Microscopy

Materials

Video microscope consisting of CK2 Olympus (Norwood, MA) inverted microscope with ×10 bright-field objective, monocular phototube, ×2.5 phototube ocular, Insight Vision series 75 camera with extended red Newvicon tube (Insight Vision Systems, Malvern, UK), progressive scan charge-coupled device (CCD) camera (Total Turnkey Solutions, Mona Vale, New South Wales, Australia), PC with

[18] H. C. Wilson and N. C. Milos, *In Vitro Cell Dev. Biol.* **23,** 323 (1987).

at least 1 GB of memory, Optimas image analysis software (Meyer Instruments, Houston, TX). The power to the condenser light of the microscope is connected to an automatic timer. There is an infrared filter placed in the condenser light path (Oriel, Stratford, CT).

Digital photometer/radiometer with radiometer probe (Tektronix, Beaverton, OR) Neutral density filters (Oriel)

LPx: LP (prepared as described above) with 1× GIBCO insulin–transferrin–selenium–X supplement (Life Technologies, Grand Island, NY)

LPx/RAL: Prepare a 10 mM all-*trans*-retinaldehyde stock solution in ethanol. Add 10 μl of the ethanolic 10 mM retinal to 5 ml of sterile 0.2% aqueous bovine serum albumin (BSA) to prepare a stock retinal–BSA solution. Add 100 μl of the stock retinal–BSA solution to 20 ml of LPx

Melanophores in 25-cm^2 culture flask

Mel-Gel 400 (400 nM melatonin in 0.1% gelatin): Add 10 μl of a 1-mg/ml ethanolic melatonin solution to 100 ml of sterile 0.1% gelatin in 0.6% NaCl. Add 2.5 μl of Mel-Gel 400 per milliliter of medium to obtain a 1 nM melatonin solution

Assay Procedure

Melanophores are adapted to darkness in a serum-free medium (LPx/RAL) for 1 week to obtain a reproducible response to light. The slow division rate of fully differentiated melanophores, which can be frustrating when trying to increase cell number, allows one to leave flasks undisturbed for prolonged periods without concern. On the day before light treatment, assemble the video microscope in a light-tight box with as much of the electrical gear as possible placed outside the light box to avoid heat buildup. Depending on the characteristics of the system, procedures to control temperature need to be devised, particularly when using CCD cameras in an enclosed space. The microscope condenser with a filter assembly that accommodates three neutral density filters between the condenser and the tissue culture flask is used to generate the light for stimulating the cells. Care must be taken to ensure that the cells cannot be exposed to extraneous light from the condenser assembly.

Before experimentation, place a flask of melanophores on the microscope stage and optimize microscope optics, camera configurations, and image analysis settings to identify the lowest possible condenser light intensity that results in a reliable image. We generally use a ×10 objective with a ×2.5 phototube ocular to obtain an image with sufficient magnification and resolution for melanophore index readings. Measure the light intensities obtained with various filter configurations, condenser light settings, and camera configurations using a photometer/radiometer. Because of the limited dynamic ranges of most video cameras, we use two cameras: one with high sensitivity (i.e., with a Newvicon photomultiplier tube) and one with relatively low sensitivity (i.e., a CCD camera). We can reliably observe melanophores with 0.03-μW/cm^2 condenser light, using Optimas image analysis software and a camera with a Newvicon extended red photomultiplier tube. When the light intensity exceeds 50,000 μW/cm^2 our CCD camera becomes saturated.

On the day of the experiment, exchange the medium with LPx/RAL containing 1 nM melatonin for 1 hr, and then replace the medium with LPx/RAL without melatonin. Melatonin aggregates the melanosomes in the cells to stabilize the baseline response, but it also affects the fluence response curve, shifting it so that the cells are not as sensitive to light. Place the cells on the microscope stage for 1 hr to equilibrate in darkness. Set the computer to acquire images automatically at 2-min intervals for the next 13–14 hr. Put the microscope on a timing circuit so that the lights are on for 0.5 hr and off for 1.5 hr. During the dark intervals sequentially exchange the neutral density filters for the desired light intensity of the next exposure. When finished, exchange the medium with standard culture medium (LFMP) so that the cells can be tested at a later time if necessary. Calculate the melanophore index in the first image obtained after the lights came on and in the last image before the lights turn off. Control cells should have a melanophore index value of 1.0–1.5 at lights on and a value of 4.0–5.0 after exposure to light greater than 1000 μW/cm^2 for 30 min. If this dynamic range is not observed, there may be residual serum or MSH that was not removed when the medium was changed to LPx/RAL, or the cells might not be healthy.

Example of Transgenic Melanophore Response to Light

The photic response of melanophores expressing the melanopsin sense transgene is shown in Fig. 4. The control melanophores contain a transgene for universal expression of green fluorescent protein instead of melanopsin. The initial images, labeled 0 μW/cm^2, were collected during the first minute after lights were turned on. As can be seen in Fig. 4, the melanosomes in the dark condition are aggregated in the perinuclear region of the cell. When exposed to 100-μW/cm^2 light, the control melanophores remain aggregated, whereas those overexpressing melanopsin have responded to the light and dispersed their pigment throughout the cytoplasm. A 10,000-μW/cm^2 light exposure results in photodispersion of pigment in both the control and melanopsin-expressing melanophores.

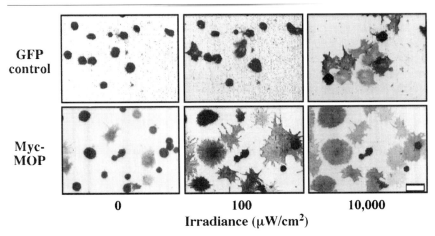

FIG. 4. Transgenic melanophores expressing sense melanopsin with an added Myc epitope (Myc-MOP) or expressing green fluorescent protein (GFP) were prepared as described in text. The melanosome distributions within the treated melanophores are shown for selected photic irradiances. These are the same melanophores in each of the three frames for each treatment; cell number has changed owing to migration of cells out of the field of view during the 16-hr experimental interval. Bar: 25 μm.

A more complete representation of the transgenic melanophore responses to light is shown in Fig. 5. A total of 10,000 $\mu W/cm^2$ is approximately the irradiation received outside on a cloudy day; 100 $\mu W/cm^2$ approximates normal room light in our laboratory; 1 $\mu W/cm^2$ can be detected on a moonlit night. The median effective dose (ED_{50}) for the effects of light on melanopsin-expressing melanophores is about 3 $\mu W/cm^2$. The ED_{50} for the GFP-expressing melanophores is about 300 $\mu W/cm^2$. Note that the GFP transgenics have some expression of melanopsin by the endogenous gene.

Analysis of Melanophore Responses, Using Microtiter Plate Reader

Materials

Flat-bottom tissue culture plates, 96-well (Nunclon; Nunc A/S, Denmark)

Microtiter plate reader (e.g., Bio-Tek, Anachem, UK)

Melanophores: Grow from primary cultures as described above, or obtain from another laboratory. In this case, the initial melanophore cultures are a generous gift from M. Lerner; material transfer agreements can be obtained from E. A. Lerner, CBRC, Charlestown, MA)

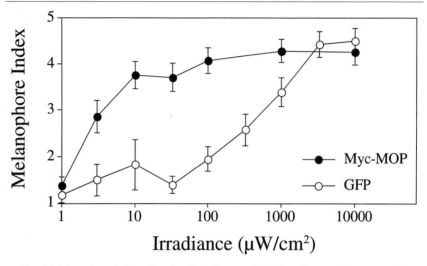

FIG. 5. Melanophore index values for the full range of light irradiances of the melanophores with a melanopsin transgene. These are the numerical scores for the cells shown in Fig. 4. Values represent the melanophore indices (means ± SEM) for 25 (GFP) and 32 (Myc-MOP) melanophores, respectively.

Growth medium: 70% Leibovitz L-15 medium with 20% fetal bovine serum, penicillin G (100 U/ml), and streptomycin (100 µg/ml)

Assay medium: 70% Leibovitz L-15 medium with penicillin G (100 U/ml) and streptomycin (100 µg/ml)

Microtiter Plate Assay Procedure

The use of a microtiter plate reader to measure melanophore responses was developed and popularized by M. R. Lerner and co-workers.[5] It allows one to obtain repeated measurements from multiple wells for kinetic analyses. When using repeated spectroscopic measurements to study the effects of light, one must be concerned by possible artifacts introduced by the spectrophotometer illumination. It has been shown, for example, that tailfin melanophores exposed to white light for 10 sec display pigment aggregation after a 3-min delay and that the response persists for about 20 min even though the cells are maintained in the dark.[19] In Fig. 6 we show that artifacts from repeated evaluations with a microtiter plate reader are not present when using 630-nm wavelength light to obtain absorbance values from cultured melanophores.

[19] A. C. J. Burgers and G. J. van Oordt, *Gen. Comp. Endocrinol. Suppl.* **1,** 99 (1962).

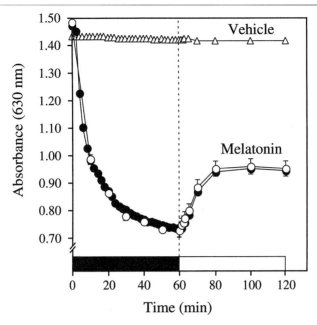

FIG. 6. Repeated measurement of melanophore absorbance at 630 nm, using a microtiter plate reader, does not inhibit or slow *Xenopus* pigment aggregation induced by a low concentration of melatonin. Melatonin (0.1 nM) or vehicle (0.0001% methanol) was added at time 0. Data shown are means ± SEM. See text for explanation.

To use the microtiter plate reader assay system as described by McClintock and Lerner,[20] suspend melanophores from a seed culture in a 75-cm^2 flask, using the trypsinization procedure described above. Plate the cells at a density of 2000 to 5000 melanophores/well into a 96-well tissue culture plate and let grow until confluent. At least 1 hr before experimentation, replace the growth medium with 100 μl of assay medium that does not contain serum or other melanophore-stimulating agents. To maximize responses to light one can pretreat the cells with all-*trans*-retinaldehyde for 1 week; however, growth in 20% fetal calf serum provides nearly all, if not all, of the retinoid requirement.[3] Using a 630-nm filter in the microtiter plate reader, measure the absorbance of the experimental wells at repeated intervals. In the example described here (Fig. 6), diluent (0.0001% methanol, triangles in Fig. 6) was added to two rows of cells (24 wells total) and 0.1 nM melatonin was added to the six other rows of cells (72 wells total, circles in Fig. 6). Black cardboard was taped to the bottom of the dish to prevent

---

[20] T. S. McClintock and M. R. Lerner, *Brain Res. Brain Res. Protocol* **2**, 59 (1997).

light transmittance through one row of diluent-treated cells and three rows of melatonin-treated cells. Immediately after drug addition, the microtiter plate was transferred to the microtiter plate reader and measurements were obtained at 2-min intervals for 1 hr (closed circles in Fig. 6). The cells were in the plate reader in a dark room at all times. The only light exposure was that generated by the plate reader during absorbance measurement. During this experiment, the black cardboard backing from two columns of cells (i.e., two diluent controls and six melatonin-treated wells) was removed every 10 min. The absorbance values for a single microtiter plate pass was determined for these wells (open circles in Fig. 6). In all wells treated with vehicle, absorbance values remained high throughout the experiment, reflecting full dispersion of pigment. The results of the experiment demonstrate that repeated exposure to the 630-nm light used by the microtiter plate reader did not have any significantly effect on either the rate or degree of pigment aggregation induced by melatonin. The concentration of melatonin used in this experiment was low (0.1 nM) to allow the best opportunity to reveal any potential pigment dispersion triggered by repeated absorbance measurement.

After the 1 hr of repeated measurement, the room lights were turned on (60-μW/cm^2 irradiance) and the plate ejected from the reader. The photoinduced change in absorbance was measured at 10-min intervals. The reversal of melatonin-induced pigment aggregation by light indicates that these melanophores were indeed photosensitive (incomplete reversal of aggregation is expected for a 60-μW/cm^2 irradiance). Thus, use of a microtiter plate reader provides an efficient method for studying the kinetics of pigment cell responses to light and the effects of drugs on the light response.

Conclusion

Model systems are valuable to the extent that they are simple to use and relevant to the issue of primary interest. Simplicity is a major strength of the amphibian melanophore model system. Large numbers of transgenic *Xenopus* embryos can be produced daily at modest cost. Melanophores expressing the transgene are obtained from first-generation embryos early in development. Using pigment granule translocation as a response parameter, repeated measurements show the time course and reversibility of treatment effects. Pigment granule translocation occurs within minutes after light exposure and can be measured in single cells at room temperature in normal atmospheric environments. One can use video microscopy techniques to collect repeated measurements from single cells on a millisecond time scale. Alternatively, a microtiter plate reader can be used to obtain measurements simultaneously from replicate melanophore populations on a minute-to-minute basis.

General principles governing the expression and regulation of melanopsin in amphibian melanophores are likely to apply to other species and cell types expressing melanopsin. Even if this expectation is not realized, knowledge concerning the function of melanopsin in amphibian melanophores is important in its own right. Amphibian melanophore responses to light protect vital organs against ultraviolet irradiation and produce rapid color changes needed for camouflage responses and animal survival. From a technical standpoint, it is important to understand the regulation and expression of melanopsin to avoid misinterpreting the actions of other agents on amphibian melanophores studied in lighted environments. Developmental expression of melanopsin in amphibian melanophores could serve as a cell-specific marker when studying neural crest development.

In this chapter we have described methods for monitoring the effects of melanopsin gene overexpression on the photobiological response of cultured transgenic melanophores. We have relied heavily on the pioneering work of Amaya and Kroll for the initial development of methods to produce transgenic frogs, and on the work of Lerner and co-workers for application of microtiter plate reader technology in the measurement of melanophore responses to stimulating agents. As computerized image analysis techniques for easy and reliable automated measurement of pigment dispersion are developed, we anticipate that assay technologies in the melanophore model system will become more sophisticated. Future transgenic studies are likely to use *Xenopus tropicalis* instead of *Xenopus laevis* as a melanophore source to avoid the complications of pseudotetraploidy. It is expected that functional knockouts resulting from either antisense expression or expression of dominant negative mutations will be developed for study of melanophore photic responses. The ease with which transgenic frogs are generated is of great value to insertional mutagenesis schemes, including gene-trap or enhancer-trap screens. Insertional mutagenesis provides an opportunity for unexpected advances in our understanding of extraretinal photoreception by melanophores. We expect the amphibian melanophore model system to be an important tool in our efforts to elucidate the regulation and functional significance of melanopsin expression in vertebrate cells.

Acknowledgments

The techniques and research described in this chapter were supported by NSF Grant IBN 9809916 (M.D.R.) and NIH Grant R01 EY11489-03 (C.B.G.). We are grateful to Dr. Michael R. Lerner (Department of Dermatology, University of Texas Southwestern Medical Center at Dallas, Dallas, TX) for supplying the clonal *Xenopus laevis* melanophore cell line used to construct Fig. 6.

Section IV

Enzymes of the Visual Cycle

[19] High-Performance Liquid Chromatography Analysis of Visual Cycle Retinoids

By GREGORY G. GARWIN and JOHN C. SAARI

Introduction

The retina is rich in retinoids. 11-*cis*-Retinal or a closely related derivative is the chromophore of all known visual pigments, and photoisomerization and regeneration involve a cycle of retinoid metabolism in which all known naturally occurring derivatives of vitamin A, except retinoic acid and retinoyl glucuronide, are found.[1,2] Analysis of these retinoids in the retina presents special challenges. First, 11-*cis*-retinal is light sensitive, especially when attached to opsin, necessitating a dim, red light environment during all stages of analysis. Second, retinal and retinol exist in all-*trans* and 11-*cis* isomeric forms in the retina, and trace amounts of other isomers may form during workup of samples. Third, because 11-*cis*- and all-*trans*-retinal are chemically bonded to opsin through a Schiff base linkage, their quantitative extraction requires the use of reagents that will disrupt this bond. Fourth, the reagents commonly used to break the Schiff base bond (hydroxylamine and derivatives) result in the formation of pairs of *syn* and *anti* isomers, generating a complex mixture of isomers that can be difficult to resolve. Finally, the presence of retinoids with widely different aqueous solubilities (e.g., retinyl palmitate and retinol) limits the analytical systems that can be used.

Analysis of visual cycle retinoids under various experimental conditions has provided a wealth of information regarding the fundamentals of the visual process.[1,3–6] A number of articles reviewing retinoid separations and properties have appeared.[7–12] However, advances in column packings and

---

[1] J. E. Dowling, *Nature* (*London*) **188,** 114 (1960).
[2] G. Wald, *Science* **162,** 230 (1968).
[3] C. D. B. Bridges, *Exp. Eye Res.* **22,** 435 (1976).
[4] K. Palczewski, J. P. Van Hooser, J. Chen, G. I. Liou, and J. C. Saari, *Biochemistry,* **38,** 12012 (1999).
[5] J. C. Saari, G. G. Garwin, J. P. Van Hooser, and K. Palczewski, *Vision Res.* **38,** 1325 (1998).
[6] W. F. Zimmerman, *Vision Res.* **14,** 795 (1974).
[7] C. C. B. Bridges, *Methods Enzymol.* **189,** 60 (1990).
[8] C. D. B. Bridges and R. A. Alvarez, *Methods Enzymol.* **81,** 463 (1982).
[9] G. W. T. Groenedijk, P. A. A. Jansen, S. L. Bonting, and F. J. M. Daemen, *Methods Enzymol.* **67,** 203 (1980).
[10] R. Hubbard, P. K. Brown, and D. Bownds, *Methods Enzymol.* **18,** 615 (1971).

quality and in instrument design have resulted in considerable improvements in separations and sensitivities, and the topic of high-performance liquid chromatography (HPLC) analysis of visual cycle retinoids deserves a revisit at this time. In addition, several of the common chemical procedures employed during retinoid analysis are described here on a micromole scale.

Methods

Chemical Procedures

General. All procedures involving retinoids are conducted under dim, red illumination or in aluminum foil-covered vials to minimize photoisomerization and photodecomposition. Reactions are carried out in screwcap vials fitted with Teflon cap liners. The air in these vials is displaced with argon before sealing. The volume of organic solvents is reduced at 37° (water bath) with a gentle flow of argon directed to the surface of the liquid with a Pasteur pipette.

Sources of Retinoids. all-*trans*-Retinol, all-*trans*-retinal, 9-*cis*-retinal, 13-*cis*-retinal, all-*trans*-retinyl acetate, and all-*trans*-retinyl palmitate are purchased from Sigma (St. Louis, MO) and purified by HPLC before use. 11-*cis*-Retinal is obtained from R. Crouch through the generosity of the National Eye Institute. 9-*cis*-, 11-*cis*-, and 13-*cis*-Retinols are generated by reduction of the corresponding retinals with $NaBH_4$ (see below). Vitamin A_2 is obtained from Distillation Products (Rochester, NY). The concentrations of retinoids are determined from their absorptions and extinction coefficients (Table I).

Extraction of Retinoids from Reaction Mixtures. Five milliliters of hexane followed by 2 ml of water are added to a 2-ml solution of retinol, retinal, or retinyl ester in alcohol (usually ethanol). The upper phase is passed through the lower phase several times with a Pasteur pipette, taking care to avoid generating an aerosol, which promotes oxidation. After a brief centrifugation (1 min, 1000g, 4°) to obtain a clean separation of phases, the upper phase is removed, a fresh portion of hexane added, and the extraction repeated. The combined upper phases from three extractions are back-extracted with water to remove alcohol dissolved in the hexane and reduced to a small volume or dried with flowing argon.

Reduction of Retinals. To reduce retinal,[10] dissolve retinal (1 mg, ~3.5 μmol) in 1 ml of ethanol and add 10 mg of $NaBH_4$. The capped tube is

---

[11] G. M. Landers, *Methods Enzymol.* **189,** 70 (1990).
[12] A. M. McCormick, J. L. Napoli, and H. F. DeLuca, *Methods Enzymol.* **67,** 220 (1980).

TABLE I
Absorption Properties of Stereoisomers of Vitamin A Compounds

Compound	Molecular weight	Ethanol		Hexane	
		λ_{max} (nm)	ε_M (M cm^{-1})[a]	λ_{max} (nm)	ε_M (M cm^{-1})[a]
Retinol	286				
9-*cis*-		323	42,300[b]	322	42,200[c]
11-*cis*-		319	34,892[d]	318	34,320[d]
13-*cis*-		328	48,300[b]	328	48,500[c]
all-*trans*-		325	52,800[b]	325	51,800[e]
Retinal	284				
9-*cis*-		373	36,068[f]	363	37,658[c,g]
11-*cis*-		376.5	24,935[d]	362.5	26,355[d]
13-*cis*-		375	35,500[f]	363	38,766[c,g]
all-*trans*-		383	42,900[h]	368	47,996[c,g]
Retinal oxime	299				
9-*cis*-		354	56,212[c]	351.5	54,119[c,g]
11-*cis*-		356	37,076[c]	347 (*syn*)	35,900[i] (*syn*)
				351 (*anti*)	30,000[i] (*anti*)
13-*cis*-		356	54,867[c]	352	52,774[c,g]
					49,000[j] (*syn*)
					52,100[j] (*anti*)
all-*trans*-		355	60,398[h]	357 (*syn*)	55,500[i] (*syn*)
				361 (*anti*)	51,700[i] (*anti*)
Retinyl palmitate	524				
9-*cis*-				322[i]	
11-*cis*-				318[k]	
13-*cis*-				328[i]	
all-*trans*-		325	49,256[l]	326[k]	
Retinyl acetate	328				
all-*trans*-		325	49,528[l]	326	51,000[e]

[a] Determined at λ_{max} unless otherwise indicated.
[b] C. D. Robeson, J. D. Cawley, L. Weisler, M. H. Stern, C. C. Eddinger, and A. J. Chechnak, *J. Am. Chem. Soc.* **77,** 4111 (1955).
[c] R. Hubbard, *J. Am. Chem. Soc.* **78,** 4662 (1956).
[d] P. K. Brown and G. Wald, *J. Biol. Chem.* **222,** 865 (1956).
[e] H. R. Cama, F. D. Collins, and R. A. Morton, *Biochem. J.* **50,** 48 (1951).
[f] C. D. Robeson, W. P. Blum, J. M. Dieterle, J. D. Cawley, and J. G. Baxter, *J. Am. Chem. Soc.* **77,** 4120 (1955).
[g] R. Hubbard, P. K. Brown, and D. Bownds, *Methods Enzymol.* **18C,** 615 (1971).
[h] G. Wald and P. K. Brown, *J. Gen. Physiol.* **37,** 189 (1953–1954).
[i] G. W. T. Groenendijk, P. A. A. Jansen, S. L. Bonting, and F. J. M. Daemen, *Methods Enzymol.* **67F,** 203 (1980).
[j] K. Ozaki, A. Terakita, R. Hara, and T. Hara, *Vision Res.* **26,** 691 (1986). Determined at λ 360 nm.
[k] C. D. B. Bridges, *Exp. Eye Res.* **22,** 435 (1976).
[l] J. G. Baxter and C. D. Robeson, *J. Am. Chem. Soc.* **64,** 2407 and 2411 (1942).

mixed at room temperature for 15 min. The course of the reaction is followed by measuring the loss of absorbance of the retinal at ~360 nm. The product is extracted into hexane and purified by HPLC if necessary. The yield of this procedure is essentially quantitative with retention of isomeric composition.

Reduction of 11-cis-Retinal with NaB^3H_4. This reaction must be run with an appropriate trap because tritium gas is released during the reaction. NaB^3H_4 (5 mCi, 10.2 μmol; Du Pont-New England Nuclear, Boston, MA) is dissolved in 400 μl of 50 mM NaOH and the solution added to 4 mg (14 μmol) of 11-*cis*-retinal dissolved in 2 ml of ethanol. The capped tube is mixed at room temperature for 15 min. After this reaction, small aliquots (20 μl) of a 10-mg/ml solution of $NaBH_4$ in 50 mM NaOH are added until the solution is colorless. The tritiated product is extracted into hexane and purified by HPLC. The product obtained by this procedure typically has a specific radioactivity of 300,000–400,000 dpm/nmol, which is reduced to the desired value by addition of 11-*cis*-retinol.

Oxidation of Retinol to Retinal. To oxidize retinol to retinal,[10] dissolve retinol (1 mg, 3.5 μmol) in 5 ml of chilled hexane. After addition of 15 mg of MnO_2 (Aldrich, Milwaukee, WI), the tube is flushed with flowing argon and sealed. The tube is suspended in a 37° water bath, and the contents are stirred continuously for 10 min (small magnetic stirring bar). The tube is again chilled on ice and the retinoids are extracted three times with hexane. The combined hexane phases are dried with flowing argon and taken up in hexane for purification by normal-phase HPLC (see below). The yield of this procedure is ~65%.

Esterification of Retinols. To esterify retinols,[13] dissolve the desired retinol (1 mg, 3.5 μmol) and triethylamine in 1 ml of dichloromethane to give final concentrations of 3.5 and 200 mM, respectively. The desired fatty acyl chloride (Sigma) is dissolved in dichloromethane (1 mg/ml), and 1 ml is added to the retinol solution. After incubation on ice for 30 min, retinoids are extracted three times with diethyl ether, and the combined diethyl ether extracts are dried with anhydrous Na_2SO_4. The extracts are then dried with flowing argon, and the residue is taken up in 2 ml of hexane. Retinyl esters are analyzed by HPLC and purified if necessary. The average yield with four different fatty acyl chlorides (palmitate, oleate, linoleate, and stearate) is 75%.

Saponification. For saponification,[14] KOH (0.12 ml, 33%) is added to retinyl esters (~1 μmol) dissolved in 1.9 ml of ethanol to give a final

---

[13] R. A. Alvarez, C. D. B. Bridges, and S.-L. Fong, *Invest. Ophthalmol. Visual Sci.* **20**, 304 (1981).

[14] S. Futterman and J. S. Andrews, *J. Biol. Chem.* **239**, 81 (1964).

concentration of 0.35 N KOH. The capped vial is heated at 55° for 30 min. After cooling on ice, the retinols are extracted with hexane. A yield of ~70% is obtained by this procedure with retention of the isomeric configuration.

Esterification of Retinoic Acids. Methyl esters of retinoic acid are generated in a commercial diazomethane generator (Wheaton Scientific, Millville, NJ).[15] Diethyl ether (1 ml) containing ~1 μmol or less of retinoic acid is placed in the outside tube of the generator. The inside tube contains 33 mg of N-methyl-N'-nitro-N-nitrosoguanidine (Aldrich) and 0.1 ml of water. The apparatus is sealed, immersed in ice–water, and 0.15 ml of 5 N NaOH is injected through the septum into the inner tube. After 1 hr on ice, the diethyl ether is collected and dried with flowing argon and the retinoyl esters purified by reversed-phase HPLC.

Oxime Formation. To form oximes,[10] the appropriate isomers of retinal (0.5 μmol) are dissolved in 1 ml of ethanol, and 0.1 ml of a 0.1 M solution of NH_2OH in 0.1 M MOPS, pH 6.5, is added. After 30 min of reaction at room temperature, the retinoids are extracted with hexane, and the oxime purified by HPLC if necessary. Substitution of O-methyl- or O-ethylhydroxylamine (Fluka Chemical, Milwaukee, WI) for NH_2OH in the procedure yields retinal (O-methyl)oxime or retinal (O-ethyl)oxime, respectively.[16] Note that NH_2OH is unstable at neutral pH. The hydrochloride salt ($NH_2OH \cdot HCl$) should be neutralized and discarded after use.

Extraction of Retinoids from Tissues

General. All procedures are carried out under dim, red light or in aluminum foil-covered, screwcap vials. Tissues are carefully homogenized to avoid oxygenation of the sample.

Extraction of Retinoids from Tissues. Two to four frozen mouse eye backs are placed in a Duall glass–glass homogenizer in 2 ml of 50 mM MOPS [3-(N-morpholino)propanesulfonic acid], pH 6.5, 10 mM NH_2OH (freshly neutralized $NH_2OH \cdot HCl$), 50% in ethanol, and homogenized with six passes of the pestle. Homogenates are incubated at room temperature for 30 min to allow formation of retinal oximes. Retinyl acetate in ethanol is added (1.5 to 2 nmol per tube) as an external standard for quantitative analysis. After addition of 4 ml of hexane, the homogenate is drawn up and down in a Pasteur pipette and centrifuged in the homogenizer tube at 900g for 2 min at 4° to break the phases. The upper phase is removed and the extraction repeated after addition of another 4 ml of hexane. The combined upper phases from three extractions are dried with flowing argon and dissolved in 200 μl of hexane (for normal-phase HPLC analysis), or

[15] H. M. Fales and T. M. Jaouni, *Anal. Chem.* **45**, 2302 (1973).
[16] F. J. G. M. Van Kuijk, G. J. Handelman, and E. A. Dratz, *J. Chromatogr.* **348**, 241 (1985).

in aqueous acetonitrile (for reversed-phase HPLC analysis). The recovery of all-*trans*-retinyl acetate is used to correct for losses during extraction. Substitution of *O*-methyl- or *O*-ethylhydroxylamine for NH_2OH in the procedure yields retinal (*O*-methyl)oxime or retinal (*O*-ethyl)oxime, respectively. Retinal oximes coelute with retinols on reversed-phase chromatography, whereas the (*O*-ethyl)oximes elute later in the profile.[16]

Retinoic acids are poorly extracted from tissues by the preceding procedure because of the limited solubility of the carboxylate anion in hexane. The yield can be improved by acidifying the homogenate with 0.1 vol of 1 *M* sodium phosphate, pH 5, before the extraction.[17]

High-Performance Liquid Chromatography Analysis

Equipment. HPLC analyses shown here are obtained with a Hewlett Packard (Palo Alto, CA) 1050 instrument equipped with a solvent conditioning module, quaternary pump, and a diode array detector. The latter feature allows the acquisition of spectra "on the fly" and simultaneous measurement of two wavelengths during elution. Chemstation A.04.02 (Hewlett Packard) is employed to drive the instrument and to store and process the data. Normal-phase chromatography employs a Supelcosil LC-Si analytical column (4.6 × 150 mm, 3-μm particle size) or a Supelcosil LC-Si narrow-bore column (2.1 × 150 mm, 3-μm particle size) (Supelco, Bellefonte, CA). Reversed-phase chromatography employs a Vydac C_{18} narrow-bore column (2.1 × 150 mm, 5-μm particle size; The Separations Group, Hesperia, CA).

Normal-Phase HPLC. The analytical silica column is equilibrated with hexane. After injection of the sample, the column is pumped with the following ratios of solvent (hexane–ethyl acetate) for the indicated times at a flow rate of 1 ml/min: 0 to 10 min, 100:0 (v/v); 10 to 20 min, 99.5:0.5 (v/v); 20 to 35 min, 90:10 (v/v); and 35 to 45 min, 100:0 (v/v). The effluent is monitored simultaneously at 325 nm (retinols and retinyl esters) and 350 nm (retinal oximes). Spectra are obtained from various components on demand.

Retinols and retinals, including the 9-*cis,* 11-*cis,* 13-*cis,* and all-*trans* isomers, are resolved by normal-phase chromatography employing dioxane–hexane at a 2:98 (v/v) ratio for the retinols and at a 6:94 (v/v) ratio for the retinals[9] (results not shown).

Reversed-Phase HPLC. The narrow-bore reversed phase column is equilibrated at a flow rate of 0.7 ml/min with 72% aqueous acetonitrile,

---

[17] J. L. Napoli, *Methods Enzymol.* **123**, 112 (1986).

0.015 M in ammonium acetate, pH 5.5. After injection of the sample in aqueous acetonitrile, retinoids are eluted with the equilibration buffer.

Isomeric Composition of Retinyl Esters. The isomeric composition of retinyl esters is determined by HPLC analysis of the retinols generated by saponification.

Sensitivity. The lower limit of sensitivity of the HPLC detection system is approximately 25 pmol, or 7 ng.

Calculation of Results. Graphs relating areas to injected amounts of retinoid are constructed for all-*trans*- and 11-*cis*-retinols at 325 nm, for all-*trans*- and 11-*cis*-retinal oximes at 350 nm, and for retinyl acetate and palmitate at 325 nm. Values (picomoles per unit area) are derived over the linear portion of the relationship and used with the extraction yield to determine the absolute amount of retinoid in tissue extracts.

Stability. Retinoids are extremely susceptible to oxidative decomposition, especially in aqueous solutions, with decay times varying from minutes to hours.[18,19] Extracted retinoids should be stored in hexane under argon at $-80°$ until analyzed. Addition of α-tocopherol (vitamin E, 1 mg/ml) will increase the stability of retinoids in solution.

Discussion

Separation of visual cycle retinoids with normal-phase silica columns and hexane–ethyl acetate eluents offers two distinct advantages over separations with reversed-phase columns. First, retinyl esters are readily soluble in hexane, the main component of the eluent, and only marginally soluble in the aqueous acetonitrile or methanol solutions employed in reversed-phase columns. Second, all visual cycle retinoids are well resolved on normal-phase silica columns, with the exception of retinyl esters, which elute as a mixture of all-*trans*- and 11-*cis*-retinyl isomers esterified to a variety of fatty acids. A typical chromatogram obtained by normal-phase chromatography and the solvent system described is shown in Fig. 1. Baseline resolution of all the visual cycle retinoids and retinal oxime derivatives was obtained with this system. 11-*cis*-Retinol and 13-*cis*-retinol were not resolved, nor were 9-*cis*- and 11-*cis*-retinal oximes. Retinoic acids elute well after the components shown in Fig. 1, but they are relatively minor components compared with visual cycle retinoids.[20]

Retinoic acids, retinols, and free retinals are well resolved on reversed-

---

[18] S. Futterman and J. Heller, *J. Biol. Chem.* **247**, 5168 (1972).
[19] R. Crouch, E. Hazard, T. Lind, B. Wiggert, G. Chader, and D. Corson, *Photochem. Photobiol.* **56**, 251 (1992).
[20] P. McCaffery, J. Mey, and U. C. Drager, *Proc. Natl. Acad. Sci. U.S.A.* **93**, 12570 (1996).

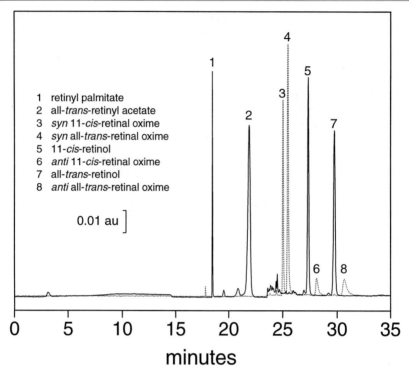

FIG. 1. Elution profile of retinoids separated by normal-phase HPLC. The solid trace was obtained at 325 nm (retinols and retinyl esters) and the dotted trace at 350 nm (retinal oximes).

phase C_{18} columns (Fig. 2). If a total retinoid analysis is not required, reversed-phase columns, with their water-based eluants, offer the advantage of more reproducible elution times and rapid column conditioning. However, as already mentioned, retinals are poorly extracted from retina unless a derivatizing agent (NH_2OH) is added, and the resulting retinal oximes coelute with retinols. Retinal (*O*-ethyl)oximes elute much later than retinols[16]; however, we were unable to completely resolve a mixture of 11-*cis*- and all-*trans*-retinal (*O*-ethyl)oximes on a C_{18} reversed-phase HPLC column. Examples of spectra obtained during a chromatographic analysis are shown in Fig. 3.

Retinyl esters are soluble in hexane; paradoxically, they are extracted in lower yield from tissue homogenates than polar retinoids. The reason

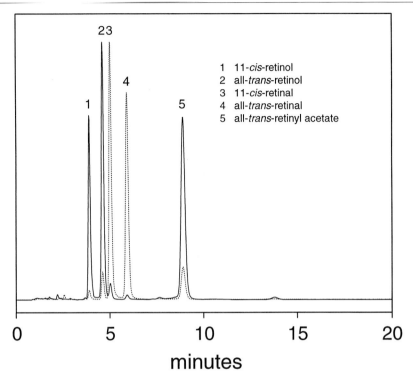

FIG. 2. Elution profile of retinoids separated by reversed-phase HPLC. The solid trace was obtained at 325 nm and the dotted trace at 383 nm. Retinal (*O*-ethyl)oximes (11-*cis*-, all-*trans*-, *syn*- and *anti*-isomers) elute between 12 and 16 min and are incompletely resolved (not shown).

for this appears to be related to their solubility in the initial aqueous–alcohol homogenate.

The physical chemical basis for the fluorescence properties of retinoids has been examined in depth.[21–23] In general, their fluorescence depends strongly on the polarity of the solvent, varying from weak fluorescence in polar solvents to strong fluorescence in nonpolar solvents such as hexane or pentane. Binding to proteins generally results in a large enhancement of fluorescence.[18] Retinals and retinoic acids are weakly fluorescent at room

---

[21] J. Kahan, *Methods Enzymol.* **28C,** 574 (1971).
[22] U. Swieter and O. Isler, *in* "The Vitamins" (W. H. Sebrell, Jr., and R. S. Harris, eds.), p. 5. Academic Press, New York, 1967.
[23] A. J. Thomson, *J. Chem. Phys.* **51,** 4106 (1969).

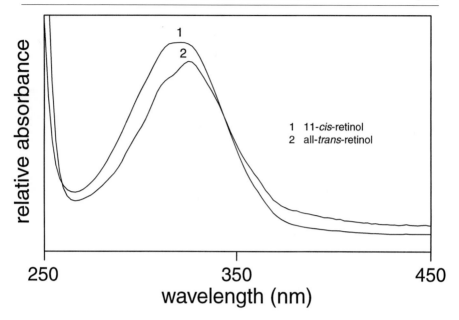

Fig. 3. Spectra of retinoids obtained during their separation by normal-phase HPLC. 1, 11-*cis*-Retinol; 2, all-*trans*-retinol.

temperature but show fluorescence at 77 K.[23,24] The fluorescence of retinols and retinyl esters can be used in their detection during chromatographic procedures and for their identification (see discussion below). Examples of emission and excitation spectra of retinols and retinyl palmitate are shown in Fig. 4.

The identity of retinoids extracted from tissues should always be established according to criteria in addition to their coelution with authentic standards. It is not always possible to submit samples for mass and/or nuclear magnetic resonance analyses because of the limited amounts available; however, a combination of chemical modification and rechromatography, spectral analysis, and coelution is usually sufficient to identify a common retinoid. Retinals can be identified by their lack of fluorescence, by their absorption maxima near 360 nm, by the characteristic shape of their absorption spectra,[25] and by their conversion to a retinol on reduction with $NaBH_4$ or to an oxime by treatment with NH_2OH. Retinols are highly fluorescent, have an absorption maximum near 320 nm, have distinctive absorption spectra,[25] and can be converted to retinals by oxidation with

[24] R. L. Christensen and B. E. Kohler, *Photochem. Photobiol.* **19**, 401 (1974).
[25] R. Hubbard, *J. Am. Chem. Soc.* **78**, 4662 (1956).

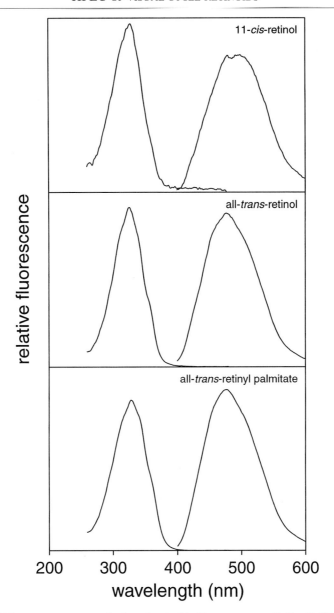

FIG. 4. Fluorescence spectra of selected retinoids. Fluorescence excitation (left) and emission (right) spectra were obtained in ethanol at room temperature. *Top:* 11-*cis*-Retinol. *Middle:* all-*trans*-Retinol. *Bottom:* all-*trans*-Retinyl acetate.

MnO$_2$ or to a retinyl ester after treatment with an acyl chloride. Retinyl esters are fluorescent, with absorption spectra similar to the retinols, and can be converted to retinols by saponification. Retinoic acids have characteristic absorption spectra, which are pH dependent (335 nm, base, 355 nm, acid for all-*trans*-retinoic acid) and can be converted to their methyl esters with diazomethane.

Acknowledgments

The methods reported here were developed or utilized in the laboratory of the authors with support from The National Eye Institute (EY02317, EY01730) and from Research to Prevent Blindness, Inc. (RPB). J.C.S. is a Senior Scientific Investigator of RBP, Inc. The authors are grateful to Dr. Yashushi Imamoto, Osaka University, for the fluorescence spectra. Conversations with Krzysztof Palczewski were helpful during the development of these methods.

[20] Quantitative Measurements of Isomerohydrolase Activity

By ANETTE WINSTON *and* ROBERT R. RANDO

Introduction

The absorption of light by rhodopsin leads to the *cis*-to-*trans* isomerization of the 11-*cis*-retinal Schiff base chromophore to generate all-*trans*-retinal. For vision to continue, the all-*trans*-retinal must be converted back into the visual chromophore 11-*cis*-retinal in a series of enzymatic reactions that compose the visual cycle in vertebrates. The liberated all-*trans*-retinal is reduced in the photoreceptors, to generate all-*trans*-retinol (vitamin A),[1] which is then shuttled across the interphotoreceptor space by binding proteins and delivered to the retinal pigment epithelium (RPE).[2-4] In the mammalian eye, the RPE plays an essential role in the visual cycle. The known reactions of the visual cycle that occur in the RPE are catalyzed by enzymes located in the microsomal membranes of the RPE. The cloned

---

[1] G. Wald and R. Hubbard, *J. Gen. Physiol.* **32,** 367 (1949).
[2] Y.-L. Lai, B. Wiggert, Y. P. Liu, and G. J. Chader, *Nature (London)* **298,** 848 (1982).
[3] A. J. Adler and K. J. Martin, *Biochem. Biophys. Res. Commun.* **108,** 1601 (1982).
[4] G. J. Liou, C. D. Bridges, S.-L. Fong, R. A. Alvarez, and F. Gonzalez-Fernandez, *Vision Res.* **22,** 1457 (1982).

lecithin:retinol acyltransferase (LRAT)[5] catalyzes the esterification of all-*trans*-retinol into all-*trans*-retinyl esters, using lecithin as an acyl donor. The all-*trans*-retinyl esters are then processed, with hydrolysis and double-bond isomerization, by an isomerohydrolase enzyme to generate 11-*cis*-retinol. The critical *trans*-to-*cis* reisomerization reaction of the visual cycle occurs in this latter step.

Here we report on a quantitative assay for the isomerohydrolase in a preparation using microsomal membranes harvested from bovine retinal pigment epithelia. Using this quantitative assay system, it is demonstrated that the isomerohydrolase is powerfully inhibited by added 11-*cis*-retinol. This suggests product inhibition as a mechanism for regulating isomerohydrolase activity.

Principle

The isomerohydrolase processes all-*trans*-retinyl esters into 11-*cis*-retinol.[6,7] However, because of solubility problems, retinyl esters cannot be effectively added to buffer solutions. In the current assay, all-*trans*-[^{3}H]retinyl esters are preformed *in situ* from added all-*trans*-[^{3}H]retinol, which is then processed to form 11-*cis*-[^{3}H]retinol. This is possible because rates of conversion by LRAT are substantially greater than isomerohydrolase rates.[8] Two versions of this assay are presented. In the first method, which is more time consuming and more costly than the second, high concentrations ($>K_m$ of LRAT) of [^{3}H]retinol are added. The second, abbreviated method uses tracer levels of all-*trans*-[^{3}H]retinol, and is useful for qualitative measurements.

Materials and Methods

Materials. Frozen eyecups devoid of retinas are obtained from J. A. and W. L. Lawson Company (Lincoln, NE). all-*trans*-[11,12-^{3}H]Retinol (specific activity, 31.4 Ci/mmol) is purchased from Du Pont–New England Nuclear (Boston, MA). 13-*cis*-Retinol and fatty acid-free bovine serum albumin (BSA) are purchased from Sigma (St. Louis, MO). 11-*cis*-Retinal is obtained from the National Eye Institute (Bethesda, MD). High-performance liquid chromatography (HPLC)-grade solvents are from J. T. Baker (Phillipsburg, NJ).

---

[5] A. Ruiz, A. Winston, Y.-H. Lim, B. A. Gilbert, R. R. Rando, and D. Bok, *J. Biol. Chem.*, **274**, 3834 (1999).
[6] P. S. Bernstein, W. C. Law, and R. R. Rando, *J. Biol. Chem.* **262**, 16848 (1987).
[7] P. S. Bernstein, W. C. Law, and R. R. Rando, *Proc. Natl. Acad. Sci. U.S.A.* **84**, 1849 (1987).
[8] A. Winston and R. R. Rando, *Biochemistry* **37**, 2044 (1998).

Preparation of Enzyme Source. The procedure for the preparation of bovine retinal pigment epithelium membranes is described elsewhere.[9] A typical preparation yields about 230 μg of protein per eyecup. The membranes can be stored in 100 mM phosphate buffer, pH 7.5, at $-80°$ for up to 3 months without loss of activity. Protein determinations are carried out by the amido black protein assay.[10]

Procedure

The isomerohydrolase assay is carried out in two steps. In the first step, all-*trans*-[^{3}H]retinol is added to membranes to allow for the LRAT-catalyzed formation of all-*trans*-[^{3}H]retinyl esters. The retinyl esters formed *in situ* are primarily located in the membrane fraction. In the second step, the membranes are incubated in the presence of BSA for assaying isomerohydrolase. Retinoid-binding proteins, such as BSA, have been shown to stimulate the formation of 11-*cis*-retinol.[8] Isomerohydrolase activity is followed by monitoring the formation of 11-*cis*-[^{3}H]retinol by HPLC methods.

Unless otherwise mentioned, all procedures are performed under dim red light with samples kept on ice. Prior to use, the membranes are irradiated with UV light (365 nm) for 5 min in order to destroy endogenous retinoids that might interfere with the formation of 11-*cis*-[^{3}H]retinol.[11] all-*trans*-[^{3}H]Retinol (3.6 nmol) in ethanol is dried under nitrogen on the bottom of a glass vial and the retinol is redissolved in 15 μl of methanol. In our experiments methanol proved to be the solvent that least inhibited LRAT and isomerohydrolase activity.[12] The methanolic solution of all-*trans*-[^{3}H]retinol is added to 1425 μl of the pigment epithelium membranes [180 μg of protein, 100 mM Tris-HCl (pH 9.0), 2.5 μM final retinol concentration]. The membranes are incubated on a shaker for 60 min at room temperature, allowing the formation of [^{3}H]retinyl ester. BSA is added [final concentration, 1% (w/v)] to solubilize the unconverted all-*trans*-retinol. The retinyl ester-containing membranes are collected by centrifugation (100,000g for 30 min at 4°), washed twice, and redissolved in 100 mM Tris-HCl, pH 8.0. The solution is sonicated briefly in ice water. At this point, the membranes usually contain between 5 and 6 nmol of retinoids per milligram of protein, the majority of which is found in the form of all-*trans*-retinyl esters (70% of total retinoids). The membranes also contain a substantial quantity of retinol isomers, which dissolve in the membranes and, therefore, are not completely removed by centrifugation. all-*trans*-

---

[9] B. S. Fulton and R. R. Rando, *Biochemistry* **26**, 7938 (1987).
[10] W. Schaffer and C. Weissmann, *Anal. Biochem.* **56**, 502 (1973).
[11] P. S. Deigner, W. C. Law, F. J. Cañada, and R. R. Rando, *Science* **244**, 968 (1989).
[12] A. Winston and R. R. Rando, unpublished observation (1997).

Retinol accounts for 15%, 11-*cis*-retinol for 10%, and 13-*cis*-retinol for 5% of total retinoids.

BSA is added to the membrane solution to a final concentration of 5% (w/v). At this point the suspension is divided into aliquots (5 μg of protein per sample) and any compounds to be tested for their effects on isomerohydrolase activity are added. The samples are incubated for 2 hr in a shaking water bath at 37° [100 mM Tris-HCl (pH 8.0), 100-μl total volume]. The reactions are quenched with methanol (500 μl per sample), 100 μl of H_2O is added, and 500 μl of hexane (containing butylated hydroxytoluene at 1 mg/ml) is used for extraction of the retinoids.

The isomeric retinols are analyzed on a 5-μm PVA-Sil column (250 × 4.00 mm; YMC, Wilmington, NC). The eluent of choice is 7% dioxane in hexane, used at a flow rate of 1.5 ml/min. The isomers are identified through coelution of standard mixtures of isomeric retinols (monitored at 325 nm) prepared as described previously.[13] Radioactivity is counted with an on-line Bertold LB 506-C HPLC radioactive monitor interfaced with an IBM computer. Calibration of the system is done by extracting and injecting defined amounts of all-*trans*-[11,12-^{3}H]retinol.

In a typical isomerohydrolase assay, samples optimized for 11-*cis*-retinol formation contain on average 32% 11-*cis*-retinol, 11% 13-*cis*-retinol, 24% all-*trans*-retinol, and 33% retinyl ester. Control samples are taken out in triplicate before the isomerohydrolase reaction and analyzed for their retinoid content. The amount of 11-*cis*-retinol formed during the second reaction step is determined by subtracting the average amount of 11-*cis*-retinol found in these three aliquots (analyzed after the first reaction step involving retinyl ester formation) from the amount of 11-*cis*-retinol measured at the end of the isomerohydrolase reaction. Unless indicated otherwise, the amounts of 11-*cis*-retinol obtained are given as a percentage of total retinoids or as picomoles per milligram of microsomal protein.

In the assay procedure described above, yields of about 1 nmol of 11-*cis*-retinol per milligram of protein are obtained. This closely approximates retinoid turnover under physiological conditions.[8] Moreover, sufficient concentrations of all-*trans*-retinol can be added that are saturating for LRAT. Alternatively, the isomerohydrolase assay can be carried out with tracer concentrations of substrate. The assay procedure with small amounts of all-*trans*-[^{3}H]retinol is simpler because centrifugation, washing, and sonication are not necessary, since 100% ester is formed during the first reaction step. Furthermore, the abbreviated method takes far less time (approximately 2 hr less) than the expanded method. The simplified procedure described below is sufficient for most preliminary experiments, for example,

---

[13] C. D. B. Bridges and R. A. Alvarez, *Methods Enzymol.* **81**, 463 (1982).

when checking for isomerohydrolase activity in enzyme solubilization or purification procedures. all-*trans*-[^{3}H]Retinol (1 nmol in ethanol) is dried under nitrogen and 40 μl of BSA (10%, w/v) is added to solubilize retinol. The use of BSA generally results in poor yields of solubilized all-*trans*-retinol (about 5%),[12] but circumvents the use of organic solvents, which inhibit isomerohydrolase. The amount of all-*trans*-retinol solubilized by BSA, however, is still sufficient to detect formation of 11-*cis*-retinol. BSA-solubilized all-*trans*-[^{3}H]retinol is added to 1400 μl of the pigment epithelium membranes [180 μg of protein, 100 mM Tris-HCl (pH 9.0), 0.3% final BSA concentration] and incubated for 15 min at room temperature. In this incubation the all-*trans*-retinol is almost quantitatively esterified, resulting in the formation 95–100% all-*trans*-retinyl ester. The sample is divided into aliquots (5 μg of protein per sample) and BSA is added to a final concentration of 5% (w/v). The isomerohydrolase reaction is carried out at 37° for 2 hr [100 mM Tris-HCl (pH 8.0), 100-μl total volume] and the reactions are quenched and analyzed as described above.

Applications

Activity of Isomerohydrolase Inhibited by Product 11-cis-Retinol

In the assay described above, LRAT is used to preform radioactive all-*trans*-retinyl esters, which are then processed into 11-*cis*-retinol by isomerohydrolase. This method obviates the need to directly add all-*trans*-retinyl esters as isomerohydrolase substrates, an important issue given the low solubility of retinyl esters. The assay also allows one to quantitatively determine the effects of various inhibitors that specifically act on the isomerohydrolase, without being concerned about the possible inhibition of LRAT. Clearly, inhibitors of LRAT are expected to inhibit the formation of 11-*cis*-retinol from added all-*trans*-retinol because the latter must first be converted into an ester prior to the isomerization reaction. The isomerohydrolase assay is used here to uncover the inhibition of isomerohydrolase caused by 11-*cis*-retinol.

The importance of soluble retinoid-binding proteins in promoting 11-*cis*-retinol formation suggests that the activity of isomerohydrolase may be regulated by product inhibition. In this scenario, the addition of BSA leads to product removal by partitioning 11-*cis*-retinol from the membrane into the soluble phase. This can be tested by measuring isomerohydrolase activity in the presence of various amounts of nonradioactive 11-*cis*-retinol. As negative control we used nonradioactive 13-*cis*-retinol. Isomerohydrolase

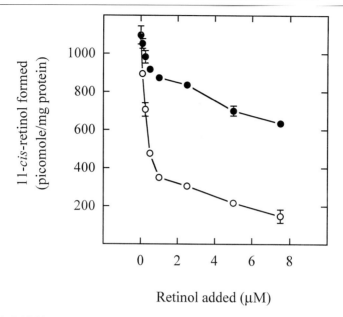

FIG. 1. Inhibition of isomerohydrolase activity in the presence of 11-*cis*-retinol or 13-*cis*-retinol. Samples were prepared for measuring 11-*cis*-retinol formation as described in Materials and Methods. The samples were incubated at 37° for 2 hr [100 mM Tris-HCl (pH 8.0), 100-μl total volume, 5% BSA] with nonradioactive 11-*cis*-retinol (open circles) or 13-*cis*-retinol (filled circles), respectively, added to a final concentration of 0, 0.1, 0.25, 0.5, 1.0, 2.5, 5.0, or 7.5 μM. Each measured point represents the results of duplicate determinations. Where no error bars are given, the error bars were smaller than the symbol used. The membranes contained 6.0 nmol of total retinoids per milligram of protein (SD = 0.49, n = 35).

is specific for all-*trans*-retinyl ester as substrate,[14] and is, therefore, not expected to have high affinity for 13-*cis*-retinol. Also, 13-*cis*-retinyl ester formed by LRAT activity is not a precursor of 11-*cis*-retinol.

11-*cis*-Retinol is a significantly better inhibitor of isomerohydrolase than its 13-*cis* congener. The formation of 11-*cis*-[11,12-^{3}H]retinol in the presence of added 11-*cis*-retinol is significantly (p < 0.05) lower than in the presence of added 13-*cis*-retinol (Fig. 1). 13-*cis*-Retinol is therefore not an effective inhibitor of isomerohydrolase. With 11-*cis*-retinol, isomerohydrolase activity reaches half of its maximum value (1096 pmol/mg of protein, SD = 46.9, n = 2) at about 400 nM added cold 11-*cis*-retinol. This low median

[14] W. C. Law, R. R. Rando, S. Canonica, F. Derguini, and K. Nakanishi, *J. Am. Chem. Soc.* **110**, 5915 (1988).

inhibitory concentration (IC_{50}) value indicates that 11-*cis*-retinol may indeed act as a specific inhibitor of isomerohydrolase.

Because added retinols are esterified by endogenous LRAT, it is unclear whether the inhibition observed with 11-*cis*-retinol is due to this molecule, or to retinyl esters formed from it. Studies with 11-*cis*-retinal as putative inhibitor decide this issue. Indeed, 11-*cis*-retinal is approximately equipotent with 11-*cis*-retinol with respect to isomerohydrolase inhibition.[15] The specificity of the 11-*cis*-retinol(al)-mediated inhibition is illustrated by the observation that neither 11-*cis*-retinal oxime nor aniline Schiff base proved to be inhibitors of the enzyme.[15] The oxime and aniline Schiff bases of all-*trans*-retinal were also completely inert as inhibitors of the isomerohydrolase.[15] Although retinyl esters could not be directly studied because of insolubility problems as well as retinyl ester hydrolytic activity present in the membrane system used, results with the oxime and Schiff base render it unlikely that 11-*cis*-retinyl esters would be capable of potently inhibiting the isomerohydrolase.

Isomerohydrolase is situated at a branch in the visual cycle, its activity determining whether the retinoids are stored in the microsomal membranes of the RPE as retinyl esters, or processed into the 11-*cis* form that is transported back to the retina for rhodopsin regeneration. The feedback inhibition of isomerohydrolase by 11-*cis*-retinol(al) may be the critical mechanism for regulating the visual cycle.

Acknowledgments

The work reported here was funded by the U.S. Public Health Service National Institutes of Health Grant EY-04096. A.W. was supported by Deutsche Forschungsgemeinschaft Grant PA 222/1-1.

[15] E. Cheung and R. R. Rando, unpublished experiments (1998).

[21] Multienzyme Analysis of Visual Cycle

By HARTMUT STECHER and KRZYSZTOF PALCZEWSKI

Introduction

The visual cycle is a fundamental chemical transformation that generates 11-*cis*-retinal from all-*trans*-retinal. The chromophore 11-*cis*-retinal recombines with apoprotein opsin to form rhodopsin, the light-sensitive receptor of the visual transduction cascade.[1] Several reactions are required for this transformation. First, all-*trans*-retinal, the bleaching product, is reduced by all-*trans*-retinol dehydrogenase (RDH) in the rod outer segment (ROS). Next, all-*trans*-retinol diffuses to the retinal pigment epithelium cells (RPE). Within RPE further reactions take place: (1) all-*trans*-retinol is isomerized to 11-*cis*-retinol, and (2) 11-*cis*-retinol is oxidized to 11-*cis*-retinal by 11-*cis*-retinol dehydrogenase (11-*cis*-RDH[2]). Last, 11-*cis*-retinal diffuses back to the ROS, where rhodopsin is regenerated. Both diffusion of all-*trans*-retinol to the RPE and uptake of 11-*cis*-retinal by photoreceptors remain poorly understood.

11-*cis*-Retinol is produced enzymatically by an isomerase, which still has defied purification and molecular characterization. The proposed substrates for this reaction are all-*trans*-retinyl carboxylic esters (predominantly palmitoyl), which undergo all-*trans* to 11-*cis* isomerization coupled to ester hydrolysis.[3] Retinoid-binding proteins, such as cellular retinaldehyde-binding protein (CRALBP) or albumin (BSA), enhance the formation of 11-*cis*-retinol.[4] Several puzzling features, however, are inconsistent with this model, in which all-*trans*-retinyl esters are substrates for a putative isomerohydrolase.[5]

Phosphatidylcholine is an acyl donor for all-*trans*-retinyl ester synthesis. This reaction is catalyzed by lecithin : retinol acyltransferase (LRAT, EC 2.3.1.135) in RPE.[6] LRAT is also responsible for the formation of 11-*cis*-retinyl esters.[7] 11-*cis*-Retinyl esters are substrates for an 11-*cis*-retinyl ester-specific hydrolase, and are another source for 11-*cis*-retinol.[8] Storage and time-dependent release of 11-*cis*-retinol from the ester pool could be important for the regulation of the visual cycle. To understand the production of 11-*cis*-retinal in RPE, an assay system is required for the simultaneous detection of LRAT, retinol isomerase, and retinyl ester hydrolase activities. Techniques designed to study these enzymatic reactions in RPE are summarized below.

---

[1] A. Polans, W. Baehr, and K. Palczewski, *Trends Neurosci.* **19**, 547 (1996).
[2] A. Simon, U. Hellman, C. Wernstedt, and U. Eriksson, *J. Biol. Chem.* **270**, 1107 (1995).
[3] P. S. Deigner, W. C. Law, F. J. Canada, and R. R. Rando, *Science* **244**, 968 (1989).
[4] A. Winston and R. R. Rando, *Biochemistry* **37**, 2044 (1998).
[5] H. Stecher, M. Gelb, J. Saari, and K. Palczewski, *J. Biol. Chem.* **274**, 8577 (1999).
[6] A. Ruiz, A. Winston, Y.-H. Lim, B. A. Gilbert, R. R. Rando, and D. Bok, *J. Biol. Chem.* **274**, 3834 (1999).
[7] J. C. Saari, and D. L. Bredberg, *J. Biol. Chem.* **264**, 8630 (1989).
[8] N. L. Mata, A. T. C. Tsin, and J. P. Chambers, *J. Biol. Chem.* **267**, 9794 (1992).

Materials and Methods

General

All procedures involving retinoids are performed under dim red illumination to prevent photoisomerization and photodecomposition. Retinoids are stored under argon at $-80°$. Bovine RPE microsomes are the source for LRAT, isomerase, and retinyl ester hydrolase. Native RPE microsomes contain endogenous retinoids (mostly retinyl esters). In some experiments, these retinoids are destroyed by UV light treatment.[3]

Source of Bovine Eyes and all-trans-Retinol

Fresh bovine eyes are obtained from a local slaughterhouse (Schenk Packing Company, Stanwood, WA). all-*trans*-Retinol is purchased from Sigma (St. Louis, Mo). all-*trans*-[11,12-^{3}H(N)]retinol is purchased from NEN Life Science Products (Boston, MA).

Preparation of Retinal Pigment Epithelium Microsomes

A microsomal membrane fraction is obtained from fresh bovine RPE as described previously.[9] The microsomal fraction is resuspended in 10 mM 3-(N-morpholino) propanesulfonic acid (MOPS), pH 7.0, containing 1 μM leupeptin and 1 mM dithiothreitol (DTT) to a final protein concentration of 2.3 mg/ml according to the Bradford method,[10] and stored in small aliquots at $-80°$.

Ultraviolet Treatment

To destroy endogenous retinoids, RPE microsomes (200-μl aliquots) are irradiated in a quartz cuvette for 5 min at $0°$, using a Chromato UVE-transilluminator (UVP, Upland, CA). UV treatment produces RPE microsomes without detectable amounts of all retinoids (Fig. 2B, top, inset).

Aporecombinant Cellular Retinaldehyde-Binding Protein

Aporecombinant CRALBP (apo-rCRALBP) is expressed in *Escherichia coli* and purified to apparent homogeneity by Ni^{2+}-NTA affinity chromatography.[11] Two fractions with the highest protein content [checked

---

[9] J. C. Saari, and D. L. Bredberg, *Methods Enzymol.* **190**, 156 (1990).
[10] M. M. Bradford, *Anal. Biochem.* **72**, 248 (1976).
[11] J. W. Crabb, Y. Chen, S. Goldflam, K. West, and J. Kapron, in "Methods in Molecular Biology: Retinoid Protocols" (C. P. F. Redfern, ed.), Vol. 89, pp. 91–104. Humana Press, Totowa, New Jersey, 1998.

by sodium dodecyl sulfate–polyacrylamide gel electrophoresis (SDS–PAGE)] are dialyzed overnight against 10 mM 1,3-bis[tris(hydroxymethyl)-methylamino]propane (BTP), 250 mM NaCl. Fractions of freshly purified and dialyzed apo-rCRALBP are stored at 4° in the presence of 0.02% (w/v) NaN$_3$ and used within 1 week.

Preparation of all-trans-Retinol as Substrate for Enzyme Reaction

all-*trans*-[11,12-^{3}H(N)] Retinol is diluted with all-*trans*-retinol to give the desired specific radioactivity (550,000 dpm/nmol), and purified on a normal-phase high-performance liquid chromatography (HPLC) column [Ultrasphere-Si 5u, 4.6 × 250 mm (Altex), flow rate 1.4 ml/min, 10% (v/v) ethyl acetate in hexane] according to modified procedure published previously.[12] Aliquots of the freshly purified all-*trans*-retinol in 10% (v/v) ethyl acetate–hexane are transferred to 1.5-ml polypropylene tubes (containing between 0.5 and 10.0 nmol of all-*trans*-retinol) and dried under argon. For this procedure, the tubes are placed in a water bath at 37°. A gentle flow of argon is directed to the surface of the liquid with a Pasteur pipette. This evaporation procedure does not last longer than 2 min. Purified and dried all-*trans*-retinol is stored (0.5–10 nmol per vial at −80°) for up to 3 months.

Assay Conditions for Lecithin : Retinol Acyltransferase, Isomerase, and Hydrolase

The 1.5-ml polypropylene tube with purified and dried substrate, all-*trans*-[^{3}H] retinol (0.5–10.0 nmol, 550,000 dpm/nmol) is used as the reaction vial. Twenty microliters of 5% BSA in 10 mM BTP, pH 7.0, is added, followed by 30–40 μl of apo-rCRALBP in 10 mM BTP, pH 7.5, containing 250 mM NaCl to give a final concentration of 25 μM. Next, 10 mM BTP, pH 7.0, is added to bring the final volume to 165 μl. For some experiments, additional compounds (i.e., phosphate-containing compounds, alcohols), solubilized in 10 mM BTP, pH 7.0, are added. Finally, 35 μl of RPE microsomes (~80 μg of protein) is added to this mixture, and the reactions are incubated at 37° for the indicated times.

Extraction of Retinoids from Reaction Mixtures

The reaction mixture (180 μl of 200 μl) is transferred to a new vial containing 300 μl of ice-cold methanol, and 300 μl of hexane is added. The sample is vortexed for 2 min and centrifuged for 4 min at 14,000g

---

[12] G. M. Landers, and J. A. Olson, *J. Chromatogr.* **438**, 383 (1988).

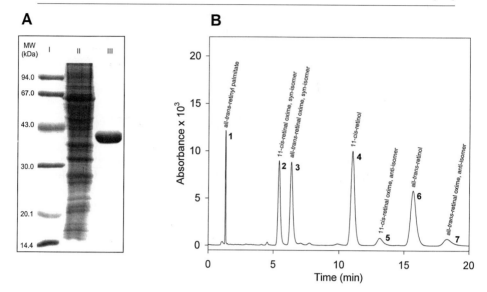

FIG. 1. Separation of retinoids and characterization of RPE microsomes and apo-rCRALBP. (A) SDS–PAGE analysis of RPE microsomes (lane II) and apo-rCRALBP (lane III). Molecular weight markers are shown in lane I. (B) HPLC separation of retinyl esters (1), *syn* 11-*cis*-retinal oxime (2), *syn* all-*trans*-retinal oxime (3), 11-*cis*-retinol (4), *anti* 11-*cis*-retinal oxime (5), all-*trans*-retinol (6), and anti all-*trans*-retinal oxime (7).

to separate organic and aqueous phases. all-*trans*-Retinol and 11-*cis*-retinol are extracted with hexane in 75–95% yield. These yields are determined with all-*trans*-[^{3}H]retinol and 11-*cis*-[^{3}H]retinol as tracers. To estimate the yield of retinyl ester extraction, all-*trans*-[^{3}H]retinol is incubated for 1 hr with RPE microsomes as a source of LRAT. Most of the all-*trans*-[^{3}H]retinol is converted to more hydrophobic all-*trans*-[^{3}H]retinyl esters, which can be extracted in ~60% yield.

High-Performance Liquid Chromatography Separation of Retinoids

Retinoids are separated using a Hewlett Packard, (HP; Palo Alto, CA) HP1100 HPLC with a diode-array detector and HP Chemstation A.04.05 software. The latter feature allows the online recording of UV spectra and identification of retinoid isomers according to their specific absorption maxima between 280 and 400 nm. A normal-phase, narrow-bore column [Alltech (Deerfield, IL) Silica 5μ Solvent Miser, 2.1 × 250 mm] and an isocratic solvent composed of 4% (v/v) ethyl acetate in hexane at a flow rate of 0.7 ml/min are used to separate retinyl esters from 11-*cis*-retinal (as oximes[5]), all-*trans*-retinal (as oximes[5]), and 11-*cis*-retinol and all-*trans*-retinol (Fig. 1B). Thirty microliters of the hexane extract is injected onto the HPLC column. Because retinals are not present, except for trace amounts in the case of native RPE microsomes (<0.1 nmol/mg of RPE protein), a similar isocratic solvent composed of 10% (v/v) ethyl acetate in hexane at a flow rate of 0.3 ml/min is used. The separation with this isocratic system can be done faster and is sufficient for the separation of retinyl esters from 11-*cis*-retinol and all-*trans*-retinol (traces in Fig. 2).

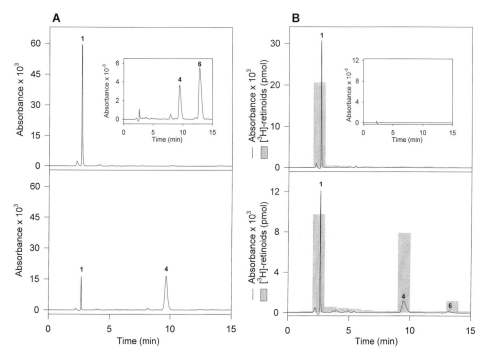

FIG. 2. Influence of apo-rCRALBP on the formation of 11-*cis*-retinol in native (A) and UV-treated (B) RPE microsomes. RPE microsomes (35 μl; 2.3 mg/ml) were incubated for 60 min at 37° in 10 mM BTP, pH 7.0, containing 1% BSA. After quenching the reaction with methanol, retinoids were extracted with hexane and one-tenth of the extract was analyzed by HPLC. (A) *Top:* HPLC traces before and after saponification (inset). *Bottom:* RPE microsomes were incubated in the presence of 25 μM apo-rCRALBP. (B) *Top:* UV-treated RPE microsomes were incubated with 2.5 μM all-*trans*-[³H]retinol (550,000 dpm/nmol). UV-treated RPE microsomes did not contain endogenous retinoids (inset). *Bottom:* UV-treated RPE microsomes were incubated with 2.5 μM all-*trans*-[³H]retinol in the presence of 25 μM apo-rCRALBP. HPLC fractions were collected and radioactivity was counted with a Beckman LS 3801. For peak identification, see Fig. 1.

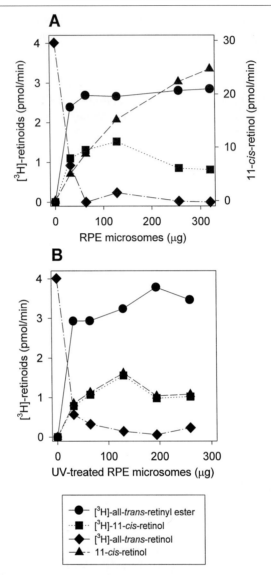

Fig. 3. Formation of all-*trans*-[³H]retinyl esters and 11-*cis*-retinol (total and ³H-labeled) as a function of RPE microsome concentrations. Native RPE microsomes (A) and UV-treated RPE microsomes (B) (in both cases 35 μl; 2.3 mg/ml) were incubated with 2.5 μM all-*trans*-[³H]retinol (550,000 dpm/nmol) in the presence of 25 μM apo-rCRALBP for 60 min at 37° in 10 mM BTP, pH 7.0, containing 1% BSA. After quenching the reaction with methanol, retinoids were extracted with hexane and one-tenth of the extract was analyzed by HPLC. HPLC fractions were collected and radioactivity was counted with a Beckman LS 3801.

Hydrolysis of Retinyl Esters

Hexane from the extracts containing retinyl esters or from the HPLC-purified retinyl ester fractions (typically 200 μl) is evaporated under argon as described above. The retinyl esters are dissolved in 230 μl of absolute ethanol and hydrolyzed with 20 μl of 6 M KOH for 30 min at 55°. To extract the products, the sample is diluted with 100 μl of water, chilled on ice for 2 min, and extracted with 300 μl of hexane. The retinoids in hexane are analyzed directly by HPLC.

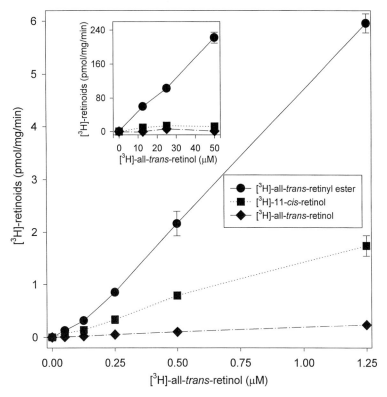

FIG. 4. Formation of all-trans-[^{3}H]retinyl esters and 11-cis-[^{3}H]retinol as a function of all-trans-[^{3}H]retinol concentrations. RPE microsomes (35 μl; 2.3 mg/ml) were incubated with different concentrations of all-trans-[^{3}H]retinol in the presence of 25 μM apo-rCRALBP for 60 min at 37° in 10 mM BTP, pH 7.0, containing 1% BSA. After quenching the reaction with methanol, retinoids were extracted with hexane and one-tenth of the extract was analyzed by HPLC. HPLC fractions were collected and radioactivity was counted with a Beckman LS 3801. Inset: A high range of the substrate concentration.

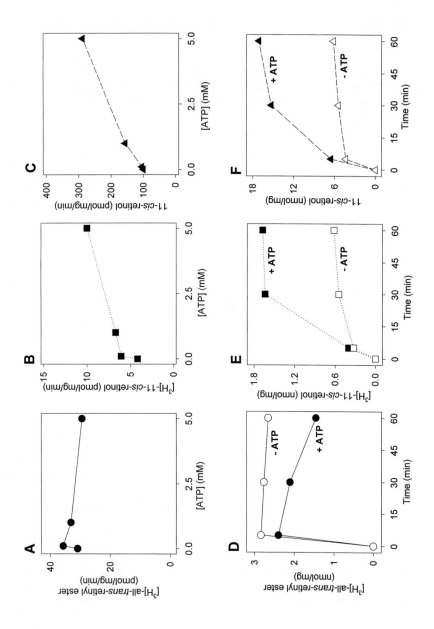

Sodium Dodecyl Sulfate–Polyacrylamide Gel Electrophoresis

SDS–PAGE is performed according to Laemmli[13] using 12% SDS–acrylamide gels in a Hoefer (San Francisco, CA) minigel apparatus and low molecular weight markers from Pharmacia Biotech (Piscataway, NJ). The gels are stained with Coomassie Brilliant Blue R-250 and destained with 50% (v/v) methanol and 7% (v/v) acetic acid.

Calculation of Results

The HPLC separation system is calibrated by injection of various amounts of retinoids. Graphs relating HPLC areas to different injected amounts of retinoids are constructed for all-*trans*-retinyl palmitate, 11-*cis*-retinol, and all-*trans*-retinol at 325 nm. Values (picomoles per unit area) are calculated over the linear portion of this relationship. all-*trans*-Retinyl palmitate and all-*trans*-retinol concentrations in hexane are determined spectrophotometrically at 325 nm with ε values of 51,770 and 49,260 M^{-1} cm^{-1}, respectively.[14,15] Radioactive HPLC fractions are collected and picomole values are calculated according to the specific radioactivity (550,000 dpm/nmol). The apo-rCRALBP concentration in buffer (10 mM BTP, 250 mM NaCl) is determined spectrophotometrically at 280 nm with ε of 9.7 ml mg^{-1} cm^{-1}.[16]

Comments

Stability of Proteins

LRAT activity was stable for 6 months, but significant variations in ester hydrolase and isomerase activities of RPE microsome preparations

---

[13] U. K. Laemmli, *Nature* (*London*) **227**, 680 (1970).
[14] R. Hubbard, P. K. Brown, and D. Bownds, *Methods Enzymol.* **18**, 615 (1971).
[15] U. Schwieter, and O. Isler, in "The Vitamins" (W. H. Sebrell, Jr. and R. S. Harris, eds.), Vol. 1, p. 5. Academic Press, New York, 1967.
[16] G. W. Stubbs, J. C. Saari, and S. Futterman, *J. Biol. Chem.* **254**, 8529 (1979).

FIG. 5. The effect of ATP on the formation of all-*trans*-[³H]retinyl esters and 11-*cis*-retinol (total and ³H-labeled). Native RPE microsomes (35 μl; 2.3 mg/ml) were incubated with all-*trans*-[³H]retinol in the presence of 25 μM apo-rCRALBP for 60 min at 37° in 10 mM BPT, pH 7.0, containing 1% BSA. After quenching the reaction with methanol, retinoids were extracted with hexane and one-tenth of the extract was analyzed by HPLC. HPLC fractions were collected and radioactivity was counted with a Beckman LS 3801. *Top:* Formation of all-*trans*-[³H]retinyl ester (A), 11-*cis*-[³H]retinol (B), and total 11-*cis*-retinol (C) as a function of ATP concentration. *Bottom:* Time-dependent formation with and without 5 mM ATP of all-*trans*-[³H]retinyl ester (D) 11-*cis*-[³H]retinol (E), and total 11-*cis*-retinol (F).

were observed depending on length of storage at −80°. The activities of isomerase and hydrolase declined by ∼50% over 3 months. Therefore, less than 2-month-old preparations were used for all studies. SDS–PAGE analysis showed a typical composition of proteins (Fig. 1A, lane II) as observed by others.[2,17] The protein pattern did not vary significantly between preparations, for RPE microsomal protein and apo-rCRALBP (Fig. 1A, lane III). The latter protein was used within 1 week of purification.

High-Performance Liquid Chromatography Separation

all-*trans*-Retinyl esters and 11-*cis*-retinyl esters were eluted 0.5 min after the solvent front (peak 1 in Figs. 1B and 2), followed by 11-*cis*-retinol (peak 4) and all-*trans*-retinol (peak 6), all with a chromatographic yield of >95%. 9-*cis*-Retinol eluted ∼1 min earlier than all-*trans*-retinol, while 13-*cis*-retinol eluted on the descending side of the 11-*cis*-retinol peak (data not shown).

Enzyme Assay

Retinoid analysis of native RPE microsomes, which were incubated without additional compounds, extracted with hexane, and separated by HPLC, showed only the presence of retinyl esters (Fig. 2A, top). Native RPE microsomes contained 0.0–0.9 nmol of all-*trans*-retinol per milligram of protein, 0.0–0.3 nmol of 11-*cis*-retinol, 8.6 ± 0.9 nmol of all-*trans*-retinyl esters, and 9.6 ± 1.4 nmol of 11-*cis*-retinyl esters. Retinyl esters were in 11-*cis* and all-*trans* configurations (Fig. 2A, top, inset). These isomers could be demonstrated via the following procedures: (1) HPLC separation and collection of the ester fraction, (2) hydrolysis of the esters, and (3) another round of HPLC separation to quantify the two retinol isomers. Different preparations of RPE microsomes had similar trace amounts of retinols and retinals; however, they differed in amounts of endogenous esters (with similar ratios between 11-*cis*- and all-*trans*-retinyl esters). In some cases, the ester pool was as high as ∼60 nmol/mg of protein. 11-*cis*-Retinol was released from 11-*cis*-retinyl esters by a hydrolase and accumulated when native RPE microsomes (with a high endogenous amount of 11-*cis*-retinyl esters) were incubated in the presence of apo-rCRALBP (Fig. 2A, bottom). Treatment of RPE microsomes with UV light destroyed all the endogenous retinoids (Fig. 2B, top, inset). Incubation of UV-treated RPE microsomes with all-*trans*-[³H]retinol in the absence of apo-rCRALBP led to esterifica-

---

[17] R. J. Barry, F. J. Canada, and R. R. Rando, *J. Biol. Chem.* **264,** 9231 (1989).

tion of the substrate by LRAT (Fig. 2B, top). When apo-rCRALBP was present during the incubation period, all-*trans*-retinol was both esterified by LRAT and converted to 11-*cis*-retinol by the isomerase (Fig. 2B, bottom). In our studies, apo-rCRALBP could not be substituted by albumin, as observed by others.[4]

Kinetics and Further Tests

This reliable enzyme assay allows us to measure the kinetic parameters of these reactions, as well as the influence of different compounds on the activities of LRAT, retinyl ester hydrolase, and retinol isomerase. The enzymatic activities of all three enzymes were a function of the concentration of native RPE microsomes (Fig. 3A) or UV-treated RPE microsomes (Fig. 3B). The influence of different substrate concentrations on LRAT and isomerase in UV-treated RPE microsomes is shown in Fig. 4. Using

TABLE I
EFFECT OF PHOSPHATE-CONTAINING COMPOUNDS ON ISOMERASE AND HYDROLASE ACTIVITY

Compounds	Concentration (mM)	11-*cis*-[^{3}H]Retinol (pmol/mg/min) (a)	11-*cis*-Retinol (pmol/mg/min) (b)	Ratio (b/a)
—	—	3.24	51.84	16.0
AMP	5	5.41	77.92	14.4
ADP	5	9.96	155.84	15.6
ATP	5	14.28	207.79	14.5
cAMP	5	4.11	60.61	14.7
Adenosine (2',5'- and 3',5'-diphosphate	5	8.87	160.17	18.0
ATPγS	5	11.25	194.80	17.3
CTP	5	14.50	203.46	14.0
GMP	5	6.06	95.24	15.7
GTP	5	15.58	168.83	10.8
cGMP	5	3.46	51.95	15.0
GTPγS	5	9.09	168.83	18.6
Sodium phosphate	10	7.78	133.92	17.2
(P_i)	100	9.07	151.2	16.6
Tetrasodium	10	15.34	211.68	13.8
pyrophosphate	20	15.80	216.45	13.7
(PP_i)	100	9.72	194.4	20.0
Tripolyphosphate (PPP_i)	25	14.07	199.13	14.1
Imidodiphosphate	10	6.26	129.6	20.7
(PNP_i)	100	10.37	168.48	16.2

TABLE II
EFFECT OF ATP ON ISOMERASE ACTIVITY

	Preincubation with all-*trans*-retinol (2.5 μM)					
	all-*trans*-Retinyl ester (nmol/mg)		11-*cis*-Retinol (nmol/mg)		all-*trans*-Retinol (nmol/mg)	
Time (min)	+ATP	−ATP	+ATP	−ATP	+ATP	−ATP
0	2.40 ± 0.01	2.30 ± 0.01	0.00	0.00	0.00	0.00
15	1.44 ± 0.20	1.64 ± 0.06	0.60 ± 0.21	0.42 ± 0.15	0.15 ± 0.02	0.00
30	1.74 ± 0.18	1.70 ± 0.03	0.90 ± 0.18	0.52 ± 0.16	0.13 ± 0.01	0.00

this assay, we tested how different compounds, such as phosphate-containing compounds or alcohols, influenced the activities of visual cycle enzymes. ATP, for example, strongly influenced the activities of isomerase and hydrolase. Their activities are increased by a factor of ~3 when ATP is present. ATP had initially no influence on LRAT activity. The apparent decrease in the ester levels after longer incubation resulted from high production and removal of 11-*cis*-retinol (Fig. 5). The ATP effect was not specific only for high-energy compounds, because a variety of phosphate-containing compounds also stimulated isomerase and hydrolase (Tables I–III). Alcohols, on the other hand, are strong inhibitors of enzymes from the visual cycle. Branched alcohols, in particular, such as 2-propanol, isobutanol, or isopentanol, profoundly inhibited isomerase activity but had little effect on LRAT activity (Table IV). The mechanistic explanation of these observations awaits further investigation.

TABLE III
EFFECT OF ATP ON HYDROLASE ACTIVITY

	Preincubation with 11-*cis*-retinol (2.5 μM)					
	all-*trans*-Retinyl ester (nmol/mg)		11-*cis*-Retinol (nmol/mg)		all-*trans*-Retinol (nmol/mg)	
Time (min)	+ATP	−ATP	+ATP	−ATP	+ATP	−ATP
0	1.84 ± 0.01	1.80 ± 0.01	0.00	0.00	0.00	0.00
5	1.53 ± 0.09	1.56 ± 0.10	0.33 ± 0.03	0.31 ± 0.01	0.00	0.00
10	1.73 ± 0.01	1.44 ± 0.06	0.63 ± 0.04	0.54 ± 0.03	0.00	0.00

TABLE IV
EFFECT OF ALCOHOLS ON LRAT AND ISOMERASE ACTIVITY

Alcohol	Structure	all-*trans*[³H] Retinyl ester (pmol/mg/min) (a)	11-*cis*-[³H]Retinol (pmol/mg/min) (b)	Ratio (a/b)
—		9.49	12.72	0.7
Methanol	CH_3OH	12.00	7.71	1.5
Ethanol	CH_3CH_2OH	10.15	3.84	2.6
n-Propanol	$CH_3CH_2CH_2OH$	6.72	2.64	2.5
2-Propanol	$H_3C{>}CHOH$ (H_3C)	12.27	0.89	13.8
Isobutanol	$H_3C{>}CHCH_2OH$ (H_3C)	10.52	0.72	14.6
Isopentanol	$H_3C{>}CHCH_2CH_2OH$ (H_3C)	14.16	1.30	10.9
2,2,2-Trifluoroethanol	F_3CCH_2OH	11.86	2.67	4.4
Benzyl alcohol	C₆H₅–CH₂OH	3.05	0.89	3.4
2-(2-Ethoxyethoxy)ethanol	$CH_3H_2C-O-CH_2H_2C-O-CH_2CH_2OH$	9.19	1.37	6.7
Geraniol	$H_3C{>}C=CHCH_2CH_2C=CHCH_2OH$ (H_3C, CH_3)	2.81	1.20	2.3

Acknowledgments

We thank Dr. J. C. Saari, Dr. M. H. Gelb, and J. Preston Van Hoosier for help during the course of these studies. This research was supported by United States Public Health Service Research Grant EY08061, and by a grant from Research to Prevent Blindness for the University of Washington, Department of Ophthalmology. K.P. is the recipient of a Jules and Doris Stein Research to Prevent Blindness Professorship.

[22] Analyzing Membrane Topology of 11-cis-Retinol Dehydrogenase

By ANDRÁS SIMON, ANNA ROMERT, and ULF ERIKSSON

Introduction

Opsins and 11-*cis*-retinaldehyde (RAL) form the visual pigments in the photoreceptor cells of the eye. The photoreceptor cells are mainly supplied with 11-*cis*-RAL by the underlying retinal pigment epithelium (RPE). The RPE obtains all-*trans*-retinol (ROH) from two main sources: the choroid circulation, and the photoreceptor cells after bleaching of the visual pigments and subsequent reduction of generated all-*trans*-RAL to all-*trans*-ROH. In the RPE, isomerization to 11-*cis*-ROH is followed by the oxidation of 11-*cis*-ROH to 11-*cis*-RAL (reviewed in Ref. 1).

The enzymatic activity responsible for the oxidation of 11-*cis*-ROH has been shown to be microsomal.[2] We have previously identified a bovine microsomal stereospecific 11-*cis*-retinol dehydrogenase (11-*cis*-RDH) and cloned the corresponding cDNA.[3] By comparing a number of bovine tissues, we showed that 11-*cis*-RDH is abundantly expressed in RPE. Thus, 11-*cis*-RDH is likely to be responsible for the generation and/or regeneration of 11-*cis*-RAL during the visual cycle.

To delineate the complex retinoid-metabolizing pathways in the RPE, and the interplay between the RPE and the adjacent photoreceptor cells, it is important to investigate the cell biology of retinoid processing. This consideration was the main motif of our studies of the membrane topology of 11-*cis*-RDH.

Hydropathy analysis of the primary structure of bovine 11-*cis*-RDH

---

[1] J. C. Saari, *in* "The Retinoids; Biology, Chemistry and Medicine" (M. B. Sporn, A. B. Roberts, and D. S. Goodman, eds.), p. 351. Raven Press, New York, 1994.
[2] F. Lion, J. P. Rotmans, F. J. M. Daemen, and S. L. Bonting, *Biochim. Biophys. Acta* **384**, 283 (1975).
[3] A. Simon, U. Hellman, C. Wernstedt, and U. Eriksson, *J. Biol. Chem.* **270**, 1107 (1995).

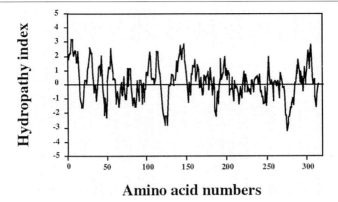

FIG. 1. Hydropathy analysis of bovine 11-*cis*-RDH. The amino acid sequence was analyzed according to Kyte and Dolittle,[3a] using a window of nine amino acid residues.

revealed the existence of an N-terminal hydrophobic putative signal sequence, which could mediate a cotranslational translocation of parts of the enzyme into the lumen of the endoplasmic reticulum (ER) (Fig. 1[3a]). To determine the membrane topology of 11-*cis*-RDH we have used several classic techniques to study the topology of membrane proteins.

Experiments were conducted by applying a protease protection technique to both RPE-derived microsomes and heterologous microsomes containing 11-*cis*-RDH generated by *in vitro* translation. As an independent method to determine the membrane topology of 11-*cis*-RDH, we have also examined the ability of mutants of 11-*cis*-RDH, containing consensus acceptor sites for N-linked glycosylation, to become glycosylated. Here we describe these methods and discuss the implications of the obtained results.

Principles of Methods

The protease protection assay is a convenient and widely used method for determining the membrane topology of membrane-associated proteins. Peptide segments of transmembrane proteins, or peripheral membrane proteins, that are present in the lumen of intact microsomes are protected from exogenously added proteases by the lipid bilayer under appropriate experimental conditions, because the protease cannot diffuse through the microsomal membrane. In contrast, peptide segments that are present in the outer aspect of the lipid bilayer of microsomes are available to exogenously

[3a] J. Kyte and R. F. Doolittle, *J. Mol. Biol.* **157,** 105 (1982).

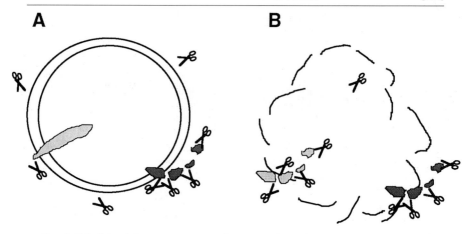

FIG. 2. Principle of the protease protection assay of membrane-associated proteins. (A) In intact microsomes, the lipid bilayer protects lumenal peptide segments, and only peptide segments exposed on the outer, cytosolic side are accessible to added proteases (represented by the scissors). (B) In detergent-solubilized microsomes, the lipid bilayer is disrupted and added proteases are able to digest peptide segments exposed on both the lumenal and the cytosolic aspect of the membrane.

added proteases, and consequently they become accessible to proteolytic digestion. It is recommended that proteases with broad specificities be used in these analyses. Most commonly used is proteinase K, a bacterial endo- and exoprotease. As a control for the ability of the protease to digest the protein, digestions should also be carried out in detergent-solubilized microsomes. The presence of detergents, preferably the nondenaturing detergent Triton X-100, disrupts the integrity of the lipid bilayer and will make the lumenal peptide segments accessible to the protease. The principle of the protease protection assay is outlined in Fig. 2.

As an independent method to control the results of the protease protection assays, mutant proteins containing ectopic glycosylation acceptor sites and/or specific antibody epitopes can be analyzed. Introducing ectopic glycosylation acceptor sites can function as reporters of lumenal orientation on synthesis and membrane translocation. The cotranslational core glycosylation reaction, whereby the first high-mannose glycan is linked to the growing polypeptide chain, occurs exclusively in the lumen of the ER.[4,5] Hence, proteins containing lumenally oriented glycosylation acceptor sites might undergo glycosylation on *in vitro* translation and cotranslational translocation into ER-derived microsomes. Glycosylation is detected by the in-

[4] C. Abeijon and C. B. Hirschberg, *Trends Biochem. Sci.* **17**, 32 (1992).
[5] R. Kornfeld and S. Kornfeld, *Annu. Rev. Biochem.* **54**, 631 (1985).

crease in the apparent molecular weight of the glycosylated protein in sodium dodecyl sulfate–polyacrylamide gel electrophoresis (SDS–PAGE) analysis. Furthermore, enzymatic digestion of the core-glycosylated protein with endoglycosidase H, or other suitable enzymes, which are able to digest the high-mannose glycans, generates a protein that migrates like unmodified protein in SDS–PAGE. The construction of glycosylation mutants is achieved by site-directed mutagenesis, using primers carrying the sequences for N-linked glycosylation acceptor sites (amino acids N–X–T/S, where N is asparagine, X is any amino acid except proline, and T/S is either threonine or serine). It should be noted that an unmodified glycosylation acceptor site should not necessarily be interpreted as having a cytosolic location, because the efficiency of the glycosylation reaction is dependent on the structural context of the ectopic site. Both the secondary structure and the folding of the polypeptide in three dimensions might affect the ability of the ectopic site to be glycosylated.[6] In particular, glycosylation sites introduced into peptide segments folded into α helices might not be easily accessible to the enzymes responsible for making the glycosylation reaction occur, because α helices can form almost immediately after synthesis of the polypeptide chain. Therefore, it might be necessary to find the optimal location of the introduced glycosylation acceptor site by "scanning" several stretches in the protein under study. The ectopic glycosylation acceptor site should not be introduced in, or near, potential membrane-spanning segments, because a minimal distance from the ER membrane is required for glycosylation to occur.[7]

Another way to generate mutants of proteins suitable for analysis of the membrane topology is to introduce ectopic antigenic epitopes into proteins by mutagenesis. By placing the ectopic epitopes in various positions in the protein under study, and then analyzing the antigenic properties of the protein in intact cells, or in isolated microsomes, information about the folding of the protein in membranes can be achieved. These types of studies can also be combined with protease protection analyses to reveal whether the epitope is accessible to proteolytic digestion. The most commonly used ectopic epitopes are the c-Myc epitope,[8] the hemagglutinin (HA) epitope from influenza virus,[9] and the FLAG epitope (Kodak, Rochester, NY). Monoclonal antibodies to these epitopes are available from several biotechnology companies. In this chapter we do not comment any further on the

---

[6] B. Holst, A. W. Bruun, M. C. Kielland-Brandt, and J. R. Winther, *EMBO J.* **15**, 3538 (1996).
[7] I. M. Nilsson and G. von Heijne, *J. Biol. Chem.* **268**, 5798 (1993).
[8] G. I. Evan, G. K. Lewis, G. Ramsay, and J. M. Bishop, *Mol. Cell. Biol.* **5**, 3610 (1985).
[9] I. A. Wilson, H. L. Niman, R. A. Houghten, A. R. Cherenson, M. L. Connolly, and R. A. Lerner, *Cell* **37**, 767 (1984).

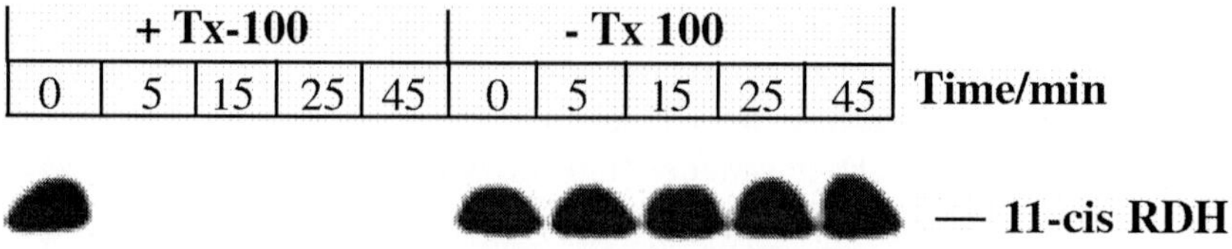

+ Tx-100					− Tx 100					
0	5	15	25	45	0	5	15	25	45	Time/min

— 11-cis RDH

FIG. 3. Protease protection analysis of 11-*cis*-RDH in RPE-derived microsomes. The protein was detected by immunoblotting after protease digestions of intact (−Tx-100) and Triton X-100-solubilized (+Tx-100) microsomes. [Modified from Simon *et al.* with permission.[10]]

use of epitope tagging, but instead focus on the use of protease protection analysis and the analysis of mutants carrying ectopic glycosylation acceptor sites in determining the membrane topology of 11-*cis*-RDH.

Protease Protection Analysis of 11-*cis*-Retinol Dehydrogenase in Retinal Pigment Epithelium-Derived Microsomes

Protease protection studies performed on microsomes isolated from bovine RPE revealed the membrane topology of 11-*cis*-RDH *in vivo*.[10] In intact microsomes, 11-*cis*-RDH appeared completely protected from the action of added proteinase K, indicated in the immunoblot analysis of the samples incubated for up to 45 min. In contrast, in detergent-solubilized microsomes, 11-*cis*-RDH was rapidly digested by the protease (Fig. 3).

Preparation of Bovine Retinal Pigment Epithelium-Derived Microsomes

Materials

Tris-HCl, pH 7.5 (1 M)
EDTA, 0.5 M
Phenylmethylsulfonyl fluoride (PMSF), 100 mM: Dissolve in 2-propanol
Aprotinin (10,000 IEK/ml; Bayer)
Sucrose, 2 M in distilled H_2O
Benzamidine, 0.5 M

RPE cells are obtained from bovine eyes. It is important that the bovine eyes be freshly isolated, and placed on ice directly after removal from the animals. RPE cells are isolated as described earlier.[11]

From about 50 eyes as starting material, RPE cells are isolated and resuspended in 20 volumes (approximately 30 ml) of swelling buffer [10

---

[10] A. Simon, A. Romert, A.-L. Gustafsson, J. M. McCaffrey, and U. Eriksson, *J. Cell Sci.* **112**, 549 (1999).
[11] C.-O. Båvik, U. Eriksson, R. A. Allen, and P. A. Peterson, *J. Biol. Chem.* **266**, 14978 (1991).

mM Tris-HCl (pH 7.5), 1 mM EDTA, 1 mM PMSF, and aprotinin (10 IU/ml)] in a 50-ml Falcon tube (Becton Dickinson Labware, Lincoln Park, NJ). After 10 min of swelling, the cells are homogenized by 15 to 20 up-and-down strokes in a glass Dounce homogenizer and transferred to a centrifuge tube. The sucrose concentration is adjusted to 0.25 M by addition of 2 M sucrose containing 10 mM Tris-HCl, pH 7.5, and 5 mM benzamidine. Cell debris and nuclei are pelleted at 1000g for 10 min at 4°. The supernatant is transferred to a new centrifuge tube, and the mitochondrial fraction is pelleted at 7700g for 15 min at 4°. The resulting postmitochondrial supernatant is loaded on a two-step sucrose gradient in an ultraclear centrifuge tube of an SW 41 rotor (Beckman, Fullerton, CA) as follows. Two milliliters of 2 M sucrose and 4.5 ml of 0.35 M sucrose are overlaid with 3.5 ml of the postmitochondrial supernatant. After centrifugation for 1.5 hr at 100,000g and 4°, the microsomes are harvested at the 0.35–2 M sucrose interface. The sucrose concentration of total RPE microsomes is diluted to 0.25 M (isotonic concentration) and the microsomes are pelleted by centrifugation at 100,000g for 60 min at 4°. This fraction is used directly or snap frozen in liquid nitrogen and stored at −80° until further processed. With this method, mostly intact right-side-out vesicles are prepared from RPE cells. This preparation is suitable for the subsequent protease protection assay.

Protease Protection Assay of Retinal Pigment Epithelium-Derived Microsomes and Immunoblotting Analysis

Materials

Phosphate-buffered saline (PBS), pH 7.2
Triton X-100, 10% in distilled H$_2$O
PMSF, 100 mM
Proteinase K in distilled H$_2$O (100 μg/ml)
PBS–0.1% Tween 20 (PBS-T)
Nonfat dry milk
Antibodies to 11-*cis*-RDH
Equipment for SDS–PAGE and immunoblotting

The entire protease protection assay is carried out on ice in 1.5-ml Eppendorf tubes. RPE-derived microsomes (20 μg of total protein) are suspended in 200 μl of PBS by as little pipetting as possible, and without vortexing. The samples are divided into two equal fractions of 90 μl. One of the fractions is diluted by the addition of 10 μl of PBS (−Triton X-100), and the other is solubilized by addition of 10 μl of 10% Triton X-100 to a final concentration of 1% (+Triton X-100). After 15 min of solubilization, 20 μl from each fraction is withdrawn to separate Eppendorf tubes containing 0.5 μl of 100 mM PMSF. To the remaining samples, 10 μl of proteinase K stock solution is added, yielding a final protease concentration of 10 μg/ml. After incubations of 5, 15, 25, and 45 min, 20-μl aliquots are withdrawn and transferred to new tubes. To stop the digestions, 0.5 μl of 100 mM PMSF is added to each tube. The samples are then subjected to SDS–PAGE and prepared for immunoblotting analysis. 11-cis-RDH is detected by incubating the filters with affinity-purified immunoglobulin to 11-cis-RDH (1 μg/ml) for 2 hr in PBS-T containing 5% nonfat dry milk. After extensive washing in PBS-T, the filters are incubated with horseradish peroxidase-labeled anti-rabbit immunoglobulin for 1 hr, washed extensively in PBS-T, and incubated with the substrate according to the recommendations of the manufacturer [ECL (enhanced chemiluminescence) protocols; Amersham, Arlington Heights, IL]. The filters are then exposed to Kodak XR film.

Analysis of in Vitro-Translated Wild-Type and Mutant 11-cis-Retinol Dehydrogenase

It is often of interest to determine the topology of an in vitro-generated membrane protein by protease protection. The recombinant protein can be expressed in transfected cells followed by the isolation of intact microsomes as described above, or it can be expressed by in vitro translation systems, where the recombinant protein is inserted into microsomes added to the reaction. The RPE-derived and in vitro-generated 11-cis-RDH can be detected by immunoblotting with specific antibodies or radiolabeled by biosynthetic incorporation of radiolabeled amino acids, normally [^{35}S]methionine or [^{35}S]cysteine. Here we describe procedures used by us to generate and analyze 11-cis-RDH, using in vitro translation in a rabbit reticulocyte lysate system. The results indicate that in vitro-translated 11-cis-RDH acquires the same membrane topology as in RPE-derived microsomes (Fig. 4).

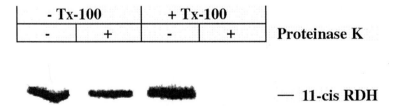

FIG. 4. Protease protection analysis of 11-cis-RDH in heterologous microsomes. Radiolabeled 11-cis-RDH was generated by in vitro translation in the presence of [^{35}S]methionine. The protein was detected by autoradiography after protease digestion of intact (−Tx-100) and Triton X-100-solubilized (+Tx-100) microsomes. [Modified from Simon et al. with permission.[10]]

In Vitro Transcription of 11-cis-Retinol Dehydrogenase mRNA

Materials

Plasmid DNA
HEPES–KOH, pH 7.4 (0.4 M)
RNA polymerase (T3, T7, or SP6), 50 U/μl
RNase-free DNase I, 10 U/μl
Nuclease-free distilled H_2O
Sodium acetate, pH 5.2 (3 M)
RNasin, 20 U/μl
mRNA cap analog [$m^7G(5')ppp(5')G$], 10 mM
Stock solutions (5 mM) of ATP, GTP, CTP, UTP
Magnesium acetate, 60 mM
Spermidine hydrochloride, 40 mM
1,4-Dithiothreitol (DTT), 1 M
Ethanol, 95%
Materials for agarose gel electrophoresis

The cDNA encoding 11-cis-RDH in pBluescript SK (Stratagene, La Jolla, CA) is linearized 3' of the insert with BamHI, and the plasmid DNA is purified by phenol extraction and precipitation. In an Eppendorf tube, 1 μg of linearized plasmid DNA is added to a buffer containing 40 mM HEPES–KOH (pH 7.4), 6 mM magnesium acetate, 4 mM spermidine hydrochloride, 20 mM DTT, 1 mM mRNA cap analog, 0.1 mM GTP, 0.5 mM ATP, 0.5 mM UTP, 0.5 mM CTP, 1 μl of RNasin, and 1 μl of T7 RNA polymerase, in a total reaction volume of 47 μl. After a 15-min incubation at 37°, 2 μl of 10 mM GTP is added. After a 45-min incubation, 1 μl of T7 polymerase is added and the sample is incubated for an additional 1 hr at 37°. Subsequently, the plasmid DNA is digested with 5 μl of RNase-free DNase at 25° for 15 min, and the in vitro-transcribed RNA is precipitated with 5.5 μl of 3 M sodium acetate and 137 μl of 95% ethanol. The RNA is resolved in 30 μl of nuclease-free distilled H_2O containing 1 μl of RNasin. The quality of the RNA is analyzed on agarose gels and stained with ethidium bromide.

In Vitro Translation of 11-cis-Retinol Dehydrogenase and Preparation of Microsomes

Materials

Synthetic mRNA
Rabbit reticulocyte lysate (RRL), microsomal (Promega, Madison, WI)
Methionine-free amino acid mixture, 1 mM

RNasin, 20 U/μl
Nuclease-free distilled H_2O
[^{35}S]Methionine, 1000 Ci/mmol
Canine pancreatic membranes, 2 Eq/μl (Promega)
PBS

The *in vitro* translation reactions are carried out at 30° for 90 min. Three microliters of synthetic mRNA (50–200 ng of RNA) from the *in vitro* transcription reaction, 17.5 μl of RRL, 2.5 μl of the methionine-free amino acid mixture, 1 μl of RNasin, 2 μl of [^{35}S]methionine (20 μCi), and 3 μl of canine pancreatic microsomal membranes are mixed in an Eppendorf tube. After the incubation, the microsomes are pelleted in an Eppendorf centrifuge at 14,000g for 20 min at 4°. The microsomes are then washed once in PBS, and resuspended in 80 μl of PBS with as little pipetting as possible and without vortexing, and subjected to protease protection studies as follows. It should be noted that when the efficiency of the *in vitro* transcription is low, the amounts of RNA added in the reactions might be increased. For optimal result, the amounts of RNA might need to be titrated for individual RNA species.

*Proteinase K Treatment of in Vitro-Translated
11-cis-Retinol Dehydrogenase*

Materials

Microsomes containing *in vitro*-translated 11-*cis*-RDH
Triton X-100, 10% in distilled H_2O
PBS
Proteinase K, 100 μg/ml in distilled H_2O
PMSF, 100 mM
Materials for SDS–PAGE and autoradiography

As for RPE-derived microsomes, the assays are carried out on ice. The resuspended microsomes are split into two Eppendorf tubes, 36 μl each. Four microliters of PBS is added to one of the tubes (intact microsomes) and 4 μl of 10% Triton X-100 to the other (solubilized microsomes). The samples are incubated on ice for 15 min to allow solubilization. Subsequently, each of the microsomal fractions is split into two additional Eppendorf tubes with 18 μl each (total of four tubes). Two microliters of PBS is added to one of them (− proteinase K), and 2 μl of the proteinase K stock solution is added to the other (+ proteinase K). The samples are incubated for 15 min and the digestions are stopped by the addition of 0.5 μl of 100 mM PMSF to each tube.

The samples are prepared for SDS–PAGE and analyzed under reducing conditions. The gels are fixed in 10% acetic acid, dried, and exposed to Kodak XR film.

Generation and Analysis of 11-cis-Retinol Dehydrogenase Mutants Containing Ectopic Glycosylation Acceptor Sites

Construction of Glycosylation Mutant 11-cis-Retionol Dehydrogenase GM71–73

Solutions and Materials

Plasmid DNA
Tris-HCl, pH 8.0 (1 M)
$MgCl_2$, 2 M
DTT, 1 M
T4 polynucleotide kinase (4.5 U/μl)
Synthetic oligonucleotide, 10 μM
E. coli dut$^-$ung$^-$, RZ 1032
LB–ampicillin medium (50 μg of ampicillin per milliliter)
Vitamin B_1, 1% (w/v) in distilled H_2O
Phage M13K07
Kanamycin solution, 50 mg/ml
Polyethylene (PEG) 8000, 20% (w/v) in 2.5 M NaCl
NaCl, 5 M
TE buffer: 10 mM Tris (pH 8.0), 0.1 mM EDTA
Phenol (saturated in TE buffer)
Chloroform
Sodium acetate, pH 5.2 (3 M)
Ethanol, 95%
Tris-HCl, pH 7.4 (1 M)
Stock solutions (50 mM) of dATP, dTTP, dGTP, dTTP
T4 DNA polymerase, 1 U/μl
T4 DNA ligase, 3 U/μl
EDTA, 0.5 M
Escherichia coli DH5α
Materials for agarose gel electrophoresis

The single-strand mutagenesis protocol is adapted from previously published methods.[12,13]

---

[12] J. Vieira and J. Messing, Methods Enzymol. **153**, 3 (1987).
[13] T. A. Kunkel, J. D. Roberts, and R. A. Zakour, Methods Enzymol. **154**, 367 (1987).

A sense oligonucleotide, 5' GACCTCCAGCGGAACATCACCTCC-CGCCTCCAC, is synthesized. The oligonucleotide encodes amino acids 67–77 in bovine 11-*cis*-RDH, with the exception that amino acids 71–73 are replaced by a sequence encoding a consensus glycosylation acceptor site, N–I–T. The oligonucleotide is phosphorylated in a 30-μl reaction containing 100 mM Tris-HCl (pH 8.0), 10 mM MgCl$_2$, 5 mM DTT, 0.4 mM ATP, 4.5 units of T4 polynucleotide kinase, and 200 pmol of the oligonucleotide. The mixture is incubated at 37° for 45 min and the reaction is terminated by incubation at 65° for 10 min. The phosphorylated oligonucleotide is then annealed to the antisense single-stranded DNA encoding 11-*cis*-RDH, which is prepared as follows.

The plasmid (pSG5) encoding bovine 11-*cis*-RDH is transformed into the *Escherichia coli* dut$^-$ung$^-$ double-mutant strain RZ 1032, and a transformant is grown overnight in LB–ampicillin medium. Ten microliters of an overnight culture is added to 1 ml of LB–ampicillin containing 1 × 10^{-3}% vitamin B$_1$ and grown at 37° until the OD$_{600}$ is 0.2. M13K07 helper phage (1 × 10^9 pfu) is added and further incubated until the OD$_{600}$ is 0.8. Three milliliters of prewarmed LB–ampicillin medium containing 1 × 10^{-3}% vitamin B$_1$ and kanamycin (70 μg/ml) is added and incubation continued for 14–18 hr with vigorous shaking at 37°. One and one-half milliliters of the overnight culture is transferred to an Eppendorf tube, centrifuged at 7000g for 10 min at 4°, and 1.3 ml of the supernatant is transferred to a fresh Eppendorf tube. The phage particles are precipitated at 4° for 2 hr with 300 μl of 20% PEG 8000 dissolved in 2.5 M NaCl buffer. The precipitate is collected by centrifugation at 7000g for 10 min and 4° and the pellet is resuspended in 200 μl of TE buffer. Single-stranded DNA is extracted from the resuspended phages by the addition of 200 μl of phenol and 50 μl of chloroform, and the aqueous phase is transferred to a new Eppendorf tube. The DNA is precipitated by adding 20 μl of 3 M sodium acetate, pH 5.2, and 500 μl of 95% ethanol. The single-stranded DNA is dissolved in 40 μl of TE buffer. The yield of DNA is quantified by spectrophotometry at 260 nm and the quality is analyzed by agarose gel electrophoresis.

The phosphorylated oligonucleotide (3 pmol) is added to a 10-μl reaction mixture containing 200 ng of the single-stranded DNA in 20 mM Tris-HCl (pH 7.4), 2 mM MgCl$_2$, and 50 mM NaCl in an Eppendorf tube. The tube is placed in a water bath at 90° and incubated for a few minutes before the water bath is turned off and the reaction mixture allowed to reach 30° (approximately 1 hr). Finally, the sample is put on ice.

After the annealing reaction, double-stranded DNA is synthesized by the addition of 1 μl of a buffer containing 5 mM dATP, 5 mM dTTP, 5 mM dGTP, 5 mM dCTP, 10 mM ATP, 100 mM Tris-HCl (pH 7.4), 50 mM MgCl$_2$, 20 mM DTT, and 1 μl of T4 DNA polymerase, and 1 μl of T4 DNA ligase. The reaction mixture is incubated on ice for 5 min, warmed to 25° for 5 min, and finally incubated at 37° for 90 min. The reaction is stopped by the addition of 0.5 μl of 0.5 M EDTA. Aliquots of the DNA are used to transform *E. coli* DH5α cells. Ampillicin-resistant colonies are inoculated in LB–ampicillin medium and grown overnight. DNA is prepared and mutant clones are identified by nucleotide sequencing. The DNA encoding the mutant is then used for *in vitro* transcription and subsequent *in vitro* translation as described previously.

Analysis of Glycosylation Pattern of Wild-Type 11-cis-Retinol Dehydrogenase and 11-cis-Retinol Dehydrogenase GM71–73

Solutions and Materials

Microsomes containing *in vitro*-translated
PBS
11-*cis*-RDH and 11-*cis*-RDH GM71–73
Sodium dodecyl sulfate (SDS), 20%
Sodium acetate (pH 5.2), 500 mM
PMSF, 100 mM in 2-propanol
Endoglycosidase H (endo H), 1 mU/μl
Materials for SDS–PAGE

In vitro transcription and translation of 11-*cis*-RDH GM71–73 are carried out as described above. Washed microsomes containing either *in vitro*-translated wild-type 11-*cis*-RDH or 11-*cis*-RDH GM71–73 are washed in PBS and solubilized at room temperature for 15 min in 50 μl of 50 mM sodium acetate, pH 5.2, containing 0.3% SDS and 1 mM PMSF. The samples are subsequently heated for 2 min at 98°. Debris is removed by centrifugation at 7000g for 2 min at room temperature. The samples are divided into two fractions of 20 μl each. Two microliters of endoglycosidase H is added to one tube (+ endo H) while 2 μl of acetate buffer is added to the other (− endo H). The samples are incubated for 12–14 hr at 34°. After the incubation, 1 μl of the enzyme is added to the endo H-treated sample and incubated for an additional 3–4 hr. The samples are then subjected to SDS–PAGE analysis and autoradiography. The increased apparent molecular weight of the mutant, 11-*cis*-RDH GM71–73, and the decrease in the apparent molecular weight after endo H treatment, show that the glycosylation acceptor site is modified by the addition of the high-mannose glycan (Fig. 5). This suggests that the region containing the glycosylation site has a lumenal orientation.

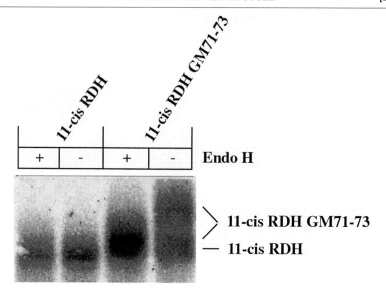

FIG. 5. Endoglycosidase H digestions of wild-type 11-*cis*-RDH and a mutant 11-*cis*-RDH, 11-*cis*-RDH GM71–73, containing an ectopic glycosylation acceptor site at positions 71–73 of the bovine protein. The increase in apparent molecular weight, and the decreased apparent molecular weight after endoglycosidase H (endo H) digestion, show that the mutant is modified by N-linked high-mannose glycans. Wild-type 11-*cis*-RDH is not glycosylated because no shift in migration in SDS–PAGE is seen on endo H treatment. [Modified from Simon *et al.* with permission.[10]]

Discussion

The results of the protease protection assay of RPE-derived microsomes show that 11-*cis*-RDH is protected in intact microsomes, because no shift in the apparent molecular weight of 11-*cis*-RDH, as revealed by SDS–PAGE, could be observed. Conversely, in the presence of Triton X-100, the enzyme was efficiently degraded, demonstrating that 11-*cis*-RDH is not protease resistant per se. This suggests that a major part of the enzyme is confined to the lumenal aspects of the microsomal membranes (Fig. 3). It can be argued that solubilization of membranes would more efficiently expose 11-*cis*-RDH to the protease, and that Triton X-100 treatment of microsomes denatures the enzyme, and thereby would make it more prone to proteolytic degradation. While these possibilities are valid, it is also well known that Triton X-100 is a nondenaturing detergent. Furthermore, the use of proteinase K, instead of a protease with a more restricted specificity, would limit the possibility that the vicinity of membranes does not allow the protease to digest the enzyme efficiently. We have investigated the ability of proteinase K to digest 11-*cis*-RDH in intact membranes, by careful sonication of the otherwise sealed microsomes in the presence of proteinase K. These studies revealed that 11-*cis*-RDH was readily digested by the protease even in the absence of Triton X-100, and also when integrated into microsomal membranes (our unpublished observations, 1998).

There are several variables that can be altered in the protease protection assay. We have obtained similar results, using a broad range of the protease (from 1 to 100 μg of proteinase K per milliliter), and similar results were also obtained using trypsin (10 μg/ml) at 37°.

When performing the protease protection assay with heterologous microsomes containing radiolabeled 11-*cis*-RDH, generated by *in vitro* translation, the protein was also protected in the intact microsomes, and readily degraded when the integrity of the microsomes was disrupted by the addition of detergent (Fig. 4). These data confirmed and extended the results of the protease protection assay performed on RPE-derived microsomes, and indicated that the membrane topology of 11-*cis*-RDH is an intrinsic property of the enzyme and is not dependent on cell-specific factors in RPE cells.

The use of ectopic N-linked glycosylation acceptor sites offers an additional method for exploring the membrane topology of 11-*cis*-RDH because this glycosylation reaction occurs exclusively in the lumen of the ER. This method is based on the assumption that the signals required for proper translation and translocation are intact in the glycosylation acceptor site mutants. An N-terminal hydrophobic signal sequence is the most common feature of proteins undergoing cotranslational translocation into the ER. In the glycosylation acceptor site mutants we have analyzed, the signals for membrane translocation have not been modified and, consequently, membrane translocation is not affected. Analysis of the *in vitro*-translated glycosylation acceptor site mutant, 11-*cis*-RDH GM71-73, by SDS-PAGE revealed two distinct bands, one of which migrated significantly slower than the wild-type protein. Treatments with endo H increased the mobility of the upper band of the mutant protein, but not of the lower band, nor did it affect the mobility of the wild-type protein (Fig. 5). These observations showed that the mutant is translocated into the lumen of the ER and becomes modified by glycosylation. The results confirmed the protease protection experiments, suggesting a lumenal orientation of the catalytic domain of the enzyme. These data also show that wild-type 11-*cis*-RDH is not glycosylated, despite the presence of a glycosylation acceptor site at positions 160-163 (amino acid residues N-I-T).

In addition to the observation that the catalytic domain of 11-*cis*-RDH appears to be confined to the lumen of the ER, we also mapped the mode of membrane insertion of the enzyme by protease protection analyses performed on different mutant, *in vitro*-translated 11-*cis*-RDHs. Those studies revealed that 11-*cis*-RDH is anchored to the lipid bilayer by an N-terminal uncleaved signal sequence and a C-terminal transmembrane region that leaves the last seven or eight amino acid residues exposed to the cytosol.[10]

The most important finding in our study of the cell biological properties of 11-*cis*-RDH is the lumenal orientation of its catalytic domain. Assuming that 11-*cis*-RDH represents the only (or the predominant) enzyme generating 11-*cis*-RAL from 11-*cis*-ROH in the RPE, our observation suggests that synthesis of 11-*cis*-RAL is a compartmentalized process in the RPE. This result might contrast with the previously proposed model that the cytosolic cellular retinaldehyde-binding protein (CRALBP) acts as a substrate carrier in the oxidation of 11-*cis*-RDH,[14,15] because the catalytic domain of the enzyme and the binding protein for the substrate apparently are present in two different subcellular compartments. One direct implication of a lumenal orientation of 11-*cis*-RDH is that the substrate, 11-*cis*-ROH, and the product, 11-*cis*-RAL, at some point are present in lumenal compartments in the RPE. Similarly, NAD(H) must also be present in the lumen of the ER for the enzyme to carry out the catalysis. A key issue in the understanding of the cell biology of retinoid processing in the RPE is how retinoids and NAD(H) are imported to, and handled in, the lumen of the ER.

The molecular and cell biological mechanisms that guide accumulation, metabolism, and transcytosis of retinoids in the RPE remain obscure. Further insights into these processes might be gained once the identity and the cell biological properties of the enzymes that are responsible for isomerization and esterification of retinol are revealed, and how the flow of substrates between these enzymes is carried out and how it is regulated.

---

[14] J. C. Saari and D. L. Bredberg, *Biochim. Biophys. Acta* **716**, 266 (1982).
[15] J. C. Saari, D. L. Bredberg, and N. Noy, *Biochemistry* **33**, 3106 (1994).

[23] Phase Partition and High-Performance Liquid Chromatography Assays of Retinoid Dehydrogenases

By JOHN C. SAARI, GREGORY G. GARWIN, FRANÇOISE HAESELEER, GEENG-FU JANG, and KRZYSZTOF PALCZEWSKI

Introduction

Photoisomerization triggers mammalian vision, converting the chromophore of rhodopsin and cone visual pigments, 11-*cis*-retinal, to all-*trans*-retinal. Regeneration of 11-*cis*-retinal for continued vision occurs in neighboring retinal pigment epithelial (RPE) cells and involves several enzymatic reactions and transcellular diffusion of the retinoids. The overall process is called the *visual cycle*.[1,2] Developments in our understanding of these reactions have been reviewed.[3–6] At least two different retinoid dehydrogenases play important roles. In photoreceptor cells, an all-*trans*-specific retinol dehydrogenase catalyzes the reduction of all-*trans*-retinal by NADPH.[7–9] This reaction is the ultimate step in quenching the reactivity of photoactivated rhodopsin[10] and its rate in mouse retina is equal to the rate of appearance of 11-*cis*-retinal in rhodopsin.[11,12] In RPE cells a *cis*-specific retinol dehydrogenase catalyzes the oxidation of 11-*cis*-retinol to 11-*cis*-retinal by NAD/NADP.[8,13,14] Both dehydrogenases are members of

---

[1] J. E. Dowling, *Nature* (*London*) **168,** 114 (1960).
[2] G. Wald, *Science* **162,** 230 (1968).
[3] D. Bok, *J. Cell Sci. Suppl.* **17,** 189 (1993).
[4] R. K. Crouch, G. J. Chader, B. Wiggert, and D. R. Pepperberg, *Photochem. Photobiol.* **64,** 613 (1996).
[5] R. R. Rando, *Chem. Biol.* **3,** 255 (1996).
[6] J. C. Saari, *in* "The Retinoids: Biology, Chemistry, and Medicine," 2nd Ed. (M. B. Sporn, A. B. Roberts, and D. S. Goodman, eds.), p. 351. Raven Press, New York, 1994.
[7] F. Haeseleer, J. Huang, L. Lebioda, J. C. Saari, and K. Palczewski, *J. Biol. Chem.* **273,** 21790 (1998).
[8] F. Lion, J. P. Rotmans, F. J. M. Daemen, and S. L. Bonting, *Biochim. Biophys. Acta* **384,** 283 (1975).
[9] G. Wald and R. Hubbard, *J. Gen. Physiol.* **32,** 367 (1949).
[10] K. P. Hofmann, A. Pulvermüller, J. Buczylko, P. Van Hooser, and K. Palczewski, *J. Biol. Chem.* **267,** 15701 (1992).
[11] K. Palczewski, J. P. Van Hooser, G. G. Garwin, J. Chen, G. I. Liou, and J. C. Saari, *Biochemistry* **37,** 12012 (1999).
[12] J. C. Saari, G. G. Garwin, J. P. Van Hooser, and K. Palczewski, *Vision Res.* **38,** 1325 (1998).
[13] C. A. G. G. Driessen, B. P. M. Janssen, H. J. Winkens, A. H. M. van Vugt, T. L. M. de Leeuw, and J. J. M. Janssen, *Invest. Ophthalmol. Visual Sci.* **36,** 1988 (1995).
[14] A. Simons, U. Hellman, C. Wernstedt, and U. Eriksson, *J. Biol. Chem.* **270,** 1107 (1995).

the short-chain dehydrogenase/reductase superfamily of oxidoreductases.[7,14] Additional *cis*- and *trans*-specific dehydrogenases may be involved in retinoid metabolism in RPE cells[15] (see discussion below).

Spectroscopic methods for assaying pyridine nucleotide-dependent dehydrogenases rely on the change in absorbance or fluorescence of the coenzyme. These methods have not been useful with retinoid dehydrogenases because of the relatively low extinction coefficient of the reduced pyridine nucleotides, the turbidity of membrane-associated enzymes, and the fluorescence of retinols and retinyl esters. We developed a phase partition assay for retinoid dehydrogenases that is rapid, accurate, and sensitive.[16] The assay relies on the transfer of tritium from water-soluble [^{3}H]NADPH or [^{3}H]NADH to water-insoluble retinoids. Thus, the activity can be detected by measuring the amount of tritium appearing in a hexane extract of the reaction mixture. Alternatively, the reaction can be followed with [15-^{3}H]retinoid substrate by measuring the amount of tritium that appears in the aqueous phase as [^{3}H]NADPH or [^{3}H]NADH after hexane extraction. These methods are suitable for dehydrogenase reactions involving water-insoluble reactants and products, such as retinoid, steroid, and eicosanoid dehydrogenases.

In this chapter we present methods for the assay of retinoid dehydrogenases by phase partition and high-performance liquid chromatography (HPLC). Examples of the utility and limitation of these assays are provided from studies of retinoid metabolism in the visual cycle. In principle, dehydrogenases that interconvert alcohols and carbonyls can be assayed either in the direction of reduction or oxidation. Pyridine nucleotide-dependent dehydrogenase reactions are pH dependent (see reaction below): low pH favors reduction and high pH favors oxidation.

$$RH_2 + NAD(P)^+ \rightleftharpoons R + NAD(P)H + H^+$$

In general, we have used pH 5.5 for phase partition assays that depend on the transfer of tritium to the hydrophobic substrate (reduction). However, we also give examples of conditions at higher pH for the phase partition assay in which tritium is transferred from the hydrophobic substrate to the pyridine nucleotide (oxidation).

---

[15] C. A. G. G. Driessen, H. J. Winkens, E. D. Kuhlmann, A. P. M. Janssen, A. H. M. van Vugt, A. F. Deutman, and J. J. M. Janssen, *FEBS Lett.* **428**, 135 (1998).

[16] J. C. Saari, D. L. Bredberg, G. G. Garwin, J. Buczylko, T. Wheeler, and K. Palczewski, *Anal Biochem.* **213**, 128 (1993).

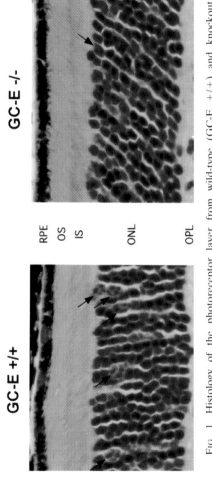

FIG. 1. Histology of the photoreceptor layer from wild-type (GC-E +/+) and knockout (GC-E −/−) mice. Light micrographs of 3-μm sections taken from 5-week-old animals and stained with hematoxylin and eosin. Original magnification: ×300. RPE, Retinal pigment epithelium; OS, outer segments; IS, inner segments; ONL, outer nuclear layer; OPL, outer plexiform layer. Cone cell nuclei are clearly identifiable in the wild-type animal (arrows); however, only a single cone is evident in the GC-E −/− littermate.

Methods

Chemical Procedures

General. All procedures involving retinoids are conducted under dim, red illumination or in aluminum foil-covered vials to minimize photoisomerization and photodecomposition. Chemical reactions are carried out in screwcap vials fitted with Teflon cap liners. Buffers are bubbled with argon and stored under argon.

Materials. 9-*cis*-, 13-*cis*-, and all-*trans*-Retinal are purchased from Sigma (St. Louis, MO) and purified by HPLC. 11-*cis*-Retinal is obtained from R. Crouch (Medical University of South Carolina, Charleston, SC) and the National Eye Institute (Bethesda, MD).

Preparation of [^{3}H]NADH and [^{3}H]NADPH. NaB^{3}H$_4$ (5 mCi; Du Pont New England Nuclear, Boston, MA) is dissolved in 0.1 ml of H$_2$O and immediately added to 5 mg of NADP or NAD (Sigma) in 0.5 ml of ice-cold 80 mM Tris buffer, pH 8.5. After 10 min of reaction at 0°, the reaction products are separated by HPLC with a column of Nucleosil SB 10 (4.6 × 250 mm) (Macherey-Nagel, Dueren, Germany) equilibrated in 10 mM KH$_2$PO$_4$, pH 7.

Preparation of pro-S-[^{3}H]NADH and pro-S-[^{3}H]NADPH. L-Glutamate dehydrogenase (400 μl; Sigma) is extensively dialyzed against 0.1 M NaCl and 10 mM Bis–Tris propane {1,3-bis[tris(hydroxymethyl)methylamino]-propane, BTP, pH 7.5 at 4°} and mixed with 0.25 mCi of L-[2,3-^{3}H]glutamic acid (24 Ci/mmol; Du Pont New–England Nuclear), 50 μl of 0.1 M L-glutamic acid, 50 μl of 27 mM ADP, and 75 μl of 0.1 M β-NAD, dissolved in H$_2$O; Sigma) or 200 μl of 0.1 M β-NADP (sodium salt, dissolved in H$_2$O; Sigma), in a final volume of 1 ml in a 1.5-ml polypropylene tube. The pH of the reaction mixture is checked with pH paper and adjusted to pH 7–8 with 1 M Tris-HCl, pH 8.4, if necessary. The tube is wrapped in aluminum foil and held at room temperature overnight with occasional mixing. Insoluble material is pelleted by centrifugation, and the pellet is washed twice with 1 ml of 10 mM BTP, pH 7.4. The supernatant is combined in a 15-ml tube and diluted with 10 mM BTP, pH 7.4, to a final volume of 9.2 ml. Half of the solution is applied to a Mono Q HR 5/5 column (Pharmacia, Piscataway, NJ) equilibrated with 10 mM BTP, pH 7.4. The column is developed with a linear gradient from 0 M NaCl in 10 mM BTP to 0.5 M NaCl over 60 min at a flow rate of 1 ml/min. The absorbance is monitored at 340 nm. The elution of radiolabeled NADH and NADPH and nonlabeled NAD and NADP is shown in Fig. 1.

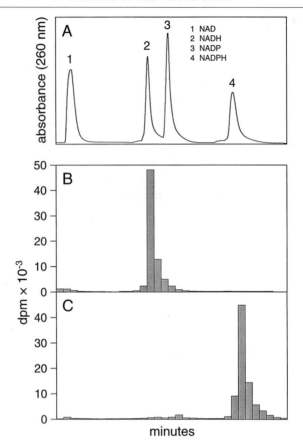

FIG. 1. Separation of pyridine nucleotides by Mono Q Sepharose ion-exchange chromatography. (A) Profile obtained by measuring the absorbance of the eluate. (B) Preparation of [^{3}H]NADH. (C) Preparation of [^{3}H]NADPH. NADP and NAD were reduced with L-[2,3-^{3}H]glutamate and glutamate dehydrogenase as described in text. The profile was obtained by following the radioactivity of the fractions.

Preparation of all-trans-[15-^{3}H]Retinal. all-*trans*-[15-^{3}H]Retinal is prepared by oxidizing all-*trans*-[15-^{3}H]retinol (Du Pont–New England Nuclear) to all-*trans*-[15-^{3}H]retinal with MnO$_2$ as described in [19] in this volume.[17] The specific radioactivity is adjusted to ~40,000 dpm/nmol by the addition of all-*trans*-retinal.

[17] G. G. Garwin and J. C. Saari, *Methods Enzymol.* **316**, [19], 1999 (this volume).

Preparation of Membranes and Enzymes

Photoreceptor Membranes. Photoreceptor membranes are prepared as described previously[18] in the dark or under red illumination for studies of all-*trans*-retinol dehydrogenase with endogenous substrate. The membranes are bleached in the presence of NH_2OH and extracted with petroleum ether when exogenous substrate is employed.

Retinal Pigment Epithelium Microsomes. Microsomes are isolated as described previously.[19]

Insect Cell Membranes. Insect cells (Sf9, *Spodoptera frugiperda* ovary cells) are transfected with a baculovirus shuttle vector (bacmid) carrying the cloned DNA of interest. The empty vector is used as a control. Insect cells are cultured at 27° in Sf-900 II SFM (Life Technologies, Gaithersburg, MD) and harvested 72–96 hr after infection by centrifugation for 15 min at 1200g. The cells are homogenized in water containing 2 mM benzamidine, 0.1 mM NADP, 0.5 mM dithiothreitol (DTT) and centrifuged at 100,000g for 30 min at 4°. The membrane fraction is washed in the same solution, pelleted by centrifugation at 100,000g for 5 min at 4°, and resuspended in five times the volume of the pellet.

Retinal Dehydrogenase. A supernatant prepared by homogenization of bovine retina is used as the source of retinal dehydrogenase.[20]

Phase Partition Assays

all-trans-Retinol Dehydrogenase with [³H]NADPH and Endogenous Substrate. In this assay,[18] the conditions described employ a pH close to neutrality to mimic more closely the *in vivo* milieu of the photoreceptor outer segment. More rapid reaction rates are obtained at pH 5.5. The reaction mixture contains 10 mM HEPES (pH 7.5) and 20 μM [³H]NADPH, in a final volume of 60 μl. Reactions are initiated by the addition of rod outer segment (ROS) membranes (0.5 to 1.0 mg/ml, unbleached) under red illumination and are either incubated in the dark (control) or flashed to produce from 0.1 to 30% bleach of the visual pigment (Auto 322 Thyrister; SunPak, Newark, NJ). After the flash, reaction mixtures are incubated for 10 min at 37°. Reactions are stopped and extracted as described below.

all-trans-Retinol Dehydrogenase with [³H]NADPH and Exogenous Substrate. The reaction mixture contains 40–50 mM 2-(*N*-morpholino)ethane-

---

[18] K. Palczewski, S. Jager, J. Buczylko, R. K. Crouch, D. L. Bredberg, K. P. Hofmann, M. A. Asson-Batres, and J. C. Saari, *Biochemistry* **33,** 13741 (1994).
[19] J. C. Saari and D. L. Bredberg, *Methods Enzymol.* **190,** 156 (1990).
[20] J. C. Saari, R. J. Champer, M. A. Asson-Batres, G. G. Garwin, J. Huang, J. W. Crabb, and A. H. Milam, *Visual Neurosci.* **12,** 263 (1995).

sulfonic acid (MES, pH 5.5), 6–10 μM [^{3}H]NADPH (20,000 dpm/nmol), and membranes from insect cells (final concentration of 2.5 to 3 mg/ml) or bleached rod outer segments (final concentration of 0.1 mg/ml) in a final volume of 100–150 μl. all-*trans*-Retinal in dimethylformamide (DMF) is added to give a final concentration of 100–140 μM (final concentration of DMF, <0.1%). After 5 to 10 min at 30°, the reaction is quenched with 400 μl of methanol, 50 μl of 100 mM NH$_2$OH (pH 7), and 50 μl of 1 M NaCl. After 2–3 min of vortexing, the radioactive product is extracted with 500 μl of hexane. Radioactivity is determined in 250–300 μl of the organic phase.

11-cis-Retinol Dehydrogenase with [^{3}H]NADPH and Exogenous Substrate. Reaction and extraction conditions employed are similar to those described for all-*trans*-retinol dehydrogenase with exogenous substrate (above). 11-*cis*-Retinal, or an isomer of interest, is added at 20 μM final concentration. The reaction is initiated by addition of RPE microsomal membranes (100 μg/ml) or insect cell membranes (2.5 to 3 mg/ml).

11-cis-Retinol Dehydrogenase with 11-cis-[15-^{3}H]Retinol. In this assay,[16] the reaction conditions employ a pH close to neutrality. More rapid oxidation is observed at pH 8.5. The reaction (200-μl final volume) is performed in plastic, 1.5-ml centrifuge tubes at 37° for 20 to 30 min. Conditions include 50 mM Tris (pH 7.2), 30 μM bovine serum albumin (BSA), 1 mM dithiothreitol (DTT), 60 μM NAD, and 10 μM 11-*cis*-[15-^{3}H]retinol (added as a solution in ethanol; final ethanol concentration <1%). The reaction is initiated by the addition of 5 μl of RPE microsomal suspension (~20 μg of protein). The reaction is stopped by the addition of 400 μl of cold methanol and 20 μl of neutralized NH$_2$OH, followed by incubation for 5 min at 37° to allow retinal oximes to form. After addition of 180 μl of 1 M NaCl, the tubes are placed on ice and 400 μl of dichloromethane is added with vortexing. A brief centrifugation is used to separate the phases. After removal of the lower (organic) phase with a Hamilton syringe, the upper phase is extracted twice more with 400 μl of dichloromethane. Finally, radioactivity is determined in 400 μl of the upper phase.

Retinaldehyde Dehydrogenase. all-*trans*-[15-^{3}H]Retinal (10 μM final concentration) is dissolved in 20 mM 3-(*N*-morpholino)propanesulfonic acid (MOPS, pH 7), containing 150 mM KCl, 20 μM BSA, 2 mM DTT, and 60 μM NAD, in a final reaction volume of 200 μl.[16,20] The reaction is initiated by the addition of retinal supernatant (0.1 to 0.2 mg of protein per milliliter). After 30 min at 37°, 920 μl of dichloromethane–methanol (1:1, v/v) is added, followed by 22 μl of freshly neutralized 1 M NH$_2$OH (final concentration 40 mM). After an incubation of 15 min to allow formation of retinal oximes, 46 μl of 1 M NaCl is added. After vortexing, the phases are separated by centrifugation (1000*g*, 1–2 min, 4°) and the upper (aqueous) phase is removed. The remaining aqueous phase is extracted again with another addition of 460 μl each of methanol and 1 M NaCl and radioactivity is determined in the combined upper phases.

Sterol Dehydrogenases. In this assay,[7,21] sterol substrates dissolved in ethanol are added to a reaction mixture containing 25 mM MES (pH 5.5) and 20 μM [^{3}H]NADPH to give a final concentration of 0.5 mM. Membranes (typically bacterial or insect cell membranes) are added to give a final concentration of 1 mg/ml. After 10 min at 30°, steroids are extracted with a 10-fold volume of dichloromethane and radioactivity is determined. The reaction product can be further identified by fluorography and thin-layer chromatography on silica with chloroform–ethyl acetate (3:1, v/v). As a positive control, the reduction of 4-androstene-3,17-dione to testosterone by rat liver microsomes is used. The following substrates are used for assays of 13α-, 11β-, or 17β-hydroxysteroid dehydrogenase activities: androsterone (5α-androstan-3α-ol-17-one), 5α-androstan-17β-ol-3-one, dihydroandrosterone (5α-androstane-3α,17β-diol), 11-dehydrocorticosterone (4-pregnen-21-ol-3,11,20-trione), corticosterone (4-pregnene-11β,21diol-3,20-dione), esterone (1,3,5[10]-estratrien-3-ol-17-one), β-estradiol (1,3,5[10]-estratrien-3,17β-diol), testosterone (4-androsten-17β-ol-3-one), 4-androstene-3,17-dione, and 5α-androstane-3,17-dione. The steroids are dissolved in ethanol to give 15–50 mM stock solutions.

High-Performance Liquid Chromatography Assays

all-trans-Retinol Dehydrogenase with Exogenous Substrate. In this assay,[18] reaction conditions are as described for the phase partition assay (see above). Reactions are stopped by the addition of 1 ml of cold ethanol and 0.1 ml of neutralized *O*-ethylhydroxylamine (Fluka, Milwaukee, WI) (final concentration 40 mM). After 10 min at room temperature, an additional 1 ml of ethanol and 5 ml of hexane are added and mixed with a Pasteur pipette. Two milliliters of H_2O is added, and the phases are separated by centrifugation. The upper (hexane) phase is removed, dried in a stream of argon, and taken up in acetonitrile, and retinoids are separated by isocratic reversed-phase HPLC (see [19] in this volume[17]). The addition of *O*-ethylhydroxylamine results in conversion of the residual retinal substrate to *O*-ethyl oximes, which are separated from the retinols on reversed-phase HPLC.[22]

11-cis-Retinol Dehydrogenase. In this assay,[16] 11-*cis*-retinal (or another isomer), dissolved in ethanol, is added to a reaction mixture containing

---

[21] M. G. Biswas and D. W. Russell, *J. Biol. Chem.* **272**, 15959 (1997).
[22] F. J. G. M. Van Kuijk, G. J. Handelman, and E. A. Dratz, *J. Chromatogr.* **348**, 241 (1985).

50 mM MOPS, (pH 5.5), 30 μM BSA, 60 μM NADPH, and 1 mM DTT (final concentration of ethanol, <0.1%). After a brief incubation at 37°, a suspension of RPE microsomes is added to give a final microsomal protein concentration of 30 μg/ml. After an appropriate time (usually 10 min) a 1-ml portion is removed and the reaction quenched by addition of 1 ml of ice-cold ethanol. After addition of 4 ml of hexane, the upper phase is repeatedly introduced into the lower phase with a Pasteur pipette, taking care not to introduce air into the sample. The upper hexane phase is removed, an additional 4 ml of hexane is introduced, and the extraction is repeated. The combined upper phases from three extractions are back-extracted with H_2O to remove dissolved ethanol, and the hexane is evaporated to dryness with flowing argon, taken up in a small volume of hexane (<200 μl), and introduced into an HPLC column. Retinoids are resolved by normal-phase HPLC as described in [19] in this volume.[17]

Discussion

The methods reported here for assay of visual cycle and other dehydrogenases can be employed advantageously for a variety of specialized purposes, or in combination to obtain further information.

Phase-Partition Assay

Improvements in the separation of pyridine nucleotides on ion-exchange resins has simplified the preparation of [^{3}H]NADPH and [^{3}H]NADH for use with the phase partition assay (Fig. 1). In addition, the use of stereospecific enzymes for the production of pro-4S and pro-4R forms of the pyridine nucleotides further extends the utility of the method. As an example, we used the phase partition assay to determine that all-*trans*-retinol dehydrogenase (RDH) from photoreceptor outer segments catalyzed transfer of the pro-4S proton of NADPH to retinol.[7] The phase partition assay is particularly useful when multiple determinations are required; for instance, during the development of purification procedures[20] and for kinetic analysis.[18] Reaction products should always be verified by HPLC analysis before routine use of the phase partition assay, especially if the substrate is readily isomerized or if additional, competing enzymes are present.

Chemical reduction of NADP is not specific to the 4-position of the substituted pyridine ring and the isomers generated are incompletely separated from each other by Nucleosil chromatography. This has not proved to be a problem because of the strict specificity of the dehydrogenases for proton transfer from the 4-position of the pyridine ring. Comparison of the results of the phase partition assay with those of the HPLC assay for all-

trans-retinol dehydrogenase of ROS has not revealed major discrepancies.[18] Reduction with enzymes and substrates specific for either the pro-R or pro-S forms of the pyridine nucleotide avoids this problem completely and provides further information about the specificity of the dehydrogenase under study. [^{3}H]NADH or [^{3}H]NADPH could be stored over a 3-month period in the dark at $-80°$ with no noticeable decomposition.

The phase partition assay was used to compare the retinoid–dehydrogenase activities of RPE microsomes with those of 11-*cis*-RDH expressed in insect cells (r11-*cis*-RDH). The recombinant enzyme showed a marked preference for 11-*cis*- and 13-*cis*-isomers of retinal (Fig. 2A and C), which was not apparent with RPE microsomes (Fig. 2B and D). The apparent activity observed with all-*trans*-retinal resulted from isomerization of the substrate (confirmed by HPLC assay). In addition, r11-*cis*-RDH was completely inhibited by NADH, whereas the 11-*cis*-RDH activity of RPE

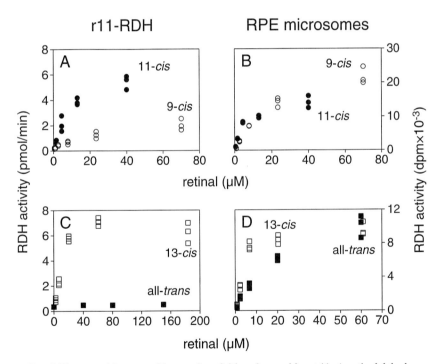

FIG. 2. Phase partition assay. Enzymatic activities of recombinant 11-*cis*-retinol dehydrogenase (r11-*cis*-RDH) (A and C) and RPE microsomes (B and D). (A)–(D) depict the results obtained when all-*trans*-, 9-*cis*-, or 13-*cis*-retinal and [^{3}H]NADPH were incubated with the indicated source of the enzyme. The large difference in the ratio of 11-*cis*-retinol to 9-*cis*-retinol dehydrogenase activities displayed by the two preparations suggests that the activities of RPE microsomes are not accounted for by r11-*cis*-RDH.

microsomes was only partially inhibited by this nucleotide (Fig. 3). The results suggest that r11-*cis*-RDH does not account for all the retinoid dehydrogenase activities of RPE microsomes and that other enzymes remain to be characterized.[15]

The phase partition assay is particularly useful for determining the enzymatic activities of expressed cDNAs encoding short-chain dehydrogenase/reductases (SDRs) thought to be involved in retinoid metabolism. Some of these enzymes also accept steroid substrates,[21] and the phase partition assay with appropriate steroid substrates can be used to assess this parameter.[7] Expression in insect cells proved to be superior to expression in bacteria and addition of NADP to the insect cell homogenate increased the stability of the dehydrogenase.

High-Performance Liquid Chromatography Assay

The HPLC assay, while less amenable to multiple samples, provides a positive identification of the product and is useful when further metabolism of the product occurs or when isomerization of the substrate is likely. These two factors are illustrated by the metabolism of various isomers of retinal

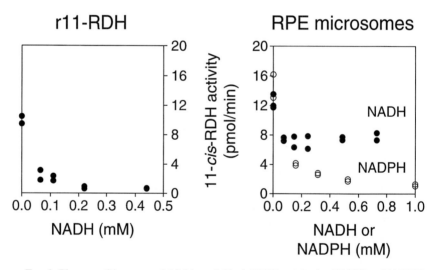

FIG. 3. Phase partition assay: inhibition of 11-*cis*-RDH activity by NADH or NADPH. 11-*cis*-RDH activity was measured at 37° for 15 min with 11-*cis*-retinal and [³H]NADPH and with RPE microsomes or recombinant 11-*cis*-RDH (r11-*cis*-RDH). Inhibition by varying concentrations of NADH or NADPH was measured. However, NADH only partially inhibits the 11-*cis*-RDH activity of RPE microsomes whereas r11-*cis*-RDH activity is inhibited completely.

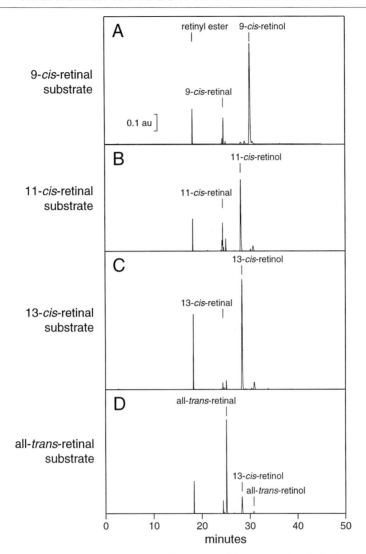

FIG. 4. HPLC assay: activity of RPE microsomes with isomers of retinal. The retinoids were incubated at 10 μM with RPE microsomes and 200 μM NADH as described in the text. After 10 min of reaction, the retinoids were extracted and analyzed by normal-phase HPLC as described. (A–C) Reduction of each of the *cis*-isomers to the corresponding retinol is evident. The retinyl ester peak indicates that the product retinol was further esterified. (D) No activity was observed with all-*trans*-retinal as the substrate. The 13-*cis*-retinol and retinyl ester observed are likely to have resulted from reduction of 13-*cis*-retinal, formed by nonenzymatic isomerization during the incubation.

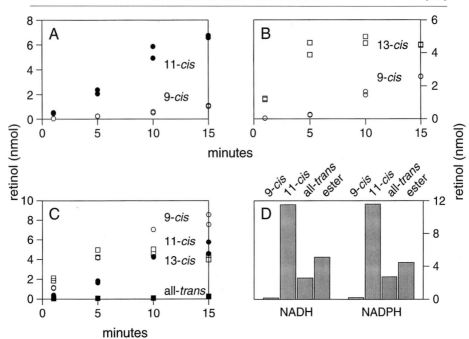

FIG. 5. HPLC assay: activity of RPE microsomes with pairs of retinal substrates. RDH activity with pairs of retinal substrates was measured at 37° with 10 μM retinoids and 200 μM NADH. (A) 11-cis-Retinal is reduced in preference to 9-cis-retinal. (B) Reduction of 9-cis-retinal is delayed until reduction of 13-cis-retinal plateaus. (C) Reduction of retinal substrates presented individually. 9-cis-Retinal is readily reduced when it is the only substrate. (D) Comparable amounts of reduction of 11-cis-retinal are obtained with NADH or NADPH. (D) also depicts the amount of nonenzymatic isomerization of 11-cis-retinal with this preparation and the amount of retinyl ester produced by subsequent esterification of the retinol.

by RPE microsomes in the presence of NADH (Fig. 4). Addition of each of the cis-isomers of retinal resulted in the appearance of the corresponding alcohol and varying amounts of retinyl ester (Fig. 4A–C). Thus, the retinol produced by the dehydrogenase reaction was esterified by the powerful lecithin : retinol acyltransferase (LRAT, EC 2.3.1.135) activity known to be present in RPE microsomes.[23,24] In contrast, addition of all-trans-retinal (Fig. 4) did not result in the appearance of all-trans-retinol, illustrating the stereospecificity of the reaction. Some substrate is converted to 13-cis-retinol, probably because of isomerization to 13-cis-retinal followed by reduction.

[23] R. J. Barry, F. J. Cañada, and R. R. Rando, J. Biol. Chem. **264,** 9231 (1989).
[24] J. C. Saari and D. L. Bredberg, J. Biol. Chem. **264,** 8636 (1989).

The HPLC assay is also useful when pairs of competing substrates are presented to an enzyme. For instance, a *cis*-specific RDH activity of RPE microsomes preferentially metabolizes 11-*cis*- or 13-*cis*-retinals when the substrates are presented paired with 9-*cis*-retinal (Fig. 5A). Others have suggested that this 11-*cis*-RDH of RPE microsomes catalyzes the first step in the biosynthesis of 9-*cis*-retinoic acid.[25,26] In RPE, where 11-*cis*-retinoids are relatively abundant, it appears unlikely that this enzyme could play a major role in the production of 9-*cis*-retinoic acid unless the reaction occurs in a defined intracellular compartment.

Perspective

Study of the physiologic roles of SDRs is complicated by the general lack of substrate specificity exhibited by these enzymes. For instance, all-*trans*-retinol dehydrogenases from liver[27] have been shown to catalyze the efficient oxidation of hydroxysteroids, raising the question of the identity of their physiological substrates.[21] A similar problem exists with defining the physiologic role of 11-*cis*-retinol dehydrogenase (11-*cis*-RDH), which has been suggested to catalyze the oxidation of 9-*cis*-retinol to 9-*cis*-retinal during the synthesis of 9-*cis*-retinoic acid.[25] Rigorous identification of a physiological substrate must include the presence of both the enzyme and its substrate in the same cellular environment. The genome of *Caenorhabditis elegans* contains ~80 genes encoding proteins with sequences containing SDR motifs (http://www.sanger.ac.uk/Projects/C_elegans/). The mammalian genome is likely to contain many more SDR-related sequences and it seems certain that assigning the physiological roles of these enzymes will be complicated. Furthermore, the presence of several enzymes with overlapping substrate specificities makes it likely that knockout animals will have a subtle phenotype.

Acknowledgments

The methods reported here were developed or utilized in the authors' laboratories with support from the National Eye Institute (EY02317, 01730, 09339), and by an unrestricted grant to the Department of Ophthalmology Research from Research to Prevent Blindness, Inc. (RBP). J.C.S. is a Senior Scientific Investigator of RBP and K.P. is a Jules and Doris Stein Professor of RBP.

[25] J. R. Mertz, E. Shang, R. Piantedosi, S. Wei, D. J. Wolgemuth, and W. S. Blaner, *J. Biol. Chem.* **272,** 11744 (1997).
[26] A. Romert, P. Tuvendal, A. Simon, L. Dencker, and U. Uriksson, *Proc. Natl. Acad. Sci. U.S.A.* **95,** 4404 (1998).
[27] M. H. E. M. Boerman and J. L. Napoli, *Biochemistry* **34,** 1027 (1995).

[24] Short-Chain Dehydrogenases/Reductases in Retina

By FRANÇOISE HAESELEER *and* KRZYSZTOF PALCZEWSKI

Introduction

The short-chain dehydrogenases/reductases (SDRs) are ubiquitously expressed enzymes present from bacteria to mammals. SDRs catalyze the oxidation/reduction of a wide range of substrates[1–3] including retinoids and steroids. Retinol dehydrogenases (RODHs)[4–6] are members of the SDR superfamily involved in the biosynthesis of all-*trans*-retinoic acid, a potent transcriptional activator.[7] The SDR superfamily also comprises enzymes that are important components of the visual cycle, including 11-*cis*-retinol dehydrogenase expressed in the retinal pigment epithelium cells (RPE) and all-*trans*-retinol dehydrogenase in photoreceptor cells. 11-*cis*-Retinol dehydrogenase oxidizes 11-*cis*-retinol to 11-*cis*-retinal, the last step in the regeneration of the chromophore of the visual pigments.[8] Photoreceptor all-*trans*-retinol dehydrogenase catalyzes the reduction of all-*trans*-retinal to all-*trans*-retinol, the final step in the quenching of photoactivated rhodopsin during phototransduction.[9]

The molecular mass of SDR superfamily members ranges from ~25 to 35 kDa. They share a common domain composed of specifically assembled α helixes and β sheets named the Rossman fold[10] and two sequence signatures: GlyXXXGlyXGly in the nucleotide-binding site and TyrXXXLys in the active site (see Fig. 1). The reducing enzymes typically employ NADPH as a cofactor while the oxidizing enzymes most often use NAD. SDRs show a stereospecificity for the cofactor, transferring specifically the pro-4S hydrogen of NAD(P)H.[1]

This chapter presents a procedure useful in searching for and cloning

---

[1] Z. Krozowski, *J. Steroid Biochem. Mol. Biol.* **51,** 125 (1994).
[2] H. Jörnvall, B. Persson, M. Krook, S. Atrian, R. Gonzalez-Duarte, J. Jeffery, and D. Ghosh, *Biochemistry* **34,** 6003 (1995).
[3] B. Persson, M. Krook, and H. Jörnvall, *Eur. J. Biochem.* **200,** 537 (1991).
[4] X. Chai, M. H. Boerman, Y. Zhai, and J. L. Napoli, *J. Biol. Chem.* **270,** 3900 (1995).
[5] X. Chai, Y. Zhai, G. Popescu, and J. L. Napoli, *J. Biol. Chem.* **270,** 28408 (1995).
[6] X. Chai, Y. Zhai, and J. L. Napoli, *Gene* **169,** 219 (1996).
[7] J. L. Napoli, *FASEB J.* **10,** 993 (1996).
[8] A. Simon, U. Hellman, C. Wernstedt, and U. Eriksson, *J. Biol. Chem.* **270,** 1107 (1995).
[9] K. Palczewski, S. Jager, J. Buczylko, R. K. Crouch, D. L. Bredberg, K. P. Hofmann, M. A. Asson-Batres, and J. C. Saari, *Biochemistry* **33,** 13741 (1994).
[10] M. G. Rossmann and P. Argos, *Mol. Cell. Biochem.* **21,** 161 (1978).

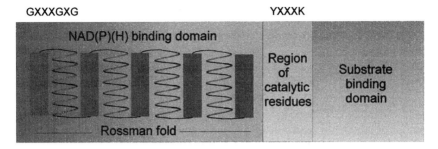

FIG. 1. Schematic representation of domains of SDRs. The cofactor-binding site consists of parallel β sheets and α helices that constitute the Rossman fold.[10] A region of catalytic residues links the nucleotide-binding domain to the substrate-binding domain. Catalytic residues and residues important for NAD(P)(H) binding are shown at the top of the scheme.

novel retinal SDR genes. We show examples of cloning of novel SDRs (retSDR2, -3, and -4)* from the retina and RPE. The cloning and characterization of retSDR1 has been described elsewhere.[11]

Methods and Results

Expressed Sequence Tag Database Search

Cloning of novel short-chain dehydrogenases/reductases is done through diverse approaches such as screening libraries under low-stringency hybridization conditions, using conserved fragments or degenerate oligonucleotides comprising conserved amino acids as probes, or by polymerase chain reaction (PCR), using two degenerate oligonucleotides designed in two conserved regions of SDRs. The generation of the expressed sequence tag (EST) database allowed us to develop a faster method to identify novel SDR genes. This method involves database searching for SDR conserved sequence motifs, cloning of full-length cDNAs, and functional analysis.

The amino acid sequence homology among SDRs is ~25–35%. Only a few amino acids important for the cofactor specificity and/or the conformation of the protein are largely conserved among SDRs, but are dispersed throughout the protein sequence and, therefore, are not useful to design primers to search for homologous proteins. On the other hand, the two

---

* The sequences reported in this chapter have been submitted to GenBank with accession numbers AF061741, AF126780, AF26781, and AF126782 for retSDR1, retSDR2, retSDR3, and retSDR4, respectively.

[11] F. Haeseleer, J. Huang, L. Lebioda, J. C. Saari, and K. Palczewski, *J. Biol. Chem.* **273**, 21790 (1998).

conserved sequence motifs, GlyXXXGlyXGly and TyrXXXLys, are more appropriate to design degenerate oligonucleotides or to search the EST database.

Two types of "probes" are used to search the EST database for new SDR genes:

1. Primers designed in SDR conserved regions: primers are designed on the basis of, for example, the sequence of an SDR with an activity similar to the one of interest. We used three primers from the RODHI sequence (GenEMBL accession number U18762): oligo I (5' TCTTCAT CACGGGCTGTGACTCGGGCTTTGG 3'; designed in the cofactor-binding site), oligo II (5' GGCCTGGTCAACAATGCTGG 3', in domainB[1]), and oligo III (5' TACTGCATCTCCAAGTTTGG 3', designed within the active site).

2. The full coding sequences of known SDRs: in our searches, we used typically 11-*cis*-RDH sequence (GenEMBL accession number Q92781), retSDR1, retSDR2, and retSDR3. These query sequences are used in FastA or tFastA softwares, which employ the method of Pearson and Lipman[12] to search for similarity between a query sequence and the EST database sequences (nucleic acid or protein).

The source of the ESTs identified through this procedure is checked first. In our case, we analyzed ESTs from retina and pineal gland. A new search is done with the selected ESTs (1) in GenEMBL to confirm that the EST is a new sequence, (2) in the EST database to obtain possibly overlapping ESTs that can be assembled in a larger sequence. The novel SDR partial sequences are then translated and checked for the presence of the two conserved motifs and highly conserved amino acids. Sequences that show an SDR conserved amino acid pattern are selected for further analysis. This EST database search is run as often as the database is updated. The EST accession numbers are indexed in a table that is constantly updated with homologous and new EST accession numbers in order to facilitate the selection of novel SDR candidates during the following searches.

Only the EST W22782 (generated by J. Nathans, Johns Hopkins School of Medicine, Baltimore, MD), helpful in cloning of retSDR1, has been identified by using oligonucleotides as query sequence. Better results were obtained by searching databases with full-length SDRs. EST H38614 (retSDR2) and EST H92520 (retSDR3) (generated by R. K. Wilson, Washington University School of Medicine, St. Louis, MO) were selected in searching databases with retSDR1. EST W26871 (retSDR4) (from J. Nathans) was selected with retSDR2 and retSDR3 database searches. The

---

[12] W. R. Pearson and D. J. Lipman, *Proc. Natl. Acad. Sci. U.S.A.* **85**, 2444 (1988).

amino acid sequence covered by these ESTs is in italic in the full-length sequence of the SDRs shown in Fig. 2. At least one SDR conserved sequence motif and amino acid is present in all selected ESTs.

It should be noted that PCRs using oligo I, II, and III (described above), or degenerated oligo I, II, and III designed in conserved motifs, have also

```
              1                                         .......βA.......       50
   retSDR1    MAWKRLGALV MFPLQMIYLV VKAAVGLVLP AKLRDLSREN VLITGGGRGI
   retSDR2    ~~MKFLLDIL LLLPLLIVCS LESFVKLFIP KRRKSVTGEI VLITGAGHGI
   retSDR3    ~~~~~~~~~~ ~~~~~~~~~~ ~~~~~~~~~M ATGTRYAGKV VVVTGGGRGI
   retSDR4    ~~~~~~~~~~ ~~~~~~~~~~ ~~~~~~~~~~ ~~~~~~~~MV VWVTGASSGI
                                                                  GXXXGX
              51..αB........  ....βB....   ......αC........                    100
   retSDR1    GRQLAREFAE RGARKIVLWG RTEKCLKETT EEIRQMG... .TECHYFICD
   retSDR2    GRLTAYEFAK LKS.KLVLWD INKHGLEETA AKCKGLG... .AKVHTFVVD
   retSDR3    GAGIVRAFVN SGAR.VVICD KDESGGRALE QELPG..... ...AVFILCD
   retSDR4    GEELAYQLSK LGV.SLVLSA RRVHELERVK RRCLENGNLK EKDILVLPLD
              G
              101                                                              150
   retSDR1    VGNREEVYQT AKAVREKVGD ITILVNNAAV .VHGKSLMDS DDDALLKSQH
   retSDR2    CSNREDIYSS AKKVKAEIGD VSILVNNAGV .VYTSDLFAT QDPQIEKTFE
   retSDR3    VTQEDDVKTL VSETIRRFGR LDCVVNNAGH HPPPQRPEET SAQGFRQLLE
   retSDR4    LTDTGSHEAA TKAVLQEFGR IDILVNNGGM .SQRSLCMDT SLDVYRKLIE
              151                                                              200
   retSDR1    INTLGQFWTT KAFLPRMLEL QNGHIVCLNS VLALSAIPGA IDYCTSKASA
   retSDR2    VNVLAHFWTT KAFLPAMTKN NHGHIVTVAS AAGHVSVPFL LAYCSSKFAA
   retSDR3    LNLLGTYTLT KLALPYL.RK SQCNVINISS LVGAIGQAQA VPYYVATKGAV
   retSDR4    LNYLGTVSLT KCVLPHMIER KCGKIVTVNS ILGIISVPLS IGYCASKHAL
                                                                  YXXXK
              201                                                              250
   retSDR1    FAFMESLT.. LGLLDCPGVS ATTVLPFHTS TEMFQGMRVR FPNLFPPLKP
   retSDR2    VGFHKTLTDE LAALQITGVK TTCLCPNFVN TGFIKNPST. ...SLGPTLEP
   retSDR3    TAMTKALALD ESPY...GVR VNCISPGNIW TPLWEELAAL MPDPRASIRE
   retSDR4    RGFFNGLRTE LATY..PGII VSNICPGPVQ SNIVENSLAG EVTKTIGNNG
              251                                                              300
   retSDR1    ETVAR...RT VEAVQLNQAL LLLPWTMHAL VILKSILPQA ALEEI.HKFS
   retSDR2    EEVVN...RL MHGILTEQKM IFIPSSIAFL TTLERILPER FLAVLKRKIS
   retSDR3    GMLAQPLGRM GQPAEVGAAA VFLASEANFC TGIELLVTGG AELGYGCKAS
   retSDR4    DQSHK...MT TSRCVRLMLI SMANDLKEVV ISEQPFLLVT YLWQYMPTWA
              301                   335
   retSDR1    GTYTCMNTFK GRT*~~~~~~ ~~~~~~~~~~ ~~~~~
   retSDR2    VKFDAVIGYK MKAQ*~~~~~ ~~~~~~~~~~ ~~~~~
   retSDR3    RSTPVDAPDI PS*~~~~~~~ ~~~~~~~~~~ ~~~~~
   retSDR4    WWITNKMGKK RIENFKSGVD ADSSYFKIFK TKHD*
```

FIG. 2. Amino acid alignment of human retinal short-chain dehydrogenases/reductases. Residues covered by the ESTs are shown in italic. The amino acids strictly conserved among SDRs are boxed in black. The residues conserved among at least 70% of the SDRs are boxed in gray. The two conserved motifs are shown under the sequences. The α helices and β sheets present in domains important for cofactor specificity are shown above the sequences. GenBank accession numbers: *retSDR1*, AF061741; *retSDR2*, AF126780; *retSDR3*, AF126781; and *retSDR4* AF126782.

been tried. This approach led to the multiple cloning of known SDRs; however, no new SDR genes were obtained.

Retinal Short-Chain Dehydrogenase/Reductase cDNA Cloning

The EST database is generated from "single-shot sequencing" and, therefore, contains only partial cDNA sequences. To clone the full-length cDNA for these novel SDRs, we have used the Marathon cDNA amplification system.[13] This is a unified method for the 5' and 3' rapid amplification of cDNA ends (RACE)[14] technique, which is used to rapidly generate full-length cDNA clones after a partial cDNA has been obtained. The procedure starts with the generation of a double-strand (ds) cDNA from an RNA sample followed by the ligation of adaptors to the ds cDNA. The cDNA of interest is amplified from this "ds cDNA-adaptor" by PCR, using a primer designed on the adaptor and a primer specific to the partially known sequence. Because multiple bands are usually generated at the first PCR, we perform a second PCR with a nested specific primer. To confirm that we have obtained correct products, we also probe a Southern blot of our RACE reactions, using the EST as a probe. This probe is generated by PCR with two EST specific primers. This is particularly useful when multiple bands are still obtained after the nested PCR.

We have used the Marathon cDNA amplification kit (Clontech, Palo Alto, CA) to clone the novel SDRs in two overlapping fragments. Human retina double-stranded cDNA is prepared from 20 μg of total RNA isolated from retinal tissue and ligated to the Marathon cDNA adaptor. Amplification of cDNA ends is performed using the Expand High Fidelity PCR System (Boehringer Mannheim, Indianapolis, IN) with the Marathon adaptor primer (AP1) and an internal gene-specific primer (GS1) as follows: after denaturation at 94° for 5 min, samples are amplified for 35 cycles at 94° for 30 sec and 68° for 4 min. A nested PCR is then performed with the nested AP2 Marathon adaptor primer and a nested gene-specific primer (GS2), following the same PCR conditions. The amplification products are cloned into a vector (pCRII vector; InVitrogen, San Diego, CA) for sequencing (by dideoxy terminator sequencing, ABI-Prism; Perkin-Elmer, Norwalk, CT).

We generally design about 23-base-long gene-specific primers that can hybridize at the polymerization temperature. The GS1 and GS2 gene-

---

[13] A. Chenchik, L. Diachenko, F. Moqadam, S. Tarabykin, S. Lukyanov, and P. D. Siebert, *BioTechniques* **21,** 526 (1996).

[14] B. C. Schaefer, *Anal. Biochem.* **227,** 255 (1995).

specific primers used to clone the 5' and 3' end of retSDRs are shown in Table I.

Alternatively, a similar procedure is followed by using a retina cDNA library as the cDNA source. One primer is then designed on the arm of the vector. However, cDNA libraries often contain partial products owing to the procedure followed to construct them. Indeed, in the procedure to construct cDNA libraries, the cDNA is often digested with one restriction enzyme (e.g., *Eco*RI, *Xho*I) prior to cloning in a vector. If the cDNA of interest contains the same restriction site, only partial clones will be present in the library.

Sequence Analysis

Once the complete amino acid sequence is obtained, we look for the criteria shared by the members of the SDR superfamily:

1. Molecular mass of ~25–35 kDa
2. Conserved motifs and highly conserved residues: In addition to the two highly conserved sequence motifs (GXXXGXG and YXXXK), SDRs have a well-conserved LXNNAG sequence of unknown function. Furthermore, ~20 additional amino acids, dispersed throughout the sequence, are conserved in at least 70% of SDRs (shown in gray boxes in Fig. 2)
3. Type of cofactor preferred: Amino acid sequence alignments and three-dimensional structure analysis of various NADP(H) and NAD(H)-preferring enzymes reveal residues conserved among NADPH and NADH enzymes. SDRs contain a motif at the amino terminus consisting of β-strand A, α-helix B, β-strand B, and α-helix C (part of the βA–αB–βB–αC–βC–αD–βD that forms the Rossman fold), which interact with the adenosine phosphate moiety of the cofactor. The residues present at the junctions βA–αB and βB–αC are thought to be important for selectivity

TABLE I
GENE-SPECIFIC PRIMERS USED FOR AMPLIFICATION OF 5' AND 3' ENDS OF retSDR2, 3, AND 4

Site	Gene-specific primer 1	Gene-specific primer 2
retSDR2		
5' end	5' CTTCTGGACATCCTCCTGCTTCTC 3'	5' CGTTACTGATCGTCTGCTCCCTAG 3'
3' end	5' CTTCTCGGTTGCTGCAGTCTACC 3'	5' GCCTTCAACCTTCTTTGCAGAGCT 3'
retSDR3		
5' end	5' ATGGCTACGGGAACGCGCTATG 3'	5' AGGTGGTGGTCGTGACCGGG 3'
3' end	5' GGGTGGGTGGTGGCCAGCGT 3'	5' TTGACAACACAATCCAGGCGGCC 3'
retSDR4		
5' end	5' GACCTGACCGACACTGGTTCCC 3'	5' CCAAAGCTGTTCTCCAGGAGTTTG 3'
3' end	5' AAAACCCCGGAGAGCATGCTTGC 3'	5' CACAGTATCCAATGGAAAGAGGTAC 3'

of NAD(H) or NADP(H). For a favorable interaction with NADP(H), positively charged residues are present at the $\beta A-\alpha B$ junction (glycine-rich motif) and/or at the beginning of α-helix C.[15,16] The presence of a negative charge at the $\beta B-\alpha C$ junction would favor binding of NAD(H).

When we have analyzed each sequence for those three criteria, we found that retSDR1 has (1) a calculated molecular mass of 33,520 kDa (typical of SDRs) (2) 22 highly conserved amino acids, and (3) the presence of RTEK in α-helix C (like RODHI[4]) and an arginine in the glycine-rich motif, which suggests NADP(H) preference as a cofactor (see Fig. 2). retSDR2 has (1) a calculated molecular mass of 32,963 kDa, (2) 21 highly conserved amino acids, and (3) a positively charged amino acid before the second glycine of the glycine-rich motif but also an isoleucine at the beginning of α-helix C, together with some positively charged residues. So, both NAD(H) and NADP(H) are possible cofactors for this enzyme. retSDR3 has (1) a calculated molecular mass of 28,030 kDa, (2) 21 highly conserved amino acids, and (3) positively charged residues in the glycine-rich motif and at the beginning of α-helix C, suggesting a preference for NADP(H). retSDR4 has (1) a calculated molecular mass of 32,267 kDa, (2) 21 highly conserved amino acids, and (3) positively charged residues only at the beginning of α-helix C, so either cofactor might be possible.

Phylogenetic Analysis

A phylogenetic analysis of the new proteins can give hints about the function of these SDRs by extrapolation from their closest homologs, as some structure–function can be maintained during evolution. It can also serve as a tool to classify new sequences among known members.

The amino acid sequences are first aligned by PileUp from the GCG package (Wisconsin Package Version 10.0; http://www.gcg.com/attractions/v10.html), which creates a multiple sequence alignment by a simplification of the progressive alignment method of Feng and Doolittle.[17] Starting with a set of aligned sequences, programs such as PAUP (phylogenetic analysis using parsimony) search for optimal trees. Growtree reconstructs a phylogenetic tree, using a distance matrix created by Distances or as described in [29] in this volume.[18] PAUPSearch (PAUP GCG interface), Distances, and Growtree are from the GCG package.

Phylogenetic trees built by those diverse softwares give similar results

---

[15] N. Tanaka, T. Nonaka, M. Nakanishi, Y. Deyashiki, A. Hara, and Y. Mitsui, *Structure* **4**, 33 (1996).
[16] I. Tsigelny and M. E. Baker, *Biochem. Biophys. Res. Commun.* **226**, 118 (1996).
[17] D. F. Feng and R. F. Doolittle, *J. Mol. Evol.* **25**, 351 (1987).
[18] S. Yokoyama, *Methods Enzymol.* **316**, Chap. 29, 2000 (this volume).

for retSDR1–retSDR4. Figure 3 shows an example of a tree built in PAUP-Search from a bootstrap analysis using neighbor-joining distance with maximum parsimony. The tree clusters retSDR1 with retSDR2 close to retSDR4 and corticosteroid 11β-dehydrogenase isozyme 1 (11βHSD1). retSDR3 is close to mouse lung carbonyl reductase (MLRC). Phylogenetic analysis classifies retSDR2, -3, and -4 closer to steroid dehydrogenases than to retinoid dehydrogenases.

Tissue Distribution of Short-Chain Dehydrogenases/Reductases

A Northern blot analysis is performed to determine the tissue distribution of SDRs. Total RNA is isolated from human or bovine retinal tissue with the Ultraspec RNA reagent (Biotecx, Houston, TX), which is a mixture

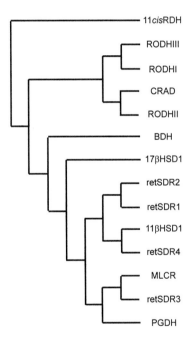

FIG. 3. Phylogenetic tree. The tree was built by a bootstrap analysis of neighbor-joining distance, using PauPSearch in GCG. Abbreviations: 11 βHSD1, corticosteroid 11 β-dehydrogenase isozyme 1 (P28845); RODHI, RODHII, and RODHIII, retinol dehydrogenase types 1 (U18762), 2 (U33500), and 3 (U33501), respectively; CRAD, *cis*-retinol/androgen dehydrogenase (AF030513); 11*cis*RDH, 11-*cis*-retinol dehydrogenase (Q92781); BDH, D-β-hydroxybutyrate dehydrogenase precursor (Q02338); 17βHSD1, estradiol 17β-dehydrogenase 1 (P14061); PGDH, 15-hydroxyprostaglandin dehydrogenase (P15428); MLCR, mouse lung carbonyl reductase (D26123); retSDR1, retinal short-chain dehydrogenase/reductase 1 (AF061741); retSDR2 (AF126780), retSDR3 (AF126781), and retSDR4 (AF126782) cloned in this study.

of guanidium salts, urea, and phenol. Poly(A)$^+$ RNAs are purified from total RNA, using a column containing oligo(dT)–cellulose (mRNA purification kit; Pharmacia-Amersham, Piscataway, NJ). Two micrograms of poly(A)$^+$-enriched RNA are resolved by agarose gel electrophoresis as described by Davis[19] in the presence of 0.66 M formaldehyde and transferred to nylon membrane. Hybridization with a ^{32}P-labeled full-length retSDR cDNA is performed in 40% (v/v) formamide, 10% (w/v) dextran sulfate, 1% (w/v) sodium dodecyl sulfate (SDS), 1 M NaCl, 50 mM Tris (pH 7.5), and herring sperm DNA (25 μg/ml). The membrane is washed at 50° in 0.1× SSC (1× SSC is 0.15 M NaCl plus 0.015 M sodium citrate) until the background monitored with a Geiger counter is low. To analyze the expression of retSDRs in other tissues, we use a human multiple tissue Northern blot containing 2 μg of poly(A)$^+$ RNA from various tissues (Clontech).

The new SDRs are cloned on the basis of ESTs from retina and, as expected, retSDR mRNAs are detected in retina (Fig. 4), although not exclusively. retSDR2 is the most widely expressed, whereas retSDR3 is almost exclusively expressed in retina. Only a low level of expression is detected in kidney. retSDR4 is expressed at similar levels in the retina, skeletal muscle, and heart.

The cellular localization of these SDR proteins in retina will be obtained by immunohistochemical methods when antibodies become available.

Expression in Insect Cells

The availability of large quantities of biologically active proteins is a prerequisite for biochemical studies of these new proteins. We have used insect cells to express functional SDRs because this system produces high levels of recombinant proteins. They usually maintain biological activity and undergo many posttranslational modifications similar to those that take place in mammalian cells. Moreover, owing to the constant development of new systems, the construction of recombinant virus has become a fast and easy procedure. We have used the baculovirus expression system developed by Luckow *et al.*[20] and commercialized by Life Technologies (Gaithersburg, MD) (Bac-to-Bac baculovirus Expression system). It is based on site-specific transposition of an expression cassette into a baculovirus shuttle vector (bacmid) propagated in *Escherichia coli*. The advantage of this system is that recombinant virus is selected in *E. coli*. Therefore,

---

[19] L. G. Davis, *in* "Basic Methods in Molecular Biology," 2nd Ed. (L. G. Davis, W. M. Kuehl, and J. F. Battey, eds.), pp. 223–350. Appleton & Lange, Norwalk, Connecticut, 1994.
[20] V. A. Luckow, S. C. Lee, G. F. Barry, and P. O. Olins, *J. Virol.* **67,** 4566 (1993).

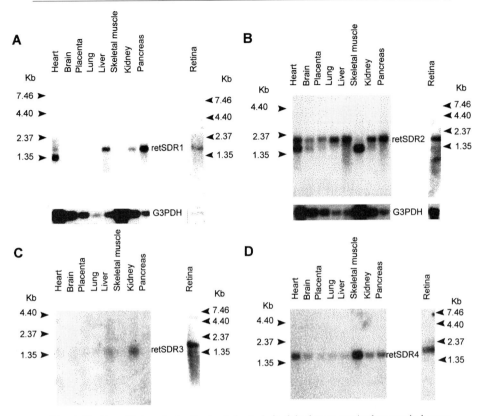

FIG. 4. Northern blot analysis of retinal short-chain dehydrogenases/reductases in human tissues and bovine retina. Poly(A)$^+$-enriched RNA from various human tissues and from bovine retina was probed with a ^{32}P-labeled retSDR1 in (A), retSDR2 in (B), retSDR3 in (C), and retSDR4 in (D). RNA molecular size standards are shown on both sides. Control hybridization with a ^{32}P-labeled glyceraldehyde-3-phosphate dehydrogenase probe is shown at the bottom of (A) and (B). The control for (C) and (D) is the same as in (B).

there is no need for long purifications of recombinant virus by plaque assay in insect cells, which can take as long as 4 to 6 weeks.

The coding sequence for retSDRs is amplified by PCR, using a sense primer that hybridizes on the initiation methionine and an antisense primer covering the stop codon as follows:

Sense retSDR2: 5' CATATGAAATTTCTTCTGGACATCCTC 3'
Antisense retSDR2: 5' TTATTGCGCTTTCATTTTATATCCAA 3'
Sense retSDR3: 5' GGTACCATGGCTACGGGAACGCGCTATG 3'
Antisense retSDR3: 5' TCACGAGCAGTTCAATGCCCGTG 3'

Sense retSDR4: 5' GGTACCATGGTGGTGGTGGGTGACTGGAG 3'
Antisense retSDR4: 5' TCAGTCATGTTTTGTCTTAAAGATTTT 3'.
The PCR conditions are the same as described previously. The purified fragment is cloned in a vector (pCRII-TOPO; Invitrogen) for sequencing. An *Xba*I–*Hin*dIII fragment (restriction sites of the pCRII-TOPO vector) covering the SDR coding sequence is then cloned between the *Xba*I and *Hin*dIII sites of the pFastBAc baculovirus expression vector (Bac-to-Bac baculovirus expression system; Life Technologies). The expression cassette is then transferred into the baculovirus shuttle vector (bacmid) by transposition (a detailed protocol can be obtained at the Life Technologies Web site: http://www.lifetech.com). Insect cells are then transfected with the recombinant bacmid, using cationic liposome-mediated transfection (Cell-Fectin reagent; Life Technologies). Four days after transfection, fresh insect cell cultures are inoculated with the supernatant to amplify the recombinant virus.

SF9 and HighFive insect cells have been tested for the expression of retSDR1. Because retSDR1 expression was detected only in SF9 cells, SF9 cells have been chosen to express retSDR1, -2, -3, and -4.

Retinoid and Steroid Dehydrogenase Activity Assay

The dehydrogenase activity of these novel retSDRs is measured by following the transfer of ^{3}H from [^{3}H]NADPH to the substrate. The assays are performed with extracts from insect cells carrying the retSDR constructs as described in [23] in this book.[21]

Membranes from insect cells expressing retSDR1 reduce specifically all-*trans*-retinal as substrate (as described in Ref. 11) but no steroid or retinol dehydrogenase activity has been detected so far for retSDR2, retSDR3, and retSDR4.

Discussion

SDRs catalyze the production of important signaling molecules in vision-related and many other tissues. In this chapter, we described a strategy to clone new SDRs from the retina and RPE. This method is applicable for the identification of new members of any superfamily as long as they share some conserved motifs. This approach takes advantage of the increasing amount of available sequences in the EST database and allows a relatively

---

[21] J. C. Saari, G. G. Garwin, F. Haeseleer, G. Jang, and K. Palczewski, *Methods Enzymol.* **316**, Chap. 23, 2000 (this volume).

fast selection of putative new members. The most difficult step is to determine the function of these enzymes. No activity has been detected so far for retSDR2, retSDR3, and retSDR4. It is possible that these recombinant proteins were not biologically active in insect cells. These enzymes might need the coexpression of other proteins to be active. Therefore, it is also advantageous to choose insect cells as an expression system because it allows simultaneous expression of multiple foreign genes in a single cell.

This method has proved successful in cloning new retinal SDRs, but this is only the first step in the characterization of these interesting enzymes. It is important to characterize functionally each member of the superfamily in order to determine which residues are essential for the substrate specificity and for the enzyme function. Both the extension of the superfamily and the full characterization of these novel enzymes are needed to define criteria necessary to search and analyze databases. This is particularly important considering that the human genome should be completely sequenced by the end of 2003.[22] At least 80 SDRs are present in the genome of *Caenorhabditis elegans*,[23] the first multicellular eukaryote to be sequenced. We can extrapolate that many more than 80 SDRs will be present in the human genome. Therefore, the experimental characterization of many SDRs is necessary to collect information to allow the analysis of the entire human genome. This is particularly important for SDRs, as the localization and substrate availability will dictate the function of these enzymes *in vivo* rather than their catalytic preferences in *in vitro* assays.

Acknowledgments

We thank Dr. J. C. Saari for advice during this project. We also thank Dr. G. Yang for help with the enzymatic assays and J. P. Van Hooser for critical reading of the manuscript. The method described was developed with support from National Institute of Health NEI Grant EY09339 and, in part, by an unrestricted grant from Research to Prevent Blindness, New York, New York. K.P. is a recipient of a Jules and Doris Stein Professorship.

---

[22] F. S. Collins, A. Patrinos, E. Jordan, A. Chakravati, R. Gesteland, and L. Walters, *Science* **282,** 682 (1998).
[23] The *C. elegans* sequencing consortium. *Science* **282,** 2012 (1998).

[25] Substrate Specificities of Retinyl Ester Hydrolases in Retinal Pigment Epithelium

By ANDREW T. C. TSIN, NATHAN L. MATA, JENNIFER A. RAY, and ELIA T. VILLAZANA

Introduction

Previous studies have shown that retinyl esters are located in the retinal pigment epithelium (RPE) of many vertebrate species from fishes, amphibians, reptilians, and birds to mammals including humans.[1-8] Rodents such as laboratory rats are somewhat different in that retinyl esters accumulate in their RPE only on light exposure.[9,10] During dark adaptation, rhodopsin regeneration takes place in the retina at the expense of retinyl esters in the RPE.[10] High-performance liquid chromatography (HPLC) has been used to show that 11-*cis*-retinyl esters were selectively depleted (over all-*trans*-retinyl esters) during visual pigment regeneration.[11]

To investigate the role of retinyl esters in the visual process, we proposed a simple hypothesis that in the RPE, there exists a substrate-specific retinyl ester hydrolase relevant to the visual cycle. In the following sections we present biochemical methods employed in our laboratory to investigate retinyl ester hydrolysis in the RPE, with an emphasis on substrate specificity.

Enzyme Assay

In the course of our study of retinyl ester hydrolysis in RPE, we have conducted retinyl ester hydrolase assays with labeled (radiometric) or non-labeled (nonradiometric) substrate. Enzyme activity is recorded by measuring product formation (by HPLC or by liquid scintillation counting). Only

---

[1] G. Wald, *J. Gen. Physiol.* **22,** 391 (1939).
[2] R. Hubbard and J. E. Dowling, *Nature (London)* **193,** 341 (1962).
[3] E. R. Berman, N. Segal, and L. Feeney, *Biochim. Biophys. Acta* **572,** 167 (1979).
[4] A. T. C. Tsin and D. D. Beatty, *Exp. Eye Res.* **29,** 15 (1979).
[5] A. T. C. Tsin and D. D. Beatty, *Exp. Eye Res.* **30,** 143 (1980).
[6] C. D. B. Bridges, R. A. Alvarez, and S. L. Fong, *Invest. Ophthal. Vis. Sci.* **22,** 706 (1982).
[7] K. A. Rodriguez and A. T. C. Tsin, *Am. J. Physiol.* **256,** R255 (1989).
[8] A. Bongiorno, L. Tesoriere, M. A. Livera, and L. Pandolpho, *Vision Res.* **39,** 1099 (1991).
[9] J. E. Dowling, *Nature (London)* **188,** 114 (1960).
[10] W. F. Zimmerman, *Vision Res.* **14,** 795 (1974).
[11] C. D. B. Bridges, *Exp. Eye Res.* **22,** 435 (1976).

Synthesis of Labeled Substrates

To carry out the radiometric assay method (modified from Prystowsky et al.[12] and Blaner et al.[13]), the labeled substrate of the enzyme reaction is synthesized as follows. Five millicuries of tritium-labeled palmitic acid (9,10-^{3}H; specificity activity, 30 Ci/mmol) in ethanol is dried under a stream of nitrogen. Dicyclohexylcarbodiimide (10 μmol in 10 μl of pyridine) is added, followed by the addition of 90 μl of pyridine. This mixture is incubated at room temperature for 30 min under argon. Cold palmitic acid (30 μmol in 300 μl of pyridine) is then added to the reaction mixture and then incubation is resumed for 1 hr. The resulting solution containing palmitic anhydride is evaporated by nitrogen, 10 μmol of retinol (in 500 μl of pyridine) is added, and the mixture is incubated overnight under argon. The synthesized retinyl palmitate is then separated from palmitic acid by passage through a 5% water-deactivated alumina column and then purified by HPLC. The specific activity of this retinyl palmitate is adjusted to about 60,000 disintegrations per minute (dpm)/nmol by the addition of cold retinyl palmitate. All procedures are carried out in dim red light.

Retinyl Ester Hydrolase Assay

Assay of enzyme activity is carried out as follows. Two nanomoles of labeled retinyl palmitate is delivered in 10 μl of ethanol to preincubated (1–2 min at 37°) reaction tubes containing 50 mM Tris–acetate (pH 8) and 1–5 μg of protein (final volume of assay mixture is 200 μl). After a 30-min incubation at 37°, the reaction is terminated by the addition of 3 ml of quenching solution (methanol–chloroform–heptane, 1.41:1.25:1, v/v/v), followed by 1 ml of 0.05 M potassium carbonate buffer, pH 10. After vortex mixing and centrifugation (1000 rpm for 15 min at 4°), 1 ml of the aqueous (upper) phase is removed for liquid scintillation counting. Retinyl ester hydrolase (REH) activity is calculated as the picomolar quantity of palmitic acid liberated per milligram of protein per minute. Background correction is carried out by incubating substrate with denatured protein.

Neutral, Bile Salt-Independent Retinyl Ester Hydrolases in Microsomes

It has been noted that the inclusion of bile salt in the assay medium significantly influences REH enzyme activity.[13] Gad and Harrison[14] showed

---

[12] J. H. Prystowsky, J. E. Smith, D. S. Goodman, *J. Biol. Chem.* **256,** 4498 (1981).
[13] W. S. Blaner, S. R. Das, P. Gour, and M. T. Flood, *J. Biol. Chem.* **262,** 53 (1987).
[14] M. Z. Gad and E. H. Harrison, *J. Lipid Res.* **32,** 685 (1991).

that in the liver, the neutral, bile salt-independent REH is relatively specific for the hydrolysis of retinyl esters while the bile salt-dependent REH activity is nonspecific for carboxyl esters and cholesteryl esters. Using bovine RPE, we have similarly concluded that in the RPE (as in the liver), the neutral, bile salt-independent REH activity in microsomes is specific for retinyl ester metabolism.[15]

Hydrolytic Reaction

Hydrolysis of retinyl esters results in the formation of equal quantities of retinol and fatty acid. We have used HPLC and liquid scintillation counting to verify that in our assay mixture, equal molar amounts of retinol and palmitic acid are released after incubating retinyl palmitate with various concentrations of microsomal protein.[16] The addition of an acyl recipient (L-α-lysophosphatidylcholine) does not increase the level of hydrolysis, and the released (labeled) fatty acid is not transferred to the acyl recipient. This confirms that *hydrolysis* rather than *acyl transfer* takes place in the enzyme assays.

Hydrolysis of Retinyl Esters by Retinal Pigment Epithelium Microsomes

Blaner *et al.*[13] first reported that hydrolysis of 11-*cis*-retinyl palmitate in human RPE homogenates takes place 20 times faster than that of the all-*trans* isomer. These results suggest that the hydrolysis of retinyl esters by RPE may be specific to the geometry of the retinol isomer. We subsequently used bovine RPE microsomes to show that hydrolysis of 11-*cis*-retinyl palmitate also takes place faster than the hydrolysis of the all-*trans* isomer.[16] However, it is not clear if this increased rate of substrate hydrolysis of the 11-*cis* isomer is due to stereospecificity of the catalytic site for the substrate or to other mechanisms.

Retinol and Fatty Acid Specificities

To investigate further the specificity of REH for the retinol moiety of the retinyl ester substrate, four [^{3}H]retinyl palmitate isomers (i.e., 9-*cis*-, 11-*cis*-, 13-*cis*-, and all-*trans*-[^{3}H]retinyl palmitate) are synthesized by reacting [9,10-^{3}H]palmitic acid anhydride with either 9-*cis*-, 11-*cis*-, 13-*cis*-, or all-*trans*-retinol, using the procedure previously described. Purification of the [^{3}H]retinyl ester products is performed by HPLC after separation from

---

[15] A. T. C. Tsin and D. W. Malsbury, *Brain Res. Bull.* **28**, 121 (1992).
[16] N. L. Mata, A. T. C. Tsin, and J. P. Chambers, *J. Biol. Chem.* **267**, 9794 (1992).

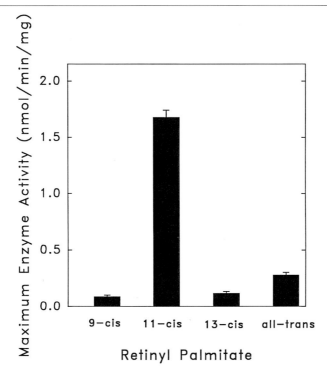

Fig. 1. Retinyl ester hydrolase activity (in RPE microsomes) for various isomers of retinyl palmitate.

the unreacted species. REH activity is quantified according to the routine radiometric assay, in which the amount of [^{3}H]palmitic acid released from a particular substrate (dpm) is related to the specific activity of that substrate (dpm/pmol). REH activities are measured with fixed protein (5 μg) and increasing concentrations of each substrate (0–100 μM). In some analyses, the released, unlabeled retinol is also measured by HPLC. As shown in Fig. 1, the highest V_{max}^{app} is associated with the 11-*cis* isomer (~1.7 versus 0.1–0.3 nmol/min/mg for the other isomers).[17] It is clear from this analysis that hydrolysis of retinyl palmitate by RPE microsomes proceeds with the greatest efficiency against the *cis* isomer, in which the bend in the conjugated double-bond system is at C-11.

The principal fatty acids present in the eye are palmitate (57%), stearate (25%), oleate (10%), myristate (1.1%), and linoleate (0.6%).[18] Retinyl palmi-

---

[17] N. L. Mata, J. R. Mata, and A. T. C. Tsin, *J. Lipid Res.* **37,** 1947 (1997).
[18] S. Futterman and J. S. Andrews, *J. Biol. Chem.* **239,** 81 (1964).

tate comprises the largest fraction of the retinyl ester pool and, in fact, is the preferred fatty acyl group for retinyl ester synthesis via lecithin : retinol acyltransferase (LRAT, EC 2.3.1.135) activity.[19] Although it is now clear that the hydrolysis of 11-*cis*-retinyl palmitate is faster than that of other isomers of retinyl palmitate, the hydrolytic rate for retinyl esters of different acyl chain lengths (or degree of saturation) has not been examined. Therefore an investigation has been undertaken to determine whether RPE microsomes exhibit different REH activities for various retinyl esters. Methods for these experiments are as follows. Unlabeled all-*trans*- and 11-*cis*-retinyl esters are synthesized by reacting each of the five principal fatty acids (as acyl chlorides) with either all-*trans*- or 11-*cis*-retinol in a standard esterification reaction.[20] Products of the reaction are separated from the unreacted species on disposable alumina columns and the retinyl ester fractions are further purified by HPLC. Thus, five different all-*trans*-retinyl esters and five different 11-*cis*-retinyl esters are synthesized. The extent to which each of the substrates can be hydrolyzed by RPE microsomal protein is then quantified in nonradiometric, single-substrate assays in which the released retinol product is measured by HPLC. Analysis of retinoids is routinely performed by normal-phase HPLC, using a 250 × 4.6 mm silica column equilibrated with either 10 or 0.2% dioxane in *n*-hexane (for analysis of retinols or retinyl esters, respectively) at a flow rate of 2 ml/min. Separation of *cis* and *trans* isomers under these conditions is typically >1.5 min. Results from this experiment have revealed a trend that approximates the retinyl ester–fatty acid composition in RPE cells. Thus, the preference of 11-*cis*-REH activity for different fatty acyl groups is as follows: palmitate > oleate = linoleate > myristate = stearate; the preference for all-*trans*-retinyl esters is as follows: palmitate > oleate = linoleate = myristate = stearate (see Mata *et al.*[20a]). Interestingly, higher rates of hydrolysis are associated with unsaturated 11-*cis*-retinyl esters compared with saturated 11-*cis*-retinyl esters (except for 11-*cis*-retinyl palmitate, which has the highest rate). It is known that the presence of double bonds in (unsaturated) fatty acids reduces the overall length of the molecule (in comparison with saturated fatty acids[21,22] with the same number of carbons). Therefore, the unsaturated, 18-carbon *cis* esters may closely approximate the length, and dimension in space, of the preferred 16-carbon, 11-*cis*-retinyl palmitate. This facilitates binding of the substrate to the active site,

---

[19] J. C. Saari and D. L. Bredberg, *J. Biol. Chem.* **264,** 8636 (1989).
[20] C. D. B. Bridges and R. A. Alvarez, *Methods Enzymol.* **81,** 463 (1982).
[20a] J. R. Mata, N. L. Mata, and A. T. C. Tsin, *J. Lipid Res.* **39,** 604 (1998).
[21] M. H. Bickel and J. Romer, *Experientia* **34,** 1047 (1978).
[22] G. Wang, H. N. Lin, S. Li, and C. H. Huang, *J. Biol. Chem.* **270,** 22738 (1995).

leading to increased hydrolysis. Therefore, these studies provide the first indication of enzyme specificity as a result of selective binding of substrates.

Substrate Specificity: Preferential Binding

It is apparent, on the basis of the data presented above, that RPE microsomes possess a higher hydrolytic activity for 11-*cis*-retinyl palmitate over all other retinyl esters. However, is it possible that this increased hydrolytic activity for 11-*cis*-retinyl palmitate is the result of this substrate being presented more "effectively" to nonspecific hydrolytic enzyme(s), rather than the result of the presence of a substrate-specific 11-*cis*-retinyl ester hydrolase (see review by Harrison[23])? By definition, enzyme specificity is "the enzyme's discrimination between several substrates competing for an active site" and the desired substrate must exhibit preferential binding as well as higher catalysis over other substrates.[24] To demonstrate enzyme specificity, we performed additional experiments to illustrate both preferential binding as well as selective catalysis.

Substrate Competition Studies

Because pure protein is not available at this time, it is not possible to obtain reliable substrate-binding data. In fact, the apparent K_m obtained from different retinyl ester substrate saturation plots are all in the micromolar range, making it difficult to discriminate binding affinities of substrates by kinetic parameters. Consequently, we have used a series of substrate competition experiments to approach the issue of substrate binding. In these studies, hydrolysis of all-*trans*-, 11-*cis*-, 13-*cis*-, and 9-*cis*-[^{3}H]retinyl palmitate is measured over time (0–10 min) in the absence and presence of an equimolar amount of unlabeled isomers of retinyl palmitate. Specifically, 10 μM labeled retinyl palmitate and 10 μM unlabeled retinyl palmitate are added (in 10 μl of ethanol) to the routine REH assay mixture and allowed to incubate for 2 min at 37°. Reactions are then initiated by the addition of 5 μg of microsomal protein (control reactions received 5 μg of heat-denatured microsomal protein) and incubation is resumed. At the indicated times representative sample and control tubes are removed, the REH reactions are quenched, and the radiolabeled free fatty acid products are partitioned into the aqueous phase as described previously. Additional experiments are also carried out in which the hydrolyses of 11-*cis*- and all-*trans*-retinyl palmitate are "challenged" by the addition of increasing

[23] E. H. Harrison, *Biochim. Biophys. Acta* **1170,** 99 (1993).
[24] A. Fresht, "Enzyme Structure and Mechanism," 2nd Ed. W. H. Freeman, New York, 1985.

concentrations of competing substrates or by the addition of [^{14}C]cholesteryl oleate. These results, as summarized in Table I, indicate the presence of an 11-*cis* specific substrate-binding site for retinyl palmitate hydrolysis in RPE microsomes because the addition of other retinyl ester isomers or cholesteryl oleate to the incubation mixture did not reduce the hydrolysis of 11-*cis*-retinyl palmitate.[20a] In contrast, hydrolysis of other retinyl ester isomers by RPE microsomes, particularly that of all-*trans*-retinyl palmitate, is significantly reduced in the presence of competing substrates including cholesteryl oleate. These data provide strong evidence of several retinyl ester hydrolases in the RPE membrane as well as the presence of a substrate-specific 11-*cis*-retinyl ester hydrolase in the RPE.

Inhibition by Chemical Modification

Efforts have been made to analyze the biochemical nature of the REH enzyme binding of different retinyl ester substrates. It is reasoned that if the hydrolysis of all-*trans*- and 11-*cis*-retinyl ester substrates are catalyzed by the same active sites, or at sites with similar amino acid composition, then modification of a specific amino acid residue within this site would alter both all-*trans*- and 11-*cis*-REH apparent affinity constants (K_m^{app}). In addition, apparent inhibitor constants (K_i^{app}) would be expected to be similar. The experimental approach involves selectively modifying either cysteine, histidine, or serine residues in RPE microsomal proteins with *N*-ethylmaleimide (NEM), L-tosylamino-2-phenylethyl chloromethyl ketone (TPCK), or phenylmethylsulfonyl fluoride (PMSF), respectively. The effect of each inhibitor on the K_m^{app} and K_i^{app} for all-*trans*- and 11-*cis*-REH activities would then be determined by substrate saturation analysis. Briefly, each inhibitor is delivered separately (in dimethyl sulfoxide, DMSO) to REH

TABLE I
SUBSTRATE COMPETITION STUDIES OF RETINYL ESTER HYDROLYSIS[a]

Competing substrate	Substrate of enzyme assay			
	all-*trans*-RP	9-*cis*-RP	11-*cis*-RP	13-*cis*-RP
all-*trans*-RP	NA	+	−	+
9-*cis*-RP	+	NA	−	+
11-*cis*-RP	+	+	NA	+
13-*cis*-RP	+	+	−	NA
Cholesteryl oleate	+	NA	−	NA

[a] The addition of competing substrate to the enzyme assay significantly (Student's *t* test; α = 0.05) reduced (+) or did not affect (−) substrate hydrolysis by RPE microsomes. For details, see text. RP, Retinyl palmitate; NA, not applicable.

reaction mixtures containing REH assay buffer and 5 μg of protein to give a final inhibitor concentration of 0.025–1 mM. Control samples containing DMSO without inhibitor are analyzed concurrently. Nonenzymatic hydrolysis of the substrates is assessed in each analysis, using heat-denatured microsomal protein. Samples are preincubated for 2 min at 37° prior to addition of substrate (1 to 20 nmol of either all-*trans*- or 11-*cis*-[^{3}H]retinyl palmitate in 10 μl of ethanol). Incubation is then resumed for 20 min at 37°. REH activities are quantified and K_m^{app} and K_i^{app} values are determined following Lineweaver–Burk transformation and Dixon plot of the data.[25] In general, K_m^{app} and K_i^{app} values determined for 11-*cis*-REH in RPE are distinct from those associated with the other REH activities for each inhibitor tested. The data show that the K_m^{app} for all-*trans*-REH activity in RPE (K_i^{app} = 100–200 μM) is altered in the presence of NEM. However, the K_m^{app} of 11-*cis*-REH in RPE is relatively unaffected by NEM (K_i^{app} = 480 μM) and therefore does not appear to rely on cysteine residues for substrate binding.[26]

Substrate Specificity: Selective Catalysis

Enzyme activity is a function of substrate binding and its subsequent catalysis to reaction products. In the following sections we delineate experiments to show that catalytic mechanisms of 11-*cis*- versus all-*trans*-retinyl ester hydrolysis are also distinct.

Inhibition of all-trans-Retinyl Ester Hydrolases by Bis(p-nitrophenyl) Phosphate

It has become clear from various characterizations of REH activities that the observed retinyl ester hydrolysis could be due, in large part, to nonspecific esterases. Consideration of the nature of the enzyme source used in these studies (i.e., microsomal protein) has led logically to the idea that microsomal carboxylesterases could be catalyzing the hydrolysis of the retinyl ester substrates. In efforts to determine the extent of the contribution of nonspecific carboxylesterases to the observed REH activities, microsomal proteins are pretreated (20 min at 4°) with increasing concentrations of an irreversible inhibitor of microsomal carboxylesterases, bis(*p*-nitrophenyl) phosphate (pNPP; 1–5 mM in DMSO) prior to REH assay. pNPP acts by irreversibly binding to prone catalytic sites, thereby precluding subsequent binding and hydrolysis of alternative substrates. However, this inhibitor can

[25] T. Palmer, "Understanding Enzymes," 4th Ed. Prentice Hall, London, 1995.
[26] N. L. Mata, Doctoral dissertation. University of Texas at San Antonio, Texas, 1997.

be hydrolyzed by nonsusceptible hydrolytic enzymes; therefore, it became important to remove excess pNPP by extensively washing (homogenization in 10 mM Tris–acetic acid, pH 7) and repelleting (150,000g, 60 min, 4°) the pNPP-treated protein. Control samples containing only the inhibitor vehicle are treated similarly. pNPP treatment causes marked dose-dependent inhibition of all-*trans*-REH activities (17–56%, 1–5 mM). In contrast, 11-*cis*-REH activity is relatively unaffected throughout the examined concentration range (17% inhibition at 5 mM).[26]

Inhibition of all-trans-Retinyl Ester Hydrolase Activity by Phospholipase A_2

During our study of the properties of 11-*cis*-REH in RPE, it was noted that this enzyme is largely associated with plasma membrane (PM)-enriched fractions (S2) of sucrose gradients prepared from RPE microsomes (see below). This suggested the possibility that this activity may be dependent on phospholipid in the membrane milieu. In fact, the integrity of phospholipids in this compartment would be crucial if the observed 11-*cis*-REH activity were actually due to a transesterification reaction, as the LRAT mechanism appears to rely on the acyl group in the *sn*-2 position of phospholipids for substrate recognition.[19] In addition, information regarding the phospholipid dependence of 11-*cis*-REH activity would be useful during purification protocols. With this impetus, the effects of three different phospholipases (A_2, C, and D) on REH activity in S2 membranes have been examined. The phospholipases are added separately to aliquots of S2 membranes and thoroughly mixed. After a 20-min incubation on ice, the mixture is centrifuged (150,000g, 60 min, 4°) and the resulting pellet washed with a low ionic strength buffer to remove excess PLA_2. To ensure that the washing procedure is sufficient to remove excess PLA_2, the hydrolysis of [^{14}C]phosphatidylcholine by the pooled supernatants and the PLA_2-treated S2 membrane fraction is measured. The data show that the washing procedure is indeed effective, as the initial PLA_2 activity is nearly quantitatively recovered in the supernatant fractions and the PLA_2-treated membrane fraction demonstrates no hydrolysis of [^{14}C]phosphatidylcholine. Hydrolysis of 11-*cis*- and all-*trans*-[^{3}H]retinyl palmitate is then determined with PLA_2-treated S2 membrane fractions. Treatment of S2 protein with the examined phospholipases (up to 14 mg of phospholipase per milligram of S2) does not result in significant inhibition of 11-*cis*-REH activity. However, PLA_2 treatment does produce a profound, dose-dependent inhibition of the all-*trans*-REH component in S2 (Fig. 2).[26] Thus, all-*trans*-REH activity is completely abolished when the PLA_2 concentration is greater than 10 mg/mg of S2 protein; inhibition of 11-*cis*-REH activity is less than

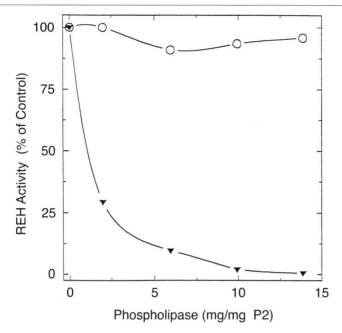

FIG. 2. Effect of phospholipase A$_2$ on 11-*cis*- and all-*trans*-REH activities in plasma membrane-enriched (S2) membranes. Values are indicated as a percentage of control (prior to PLA$_2$ treatment). (▲) all-*trans*-REH activity; (○) 11-*cis*-REH activity.

5% in this concentration range. [It is noteworthy that although the PLA$_2$ preparation employed in our studies does contain detectable protease activity (assayed in a colorimetric assay in which primary amines, which result from hydrolysis of succinylated casein, react with trinitrobenzenesulfonic acid to produce a 450 nm-absorbing conjugate; Pierce, Rockford, IL), the concentrations used in the present study cannot account for the pronounced inhibition observed for all-*trans*-REH activity in S2.] These results show that the catalysis of 11-*cis*-retinyl ester proceeds distinctly from that of the all-*trans* isomer.

Properties of 11-*cis*-Retinyl Ester Hydrolases in Retinal Pigment Epithelium

Colocalization of 11-cis-Retinyl Ester Hydrolases and 11-cis-Retinyl Esters in Retinal Pigment Epithelium Plasma Membrane

To evaluate the physiological relevance of 11-*cis*-REH in the visual pathway, RPE membranes are fractionated in a sucrose gradient and the distribution of recovered protein, enzyme markers, and 11-*cis*- and all-*trans*-REH activities is recorded.[27] Enzyme markers for endoplasmic reticulum (ER, glucose-6-phosphatase and NADPH-cytochrome *c* reductase) and plasma membrane (PM, alkaline phosphatase and ouabain-sensitive Na^+,K^+-ATPase) are used to verify the enrichment of these membranes in each sucrose gradient fraction. Analysis of REH activities reveals that 11-*cis*-REH activity is primarily associated with the PM-enriched fraction, while the greatest percentage of all-*trans*-REH activity is recovered in the ER-enriched fraction.[27] It is noteworthy that the specific activity of 11-*cis*-retinyl ester hydrolysis is increased (about twofold) in PM-enriched membrane relative to that of microsomal membrane. These results confirm again the multiplicity of hydrolytic enzymes in the RPE membrane and that 11-*cis*-REH is a separate protein from all-*trans*-REH. Additional experimental results (see below) provide further support for this suggestion.

Retinoids are extracted from RPE microsomal protein in sucrose gradient fractions. Protein samples (5–10 mg of protein in a total volume of 1 ml) are incubated at room temperature for 5 min in the presence of 2 volumes of absolute ethanol in order to liberate membrane- and protein-associated retinoids. Retinoids are then partitioned into 15 ml of petroleum ether (three 5-ml extractions). The petroleum ether extract is evaporated to dryness under a stream of N_2 and the residue is resuspended in 1 ml of hexane and applied to an alumina column in order to separate retinyl esters (eluted with two 5-ml volumes of 0.5% dioxane–*n*-hexane) and retinols (eluted with two 5-ml volumes of 10% dioxane–*n*-hexane). The retinyl ester and retinol fractions are taken to dryness under N_2 and then resuspended in either 0.2 or 10% dioxane–*n*-hexane for isocratic, normal-phase analysis of retinyl esters and retinols. The identity of the retinyl esters found in these proteins is established from photodiode array absorption spectra and also from coelution with authentic retinoid standards. These results reveal that 11-*cis*- and all-*trans*-retinyl esters are localized in different membrane compartments within the RPE. Using the microsomal 11-*cis*- and all-*trans*-retinyl ester concentrations as a reference (0.70 and 1.10 nmol/mg, respectively), 67% of the 11-*cis*-retinyl esters are recovered in the plasma membrane (S2)-enriched fraction and 70% of the all-*trans*-retinyl esters are recovered in the endoplasmic reticulum (S4) fraction.[27] Thus, 11-*cis*-retinyl esters are localized in a compartment in which recovery of 11-*cis*-REH activity is also greatest. A similar colocalization is also observed for all-*trans*-retinyl esters and all-*trans*-REH activity. [The remaining gradient fractions demonstrate relatively insignificant recoveries (<5%) of retinyl ester. In addition, neither 11-*cis*- nor all-*trans*-retinols are detected in any

---

[27] N. L. Mata, E. T. Villazana, and A. T. C. Tsin, *Invest. Ophthalmol. Vis. Sci.* **39,** 1312 (1998).

of the gradient fractions.] These results provide strong support for the physiological relevance of 11-*cis*-REH in the RPE.

In Situ Hydrolysis of 11-cis-Retinyl Ester in Retinal Pigment Epithelium Plasma Membrane

The colocalization of 11-*cis*-REH activity and 11-*cis*-retinyl esters led to the examination of the functional/physiological relevance of this association.[27] It is clear that plasma membrane-enriched fraction (S2) will hydrolyze 11-*cis*-retinyl palmitate at a lipid–water interface in our *in vitro* assay system. However, if the association of 11-*cis*-REH activity and 11-*cis*-retinyl esters contributes to the production of visual chromophore *in vivo*, we would expect 11-*cis*-retinol to be liberated from the endogenous S2 membrane-associated 11-*cis*-retinyl esters. Initial efforts to measure 11-*cis*-retinol liberated in this manner were hampered by the relatively low concentrations of 11-*cis*-retinyl esters in S2 ($\sim$0.47 nmol/mg) and the relatively high K_m^{app} of 11-*cis*-REH activity ($\sim$66 μM).[16] It was reasoned that in order to establish a substrate-dependent relationship of 11-*cis*-REH activity and membrane-associated 11-*cis*-retinyl esters, exogenous 11-*cis*-retinyl esters would first have to be incorporated into S2 membranes. The capacity of phospholipid bilayers, and unilamellar vesicles, to accommodate retinoids has been determined to be in the range of 1–10 membrane mol%.[28,29] Quantitation of the phosphorus content (P_i) of S2 membranes gives 0.015 μg of P_i per microgram of protein. According to the phospholipid–inorganic phosphate relationship, 25 μg of phospholipid per microgram of P_i,[30] 1 mg of S2 would contain $\sim$0.4 mg of phospholipid. Because we intended to deliver the 11-*cis*-[^{3}H]retinyl palmitate to S2 membranes in a phospholipid suspension [i.e., phosphatidylcholine (PC) liposomes], we could adjust the amount of PC to facilitate incorporation into a fixed amount of S2 membranes without exceeding the limitations of the membrane to accommodate the exogenous retinyl ester (10 mol%). In a typical REH rate analysis, up to 20 nmol of retinyl ester substrate is present in the reaction mixture (reaction volume, 200 μl). Incorporation of 20 nmol of retinyl ester into 500 nmol of PC gives 4 mol% retinyl ester. Therefore, approximately 500 nmol of PC dissolved in chloroform is added to dried residues of 11-*cis*-[^{3}H]retinyl palmitate (eight separate samples are prepared from 2 to 20 nmol). The samples are thoroughly mixed and then evaporated to dryness under N_2. One milliliter of 0.2 M KH_2PO_4, pH 7.2, containing 9.5 mg of Triton WR-1339, is added to each of the dried samples. The

---

[28] M.-T. P. Ho, H. J. Pownall, and J. G. Hollyfield, *J. Biol. Chem.* **264**, 17759 (1989).
[29] W. Stillwell and I. Wassall, *Methods Enzymol.* **189**, 373 (1990).
[30] L. Sokoloff and G. H. Rothblat, *Proc. Soc. Exp. Biol. Med.* **146**, 1166 (1974).

detergent–lipid suspensions are dispersed by sonication on ice. Eight separate aliquots of the RPE plasma membrane fraction (S2; ~5 mg/ml) are diluted 1:5 with one of the eight detergent–lipid preparations; the combined mixtures are briefly sonicated and then incubated at room temperature for 20 min. After incubation, the mixtures are subjected to ultracentrifugation (150,000g, 60 min, 4°). The resulting protein is washed twice in a low ionic strength buffer to remove residual 11-*cis*-[^{3}H] retinyl palmitate that is not incorporated into S2 membranes. The washed S2 pellets are resuspended in REH assay buffer to give a final protein concentration of ~1 mg/ml. The amount of 11-*cis*-[^{3}H]retinyl palmitate that is associated with the pellets is determined after extraction and HPLC analysis of the incorporated 11-*cis*-[^{3}H]retinyl palmitate. An aliquot of each sample is then incubated at 37° and the liberated [^{3}H]palmitic acid and 11-*cis*-retinol are quantified by scintillation counting and HPLC, respectively. Control reactions for this experiment are performed in an identical fashion with either heat-denatured or chemically (2-mercaptoethanol) denatured protein. The data shown in Fig. 3 clearly demonstrate a substrate-dependent hydrolysis of the membrane-associated 11-*cis*-[^{3}H]retinyl palmitate.[26] In addition, the

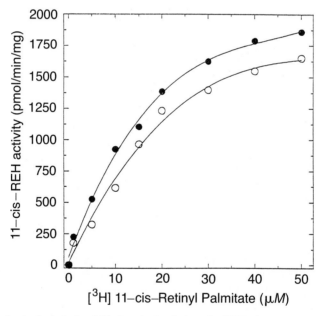

FIG. 3. *In situ* hydrolysis of 11-*cis*-retinyl palmitate by RPE plasma membrane. 11-*cis*-REH activity was measured by the released labeled fatty acid (liquid scintillation counting, ●) and 11-*cis*-retinol (HPLC, ○).

fact that both reaction products, [³H]palmitic acid and 11-*cis*-retinol, are liberated in essentially equimolar stoichiometry indicates the mechanism under study is a true ester hydrolysis rather than a reversal of the esterification (LRAT) reaction. If, in fact, a transesterification reaction is responsible for the observed hydrolysis, then the amount of liberated 11-*cis*-retinol would be expected to exceed that of palmitic acid, as the latter product would be incorporated as PC and therefore not be extracted into the free fatty acid phase of the reaction mixture. In separate experiments, we have also determined that 11-*cis*-LRAT activity is not present in the S2 membranes.[31]

Isolation of 11-cis-Retinyl Ester Hydrolase Enzyme by Percoll Gradient Separation of S2 Membranes

To isolate 11-*cis*-REH activity from other nonspecific hydrolases including all-*trans*-REH, plasma membrane-enriched fraction S2 is subfractionated on self-forming gradients of Percoll. The procedure followed was first described by Ottonello and Mariani.[32] Briefly, S2 membranes collected from the sucrose gradient are pelleted by ultracentrifugation (150,000g, 60 min, 4°) and resuspended in 10 mM Tris–acetate, pH 7.5. The protein suspension is then mixed with a Tris-buffered Percoll solution (pH 7.5) to achieve a density of 1.045. Preliminary experiments have shown that this density is absolutely critical for the optimal resolution of enzyme activities in the S2 protein. The Percoll–protein solution (volume, 36 ml) is then centrifuged for 20 min at 10,000 rpm (Sorvall SS-34). Two hundred-microliter fractions are removed from the centrifuge tube, from top to bottom, by careful pipetting. The recovered fractions (designated P1–P18) are assayed for protein concentration, 5'-nucleotidase, carboxylesterase, ouabain-sensitive Na$^+$,K$^+$-ATPase, alkaline phosphatase, and 11-*cis*- and all-*trans*-REH activities. Recovery of 11-*cis*-REH activity is greatest in fraction P2 ($\rho 5° = 1.042$). Specific activities of the PM enzyme markers are also markedly increased in P2. Enrichment of the enzyme marker activities, relative to values obtained with S2 membranes, is as follows: 5'-nucleotidase (5-fold), Na$^+$,K$^+$-ATPase (15-fold), and alkaline phosphatase (10-fold). The enrichment factors of Na$^+$,K$^+$-ATPase and alkaline phosphatase and the fact that, in RPE, Na$^+$,K$^+$-ATPase activity is localized predominantly in the apical surface of the RPE,[33,34] indicate that 11-*cis*-REH activity may

---

[31] N. L. Mata and A. T. C. Tsin, *Biochim. Biophys. Acta* **1394**, 16 (1998).

[32] S. Ottonello and G. Mariani, *Curr. Eye Res.* **3**, 1085 (1984).

[33] T. Okami, A. Yamamoto, K. Omori, T. Takada, M. Uyama, and Y. Tashiro, *J. Histochem. Cytochem.* **38**, 1267 (1990).

[34] A. K. Mircheff, S. S. Miller, D. B. Farber, M. E. Bradley, W. T. O'Day, and D. Bok, *Invest. Ophthalmol. Vis. Sci.* **31**, 863 (1990).

be associated primarily with the apical PM region. Further biochemical characterization of P2 indicates relatively low carboxylesterase activity and no detectable all-*trans*-REH activity.[26] This latter result is particularly significant as it represents our first physical isolation of 11-*cis*-REH activity from that of all-*trans*-REH.

Isolation of 11-cis-Retinyl Ester Hydrolase Enzyme by Liquid Chromatography

We have developed a method to solubilize 11-*cis*-REH activity from RPE microsomes. This involves the use of a neutral detergent, CHAPS, subsequent to which the enzyme activity in the protein extract is stable for weeks at 4°. We have also introduced this protein extract to a hydrophobic interaction column (HIC), which isolates the 11-*cis*-REH enzyme from the all-*trans*-REH and from nonspecific carboxylesterase. The methods used in these experiments are as follows.[35] A typical experiment involves loading 2–4 ml of protein extract prepared from bovine RPE microsomes [in 10 mM Tris-HCl (pH 7.0), 20% (v/v) glycerol, and 1% (w/v) CHAPS; about 10 mg of protein and 10,000 pmol/min/mg of 11-*cis*-REH activity]. Figure 4 shows an HIC eluted with a step gradient of 20 to 0% ammonium sulfate in 10 mM Tris, 20% (v/v) glycerol, and 1% CHAPS (w/v) at a flow rate of 0.5 ml/min. Two milliliter fractions are collected and analyzed for protein concentration, 11-*cis*-REH activity (Fig. 4A, open and solid symbols, respectively), all-*trans*-REH activity, and carboxylesterase activity (Fig. 4B, open and solid symbols, respectively). 11-*cis*-REH activity was recovered in two major peaks (peaks 1 and 2). Proteins were also fully (100%) recovered from the HIC column and they were mainly eluted with peak 1. 11-*cis*-REH enzyme activity associated with peak 2 was about five times higher than that of peak 1 because peak 2 contained significantly less protein. Therefore, a fivefold protein purification of the 11-*cis*-REH was achieved by HIC. Moreover, Fig. 4B shows that both nonspecific carboxylesterase and all-*trans*-REH activities were associated mainly with peak 1, indicating that this procedure was effective to achieve separation of 11-*cis*-REH enzyme from other esterases or hydrolases including the all-*trans*-REH enzyme.

Concluding Statements

On the basis of results reported in the literature and on those from our laboratory, the location and role of 11-*cis*-REH seem apparent. Starting

[35] E. T. Villazana, J. E. Salinas, C. Contreras, A. Biswas, D. A. Sierra, and A. T. C. Tsin, *Invest. Ophthal. Vis. Sci.* **39**, S40 (1998).

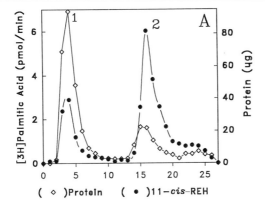

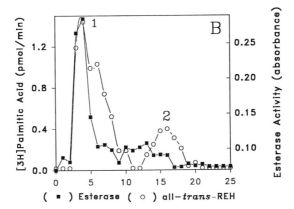

FIG. 4. Hydrophobic interaction column (HIC) isolation of 11-*cis*-REH enzyme. Ten milligrams of RPE microsomes was solubilized in 12% CHAPS and introduced into the HIC column equilibrated with 20% ammonium sulfate. The gradient was run from 20 to 0% sodium sulfate at 0.5 ml/min. Fractions (2 ml) were collected for 11-*cis*-REH activity (●, A), protein concentration (◇, A), nonspecific carboxylesterase activity (■, B) and all-*trans*-REH activity (○, B).

from the branch point where 11-*cis*-retinyl esters are synthesized at the ER, they are somehow relocated to the plasma membrane, where they are stored. Hydrolysis of this pool of retinyl esters is carried out by a substrate-specific 11-*cis*-retinyl ester hydrolase and the product of the reaction, 11-*cis*-retinol, is then oxidized by the 11-*cis*-RD at the PM.[31] Because 11-*cis*-LRAT activity is found only in the ER and not in the PM,[31] there is no bifurcation point for the 11-*cis*-retinol at this location and 11-*cis*-retinal can be readily formed. Because of the PM location, the retinal product will be readily available for transfer to the retina via IRBP without the need for intracellular delivery by CRALBP. This provides a novel branch of the visual pathway yet to be explored by scientists studying enzymes of the visual cycle.[27]

Acknowledgment

We thank the NIH (NEI and MBRS) for financial support of this work. We also thank Drs. Lafer and Prasad for advice on HIC column isolation of 11-cis-REH enzyme.

[26] Molecular Characterization of Lecithin–Retinol Acyltransferase

By ALBERTO RUIZ and DEAN BOK

Introduction

The retinal pigment epithelium (RPE) of the eye is a multifunctional component of the vertebrate retina. Anatomically, this epithelium is composed of a monolayer of cuboidal cells and is located between the large-bore capillary bed of the choroidal layer and the rod and cone photoreceptors of the neurosensory retina.[1] One of the most important functions of the RPE in vision is the uptake of vitamin A (retinol) and the processing and transport of its derivatives (retinoids) in a complex phenomenon known as the *visual cycle*.[2,3] Several enzymatic steps are involved in this process and the product of each reaction is an essential chemical substrate for the appropriate continuation of the cycle.[1,4,5] Although most of the enzymes participating in this cycle have been biochemically characterized and their activity has been well documented, the molecular characterization of proteins such as lecithin–retinol acyltransferase (LRAT, EC 2.3.1.135) and isomerohydrolase still remains unresolved. In this study we present some of the

[1] D. Bok, *J. Cell Sci. Suppl.* **17**, 189 (1993).
[2] D. S. Goodman and W. S. Blaner, *in* "The Retinoids" (M. B. Sporn, A. B. Roberts, and D. S. Goodman, eds.), p. 2. Academic Press, New York, 1984.
[3] D. S. Goodman, *in* "The Retinoids" (M. B. Sporn, A. B. Roberts, and D. S. Goodman, eds.), p. 42. Academic Press, New York, 1984.
[4] A. C. Ross and V. L. Hammerling, *in* "The Retinoids: Biology, Chemistry, and Medicine" (M. B. Sporn, A. B. Roberts, and D. S. Goodman, eds.), p. 521. Raven, New York, 1994.
[5] J. C. Saari, *in* "The Retinoids: Biology, Chemistry, and Medicine" (M. B. Sporn, A. B. Roberts, and D. S. Goodman, eds.), p. 351. Raven, New York, 1994.

methodology utilized for the characterization of LRAT at the molecular level.

Biochemical Characteristics of Lecithin–Retinol Acyltransferase

LRAT is a membrane-bound protein presumed to be located in the smooth endoplasmic reticulum and is responsible for the transfer of fatty acyl groups in a regiospecific manner from the *sn*-1 position of membrane phospholipids, specifically phosphatidylcholine (lecithin) to all-*trans*-retinol for the generation of the all-*trans*-retinyl esters.[6,7] This enzymatic reaction has been shown to take place in several organs known to be involved in the processing of vitamin A such as liver, testis, and the intestine.[6,8,9] In the RPE of the eye, LRAT plays a critical role in visual pigment regeneration because it supplies retinyl esters to the isomerohydrolase for the formation of 11-*cis*-retinol.[10,11] It is known that for most membrane-bound proteins, enzyme denaturation occurs as purification proceeds. Likewise the process of purification of LRAT to homogeneity has been extremely difficult. However, LRAT protein has been partially purified and vital information has been generated, such as its substrate specificity, the kinetics of its mechanism of action, and other biochemical properties that have helped to clarify its specific role in the visual cycle.[7,12,13] An important step forward in the characterization of this important enzyme has been the generation by Rando *et al.* of a series of sulfhydryl-directed reagents that inhibit LRAT enzymatic activity in a highly specific manner.[14,15] These reagents were radioactively labeled and used as probes to search for targeted proteins. By applying the radiolabeled inhibitor all-*trans*-[^{3}H]retinyl-bromoacetate to bovine RPE membranes, a protein product of about 25 kDa was labeled and resolved in acrylamide gels.[13] Another reagent, *N-tert*-butoxycarbonyl-L-biocytin-11-aminoundecanochloromethyl ketone (BACMK), which binds irreversibly to LRAT, was used to confirm a putative mass of 25 kDa for LRAT[16] and led to its isolation and partial amino

---

[6] J. C. Saari and D. L. Bredberg, *J. Biol. Chem.* **264**, 8636 (1989).
[7] R. J. Barry, F. J. Canada, and R. R. Rando, *J. Biol. Chem.* **16**, 9231 (1989).
[8] P. N. MacDonald and D. E. Ong, *Biochem. Biophys. Res. Commun.* **156**, 157 (1988).
[9] M. C. Schmitt and D. E. Ong, *Biol. Reprod.* **49**, 972 (1993).
[10] P. S. Bernstein, W. C. Law, and R. Rando, *Proc. Natl. Acad. Sci. U.S.A.* **84**, 1849 (1987).
[11] P. S. Deigner, W. C. Law, F. J. Canada, and R. R. Rando, *Science* **244**, 968 (1989).
[12] Y-Q. Shi, I. Hubacek, and RT. R. Rando, *Biochemistry* **32**, 1257 (1993).
[13] Y-Q. Shi, S. Furuyoshi, I. Hubacek, and R. R. Rando, *Biochemistry* **32**, 3077 (1993).
[14] B. S. Fulton and R. R. Rando, *Biochemistry* **26**, 7938 (1987).
[15] A. Trehan, F. J. Canada, and R. R. Rando, *Biochemistry* **29**, 309 (1990).
[16] A. Ruiz, A. Winston, Y.-H. Lim, B. A. Gilbert, R. R. Rando, and D. Bok, *J. Biol. Chem.* **274**, 3834 (1999).

acid sequencing. The data generated by these studies have been used for the design of degenerate primers for molecular cloning and elucidation of LRAT primary structure as described below.

Assay Methods

Preparation of Retinal Pigment Epithelial Cells

We have used normal human RPE derived by dissection or amplification by cell culture. Native RPE cells are prepared by dissection of fetal or adult donor eyes under a stereo microscope. Using enucleated fetal eyes from abortuses (gestation week 15–24), sheets of RPE cells can be carefully peeled away from their choroidal attachment with fine forceps. In fetal eyes that have been frozen, or in any adult eyes, RPE cells must be isolated in a different manner. The anterior segment, lens, and retina are first removed. The eyecup is then rinsed in calcium- and magnesium-free Hanks' balanced salt solution (HBSS). RPE cells in a minimal volume of HBSS are lightly scraped free from their underlying attachment to Bruch's membrane with a No. 15 scalpel blade. Cells thus collected are transferred with a dropper pipette to fresh HBSS on ice. The procedure for culturing human fetal RPE in our laboratory has been described previously.[17–19] Confluent RPE monolayers are maintained in culture until they are morphologically well differentiated. Cultures are rinsed with HBSS, collected by scraping with a Teflon policeman, and transferred to fresh HBSS on ice. RPE cells may be used immediately or stored temporarily at $-80°$.

Isolation of mRNA

Extraction of mRNA is performed either on RPE cells or from transfected and nontransfected HEK-293 cultured cells. Cells are added into a lysis buffer containing 200 mM NaCl, 200 mM Tris (pH 7.5), 1.5 mM MgCl$_2$, and 2% (w/v) sodium dodecyl sulfate (SDS) and protein/RNase degrader. When necessary the cell lysate is passed two to four times through a sterile plastic syringe fitted with an 18 to 21-gauge needle, before incubation at 45°. During incubation the tube is rocked gently for 20 min to assure complete digestion of proteins and ribonucleases. By this treatment RNA molecules are released from broken cells into the buffer solution. By in-

---

[17] J. G. Flannery, W. O'Day, B. A. Pfeffer, J. Horwitz, and D. Bok, *Exp. Eye Res.* **51,** 717 (1990).

[18] D. A. Frambach, G. L. Fain, D. B. Farber, and D. Bok, *Invest. Ophthalmol. Vis. Sci.* **31,** 1767 (1990).

[19] A. K. Mircheff, S. S. Miller, D. B. Farber, M. E. Bradley, W. T. O'Day, and D. Bok, *Invest. Ophthalmol. Vis. Sci.* **31,** 863 (1990).

creasing the salt concentration in the buffer (NaCl, 0.5 M final concentration) the poly(A)$^+$ fraction of RNA can be separated by adding lyophilized oligo(dT) coupled to cellulose resin (InVitrogen, San Diego, CA) followed by incubation of the mixture at room temperature for 20 min. The resin is then sedimented at 3000g for 5 min at 4°. After washing the resin several times with a binding buffer containing 500 mM NaCl, 10 mM Tris-HCl (pH 7.5) in water treated with diethyl pyrocarbonate (DEPC; Sigma, St. Louis, MO) followed by washes with low-salt buffer containing 250 mM NaCl, 10 mM Tris-HCl (pH 7.5) in DEPC-treated water, the poly(A)$^+$ RNA fraction can be recovered from the resin by addition of an eluent buffer containing 10 mM Tris-HCl, pH 7.5, in DEPC-treated water.

Reverse Transcriptase-Polymerase Chain Reaction Analysis

Using BACMK, which specifically inhibits LRAT activity in bovine retinal RPE membranes, a single protein band of about 25 kDa is affinity labeled, gel purified, digested with trypsin, and microsequenced as described previously.[16] Two peptides with the sequence MKNPMLEAVSLVLEKLL-FISYFKF and HLTHYGIYLGDNR were obtained. On the basis of these peptide sequences, theoretically corresponding to the N-terminal region and an internal sequence of the putative bovine LRAT, respectively, several sets of forward and reverse degenerate primers are analyzed. The following pair of degenerate primers has been designed: 5' ATGAARAAYCCNAT-GCTNGARGC 3' (forward) and 5' DATNCCRTARTGNGTNARRTG 3' (reverse). Reverse transcriptase-polymerase chain reaction (RT-PCR) experiments are performed on poly(A)$^+$ RNA from human RPE cells, according to the vendor recommendations in the RNA-PCR kit (Perkin-Elmer, Norwalk, CT). In 0.5-ml microcentrifuge tubes, 200 ng of poly(A)$^+$ RNA extracted from RPE cells is reverse transcribed in a mixed reaction containing a final concentration of 5 mM MgCl$_2$, 1× PCR buffer, 1 mM dNTPs, RNase inhibitor (1 U/μl), reverse transcriptase (2.5 U/μl), and a 2.5 μM concentration of degenerate reverse primer. The total 20-μl reaction is overlaid with 50 μl of mineral oil and incubated at 42° for 15 min. The enzymatic activity is inactivated by subsequent incubation at 94° for 5 min and single-strand cDNA obtained after reverse transcription is then recovered by using a fine-tipped Pasteur pipette. Seventy-eight microliters of a mixture containing a final concentration of 2.5 μM degenerate forward primer, 2 mM MgCl$_2$, 1× PCR buffer, and 65.5 μl of sterilized water is mixed with the 20-μl reverse transcription reaction.

PCR experiments are performed with a Robocycler 40 apparatus (Stratagene, La Jolla, CA) under the following conditions: 2.5 U of Ampli-Taq DNA polymerase is added after 2 min of heat shock followed by 35 amplification cycles of denaturing at 94° for 1 min; annealing at 55 or 60° for 1 min; and elongation at 72° for 2 min. A final incubation at 72° for 10 min is used to complete the polymerase activity on the final PCR products. As shown in Fig. 1A, a predominant ~200-base pair (bp) product is observed when using an annealing temperature of either 55 or 60°. The higher molecular weight background, which is probably due to degeneration of the primers, is notably reduced by the increase in the annealing temperature to 60°. The ~200-bp PCR product is excised from low melting point agarose and purified by standard methods. The product is cloned into the TA pCR II vector (InVitrogen), and sequenced by the dideoxy chain termination method (Amersham, Cleveland, OH). DNA primers complementary to the M13 and T7 promoters present in the vector and flanking the PCR insert are used for this purpose. A fragment of 189 bp is further characterized and subsequently labeled with [α-^{32}P]dCTP (Amersham, Arlington Heights, IL) by the nick translation technique (GIBCO-BRL-

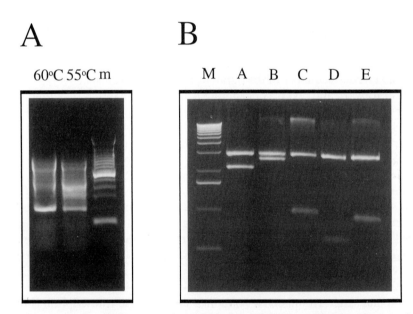

Fig. 1. Analysis of DNAs encoding LRAT. (A) RT-PCR amplification of a putative DNA fragment (~200 bp) encoding LRAT by using degenerate primers under two different annealing temperatures, 55 and 60°. m, 100-bp DNA ladder (GIBCO-BRL). (B) LRAT cDNA clones A, B, C, D, and E isolated from a human RPE cDNA library. Digestion with *Xba*I/*Xho*I restriction enzymes was performed to determine the insert size for each clone. The band at about 3.0 kb, present in all clones, is the linear pBluescript vector, whereas each single band below the vector represents the DNA insert. M, 1-kb DNA ladder (GIBCO-BRL)

Life Technologies, Gaithersburg, MD) for use as a probe in the screening of a cDNA library.

Screening of Human Retinal Pigment Epithelium cDNA Library

A human fetal RPE cDNA library was constructed in the Uni-ZAP XR vector (Stratagene) by using poly(A)$^+$ RNA extracted from cultured fetal human RPE cells obtained from abortuses (gestation week 15–24) and maintained and harvested as described above. The DNA inserts are unidirectionally placed between the unique *Eco*RI and *Xho*I restriction sites of the pBluescript vector. The cDNA library is screened with the 189-bp DNA fragment labeled with [α-^{32}P]dCTP (1.0 × 10^8 cpm). Replicas from a set of Luria broth (LB)–tetracycline agar plates (100 mm) containing approximately 2.5 × 10^5 plaques from the cDNA library are transferred in duplicate to Nylon membrane disks (Amersham), UV cross-linked at 120,000 μJ of energy in a Stratalinker apparatus (Stratagene), and hybridized with a solution containing 2.5% (w/v) dextran sulfate, 0.5% (w/v) SDS, 5× Denhardt's reagent, 5× SSC (1× SSC is 0.15 M NaCl plus 0.015 M sodium citrate), 10% (v/v) salmon sperm DNA (10 mg/ml), and 50% (v/v) formamide at 42°.

After overnight hybridization the filters are washed with 2× SSC–0.1% (w/v) SDS twice at room temperature for 15 min, followed by high-stringency washes in 0.1× SSC–0.1% (w/v) SDS at 50° for two rounds of 30 min. Filters are exposed to X-ray film, using an intensifying screen at −80°. In our experiment, five cDNA clones were positively identified. Clones are removed from the ZAP phage vehicle by *in vivo* excision, using SOLR cells and the Exassist helper phage (Stratagene). The phagemids (pBluescript vector containing a cDNA insert) are rescued by heating the *in vivo* excision mixture at 75° for 20 min. After centrifugation at 5000g for 10 min at 4°, the supernatant is mixed with XL1-Blue MR cells (Stratagene), incubated at 37° for 15 min, and plated on LB–ampicillin agar plates. Colonies obtained after overnight incubation at 37° are grown in liquid LB–ampicillin medium and their DNA extracted by resuspending the pellet in 50 mM Tris-HCl (pH 7.5), 10 mM EDTA, and RNase A (100 μg/ml) and lysing the cells with 0.2 M NaOH and 1% (w/v) SDS. After separation from cell debris by centrifugation at 10,000g for 5 min at 4°, the DNA solution is passed through a silica column (Promega, Madison, WI) and eluted with warm water. The size of the insert in each clone is determined by digestion with *Xba*I and *Xho*I as shown in Fig. 1B. All clones are further characterized by sequencing; the size and arrangement of our five clones are displayed in Fig. 2.

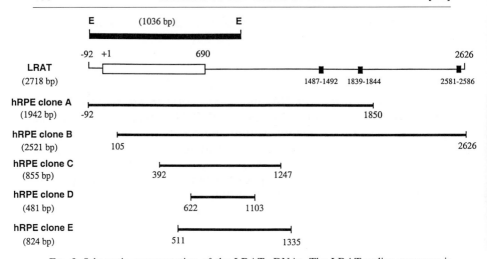

FIG. 2. Schematic representation of the LRAT cDNAs. The LRAT-coding sequence is represented by an empty box flanked on the left by negative numbers corresponding to the 5' UTR and on the right by a long 3' UTR. Small solid boxes in the 3' UTR indicate the position of three alternative polyadenylation signals. The size and regional position of each cDNA clone are indicated at the bottom. *Top:* The solid box flanked by *Eco*RI sites (E) represents a DNA fragment used for transfection experiments.

Characterization of Human Lecithin–Retinol Acyltransferase cDNAs

cDNA clones with sizes of 1942 bp (clone A), 2521 bp (clone B), 855 bp (clone C), 481 bp (clone D), and 824 bp (clone E) have been characterized. The overlapping nucleotide sequences of the five clones span 2718 bp. This sequence contains 92 nucleotides of 5' untranslated region (UTR) and an open reading frame of 690 bp encoding a deduced 230-amino acid polypeptide with a calculated mass of 25.3 kDa. The 3' UTR contains 1936 nucleotides, including a poly(A) tail of 20 residues (GenBank accession number AF071510). The methionine initiation codon ATG at nucleotides +1 to +3 is presumed to initiate the transcription process of the LRAT protein and is in reasonable agreement with consensus sequences for the translation initiation site of eukaryotic mRNAs.[20] The amino acid sequence predicts a potential glycosylation site, N–X–T, at positions 21–23. Whether this site is glycosylated *in vivo* has not been experimentally determined. However the site is not present in the bovine LRAT sequence[21] and therefore it is unlikely to be an active glycosylation site. In addition, sharp bands have been consistently observed in Western blot analyses

---

[20] M. Kozak, *Mammalian Genome* **7**, 563 (1996).
[21] A. Ruiz and D. Bok, unpublished results (1998).

and no diffuse bands commonly seen for typical glycoproteins have been observed. At least three alternative polyadenylation sites have been found in the 3' UTR sequence and are indicated by small solid boxes in Fig. 2. Consensus polyadenylation signals have been found at positions 1487–1492 and 1839–1844. An atypical polyadenylation signal, ATTAAA, has been localized 20 nucleotides upstream from the poly(A) tail at positions 2581–2586.

Subcloning of Lecithin–Retinol Acyltransferase into Expression Vector

A full restriction map of the nucleotide sequence of LRAT reveals the presence of two internal *Eco*RI sites at positions 939–944 and 1475–1480 of the 3' UTR sequence. Because these clones were obtained from a unidirectional cDNA library it was known to us that all of the inserts would be flanked by an *Eco*RI site in the upstream region and by an *Xho*I site in the downstream region. Both of these restriction sites are part of the multicloning site of the pBluescript vector. Therefore by digesting clone A with *Eco*RI, a 1036-bp fragment containing 92 bp of the 5' UTR, the entire coding sequence and 254 bp of the 3' UTR are removed as schematically represented in the upper part of Fig. 2. This LRAT/*Eco*RI fragment has been gel purified and subcloned into the *Eco*RI site in the multicloning region of the mammalian expression vector, pcDNA3, in which expression is driven by the human cytomegalovirus promoter (InVitrogen). The presence of the cloned insert in the pcDNA3 vector has been confirmed by digestion with *Eco*RI and analysis on an agarose gel. The correct orientation of the LRAT/*Eco*RI cDNA fragment has been confirmed by sequencing analysis.

Transfection of HEK-293 Cells

Human embryonic kidney 293 cells (HEK-293 cells, a kind gift of M. Simon, California Institute of Technology, Pasadena, CA) were initially analyzed by RT-PCR with sets of specific primers to confirm the absence of possible transcripts encoding LRAT. On this basis HEK-293 cells were selected for the induction of specific functional LRAT activity. Cells with a passage number of 25 are grown to confluence on 100-mm culture plates for 3 to 4 days at 37°, 5% (v/v) CO_2. The culture growth medium contains Dulbecco's modified Eagle's medium (DMEM) with high glucose, 10% (v/v) fetal bovine serum, 1 mM sodium pyruvate (Irvine Scientific, Santa Ana, CA), plus 1% L-glutamine (Sigma). Normally, a 100-mm plate is split into four plates and, after the monolayers become confluent, they are split into 12 or 16 plates depending on the number of plates needed for a particular experiment. The transfections are notably more efficient if the culture is 75 to 85% confluent. Therefore, for convenience, the last passage is made 24 to 36 hr before the transfection takes place. LipofectAMINE (2 mg/ml; GIBCO-BRL, Gaithersburg, MD) is used as the transfection vehicle to deliver the LRAT construct to HEK-293 cells. Empty vector pcDNA3 (InVitrogen) is used as a negative control. For each 100-mm culture plate, about 160 μg of LipofectAMINE is diluted with 270 μl of serum-free DMEM containing 30 μg of DNA. This solution is mixed and incubated at room temperature for 45 min. The combined solution is then diluted with an additional 4.8 ml of DMEM with high glucose to give a total transfection volume of 6.4 ml. While the DNA–lipid complexes are forming, the culture plates are rinsed gently once with serum-free DMEM. Each culture plate is overlaid with 6.4 ml of transfection medium and, after 5 hr of incubation at 37°, 5% CO_2, it is replaced with 7 ml of culture growth medium. Twenty-four hours after transfection, the culture plates are gently rinsed twice with HBSS (Irvine Scientific) and collected with a cell lifter (Costar, Cambridge, MA) in HBSS and pelleted in a 15-ml tube at 1000g for 5 min at 4°. Cell pellets are frozen at −80° until further analysis.

Northern Blot Analysis

Northern blot analysis is used to accomplish two separate goals: first, to study the tissue distribution of the LRAT message in several human tissues, including the pigment epithelium, at fetal and adult stages and second, to confirm the induction of LRAT transcripts in a cell line, for the study of putative enzymatic activity. For these purposes, 2 μg of poly(A)$^+$ RNA extracted from RPE cells or HEK-293 cells transfected with LRAT cDNA is electrophoretically separated in a 1.2% (w/v) agarose–formaldehyde gel and blotted onto nylon membranes according to a previously published method.[22] Filter membranes are hybridized at 42° overnight in the presence of 50% (v/v) formamide with a gel-purified *Xba*I/*Xho*I DNA insert from LRAT cDNA clone A when analyzing HEK-293 cells or the insert from clone B when poly(A)$^+$ RNA from other human tissues is analyzed. Either DNA fragment is labeled with [α-^{32}P]dCTP (Amersham) by nick translation and used as a probe for the detection of RNA transcripts specific for the LRAT gene. After hybridization, the filters are washed with 2× SSC–0.1% (w/v) SDS twice at room temperature for 15 min, and by high-stringency washes in 0.1× SSC–0.1% (w/v) SDS at 50° for two rounds of 30 min. Filters are exposed to X-ray film, using an intensifying screen at −80°.

---

[22] J. Sambrook, E. F. Fritsch, and T. Maniatis, p. 7.43. Molecular Cloning: A Laboratory Manual. Cold Spring Harbor Press, Cold Spring Harbor, New York, 1989.

As shown in Fig. 3, which depicts the analysis of poly(A)$^+$ RNA from transfected HEK-293 cells, a single RNA transcript of about 1.5 kb is observed. This transcript includes the LRAT insert plus parts of the upstream promoter and termination signal of the expression vector and is found only in samples from HEK-293 transfected with LRAT cDNA. No transcripts are observed in the other two samples. This result confirms the successful induction of LRAT mRNA, and the absence of endogenous mRNA encoding LRAT in control HEK-293 cells transfected with empty vector or in nontransfected HEK-293 cells. When using the DNA insert from clone B as a probe, a major RNA transcript of 5.0 kb is observed in tissues that are known for their high vitamin A-processing activity (see Ref. 16). In fetal tissues, the specific message is expressed in RPE, liver, and barely in brain. In adult tissues, the highest level of expression is observed in testis and liver, followed by the RPE, small intestine, prostate,

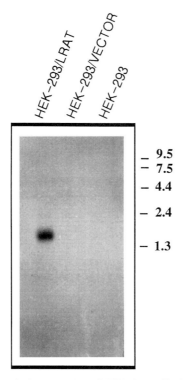

FIG. 3. Northern blot analysis of transfected HEK-293 cells. Poly(A)$^+$ RNA hybridized with a DNA insert from clone A confirmed the induction of an RNA transcript in cells transfected with LRAT, which was absent in cells transfected with empty vector and nontransfected cells.

pancreas, colon, and brain (low expression). Additional smaller messages, which could represent polyadenylation variants, were also detected in these tissues and others such as adult skeletal muscle, spleen, thymus, and uterus. A 470-bp DNA fragment corresponding to positions 1996–2465 near the end of the LRAT 3' UTR nucleotide sequence hybridize exclusively with the 5.0-kb message (data not shown), eliminating the lower molecular weight bands and suggesting that they represent polyadenylation variants.

Induced LRAT activity in HEK-293 cells transfected with LRAT cDNA is demonstrated by providing [^{3}H]retinol to cell membranes.[16] Esterification of all trans-[^{3}H]retinol into all-trans-retinyl palmitate is detected by high-performance liquid chromatography (HPLC) by virtue of its coelution with authentic nonradioactive retinyl palmitate. Further characterization of this radioactive product by I_2-catalyzed isomerization yields an isomeric mixture containing predominantly all-trans-retinyl esters.[16]

Western Blot Analysis

Polyclonal antibodies against a mixture of two peptide sequences GAAGKDKGRNSFYETSS and HLDESLQKKALLNEEVARRAE corresponding to positions 28–44 and 126–146 of the LRAT polypeptide were generated in rabbits by contract with Alpha Diagnostics International (San Antonio, TX). The specificity of these antibodies has been shown previously (see Ref. 16). Native human RPE cells, nontransfected HEK-293T cells (a variant of HEK-293, carrying the T antigen from SV40 for higher expression), cells transfected with empty plasmid, and cells transfected with LRAT cDNA are collected 24 hr after transfection. For this purpose, 10 μg of microsomal protein from each sample in buffer containing 1% (v/v) 2-mercaptoethanol is boiled for 2 min and loaded onto a 12.5% (w/v) SDS–polyacrylamide gel. Blot analysis is performed on nitrocellulose filters according to a published method,[23] using antiserum diluted to 1:1000 for the identification of LRAT. Protein bands are detected by enhanced chemiluminescence, using the ECL system (Amersham). The antiserum reacts specifically with a single protein band of about 25–26 kDa in human RPE cells. A band with the same mobility is also observed in HEK-293T cells transfected with LRAT cDNA. In contrast, this product is absent in cells that are nontransfected or transfected with the empty plasmid.[16]

Analysis of Lecithin–Retinol Acyltransferase Sequences

Further analysis of the entire nucleotide and amino acid sequence encoding LRAT is performed with sequences available in the GenBank database.

[23] H. Towbin, T. Staehlin, and J. Gordon, *Proc. Natl. Acad. Sci. U.S.A.* **76**, 4350 (1979).

H-LRAT	160 NCEHFVTYCRYG	171
B-LRAT	160 NCEHFVTYCRYG	171
M-LRAT	160 NCEHFVTYCRYG	171
H-TIG3	112 NCEHFVAQLRYG	123
H-rev 107	112 NCEHFVNELRYG	123
R-rev 107	110 NCEHFVNELRYG	121

FIG. 4. Comparison of a short stretch of amino acids with high homology to LRAT sequences: H-LRAT (human-LRAT), B-LRAT (bovine-LRAT), M-LRAT (mouse-LRAT), H-*rev* 107 (human H-*rev* 107), R-*rev* 107 (rat H-*rev* 107), and H-TIG3 (human-TIG3). Numbers flanking the amino acid sequences represent original position of residues along the entire polypeptide.

At the amino acid level LRAT does not show close homology with previously reported acyltransferases, such as lecithin–cholesterol acyltransferase (LCAT).[24,25] In a more recent database analysis, a low homology has been found with proteins of 18-kDa H-*rev* 107 from human and rat (accession numbers ×92814 and ×76453, respectively) and human TIG-3 (accession number AF060228). Both of these proteins have been associated with tumor suppressor activities and were initially identified in a phenotypic revertant of H-*ras*-transformed 208F rat fibroblasts in the case of H-*rev* 107[26] and more recently in human keratinocytes cells induced by a synthetic retinoid, Tazarotene, in the case of TIG-3.[27] Interestingly, a small stretch of residues shown in Fig. 4 is highly conserved among these three proteins. Nine of 12 amino acids were identical. Comparison of this region shows that two of the four cysteine residues present in the LRAT polypeptide, namely, Cys-161 and Cys-168, are conserved in human, bovine, and mouse LRAT sequences. Cys-161 in LRAT is conserved at position 113 of human H-*rev* 107 and human TIG3, and at position 111 of rat H-*rev* 107, respectively. In contrast, Cys-168 is present only in LRAT sequences. On the

---

[24] J. Lean, D. Fielding, D. Drayna, H. Dieplinger, B. Baer, W. Kohr, W. Henzel, and R. Lawn, *Proc. Acad. Natl. Sci. U.S.A.* **83,** 2335 (1986).

[25] O. L. Francone and C. J. Fielding, *Proc. Natl. Acad. Sci. U.S.A.* **88,** 1716 (1991).

[26] A. Hajnal, R. Klemenz, and R. Schafer, *Oncogene* **9,** 479 (1994).

[27] D. DiSepio, C. Ghosn, R. L. Eckert, A. Deucher, N. Robinson, M. Duvic, R. A. S. Chandraratna, and S. Nagpal, *Proc. Natl. Acad. Sci. U.S.A.* **95,** 1481 (1998).

basis of this observation it is tempting to hypothesize that Cys-168 could play an important role in the esterifying activity of LRAT because this enzyme uses an ordered ping–pong Bi Bi mechanism whereby palmitic acid could bind initially to this residue protein and subsequently to retinol to form retinyl palmitate. Site-directed mutagenesis analysis on these residues should shed some light regarding the putative active site of the LRAT enzyme.

At the nucleotide level, as shown in Fig. 5, a number of unidentified expressed sequence tag (EST) clones with high homology to LRAT have been found. The full size of each insert is denoted by the number in parentheses and the fraction of nucleotide sequences available in the database for each clone is represented as solid black boxes. Their specific position relative to the schematic representation of the LRAT nucleotide sequence is indicated. Because these clones were generated from a cultured cell line (NT2) and other human tissues it is of interest to us to identify potentially important variations at the nucleotide or deduced amino acid level with respect to the LRAT clones from RPE. Importantly, an additional 119 nucleotides upstream from the -92 position of RPE clone A is apparent from EST clone 664435. An analysis on these clones shows a few nucleotide changes in some of the EST clones compared with RPE LRAT sequenced in our laboratory. These changes are likely to represent sequencing errors

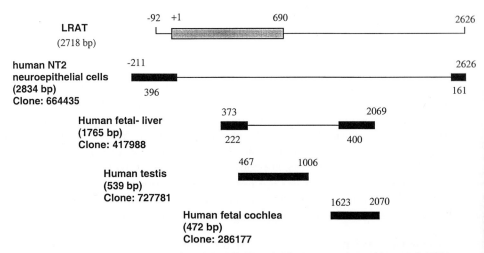

FIG. 5. Schematic representation of the LRAT, as in Fig. 2: comparison with several cDNA (EST) clones previously deposited in the GenBank database with high homology sequence to LRAT. Tissue origin of each clone, full size of the insert (numbers in parentheses), clone number according to the database, and mapping localization with respect to LRAT are indicated.

in ESTs as a function of the massive sequencing activity inherent in this methodology. At least in these clones, no relevant sequence variation has been found that would account for differences in the LRAT gene or tissue specificity. Although more studies at the biochemical and molecular level need to be performed for LRAT, the possibility of the existence of LRAT isoforms, closely related proteins, or even larger versions of the LRAT sequence reported here, should not be discarded.

Acknowledgments

We thank Dr. William O'Day for the culture and dissection of RPE cells, Jane Hu for transfection of HEK-293 cells, and Alice Van Dyke for expert photographic assistance. This research was supported by NIH Grants EY00444 and EY00331 and by a Center Grant from the National Retinitis Pigmentosa Foundation Fighting Blindness Inc. D.B. is the Dolly Green Professor of Ophthalmology at UCLA and a Research to Prevent Blindness Senior Scientific Investigator.

[27] Analysis of Chromophore of RGR: Retinal G-Protein-Coupled Receptor from Pigment Epithelium

By WENSHAN HAO, PU CHEN, and HENRY K. W. FONG

Introduction

The retinal pigment epithelium (RPE) and Müller cells contain an abundant opsin that is distinct from rhodopsin and cone visual pigments. The RPE and Müller cell opsin, first referred to as RPE retinal G-protein-coupled receptor (RGR), represents a distant evolutionary branch of vertebrate opsins that are most closely related in amino acid sequence to invertebrate visual pigments and retinochrome, a photoisomerase that catalyzes the conversion of all-*trans*- to 11-*cis*-retinal in squid photoreceptors.[1] Whereas vertebrate rhodopsin and the cone pigments are ~45% identical in amino acid sequence, the amino acid sequences of rhodopsin and RGR are only ~25% identical.[2] Moreover, various characteristics of RGR contrast markedly with those of visual pigments, if not with those of the G-protein-coupled receptors. RGR is localized to intracellular membranes in RPE and Müller cells, rather than at the cell surface. It is capable

---

[1] T. Hara and R. Hara, *in* "Retinal Proteins" (Y. A. Ovchinnikov, ed.), p. 457. VNU Science Press, Utrecht, The Netherlands, 1987.
[2] M. Jiang, S. Pandey, and H. K. W. Fong, *Invest. Ophthal. Vis. Sci.* **34**, 3669 (1993).

of forming a stable photopigment with bound all-*trans*-retinal. Analysis of purified RGR indicates that the endogenous chromophore of RGR is predominantly all-*trans*-retinal.[3] The absorption spectrum of bovine RGR bound to all-*trans*-retinal contains two pH-dependent absorption peaks with maxima in the blue ($\lambda_{max} \sim 469$ nm) and near-ultraviolet ($\lambda_{max} \sim 370$ nm) regions of light.[4] Irradiation of RGR with 470-nm monochromatic or near-UV light results in stereospecific isomerization of the bound all-*trans*-retinal to an 11-*cis* configuration.[3]

Comparison of bovine, human, and mouse RGR shows that the three proteins are 78–86% identical in amino acid sequence.[5] Lys-[255] and His-[91] are conserved in RGR from all three species, and their positions correspond to those of two critical amino acids in visual pigments. Lys-[255] is homologous with the conserved lysine residue that serves as the retinaldehyde attachment site in visual pigments. His-[91] corresponds to the position of the retinylidene Schiff base counterion in bovine rhodopsin. RGR differs from the visual pigments by the absence of a negatively charged amino acid residue that acts as a stabilizing counterion to its protonated Schiff base. The pK_a (~ 6.5) of its retinylidene Schiff base chromophore is far lower than that of the visual pigments. With few exceptions, the conserved amino acid sequence of mammalian RGR contains the consensus sequence motif of G-protein-coupled receptors.[6]

Despite the unique features of RGR, many established methods are directly applicable to the characterization of this novel opsin. To elude various limitations in accessibility, abundance, and biochemical properties, the following procedures have been used to analyze the chromophore of RGR.

Procedures

Preparation of Retinal Pigment Epithelium Microsomes

Fresh bovine eyes are dark adapted over ice in transit from the local abattoir, and RPE cells are isolated within 4 hr of enucleation. Dissection of the tissue is carried out under dim yellow light. The anterior segments, lens, vitreous, and neural retina are removed from the eyecup. The eyecup is rinsed and half-filled with phosphate-buffered saline (PBS), and the RPE

---

[3] W. Hao and H. K. W. Fong, *J. Biol. Chem.*, **274**, 6085 (1999).
[4] W. Hao and H. K. W. Fong, *Biochemistry* **35**, 6251 (1996).
[5] L. Tao, D. Shen, S. Pandey, W. Hao, K. A. Rich, and H. K. W. Fong, *Mol. Vis.* **4**, 25 (1998). ⟨http://www.molvis.org/molvis/v4/p25⟩
[6] J. M. Baldwin, *EMBO J.* **12**, 1693 (1993).

cells are removed by gently scraping the cell monolayer with a spatula. The cells are collected by low-speed centrifugation, resuspended, and washed twice in 5–8 ml of ice-cold homogenization buffer containing 30 mM sodium phosphate (pH 6.5) and 250 mM sucrose. The cells are homogenized with a Dounce glass homogenizer, applying ~10 strokes of the tight pestle. The homogenate is centrifuged at 700g for 6 min at 4° to remove nuclei and unbroken cells. The pellet is resuspended, and the homogenization and centrifugation steps are repeated four times. The supernatants from the homogenization steps are pooled and centrifuged in a Sorvall (Newtown, CT) SS-34 rotor at 15,000g for 20 min at 4°. The 15,000g supernatant is recovered and centrifuged in a Beckman (Fullerton, CA) 70 Ti rotor at 150,000g for 1 hr at 4°. The RPE microsomal membranes are then used immediately or stored at $-80°$. The preparation of RPE microsomes results in enrichment of RGR and preferential depletion of contaminant rhodopsin (Fig. 1).

Isolation of RGR by Immunoaffinity Purification

All steps in the isolation of RGR from RPE microsomes are conducted under dim red light or in the dark. The buffers used in the purification are

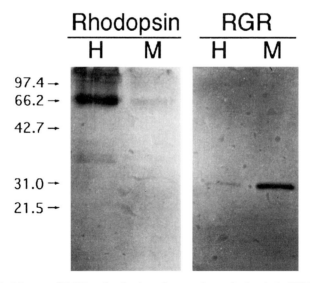

FIG. 1. Enrichment of RGR and reduction of contaminant rhodopsin in RPE microsomal membranes. RGR and rhodopsin were detected by immunoblot analysis of proteins (5 μg) in the microsomal membrane preparation (M) and post-700g homogenate (H) of bovine RPE cells. Duplicate blots were made after SDS–PAGE and probed with monoclonal antibodies directed against bovine RGR and rhodopsin.

filtered through sterile membranes of 0.2-μm pore size. The microsomal membranes are extracted three times for 1 hr at 4° in 1.2% (w/v) digitonin in 10 mM sodium phosphate buffer (pH 6.5), 150 mM NaCl, and 0.5 mM EDTA. After centrifugation of the extracts at 100,000g for 20 min at 4°, all supernatants are combined and mixed for 2 hr at 4° with Affi-Gel 10 resin (Bio-Rad, Hercules, CA) conjugated to anti-bovine RGR monoclonal antibody 2F4.[7] The immunoaffinity resin is transferred to a small plastic column for washing with 25 bed volumes of 10 mM sodium phosphate buffer, pH 6.5, containing 0.1% (w/v) digitonin, 150 mM NaCl, and 0.5 mM EDTA. To elute RGR, the column is loaded 10 times with a 0.5 bed volume of the wash buffer containing 50 μM bovine RGR carboxyl terminal peptide (CLSPQRREHSREQ). Each fraction is loaded and cycled through the column four times. The eluates are pooled and concentrated approximately fourfold with a Centricon-3 concentrator (Amicon, Beverly, MA). The purified fraction contains a major ~31-kDa protein that reacts specifically to a bovine RGR antibody (Fig. 2).

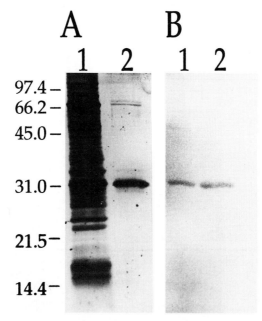

FIG. 2. Isolation of RGR from the bovine retinal pigment epithelium. RGR was isolated from a digitonin extract of microsomal proteins by an immunoaffinity procedure and electrophoresed in a 12% SDS–polyacrylamide gel. The protein extract (14 μg) in lane 1 and purified RGR (0.7 μg) in lane 2 were analyzed by (A) protein silver staining and (B) immunoblot analysis, using monoclonal antibody 2F4. [Reproduced with permission from Hao and Fong.[4]]

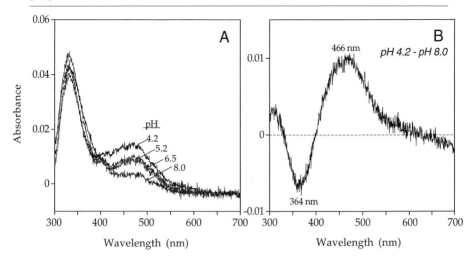

FIG. 3. Absorption spectra of RGR. RGR from bovine RPE was purified in 10 mM sodium phosphate buffer, pH 6.5, containing 150 mM NaCl, 0.5 mM EDTA, and 0.1% (w/v) digitonin. (A) The spectrum of RGR was determined at pH 6.5, 5.2, 4.2, and 8.0. The absolute curves were superimposed at 660 to 680 nm to equalize the baselines. (B) The spectrum at pH 8.0 was subtracted from the spectrum at pH 4.2. The maximum and minimum for the absorption peaks lie at approximately 466 and 364 nm, respectively. [Reproduced with permission from Hao and Fong.[4]]

The immunoaffinity resin is prepared by conjugation of Affi-Gel 10 resin with monoclonal antibody 2F4, which is an IgM molecule that specifically recognizes RGR in a crude extract of bovine RPE microsomes.[7] The antibody is added to activated Affi-Gel 10 resin (Bio-Rad) in 0.1 M morpholine-propanesulfonic acid (MOPS, pH 7.5), and the suspension is agitated gently for 4 hr at 4°. After the coupling reaction, the resin is incubated for 1 hr in 0.1 M ethanolamine to block the remaining reactive sites. The immunoaffinity gel is then washed with water, equilibrated with binding buffer [10 mM sodium phosphate (pH 7.2), and 500 mM NaCl], and stored at 4° until use. After use, the immunoaffinity resin is regenerated and bound peptide is removed in a solution of 5 M LiCl and 0.2% (w/v) digitonin.

Spectroscopic Measurements

The UV–visible absorption spectra of purified RGR may be recorded with a scanning spectrophotometer at room temperature. The reference sample consists of the elution buffer—10 mM sodium phosphate (pH 6.5),

---

[7] D. Shen, M. Jiang, W. Hao, L. Tao, M. Salazar, and H. K. W. Fong, *Biochemistry* **33**, 13117 (1994).

150 mM NaCl, 0.5 mM EDTA, 0.1% (w/v) digitonin, and 50 μM carboxyl-terminal peptide. The pH of the sample may be raised by addition of 1 M Na$_2$HPO$_4$, or lowered by addition of 1 M NaH$_2$PO$_4$ or 1 M citrate buffer, pH 3.8. The absorption spectrum of RGR is pH sensitive and shows an apparent absorption peak in the blue region of light. The difference spectrum of RGR at pH 4.2 and 8.0 reveals absorption peaks with maxima at ~466 and ~364 nm (Fig. 3).

Extraction of Chromophore of RGR by Hydroxylamine Derivatization

The retinal chromophore of purified bovine RGR is extracted under dim red light and analyzed by the method of hydroxylamine derivatization,

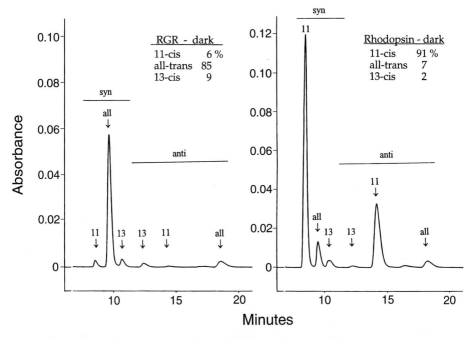

FIG. 4. The endogenous chromophore of RGR and rhodopsin. The bound chromophores of RGR (left) and rhodopsin in ROS membranes (right) were reacted with hydroxylamine and analyzed by HPLC. Absorbance was measured at 360 nm, and the amounts of both *syn* and *anti* isomers of retinaloxime were added to determine the percent quantity of each retinal isomer, all, all-*trans*-retinal oxime; 13, 13-*cis*-retinal oxime; 11, 11-*cis*-retinal oxime. [Reproduced with permission from Hao and Fong.[3]]

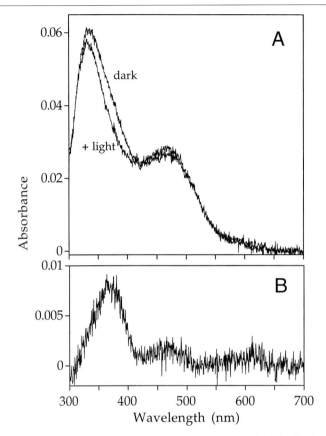

FIG. 5. The effect of light on the absorption spectrum of RGR. (A) Absolute spectrum of RGR before and after illumination. (B) Difference between dark and postillumination spectra shown in (A). [Reproduced with permission from Hao and Fong.[4]]

as described by Groenendijk et al.[8,9] This method has been used successfully to extract retinal from ROS membranes and has given quantitative recovery and complete retention of the geometric structure of retinal isomers.[9] In a typical extraction procedure, 100 to 300 μl of purified RGR is mixed with 0.1 volume of 2 M hydroxylamine, pH 6.5, followed by 300 μl of methanol and 300 μl of dichloromethane. Sodium phosphate buffer, pH 6.5, is added to bring the sample volume to 900 μl. The extraction with dichloromethane

[8] G. W. T. Groenendijk, W. J. De Grip, and F. J. M. Daemen, *Anal. Biochem.* **99**, 304 (1979).
[9] G. W. T. Groenendijk, W. J. De Grip, and F. J. M. Daemen, *Biochim. Biophys. Acta* **617**, 430 (1980).

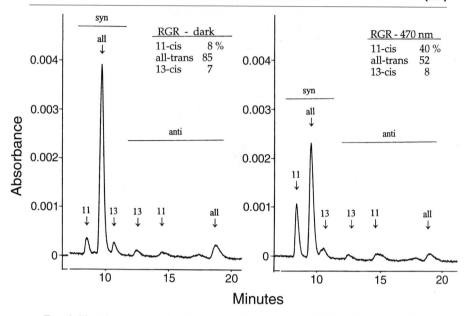

FIG. 6. Photoisomerization of the retinal chromophore of RGR. The chromophore of purified RGR was analyzed after incubation in darkness (left) or after irradiation with 470-nm monochromatic light at an illuminance of 410 lux (right). The protein at pH 6.5 was irradiated for 3 min at room temperature in a quartz cuvette positioned 60 cm from the lamp. [Reproduced with permission from Hao and Fong.[3]]

is performed by vortexing the mixture (aqueous buffer–methanol–dichloromethane, 1:1:1 by volume) for 30 sec, followed by centrifugation in a microcentrifuge at 14,000 rpm for 1 min at 25°C. The lower organic phase is removed carefully, and the upper phase is extracted twice more with 300 μl of dichloromethane. The organic layers are pooled and dried down under a nitrogen stream. The extracted retinal oximes are then solubilized in 100 μl of hexane, and the solution is filtered through glass wool held in a pipette tip. The tube is rinsed with 100 μl of hexane four more times. The filtered solutions of retinal oximes are combined and dried again under nitrogen. The residue is either stored in darkness under nitrogen at −80° or analyzed immediately by high-performance liquid chromatography (HPLC).

High-Performance Liquid Chromatography Analysis of Retinal Oximes

The isomers of retinal oximes are analyzed by HPLC, as described previously.[8–10] The extracted retinal oximes are dissolved in hexane

[10] K. Ozaki, A. Terakita, R. Hara, and T. Hara, *Vision Res.* **26**, 691 (1986).

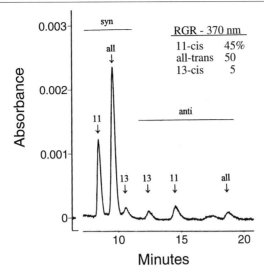

FIG. 7. Irradiation of RGR by near-UV light. The chromophore of RGR was analyzed after irradiation of the protein at pH 6.5 for 3 min at room temperature with near-UV light passing through an Oriel 370-nm interference filter. [Reproduced with permission from Hao and Fong.[3]]

and applied to a LiChrosorb RT Si60 silica column (4 × 250 mm, 5 μm; E. Merck, Darmstadt, Germany) in an HPLC system equipped with solvent module and UV–visible detector. The samples are injected in a volume of 20 μl, and the components are resolved in a running solvent consisting of hexane supplemented with 8% (v/v) diethyl ether and 0.33% (v/v) ethanol. The HPLC column is precalibrated with the reaction products of hydroxylamine and purified all-*trans*-retinal or 11-*cis*-retinal standards. Identification of the retinal oxime isomers is based on the retention times of the known retinal oxime products. Absorbance is measured at 360 nm, and the absorbance peaks from the chromatograph are analyzed by HPLC system software. The proportion of each isomer in the loading sample is determined from the total peak areas of both its *syn*- and *anti*-retinal oxime and is corrected according to the following extinction coefficients (ε_{360}, in hexane): all-*trans*-*syn*, 54,900; all-*trans*-*anti*, 51,600; 11-*cis*-*syn*, 35,000; 11-*cis*-*anti*, 29,600; 13-*cis*-*syn*, 49,000, and 13-*cis*-*anti*, 52,100.[8,10] Analyses of the endogenous chromophore of RGR and rhodopsin reproducibly demonstrate that the predominant isomer in the dark is all-*trans*-retinal for RGR and, as expected, 11-*cis*-retinal for rhodopsin (Fig. 4).

Irradiation of RGR

RGR is purified, as described, and may be irradiated at room temperature by a 30-W fiber optic light source for 5 min at a distance of 10 cm. UV–visible absorption spectra are determined before and immediately after illumination. Under these conditions, irradiation of RGR does not lead to significant bleaching of RGR_{469}, the blue light-absorbing form of RGR (Fig. 5).

RGR may be irradiated with monchromatic light, using an Oriel (Stratford, CT) light source equipped with a 150-W xenon arc lamp. The lamp produces uniform irradiance from 300 to 800 nm. Monochromatic light beams at 370 or 470 nm are formed by passing the light through a 370-nm interference filter, or both a 470-nm interference filter and 455-nm longpass filter, respectively. The protein samples are held at room temperature in a quartz cuvette positioned 60 cm from the lamp. After delivery of the intended amount of light, the retinal chromophores are extracted by hydroxylamine derivatization and analyzed by HPLC, as described previously. Irradiation of the all-*trans*-retinal–RGR complex modifies the relative amounts of the two physiologically relevant isomers only, i.e., all-*trans*- and 11-*cis*-retinal. When irradiated with monochromatic light at 470 nm (Fig. 6) or 370 nm (Fig. 7), all-*trans*-retinal in RGR isomerizes stereospecifically to 11-*cis*-retinal.

Hydroxylamine Reactivity of RGR

The reaction of RGR and hydroxylamine may be followed by measurement of the absorbance of RGR at various times after the addition of hydroxylamine. Difference spectra are obtained by subtraction of the absorption spectrum at completion from the spectrum obtained at each time point and before the addition of 80 μM hydroxylamine.[4] The difference spectra indicate the disappearance of one pigment (λ_{max} ~470 nm) and formation of a new pigment (λ_{max} ~364 nm), which is consistent with retinaldehyde oxime. From the difference spectra and the known extinction coefficient of all-*trans*-retinal oxime, the molar extinction coefficient of RGR_{469} is ~62,800 $cm^{-1} M^{-1}$. The extinction coefficient of RGR_{370} is ~66,100 $cm^{-1} M^{-1}$, as can be determined from the pH difference spectrum of RGR between pH 8.0 and pH 4.2.[4] The extinction coefficients of RGR_{469} and RGR_{370} are slightly higher than that of retinochrome (60,800)[11] and 1.5- and 1.6-fold higher than that of bovine rhodopsin (40,600),[12] respectively.

---

[11] T. Hara and R. Hara, *Methods Enzymol.* **81**, 190 (1982).
[12] R. Hubbard, P. K. Brown, and D. Bownds, *Methods Enzymol.* **XVIII C**, 615 (1971).

Section V

Posttranslational and Chemical Modifications

[28] Low-Temperature Photoaffinity Labeling of Rhodopsin and Intermediates along Transduction Path

By María L. Souto, Babak Borhan, and Koji Nakanishi

Introduction

An understanding of the tertiary structure of rhodopsin, including the site and orientation of the 11-*cis*-retinal chromophore, is essential in determining the visual transduction mechanism. Also critical is an understanding of the short-lived 11-*cis*-retinal to all-*trans* intermediates along the isomerization pathway, which have been identified by low-temperature spectroscopy (Scheme 1).[1–7] However, owing to difficulties in securing crystals suited for X-ray and in applying solution nuclear magnetic resonance (NMR) techniques to rhodopsin, high-resolution structural data are lacking. This has led to the development of alternate biophysical techniques to clarify these crucial aspects.[8–10] Photoaffinity labeling of rhodopsin, after many years of coping with inherent difficulties associated with membrane proteins as well as the photolabile nature of the pigment itself, has yielded invaluable information regarding the site and orientation of the chromophore within the binding site prior to isomerization.[11,12] The major hindering factors associated with these studies were difficulties in separating the sticky peptide fragments after proteolysis of the photo-cross-linked rhodopsin for sequencing the cross-linked sites,[13] and scrambling of cross-linked sites on

---

[1] W. J. DeGrip, D. Gray, J. Gillespie, P. H. Bovee, E. M. Van den Berg, J. Lugtenburg, and K. J. Rothschild, *Photochem. Photobiol.* **48,** 497 (1988).
[2] B. Konig, W. Welte, and K. P. Hofmann, *FEBS Lett.* **257,** 163 (1989).
[3] R. R. Birge, *Annu. Rev. Biophys. Bioeng.* **10,** 315 (1981).
[4] R. Mathies, *Methods Enzymol.* **88,** 633 (1982).
[5] A. Cooper, *Nature (London)* **282,** 531 (1979).
[6] D. Emeis, H. Kuhn, J. Reichert, and K. P. Hofmann, *FEBS Lett.* **143,** 29 (1982).
[7] N. Bennett, M. Michel-Villaz, and H. Kuhn, *Eur. J. Biochem.* **127,** 97 (1982).
[8] D. D. Thomas and L. Stryer, *J. Mol. Biol.* **154,** 145 (1982).
[9] T. Kawaguchi, T. Hamanaka, Y. Kito, and T. Mitsui, *Nagoya Kogyo Daigaku Kiyo* **43,** 245 (1991).
[10] G. F. X. Schertler, C. Villa, and R. Henderson, *Nature (London)* **362,** 770 (1993).
[11] H. Z. Zhang, K. A. Lerro, T. Yamamoto, T. H. Lien, L. Sastry, M. A. Gawinowicz, and K. Nakanishi, *J. Am. Chem. Soc.* **116,** 10165 (1994).
[12] K. Nakanishi, H. Zhang, K. A. Lerro, S. Takekuma, T. Yamamoto, T. H. Lien, L. Sastry, D. J. Baek, C. Moquin-Pattey, M. F. Boehm, and K. Nakanishi, *Biophys. Chem.* **56,** 13 (1995).
[13] K. Nakanishi and R. Crouch, *Israel J. Chem.* **35,** 253 (1995).

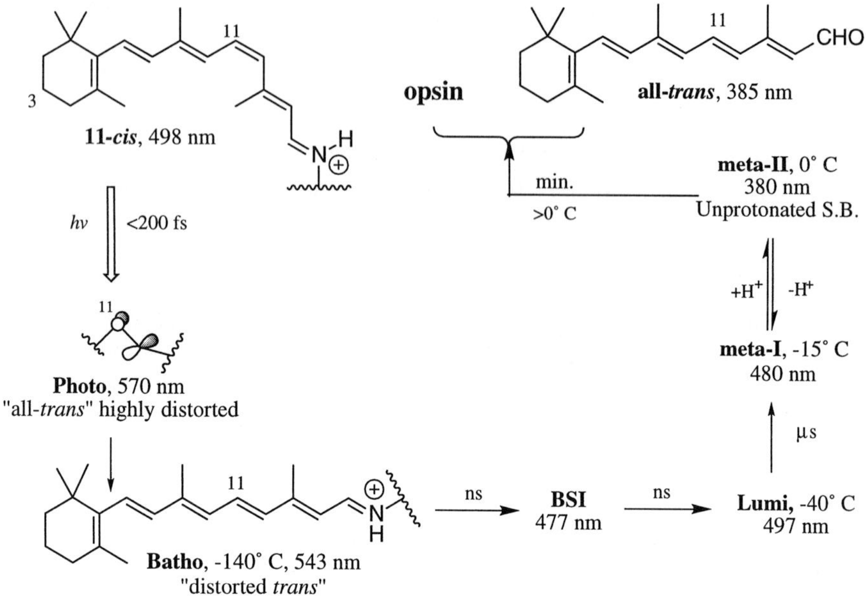

SCHEME 1. The visual transduction intermediates of rhodopsin.

photoactivation of the photolabel owing to isomerization of the 11-*cis*-ene to *trans*-ene.[14]

In the following, an overview of photoaffinity experiments performed with retinal analogs **1–4** is given to describe the general protocol, cross-linking results involving key intermediates along the visual transduction pathway, and the major difficulties involved in these studies.

Photoaffinity labeling of rhodopsin was carried out according to the following sequence: (1) design and synthesis of photoaffinity labeling retinal analogs, (2) regeneration of pigment analogs and verification of "wild-type" binding via biophysical methods, (3) photoactivation of the photolabel leading to photo-cross-linked rhodopsin and its photointermediates, and (4) protein cleavage and sequencing of cross-linked sites.

Synthesis of Photolabeled Chromophore Analogs

Retinal analogs designed as photoaffinity labels for rhodopsin should satisfy the following requirements: (1) the labeled retinal should be able to bind readily and should not cause conformational distortions of the

---

[14] T. A. Nakayama and H. G. Khorana, *J. Biol. Chem.* **265**, 15762 (1990).

opsin; (2) the photoaffinity label should adopt the same conformation as the endogenous substrate, 11-*cis*-retinal; (3) the photolabel bearing retinal should be reasonably stable under the conditions employed in binding studies; (4) the photolabile group should undergo facile photolysis at such a wavelength that irradiation causes minimal protein damage; (5) the photolysis wavelength should not electronically excite the retinal chromophore and change its orientation relative to the protein; (6) the intermediate, e.g., carbene, generated by light should react rapidly and indiscriminately with its immediate environment; and (7) a reasonably simple route should be available not only for the synthesis of the photoaffinity label-bearing retinal but also for its ^{3}H- or ^{14}C-labeled analog at high levels of radioactive specific activity.

The photolabile groups chosen for our studies with rhodopsin were α-diazoacetyl and α-diazoketone groups. The use of these groups appeared to have several advantages, one of which is good stability in the experimental pH range. Also, these photolabels require activation at 254 nm, which should not interfere with the photochemically active β band of rhodopsin at 350 nm.

The first photolabeled analogs synthesized were racemic 9-*cis*-3-[^{14}C$_2$] diazoacetoxyretinal **1**,[15] which was subsequently prepared in optically pure forms 3*S* and 3*R*,[16] and 9-*cis*-retinal with 9-^{14}CH$_2$O-^{14}CO-CHN$_2$ residue,

[15] R. Sen, J. D. Carriker, V. Balogh-Nair, and K. Nakanishi, *J. Am. Chem. Soc.* **104**, 3214 (1982).
[16] H. Ok, C. Caldwell, D. R. Schroeder, A. K. Singh, and K. Nakanishi, *Tetrahedron Lett.* **29**, 2275 (1988).

2.[17] The 9-*cis* geometry was chosen rather than the 11-*cis* because it was synthetically more accessible and because the ring binding site in isorhodopsin (9-*cis*-retinal) was not expected to differ much from that in rhodopsin. The photoaffinity radiolabel was introduced by esterification of the hydroxy group of the corresponding precursors with the tosylhydrazone of glyoxylic acid by slight modification of conventional procedures.[15] Diazoester **2** was extremely unstable [1% yield after high-performance liquid chromatography (HPLC) purification] and hence, although binding studies and photolysis were performed, the sequencing was abandoned. Despite the fact that 3*S* and 3*R* enantiomers were found to bind to opsin in a stereo-discriminating manner (only the 3*S* enantiomer binds)[15] and that the extent of cross-linking was at a satisfactory level of 10–15%,[17] further studies were discontinued because of the following development.

Satisfactory photo-cross-linking results were finally secured by incorporating the photolabel into an analog in which the 11-*cis* → *trans* isomerization is blocked. Namely, 3-diazo-4-oxo-10,13-ethano-11-*cis*-[15-^{3}H]retinal analog **3**[11,12] was prepared by modification of the known synthesis of 10,13-ethano-11-*cis*-retinal,[18] which adopts a conformation similar to that of native 11-*cis*-retinal within the protein pocket.[19] Tritium label was used instead of ^{14}C because of its higher radioactive specific activity; this was introduced by selective reduction of the aldehyde with NaBT$_4$ followed by MnO$_2$ oxidation.

Subsequent studies with 3-diazo-4-oxo-11-*cis*-[15-^{3}H]retinal **4** under controlled conditions have disclosed the sequential changes occurring in retinal–receptor interactions. Thus analog **4** without the locked ring was designed to follow the isomerization pathway in a temperature-resolved manner. However, difficulties were encountered in the chemical synthesis owing to the presence of the inherently unstable 11-*cis* geometry as well as the diazo-ketone photolabel. In the present case the problem is magnified because functional groups must be introduced after formation of the 11-*cis* double bond. Most schemes geared toward introducing the 11-*cis* geometry close to the end of the synthesis have either failed or resulted in low yields of complex isomeric mixtures.[20–26] In contrast, synthesis of the more stable

---

[17] R. Sen, A. K. Singh, V. Balogh-Nair, and K. Nakanishi, *Tetrahedron* **40**, 493 (1884).

[18] R. van der Steen, M. Groesbeck, L. J. P. van Amsterdam, J. Lugtenburg, J. van Oostrum, and W. J. DeGrip, *Recl. Trav. Chim. Pays-Bas* **108**, 20 (1989).

[19] S. Bhattacharya, K. D. Ridge, B. E. Knox, and H. G. Khorana, *J. Biol. Chem.* **267**, 6763 (1992).

[20] E. I. Negishi and Z. Owczarczyk, *Tetrahedron Lett.* **32**, 6683 (1991).

[21] A. Wada, Y. Tanaka, N. Fujioka, and M. Ito, *Bioorg. Med. Chem. Lett.* **6**, 2049 (1996).

all-*trans* analogs are more straightforward. It would thus be advantageous to synthesize all-*trans* isomers and then introduce the 11-*cis* geometry at the end of the synthesis. This was accomplished via chemoenzymatic synthesis, in which the 11-*cis* geometry is established as the last step in the reaction sequence by isomerization of the all-*trans*-retinal analog by retinochrome, a photosensitive isomerase isolated from the visual cells of cephalopods.[27] This methodology, although efficient, is unsuited for securing milligram quantities of compound. Therefore, after considerable difficulty, we developed a chemical synthesis of an efficient, general, and mild preparation of 11-*cis*-retinoids, e.g., **4**, via semihydrogenation of 11-yne-retinoid precursors with Cu/Ag-activated zinc dust.[28]

Reconstitution of Pigment Analogs

Rhodopsin analogs were regenerated by combining the HPLC-purified retinal analogs with bleached rod outer segment (ROS) membranes. Rod outer segments were isolated from bovine retinas by following a standard procedure[29] with minor modifications.[30] The bleaching was performed after resuspension in phosphate buffer (67 mM, pH 7.0) containing hydroxylamine (100 mM) at room light and ice-bath temperature. Reconstitution was effected by the addition of an ethanolic analog stock solution to opsin in 10 mM CHAPSO/10 mM HEPES (pH 7.0) at a 0.6–1.5 molar ratio equivalent of retinal. The incorporation into apoprotein was monitored spectrophotometrically. The regeneration in ROS suspension (without detergent) is also possible, but the efficiency is much lower and requires the use of a severalfold excess of retinal analog. This could lead to random cross-linking by unincorporated chromophore during photolysis.

---

[22] M. B. Sporn, A. B. Roberts, and D. S. Goodman, "The Retinoids: Biology, Chemistry, and Medicine," 2nd Ed. Raven, New York, 1994.
[23] A. Hosoda, T. Taguchi, and Y. Kobayashi, *Tetrahedron Lett.* **28**, 65 (1987).
[24] D. Mead, A. E. Asato, M. Denny, R. S. H. Liu, Y. Hanzawa, T. Taguchi, A. Yamada, N. Kobayashi, A. Hosoda, and Y. Kobayashi, *Tetrahedron Lett.* **28**, 259 (1987).
[25] A. Trehan and R. S. H. Liu, *Tetrahedron Lett.* **29**, 419 (1988).
[26] C. G. Knudsen, R. A. S. Chandraratna, L. P. Walkeapaa, Y. S. Chauhan, S. C. Carey, T. M. Cooper, R. R. Birge, and W. H. Okamura, *J. Am. Chem. Soc.* **105**, 1626 (1983).
[27] B. Borhan, R. Kunz, A. Y. Wang, K. Nakanishi, N. Bojkova, and K. Yoshihara, *J. Am. Chem. Soc.* **119**, 5758 (1997).
[28] B. Borhan, M. L. Souto, J. M. Um, B. Zhou, and K. Nakanishi, *Chem. Eur. J.* **5**, 1172 (1999).
[29] D. S. Papermaster, *Methods Enzymol.* **81**, 48 (1982).
[30] S. Hu, P. J. Franklin, J. Wang, B. E. Ruiz Silva, F. Derguini, and K. Nakanishi, *Biochemistry* **33**, 408 (1994).

3-Diazoacetoxy-9-*cis*-rhodopsin was formed from retinal **1** in bleached ROS suspension and absorbs at 465 nm in 2% (w/v) digitonin solution. It was demonstrated that the chromophore of this pigment analog occupies the natural binding site and that the opsin binds only the 3*S* enantiomer.[15,17]

Despite the poor synthetic yield of retinal **2**, 9-methylenediazoacetoxy-9-*cis*-rhodopsin was obtained in 90% yield when **2** was combined with detergent solution of bleached ROS for 22 hr. The pigment had an absorption maximum at 460 nm and displayed circular dichroic (CD) extrema at 467 and 360 nm characteristic of visual pigments. This highly modified retinal was tightly bound to the natural binding site and could not be displaced by addition of 11-*cis*-retinal to the pigment. Thus, this system should be suited for investigating the region of the binding site surrounding the polyene chain.

Analog **3** was incorporated into bovine opsin (0.65:1 ratio) to yield the rhodopsin that absorbed maximally at 483 nm. Reconstitution was essentially complete after 1 day, with a reconstitution yield of ~50% based on apoprotein as determined spectrophotometrically. The CD spectrum of the reconstituted pigment showed a strong positive Cotton effect (CE) at 308 nm, in contrast to that of native rhodopsin with two positive CEs at 340 nm (β band) and 490 nm (α band). This lack of the α band is in line with the CD spectrum of rhodopsin reconstituted from the analog containing a C-10/C-13 etheno-bridged six-membered ring, which removes the twist around the 12-*s-cis* bond[30] and supports the assignment of the α band to the distortion around the 12–13 bond.[31]

For reconstitution, the 11-*cis*-retinal analog **4** was added from an ethanol stock solution to the opsin solution at a chromophore-to-opsin molar ratio of 1:1.2. It gave a rhodopsin analog that absorbed maximally at 483 nm. The UV profile indicated full regeneration within 5 min and the chromophore was not displaced by addition of 11-*cis*-retinal. The CD spectrum of the regenerated chromophore exhibited two positive peaks at 306 and 462 nm, closely matching the α and β peaks exhibited by native rhodopsin. The high rate of regeneration and the biophysical data suggest **4** binds to opsin in a manner similar to that of native 11-*cis*-retinal.

Photolysis of Rhodopsin Analogs

In all cases, the pigments reconstituted from diazoacetoxy as well as diazoketone derivatives were irradiated at 254 nm to generate the carbene, to induce cross-linking to the amino acid residues close to the photolabeled

---

[31] M. Ito, A. Kodama, K. Tsukida, Y. Fukada, Y. Shichida, and T. Yoshizawa, *Chem. Pharm. Bull.* **30**, 1913 (1982).

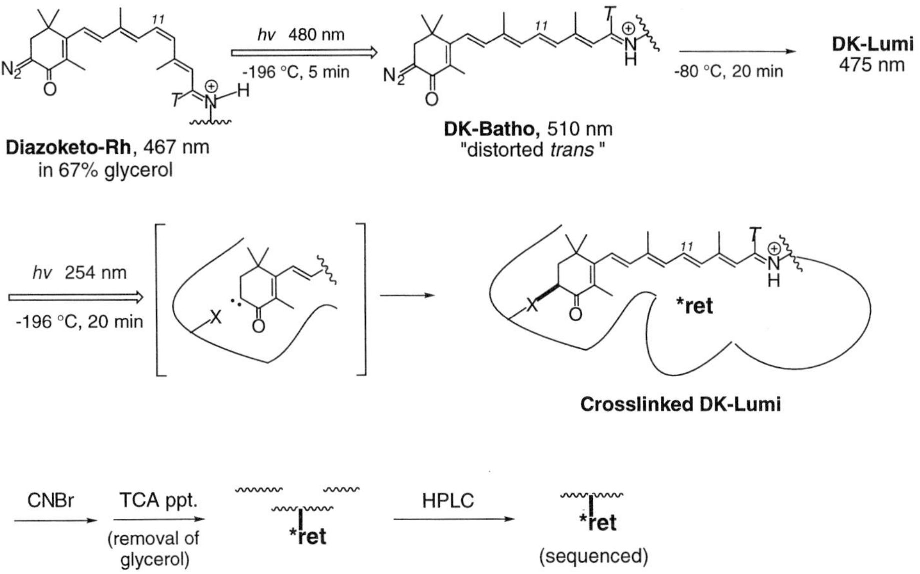

FIG. 1. Protocol for sequencing transduction intermediates resulting from DK-retinal **4**.

positions. The various conditions tested to activate the label[17] resulted in the following observations: (1) shorter irradiation time is necessary when the samples are in detergent solution (CHAPS or CHAPSO) than in suspension; (2) a medium-pressure Hg lamp (450 W) equipped with a narrow-band interference filter at 254 nm as well as an excimer laser tuned to emit at 249 nm did not succeed in activating all diazoacetoxy groups at room temperature but caused extensive damage to the protein even in short periods of irradiation; (3) irradiation with a low-pressure mercury lamp (Southern New England Ultraviolet Co., Branford, CT) with a 254-nm narrow-band emission activated diazoacetoxy groups at 4° and diazoketone groups even at liquid nitrogen temperature (see below) with no significant damage to the protein.

The temperatures in which the respective bleaching intermediates incorporating diazoketo (DK) analog **4** attain maximal population is parallel to those of native rhodopsin (Rh), but lower. DK (diazoketo)bathorhodopsin ($-196°$ versus $-140°$ for native) → DK-lumi-Rh($-80°$ versus $-40°$) → DK-meta I-Rh($-40°$ versus $-15°$) → DK-meta II-Rh ($0°$).[32] The bleaching of 3-diazo-4-oxo-11-*cis*-rhodopsin was performed at $-196°$ in 67% glycerol,

[32] B. Borhan, M. L. Souto, H. Imai, Y. Shichida, and K. Nakanishi, unpublished results (1998).

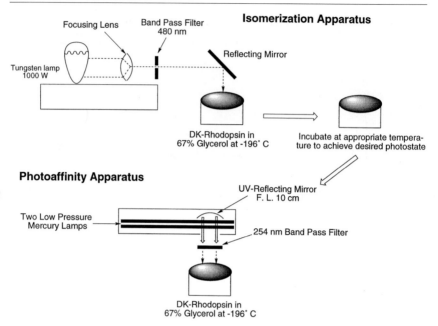

FIG. 2. Photoisomerization to the bathorhodopsin state was achieved by irradiating the DK-Rh in 67% glycerol at −196° with a 1000-W tungsten lamp. The light was focused with a lens through a 480-nm bandpass filter and reflected onto the sample with a reflecting mirror. After 10 min of irradiation the sample was incubated at the appropriate temperature for 20 min in total darkness. The photoaffinity labeling was initiated by cooling the sample to −196° and irradiating it with two low-pressure mercury UV lamps fitted with a 254-nm bandpass filter for 20 min. The UV light was focused with a UV-reflecting mirror onto the sample.

which was followed by photolysis performed at the temperatures corresponding to the respective intermediate.

For example, the photo-cross-linking to DK-lumi-Rh (−80°) was performed in the following steps (Fig. 1): (1) irradiate DK-Rh (λ_{max} of 467 nm in 67% glycerol) with 480-nm light at −196°; (2) warm the system to −80°, or the temperature for optimal DK-lumi production (λ_{max} of 475 nm), for 20 min; (3) lower the temperature to −196° to freeze the protein in the lumi state, and then irradiate at 254 nm to activate photolabel; (4) perform cyanogen bromide cleavage at room temperature; (5) remove glycerol from the system by precipitating the protein with trichloroacetic acid; (6) separate peptidic fragments by HPLC; and (7) sequence the radioactive fragment that contains the photo-cross-linked DK analog. The isomerization apparatus is shown in Fig. 2.

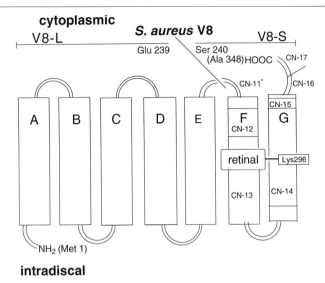

FIG. 3. Cleavage of rhodopsin incorporating retinal analog DK-ret6 **3** by V8 protease.

Proteolysis and Sequencing

Because rhodopsin is a membrane-bound protein, the hydrophobic and "sticky" nature of the cleaved peptidic fragments complicates their purification and sequencing. As indicated above, this factor was a major obstacle in performing photoaffinity labeling studies of this membrane protein and the main cause of the unsuccessful results with analog **1**, whose cross-linked amino acids could not be identified despite the cross-linking yield of 20–30%.[16,17]

However, with analog **3** bearing the 11-*cis*-locked side chain, diazo-4-oxo-10,13-ethano-11-*cis*-[5-^{3}H]retinal, irradiation selectively activated the diazoketo function and gave rise to clear-cut results for the first time. The cross-linked apoprotein in membrane suspension was cleaved by V8 protease between Glu-239 and Ser-240 into large (V8-L) and small (V8-S) fragments (cleavage yield, ca. 50%). SDS–PAGE showed all radioactivity was contained in V8-S, indicating that the cross-linking site resided in helix F or G (Fig. 3). The V8-S polypeptide separated by HPLC was further cleaved by CNBr into seven fragments. It was fortunate that radioactivity was associated with the V8-S HPLC peak because this was the only peak that was baseline separated from an inseparable mixture of V8-L, uncleaved rhodopsin, and aggregated rhodopsin. The CNBr-cleaved mixture was separated by SDS–PAGE and the labeled peptide was identified by autoradiography. The major radioactive band was characterized as CNBr peptide 13

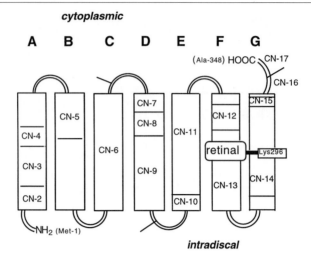

FIG. 4. The separated cyanogen bromide fragments from rhodopsin cross-linked to DK-retinal **4**.

(CN-13, Fig. 3) by blotting on polyvinylidene difluoride (PVDF) membrane and amino acid sequencing. Edman degradation of the CN-13 fragment revealed locked analog **3** to be almost exclusively linked to Trp-265 and Leu-266, revealing that the β-ionone C-3 is close to helix F. Moreover, because these labeled amino acids are in the middle of helix F, while the Schiff base linkage to Lys-296 at the other terminus of the chromophore is also close to the middle of helix G, the chromophore lies horizontally near the center of the lipid bilayer.

Determination of the sites of cross-linking in the intermediates along the transduction pathway using the pigment incorporating DK-retinal analog **4** requires a large number of photo-cross-linking and sequencing experiments (for each intermediate). It is imperative to optimize the cleavage of rhodopsin and purification of its fragments. Also, in addition to the difficulties usually encountered in cross-linking studies, we had to establish protocols to recover the cross-linked protein from the glycerol solution. Various proteolytic systems, both enzymatic and nonenzymatic, were investigated. We finally succeeded by adapting the protocol developed by Knapp et al. in their cyanogen bromide cleavages studies.[33] Cysteine residues were pyridyethylated in the glycerol solution prior to separating the protein from the membrane. After removal of glycerol and reducing agents by

[33] L. E. Ball, J. E. J. Oatis, K. Dharmasiri, M. Busman, J. Wang, L. B. Cowden, A. Galijatovic, N. Chen, R. K. Crouch, and D. R. Knapp, *Protein Sci.* **7**, 758 (1998).

trichloroacetic acid precipitation, the reduced apoprotein was cleaved with cyanogen bromide. The cleaved peptidic fragments were all separated by reversed-phase HPLC and identified by Edman degradation (Fig. 4). It was found that in bathorhodopsin the labeled amino acid was Trp-265 in fragment CN-13 (Val-258 to Phe-287) helix F (Fig. 3). This is the same amino acid that was cross-linked with photoaffinity label **3**, suggesting that the β-ionone ring does not move from its original position at the batho state.[11] In lumi, meta I, and meta II the cross-linking site now resides in helix D, fragment CN-9 (Ala-164 to Gly-182). Thus, a change in the ligand–receptor complex is seen between the intermediates batho-Rh and lumi-Rh. Namely, in the batho state the β-ionone ring is still in close proximity to helix F, whereas in lumi-Rh and subsequent intermediates, the C-3 of the chromophore flips over toward helix D.[34]

Acknowledgments

These studies were supported by NIH Grant GM 34509 (to K.N.) and by a grant from the Ministry of Education and Culture of Spain (to M.L.S.).

[34] J. M. Baldwin, G. F. Schertler, and V. M. Unger, *J. Mol. Biol.* **272,** 144 (1997).

[29] Structural Analysis of Protein Prenyl Groups and Associated C-Terminal Modifications

By Mark E. Whitten, Kohei Yokoyama, David Schieltz, Farideh Ghomashchi, Derek Lam, John R. Yates III, Krzysztof Palczewski, and Michael H. Gelb

Introduction

The prenylation of proteins, including several in the visual transduction pathway, is one of the most recently discovered modifications of eukaryotic cell proteins.[1-4] Some prenylated proteins, including rhodopsin kinase, the γ subunit of transducin, the α subunit of retinal cGMP phosphodiesterase, Ras, and lamin B, are initially produced with a C-terminal sequence motif CaaX (C is cysteine, a is usually but not necessarily an aliphatic residue, and X is typically M, S, or Q). Prenylation occurs by enzyme-catalyzed attachment of a 15-carbon farnesyl group to cysteine via a thioether linkage. After prenylation, aaX is released by a membrane-bound endoprotease,[5-8] and the newly exposed S-farnesylcysteine is methylated on its α-carboxyl group by a membrane-bound methyltransferase.[9,10] Other prenylated proteins, including the γ subunits of many heterotrimeric G proteins, the β subunit of retinal cGMP phosphodiesterase, and a subset of small GTP-binding proteins, contain a CaaX motif (X = L, F) and are modified by the attachment of the 20-carbon geranylgeranyl group. Removal of aaX and C-terminal methylation occurs as for farnesylated proteins. Finally, the subset of GTP-binding proteins termed Rab contains two cysteines near or at the C terminus (CCXX, XXCC, CXC), and both cysteine sulfhydryls contain a thioether-linked geranylgeranyl group.[11] The attachment of prenyl

---

[1] J. A. Glomset, M. H. Gelb, and C. C. Farnsworth, *Trends Biochem. Sci.* **15,** 139 (1990).
[2] W. A. Maltese, *FASEB J.* **4,** 3319 (1990).
[3] S. Clarke, *Annu. Rev. Biochem.* **61,** 355 (1992).
[4] F. L. Zhang and P. J. Casey, *Annu. Rev. Biochem.* **65,** 241 (1996).
[5] Y.-T. Ma and R. R. Rando, *Proc. Natl. Acad. Sci. U.S.A.* **89,** 6275 (1992).
[6] G. F. Jang, K. Yokoyama, and M. H. Gelb, *Biochemistry* **32,** 9500 (1993).
[7] V. L. Boyartchuck, M. N. Ashby, and J. Rine, *Science* **275,** 1796 (1997).
[8] W. K. Schmidt, A. Tam, K. Fujimura-Kamada, and S. Michaelis, *Proc. Natl. Acad. Sci. U.S.A.* **95,** 11175 (1998).
[9] C. A. Hrycyna, S. K. Sapperstein, S. Clarke, and S. Michaelis, *EMBO J.* **10,** 1699 (1991).
[10] J. D. Romano, W. K. Schmidt, and S. Michaelis, *Mol. Biol. Cell* **9,** 2231 (1998).
[11] C. C. Farnsworth, M. Kawata, Y. Yoshida, Y. Takai, M. H. Gelb, and J. A. Glomset, *Proc. Natl. Acad. Sci. U.S.A.* **88,** 6196 (1991).

groups to these three classes of proteins is catalyzed by three distinct protein prenyltranferases.[12,13]

Structures of protein prenyl groups have been rigorously established by releasing the protein-bound lipids with Raney nickel (to cleave reductively the bond between cysteine S and C-1 of the prenyl group) and analyzing the released hydrocarbons by combined gas chromatography–mass spectrometry versus authentic standards.[14-17] Protein prenyl groups can also be released by treatment with methyl iodide to produce farnesol and geranylgeraniol along with isomerized prenols.[16,18] This method has been used when relatively small amounts of prenylated protein are available, typically from tissue culture cells that have been grown in the presence of radiolabeled mevalonic acid (the precursor of prenyl groups) or its lactone in the presence of statins, which block the production of endogenous mevalonic acid. The radiolabeled prenol is analyzed versus standards by reversed phase high-performance liquid chromatography (HPLC).[16] While this approach does not establish the chemical structure of the prenyl group, it provides strong circumstantial evidence for the presence of protein-bound farnesyl and geranylgeranyl groups. However, in our hands, we have experienced difficulties with methyl iodide cleavage, in that yields are highly variable and can be quite low in some cases (<10% cleavage). Work in our laboratory has shown that the Raney nickel cleavage method is superior to methyl iodide treatment, and the details of this procedure, combined with HPLC analysis of the released radiolabeled hydrocarbons, are described in this chapter. This method has been used to analyze protein prenyl groups from protein extracted from sodium dodecyl sulfate (SDS)–polyacrylamide gel slices. Also described here is the use of Raney nickel cleavage followed by combined gas chromatography–mass spectrometry to determine the structure of prenyl groups released from delipidated total cell protein.

The rigorous structural analysis of aaX removal and C-terminal methylation has been carried out by fast atom bombardment mass spectrometry

---

[12] K. Yokoyama, G. W. Goodwin, F. Ghomashchi, J. Glomset, and M. H. Gelb, *Biochem. Soc. Trans.* **20,** 479 (1992).

[13] P. J. Casey and M. C. Seabra, *J. Biol. Chem.* **271,** 5289 (1996).

[14] C. C. Farnsworth, S. L. Wolda, M. H. Gelb, and J. A. Glomset, *J. Biol. Chem.* **264,** 20422 (1989).

[15] C. C. Farnsworth, M. H. Gelb, and J. A. Glomset, *Science* **247,** 320 (1990).

[16] C. C. Farnsworth, P. J. Casey, W. N. Howald, J. A. Glomset, and M. H. Gelb, *Methods* **1,** 231 (1990).

[17] M. H. Gelb, C. C. Farnsworth, and J. A. Glomset, *in* "Lipidation of Proteins: A Practical Approach" (A. J. Turner), p. 231. IRL Press, Oxford, 1992.

[18] Y. Sakagami, A. Isogai, A. Suzuki, S. Tamura, C. Kitada, and M. Fujino, *Agric. Biol. Chem.* **43,** 2643 (1979).

of C-terminal peptides from prenylated proteins.[11,19] A less rigorous approach has been used to explore C-terminal methylation: HPLC analysis of the product of oxidative scission of the prenyl–cysteine linkage combined with amide bond hydrolysis to yield cysteic acid methyl ester.[20] In the present chapter, a method for electrospray mass spectrometric analysis of protein C-terminal peptides is presented, which has been applied to the analysis of prenylated peptides in a complex mixture (a tryptic digest of partially purified bovine rhodopsin kinase). This method is facilitated by the availability of prenylated peptide standards, and a convenient synthesis of peptides and peptide methyl esters radiolabeled with high specific activity farnesyl and geranylgeranyl groups is described.

Analysis of Protein Prenyl Groups with Raney Nickel Cleavage

Preparation of Radioprenylated Proteins, Protein Delipidation, and Sodium Dodecyl Sulfate–Polyacrylamide Gel Electrophoresis

To radiolabel protein prenyl groups, cells are cultured with ^{3}H- or ^{14}C-labeled mevalonolactone or mevalonic acid [New England Nuclear (Boston, MA) or American Radiolabeled Chemicals (St. Louis, MO)] in the presence of an inhibitor of hydroxymethylglutaryl-CoA reductase (such as simvastatin or lovastatin) to suppress production of endogenous mevalonic acid.[16,17] Statins such as lovastatin and simvastatin require conversion of the lactone form to the carboxylate salt as described.[17] Mevalonolactone tends to be more cell permeable than mevalonic acid but must undergo intracellular hydrolytic ring opening prior to incorporation into the isoprenoid pathway. Thus, it is best to carry out small-scale radiolabeling experiments with both mevalonic acid and its lactone to gauge the level of incorporation into proteins (determined by fluorography of SDS–polyacrylamide gels containing delipidated cell protein; see below). A disadvantage of this method for the analysis of total cell protein prenyl groups is that only proteins that become prenylated during the radiolabeling period are detected, and thus the ratio of radiofarnesylated to radiogeranylgeranylated proteins will probably not be the same as the ratio of prenyl groups present in the total pool of cellular prenylated proteins. Also, evidence from our laboratory (see below) has shown that the use of statins together with exogenous radiolabeled mevalonic acid or its lactone leads

---

[19] M. Kawata, C. C. Farnsworth, Y. Yoshida, M. H. Gelb, J. A. Glomset, and Y. Takai, *Proc. Natl. Acad. Sci. U.S.A.* **87**, 8960 (1990).

[20] H. K. Yamane, C. C. Farnsworth, H. Xie, W. Howald, B. K.-K. Fung, S. Clarke, M. H. Gelb, and J. A. Glomset, *Proc. Natl. Acad. Sci. U.S.A.* **87**, 5868 (1990).

to a distortion of the ratio of radiolabeled farnesyl versus geranylgeranyl groups attached to total cell protein, presumably owing to an effect of statins on the farnesyl pyrophosphate-to-geranylgeranyl pyrophosphate ratio. For these reasons, prenyl groups present in total cell protein are best analyzed by combined gas chromatography–mass spectrometry of Raney nickel-treated protein from cells grown in the absence of statins.

Only a small fraction of the intracellular radiolabeled mevalonic acid may be incorporated into proteins. Thus it is important to remove radioactive isoprenoids from the protein sample that are not covalently attached to protein. This can be partially accomplished by precipitating cell protein from a cell lysate with trichloroacetic acid and washing the pellet with organic solvent. However, in our hands a significant amount of non-protein-bound radiolabeled material remains, which can interfere with subsequent prenyl group analysis. Complete delipidation of protein is achieved by SDS–polyacrylamide gel electrophoresis (PAGE) followed by extraction of proteins from the gel. As described later, delipidation with organic solvent is adequate for the analysis of prenyl groups by gas chromatography–mass spectrometry.

We describe here a typical procedure that we have used to analyze prenylated proteins in mammalian cells and trypanosomatids. Bloodstream form *Trypanosoma brucei* (*T. brucei*) strain 427 (10^7 mid-log-phase cells, corresponding in protein content to $\sim 10^5$ mammalian cells) is labeled for 24 hr in a humidified atmosphere of 5% CO_2 at 37° in 1 ml of culture medium [HMI-9, 10% (v/v) fetal calf serum] containing 1 mCi of RS-[5-^{3}H]mevalonolactone (50 Ci/mmol) and 2 μM saponified simvastatin (obtained as a gift from Merck Research Laboratories, Whitehouse Station, NJ). Cells are harvested in a 1.5-ml Eppendorf tube by centrifugation and washed once with phosphate-buffered saline (PBS) in the usual way. The cell pellet is resuspended in 1 ml of ice-cold lysis buffer [20 mM Tris-HCl (pH 8.0) containing 50 mM NaCl, 1 mM dithiothreitol (DTT), 1 mM phenylmethylsulfonyl fluoride (PMSF), 30 μM N^α-tosyl-L-lysine chloromethyl ketone, 30 μM N^α-tosyl-L-phenylalanine chloromethyl ketone, pepstatin A (10 μg/ml), leupeptin (10 μg/ml), and aprotinin (10 μg/ml)]. Protease inhibitors are added fresh from properly stored stock solutions. The suspension is frozen in a −80° freezer. The frozen sample can be stored at −80° or processed further. The frozen sample is allowed to thaw on ice to lyse the cells (note, freeze–thawing may not be sufficient to lyse other types of cells, and the appropriate cell disruption technique should be used).

To the suspension of disrupted cells (cell debris including membranes should not be removed as the latter contain prenylated proteins) is added 150 μl of 100% (w/v) ice-cold trichloroacetic acid. The sample is briefly vortexed, incubated on ice for 30 min, and microcentrifuged at $\sim 10,000g$ for 10 min at 4°. The supernatant is decanted, and tissue paper is used to soak up the last bit of liquid from the lip of the inverted tube. To the pellet is added 1 ml of room temperature $CHCl_3$–methanol (2:1, v/v). The sample is vortexed at room temperature and microcentrifuged for 10 min at 4°. After decantation, the pellet is washed once more with organic solvent as above. To the pellet is added 1 ml of ice-cold reagent-grade acetone, and the sample is vortexed and microcentrifuged as described above. The acetone is decanted, and the pellet is allowed to air dry at room temperature. The protein pellet can be stored at $-80°$ or processed further for SDS–PAGE. To the pellet is added 20 μl of water, and the Eppendorf tube is placed in a bath sonicator until most of the pellet is dispersed (several minutes). Laemmli sample loading buffer (20 μl of 2× buffer) supplemented with freshly added 4% (v/v) β-mercaptoethanol is added, and the sample is boiled for 3 min.

After brief centrifugation, a small portion of the supernatant (1 μl) is subjected to scintillation counting to estimate the total amount of radiolabeled protein. Most of the sample (38 μl) is applied to two lanes of a standard Laemmli minigel [12.5% (w/v) running gel, 0.75 mm thick] for SDS–PAGE. Prestained molecular weight markers are also applied to lanes that bracket the lanes with radiolabeled protein. A second gel containing molecular weight markers and 1 μl of radiolabeled protein is run for subsequent fluorography. This gel is prepared for fluorography as described.[21] Using the fluorograph and the molecular weight markers as guides, the desired region(s) of the nonfluorographed gel containing the desired radiolabeled proteins is sliced out of the lane with a razor blade (the gel should not be fixed or dried). Slices are placed in Eppendorf tubes. The amount of gel per tube should be no more than a piece that has the width of a gel lane and a length of ~2 cm. To each Eppendorf tube is added 0.6 ml of elution buffer [50 mM ammonium bicarbonate, 0.08% (w/v) SDS, 1% (v/v) β-mercaptoethanol, made fresh] and 20 μg of bovine serum albumin (BSA) as carrier protein.

The gel slice(s) in each tube is crushed into small pieces with a small plastic pestle designed to fit conical Eppendorf tubes (typically several minutes of crushing). Another 0.6 ml of elution buffer is used to rinse the pestle, collecting the liquid into the tube with crushed gel. The sample is boiled for 5 min and rotated overnight at room temperature. The sample is microcentrifuged for a few minutes at room temperature, and most of the supernatant is transferred to a new Eppendorf tube with a Pasteur pipette. To the gel pellet is added 0.3 ml of elution buffer, and after briefly vortexing, the sample is rotated for 2 hr. After centrifugation, the super-

---

[21] P. McGeady, S. Kuroda, K. Shimizu, Y. Takai, and M. H. Gelb, *J. Biol. Chem.* **270**, 26347 (1995).

natant is combined with the first supernatant. To ensure complete removal of gel pieces, the tube of combined supernatant is microcentrifuged for a few minutes and the supernatant is carefully transferred to a new Eppendorf tube (avoid taking gel slices). The sample can be stored at $-80°$ or further processed. The remaining gel pellet is dissolved in 0.5 ml of 30% (v/v) H_2O_2 (overnight at $90°$), and the solution is subjected to scintillation counting to determine the amount of labeled protein remaining in the gel.

The sample of extracted protein is concentrated to 200 μl in a SpeedVac (Savant Instruments, Holbrook, NY), and a small aliquot is subjected to scintillation counting to determine the protein extraction yield. The final sample volume is important and is estimated by comparing the meniscus height with that of a second Eppendorf tube containing 200 μl of water. We concentrate the volume to <200 μl and then add back the appropriate amount of water. In our experience we are able to extract >90% of the radiolabeled proteins from gel slices. A high yield of gel-extracted radio-prenylated proteins is required if accurate quantification of the amount of protein prenyl groups is desired.

To the concentrate is added 50 μl of ice-cold 100% (w/v) trichloroacetic acid. After briefly vortexing, the tube is incubated on ice for 2 hr and microcentrifuged at ~10,000g for 10 min at $4°$. The supernatant is decanted, and the last bit of liquid is removed from the lip of the tube with a tissue. Ice-cold reagent-grade acetone (0.5 ml) is added. The tube is inverted twice and microcentrifuged as described above. The supernatant is removed as described above, and the last trace of acetone is removed by placing the open tube in a $37°$ water bath for several minutes. The sample is now ready for Raney nickel treatment or it can be stored at $-80°$.

Release of Radiolabeled Prenyl Groups from Proteins with Raney Nickel

To the pellet of delipidated protein (total cell protein or protein eluted from the SDS–polyacrylamide gel; see previous section) is added 0.4 ml of 8 M guanidine hydrochloride, 0.2 M sodium phosphate (pH 7.0). To dissolve the protein pellet, the sample is vortexed and sonicated briefly in a bath-type device. The solution is transferred to a 16 × 100 mm glass tube fitted with a Teflon-lined screw cap. To prevent loss of pentane (see below) due to deformation of the Teflon liner, it is important to use a screw cap that does not have a hole in the center of the cap. To the solution of dissolved protein is added 1 ml of pentane. The tube is briefly vortexed, and most of the pentane is removed and discarded with a pipette. For samples containing large amounts of protein, centrifugation is required after vortexing to separate the solvent layers. Pentane extraction is repeated once more. This pentane extraction is important for removing hydrophobic material that may be present in the protein sample and that is not thioether linked to proteins. Pentane (1 ml) and internal standards [20 μg each of N-acetyl-S-farnesyl-L-cysteine and N-acetyl-S-geranylgeranyl-L-cysteine, both from CalBiochem (La Jolla, CA), added from 1-μg/μl stock solutions in ethanol] are then added to the tube followed by ~50 mg of Raney nickel. Raney nickel is prepared as described.[17] Caution is advised because Raney nickel can ignite if allowed to dry in air. A slurry of Raney nickel from an ethanol suspension is transferred with a Pasteur pipette to a piece of weigh paper on a balance. After most of the excess ethanol has evaporated, the moist solid is weighed, and additional Raney nickel is added as needed to give ~50 mg. Most of the Raney nickel is scraped off the paper with a spatula and transferred to the reaction tube. The tube is vortexed briefly to suspend any Raney nickel from the walls of the tube above the liquid phase. A small magnetic stir bar is added (use the proper size stir bar so that rapid mixing can occur), and the tube is tightly capped and held with a clamp and ring stand in an oil bath at $100 \pm 5°$ for 15 hr with sufficient stirring to break the pentane phase into small droplets. The tube is positioned so that the pentane–water interface is at the oil level.

The tube is allowed to cool to room temperature before removing the cap. Most of the pentane is transferred to a new glass vial with a Pasteur pipette. To the aqueous phase is added 0.5 ml of pentane, the tube is vortexed briefly, and the pentane layer is combined with the first pentane extract. The extract can be stored at $-20°$ in a tightly capped vial. To concentrate the pentane solution of released hydrocarbons, a portion is transferred to a small glass tube with a conical bottom (a Pasteur pipette flame sealed at its narrow end with a Bunsen burner works well). The tube is placed in an ice–NaCl bath, and a gentle stream of N_2 is focused on the solution to remove much of the pentane, leaving about 100 μl. More pentane solution is transferred to the tube, and the evaporation process is repeated until the entire pentane extract has been concentrated. Finally, the sample is concentrated to a final volume of ~20 μl of pentane, and 100 μl of methanol is immediately added. This concentration procedure ensures that minimal amounts of radiolabeled hydrocarbons are lost (routinely <10%). Routinely, 60–80% of the radioactivity is released into pentane.

Raney nickel-released prenyl groups are analyzed by HPLC on a C_{18} reversed-phase, analytical column [Vydac 218TP52 (The Separations Group, Hesperia, CA) or similar column]. Prior to HPLC, a small aliquot of sample is subjected to scintillation counting so that yields of HPLC-eluted hydrocarbons can be obtained. The column is developed at a flow rate of 0.5 ml/min with the following solvent program: 0–5 min, 100% water; 5–10 min, 0–60% acetonitrile linear gradient; 10–70 min, 60–100%

acetonitrile linear gradient; 70–100 min, 100% acetonitrile. A UV detector set at 210 nm is used to detect the hydrocarbon products. all-*trans*-2,6,10-Trimethyl-2,6,10-dodecatriene from farnesyl groups elutes with a retention time of 58.5 min, and all-*trans*-2,6,10,14-tetramethyl-2,6,10,14-hexadecatetraene from geranylgeranyl groups elutes with a retention time of 73.7 min. Eluent fractions (0.5 ml) are collected with a fraction collector and mixed with 5 ml of scintillation fluid for counting.

Figure 1 shows a typical chromatogram of radiolabeled prenyl groups released from *T. brucei* protein by Raney nickel treatment. A major peak of radioactivity for the 15-carbon hydrocarbon is seen precisely at the elution position for hydrocarbon standard (UV detection, not shown). Although the unlabeled 20-carbon hydrocarbon elutes as a sharp peak at 73.7 min (UV detection), little if any radiolabeled material is seen in this sample, indicating that little, if any, geranylgeranylated proteins became radiolabeled. Routinely, 75% of the counts per minute applied to the HPLC

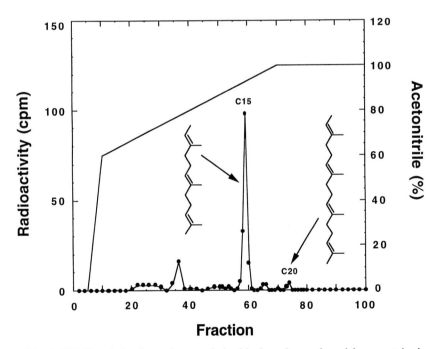

FIG. 1. HPLC analysis of prenyl group-derived hydrocarbons released from proteins by Raney nickel cleavage. HPLC conditions are given in text. Arrows show the elution position of authentic hydrocarbon standards (obtained by Raney nickel cleavage of *N*-acetyl-*S*-farnesyl-L-cysteine and *N*-acetyl-*S*-geranylgeranyl-L-cysteine).

column are eluted. In contrast to these results, methyl iodide cleavage of prenyl groups from proteins gives rise to multiple prenols derived from each type of protein prenyl group owing to multiple rearrangement pathways for the farnesyl carbocation.[22,23] A further disadvantage of the methyl iodide method is that radiolabeled prenols can also come from prenyl pyrophosphates that may be present in large amounts in protein samples. Raney nickel cleavage uniquely generates hydrocarbons rather than alcohols and is selective for carbon–sulfur bond cleavage.

Analysis of Nonradiolabeled Protein Prenyl Groups by Combined Gas Chromatography–Mass Spectrometry

The Raney nickel cleavage procedure when combined with gas chromatography–mass spectrometry can also be used to analyze prenyl groups without radiolabeling. We have been able to use this method to analyze total cell protein that has been delipidated with organic solvent only (i.e., no need for SDS–PAGE). Delipidated total cell protein (see previous section) is treated with Raney nickel as described in the previous section. The concentrated pentane extract is analyzed by combined gas chromatography–electron impact mass spectrometry exactly as described in detail.[17] In this case, internal standards other than N-acetyl-S-prenyl-L-cysteines must obviously be used. Suitable internal standards are eicosane and phytane.[17] Gas chromatogram retention times and mass spectrometry fragmentation patterns for the prenyl group-derived hydrocarbons have been published.[17]

Raney nickel-released prenyl groups from ~40 μg of total delipidated cell protein (from 10^7 *T. brucei* cells) is more than sufficient for combined gas chromatography–mass spectrometry including fragmentation pattern analysis of the prenyl group-derived hydrocarbons.[17] For example, when 39 μg of total delipidated cell protein is subjected to Raney nickel treatment to obtain the released hydrocarbons in 100 μl of methanol (see previous section), injection of 2 μl of this solution is sufficient for gas chromatography–mass spectrometry. On the basis of these numbers, it seems reasonable that gas chromatography–mass spectrometry can be carried out on <1 μg of a single prenylated protein extracted from an SDS–polyacrylamide gel slice. A more precise calculation based on the detection limits of hydrocarbons released from N-acetyl-S-prenyl-

---

[22] Y. Sakagami, A. Isogai, A. Suzuki, S. Tamura, C. Kitada, and M. Fujino, *Agric. Biol. Chem.* **42**, 1093 (1978).

[23] Y. Ishibashi, Y. Sakagami, A. Isogai, and A. Suzuki, *Biochemistry* **23**, 1399 (1984).

L-cysteine internal standards is that 30–1000 ng of prenylated protein is needed depending on the molecular weight of the protein (6000–200,000).

It is important to note that the ratio of C_{20} to C_{15} hydrocarbons in the gas chromatography analysis is much higher than the ratio in the radiometric analysis (Fig. 1). These results strongly argue that treatment of cells with statins in the presence of radiolabeled mevalonolactone leads to a distortion of the relative amounts of $C_{15,20}$-prenyl pyrophosphate in cells, which in turn distorts the relative amounts of the various prenyl groups added to newly synthesized proteins during radiolabeling. Thus, caution is advised, and gas chromatography–mass spectrometry is the method of choice for analyzing the mixture of prenyl groups attached to a collection of proteins.

Analysis of Prenyl Groups and C-Terminal Proteolysis and Methylation in Complex Peptide Mixtures by Electrospray Mass Spectrometry

Preparation of Prenylated Peptides and Peptide Methyl Esters

As is described below, electrospray mass spectrometry is a powerful method for detection of prenylated peptides and peptide methyl esters present in complex peptide mixtures such as tryptic digests of proteins. These studies are facilitated by the availability of appropriate synthetic radiolabeled and nonradiolabeled prenylated peptides and peptide methyl esters. As described in the next section, such peptides are useful for determining mass spectrometry detection limits, for providing HPLC retention times of prenylated tryptic peptides, for ensuring that prenylated peptides remain soluble, and for ensuring that prenylated peptide methyl esters have not undergone hydrolytic demethylation during sample handling.

The synthesis of nonradiolabeled prenylated peptides and prenylated peptide methyl esters by solution-phase prenylation of peptides made by solid-phase synthesis and the purification of these materials have been described in detail.[24] The same methods are used to prepare and purify radioprenylated peptides using tritiated farnesyl bromide and geranylgeranyl bromide, which are prepared as described here. There are several methods for converting prenols into their respective prenyl halides, but most are difficult to execute for the preparation of high specific activity

---

[24] L. Liu, G.-F. Jang, K. Yokoyama, F. Ghomashchi, C. C. Farnsworth, J. A. Glomset, and M. H. Gelb, *Methods Enzymol.* **250,** 189 (1994).

tritiated material on the submilligram scale. We have found that bromination with CBr_4/triphenyl phosphine (PPh_3) works well for this purpose, and is simple to execute.

A solution of [1-^{3}H]farnesol [1 mCi, 22.2 Ci/mmol (American Radiolabeled Chemicals); or prepared by reduction of farnesal with NaB^3H_4] in ethanol is added to 1 mg of unlabeled farnesol [all-*trans* (Aldrich, Milwaukee, WI)] in a 3-ml glass screwcap vial, and solvent is evaporated under a stream of dry N_2. Toluene (50 μl) is added and subsequently evaporated under N_2. This step is repeated twice to ensure that all of the ethanol is removed. Dry CH_2Cl_2 (350 μl, distilled from CaH_2) is added, followed by 2,6-lutidine (0.7 ml, 6 mmol, $\geq$99%; Eastman Kodak, Rochester, NY), and the vial is left capped on ice for 1 min. Triphenylphosphine (2.6 mg, 9.9 μmol, Aldrich) in 350 μl of dry CH_2Cl_2 is added, followed by carbon tetrabromide (3 mg, 9 μmol; Aldrich) in 350 μl of dry CH_2Cl_2. The reaction is incubated on ice for 1 hr in the dark. The solvent is evaporated under a stream of N_2, and the residue is taken up in 100 μl of acetonitrile for HPLC purification. The same procedure is used with [1-^{3}H]geranylgeraniol (1 mCi) mixed with 1.3 mg of unlabeled geranylgeraniol (both from American Radiolabeled Chemicals).

The reaction mixture in acetonitrile is injected onto a C_{18} reversed-phase semipreparative HPLC column (Vydac, 10 μm, 10 $\times$ 250 mm). The column is washed at a flow rate of 1.5 ml/min with 30% (v/v) acetonitrile in water (Milli-Q; Millipore, Bedford, MA) for 10 min, followed by a linear gradient from 30 to 100% acetonitrile over 40 min, and then holding at 100% acetonitrile for 30 min. UV detection is performed at 215 nm. The retention times for farnesyl bromide and geranylgeranyl bromide are 60 and 66 min, respectively. The radiochemical purity is verified by thin-layer chromatography on a C_{18} reversed-phase plate (20% ethyl acetate/hexane) followed by fluorography with EN^{3}HANCE (NEN, Boston, MA). Unlabeled prenol is spotted as a standard (detected with I_2 vapor staining). The reaction typically yields approximately 80 μCi of [1-^{3}H]farnesyl bromide or 120 μCi of [1-^{3}H]geranylgeranyl bromide that is >95% radiochemically pure. Unreacted farnesol (51 min) and geranylgeraniol (61 min) are also recovered (0.55 mCi of farnesol and 0.45 mCi of geranylgeraniol), which can be reused.

Figure 2 shows the HPLC trace of four variants of the C-terminal tryptic peptide of bovine rhodopsin kinase (SGMC(*S*-[1-^{3}H]farnesyl)-COOH, SGMC(*S*-[1-^{3}H]farnesyl)-COOCH$_3$, SGMC(*S*-[1-^{3}H]geranylgeranyl)-COOH, and SGMC(*S*-[1-^{3}H]geranylgeranyl)-COOCH$_3$). The peptides SGMC-COOH and SGMC-COOCH$_3$ are made by solid-phase synthesis[24] and prenylated with radiolabeled prenyl bromides (1 Ci/mol) as described.[24]

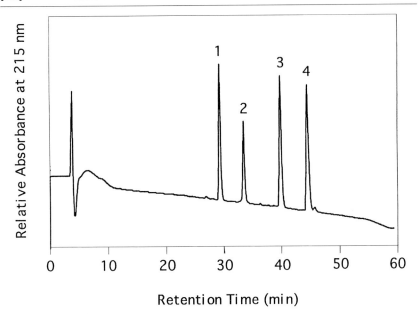

FIG. 2. Reversed-phase HPLC analysis of synthetic bovine rhodopsin kinase prenylated peptides: column, Vydac, 4.6 × 250 mm; flow rate, 1.5 ml/min; solvent gradient, 5–30% acetonitrile in water with 0.1% trifluoroacetic acid over 5 min, then 30–85% over 55 min; UV detector at 215 nm. Retention times: peak 1, 29.1 min, SGMC(S-farnesyl)-COOH; peak 2, 33.2 min, SGMC(S-farnesyl)-COOCH$_3$; peak 3, 39.9 min, SGMC(S-geranylgeranyl)-COOH; peak 4, 44.8 min, SGMC(S-geranylgeranyl)-COOCH$_3$.

Radioprenylated peptides are purified by reversed phase HPLC as shown in Fig. 2.

Trypsin Digestion of Bovine Rhodopsin Kinase

Partially purified bovine rhodopsin kinase (purified from whole retina rather than rod outer segments[25]) in 20 mM Bis–Tris propane, 1 mM dodecyl-β-maltoside (~30% pure as judged by SDS–PAGE) is concentrated to 50 μg/ml in a Centricon-30 ultrafiltration device (Amicon Danvers, MA). For trypsin digestion, 25 μg of protein containing ~7 μg of rhodopsin kinase is mixed with 0.25 μg of sequencing-grade modified trypsin (Promega, Madison, WI) in 1 M urea, 66.7 mM NH$_4$HCO$_3$, 3 mM CaCl$_2$ (pH 8.0) in a total volume of 250 μl. The sample is incubated at 30° for

[25] K. Palczewski, *Methods Neurosci.* **15**, 217 (1993).

18 hr. The digest can be stored at $-20°$ prior to analysis by mass spectrometry (see below).

With the radioprenylated tryptic peptide standards in hand, it is possible to gauge the solubility of these peptides in the buffer used for trypsin digestion and also to gauge the stability of the methyl esters under the conditions used for purification of rhodopsin kinase and for trypsinization. Such information is obviously critical for addressing by mass spectrometry (see below) the question of which C-terminal variants of rhodopsin kinase are present in native enzyme purified from retina. The synthetic peptide SGMC(S-[1-^{3}H]farnesyl)-COOCH$_3$ is incubated at $30°$ for 18 hr in the buffer used for trypsin digestion. Scintillation counting of an aliquot of the solution showed that 83% of the radiolabeled peptide is soluble, and reversed-phase HPLC analysis of the mixture shows that 60% of the methyl ester is recovered from the column with less than 20% of the peptide converted to the free acid.

Combined High-Performance Liquid Chromatography–Electrospray Mass Spectrometry

Analysis of the rhodopsin–kinase tryptic digest is performed by microcolumn microelectrospray HPLC–tandem mass spectrometry. The microcolumn is constructed from fused silica capillary (365 × 100 μm; J & W Scientific, Folsom, CA) and pulled to a tip of ∼2 μm with a laser puller.[26] Poros R2 reversed phase material (10 μm; PerSeptive Biosystems, Framingham, MA) is packed against the tip to a length of approximately 15 cm.

The tryptic digest (10 μl of 250 μl) is loaded onto the column at 150 nl/min according to the method described by Yates et al.[27,28] An HP 1100 binary HPLC pump (Hewlett-Packard, Wilmington, DE) is used to deliver the gradient to the column. A precolumn split allows a flow rate of 150 nl/min through the column to be achieved. Buffer A consists of 0.5% (v/v) acetic acid and buffer B consisted of acetonitrile–water (80:20, v/v) with 0.5% (v/v) acetic acid (Optima-grade solvents; Fisher, Pittsburgh, PA). The gradient is ramped from 2% (v/v) solvent B to 25% (v/v) solvent B in 2.5 min, followed by 25% (v/v) solvent B to 80% (v/v) solvent B in 27.5 min. Throughout the gradient, a Finnigan LCQ mass spectrometer (Finnigan, San Jose, CA) is programmed for selected ion monitoring (SIM) followed by tandem mass spectrometry (MS/MS).

[26] C. L. Gatlin, G. R. Kleemann, L. G. Hays, A. J. Link, and J. R. Yates III, *Anal. Biochem.* **263**, 93 (1998).
[27] J. R. Yates III, A. L. McCormack, J. B. Hayden, and M. P. Davey, in "Cell Biology: A Laboratory Handbook" (J. E. Celis), p. 380. Academic Press, San Diego, California, 1994.
[28] A. L. McCormack, J. K. Eng, and J. R. Yates III, *Methods* **6**, 274 (1994).

The peptide species monitored are the farnesylated nonmethylated tryptic peptide, m/z 602.4; the geranylgeranyl nonmethylated peptide, m/z 670.6; the farnesylated methylated peptide, m/z 616.3; and the geranylgeranyl methylated peptide, m/z 685.6 (structures of peptides given above). The instrument acquires three microscans of each m/z value in an SIM mode and then acquires tandem mass spectra for the m/z values. MS/MS data provide important verification that the m/z values represent the peptides. The rhodopsin kinase tryptic digest is a complex mixture of peptides. It is therefore possible that a finite amount of ion signal could be produced at the selected m/z values by multiply charged species or fragment ions from peptides other than those of interest. The MS–MS spectrum provides a fingerprint of the prenylated peptide that enables most, if not all, of the peptide sequence to be deduced from the data.

Figure 3A shows the tandem mass spectrum of the synthetic farnesylated and nonmethylated C-terminal tryptic peptide based on the bovine rhodopsin kinase C terminus [SGMC(S-farnesyl)-COOH]. The retention time for this m/z 601.4 ion is 19.23 min. From the tryptic digest of retina-derived bovine rhodopsin kinase, only the farnesylated nonmethylated peptide is observed. The retention time for this m/z 601.4 peptide ion is 19.05 min. Figure 3B shows the tandem mass spectrum for this peptide obtained from native kinase. The spectrum shows fragment ions for the y_1–y_3 and the b_3 ion. A signal is seen for the loss of the farnesyl group and the loss of the farnesyl group including SH is observed at m/z 397.0 and m/z 363.1, respectively. These fragment ions are identical to the signals observed in the tandem mass spectrum of the peptide standard. Studies with synthetic SGMC(S-farnesyl)-COOCH$_3$, and the geranylgeranylated peptide and its methyl ester, show that these compounds are readily detected by mass electrospray spectrometry (not shown). These peptides are also detected when the rhodopsin kinase trypsin digest is spiked with the synthetic standards. A final note is that given the high sensitivity of the method, a blank run should be carried out prior to the analysis of trypsin digests to be sure that there is no carryover of peptide standards from a previous analysis.

It is anticipated that the HPLC–MS/MS method for the identification of prenylated and methylated peptides in complex peptide mixtures can be applied to most tryptic digests, whether the source of protein be a partially purified preparation, an SDS–polyacrylamide gel band, or an immunoprecipitate. The method is sensitive; 0.3 μg of rhodopsin kinase is more than sufficient for trypsin digestion and tandem mass spectrometry. The protein to be analyzed does not have to be homogeneous as long as its C-terminal sequence is known. Presumably the method can be used for the analysis of doubly geranylgeranylated peptides, as these peptides elute from reversed-phase HPLC columns in good yield and have been detected

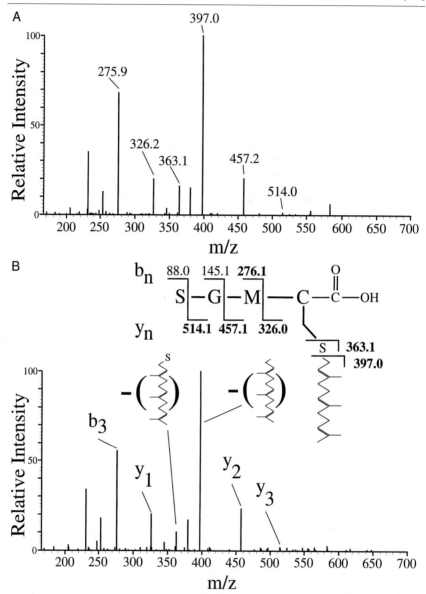

FIG. 3. (A) Tandem mass spectrum of the (M + H)⁺ farnesylated nonmethylated peptide standard ion at m/z 601.4. This peptide eluted at a retention time of 19.23 min. (B) Tandem mass spectrum of the (M + H)⁺ farnesylated nonmethylated peptide at m/z 601.4 from the tryptic digest of native bovine rhodopsin kinase. This peptide has a retention time of 19.05 min. The amino acid sequence and location of modification are shown above the spectrum. Sequence ions in boldface were found in the spectra.

by fast atom bombardment mass spectrometry.[11] Synthesis of doubly geranylgeranylated peptides has been reported.[24]

Conclusions

The methods described in this chapter for the structural analysis of prenyl groups released from proteins and of C-terminal lipidated proteolytic fragments derived from native proteins offer many advantages over methods that have been used previously. Raney nickel cleavage yields single cleavage products in high yield. These fragments can be conveniently analyzed by HPLC methods when radiolabeling is carried out or by gas chromatography–mass spectrometry when proof of structure and accurate ratios of the different length prenyl groups are desired. Tandem mass spectrometry, when combined with micro-HPLC, provides a method for characterizing the entire C-terminal structure of the protein of interest (prenylation, proteolysis, and methylation). Because of mass speciation, this method is useful for analyzing complex mixtures such as tryptic digests of native proteins. Finally, all of the methods are sufficiently sensitive to permit the analysis of relatively small amounts of proteins extracted from SDS–polyacrylamide gel bands.

Acknowledgments

Supported by National Institutes of Health Grants CA52874 (M.H.G.), 118223 (J.R.Y.), and EY08061 (K.P.) and by Molecular Biophysics Training Grant T32 GM08268 (M.E.W.).

[30] Isoprenylation/Methylation and Transducin Function

By CRAIG A. PARISH and ROBERT R. RANDO

Introduction

Members of a diverse group of proteins are posttranslationally modified at their carboxyl termini by isoprenylation/methylation.[1-4] These proteins

[1] W. R. Schafer and J. Rine, *Annu. Rev. Genet.* **26,** 209 (1992).
[2] M. Sinensky and R. J. Lutz, *BioEssays* **14,** 25 (1992).
[3] C. A. Omer and J. B. Gibbs, *Mol. Microbiol.* **11,** 219 (1994).
[4] S. Clarke, *Annu. Rev. Biochem.* **61,** 355 (1992).

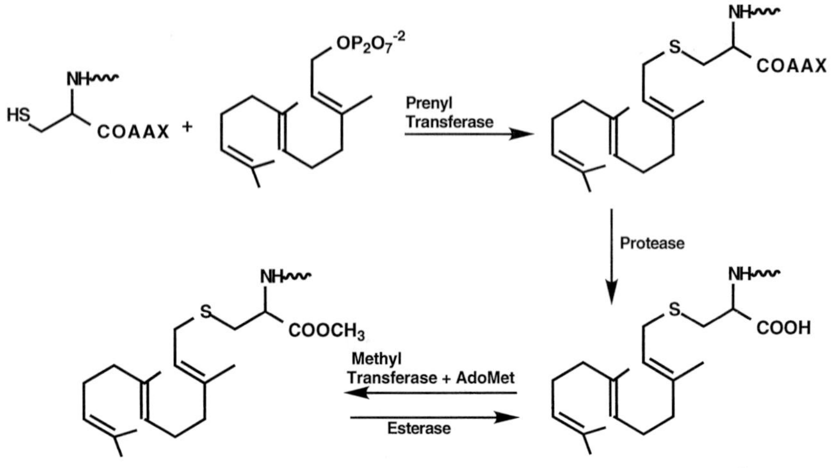

FIG. 1. Isoprenylation/methylation pathway for CAAX proteins.

include heterotrimeric G-protein γ subunits,[5–7] small G proteins,[8] protein kinases,[9] nuclear lamins,[10] and viral proteins.[11] The modifications themselves are not of a single type, and can be broadly divided into singly and doubly isoprenylated species. Geranylgeranylation (C_{20}) is by far the most prevalent modification overall. All heterotrimeric G proteins are singly geranylgeranylated, with the exception of retinal transducin, which is farnesylated (C_{15}).[6,7] The biochemical pathway for farnesylation/methylation of proteins containing a carboxyl-terminal CAAX motif (C, cysteine; A, aliphatic amino acid; X, any amino acid), such as transducin, is shown in Fig. 1.

Isoprenylation/methylation is thought to allow otherwise soluble proteins to associate with membranes, and one can readily understand how membrane association would be important in phototransduction. Because rhodopsin is an integral membrane protein, it is more efficient for it to

[5] H. K. Yamane, C. C. Farnsworth, H. Y. Xie, W. Howald, B. K. Fung, S. Clarke, M. H. Gelb, and J. A. Glomset, *Proc. Natl. Acad. Sci. U.S.A.* **87**, 5868 (1990).
[6] Y. Fukada, T. Takao, H. Ohguro, T. Yoshizawa, T. Akino, and Y. Shimonishi, *Nature (London)* **346**, 658 (1990).
[7] R. K. Lai, D. Perez-Sala, F. J. Canada, and R. R. Rando, *Proc. Natl. Acad. Sci. U.S.A.* **87**, 7673 (1990).
[8] Y. Takai, K. Kaibuchi, A. Kikuchi, and M. Kawata, *Int. Rev. Cytol.* **133**, 187 (1992).
[9] J. Inglese, W. J. Koch, M. G. Caron, and R. J. Lefkowitz, *Nature (London)* **359**, 147 (1992).
[10] C. C. Farnsworth, S. L. Wolda, M. H. Gelb, and J. A. Glomset, *J. Biol. Chem.* **264**, 20422 (1989).
[11] J. S. Glenn, J. A. Watson, C. M. Havel, and J. M. White, *Science* **256**, 1331 (1992).

interact with proteins that associate with disk membranes, rather than with proteins that are cytoplasmic. An important issue to resolve is the mechanism through which isoprenylation enhances membrane association. The simplest mechanism involves lipid–lipid-based hydrophobic interactions, in which the isoprenylated/methylated cysteine moiety is held at the surface of the membrane through direct contact with membrane phospholipids. An alternative mechanism involves specific lipid–protein interactions, in which the isoprenylated/methylated cysteine moiety is recognized by a receptor protein. This type of interaction would involve binding proteins that recognize the lipid conformation of the isoprene, as well as the carboxyl-terminal methyl group. A role for the methyl group in a specific protein-binding event presumably would lead to a significant decrease in association on demethylation owing to the exposed carboxylate negative charge. The idea that isoprenylated cysteine moieties might be involved in mediating specific protein–lipid binding may be, of course, of general interest beyond membrane docking. It is important to explore the issue of whether there might be important biochemical interactions that depend on isoprenyl–protein recognition. It is interesting to note that because methylation is the only reversible enzymatic reaction in the pathway, and it has been suggested that the activity of isoprenylated/methylated proteins may be regulated by their state of methylation.[12] These suggestions are based on analogy with what is known about the role of protein methylation in bacterial chemotaxis.[13,14] This issue can be directly addressed in studies on phototransduction.

In attempting to understand the role of isoprenylation/methylation in protein function, it is important to study an isoprenylated/methylated system that can be described quantitatively. The retinal photoreceptor is the best understood of all the heterotrimeric G-protein-coupled receptor systems, and often serves as a paradigm for the study of seven-transmembrane helical receptor signal transduction.[15] While rhodopsin is an integral membrane protein, transducin (T) is farnesylated/methylated on its γ subunit.[6,7] The target of transducin, the phosphodiesterase, is farnesylated/methylated on one subunit and geranylgeranylated on the other,[16] and rhodopsin kinase, which is involved in the downregulation of photoactivated rhodopsin (R*),

---

[12] M. R. Philips, M. H. Pillinger, R. Staud, C. Volker, M. G. Rosenfeld, G. Weissmann, and J. B. Stock, *Science* **259**, 977 (1993).
[13] M. J. Shapiro, I. Chakrabarti, and D. E. Koshland, Jr., *Proc. Natl. Acad. Sci. U.S.A.* **92**, 1053 (1995).
[14] E. N. Kort, M. F. Goy, S. H. Larsen, and J. Adler, *Proc. Natl. Acad. Sci. U.S.A.* **72**, 3939 (1975).
[15] L. Stryer, *J. Biol. Chem.* **266**, 10711 (1991).
[16] J. S. Anant, O. C. Ong, H. Y. Xie, S. Clarke, P. J. O'Brien, and B. K. Fung, *J. Biol. Chem.* **267**, 687 (1992).

is farnesylated/methylated.[9] In the visual system, R* interacts with T$\alpha\beta\gamma$ at the surface of the disk membrane and catalyzes the exchange of GDP for GTP. T$\beta\gamma$ is essential in this exchange reaction. Tα–GTP interacts with the phosphodiesterase, relieving it of an inhibitory subunit and allowing the latter enzyme to hydrolyze cyclic GMP (cGMP). As cGMP controls Na^+/Ca^{+2} conductance across the plasma membrane, the hydrolysis of cGMP results in photoreceptor hyperpolarization and the initiation of the visual response.[15] The preparation of modified protein subunits has allowed for an extensive study of the role of posttranslational modifications in heterotrimeric G-protein function.

Principle

To study the roles of isoprenylation and methylation, biochemical means to remove these two functionalities from heterotrimeric G-protein γ subunits are required and assays that determine the relative activities of the various modified proteins are needed. The enzymatic activities shown in Fig. 2 are capable of processing T$\beta\gamma$ to provide mechanistically informative fragments. A number of assays can quantitatively assess the relative functional performance of each protein.

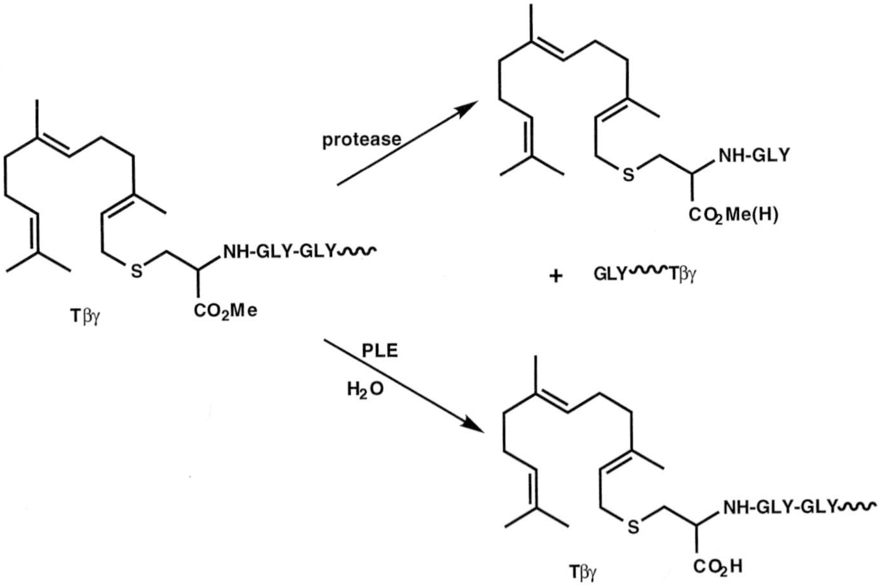

FIG. 2. Enzymatic processing of retinal transducin. Pig liver esterase (PLE) can specifically hydrolyze the methyl ester, and a rod outer segment protease cleaves off the terminal dipeptide.

In the inactive state, the αβγ subunits are associated in the heterotrimeric state and GDP is bound to the α subunit. On receptor activation, GDP is released by the α subunit and GTP is taken up. The association of βγ with α is required for GTP exchange to occur. The use of a nonhydrolyzable analog of GTP, radiolabeled [^{35}S]GTPγS, allows for the quantitative analysis of the extent of GTP binding to the α subunit under a range of experimental conditions.

After GTP binding to the heterotrimeric G protein, the GTP-bound α subunit dissociates from βγ, leading to activation of effector enzymes important in diverse signal transduction cascades. The βγ subunits are always associated, and act as a functional monomer; the two subunits can be separated only under denaturing conditions. In the case of transducin, it is Tα that binds to the inhibitory subunit of the cGMP phosphodiesterase, leading to its activation and the further propagation of the visual signal. Heterotrimeric G-protein βγ subunits can have effectors other than α subunits.[17,18] Other systems use both α and βγ subunits for the stimulation of subsequent enzymes in signaling pathways. βγ subunits can activate enzymes such as certain isoforms of phosphoinositide 3-kinase (PI3K)[19,20] and phosphatidylinositol-specific phospholipase C (PIPLC β),[21,22] adenylate cyclase,[23] and ion channels.[24,25] The role of βγ isoprenylation/methylation can be probed by assays of effector enzyme stimulation. Both PIPLC β isoforms and PI3K are activated by G protein βγ subunits and, therefore, can be used in assays to probe the role of isoprenylation/methylation in βγ function.[19,26]

The intrinsic GTPase activity of the α subunit leads to hydrolysis of the bound GTP to GDP, which is followed by heterotrimer reassociation, thus completing the G-protein cycle. Pertussis toxin-catalyzed ADP ribosylation

---

[17] D. E. Clapham and E. J. Neer, *Nature (London)* **365,** 403 (1993).

[18] P. C. Sternweis, *Curr. Opin. Cell Biol.* **6,** 198 (1994).

[19] L. Stephens, A. Smrcka, F. T. Cooke, T. R. Jackson, P. C. Sternweis, and P. T. Hawkins, *Cell* **77,** 83 (1994).

[20] P. A. Thomason, S. R. James, P. J. Casey, and C. P. Downes, *J. Biol. Chem.* **269,** 16525 (1994).

[21] A. V. Smrcka and P. C. Sternweis, *J. Biol. Chem.* **268,** 9667 (1993).

[22] A. Dietrich, M. Meister, D. Brazil, M. Camps, and P. Gierschik, *Eur. J. Biochem.* **219,** 171 (1994).

[23] W. J. Tang and A. G. Gilman, *Science* **254,** 1500 (1991).

[24] K. D. Wickman, J. A. Iniguez-Lluhl, P. A. Davenport, R. Taussig, G. B. Krapivinsky, M. E. Linder, A. G. Gilman, and D. E. Clapham, *Nature (London)* **368,** 255 (1994).

[25] E. Reuveny, P. A. Slesinger, J. Inglese, J. M. Morales, J. A. Iniguez-Lluhi, R. J. Lefkowitz, H. R. Bourne, Y. N. Jan, and L. Y. Jan, *Nature (London)* **370,** 143 (1994).

[26] N. Ueda, J. A. Iniguez-Lluhi, E. Lee, A. V. Smrcka, J. D. Robishaw, and A. G. Gilman, *J. Biol. Chem.* **269,** 4388 (1994).

of the α subunit is used to analyze the relative affinities of α for $\beta\gamma$.[27] The toxin can modify only G proteins that are in the heterotrimeric state. Therefore, this assay can indicate quantitatively the role of isoprenylation/methylation in G-protein association. In addition, the affinity of the various subunits for membranes can be investigated by mixing the modified proteins with phospholipid vesicles and comparing the relative amounts of each in membrane and cytosolic fractions.

Assays probing the role of the $\beta\gamma$ isoprenyl/methyl moiety may be sensitive to synthetic analogs that mimic the carboxyl termini of these proteins, if indeed the operative mechanism involves specific protein–lipid interactions. To this end, a variety of farnesyl cysteine-based analogs can be prepared and tested for effects in the above-described assay systems, providing further information regarding the role of the isoprenyl/methyl substituents in G-protein function.

Materials and Methods

Protein Preparation

Rhodopsin. Frozen bovine retinas are obtained from J. A. & W. L. Lawson Company (Lincoln, NE). Solubilized rhodopsin in dodecyl maltoside[28] and urea-treated rod outer segments (ROS)[29] are prepared in the dark under red light.

Transducin. The preparation of transducin is based on the GTP-dependent elution of purified ROS[30,31] and is carried out under room light. Transducin is extracted completely from the ROS membranes by two treatments with 40 μM GTP in hypotonic buffer (total volume, 75 ml) for 15 min. Further purification of transducin is accomplished by hexylagarose chromatography (see below). Fractions are analyzed for the ability to bind [^{35}S] GTPγS and those fractions with GTP exchange activity are concentrated to a small volume by microultrafiltration (Amicon, Danvers, MA). The transducin heterotrimer is separated into its purified Tα and T$\beta\gamma$ subunits by Blue Sepharose chromatography.[30–32]

---

[27] E. J. Neer, J. M. Lok, and L. G. Wolf, *J. Biol. Chem.* **259**, 14222 (1984).
[28] C. Longstaff and R. R. Rando, *Biochemistry* **24**, 8137 (1985).
[29] A. B. Fawzi and J. K. Northup, *Biochemistry* **29**, 3804 (1990).
[30] M. Wessling-Resnick and G. L. Johnson, *J. Biol. Chem.* **262**, 3697 (1987).
[31] C. A. Parish and R. R. Rando, *Biochemistry* **33**, 9986 (1994).
[32] Y. Fukada, H. Ohguro, T. Saito, T. Yoshizawa, and T. Akino, *J. Biol. Chem.* **264**, 5937 (1989).

Defarnesylated T$\beta\gamma$. The role of the farnesyl/methyl group can be studied directly in the case of transducin because a specific photoreceptor protease cleaves off the carboxyl-terminal glycyl-S-farnesylcysteinyl moiety of T$\beta\gamma$ (Fig. 2).[33] This activity has been identified in crude preparations of transducin in which the transducin heterotrimer is not stable under standard storage conditions. High-performance liquid chromatography (HPLC) analysis of Tγ indicates that that subunit is being processed to a more hydrophilic species. After purification by hexylagarose chromatography, transducin is stable for extended periods, indicating that the proteolytic activity can be separated from the G protein; preliminary characterization indicates that the proteolytic processing of Tγ is enzymatic in nature.[33]

Hexylagarose chromatography[34] (250 × 10 mm) of transducin is accomplished with a gradient of 20–200 mM NaCl in buffer [10 mM morpholine propane sulfonic acid (MOPS, pH 7.4), containing 1 mM dithiothreitol (DTT) and 1 mM EDTA; total volume of gradient, 150 ml], resulting in the separation of the proteolytic activity in a broad band as monitored by UV absorbance at 280 nm. This peak elutes from the column between 20 and 40 mM NaCl. The transducin peak, which elutes from the hexylagarose column between 50 and 80 mM NaCl, is identified by monitoring the ability of these fractions to bind GTPγS in the presence of photoactivated rhodopsin. The fractions containing the proteolytic activity are incapable of binding GTPγS. The fractions that contain protein by UV absorbance but do not bind GTPγS are combined and concentrated to a small volume with a microultrafiltration system equipped with a PM-10 membrane (Amicon). This concentrated solution is used as a source of protease. Characterization of this enzyme indicates that it is a thiol protease with a molecular weight of approximately 35,000.[33]

Demethylated T$\beta\gamma$. Removal of the carboxyl-terminal glycyl-S-farnesylcysteinyl moiety of T$\beta\gamma$ is a substantial molecular alteration, and the fact that the resulting protein is inactive (see below) may be a consequence of several influences, including the gross alteration of protein structure. Protein demethylation is a more subtle modification, and studies of demethylated T$\beta\gamma$ would be expected to yield more definitive information about the functional role of isoprenylation/methylation. If a protein receptor exists that recognizes the farnesylated/methylated cysteine moiety, then binding should be quite sensitive to the state of methylation.

As isolated, transducin exists in a partially demethylated state.[6,35] While

---

[33] H. Cheng, C. A. Parish, B. A. Gilbert, and R. R. Rando, *Biochemistry* **34,** 16662 (1995).
[34] B. K. Fung, J. B. Hurley, and L. Stryer, *Proc. Natl. Acad. Sci. U.S.A.* **78,** 152 (1981).
[35] D. Perez-Sala, E. W. Tan, F. J. Canada, and R. R. Rando, *Proc. Natl. Acad. Sci. U.S.A.* **88,** 3043 (1991).

a small amount of demethylated T$\beta\gamma$ can be obtained by chromatographic purification of this material,[36] a simple and direct protocol for obtaining large amounts of demethylated proteins would provide freshly prepared samples from protein of known bioactivity. Transducin can be demethylated by immobilized pig liver esterase (iPLE),[31] which can easily be removed from a reaction mixture. In a typical experiment to accomplish full hydrolysis of Tγ-OCH$_3$ to the corresponding free carboxylate (Tγ-OH), purified T$\beta\gamma$ subunits are treated with iPLE which has been thoroughly washed with buffer (four times). After the desired time at 4°, the supernatant is removed after centrifugation (14,000 rpm, 5 min, 4°) and the beads are further washed with buffer containing 10 mM Tris-HCl, pH 7.4/5 mM MgCl$_2$/1 mM DTT/0.1 mM EDTA/100 mM NaCl. After 1 hr, a second supernatant is removed after centrifugation.

While the hydrolysis of the intact $\beta\gamma$ subunits is accomplished, denaturing HPLC conditions allow for the analysis of the γ-subunit state of methylation. Typically, the β subunit does not elute from the column. All transducin samples are injected with added guanidinium chloride at a final concentration of greater than 3 M. Analysis of Tγ is accomplished by reversed-phase HPLC, using a C$_{18}$ column (Dynamax 300A; Rainin, Boston, MA) running a linear gradient at 0.75 ml/min. After 10 min at 5% CH$_3$CN in H$_2$O [10 mM trifluoro acetic acid (TFA)], a 40-min gradient is run to 95% CH$_3$CN in H$_2$O (10 mM TFA). At this point, the eluant is held at 95% CH$_3$CN for an additional 10 min. UV absorbance is monitored at 205 nm. The methylated and demethylated forms of the γ subunit are baseline separated, using these HPLC conditions, with Tγ-OCH$_3$ being retained on the column approximately 1 min longer than the corresponding free carboxylic acid.

To determine the role of methylation in the function of heterotrimeric G proteins, it is informative to compare the effects observed with farnesylated/methylated transducin with data obtained with another $\beta\gamma$ isoform that contains a geranylgeranylated/methylated γ subunit. $\beta_1\gamma_2$, which contains the same β subunit as transducin, but a geranylgeranylated/methylated γ isoform (γ_2), is prepared in a baculovirus expression system and purified.[37] $\beta_1\gamma_2$ is demethylated with iPLE under conditions similar to those used for T$\beta\gamma$.[38] The hydrolysis of γ_2 is quantified by reversed-phase HPLC. In the case of the more hydrophobic geranylgeranylated γ subunits, a C$_8$ column (Dynamax 300Å; Rainin) is used to separate the methylated and demethylated γ_2. $\beta_1\gamma_2$ samples are injected with 3 M guanidinium chloride.

[36] Y. Fukada, T. Matsuda, K. Kokame, T. Takao, Y. Shimonishi, T. Akino, and T. Yoshizawa, *J. Biol. Chem.* **269**, 5163 (1994).
[37] T. Kozasa and A. G. Gilman, *J. Biol. Chem.* **270**, 1734 (1995).
[38] C. A. Parish, A. V. Smrcka, and R. R. Rando, *Biochemistry* **35**, 7499 (1996).

After running the column at 5% acetonitrile in water (10 mM TFA) for 10 min, a linear gradient is run from 5% to 95% acetonitrile in water (10 mM TFA) over 40 min. At that point, the solvent is held at 95% acetonitrile in water (10 mM TFA) for 10 min. The flow rate of the gradient is 0.75 ml/min and the absorbance of each injection is monitored at 205 nm. The retention times of methylated and demethylated γ_2 are approximately 46.5 and 45.0 min, respectively. The identity of each protein species is confirmed by electrospray ionization mass spectrometry.[38]

Assays of Heterotrimeric G-Protein Function

GTP Exchange Assays. Twenty microliters of assay buffer [50 mM Tris-HCl (pH 7.4), 500 mM NaCl, 25 mM MgCl$_2$, 5 mM DTT, 0.5 mM EDTA], water (an amount that will bring the total assay volume to 100 μl), and Tα and/or T$\beta\gamma$ subunits are combined in a test tube. In the dark, either detergent-solubilized rhodopsin or urea-treated ROS is added and the tube is incubated at the desired temperature. The sample is bleached for 1 min under ordinary room light before adding [^{35}S]GTPγS (2 μM final concentration, 5000–7000 cpm/pmol). An aliquot of each experiment is removed at the desired time after [^{35}S]GTPγS addition. All assays are done in duplicate. Each aliquot is filtered through a nitrocellulose membrane (Schleicher & Schuell, Keene, NH) and immediately washed three times with ice-cold 10 mM Tris-HCl (pH 7.4)–100 mM NaCl–5 mM MgCl$_2$–0.1 mM EDTA (4 ml). Each membrane is dissolved in Filtron-X scintillation fluid (10 ml) and counted. A summary of the data obtained with methylated and demethylated T$\beta\gamma$ with urea-treated ROS is presented in Fig. 3.

Effector Enzyme Assays. The assays of phosphatidylinositol-specific phospholipase C β isoforms (PIPLC β) and phosphoinositide 3-kinase (PI3K) activity with T$\beta\gamma$ and $\beta_1\gamma_2$ have been described previously.[38,39] PIPLC β activity[21] is assayed in sonicated micelles of 50 μM phosphatidylinositol 4,5-diphosphate, 200 μM bovine brain phosphatidylethanolamine (PE), and [*inositol*-2-^{3}H(N)]phosphatidylinositol biphosphate (PIP$_2$) (4000 cpm/assay). PI3K[19] is assayed with sonicated micelles containing 600 μM bovine liver PE and 300 μM bovine liver phosphatidylinositol (PI). A summary of the data on the activation of PIPLC β is shown in Fig. 4.

Pertussis Toxin-Catalyzed ADP Ribosylation. Pertussis toxin (10 μl, 50 μg/ml) is activated by adding DTT (1.5 μl, 1 M) and sodium dodecyl sulfate (SDS, 1 μl, 2.5%) and incubating this sample for 20 min at 32°.[40] To prepare the pertussis toxin working solution, the activated toxin is diluted with water (50 μl). The assay buffer contains 50 mM Tris (pH 7.4),

[39] C. A. Parish, A. V. Smrcka, and R. R. Rando, *Biochemistry* **34,** 7722 (1995).
[40] D. J. Carty, *Methods Enzymol.* **237,** 63 (1994).

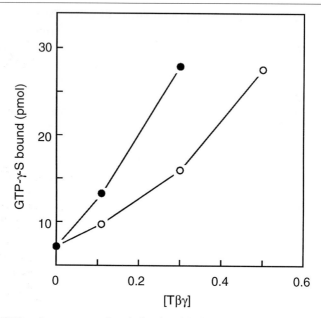

FIG. 3. GTP exchange assays of methylated and demethylated T$\beta\gamma$ with urea-washed rod outer segment membranes. The indicated quantity of GTPγS was bound to Tα after 10 min at 25°. [Tα] = 1.5 μM, [R*] = 5 nM, [GTPγS] = 2 μM. (●) T$\beta\gamma$-OCH$_3$; (○) T$\beta\gamma$-OH.

2 mM EDTA, 20 mM thymidine, 20 mM DTT, 10 μM NAD, and [*adenylate*-^{32}P]NAD (40 μCi/ml). To each assay tube is added water (8 μl), α_{i-1} [1 μl, 12 nM in 50 mM Tris (pH 7.7)–75 mM sucrose–6 mM MgCl$_2$–1 mM EDTA–1 mM DTT–0.6% Lubrol PX–10 mM NaF–10 μM AlCl$_3$–100 mM NaCl, isolated from bovine brain], and methylated or demethylated $\beta_1\gamma_2$ at varying concentrations (1 μl). Each assay tube is cooled to 0° and the pertussis toxin working solution (2.5 μl) and the radioactive assay buffer (12.5 μl) are added. The final assay volume is 25 μl. After heating to 37° for 15 min, each assay tube is immediately cooled to 0°. Cold sample buffer (12.5 μl) is added to each assay and an aliquot (10 μl) is run on a 12% SDS–polyacrylamide gel. After Coomassie blue staining of the gel, the labeled α_{i-1} subunit is observed both by autoradiography (20–40 hr) and phosphorimaging [3-hr exposure; Molecular Dynamics (Sunnyvale, CA) PhosphorImager]. The intensity of each band is determined by integrating identical areas of the gel obtained from the PhosphorImager.

Liposome Binding of $\beta\gamma$ Subunits. Azolectin vesicles can be used to determine the relative binding affinities of methylated and demethylated

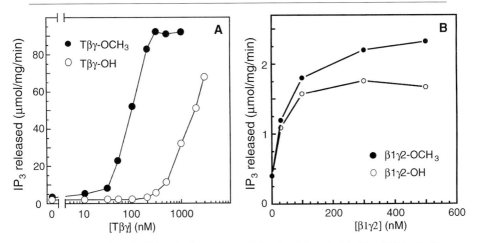

FIG. 4. Activation of PIPLC β isoforms by methylated and demethylated $\beta\gamma$. (A) Activation of PIPLC β_3 by farnesylated T$\beta\gamma$-OCH$_3$ and T$\beta\gamma$-OH. (B) Activation of PIPLC β_2 by geranylgeranylated $\beta_1\gamma_2$-OCH$_3$ and $\beta_1\gamma_2$-OH. Each assay contained 0.5 ng of purified PIPLC from rat brain.[38,39]

$\beta\gamma$ subunits for phospholipid membranes.[41] Each tube contains the appropriate $\beta\gamma$ subunit (approximately 1 μM final concentration) and the filtered azolectin vesicles (10 μl, 16 mg/ml, as determined by dry weight). An experiment with $\beta_1\gamma_2$ contains 5 μl of protein sample in 1% cholate. To compare the $\beta_1\gamma_2$ binding data with that of methylated T$\beta\gamma$, 5 μl of 1% cholate is added to experiments involving T$\beta\gamma$. Each tube is supplemented with 20 mM Tris (pH 7.5)–120 mM NaCl–1 mM MgCl$_2$ to bring the total assay volume to 50 μl. After incubating at 30° for 30 min, the lipid-bound $\beta\gamma$ is separated from the soluble fraction by centrifugation [TL100 (Beckman, Fullerton, CA), 98,000 rpm, 400,000g] for 10 min at 4°. The supernatant is removed and the pellet is washed with 50 μl of 20 mM Tris (pH 7.5)–120 mM NaCl–1 mM MgCl$_2$. After another centrifugation for 10 min at 4°, this wash is removed. The pellet and supernatant are dissolved in sample buffer and run on an SDS–12% polyacrylamide gel to determine the relative affinity of each $\beta\gamma$ for the azolectin vesicles. Each gel is stained with Coomassie blue and dried. While the γ subunit runs at the dye front under these conditions, the presence of the β subunit is used to indicate the location of the $\beta\gamma$ subunit. The percentage of each sample present in the pellet and supernatant fractions is quantified by densitometry.

[41] J. Bigay, E. Faurobert, M. Franco, and M. Chabre, *Biochemistry* **33**, 14081 (1994).

Synthetic Farnesylated Cysteine Analogs as Probes of Binding Specificity. The syntheses of a wide variety of farnesylated cysteine derivatives have been described.[42–44] Farnesylcysteine analogs can be added to any of the assays described above. In each case, a concentrated stock solution of a farnesylated derivative or other analog in dimethyl sulfoxide (DMSO) is added to the assay mixture and the response is compared with the control experiment, in which DMSO alone is added. The specificity of the response in an assay must be addressed through the use of appropriate controls. For example, a protein-binding site should exhibit stereospecificity; it would be expected, therefore, that the L enantiomer of an analog should be more potent than its D counterpart if specific binding interactions are being blocked with these compounds. The specificity of the lipid group can be addressed through modification of the hydrophobic chain. In the extreme case, farnesoic, palmitic, and oleic acids have been used to probe the specificity of the lipid group.[44]

Summary

Freshly prepared proteolyzed (deprenylated) T$\beta\gamma$[33] and material isolated from retina[45] are inert with respect to activating Tα in the presence of R* in detergent and in disk membranes. In addition, proteolyzed T$\beta\gamma$ is also incapable of supporting the pertussis toxin-catalyzed ADP ribosylation of Tα–GDP.[46] These experiments show that isoprenylation/methylation is essential for the fruitful interactions between Tα and T$\beta\gamma$ at the membrane.

When tested for its ability to support GTP-for-GDP exchange catalyzed by R*, demethylated T$\beta\gamma$ proved to be approximately 50% as active as methylated T$\beta\gamma$ in photoreceptor disk membranes (Fig. 3)[31] and in reconstituted liposomes containing rhodopsin.[36] In detergent, no difference was observed between methylated and demethylated T$\beta\gamma$, suggesting no role at all for the methyl group in functional interactions between Tα, T$\beta\gamma$,

---

[42] E. W. Tan, D. Perez-Sala, F. J. Canada, and R. R. Rando, *J. Biol. Chem.* **266**, 10719 (1991).
[43] B. A. Gilbert, E. W. Tan, D. Perez-Sala, and R. R. Rando, *J. Am. Chem. Soc.* **114**, 3966 (1992).
[44] C. A. Parish, D. P. Brazil, and R. R. Rando, *Biochemistry* **36**, 2686 (1997).
[45] H. Ohguro, Y. Fukada, T. Takao, Y. Shimonishi, T. Yoshizawa, and T. Akino, *EMBO J.* **10**, 3669 (1991).
[46] H. Ohguro, Y. Fukada, T. Yoshizawa, T. Saito, and T. Akino, *Biochem. Biophys. Res. Commun.* **167**, 1235 (1990).

and R*.[31] The twofold activity difference observed in membranes[31] can be accounted for by the twofold lessened affinity of the demethylated T$\beta\gamma$, compared with its methylated counterpart, for membranes in the presence of R* and Tα.[36] It is interesting to note that a substantially larger difference (>10-fold) in the relative binding of methylated versus demethylated T$\beta\gamma$ to membranes is observed in the absence of R* and Tα.[36] However, R* has a substantial affinity for T$\alpha\beta\gamma$, and the influence of R* and Tα greatly reduces any differences resulting from the presence or absence of a methyl group on T$\beta\gamma$.

The results from studies of demethylated T$\beta\gamma$ demonstrate that specific lipid–receptor interactions are unlikely to play a critical role in the rhodopsin–transducin system, and further show that the effect of methylation is probably due to the increased hydrophobicity of methylated T$\beta\gamma$ versus its unmethylated counterpart. These studies are, of course, relevant to heterotrimeric G proteins, and specifically to the interactions of receptor (R*) with Tα and T$\beta\gamma$. If a hydrophobic lipid–lipid mechanism is operative, the state of methylation would be expected to have a more profound effect on the membrane-associative properties of farnesylated proteins, but not on those of geranylgeranylated proteins. The increased hydrophobicity of the C_{20} geranylgeranyl group relative to the C_{15} farnesyl group will compensate for the loss of the methyl substituent.

The results obtained in the transducin–rhodopsin system can be contrasted with the effect of γ-subunit methylation on effector enzyme activation. In the case of the geranylgeranylated $\beta_1\gamma_2$, methylation proved to have only a small effect on PIPLC β activation (Fig. 4B). An approximately 25% diminution in efficacy, but not potency, was observed for the demethylated geranylgeranylated $\beta_1\gamma_2$ versus its methylated counterpart.[38] This again shows that specific lipid–protein interactions are unimportant. The effect of methylation on membrane binding would be expected to be small, given that $\beta_1\gamma_2$ is geranylgeranylated. It is of interest to compare these results with those found with methylated and unmethylated T$\beta\gamma$ as activators of PIPLC β.[39] In this instance there was a large effect noted, with methylated T$\beta\gamma$ being at least 10-fold more potent than its unmethylated counterpart with respect to activating either enzyme (Fig. 4A). This result is readily understandable in light of the role of methylation in selectively enhancing hydrophobicity of farnesylated proteins as opposed to geranylgeranylated proteins. Similar results were obtained for the activation of PI3K, further strengthening the conclusion that it is lipid–lipid interactions that direct $\beta\gamma$ subunit membrane association.

If lipid–protein interactions were involved in the association of farnesylated/methylated T$\beta\gamma$ with other proteins, farnesylated cysteine derivatives mimicking the carboxyl terminus of Tγ would be competitive inhibitors of the G-protein-mediated exchange of GTPγS for GDP.[47–49]

While these molecules can inhibit the exchange reaction in the rhodopsin-transducin pathway at high concentrations (~100 μM) of isoprenylated analogs, it is unclear if this inhibition is specific in nature. For example, high concentrations of N-acetyl-S-farnesylcysteine (AFC) partially solubilized rod outer segment membranes.[44] Moreover, the inhibitory effect of the isoprenylated peptides is nonstereospecific and also can be mimicked by nonspecific reagents such as farnesoic acid and palmitic acid.[44] Therefore, in our opinion, it appears that there is no strong evidence to support the view that specific lipid–protein interactions are involved in the transducin–rhodopsin system and that isoprenylation/methylation of heterotrimeric G protein βγ subunits is essentially a modification that operates by enhancing the hydrophobicity of the modified proteins.

In certain cases, a direct interaction between an isoprene group and a protein-binding pocket may be involved in protein association.[50] It is interesting to note that farnesyl cysteine analogs can have profound pharmacological effects on signal-transducing systems in a variety of cell types.[12,48,51–54] Whether these effects are specifically related to isoprenylated/methylated protein function is currently unknown, but this possibility must be kept in mind pending the identification of putative pharmacological targets for these analogs.

Acknowledgments

The studies reported here were funded by a grant from the National Institutes of Health (EY-03624).

---

[47] O. G. Kisselev, M. V. Ermolaeva, and N. Gautam, *J. Biol. Chem.* **269**, 21399 (1994).
[48] A. Scheer and P. Gierschik, *Biochemistry* **34**, 4952 (1995).
[49] T. Matsuda, T. Takao, Y. Shimonishi, M. Murata, T. Asano, T. Yoshizawa, and Y. Fukada, *J. Biol. Chem.* **269**, 30358 (1994).
[50] Y. Q. Gosser, T. K. Nomanbhoy, B. Aghazadeh, D. Manor, C. Combs, R. A. Cerione, and M. K. Rosen, *Nature* (*London*) **387**, 814 (1997).
[51] A. Scheer and P. Gierschik, *FEBS Lett.* **319**, 110 (1993).
[52] J. Ding, D. J. Lu, D. Perez-Sala, Y. T. Ma, J. F. Maddox, B. A. Gilbert, J. A. Badwey, and R. R. Rando, *J. Biol. Chem.* **269**, 16837 (1994).
[53] Y. T. Ma, Y. Q. Shi, Y. H. Lim, S. H. McGrail, J. A. Ware, and R. R. Rando, *Biochemistry* **33**, 5414 (1994).
[54] Huzoor-Akbar, W. Wang, R. Kornhauser, C. Volker, and J. B. Stock, *Proc. Natl. Acad. Sci. U.S.A.* **90**, 868 (1993).

[31] Functional Analysis of Farnesylation and Methylation of Transducin

By TAKAHIKO MATSUDA and YOSHITAKA FUKADA

Introduction

The γ subunit of retinal G protein, transducin, (Tγ), is farnesylated and carboxylmethylated at the C terminus.[1,2] Generally, these posttranslational modifications (isoprenylation and methylation) of G-protein γ subunits are required not only for membrane attachment of G$\beta\gamma$,[3–5] but also for its functional interactions with Gα,[4–12] receptors,[13] and effectors such as adenylyl cyclase, phospholipase C, and phosphoinositide 3-kinase.[5,14–16] To evaluate the functional significance of the modifications in the light signal transduction process, we have developed methods to prepare several T$\beta\gamma$ ($\beta_1\gamma_1$) species having various lipid modifications at the C terminus of Tγ

---

[1] Y. Fukada, T. Takao, H. Ohguro, T. Yoshizawa, T. Akino, and Y. Shimonishi, *Nature (London)* **346,** 658 (1990).

[2] R. K. Lai, D. Pérez-Sala, F. J. Cañada, and R. R. Rando, *Proc. Natl. Acad. Sci. U.S.A.* **87,** 7673 (1990).

[3] W. F. Simonds, J. E. Butrynski, N. Gautam, C. G. Unson, and A. M. Spiegel, *J. Biol. Chem.* **266,** 5363 (1991).

[4] Y. Fukada, T. Matsuda, K. Kokame, T. Takao, Y. Shimonishi, T. Akino, and T. Yoshizawa, *J. Biol. Chem.* **269,** 5163 (1994).

[5] T. Matsuda, Y. Hashimoto, H. Ueda, T. Asano, Y. Matsuura, T. Doi, T. Tako, Y. Shimonishi, and Y. Fukada, *Biochemistry* **37,** 9843 (1998).

[6] Y. Fukada, H. Ohguro, T. Saito, T. Yoshizawa, and T. Akino, *J. Biol. Chem.* **264,** 5937 (1989).

[7] H. Ohguro, Y. Fukada, T. Yoshizawa, T. Saito, and T. Akino, *Biochem. Biophys. Res. Commun.* **167,** 1235 (1990).

[8] H. Ohguro, Y. Fukada, T. Takao, Y. Shimonishi, T. Yoshizawa, and T. Akino, *EMBO J.* **10,** 3669 (1991).

[9] T. Matsuda, T. Takao, Y. Shimonishi, M. Murata, T. Asano, T. Yoshizawa, and Y. Fukada, *J. Biol. Chem.* **269,** 30358 (1994).

[10] J. A. Iñiguez-Lluhi, M. I. Simon, J. D. Robishaw, and A. G. Gilman, *J. Biol. Chem.* **267,** 23409 (1992).

[11] J. B. Higgins and P. J. Casey, *J. Biol. Chem.* **269,** 9067 (1994).

[12] M. Rahmatullah and J. D. Robishaw, *J. Biol. Chem.* **269,** 3574 (1994).

[13] O. Kisselev, M. Ermolaeva, and N. Gautam, *J. Biol. Chem.* **270,** 25356 (1995).

[14] A. Katz, D. Wu, and M. I. Simon, *Nature (London)* **360,** 686 (1992).

[15] A. Dietrich, M. Meister, D. Brazil, M. Camps, and P. Gierschik, *Eur. J. Biochem.* **219,** 171 (1994).

[16] C. A. Parish, A. V. Smrcka, and R. R. Rando, *Biochemistry* **34,** 7722 (1995).

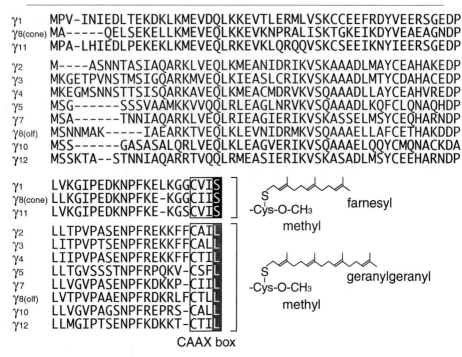

FIG. 1. Amino acid sequences and lipid modifications of G-protein γ subunits. Amino acid sequences of bovine γ_1, γ_2, γ_3, γ_5, γ_7, $\gamma_{8(cone)}$, and γ_{12}; human γ_4, γ_{10}, and γ_{11}; and rat $\gamma_{8(olf)}$ are aligned. Two forms of γ_8 are different gene products from each other: $\gamma_{8(cone)}$ is expressed in retinal cone photoreceptor cells, while $\gamma_{8(olf)}$ is expressed in olfactory and vomeronasal neuroepithelia.

(γ_1).[4-6] In this chapter, we summarize the procedures for the preparation of the T$\beta\gamma$ species and the analysis of functional roles of the modifications.

Posttranslational Modifications of G-Protein γ Subunits

The γ subunits of heterotrimeric G proteins have a C-terminal consensus sequence designated the CAAX motif, where C, A, and X represent cysteine, aliphatic amino acid, and any amino acid, respectively (Fig. 1). The X residue in the CAAX motif is a major determinant for the isoprenyl group, farnesyl (C_{15}) or geranylgeranyl (C_{20}) covalently attached to the cysteine residue. When X is serine, methionine, glutamine, or alanine, proteins are recognized by farnesyltransferase, whereas a leucine residue at this position leads to the modification by geranylgeranyltransferase I.[17]

---

[17] F. L. Zhang and P. J. Casey, *Annu. Rev. Biochem.* **65,** 241 (1996).

Among 11 subtypes of Gγ, transducin γ subunits (rod γ_1 and cone γ_8) and γ_{11} have serine at the X position, and they are farnesylated.[1,2,18,19] On the other hand, the γ subunits having leucine at the X position are geranylgeranylated[19–23] (Fig. 1). After the isoprenylation, AAX residues in the CAAX motif are removed by a specific protease, and then a newly exposed carboxyl group of the cysteine residue is methylesterified (Fig. 1). These posttranslational modifications of Tγ are indispensable for the function of transducin,[4–9] but the precise mechanism by which they contribute to the G-protein function is still controversial.

Preparation of Methylated and Nonmethylated Forms of T$\beta\gamma$

The γ subunits of G proteins are thought to be fully isoprenylated and methylated at the C terminus *in vivo*,[4] but Tγ is gradually demethylated during purification procedures. Therefore purified T$\beta\gamma$ is a mixture of methylated and nonmethylated forms,[4] while preparations of other G$\beta\gamma$ complexes do not contain detectable amount of nonmethylated γ.[22] Nonmethylated forms of G$\beta\gamma$ can be prepared by using pig liver methylesterase activity.[24] Methylated and nonmethylated forms of farnesylated T$\beta\gamma$ are separated by a gel-filtration column, Superdex 75 (Pharmacia, Piscataway, NJ),[4] possibly due to their difference in hydrophobic interaction with the column beads. A TSKgel G3000SW column (Tosoh, Tokyo, Japan) can be used as well, and other investigators reported the same result using a Superose 12 (Pharmacia) column.[25]

Preparation of Rod Outer Segments

The procedures are carried out at 4° in the dark. Dark-adapted bovine retinas (200–300) are freshly isolated and precipitated by centrifugation (10,000g, 10 min) in 100–200 ml of buffer A [10 mM morpholinepropane-

---

[18] O. C. Ong, H. K. Yamane, K. B. Phan, H. K. W. Fong, D. Bok, R. H. Lee, and B. K.-K. Fung, *J. Biol. Chem.* **270**, 8495 (1995).

[19] K. Ray, C. Kunsch, L. M. Bonner, and J. D. Robishaw, *J. Biol. Chem.* **270**, 21765 (1995).

[20] H. K. Yamane, C. C. Farnsworth, H. Xie, W. Howald, B. K.-K. Fung, S. Clarke, M. H. Gelb, and J. A. Glomset, *Proc. Natl. Acad. Sci. U.S.A.* **87**, 5868 (1990).

[21] S. M. Mumby, P. J. Casey, A. G. Gilman, S. Gutowski, and P. C. Sternweis, *Proc. Natl. Acad. Sci. U.S.A.* **87**, 5873 (1990).

[22] R. Morishita, Y. Fukada, K. Kokame, T. Yoshizawa, K. Masuda, M. Niwa, K. Kato, and T. Asano, *Eur. J. Biochem.* **210**, 1061 (1992).

[23] R. Morishita, H. Nakayama, T. Isobe, T. Matsuda, Y. Hashimoto, T. Okano, Y. Fukada, K. Mizuno, S. Ohno, O. Kozawa, K. Kato, and T. Asano, *J. Biol. Chem.* **270**, 29469 (1995).

[24] C. A. Parish, A. V. Smrcka, and R. R. Rando, *Biochemistry* **35**, 7499 (1996).

[25] J. Bigay, E. Faurobert, M. Franco, and M. Chabre, *Biochemistry* **33**, 14081 (1994).

sulfonic acid (MOPS), 2 mM MgCl$_2$, 30 mM NaCl, 60 mM KCl, 1 mM dithiothreitol (DTT), 0.1 mM phenylmethylsulfonyl fluoride (PMSF), aprotinin (4 μg/ml), leupeptin (4 μg/ml); pH 7.5]. The retinal pellet is mixed with 300 ml of buffer A containing 40% (w/v) sucrose in a capped bottle, and it is shaken vigorously by hand for 1 min. The suspension is centrifuged at 18,000g for 20 min at 4° and the supernatant is saved. The pellet is mixed again with buffer A containing 40% (w/v) sucrose, followed by centrifugation. The resultant supernatants are combined, diluted with an equal volume of buffer A, and centrifuged at 18,000g for 40 min at 4°. The pellet is suspended in 40 ml of buffer A, and overlaid on sucrose step gradients formed in six tubes [each contains two sucrose layers: 14 ml of 35% (w/v) sucrose and 10 ml of 29% (w/v) sucrose in buffer A]. They are then centrifuged in a swing-type rotor at 110,000g for 60 min. The rod outer segment (ROS) membranes are collected from the interface between 35 and 29% sucrose layers, mixed with an equal volume of buffer A, pelleted at 22,000g for 30 min at 4°, and stored at −80°.

Purification of Transducin

Transducin is purified[4,6] from bovine ROS membranes at 4°. The frozen pellet of the ROS membranes prepared from 200–300 bovine retinas as described above is thawed and homogenized in 85 ml of buffer A with a Teflon homogenizer (10 strokes). During this procedure, performed under a room light (∼30 min), rhodopsin is converted into metarhodopsin II as seen by the color change (red to orange). The homogenized ROS membranes are precipitated at 74,000g for 30 min at 4°, and the supernatant is removed. The membranes are homogenized (10 strokes) in 70 ml of buffer B [10 mM Tris-HCl, 0.5 mM MgCl$_2$, 1 mM DTT, 0.1 mM PMSF, aprotinin (4 μg/ml), leupeptin (4 μg/ml); pH 7.5] supplemented with 0.3 mM EDTA, pelleted at 74,000g for 30 min at 4°, and frozen at −80°. The frozen membranes are thawed in 70 ml of buffer B with 0.3 mM EDTA, homogenized (10 strokes), and pelleted at 74,000g for 30 min for 4°. This washing procedure is repeated three times more by using buffer B without freeze–thawing. Soluble protein-depleted ROS membranes thus prepared are homogenized (three strokes) in 70 ml of buffer B supplemented with GTP at a final concentration of 100 μM, and centrifuged at 74,000g for 30 min at 4°. The supernatant containing GTP-bound Tα and T$\beta\gamma$ is saved, and this extraction procedure is repeated three times more. The supernatants are combined, and supplemented with MgCl$_2$ at a final concentration of 2 mM to facilitate dissociation of Tα and T$\beta\gamma$ from each other. The extract (∼300 ml) is then applied at a flow rate of 20 ml/hr to a Blue Sepharose (Pharmacia) column (10 × 50 mm) which traps Tα, and the flow-through fraction containing

T$\beta\gamma$ is applied to a DEAE-Toyopearl 650S (Tosoh) column (10 × 60 mm). After application of the samples, the columns are independently washed with buffer C [10 mM MOPS, 2 mM MgCl$_2$, 1 mM DTT, 0.1 mM PMSF, aprotinin (4 μg/ml), leupeptin (4 μg/ml); pH 7.5]. Proteins bound to the two columns are eluted with 600 mM NaCl in buffer C at a flow rate of 15 ml/hr. T$\beta\gamma$ thus obtained is more than 95% pure.

Separation of Methylated and Nonmethylated T$\beta\gamma$

The T$\beta\gamma$ fraction (<10 ml) eluted from the DEAE-Toyopearl 650S column described above is applied to a Superdex 75-pg gel-filtration column (26 × 600 mm; Pharmacia) preequilibrated with buffer D (buffer C supplemented with 100 mM NaCl). Proteins are eluted at a flow rate of 2.5 ml/min, using an FPLC (fast protein liquid chromatography) system (Pharmacia). As shown in Fig. 2A, T$\beta\gamma$ forms two peaks as it elutes. The former peak (S1 fraction) contains farnesylated/nonmethylated T$\beta\gamma$, while the latter peak (S3 fraction) contains farnesylated/methylated T$\beta\gamma$ as analyzed by reversed-phase high-performance liquid chromatography (HPLC) (see below). The elution of these two forms of T$\beta\gamma$ (~45 kDa) are remarkably delayed as compared with marker proteins, probably owing to their hydrophobic interactions with the column beads. The ratio of these two peak areas varies among preparations.

We previously isolated a minor form of T$\beta\gamma$ having γ subunit truncated at the C terminus (Pro^2–Gly^{69} or Pro^2–Gly^{70}; previously designated Tγ-1).[6,26] This form is eluted from the gel-filtration column at a position (~160 ml in Fig. 2A) expected from its molecular weight. The relative content of this form in the preparation of T$\beta\gamma$ is usually low.

Analysis of Tγ by Reversed-Phase High-Performance Liquid Chromatography

To analyze the C-terminal structure (lipid modifications) of Tγ, purified T$\beta\gamma$ in fractions S1–S3 in Fig. 2A is directly injected onto a reversed-phase column (Cosmosil 5C$_{18}$-P300, 4.6 × 150 mm; Nacalai Tesque, Kyoto, Japan) equipped with an HPLC system (model 600E; Waters, Milford, MA).[4,8,26] Proteins are eluted from the column with a linear gradient (1%/min) of acetonitrile (10–80%) in 0.1% trifluoroacetic acid at a flow rate of 1 ml/min. Under these conditions, T$\beta\gamma$ is denatured and separated into each subunit. The nonmethylated and methylated forms of farnesylated Tγ are eluted at 47 and 48 min, respectively, while Tβ is adsorbed to the column (Fig. 2B). S1 and S3 fractions contain nonmethylated and methylated forms

[26] Y. Fukada, *Methods Enzymol.* **250,** 91 (1995).

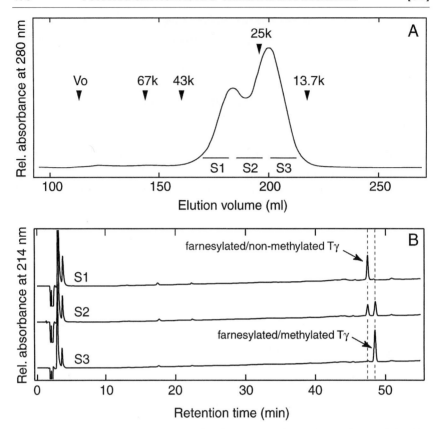

FIG. 2. Separation of Tβγ into methylated and nonmethylated forms by a gel-filtration column. (A) The fraction of Tβγ purified by the DEAE-Toyopearl 650S column was loaded onto the Superdex 75-pg gel-filtration column (26 × 600 mm; Pharmacia) preequilibrated with buffer D. Proteins are eluted at a flow rate of 2.5 ml/min, using an FPLC system (Pharmacia). Arrowheads show elution positions of standard molecules, and V_o denotes the void volume. (B) Fractions S1, S2, and S3 isolated by the gel-filtration column (A) were injected onto a reversed-phase column (Cosmosil 5C$_{18}$-P300, 4.6 × 150 mm; Nacalai Tesque) equipped with an HPLC system (model 600E; Waters). The γ subunits were eluted with a linear gradient of acetonitrile (10–80%, 1%/min) in 0.1% trifluoroacetic acid at a flow rate of 1 ml/min, and the absorbance at 214 nm of the eluate was continuously monitored. Farnesylated/nonmethylated and farnesylated/methylated Tγ species were eluted at 47 and 48 min, respectively.

TABLE I
COMPOSITIONS OF $\beta_1\gamma_1$ EXPRESSED IN INSECT CELLS AND RETINAL T$\beta\gamma^a$

Modification	Membrane $\beta_1\gamma_1$S74L (%)	Membrane $\beta_1\gamma_1$WT (%)	Cytosolic $\beta_1\gamma_1$WT (%)	Retinal T$\beta\gamma$ (%)
Farnesylated	14	33	—	100
Geranylgeranylated	86	67	—	—
Unmodified	—	—	100	—

^{a} Membrane $\beta_1\gamma_1$WT, membrane $\beta_1\gamma_1$S74L, and retinal T$\beta\gamma$ are mixtures of methylated and nonmethylated forms. The ratios of these two forms varied among preparations.

of Tγ, respectively, while Tγ in the S2 fraction is a mixture of the two forms. In this HPLC analysis, the C-terminally truncated form of Tγ (Pro^2–Gly^{69} or Pro^2–Gly^{70}) is eluted at 42 min.

Preparation of Mutant $\beta_1\gamma_1$

To prepare a recombinant of farnesylated or geranylgeranylated $\beta_1\gamma_1$ using a baculovirus–insect cell system, we employ coexpression of β_1 (Tβ) and wild-type (WT) or mutant (S74L) γ_1 (Tγ), in the latter of which the C-terminal farnesylation signal (CVIS) is replaced by the geranylgeranyl signal sequence (CVIL).[5] Unexpectedly, both γ_1WT and γ_1S74L purified from the membrane fractions of virus-infected cells are modified by a mixture of farnesyl and geranylgeranyl (Table I). Such a heterogeneous isoprenylation of γ_1 expressed in insect cells was also reported by Kalman et al.[27] But Lindorfer et al.[28] reported that γ_1 expressed in insect cells was exclusively farnesylated. At present, the reason for this is unclear.

In addition to these isoprenylated $\beta_1\gamma_1$ complexes, we can obtain an unmodified form of $\beta_1\gamma_1$ from the cytosolic fraction of baculovirus-infected insect cells (Table I). Overexpression of the γ subunit in insect cells may result in an accumulation of the unmodified form of $\beta_1\gamma_1$.

Production of Recombinant Baculoviruses

The cDNAs encoding bovine G-protein β_1 and γ_1 in the baculovirus transfer vector pVL1393, and β_1- and γ_1-recombinant baculoviruses were kindly provided by T. Doi (International Institute for Advanced Research,

[27] V. K. Kalman, R. A. Erdman, W. A. Maltese, and J. D. Robishaw, *J. Biol. Chem.* **270**, 14835 (1995).
[28] M. A. Lindorfer, N. E. Sherman, K. A. Woodfork, J. E. Fletcher, D. F. Hunt, and J. C. Garrison, *J. Biol. Chem.* **271**, 18582 (1996).

Matsushita Electric Industrial Co., Kyoto, Japan). Site-directed mutagenesis (S74L) of serine (CAT) at position 74 of γ_1 into leucine (TAT) is performed by polymerase chain reaction (PCR), using the γ_1 DNA in pVL1393 as a template with the following primers: 5'-CTAGAATTCCAC-CATGCCAGTGATC-3' and 5'-GCAGCGGCCGCTTATAAAATCA-CACAG-3'. These primers contain an *Eco*RI site and an *Not*I site, respectively. The amplified product (~250 bp) is digested with *Eco*RI and *Not*I, and ligated into the *Eco*RI/*Not*I site of pVL1393 to produce a vector termed pVL1393-γ_1S74L.

Spodoptera frugiperda cells (Sf9) can be obtained from the American Type Culture Collection (ATCC, Rockville, MD) and *Trichoplusia ni* cells (Tn5) from InVitrogen (San Diego, CA). We obtained these two cell lines from Y. Matsuura (National Institute of Infectious Diseases, Tokyo, Japan). Sf9 cells are maintained at 27° in TC100 medium (GIBCO, Grand Island, NY) supplemented with 10% (v/v) fetal calf serum (GIBCO), 0.26% (v/v) Bacto-tryptose broth (Difco, Detroit, MI), and kanamycin (100 μg/ml), and they are used to produce and amplify recombinant baculoviruses. Tn5 cells are maintained at 27° in SF900 II serum-free medium (GIBCO), and are used for large-scale protein production. Generally, the protein level expressed in Tn5 cells is higher than that in Sf9 cells.

To generate γ_1S74L-recombinant baculovirus, Sf9 cells (1×10^6 in a 35-mm dish) are transfected with a combination of 5 μg of pVL1393-γ_1S74L and 20 ng of BaculoGold viral DNA (PharMingen, San Diego, CA) using Lipofectin (GIBCO). The recombinant baculoviruses are harvested 4 days after transfection, plaque purified, and amplified [up to 5×10^7 plaque-forming units [(pfu)/ml].[29]

Expression and Purification of $\beta_1\gamma_1$ Complex

A scheme for purification of $\beta_1\gamma_1$ expressed in the insect cells is shown in Fig. 3. Prior to the experiment, Tn5 cells are grown to a density of 4×10^6 cells/ml in a 1-liter flask containing 300 ml of SF900 II medium with constant shaking at 130 rpm. To produce recombinant $\beta_1\gamma_1$, Tn5 cells (6×10^8 cells) are coinfected with β_1- and γ_1 (WT or S74L)-recombinant baculoviruses at a multiplicity of infection of 5 for each virus, and then cultured for 3 days in two 1-liter flasks, each containing 300 ml of the medium (1×10^6 cells/ml). Seventy-two hours after the infection, the cells are harvested, washed twice with PBS (10 mM sodium phosphate 140 mM

---

[29] C. D. Richardson (ed.), "Methods in Molecular Biology," Vol. 39. Humana Press, Clipton, New Jersey, 1995.

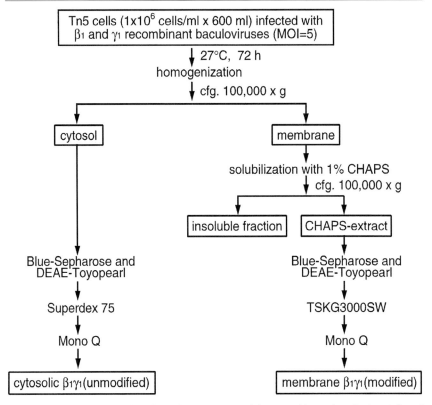

FIG. 3. Scheme for purification of $\beta_1\gamma_1$ expressed in recombinant baculovirus-infected insect cells.

NaCl, 1 mM MgCl$_2$; pH 7.4), suspended in 60 ml of buffer E [20 mM Tris-HCl, 2 mM MgCl$_2$, 1 mM DTT, 0.1 mM PMSF, aprotinin (4 μg/ml), leupeptin (4 μg/ml); pH 8.0], and homogenized with a Teflon homogenizer (10 strokes), followed by drawing/ejecting the suspension 10 times through a 25-gauge needle. The homogenate is centrifuged at 100,000g for 30 min at 4° to isolate cytosol and membrane fractions. Proteins in the membrane fraction are solubilized with 60 ml of buffer F (buffer E supplemented with 1% (w/v) 3-[(3-cholamidopropyl)-dimethyl-ammonio]-1-propanesulfonate (CHAPS)] by stirring gently for 2 hr, followed by centrifugation at 100,000g for 30 min to isolate CHAPS-extract and insoluble fractions. In both expression experiments (wild-type and mutant S74L-$\beta_1\gamma_1$), $\beta_1\gamma_1$ complex is recovered not only in the CHAPS-extract fraction but also in the cytosolic fraction.[5]

To purify $\beta_1\gamma_1$ (WT or S74L) from the CHAPS-extract fraction, the extract (60 ml) is applied at a flow rate of 22 ml/hr to a Blue Sepharose column (14 × 60 mm), to which $\beta_1\gamma_1$ does not bind. The flow-through fraction is then applied to a DEAE-Toyopearl 650S column (10 × 60 mm). After application of the sample, the column is washed with 50 ml of buffer F, and proteins are eluted with a linear gradient of NaCl (0–500 mM) in buffer F (30 ml) at a flow rate of 22 ml/hr. The peak of $\beta_1\gamma_1$ is centered at ~100 mM NaCl. Fractions containing $\beta_1\gamma_1$ are pooled (5 ml), and applied to a TSKgel-G3000SW gel-filtration column (25 × 600 mm; Tosoh), pre-equilibrated with buffer G (buffer F supplemented with 100 mM NaCl). Proteins are eluted at a flow rate of 3 ml/min, using an FPLC system. The $\beta_1\gamma_1$-rich fractions are pooled (15 ml) and applied to a Mono Q HR 5/5 (Pharmacia) column equilibrated with buffer H [20 mM Tris-HCl, 5 mM MgCl$_2$, 1 mM DTT, 1% (w/v) CHAPS; pH 7.7]. Proteins are eluted with a linear gradient of NaCl (0–500 mM) in buffer H (50 ml) at a flow rate of 1 ml/min (0.5%/min), using the FPLC system. The peak of $\beta_1\gamma_1$ is centered at ~200 mM NaCl.

Similarly, $\beta_1\gamma_1$ complex in the cytosolic fraction is purified. In this case, a Superdex 75-pg gel-filtration column is used instead of the TSKgel-G3000SW column, and CHAPS is eliminated from all the buffers. The final yields of $\beta_1\gamma_1$ purified from 600-ml cell cultures are more than 700 and 350 μg for cytosolic $\beta_1\gamma_1$ and membrane $\beta_1\gamma_1$, respectively.

Analysis of γ_1 by Reversed-Phase High-Performance Liquid Chromatography

To analyze the C-terminal structure of recombinant γ_1, purified $\beta_1\gamma_1$ is directly injected onto a reversed-phase column as described above. Proteins are eluted from the column with a linear gradient (1%/min) of acetonitrile (30–80%) in 0.1% trifluoroacetic acid at a flow rate of 1 ml/min. As shown in Fig. 4, membrane $\beta_1\gamma_1$WT (trace b) and membrane $\beta_1\gamma_1$S74L (trace c) contain four kinds of γ_1 species: Farnesylated/nonmethylated (C$_{15}$, eluted at 28 min), farnesylated/methylated (C$_{15}$-m, at 29.1 min), geranylgeranylated/nonmethylated (C$_{20}$, at 31.7 min) and geranylgeranylated/methylated (C$_{20}$-m, at 33.2 min). The ratios of farnesylated γ_1 (C$_{15}$ + C$_{15}$-m) to geranylgeranylated γ_1 (C$_{20}$ + C$_{20}$-m) are 33:67 for membrane $\beta_1\gamma_1$WT and 14:86 for membrane $\beta_1\gamma_1$S74L (Table I). The degree of methylation varies from preparation to preparation, as is observed for T$\beta\gamma$ purified from the bovine retina. In contrast, cytosolic $\beta_1\gamma_1$WT (trace d, Fig. 4) contains unmodified γ_1 (eluted at 22.8 min), which has an intact CAAX box (CVIS) at the C terminus.

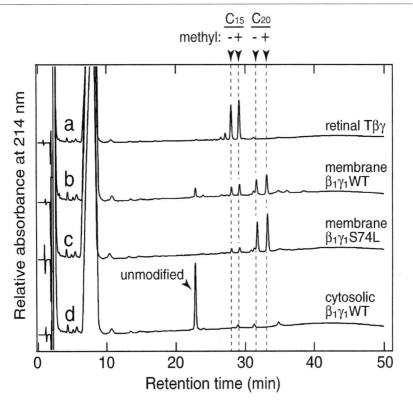

FIG. 4. Reversed-phase HPLC analysis of γ_1 expressed in insect cells. Purified $\beta_1\gamma_1$ complexes were injected onto a reversed-phase column (Cosmosil 5C$_{18}$-P300, 4.6 × 150 mm) equipped with an HPLC system (600E; Waters). The γ subunits were eluted with a linear gradient of acetonitrile (30–80%, 1%/min) in 0.1% trifluoroacetic acid at a flow rate of 1 ml/min, and the absorbance at 214 nm of the eluate was continuously monitored. Dashed vertical lines indicate the elution positions of farnesylated/nonmethylated γ_1 (28 min), farnesylated/methylated γ_1 (29.1 min), geranylgeranylated/nonmethylated γ_1 (31.7 min), and geranylgeranylated/methylated γ_1 (33.2 min), respectively. Unmodified γ_1 was eluted at 22.8 min. [Reprinted with permission of the American Chemical Society, from T. Matsuda, Y. Hashimoto, H. Ueda, T. Asano, Y. Matsuura, T. Doi, T. Tako, Y. Shimonishi, and Y. Fukada, *Biochemistry* **37,** 9843 (1998).]

Functional Analyses of $\beta_1\gamma_1$

Membrane Binding

Stripped ROS membranes are prepared as described.[8] Briefly, the ROS membranes (prepared from 200 bovine retinas) are subjected to transducin extraction with GTP (as described), and then irradiated with an orange light (>540 nm) at 4° for 5 min in the presence of 50 mM NH$_2$OH. The bleached membranes containing ~5 μmol of opsin are washed five times with buffer B, and subsequent procedures are carried out under dim red light. The pellet suspended in 60 ml of buffer B is incubated with 11-*cis*-retinal (9.0 μmol) at 25° for 3 hr, and then at 4° for 18 hr to regenerate rhodopsin. The regenerated ROS membranes are washed once with buffer B, and suspended in buffer D (termed "stripped ROS membranes"). By using its aliquot, rhodopsin concentration is determined from the difference absorbance at 498 nm before and after complete bleaching of rhodopsin in the presence of 100 mM NH$_2$OH, assuming that the molecular extinction coefficient at 498 nm is 40,600 cm^{-1} M^{-1}.[30]

Membrane binding of $\beta_1\gamma_1$ to the stripped ROS membranes thus prepared is assessed at 4° under dim red light.[4,5] $\beta_1\gamma_1$ (final concentration, 0.3–1.0 μM) is mixed with the stripped ROS membranes (final rhodopsin concentration, 20 μM) in 50 μl of buffer D, incubated for 30 min at 4°, and centrifuged at 18,000g for 30 min at 4° to sediment the membranes with bound $\beta_1\gamma_1$. The resultant clear supernatant (40 μl) containing unbound subunits is subjected to sodium dodecyl sulfate–polyacrylamide gel electrophoresis (SDS–PAGE). The amounts of β_1 and γ_1 in the gel are estimated by densitometric scanning of the Coomassie blue-stained gel or by Western blotting with antibodies against bovine T$\beta\gamma$[31] and the ECL detection system (NEN, Boston, MA). The amounts of β_1 and γ_1 bound to the membranes are calculated by subtracting the amounts of unbound subunits from the total amounts added to the tubes. As shown in Fig. 5, the membrane binding of $\beta_1\gamma_1$ exclusively requires the isoprenylation of γ subunit under these conditions, and the predominantly geranylgeranylated $\beta_1\gamma_1$ (membrane $\beta_1\gamma_1$S74L) shows much higher affinity for the ROS membranes than the purely farnesylated form (retinal T$\beta\gamma$).

GTPγS Binding to Tα Catalyzed by Metarhodopsin II

Bovine rhodopsin is extracted and purified from the stripped ROS membranes, and reconstituted in phosphatidylcholine (PC) liposomes under dim red light as described.[6] Briefly, rhodopsin (4.5 μmol) in the stripped ROS membranes is solubilized with 60 ml of buffer J [1 mM sodium phosphate, 20 mM NaCl, 1 mM MgCl$_2$, 2% (w/v) CHAPS; pH 7.4], followed by centrifugation at 105,000g for 60 min at 4°. Solubilized rhodopsin in the supernatant is applied to a concanavalin A (ConA)–Sepharose column

---

[30] G. Wald and P. K. Brown, *J. Gen. Physiol.* **37**, 189 (1953).
[31] Y. Fukada and T. Akino, *Photobiochem. Photobiophys.* **11**, 269 (1986).

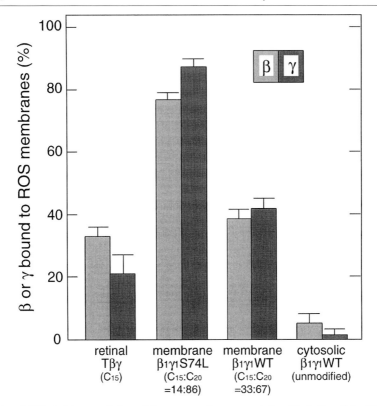

FIG. 5. Effect of isoprenylation of $\beta_1\gamma_1$ on membrane association. Each $\beta_1\gamma_1$ (0.3 μM) was mixed with nonirradiated stripped ROS membranes (20 μM rhodopsin) in the dark, and centrifuged to obtain a clear supernatant containing unbound $\beta_1\gamma_1$. This supernatant was subjected to SDS–PAGE, followed by immunoblot analysis with anti-T$\beta\gamma$ antibodies and the ECL detection system. Data represent the average ± range of variations in two independent experiments. [Reprinted with permission of the American Chemical Society, from T. Matsuda, Y. Hashimoto, H. Ueda, T. Asano, Y. Matsuura, T. Doi, T. Tako, Y. Shimonishi, and Y. Fukada, *Biochemistry* **37**, 9843 (1998).]

(22 × 150 mm; Pharmacia) at a flow rate of 33 ml/hr. After application of the sample, the column is washed with 150 ml of buffer J containing 50 mM NH$_2$OH, and proteins are eluted with buffer J containing 200 mM α-methylmannoside. The fraction (~40 ml) containing rhodopsin (300 nmol) is concentrated to 15 ml by using a stirred cell concentrator fitted with a YM30 membrane (Amicon, Danvers, MA). This solution is mixed with 50 μmol of egg phosphatidylcholine (Sigma, St. Louis, MO) and dialyzed against 600 ml of buffer D. The dialysis buffer is changed more than 10 times every 7–10 hr.

The GTPγS-binding assay is carried out as described.[4,5] Prior to the reactions, purified bovine rhodopsin in phosphatidylcholine liposomes is light activated with an orange light (>540 nm) for 1 min at 4°. The time course of the GTPγS binding to Tα is measured at 4° in a reaction mixture (150 μl of buffer D) composed of various concentrations (0.05–0.3 μM) of Tβγ, 1.0 μM Tα, 30 nM metarhodopsin II in liposomes, ovalbumin (2.5 mg/ml), and 10 μM [^{35}S]GTPγS (3 Ci/mmol). The reaction is started by adding [^{35}S]GTPγS and terminated by diluting 10-μl aliquots with 200 μl of ice-cold buffer K (100 mM Tris-HCl, 1 mM $MgCl_2$; pH 7.5) containing 2 mM GTP. [^{35}S]GTPγS-bound Tα is isolated from free [^{35}S]GTPγS by filtering the samples over 96-well cellulose membranes (type HATF; Millipore, Bedford, MA) fitted with a MultiScreen vacuum filtration manifold (Millipore). After filtration, the membranes are washed four times with 0.3 ml of buffer K, dried, and punched out manually with MultiScreen punch tips (Millipore). The radioactivities of the membranes are measured by liquid scintillation counting.

As shown in Fig. 6A, isoprenylated $β_1γ_1$ stimulates the rate of the GTPγS-binding reaction to Tα in a dose-dependent manner. The chain length of the isoprenyl group noticeably affects the function of $β_1γ_1$, as the predominantly geranylgeranylated $β_1γ_1$ (membrane $β_1γ_1$S74L) is about three times as effective in enhancing the reaction rate as the purely farnesylated form (retinal Tβγ). Unmodified $β_1γ_1$ is totally inactive, indicating that the C-terminal modifications of the γ subunit is indispensable for the activation (GDP/GTP exchange reaction) of transducin catalyzed by metarhodopsin II.

ADP Ribosylation of Tα Catalyzed by Pertussis Toxin

In solution, pertussis toxin ADP-ribosylates Tα when it is complexed with Tβγ, and therefore the rate of the ADP-ribosylation reaction quantitatively reflects the affinity between Tα and Tβγ in the absence of rhodopsin and lipid membranes. Thus, this method is suitable for evaluation of the effect of the lipid modification on the subunit interaction between Tα and Tβγ.[4,5,7,9]

The ADP-ribosylation reaction of Tα is carried out as described.[4,5,9] Prior to the reactions, pertussis toxin (Kaken-Seiyaku Co., Tokyo, Japan) is preactivated at 30° for 15 min in buffer L (72 mM HEPES–NaOH, 180 mM NaCl, 18 mM EDTA, 10 mM DTT, 3.6 mM ATP, and 0.7 mM GDP; pH 6.6). The ADP-ribosylation reactions are performed at 30° in reaction mixtures (120 μl of buffer D plus 30 μl of buffer L) containing various concentrations (0.05–0.5 μM) of Tβγ, 0.5 μM Tα, preactivated pertussis toxin (10 μg/ml), ovalbumin (2.5 mg/ml), 0.005% Lubrol-PX, and

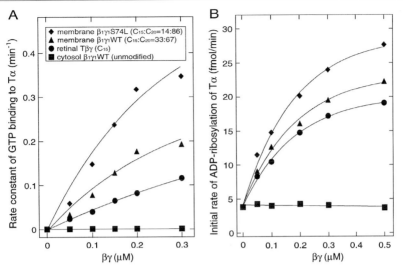

FIG. 6. Effects of isoprenylation of $\beta_1\gamma_1$ on functional interaction with Tα. (A) GTPγS binding to Tα catalyzed by metarhodopsin II. The GTPγS-binding reaction was carried out at 4° in a reaction mixture composed of various concentrations (0.05, 0.1, 0.15, 0.2, or 0.3 μM) of $\beta_1\gamma_1$, 1.0 μM Tα, 30 nM metarhodopsin II in liposomes, ovalbumin (2.5 mg/ml), 0.1% CHAPS, and 10 μM [^{35}S]GTPγS (3 Ci/mmol). At 2.5-min intervals of incubation time ($\leq$15 min), aliquots of the reaction mixture were withdrawn to determine the amounts of [^{35}S]GTPγS-bound Tα. The data points were best fitted to a single exponential equation, $B(t) = B_m [1-\exp(-kt)]$, where $B(t)$ is the amount of GTPγS-bound Tα at time t, B_m is the maximum binding at infinite time, and k is the rate constant. The rate constants calculated from the kinetic experiments were plotted against the concentration of $\beta_1\gamma_1$ in the reaction mixture. (B) ADP ribosylation of Tα catalyzed by pertussis toxin. The ADP-ribosylation reaction was carried out at 30° in a reaction mixture containing various concentrations (0.05, 0.1, 0.2, 0.3, or 0.5 μM) of $\beta_1\gamma_1$, 0.5 μM Tα, preactivated pertussis toxin (10 μg/ml), ovalbumin (2.5 mg/ml; Sigma), 0.17% CHAPS, and 10 μM [^{32}P]NAD (1 Ci/mmol). At 15-min intervals of incubation time ($\leq$60 min), aliquots of the reaction mixture were withdrawn to determine the amounts of [^{32}P]ADP-ribosylated Tα. The initial rate of the reaction was calculated from the slope of the linear fitting of the data, and plotted against the concentration of $\beta_1\gamma_1$. [Reprinted with permission of the American Chemical Society, from T. Matsuda, Y. Hashimoto, H. Ueda, T. Asano, Y. Matsuura, T. Doi, T. Tako, Y. Shimonishi, and Y. Fukada, *Biochemistry* **37**, 9843 (1998).]

10 μM [^{32}P]NAD (1 Ci/mmol). The reaction is started by the addition of preactivated pertussis toxin and terminated by diluting 20-μl aliquots with 200 μl of ice-cold buffer K containing 15% (w/v) trichloroacetic acid. [^{32}P]ADP-ribosylated Tα is isolated from free [^{32}P]NAD by filtering the samples over 96-well cellulose membranes (type HATF; Millipore). After filtration, the membranes are washed four times with 0.3 ml of buffer K containing 6% (w/v) trichloroacetic acid, dried, and punched out. The

radioactivities of the membranes are measured by liquid scintillation counting.

As is observed in the GTPγS-binding assay (Fig. 6A), the modification of the γ subunit is absolutely required for stimulating the ADP ribosylation of Tα catalyzed by pertussis toxin, and the predominantly geranylgeranylated $\beta_1\gamma_1$ (membrane $\beta_1\gamma_1$ S74L) is more effective for the stimulation than the purely farnesylated Tβγ (Fig. 6B). These results indicate that the C-terminal modification(s) of Tγ is indispensable for the interaction between Tα and Tβγ, and that geranylgeranylated $\beta_1\gamma_1$ has a higher affinity for Tα than does the farnesylated form. Such a difference in subunit interac-

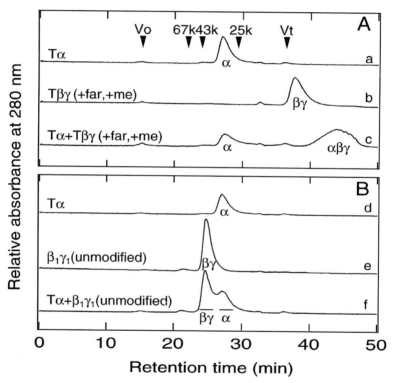

FIG. 7. Gel-filtration analysis of the interaction between Tα and Tβγ. (A) Tα (trace a; 0.5 nmol), farnesylated/methylated Tβγ (trace b; 0.5 nmol, purified from bovine retina), or their mixture (trace c) in 50 μl of buffer M was injected onto a gel-filtration column (TSKgel G3000SW, 7.5 × 300 mm; Tosoh) equipped with an FPLC system. Proteins were eluted with buffer M at a flow rate of 0.5 ml/min. Arrowheads show elution positions of standard molecules. V_o and V_t denote void and total volumes of the column, respectively. (B) Tα (trace d), unmodified $\beta_1\gamma_1$ (trace e; produced in insect cells), or their mixture (trace f) was injected onto the column, and eluted as in (A).

tion seems to account well for the functional difference among $\beta_1\gamma_1$ species shown in Fig. 6A.

Gel-Filtration Analysis

The difference in subunit interaction is more directly assessed by using a gel-filtration column. Since the subunit interaction of transducin (between Tα and T$\beta\gamma$) is much weaker than that of other G proteins, higher concentrations of transducin subunits are required for detection of $\alpha\beta\gamma$ trimer in a gel-filtration column.

Purified Tα (10 μM), T$\beta\gamma$ (10 μM), or their mixture in 50 μl of buffer M (20 mM Tris-HCl, 1 mM EDTA, 100 mM NaCl; pH 7.5) is injected onto a TSKgel G3000SW gel-filtration column (7.5 × 300 mm; Tosoh) equipped with the FPLC system, and eluted with buffer M at a flow rate of 0.5 ml/min. Peak fractions are collected and analyzed by SDS–PAGE. When farnesylated/methylated T$\beta\gamma$ is mixed with Tα (Fig. 7A), they associate with each other, and elute much later than each subunit. This retardation may reflect an increase in hydrophobicity due to the formation of the trimeric complex. In contrast with this, unmodified $\beta_1\gamma_1$ does not form a complex with Tα under these conditions (Fig. 7B), demonstrating the importance of lipid modifications of Tγ in the Tα–T$\beta\gamma$ interaction.

Conclusions

The function of T$\beta\gamma$ depends critically on the lipid modification of Tγ. This is at least partly attributed to the change in affinity for Tα observed in the absence of rhodopsin and membranes (Figs. 6B and 7). We speculate that Tα may recognize the farnesyl and methyl groups attached to the C terminus of Tγ for efficient association.[1,4,5,9] The modifications of Tγ also exert a strong effect on the membrane association of T$\beta\gamma$ (Fig. 5). These two effects of the modifications seem to contribute largely to the function of T$\beta\gamma$. Nevertheless, we do not know how the lipid modifications facilitate both the membrane-anchoring and subunit association at the same time. To address this issue, we need molecular structural information about the farnesyl and methyl groups in the trimeric complex.

Acknowledgments

This work was supported in part by Grants-in-Aid from the Japanese Ministry of Education, Science, Sports, and Culture. We thank Drs. Tomoko Doi and Yoshiharu Matsuura for providing plasmids, recombinant baculoviruses, and insect cell lines and for technical advice.

[32] Identification of Phosphorylation Sites within Vertebrate and Invertebrate Rhodopsin

By Hiroshi Ohguro

Introduction

Studies of vertebrate and invertebrate rhodopsin have been applicable to our understanding of amplification and desensitization processes of G-protein-coupled receptors in general. In vertebrate rod outer segments, photoisomerization of rhodopsin (Rho*) initiates activation of hundreds of G-protein (G_t) molecules, which in turn stimulate cGMP phosphodiesterase, resulting in closure of cGMP gated channels in response to the decrease in cytosolic cGMP concentrations. Alternatively, Rho* is phosphorylated (P-Rho*) by rhodopsin kinase (RK), and arrestin bound to P-Rho* prevents further activation of G_t, resulting in termination of the light-induced visual transduction cascade.[1] Similarly, phosphorylation of invertebrate Rho* is involved in the desensitization of visual transduction in invertebrate systems.[2] Therefore, phosphorylation of vertebrate and invertebrate Rho is a critical step in the inactivation of light-activated visual transduction pathways.

To study the enzymatic and functional roles of Rho phosphorylation, it was necessary to develop protocols for the identification of phosphorylation sites within vertebrate and invertebrate Rho. Among several possible approaches, we describe here a method that employs mass spectrometry in conjunction with specific proteolysis, and should be applicable to the study of phosphorylation in other systems.

Identification of Phosphorylation Sites within Vertebrate (Bovine) Rhodopsin

Palczewski *et al.*[3] first reported that endoproteinase Asp-N can cleave Rho at a single asparagyl residue homologous among vertebrates, generating a soluble peptide containing all potential sites of phosphorylation by RK (Fig. 1[4]). Using this simple proteolysis as a starting point, we have

[1] K. Palczewski, *Eur. J. Biochem.* **248,** 261 (1997).
[2] S. Yarfitz and J. B. Hurley, *J. Biol. Chem.* **269,** 14329 (1994).
[3] K. Palczewski, J. Buczylko, M. W. Kaplan, A. S. Polans, and J. W. Crabb, *J. Biol. Chem.* **266,** 12949 (1991).
[4] E. S. Dratz and P. A. Hargrave, *Trends Biol. Sci.* **8,** 128 (1983).

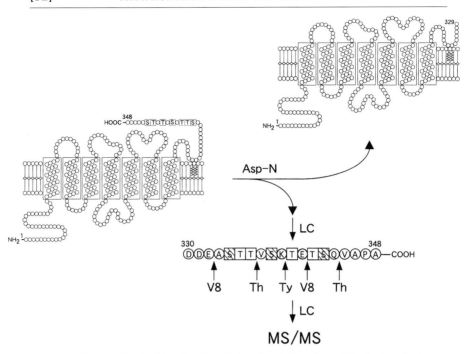

FIG. 1. Strategy for the identification of phosphorylation sites within bovine rhodopsin. The bovine rhodopsin model is a modification of the secondary structure proposed by Dratz and Hargrave.[4] Rhodopsin C-terminal peptides are isolated by endoproteinase Asp-N digestion and purified by liquid chromatography (LC), using a C_{18} reversed phase column, and the rest of the receptor is removed by centrifugation. The C-terminal peptides are subdigested by either trypsin (Ty), *S. aureus* V8 protease (V8), or thermolysin (Th). The number and sequence location of phosphorylation are determined by ES/MS and MS/MS analysis of the fragment peptides.

developed a method that identifies these sites[5,6] (a similar method has been independently described by two other groups[7,8]). The protocol is composed of the following five steps: (1) preparation of phosphorylated Rho from bovine rod outer segment (ROS) membranes, (2) washing out of soluble proteins associated with ROS membranes, (3) isolation of the C-terminal Rho peptide digested by endoproteinase Asp-N, (4) subdigestion of the C-

---

[5] H. Ohguro, K. Palczewski, L. H. Ericsson, K. A. Walsh, and R. S. Johnson, *Biochemistry* **32,** 5718 (1993).

[6] H. Ohguro, R. S. Johnson, L. H. Ericsson, K. A. Walsh, and K. Palczewski, *Biochemistry* **33,** 1023 (1994).

[7] J. H. McDowell, J. P. Nawrocki, and P. A. Hargrave, *Biochemistry* **32,** 4968 (1993).

[8] D. I. Papac, J. E. Oatis, Jr., R. K. Crouch, and D. R. Knapp, *Biochemistry* **32,** 5930 (1993).

terminal peptide, and (5) characterization of the fragment peptides by mass spectral analysis. Details of each step are described below.

Preparation of Phosphorylated Rhodopsin from Bovine Rod Outer Segment Membranes

Rho phosphorylation is carried out on ROS membranes isolated from fresh bovine retinas by discontinuous sucrose density gradient centrifugation[9] (Rho concentration is approximately 1 mg/ml) in 10 mM Bis–Tris propane (BTP) buffer, pH 7.5, containing 5 mM MgCl$_2$ and 1 mM ATP at 30° for 0–45 min under illumination from a 150-W lamp at a distance of 10 cm. The reaction is terminated by adding an equal volume of 250 mM potassium phosphate buffer, pH 7.2, containing 200 mM EDTA, 5 mM adenosine, 100 mM KF, and 200 mM NaCl to quench both rhodopsin kinase and phosphatase activities.

Washing out of Soluble Proteins Associated with Rod Outer Segment Membranes

After the preparation of phosphorylated Rho, soluble proteins weakly bound to the ROS membranes are removed by washing with 10 mM HEPES buffer, pH 7.5, containing 200 mM KCl (three times), and with 10 mM HEPES, pH 7.5 (three times) in a 1.5-ml Eppendorf tube; in each washing step, the sample is shaken vigorously for 5 min and then centrifuged at 16,000 rpm for 10 min at 4°.

Isolation of C-Terminal Rhodopsin Peptide Digested by Endoproteinase Asp-N

ROS membranes washed as described above are suspended in 0.5–1.0 ml of 10 mM HEPES, pH 7.5, and treated with endoproteinase Asp-N (at an enzyme-to-substrate ratio of about 1:10,000) overnight at room temperature. After the incubation, the extent of the cleavage is examined by sodium dodecyl sulfate–polyacrylamide gel electrophoresis (SDS–PAGE), and the unphosphorylated and phosphorylated Rho C-terminal peptides are separated from membranes by centrifugation at 16,000 rpm for 10–20 min at 4°. The peptides are purified over a reversed-phase high-performance liquid chromatography (HPLC) C$_{18}$ column (Vydac 218TP52, 2.1 × 250 mm; Separations Group, Hesperia, CA), using a linear gradient of acetonitrile (0–52%) and trifluoroacetic acid (TFA, 0.08–0.1%) over 30 min at a flow rate of 0.3 ml/min with detection at 215 nm. The Rho C-terminal

---

[9] D. S. Papermaster, *Methods Enzymol.* **81,** 48 (1982).

peptide elutes at approximately 37% acetonitrile. Electrospray mass spectrometry (ES/MS) identifies the Asp-N-generated C-terminal peptide from unphosphorylated Rho as D^{330}DEASTTVSKTETSQVAPA (observed molecular weight, 1936.2; calculated molecular weight, 1937.0). During Rho phosphorylation, mass spectra of this peptide at various times reveals $(M + 2H)^{2+}$ ions at m/z 969.1, 1009.3, 1049.3, and 1089.0, indicating non-, mono-, di-, and triphosphorylated peptide, respectively (Fig. 2).

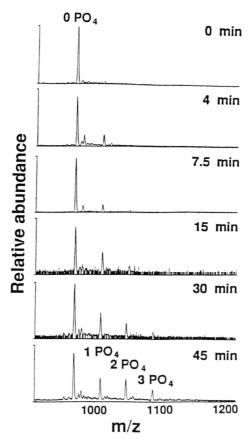

FIG. 2. Analysis of Rho C-terminal peptides by ES/MS during the time course of rhodopsin phosphorylation *in vitro*. Asp-N-generated bovine Rho C-terminal peptide (D^{330}DEASTTV-SKTETSQVAPA) obtained after ROS phosphorylation at various time points (0, 4, 7.5, 15, 30, and 45 min) was purified over a C_{18} HPLC column and analyzed by ES/MS. The ions designated as 0 PO_4, 1 PO_4, 2 PO_4, and 3 PO_4 represent the corresponding peptide with zero, one, two, and three phosphates, respectively. [Reprinted with permission from H. Ohguro, K. Palczewski, L. H. Ericsson, K. A. Walsh, and R. S. Johnson, *Biochemistry* **32**, 5718–5724 (1993). Copyright 1993 American Chemical Society.]

Phosphorylated and unphosphorylated forms of the peptide are then separated by rechromatography over a microbore C_{18} HPLC column (Vydac 218TP51, 1.0 × 250 mm) employing a linear gradient of acetonitrile (0–20%) and heptafluorobutyric acid (HFB, 0.08–0.1%) over 100 min at a flow rate of 0.12 ml/min with detection at 215 nm.[10] Non-, mono-, and multiphosphorylated bovine peptides elute at approximately 16, 14, and 12% acetonitrile, respectively. In the case of phosphorylated peptides from mouse Rho, column chromatography using HFA as a counterion gives complete separation of not only phosphorylated and unphosphorylated peptides, but also of peptides singly phosphorylated at different positions.[11] As shown in Fig. 3,[12] Rho C-terminal peptides from 100 mice kept under continuous illumination (32 foot candela) resolved into three peaks (A, B, and C), the elution positions of which correspond with those of authentic peptide monophosphorylated at Ser-338 (P338 peptide) and Ser-334 (P334 peptide), and nonphosphorylated peptide, respectively.

Subdigestion of C-Terminal Peptide

Phosphorylated forms of the peptide could, in principle, be analyzed by tandem mass spectrometry (MS/MS) to identify the specific residues phosphorylated. But because of the abundance of ions resulting from the cleavage of the Ala^{346}–Pro bond, MS/MS analysis of the $(M + 2H)^{2+}$ ions of phosphorylated and nonphosphorylated C-terminal Rho peptides (D^{330}DEASTTVSKTETSQVAPA) yields few sequence-specific fragment ions. To obtain a more complete series of fragment ions for analysis, phosphorylated forms of the peptide (approximately 2–5 nmol) are digested by trypsin–tolylsulfonyl phenylalanyl chloromethyl ketone (TPCK) (0.2 nmol; Worthington, Freehold, NJ), thermolysin (15 pmol; Boehringer Mannheim, Indianapolis, IN), or *Staphylococcus aureus* V8 protease (0.2 nmol; Boehringer Mannheim) in 0.1 ml of 100 mM Tris-HCl, pH 8.0, at 30° for 1, 16, or 2 hr, respectively. Trypsin, thermolysin, and *S. aureus* V8 protease cleave the peptide into two fragments (DDEASTTVSK and TETSQVAPA), three fragments (DDEASTT, VSKTETSQ, and VAPA), and three fragments (DDE, ASTTVSKTE, and TSQVAPA), respectively. In the case of phosphorylated peptides, subdigestion by trypsin or thermolysin is significantly inhibited by phosphorylation near the cleavage sites (trypsin, phosphoryla-

---

[10] H. Ohguro and K. Palczewski, *FEBS Lett.* **368**, 452 (1995).
[11] H. Ohguro, J. P. Van Hooser, A. H. Milam, and K. Palczewski, *J. Biol. Chem.* **270**, 14259 (1995).
[12] H. Ohguro, J. P. Van Hooser, A. H. Milam, and K. Palczewski, *J. Biol. Chem.* **270**, 14259 (1995).

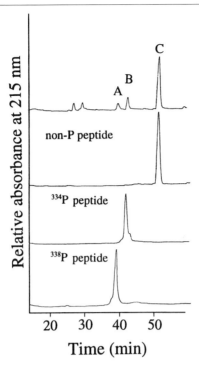

Fig. 3. Separation of synthetic phosphorylated and unphosphorylated mouse Rho peptides by HPLC, using HFB as a counterion. Asp-N-generated mouse Rho C-terminal peptides from an *in vivo* experiment with 100 mice kept under continuous illumination (32 foot candela) as described by Ohguro et al.[12] were separated into three peptides (A, B, and C) employing a linear gradient of acetonitrile (0–20%) and heptafluorobutyric acid (0.03–0.05%) over 100 min. The peptides were monitored by UV absorbance at 215 nm. The flow rate was 0.12 ml/min. The elution positions of peaks A, B, and C correspond with those of authentic monophosphorylated peptide at Ser-338 (P338 peptide) and Ser-334 (P334 peptide), and nonphosphorylated peptide, respectively.

tion at Ser-338; thermolysin, phosphorylation at Ser-338 and Ser-343), while the *S. aureus* V8 subdigestion is not inhibited by phosphorylation.

Analysis of Fragment Peptides by Mass Spectrometry

The number of phosphate groups and their sequence location are determined by electrospray mass spectrometry (ES/MS) and tandem mass spectrometry (MS/MS), using a Sciex API III triple quadrupole mass spectrometer fitted with a nebulization-assisted electrospray ionization source (PE/Sciex, Thornhill, Ontario, Canada). On the basis of MS/MS analysis of protelytic peptides from monophosphorylated Rho obtained as described above, we have found that the initial site of light-dependent phosphorylation is one of the serine residues within the C terminus (Ser-334, Ser-338, and Ser-343). Figure 4[13] shows examples of MS/MS spectra of monophosphorylated D^{330}DEASTTVSK (Fig. 4, top) and monophosphorylated T^{340}ETSQVAPA (Fig. 4, bottom) obtained by tryptic digest of Asp-N-generated C-terminal peptides from bovine Rho, identifying phosphorylation at Ser-334 and Ser-343, respectively.

Identification of the Phosphorylation Site within Invertebrate (Octopus) Rhodopsin

As yet, no proteolytic enzyme has been identified that will produce a single C-terminal peptide from invertebrate Rho that contains all potential phosphorylation sites. Consequently, we have developed a method using proteolysis by trypsin and thermolysin followed by mass spectral and automated amino acid sequence analysis to identify the octopus Rho (oRho) phosphorylation site (Fig. 5[14]). This protocol is composed of the following three steps: (1) preparation of phosphorylated oRho using octopus eye microvillar membranes, (2) proteolysis of the microvillar membranes by trypsin and thermolysin, and (3) characterization of fragment peptides by mass spectral and automated amino acid sequence analysis. Details of each step are described below.

Preparation of Phosphorylated Octopus Rhodopsin, Using Octopus Eye Microvillar Membranes

Octopus microvillar membranes[15] and oRK[16] are purified from *Octopus defreini* eyes. Microvillar membranes (1.2 mg of Rho in 600 μl), washed extensively with isotonic and hypotonic buffers, are incubated with oRK (0.1 mg) in 20 mM BTP buffer, pH 7.5, containing 5 mM MgCl$_2$ and 1 mM [γ-^{32}P]ATP (100 cpm/pmol) at 30° for 30 min under white illumination (150-W lamp). To stop phosphorylation and dephosphorylation, the reaction is quenched with 0.5 ml of 100 mM sodium phosphate, pH 7.2, containing 200 mM EDTA, 5 mM adenosine, 100 mM KF, and 200 mM KCl, following which the membranes are washed five times with 10 mM HEPES buffer, pH 7.5, containing 100 mM NaCl in a 1.5-ml Eppendorf tube; in

---

[13] K. Biemann, *Methods Enzymol.* **193**, 886 (1990).
[14] Y. A. Ovchinnikov, N. G. Abdulaev, A. S. Zolotarev, I. D. Artamonov, I. A. Bespalov, A. E. Dergachev, and M. Tsuda, *FEBS Lett.* **232**, 69 (1988).
[15] M. Tsuda, T. Tsuda, and H. Hirata, *FEBS Lett.* **257**, 38 (1989).
[16] S. Kikkawa, N. Yoshida, M. Nakagawa, T. Iwasa, and M. Tsuda, *J. Biol. Chem.* **273**, 7441 (1998).

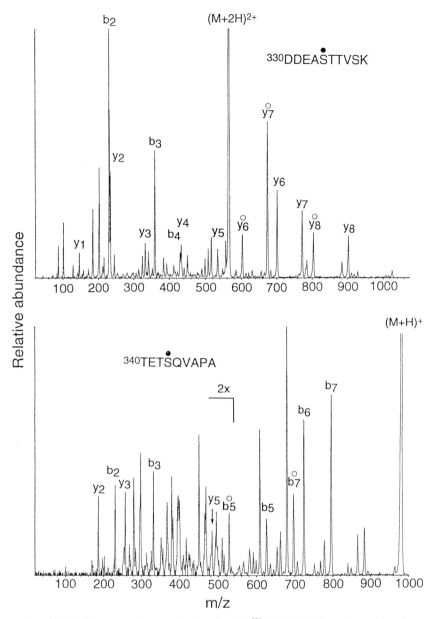

FIG. 4. MS/MS spectra of monophosphorylated D[330]DEASTTVSK and monophosphorylated T[340]ETSQVAPA from bovine Rho. *Top:* MS/MS spectrum of monophosphorylated $(M + 2H)^{2+}$ ion of m/z 566.8. *Bottom:* MS/MS spectrum of monophosphorylated $(M + H)^+$ ion of m/z 983.5. The ion nomenclature is that proposed by Biemann,[13] where the *y*- and *b*-type ions are numbered from the C and N termini, respectively. Closed circles above the sequence indicate phosphorylation sites. Those ions designated by open circles indicate additional β-elimination of a phosphate.

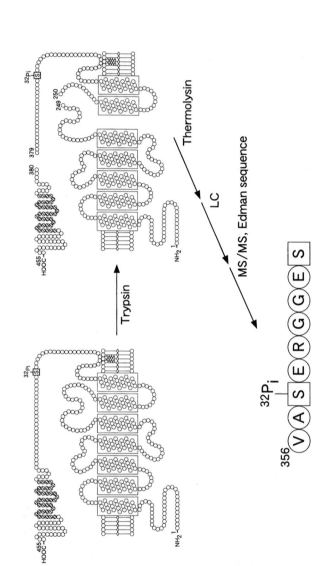

FIG. 5. Strategy of identification of phosphorylation sites within octopus rhodopsin. The model of oRho is a modification proposed by Ovchinnikov et al.[14] ^{32}P-phosphorylated oRho is first digested into three fragments: a 24-kDa fragment (M^1–K^{249}), a 16-kDa fragment (A^{250}–K^{379}), and a C-terminal fragment containing PPQYG repeated tail (M^{380}–A^{455}). The C-terminal fragment is removed by centrifugation, and the remaining fragments are subdigested with thermolysin. The radioactive proteolytic peptide purified by liquid chromatography (LC), using a C_{18} reversed-phase column, is analyzed by Edman sequencing and MS/MS analysis.

Proteolysis of Microvillar Membranes by Trypsin and Thermolysin

Phosphorylated microvillar membranes (containing 1.2 mg of oRho) are first digested with 10 μg of trypsin–TPCK (Worthington) overnight at room temperature. oRho is degraded into three fragments: a 24-kDa fragment (M^1–K^{249}), a 16-kDa fragment (A^{250}–K^{379}), and a C-terminal fragment containing PPQYG repeated tail (M^{380}–A^{455}). Autoradiography shows that radioactivity comigrates with the 16-kDa fragment, suggesting that the

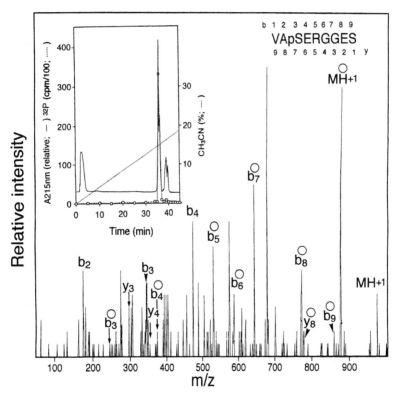

FIG. 6. MS/MS spectrum of $V^{356}S(P_i)ERGGES$ peptide Phosphorylated octopus microvillar membranes were successively treated with trypsin and thermolysin, and the radioactive $V^{356}S(P_i)ERGGES$ peptide was purified over a C_{18} reversed-phase HPLC column (*inset*). Shown is the MS/MS spectrum of a singly phosphorylated $(M + H)^+$ ion (m/z 971.5) with identified phosphorylation at the Ser-358 residue. [Reprinted with permission from H. Ohguro, N. Yoshida, H. Shidou, J. W. Crabb, K. Palczewski, and M. Tsuda, *Photochem. Photobiol.* **68**, 824–828 (1998). Copyright American Society for Photobiology.]

phosphorylation site(s) are located within the A^{250}–K^{379} fragment. To obtain smaller phosphopeptides for sequence analysis, the digest is treated with 100 mM phenylmethylsulfonyl fluoride (PMSF) to quench trypsin activity, washed five times with 10 mM HEPES buffer, pH 7.5, as described above, and subdigested with 5 μg of thermolysin (Boehringer Mannheim) for 6 hr at 30°. The digest supernatant obtained by centrifugation at 100,000g for 20 min at 4° is injected onto a reversed phase HPLC C_{18} column (Vydac 218TP52, 2.1 × 250 mm), and proteolytic peptides are separated with a linear gradient of acetonitrile (0–45%) and TFA (0.08–0.1%) over 60 min at a flow rate of 0.3 ml/min. The radioactive fractions are collected, lyophilized, and rechromatographed on the C_{18} HPLC column as described above (inset to Fig. 6[17]). Edman degradation of the phosphopeptides identifies the V^{356}ASERGGES sequence, and MS/MS analysis reveals the Ser-358 residue as the phosphorylation site (Fig. 6).

Acknowledgments

The part of the work done in the laboratory of K. Palczewski (Department of Ophthalmology, University of Washington, Seattle, WA) was supported by NIH Grant EY08061 and an unrestricted grant from RPB to the Department of Ophthalmology at the University of Washington. I especially thank Mr. Greg Garwin (Department of Ophthalmology, University of Washington) for critical reading of the manuscript.

[17] H. Ohguro, N. Yoshida, H. Shidou, J. W. Crabb, K. Palczewski, and M. Tsuda, *Photochem. Photobiol.* **68,** 824 (1998).

[33] Ocular Proteomics: Cataloging Photoreceptor Proteins by Two-Dimensional Gel Electrophoresis and Mass Spectrometry

By HIROYUKI MATSUMOTO and NAOKA KOMORI

Introduction

The identification of isolated proteins provides essential information in any biochemical inquiry. Two-dimensional (2-D) gel electrophoresis,[1–5]

[1] K. G. Kenrick and J. Margolis, *Anal. Biochem.* **33,** 204 (1970).
[2] J. Klose, *Humangenetik* **26,** 231 (1975).
[3] P. H. O'Farrell, *J. Biol. Chem.* **250,** 4007 (1975).
[4] G. A. Scheele, *J. Biol. Chem.* **250,** 5375 (1975).
[5] G. F. Ames and K. Nikaido, *Biochemistry* **10,** 616 (1976).

invented more than two decades ago, is a powerful technique for isolating proteins from a complex biological mixture. Among the original 2-D gel protocols, those described by O'Farrell[3] and by Ames and Nikaido,[5] have been used frequently. Both protocols employ an isoelectric focusing electrophoresis (IEF) in the presence of denaturants and detergents such as urea, Nonidat P-40 (NP-40) (or Triton X-100), or sodium dodecyl sulfate (SDS) in the first dimension, followed by the second dimension sodium dodecyl sulfate–polyacrylamide gel electrophoresis (SDS–PAGE) based on Laemmli.[6] In the past, however, even after an interesting protein spot was observed on a 2-D gel, it was difficult to determine the identity of the protein. Consequently, in the first several years after the invention of 2-D gel electrophoresis, the technique was regarded as analytical, to be used only to examine protein profiles for comparison. One possible way to analyze a protein separated on a 2-D gel was to electroelute[7,8] a protein of interest from multiple 2-D gel spots in order to accumulate sufficient quantities of the protein for analysis. In the late 1980s immobilizing media such as polyvinylidene difluoride (PVDF)[9] and nitrocellulose[10] were used for electroblotting followed by gas-phase Edman degradation to sequence proteins in picomolar quantities[8,9] or to determine the internal amino acid sequence after *in situ* protease digestion.[10] This innovation, combined with cDNA cloning technologies emerging at that time, resulted in the discovery of a number of new genes. Regarding the development and protocols of conventional protein microanalysis refer to previous volumes[11] of this series and/or monographs on the topic, for example, Ref. 12.

Important developments have taken place in two seemingly unrelated areas that, in combination, are likely to affect the future of molecular biosciences. The first is the initiation and maturation of "genome projects" in which all the genome information in human and in some model animals will be registered. Technological development in automated DNA sequencing has led to the completion of some genome projects for microbes and lower eukaryotes. It is expected that the entire human genome will be

---

[6] U. K. Laemmli, *Nature (London)* **227,** 680 (1970).
[7] D. Every and R. S. Green, *Anal. Biochem.* **119,** 82 (1982).
[8] N. Komori, J. Usukura, and H. Matsumoto, *J. Cell Sci.* **102,** 191 (1992).
[9] P. Matsudaira, *J. Biol. Chem.* **262,** 10035 (1987).
[10] R. H. Aebersold, J. Leavitt, R. A. Saavedra, L. E. Hood, and S. B. Kent, *Proc. Natl. Acad. Sci. U.S.A.* **84,** 6970 (1987).
[11] C. H. W. Hirs and S. N. Timasheff (eds.), *Methods Enzymol.* **91** (1983); M. P. Deutscher (ed.), *Methods Enzymol.* **182** (1990).
[12] P. Matsudaira, "A Practical Guide to Protein and Peptide Purification for Microsequencing." Academic Press, San Diego, California, 1993.

revealed by the beginning of the next millenium.[13] The second is the development in modern mass spectrometry initiated by inventions of two new ionization methods, that is, electrospray ionization and matrix-assisted laser desorption ionization.[14–17] In the fields of both DNA sequencing and mass spectrometry, it is apparent that advancement in semiconductor and microcomputer technologies has played an important role as essential infrastructure.

The genome projects have provided us with information with which to interpret the expression of proteins in a particular organism. However, at the same time, it has become apparent that in order to acknowledge the functional roles of proteins in relation to each other, one needs to know the dynamic aspect of gene expression: when, where, and to what extent will the genome information be realized in a particular cell under a particular set of biological parameters? In this context, the genome information merely predicts the potentiality of gene expression and it is necessary to study the proteins actually expressed in a cell in order to understand the biological mechanism. The situation is far more complicated in the actual biological system because proteins are modified in their main peptide chains, at their N and/or C termini, and in their side chains after they are translated.[18–20] It is apparent that (1) such biochemical reactions generally called "post translational modifications" play essential roles in every facet of biological phenomena, and (2) genome information per se often fails to predict the occurrence of posttranslational modifications.

Modern mass spectrometry assisted by complete genome information will play an essential role in the study of proteins because of its sensitivity, resolution, and the ability to analyze complex mixtures of molecules quickly. The combination of 2-D gel electrophoresis and mass spectrometry for data acquisition followed by data analysis through genome databases makes a powerful tool with which to reveal the identity of unknown proteins. This line of scientific endeavor is often called "proteome research" or "proteomics,"[21] meaning the study of proteins expressed at high throughput under a defined set of biological parameters. To make full use of these excellent

---

[13] R. Waterston and J. E. Sulston, *Science* **282**, 53 (1998).
[14] J. A. McCloskey (ed.), *Methods Enzymol.* **193** (1990).
[15] J. R. Chapman (ed.), *Methods Mol. Biol.* **61** (1996).
[16] C. Fenselau, *Anal. Chem.* **69**, 661A (1997).
[17] J. R. Yates III, *J. Mass Spectrom.* **33**, 1 (1998).
[18] R. Uy and F. Wold, *Science* **198**, 890 (1977).
[19] F. Wold, *Annu. Rev. Biochem.* **50**, 783 (1981).
[20] R. G. Krishna and F. Wold, *Adv. Enzymol. Relat. Areas Mol. Biol.* **67**, 265 (1993).
[21] M. R. Wilkins, K. L. Williams, R. D. Appel, and D. F. Hochstrasser (eds.), "Proteome Research: New Frontiers in Functional Genomics." Springer-Verlag, Berlin, 1997.

capabilities of mass spectrometry, it is important to develop a protocol for the preparation of sample.

In this chapter we describe a set of protocols to perform 2-D gel electrophoresis of proteins extracted from bovine retina, in-gel digestion of protein spots, matrix-assisted laser desorption ionization time-of-flight mass spectrometry (MALDI-TOF) to obtain peptide mass fingerprints, a search of genome databases (through the Internet) for peptide fingerprints, and analysis of the resulting data. In describing our protocols we have tried to simplify the procedures as much as possible so that one can grasp the essence of the protocols without the potential confusion of too many alternative options. In fact, many published protocols for 2-D gel electrophoresis differ only slightly from each other. Those variable methods will also work in general. It is important to establish a method optimized to each experimental system; the methods described in this chapter are expected to work for proteins extracted from retina of other animal species or from other parts of ocular systems, although some modification may be necessary for optimization.

Two-Dimensional Gel Electrophoresis

Gel Electrophoresis Apparatus and Tubes

We use an isoelectric focusing (IEF) apparatus with a jacketed beaker (10.5-cm i.d., 12.5-cm inner height) used as a lower-buffer chamber (Buchler Instrument). Unfortunately, this apparatus is no longer available commercially. However, a similar tube gel apparatus, for example, one from Bio-Rad (Hercules, CA), will serve the same purpose. For the second-dimension SDS–PAGE, we use a model V16 vertical gel electrophoresis apparatus from GIBCO-BRL (Gaithersburg, MD) with spacers 0.8 mm in thickness. Some companies such as Bio-Rad and Hoefer/Amersham Pharmacia Biotech (Piscataway, NJ) offer a 2-D gel electrophoresis system based on an electrophoresis chamber used as both the tube-gel apparatus and the slab-gel electrophoretic tank. In our experience it appears that an IEF system in which tubes are immersed in buffer during electrophoresis performs better than one in which tubes are not immersed. This is probably because the dissipation of heat from gels is more efficient and homogeneous in the former compared with the latter. The size of an electrophoresis chamber determines the volume of the buffer required in both IEF and SDS–PAGE. Therefore, it is best to avoid an unnecessarily large apparatus that requires more than 1 liter of electrophoresis buffer.

In a typical setting for IEF we use glass tubes 14–15 cm in length with 3.0-mm i.d. and 5-mm o.d. We have been making glass tubes ourselves, using a glass tube cutter. However, glass tubes suitable for IEF are commercially available, for example, from Bio-Rad or Hoefer/Amersham Pharmacia Biotech. The size of tubes may differ slightly from one company to another. The inner diameter determines the volume of the sample that can be loaded for IEF. Because the method used to prepare an IEF sample substantially affects its final volume, the tube size needs to be determined on the basis of the method of sample preparation and the apparatus. It is necessary to ensure that the tubes fit comfortably and tightly within the hole of the grommets of the IEF gel apparatus. The IEF tubes provided by one company usually fit best into the gel apparatus from the same company, and may not fit well with a similar apparatus from a different company.

Stock Solutions for Isoelectric Focusing Gel

Solution A: Acrylamide (30%, w/v)–bis-Acrylamide (1.5%, w/v)
Solution B: Riboflavin (0.004%, w/v)–N,N,N',N'-tetramethylethylenediamine (TEMED; 0.45%, v/v)
Solution C: Triton X-100 or NP-40 (20%, w/v)
Solution D: Ammonium persulfate (1.5%, w/v)
Ampholyte: Bio-Lyte 3/10, 3/5, 4/6, 5/7, and/or 8/10 (Bio-Rad), or equivalent

Isoelectric Focusing Lysis Buffer

Urea, 9.5 M
Triton X-100 or NP-40 (2.0%, w/v)
Bio-Lyte 3/10 (2.0%, w/v)
2-Mercaptoethanol (5.0%, v/v)

To make this solution, weigh an appropriate amount of urea in a small graduated cylinder (10–20 ml). Add the other ingredients and dissolve the urea. Adjust the final volume.

Isoelectric Focusing Overlay Buffer

Urea, 5.0 M
Triton X-100 or NP-40 (2.0%, w/v)
Bio-Lyte 3/10 (1.0%, w/v)
Use the same procedure as described for IEF lysis buffer.

IEF lysis and overlay buffers can be stored in aliquots (100–200 μl/tube) at $-20°$ in a non-frost-free freezer. We did not find any detectable changes in the patterns of protein spots on 2-D gels when using IEF lysis and overlay buffers stored at $-20°$ for more than 1 year.

Electrode Solutions

Top chamber (positive polarity): 20 mM H_3PO_4
Lower chamber (negative polarity): 1 M NaOH

These solutions are designed to run IEF from the acidic side to the basic side.[22,23] To run an IEF gel from the basic side one should refer to O'Farrell[3] for the concentration of buffers. We reuse 1 M NaOH solution in the lower chamber several times. Store the solutions at room temperature.

Equilibration Buffer

Tris-HCl (pH 6.8), 62.5 mM
SDS (2%, w/v)
2-Mercaptoethanol (5%, v/v)

It is convenient to make 1 liter of this solution. Store the solution at room temperature.

Agarose

Agarose, 1% (w/v) in equilibration buffer

Microwave to dissolve the agarose, make 2- to 3-ml aliquots in 15-ml tubes, and store at room temperature. One should liquefy agarose by boiling prior to use. This agarose solution is used for overlaying the first-dimension IEF gel on top of the second-dimension SDS–polyacrylamide slab gel.

Sample Preparation

IEF samples can be prepared either from a tissue block, culture cells, or a tissue/cell homogenate solution. A tissue block weighing 1–5 mg can be homogenized in ≤90 μl of IEF lysis buffer. In this case the dilution of the lysis buffer by the tissue fluid will not affect the performance of IEF significantly. After homogenization, the sample is homogenized further by sonication in a sonication bath for at least 20 min to break down chromosomal DNA. The sample is then centrifuged at 15,000g for 5 min at 25° to precipitate insoluble debris or aggregated DNA. The recovered supernatant (usually 50 to 100 μl) is loaded on top of the IEF gel. For cultured cells the procedures described by Steinberg and Coffino[24] are used with some modifications. Briefly, ~1.5 × 10^6 cells are spun down and the cell pellet is dissolved in IEF lysis buffer to make the final volume ≤90 μl. In the case of bacterial cells, a 0.5 $A_{590-600}$ equivalent is usually adequate to start

---

[22] K. Miyazaki, H. Hagiwara, M. Yokota, T. Kakuno, and T. Horio, *in* "Isoelectric Focusing and Isotachophoresis" (N. Ui and T. Horio, eds.), p. 183. Kyoritsu Shuppan, Tokyo, Japan, 1978. [In Japanese]

[23] H. Matsumoto and W. L. Pak, *Science* **223,** 184 (1984).

[24] R. A. Steinberg and P. Coffino, *Cell* **18,** 719 (1979).

with. Again, in these two cases, the dilution of lysis buffer can be disregarded.

If tissue homogenates are the starting material, mix homegenate (equivalent to 100–500 µg of total protein) with IEF lysis buffer to make the final volume ≤90 µl. One hundred micrograms of total protein is usually enough to see many protein spots by silver staining, but may not be enough for Coomassie blue staining if the protein of interest is in minor abundance. With our system, we can observe a linear increase in the size and intensity of protein spots up to 500 µg of total protein without detectable loss of small protein spots. In preparing IEF samples from culture cells or tissue homogenates, do not dilute the IEF lysis buffer too much; we have observed that a dilution of up to 30% (at least) of lysis buffer, caused by an aqueous sample, does not affect the IEF. Thus the volume of IEF buffer should be decided on the basis of the sample volume.

Isoelectric Focusing Gel Electrophoresis (First Dimension)

It appears that preparation of high-quality IEF gels is crucial for a successful IEF. By "high quality" we mean homogeneously polymerized IEF gel in clean glass tubes. The method developed by Miyazaki *et al.*[22] utilizes a combination of photoinduced polymerization by riboflavin and light-independent polymerization by ammonium persulfate of acrylamide. Such a combination appears to make high-quality IEF gels routinely.

IEF tubes must be cleaned thoroughly by soaking overnight in chromic–sulfuric acid cleaning solution (Fisher, Pittsburgh, PA) and by rinsing with hot water followed by deionized water. The tubes are dried in a vacuum oven at 80° for 1 hr. Chromic–sulfuric acid solution can be reused many times, until its brown color becomes greenish. Special care needs to be taken for the disposal of the used chromic acid solution, which makes the use of chromic acid inconvenient. We have observed that, as long as IEF tubes are kept wet, that is, not allowed to dry after each electrophoresis, cleaning them with a common laboratory detergent solution is sufficient. When tubes are clean and dry, mark them 1.5–2.0 cm from the top, seal the bottom end of each with Parafilm, and set up the tubes on a tube stand.

IEF gel solution[22,23] is made in a 125-ml filtering flask and consists of
Urea, 8.5 M
Acrylamide (4.0%, w/v) bisacrylamide (0.2%, w/v)
Riboflavin (0.0005%, w/v) TEMED (0.056%, v/v)
Triton X-100 or NP-40 (2.0%, w/v)
Bio-Lyte 3/10 (2.0%, w/v)
Ammonium persulfate (0.01%, w/v)

To make IEF gel solution the following steps should be performed. This recipe makes 12 ml of IEF gel mixture.

Step 1. Weigh 6.13 g of urea in a 125-ml filtering flask.

Step 2. Add the following ingredients in order: 1.6 ml of stock solution A; 1.5 ml of stock solution B; 1.2 ml of stock solution C; 0.6 ml of Bio-Lyte 3/10 (40%, w/v); and 2.5 ml of H_2O.

Step 3. Dissolve the urea completely and degas the mixture for 5 min. During the process we usually cover the flask with aluminum foil in order to keep riboflavin from light. It is also recommended that the ambient light be kept minimal by turning off some of the room lights until the polymerization is started at step 7.

Step 4. Add 0.08 ml of stock solution D and mix quickly by swirling the flask.

Step 5. By using a 1-ml syringe with appropriate tubing attached to the needle, fill the glass tubes with the IEF gel mixture up to the 1.5- to 2.0-cm mark.

Step 6. Overlay water to about 5 mm on top of the IEF gel solution.

Step 7. Start polymerization by illuminating the tubes with a fluorescent lamp.

Step 8. Fifteen minutes after the start of polymerization seal the top of each tube with Parafilm.

Step 9. Continue illumination for at least 4 hr or longer. Alternatively, an overnight illumination under normal room light will also polymerize IEF gels efficiently.

IEF gels can be kept at 25° for at least 1 week or as long as the bottom and top ends of the gels do not dry out. Before loading IEF gels onto a tube-gel apparatus, remove the water layer over the gels, using a Kimwipe, and also remove the Parafilm from the bottom ends. (Although we have never had an IEF gel come out of a tube during electrophoresis, it could happen; if this appears to be the case, dialysis membrane can be used to seal the bottom end in order to prevent the gel from slipping out.) Set up the tubes on a tube-gel apparatus, load samples, and slowly overlay IEF overlay buffer on the top of each sample. Electrophoresis is performed first at 100 V for 1 hr and then at 300 V for 15–18 hr. At the end of the run, we often increase the voltage to 500 V and run the apparatus for 1–2 hr. Consequently, the total voltage-hours is somewhere between 5000 and 7000. It is necessary to perform preliminary studies to determine the optimum voltage-hours for each system.

After a run, push the IEF gel out of the tube, using a syringe filled with water, and shake the gel in 20 ml of equilibration buffer for 15 min in order to equilibrate the gel in the SDS–PAGE environment. Change the equilibration buffer three times before proceeding to SDS–PAGE. If necessary, freeze the gels in the same buffer at $-20°$ until use; however, it should be kept in mind that some low molecular mass proteins (≤ 15 kDa) may diffuse out by freezing and thawing. When we freeze IEF gels, we usually perform equilibration only twice.

Sodium Dodecyl Sulfate–Polyacrylamide Gel Electrophoresis (Second Dimension)

Separating and stacking gels are prepared according to a conventional SDS–polyacrylamide gel recipe.[6] The length of our stacking gel is somewhere between 1.5 and 2.0 cm. To make a stacking gel, seal its top by means of a glass rod that is slightly shorter than the opening width of the glass plate, and pour stacking gel solution until it touches the glass rod. The glass rod will prevent acrylamide from being exposed to air and allow the gel to polymerize quickly and evenly. After the gel is polymerized, remove the glass rod and load the IEF tube gel equilibrated with equilibration buffer. Seal the space between the tube gel and the stacking gel with 1% (w/v) agarose. At this time, a well for loading molecular weight markers can be made by placing a piece of spacer (10 × 5 × 0.8 mm) over a stacking gel. After the agarose solidifies, remove the spacer and load prestained molecular weight markers (GIBCO-BRL). SDS–PAGE is carried out according to Laemmli.[6] The running front can be tracked by dropping 0.1% (w/v) bromphenol blue into the upper chamber or by using equilibration buffer containing bromphenol blue (≤0.01%, w/v).

Two-Dimensional Gel of Proteins Extracted from Bovine Retina

In Fig. 1 we show an example of 2-D gels on which various amounts of proteins extracted from bovine retina have been separated. In this experiment fresh bovine retina is homogenized with a tissue grinder in a homogenization buffer containing 320 mM sucrose, 10 mM HEPES (pH 7.5), 4 mM dithiothreitol (DTT), 0.5 mM MgSO$_4$, 2 mM EDTA, 0.1 mM sodium vanadate, 5 mM potassium fluoride, 10 mM benzamidine, phenylmethylsulfonyl fluoride (PMSF, 100 μg/ml), aprotinin (50 μg/ml), leupeptin (10 μg/ml), and pepstatin A (10 μg/ml). The amounts of proteins in the homogenate are assayed by the method of Lowry *et al.*[25] For 2-D gel electrophoresis, different amounts of the total proteins are mixed with IEF lysis buffer. The carrier ampholyte used in the first IEF dimension is Bio-Lyte 3/10 at 2% (w/v). The pH gradient is estimated by cutting a blank IEF gel into 5-mm pieces, extracting the Bio-Lyte from each piece into water, and measuring the pH. The molecular weights on the second-dimension SDS–PAGE are calibrated by the internal standards introduced

---

[25] O. H. Lowry, N. J. Rosenbrough, A. L. Farr, and R. J. Randal, *J. Biol. Chem.* **193**, 265 (1951).

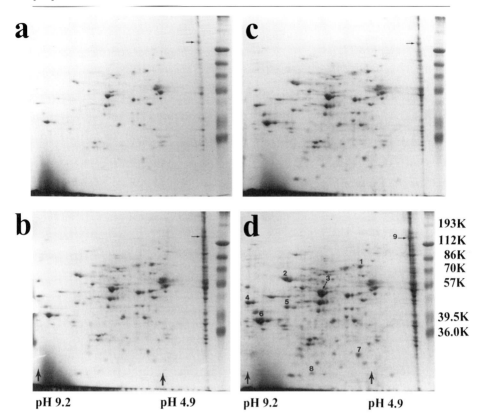

FIG. 1. Two-dimensional gels of various amounts of total protein extracted from bovine retina. Bovine retina was homogenized in the homogenization buffer and subjected to protein assay. After quantification, various amounts of proteins were loaded in the first-dimension IEF gels, followed by the second-dimension SDS–PAGE. The protein loads are as follows: (a) 100 μg, (b) 200 μg, (c) 300 μg, and (d) 400 μg. Major protein spots, such as those designated 1 through 9 in (d), are easily identifiable in all four gels [except for spot 8 in (a)]. The positions of molecular weight markers are shown to the right and the isoelectric points are shown at the bottom. All the numbered spots, 1–9, were subjected to in-gel digestion and peptide mass mapping. The MALDI-TOF spectrum for spot 7 is shown in Fig. 2. The process for peptide mass database search through MS-Fit is illustrated in Fig. 3. The identity of these nine proteins revealed by peptide mass fingerprinting is as follows: spot 1, 79-kDa heat shock cognate protein; spot 2, pyruvate kinase; spot 3, enolase; spot 4, mitochondrial aspartate aminotransferase[30]; spot 5, mitochondrial creatine kinase[30]; spot 6, glyceraldehyde-3-phosphate dehydrogenase; spot 7, Ca-binding protein P26 (recoverin); spot 8, guanylate kinase; spot (or band) 9, interphotoreceptor retinoid-binding protein.[30]

into a small well made of 1% (w/v) agarose, as described previously. The results indicate a high reproducibility of 2-D gels at total protein loads of 100 to 400 μg. As seen in Fig. 1, most of the protein spots on these gels are identifiable between each gel. This establishes the correspondence between spots on each gel unequivocally. In general, there is a linear relationship between the amount of protein loaded on the IEF gel and the appearance of the corresponding protein spot on the resulting 2-D gel. However, it should be noted that a small number of proteins do not exhibit a linear relationship between the load and the appearance. This may be partly due to aggregation of proteins at the acidic edge of the IEF gel, where we loaded the protein sample. For this analysis, the protein spots designated 1–9 in Fig. 1 were subjected to in-gel digestion and peptide mass mapping.

In-Gel Digestion and Peptide Mass Fingerprinting

Choice of Proteolytic Enzyme for Peptide Mass Fingerprinting

Peptide mass fingerprinting (mapping) is a technique for identifying proteins on the basis of the measured masses of an unknown protein subjected to defined proteolytic digestion. Therefore, the purpose of in-gel digestion is to create a set of proteolytic peptide fragments from the 2-D gel protein spot of interest. Digestion within a gel without prior extraction of the protein apparently is advantageous because of the simplicity of the whole procedure compared with other methods such as electroelution followed by digestion, or electroblotting onto a membrane followed by on-the-membrane digestion. Theoretically, any proteolytic enzymes can serve this purpose as long as digestion is highly specific, that is, dependent on the sequence of the peptide so that one can create a hypothetical digestion table. Such theoretical digestion is often called "*in silico*" digestion because the digestion table is usually created by computer simulation. Among many site-specific proteases commercially available today, trypsin is highly practical because it cleaves C-terminal to arginine or lysine, creating a set of fragments in a mass range appropriate for MALDI-TOF, that is, m/z 500–4000 Da, for most proteins. However, if trypsin generates fragments inappropriately small ($m/z \leq 500$ Da) because of its high basicity, or inappropriately large fragments ($m/z \geq 5000$ Da) because of the lack of arginine and/or lysine residues, other proteases need to be considered. These enzymes and their cleavage site(s) are as follows: endoproteinase Lys-C (C-terminal to lysine), Arg-C (C-terminal to arginine), Asp-N (N-terminal to aspartate), Glu-C in bicarbonate buffer (C-terminal to glutamate), Glu-C in phosphate buffer (C-terminal to glutamate and aspartate), and chymotrypsin

(C-terminal to bulky hydrophobic groups such as phenylalanine, tryptophan, tyrosine, leucine, and methionine).

For in-gel digestion of 2-D gel spots[26,27] we adapted the method reported by Rosenfeld et al.[28] The essence of the Rosenfeld et al. method resides in (1) soaking an excised gel in aqueous organic environment such as 50% (v/v) acetonitrile–ammonium bicarbonate buffer, and (2) dehydrating the gel piece by spontaneous evaporation of acetonitrile–water. The resulting shrunken gel will readily reabsorb the protease solution. Because trypsin is widely used and appears to be useful for virtually any protein, we describe the in-gel digestion protocol using trypsin. If another protease is to be used, it may be necessary to modify the protocol slightly, in particular the buffer system.

In-Gel Digestion by Trypsin

Stock Solutions

Ammonium carbonate (pH 8.9; pH adjusted with ammonium hydroxide), 200 mM

Acetonitrile (50%, v/v)–ammonium carbonate (pH 8.9), 200 mM

Acetonitrile (60%, v/v)–trifluoroacetic acid (TFA; 0.1%, v/v)

Tolylsulfonyl phenylalanyl chloromethyl ketone (TPCK; 0.25 μg/μl)–trypsin (Pierce, Rockford, IL) in 200 mM ammonium carbonate (pH 8.9); make fresh each time.

Procedure

Because Coomassie Brilliant Blue (CBB) interferes with mass spectrometry, it is recommended that staining gels with CBB be kept to a minimum but sufficient to identify the protein spot of interest. After destaining, gels are rinsed in deionized water several times to remove organic solvents completely. While gels are wet, a spot (2–3 × 2–3 mm) containing the protein of interest is excised with a clean blade and transferred into an Eppendorf tube. Process as many protein spots as desired, provided no mixing of samples occurs. Rinse the excised gel pieces in high-quality water for 15 min several times. After removing the water the gel pieces may be

---

[26] H. Matsumoto, B. T. Kurien, Y. Takagi, E. S. Kahn, T. Kinumi, N. Komori, T. Yamada, F. Hayashi, K. Isono, W. L. Pak, K. W. Jackson, and S. L. Tobin, *Neuron* **12,** 997 (1994).

[27] T. Kinumi, S. L. Tobin, H. Matsumoto, K. W. Jackson, and M. Ohashi, *Eur. Mass Spectrom.* **3,** 367 (1997).

[28] J. Rosenfeld, J. Capdevielle, J. C. Guillemot, and P. Ferrara, *Anal. Biochem.* **203,** 173 (1992).

stored frozen. To proceed, the gel pieces are vigorously shaken in 500 μl of 50% (v/v) acetonitrile–200 mM ammonium carbonate (pH 8.9) at room temperature to remove CBB. Proteins appear to stay within the gel unless the molecular mass of the protein is extremely small, such as several thousand daltons. Change the solution every 15–30 min and continue to shake the gel spots until CBB is no longer visible. Usually, this can be achieved after three or four changes of the buffer.

A gel piece is then transferred onto a clean glass plate and kept for 5–10 min until the appearance of the gel indicates dehydration. To the dehydrated gel, add 2 μl of TPCK (0.25 μg/μl)–trypsin in 200 mM ammonium carbonate to each side of the gel. The dehydrated gel will absorb the trypsin solution quickly. Transfer the gel piece into a 1.5-ml tube, add 150 μl of 200 mM ammonium carbonate, and incubate at 30° for 5–18 hr.

To elute tryptic peptides from the gel piece, add 150 μl of 60% (v/v) acetonitrile–0.1% (v/v) TFA to the tube and shake it at 25° for 20 min. After a brief spin in a tabletop centrifuge, collect the supernatant into a new 1.5-ml tube. Repeat the extraction two more times and combine all the supernatants. Dry the supernatant with a vacuum-concentrating system and dissolve the pellet with 5–10 μl of high-quality water for mass spectrometry. The sample can be stored frozen at least for several months.

Peptide Mass Fingerprinting by Matrix-Assisted Laser Desorption Ionization Time-of-Flight Mass Spectrometry

Mass Spectrometer for Peptide Mass Fingerprinting

To measure the mass of tryptic peptides generated by in-gel digestion, a mass spectrometer equipped with an ionization source capable of ionizing peptides[14–17] is required. There are two ionization methods commonly used for this purpose: an electrospray ionization (ESI) source and a matrix-assisted laser desorption ionization (MALDI) source. There are two essential differences between ESI and MALDI: (1) ESI takes a liquid sample, whereas MALDI takes a crystallized sample, and (2) ESI tends to generate multiply charged ions, whereas MALDI mostly generates singly charged ions. Many combinations between these ion sources and mass analyzers such as a quadrupole mass analyzer, a quadrupole ion trap mass analyzer, an ion cyclotron Fourier transform mass analyzer, and a time-of-flight mass analyzer are possible. In our opinion, MALDI interfaced with a time-of-flight mass analyzer (MALDI-TOF) is preferable for peptide mass fingerprinting for the following reasons: production of singly charged ions, high sensitivity, need for only a small amount of sample, and ease of operation of the equipment. Therefore, we show here an example of peptide mass fingerprinting by MALDI-TOF.

Equipment

We use a MALDI-TOF mass spectrometer Voyager Elite BioSpectrometry Research Station built by PE Biosystems (Framingham, MA) for a routine peptide mass mapping. However, MALDI-TOF mass spectrometers from other companies such as Bruker (Billerica, MA), Finnigan (San Jose, CA), Micromass (Beverly, MA), Kratos (Ramsey, NJ), and others can be used as well. If MALDI-TOF equipment is not available locally, mass spectrometry facilities at other universities and research institutes are available for a service fee. To obtain information about mass spectrometry facilities, conduct a web search on the Internet.

Acquisition of Matrix-Assisted Laser Desorption Ionization–Time of Flight Spectrum of Tryptic Fragments

To prepare a sample for MALDI-TOF mass spectrometry, 0.5 μl of the resulting aqueous solution (5–10 μl) of extracted peptides is mixed with a matrix solution. A typical matrix solution is α-cyano-4-hydroxycinnamic acid (10 mg/ml) in 40% (v/v) acetonitrile–0.1% (v/v) TFA for peptides up to ~3000 Da. We also often use 3-methoxy-4-hydroxycinnamic (ferulic) acid [10 mg/ml in 70% (v/v) acetonitrile–5% (v/v) formic acid] as a matrix because it generally performs better than α-cyano-4-hydroxycinnamic acid for larger peptides ($\geq$1500 Da). We usually calibrate the MALDI-TOF equipment externally. For the matrix α-cyano-4-hydroxycinnamic acid we use its dimer ($[M_2 + H]^+$ = 379.093, monoisotopic mass) and angiotensin I ($[M + H]^+$ = 1296.68, monoisotopic mass) as standards. For ferulic acid we use angiotensin I and the oxidized form of insulin B chain ($[M + H]^+$ = 3496.96, average mass) as standards.

The MALDI-TOF spectrum of the tryptic digests generated from spot 7 in Fig. 1d is shown in Fig. 2. Major peaks on the MALDI-TOF spectrum are automatically read by the Voyager operating software, which is a customized version of GRAMS/386 (Galactic, Salem, NH) and a peak table is generated. The peak table can be automatically incorporated into Microsoft (Redmond, WA) Excel and saved.

Database Search through MS-Fit

There are several database services available[21] through the Internet. We have used MS-Fit because of its user-friendly features and simplicity. We illustrate, step by step, an example of a protein database search through MS-Fit, using the tryptic fingerprint of spot 7 shown in Fig. 2.

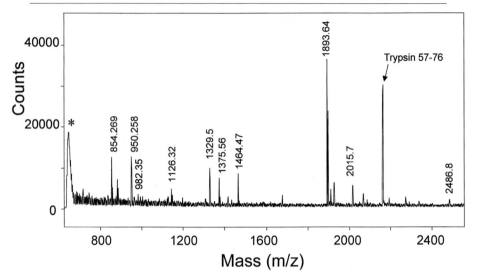

FIG. 2. MALDI-TOF spectrum of the tryptic digests generated from spot 7. The peaks indicated with numbers were unambiguously assigned as a result of the MS-Fit search shown in Fig. 3. The peak indicated as trypsin 57–76 has an observed mass of m/z 2162.77 and was assigned as amino acids 57–76 of trypsin (m/z 2163.05, monoisotopic mass). By inputting the mass numbers indicated to MS-Fit at http://prospector.ucsf.edu/, one may reproduce the results illustrated in Fig. 3. An asterisk indicates low-mass gate noise.

Step 1. Log onto ProteinProspector at http://prospector.ucsf.edu/.

Step 2. Copy the peak table in the Microsoft Excel file into Windows memory.

Step 3. Open the first page of MS-Fit as shown in Fig. 3a. Paste the peak table into the "Peptide Masses" field. Choose appropriate parameters for the database search. In the case of spot 7 we chose: Species (MAMMALS), MW of Protein (10,000–40,000), Protein pI (3–7), monoisotopic mass, with a mass tolerance of 1.0 Da. Then start the search.

Step 4. Figure 3b illustrates the output of the MS-Fit search showing "Ca-binding protein P26" as the best match. Clicking on the NCBInr accession number (132258 in this case) will link to the corresponding entry in the MEDLINE database.

Step 5. By clicking the ranking number, 1 in this case, a detailed description of the match will be displayed as shown in Fig. 3c. The displayed peptide table indicates the matched fragments and the errors. This window also displays possible posttranslational modifications of the protein.

Step 6. By further clicking the MS-Digest index number (31550 in this case), one can see the regions of the protein covered by the observed peptide fingerprints (Fig. 3d).

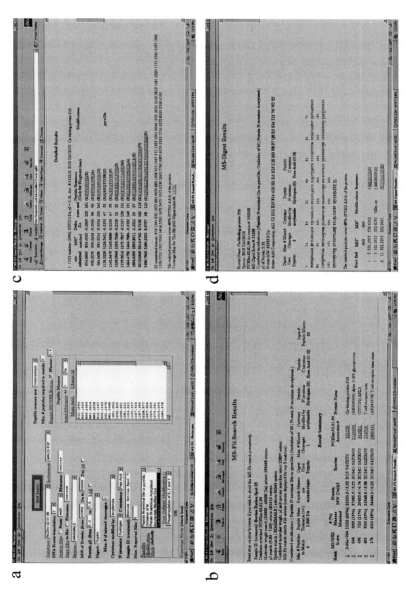

FIG. 3. MS-Fit input (a) and output (b–d) screens for mass fingerprinting search for spot 7. See text for details.

Criteria for Identification

In all nine cases of protein identification illustrated in Fig. 1d the MS-Fit search yields an unequivocal output, which is characterized by (1) high "molecular weight search (MOWSE)" scores,[29] (2) a clear gap in MOWSE score between the best score and its followers, such as $>10^4$ versus 10^2, and (3) a high proportion of coverage (usually more than 30%) of the entire amino acid sequence by the observed peptides. Moreover, when the MALDI-TOF spectrum is reexamined manually, that is, not by the automatic peak assignment function, we observed more peptide peaks that match the *in silico* digestion. Under certain circumstances it may be necessary to obtain further evidence for the assignment. Such evidence might include structural information about the peptides, determined by fragmentation using mass spectrometry. There are several techniques available for this purpose; for example, collision-induced dissociation (CID) in tandem mass spectrometry and post source decay (PSD) analysis in MALDI-TOF. For information on these techniques, refer to monographs[14,15] or reviews.[16,17] If it is necessary to pursue further biochemical evidence for tryptic fragments available at a 10 pM level, gas-phase Edman degradation after separating peptides by microbore HPLC will lead to a more convincing conclusion. Alternatively, if the N terminus of a protein is not blocked, direct Edman sequencing of a PVDF blot will yield convincing evidence. In our research we have confirmed the identity of spots 4 and 5 to be aspartate aminotransferase and creatine kinase, respectively, by Edman degradation in addition to peptide mass fingerprinting.[30]

If, unfortunately, an MS-Fit search gives an ambiguous output, several interpretations are possible. First, if a protein spot consists of multiple molecular entities, the database search may give multiple matches with high MOWSE scores. Further effort to separate the protein from the contaminants will improve the fingerprinting. Second, if the gene encoding the protein has not been registered in the database, no strong candidates will be displayed. This may happen frequently if the genome project for the animal in use is not developed enough. In such a case, expanding the search to related species or even to all species may hit a homologous protein. While using bovine retina, we often found that our MS-Fit search pinpointed the counterpart of a protein in other mammals, but not in *Bos taurus* itself. We encountered this situation with creatine kinase of bovine retina (spot 5 in Fig. 1d).[30] Theoretically, all the proteins on a 2-D gel will be identifiable by this technique after the genome database is complete. A protein spot

[29] D. J. C. Pappin, P. Hojrup, and A. J. Bleasby, *Curr. Biol.* **3**, 327 (1993).
[30] Y. Nishizawa, N. Komori, J. Usukura, K. W. Jackson, S. L. Tobin, and H. Matsumoto, *Exp. Eye Res.* **69**, 195 (1999).

of high purity (≥90%), clean proteolytic digestion, and accurate MALDI-TOF measurement will be the key to a successful identification.

Peptide Mass Fingerprinting of Protein Producing Large Fragments

We present here another example of peptide mass mapping in which tryptic digestion produces relatively large fragments (≥2000 Da). Figure 4 is the MALDI-TOF spectrum of tryptic fragments generated from spot (or band) 9 of Fig. 1d. In this case nearly half of the generated peptides were larger than m/z 2000. In the mass region $m/z \leq 1500$, the monoisotopic mass peak that represents the molecular species consisting of only ^{12}C always dominates among the cluster of various isotopic peaks. Such high resolution can easily be achieved if the MALDI-TOF instrument has a delayed extraction (DE) mode. The peak assignment function of GRAMS software, which we usually use as a first step, always picks up the monoisotopic peaks correctly in this region, that is, $m/z \leq 1500$–1800. Therefore, if the sizes of the fragments are rather small, that is, less than 2000 Da, an

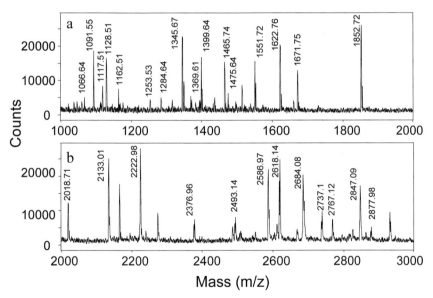

FIG. 4. (a, b) MALDI-TOF spectra of the tryptic digests generated from spot 9. Note that, in this case, relatively larger peptides are displayed in the peptide fingerprint. For the peaks in the higher mass region (≥2000; b), the monoisotopic peaks were assigned manually. If one performs an MS-Fit search at http://prospector.ucsf.edu/, using the mass numbers indicated in (a) and (b), an output displaying "interphotoreceptor retinoid-binding protein" as the first candidate will result.

MS-Fit search in the monoisotopic mass mode as shown in the case for spot 7 (Fig. 2) or the low mass region of spot 9 (Fig. 4a) will be appropriate. However, because the GRAMS software autofunction will assign an average mass for the clustered peaks in which isotopic mass peaks dominate over the monoisotopic mass peak in the higher mass region ($m/z \geq 2000$; Fig. 4b), it is necessary to manually assign the observed monoisotopic peak. The numbers in Fig. 4b represent monoisotoic masses assigned manually. For further explanation on monoisotopic mass and average mass, it is best to consult a monograph.[31]

Conclusion and Comments

It is apparent that the peptide mass fingerprinting by MALDI-TOF illustrated in this chapter, as applied to identification of retinal proteins, will be a powerful technique to guide vision researchers who are interested in understanding visual function at the protein level. The idea of peptide mass fingerprinting was first presented by W. J. Henzel[32] in a poster presentation at the Third Symposium of the Protein Society in Seattle, 1989. Several groups pursued the idea and presented the first experimental examples of peptide mass fingerprinting in various systems.[29,32–36] The technical aspects of mass spectrometers have substantially improved in two ways: (1) many excellent instruments are commercially available and (2) all are built in a user-friendly manner so that researchers new to mass spectrometry can operate the equipment. The advantage of peptide mass fingerprinting is that if a researcher identifies a protein on a gel and its identity is likely to provide information crucial to the project, it is not necessary to guess, for example, when choosing between several possible antibodies for the identification of the protein on Western blots. Instead, it is possible to pursue a rather straightforward peptide mass fingerprinting as described here. We have confirmed that archived 2-D gels that have been dried and stored at room temperature for almost two decades still serve as a source for peptide mass fingerprinting.[37] Therefore, if an intriguing protein is

[31] J. T. Watson, "Introduction to Mass Spectrometry," 3rd Ed. Lippincott-Raven, Philadelphia, 1997.
[32] P. James, M. Quadroni, E. Carafoli, and G. Gonnet, *Protein Sci.* **3,** 1347 (1994).
[33] W. Henzel, T. Billeci, J. Stults, S. Wong, C. Grimley, and C. Watanabe, *Proc. Natl. Acad. Sci. U.S.A.* **90,** 5011 (1993).
[34] P. James, M. Quadroni, E. Carafoli, and G. Gonnet, *Biochem. Biophys. Res. Commun.* **195,** 58 (1993).
[35] M. Mann, P. Hojrup, and P. Roepstorff, *Biol. Mass Spectrom.* **22,** 338 (1993).
[36] J. R. Yates III, S. Speicher, P. R. Griffin, and Hunkapillar, *Anal. Biochem.* **214,** 397 (1993).
[37] H. Matsumoto and N. Komori, *Anal. Biochem.* **270,** 176 (1999).

identified on a gel and the gel is dried and stored in a notebook, there is a good possibility that the identity of the protein can be revealed by peptide mass fingerprinting. The probability of successful identification will substantially increase after the genome database is completed. It should be noted, however, that although 2-D gel electrophoresis is a powerful technique, there is no guarantee that a gel displays all the proteins expressed in the tissue sample used. There are two challenging difficulties remaining for future development: how to study less abundant proteins and how to extract and display insoluble proteins that apparently eluded investigation by the current method. Ironically, rhodopsin is one of the most abundant proteins in the photoreceptor cell, but is not displayed on the 2-D gel in Fig. 1 because of its insolubility. Further development needs to be done to overcome these two difficulties for the preparation and processing of samples. In such efforts a well-defined fractionation of samples either anatomically and/or biochemically and a careful solubilization of proteins will be necessary. A "tissue printing" method to isolate the photoreceptor layer from bovine retina[30] will be a good starting material for this purpose. Finally, since the accuracy of mass measurement is high, there is a good possibility that peptide mass fingerprinting pinpoints posttranslational modifications such as phosphorylation[38-40] and lipidation.[41,42] Thus, the protocol described in this chapter will also serve a first step for studies of posttranslational modification of retinal proteins.

Acknowledgments

This work was supported by NIH Grant EY06595 and the Oklahoma Center for the Advancement of Science and Technology (OCAST) Awards HR3-080 to H.M. and HN5-024 to N.K. The PE Biosystems Voyager Elite laser desorption ionization time-of-flight mass spectrometer was purchased by the NSF EPSCoR Oklahoma Biotechnology Network program. Pierre Fontanille, Gaëlle Crohas, and Olivier Corre, exchange students from Blaise Pascal University at Clermont-Ferrand, France, in the summer of 1998, contributed to the advancement of peptide mass fingerprinting techniques in our group. We thank Tom Pugh and Jane Neal for technical assistance and Masaomi Matsumoto for reading this manuscript.

---

[38] D. I. Papac, J. E. Oatis, Jr., R. K. Crouch, and D. R. Knapp, *Biochemistry* **32,** 5930 (1993).
[39] H. Ohguro, J. P. Van Hooser, A. H. Milam, and K. Palczewski, *J. Biol. Chem.* **270,** 14259 (1995).
[40] J. B. Hurley, M. Spencer, and G. A. Niemi, *Vision Res.* **38,** 1341 (1998).
[41] Y. Fukada, *Methods Enzymol.* **250,** 91 (1995).
[42] R. R. Rando, *Biochem. Soc. Trans.* **24,** 682 (1996).

Section VI

Analysis of Animal Models of Retinal Diseases

[34] Animal Models of Inherited Retinal Diseases

By JEANNE FREDERICK, J. DARIN BRONSON, and WOLFGANG BAEHR

Introduction: Genes Linked to Retinitis Pigmentosa and Allied Diseases

Of several hundred inherited diseases affecting the human eye or the retina, relatively few have been characterized at a molecular genetic level. Among these, retinitis pigmentosa (RP), a large and heterogeneous group of blinding disorders, has been studied intensively at the molecular level.[1,2] Particularly useful in determining biochemical mechanisms has been animal models of hereditary degeneration that simulate human diseases. In this chapter, we summarize vertebrate animal models of retinal degeneration consisting of two groups: the naturally occurring, and the laboratory generated. Naturally occurring mutations affect a variety of genes, including genes involved in phototransduction (e.g., *rd* mouse, *rd* chicken), genes encoding structural proteins of photoreceptors (e.g., *rds* mouse), genes encoding transcription factors (e.g., **vitiligo**), or genes also expressed in other tissues (e.g., *shaker* mouse). Laboratory-generated mutants consist typically of transgenic mice or mice in which a gene of interest has been knocked out. Other vertebrate animals, for example, *Xenopus* and zebrafish, are used increasingly to generate animal models. Genes targeted include those involved in photoreception (rhodopsin), phototransduction (cGMP cascade), the visual cycle, cellular metabolism (transporters, channels, enzymes), cell structure (cytoskeleton, membrane components), or genes involved in development (transcription factors).

Naturally Occurring Animal Models of Retinitis Pigmentosa

Several models of both categories have defects in genes encoding components of the phototransduction cascade (PC) (Fig. 1[3,4]) or the visual cycle (VC). We estimate the number of genes involved in these events conservatively at about 40–60. About two dozen have been cloned and characterized by recombinant DNA techniques (Fig. 1). The number of identified PC/VC gene defects producing naturally occurring animal models

---

[1] T. P. Dryja and T. Li, *Hum. Mol. Genet.* **4**, 1739 (1995).
[2] K. Gregory-Evans and S. S. Bhattacharya, *Trends Genet.* **14**, 103 (1998).
[3] A. Polans, W. Baehr, and K. Palczewski, *Trends. Neurosci.* **19**, 547 (1996).
[4] Y. Koutalos and K. W. Yau, *Trends Neurosci.* **19**, 73 (1996).

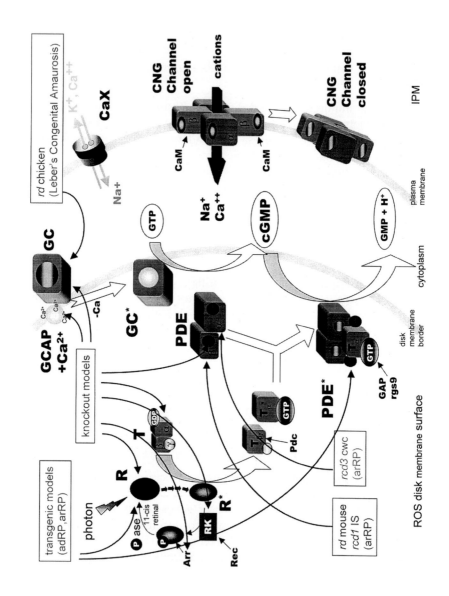

of RP is surprisingly small (four). These genes encode PDE_β (*rd* or *r* mouse, *rcd1* Irish setter), PDE_α (*rcd3* Cardigan Welsh dog), GC1 (*rd* chicken), and RPE65 (Briard dog). PDE (cGMP phosphodiesterase) is the enzyme that degrades cGMP on (indirect) activation by light, and guanylate cyclase 1 (GC1) is the enzyme that produces cGMP (Fig. 1). Because cytoplasmic cGMP levels in dark and light are carefully balanced by a number of feedback mechanisms, it is perhaps not surprising that defects in genes encoding these enzymes cause severe problems in photoreceptor metabolism. RPE65 is a component of the retinal pigment epithelium (RPE) involved in retinoid metabolism. The number of naturally occurring animal models with known gene defects affecting photoreceptor structure is even smaller (one, *rds* mouse). A number of animal models (e.g., RCS rat, *prcd* dog) have been researched thoroughly but the causative gene defects remain unknown.

Laboratory-Generated Animal Models

Owing to rapidly advancing transgenic and gene-targeting technologies, the number of laboratory-generated vertebrate animals is ever increasing (Table I). Mutations in the rhodopsin gene, responsible for a large number of dominant and recessive RP cases in human, have been used most extensively for the generation of RP models.[5] A number of genes that have been identified in photoreceptors or the inner retina have been used for generation of knockout animals, a powerful method by which to analyze the contribution of a gene product to a specific pathway. Hence the majority of animal models listed in Table I are knockout models.

[5] J. Lem and C. L. Makino, *Curr. Opin. Neurobiol.* **6,** 453 (1996).

FIG. 1. The phototransduction cascade and animal models for retinal degeneration. The cascade[3] consists of rhodopsin (R, one subunit), transducin (T, three subunits), cGMP phosphodiesterase (PDE, four subunits), and the cGMP-gated cation channel (two subunits). The task of the cascade is to rapidly hydrolyze cytoplasmic cGMP and to close CNG cation channels. Regulatory components are rhodopsin kinase (RK), arrestin (Arr), recoverin (Rec), phosducin (Pdc), RGS9 (GAP), and calmodulin (CaM), each of which consists of one subunit. The recovery phase returns activated components and cGMP/Ca^{2+} levels to their dark states. Key components of the recovery phase are guanylate cyclase (GC; two distinct GCs have been identified in photoreceptors) and guanylate cyclase-activating proteins (GCAPs; three identified to date). The cation exchanger (CaX, one subunit) is not part of the cascade but is instrumental in regulating cation concentrations. Arrows point to subunit genes whose defects have been identified in naturally occurring animal models, or to genes that have been used for targeting in knockout animal models. For reviews see Polans *et al.*[3] and Koutalos and Yau.[4]

TABLE I
VERTEBRATE ANIMAL MODELS (EXCLUDING HUMAN), GENE DEFECTS, AND PHENOTYPES[a]

Component/gene[b]		Activity/function[b]	Animal model[c]	Gene defect[d]	Phenotype	Refs.[e]
Rhodopsin (rod)	Rho	Light reception; catalyst of $G_{t\alpha} \cdot$GTP formation; disk structural protein	Transgenic mouse	P23H	Dominant RP	1
				V20G; P23H; P27L	Dominant RP	2
				T17M		3
				P347S		4–6
				Q344ter	CSNB	7
				G90D		8, 9
				K296E		10
			Transgenic pig	P347L	Dominant RP	11, 12
			Knockout mouse		Recessive RP	13–15
Transducin (rod)	$G_{t\alpha}$	Activator of rod PDE	Knockout mouse			16
Transducin (cone)		Activator of cone PDE	Transgeic mouse		Downregulation of PDE	17
cGMP phosphodiesterase (PDE)	PDE_α	Hydrolyzes cGMP	rcd3 cardigan Welsh	Exon 15 frameshift		18
	PDE_β		rd mouse	Y348ter	Recessive RP	19
			r mouse	Y348ter	Recessive RP	20
			rcd1 Irish setter	W807ter	Recessive RP	21–22
	PDE_γ		Knockout mouse	Null allele		23
			Transgenic mouse			24
Guanylate cyclase 1	GC1	Produces cGMP	rd chicken	$\Delta \times 4$–7	Recessive	25
			Knockout mouse	Null allele	LCA I	26
Guanylate cyclase-activating proteins 1 and 2	GCAP-1, -2	Mediate Ca^{2+} sensitivity of GC1 and or GC2	Double knockout mouse	Null allele		27
Recoverin (S-modulin)		Ca^{2+} sensor; blocks Rho kinase?	Knockout mouse	Null allele	None	28
Arrestin (rod)		Binds to phosphorylated Rho	Knockout mouse	Null allele	Oguchi CSNB	29, 30
Rho-kinase (rod and cone)		Phosphorylates Rho	Knockout mouse?			31
IRBP (rod and cone)		Retinoid-binding protein	Knockout mouse	Null allele		32
RPE65			CSNB Briard dog	4-bp deletion		33, 34
			Knockout mouse		LCA type II	35, 36
ABCR (rim protein)		Transporter	Knockout mouse	Null allele	Recessive Stargardt MD	37
rds/peripherin		Structural	rds mouse	Insertion		38, 39
ROM1		Structural	Knockout mouse	Null allele	Recessive RP	40
Mitf		Transcription factor	vitiligo mouse	Defect in mitf gene	Recessive RP	41, 42
?		?	RCS rat	?		43, 44
?		?	Rdy cat	?	ad rod–cone dysplasia	45

TABLE I (continued)

Component/gene[b]	Activity/function[b]	Animal model[c]	Gene defect[d]	Phenotype	Refs.[e]
Phosducin?	Regulatory protein	prcd	Missense mutation R82G?		46, 47
?	?	Rcd2		Recessive RP	48
mGluR6	Metabotropic Glu receptor, type 6	Knockout mouse	Null allele	Absence of b-wave	49, 50
?	?	Mutant fish (nba)		Dominant night blindness	51
?	?	nob mouse	?	xCSNB	52
β3 subunit gene of AP3 adaptor complex		pearl mouse		Night blindness Hermansky–Pudlak syndrome	53
REP-1	Rab escort protein	Knockout mouse	Null allele	Choroideremia	54, 55
Myosin VIIA	Structural	shaker-1 mouse	Multiple mutations	Usher syndrome type 1B	56, 57
ND		Knockout mouse	Null allele	xRP (Norrie)	58
PLCβ4	Phospholipase	Knockout mouse	Null allele	Reduction in a- and b-waves	59
Cyclin D1		Knockout mouse	Null allele	Photoreceptor degeneration	60
OAT	Ornithine δ-amino transferase	Knockout mouse	Null allele	Gyrate atrophy (chorioretinal degeneration)	61
Dystrophin dp260	?	Knockout mouse	Null allele	Prolonged implicit time of b-wave	62

[a] Rho, Rhodopsin; RP, retinitis pigmentosa; ad, autosomal dominant; ar, autosomal recessive; x, x linked; CSNB, congenital stationary night blindness; MD, macular degeneration; ND, Norrie disease; prcd, progressive rod/cone degeneration; rd, retinal degeneration; rcd, rod/cone dysplasia; OAT, ornithine aminotransferase; LCA, Leber's Congenital Amaurosis.
[b] Gene and the function of the gene product, if known.
[c] Animal model (naturally occurring or laboratory generated).
[d] Gene defect (amino acid substitutions, or nucleotide deletions), if known.
[e] Key to references: (1) J. E. Olsson, J. W. Gordon, B. S. Pawlyk et al., Neuron **9,** 815 (1992); (2) M. I. Naash, J. G. Hollyfield, M. R. Al-Ubaidi, and W. Baehr, Proc. Natl. Acad. Sci. U.S.A. **90,** 5499 (1993); (3) T. Li, M. A. Sandberg, B. S. Pawlyk et al., Proc. Natl. Acad. Sci. U.S.A. **95,** 11933 (1998); (4) T. Li, W. K. Snyder, J. E. Olsson, and T. P. Dryja, Proc. Natl. Acad. Sci. U.S.A. **93,** 14176 (1996); (5) P. C. Huang, A. E. Gaitan, Y. Hao, R. M. Petters, and F. Wong, Proc. Natl. Acad. Sci. U.S.A. **90,** 8484 (1993); (6) C.-H. Sung, C. Makino, D. Baylor, and J. Nathans, J. Neurosci. **14,** 5818 (1994); (7) C. Portera-Cailliau, C.-H. Sung, J. Nathans, and R. Adler, Proc. Natl. Acad. Sci. U.S.A. **91,** 974 (1994); (8) P. A. Sieving, J. E. Richards, F. Naarendorp, E. L. Bingham, K. Scott, and

(continued)

TABLE I (continued)

M. Alpern, *Proc. Natl. Acad. Sci. U.S.A.* **92,** 880 (1995); *(9)* V. R. Rao, G. B. Cohen, and D. D. Oprian, *Nature (London)* **367,** 639 (1994); *(10)* T. Li, W. K. Franson, J. W. Gordon, E. L. Berson, and T. P. Dryja, *Proc. Natl. Acad. Sci. U.S.A.* **92,** 3551 (1995); *(11)* Z. Y. Li, F. Wong, J. H. Chang *et al., Invest. Ophthalmol. Visual Sci.* **39,** 808 (1998); *(12)* R. M. Petters, C. A. Alexander, K. D. Wells *et al., Nature Biotechnol.* **15,** 965 (1997); *(13)* M. M. Humphries, D. Rancourt, G. J. Farrar *et al., Nature Genet.* **15,** 216 (1997); *(14)* J. Lem, N. V. Krasnoperova, P. D. Calvert *et al., Proc. Natl. Acad. Sci. U.S.A.* **96,** 736 (1999); *(15)* E. R. Weiss, Y. Hao, C. D. Dickerson *et al., Biochem. Biophys. Res. Commun.* **216,** 755 (1995); *(16)* J. Lem, R. L. Sidman, B. Kosaras *et al., Invest. Ophthalmol. Visual Sci.* **39,** S644 (1998); *(17)* C. J. Raport, J. Lem, C. Makino *et al., Invest. Ophthalmol. Visual Sci.* **35,** 2932 (1994); *(18)* S. M. Petersen-Jones and D. R. Sargan, *Invest. Ophthalmol. Visual Sci.* **40,** 1637 (1999); *(19)* S. J. Pittler and W. Baehr, *Proc. Natl. Acad. Sci. U.S.A.* **88,** 8322 (1991); *(20)* S. J. Pittler, C. E. Keeler, R. L. Sidman, and W. Baehr, *Proc. Natl. Acad. Sci. U.S.A.* **90,** 9616 (1993); *(21)* M. L. Suber, S. J. Pittler, N. Qin *et al., Proc. Natl. Acad. Sci. U.S.A.* **90,** 3968 (1993); *(22)* P. J. M. Clements, C. Y. Gregory, S. M. Peterson-Jones, D. R. Sargan, and S. S. Bhattacharya, *Curr. Eye Res.* **12,** 861 (1993); *(23)* S. H. Tsang, P. Gouras, C. K. Yamashita *et al., Science* **272,** 1026 (1996); *(24)* S. H. Tsang, M. E. Burns, P. D. Calvert *et al., Science* **282,** 117 (1998); *(25)* S. L. Semple-Rowland, N. R. Lee, J. P. Van Hooser, K. Palczewski, and W. Baehr, *Proc. Natl. Acad. Sci. U.S.A.* **95,** 1271 (1998); *(26)* D. G. Birch, R.-B. Yang, and D. L. Garbers, *Invest. Ophthalmol. Visual Sci.* **39,** S643 (1998); *(27)* A. Mendez, M. E. Burns, I. Sokal *et al., Invest. Ophthalmol. Visual Sci.* **40,** S391 (1999); *(28)* J. Chen and M. I. Simon, *Invest. Ophthalmol. Visual Sci.* **36,** S641 (1995); *(29)* J. Xu, R. L. Dodd, C. L. Makino, M. I. Simon, D. A. Baylor, and J. Chen, *Nature (London)* **389,** 505 (1997); *(30)* A. L. Lyubarski, E. N. Pugh, B. Falsini, P. Valentini and J. Chen, *Invest. Ophthalmol. Visual Sci.* **39.** S643 (1998); *(31)* A. L. Lyubarski, C.-K. Chen, M. I. Simon, and E. N. Pugh, Jr., *Invest. Ophthalmol. Visual Sci.* **40,** S390 (1999); *(32)* G. I. Liou, Y. Fei, N. S. Peachey *et al., J. Neurosci.* **18,** 4511 (1998); *(33)* G. D. Aguirre, V. Baldwin, S. Pearce-Kelling, K. Narfström, K. Ray, and G. M. Acland, *Mol. Vis.* **4,** 23 (1998); *(34)* A. Veske, S. E. Nilsson, K. Narfström, and A. Gal, *Genomics,* **57,** 57 (1999); *(35)* T. M. Redmond, S. Yu, E. Lee, D. Bok, D. Hamasaki, and K. Pfeifer, *Invest. Ophthalmol. Visual Sci.* **39,** S643 (1998); *(36)* T. M. Redmond, S. Yu, E. Lee *et al., Nature Genet.* **20,** 344 (1998); *(37)* J. Weng, S. M. Azarian, and G. H. Travis, *Invest. Ophthal. Visual Sci.* **39,** S643 (1998); *(38)* G. H. Travis, J. G. Sutcliffe, and D. Bok, *Neuron* **6,** 61 (1991); *(39)* G. H. Travis, M. B. Brennan, P. E. Danielson, C. A. Kozak, and J. G. Sutcliffe, *Nature (London)* **338,** 70 (1989); *(40)* G. A. Clarke, J. Rossant, and R. R. McInnes, *Invest. Ophthalmol. Visual Sci.* **39,** S962 (1998); *(41)* R. L. Sidman, B. Kosaras, and M. Tang, *Invest. Ophthalmol. Visual Sci.* **37,** 1097 (1996); *(42)* B. L. Evans and S. B. Smith, *Mol. Vis.* **3,** 11 (1997); *(43)* R. J. Mullen and M. M. LaVail, *Science* **192,** 799 (1976); *(44)* M. J. McLaren and G. Inana, *FEBS Lett.* **412,** 21 (1997); *(45)* R. Curtis, C. E. Barnett, and A. Leon, *Invest. Ophthalmol. Visual Sci.* **28,** 131 (1987); *(46)* G. M. Acland, K. Ray, C. S. Mellersh *et al., Proc. Natl. Acad. Sci. U.S.A.* **95,** 3048 (1998); *(47)* Q. Zhang, G. M. Acland, C. J. Parshall, J. Haskell, K. Ray, and G. D. Aguirre, *Gene* **215,** 231 (1998); *(48)* K. Ray, V. J. Baldwin, C. Zeiss, G. M. Acland, and G. D. Aguirre, *Curr. Eye Res.* **16,** 71 (1997); *(49)* A. Nomura, R. Shigemoto, Y. Nakamura, N. Okamoto, N. Mizuno, and S. Nakanishi, *Cell* **77,** 361 (1994); *(50)* M. Masu, H. Iwakabe, Y. Tagawa, *et al., Cell* **80,** 757 (1995); *(51)* L. Li and J. E. Dowling, *Proc. Natl. Acad. Sci. U.S.A.* **94,** 11645 (1997); *(52)* M. T. Pardue, M. A. McCall, M. M. LaVail, R. G. Gregg, and N. S. Peachey, *Invest. Ophthalmol. Visual Sci.* **39,** 2443 (1998); *(53)* L. Feng, A. B. Seymour, S. Jiang *et al., Hum. Mol. Genet.* **8,** 323 (1999); *(54)* J. A. van den Hurk, W. Hendriks, D. J. van de Pol *et al., Hum. Mol. Genet* **6,** 851 (1997); *(55)* J. A. van den Hurk, M. Schwartz, H. van Bokhoven *et al., Hum. Mutat.* **9,** 110 (1997); *(56)* P. Mburu, X. Z. Liu, J. Walsh *et al., Genes Funct.* **1,** 191 (1997); *(57)* F. Gibson, J. Walsh, P. Mburu *et al., Nature (London)* **374,** 62 (1995); *(58)* W. Berger, D. van de Pol, D. Bachner *et al., Hum. Mol. Genet* **5,** 51 (1996); *(59)* H. Jiang, A. Lyubarsky, R. Dodd *et al., Proc. Natl. Acad. Sci. U.S.A.* **93,** 14598 (1996); *(60)* C. Ma, D. Papermaster, and C. L. Cepko, *Proc. Natl. Acad. Sci. U.S.A.* **95,** 9938 (1998); *(61)* T. Wang, A. M. Lawler, G. Steel, I. Sipila, A. H. Milam, and D. Valle, *Nature Genet.* **11,** 185 (1995); *(62)* S. Kameya, E. Araki, A. Mizota *et al., Invest. Ophthalmol. Visual Sci.* **39,** S644 (1998).

Gene Defects Affecting Levels of cGMP

Because a number of animal models (Fig. 1) are associated with mutations in genes affecting cGMP metabolism, we summarize briefly the underlying gene defects and their consequences. The major components are cGMP phosphodiesterase (PDE) and the GC–GC-activating protein (GCAP) complex. Naturally occurring homozygous PDE_β null alleles cause retinal degeneration in the *rd* mouse,[6] and the *rcd1* Irish setter.[7] A laboratory-generated PDE_γ knockout resulted in a phenotype that was similar to that of *rd*.[8] In these animal models, PDE_β or PDE_γ was not produced, and PDE was found to be inactive, leading to elevated levels of cytoplasmic cGMP. A point mutation in PDE_γ, reintroduced into homozygous PDE_γ knockouts, impaired transducin–PDE interactions and slowed the recovery rate of the flash response in transgenic mouse rods.[9] A GC1 null mutation was linked to the retinal degeneration (*rd*) chicken[10] (and see [35] in this volume[11]), a phenotype nearly identical to that of human Leber's congenital amaurosis (LCA),[12] a severe retinopathy of the heterogeneous RP group. In the *rd* chicken, GC is inactive and cGMP is not produced to levels sufficient to sustain phototransduction. Both the GC1 gene[13] and the GCAP genes[14] have been knocked out in animal models. Preliminary results indicate that loss of GC1 function primarily affects cones, and loss of GCAP1/GCAP2 affects Ca^{2+} sensitivity of recovery.[14]

Diagnostic Tests (Genotyping) of Naturally Occurring or Transgenic/Knockout Mouse Models

To identify whether a particular mouse strain carries a known mutation (e.g., *rd*), a rapid diagnostic test is most helpful. The best diagnostic tests available are based on polymerase chain reaction (PCR) amplification of specific exons, and direct sequencing or restriction digests of amplified products. In most cases this procedure requires isolation of genomic DNA (in mice mostly from tail clippings), amplification of gene fragments or exons with specific oligonucleotide primers, digestion of the fragment with

---

[6] S. J. Pittler and W. Baehr, *Proc. Natl. Acad. Sci. U.S.A.* **88**, 8322 (1991).
[7] M. L. Suber, S. J. Pittler, N. Qin et. al., *Proc. Natl. Acad. Sci. U.S.A.* **90**, 3968 (1993).
[8] S. H. Tsang, P. Gouras, C. K. Yamashita et al., *Science* **272**, 1026 (1996).
[9] S. H. Tsang, M. E. Burns, P. D. Calvert et al. *Science* **282**, 117 (1998).
[10] S. L. Semple-Rowland, N. R. Lee, J. P. Van-Hooser, K. Palczewski, and W. Baehr, *Proc. Natl. Acad. Sci. U.S.A.* **95**, 1271 (1998).
[11] S. L. Semple-Rowland and N. R. Lee, *Methods Enzymol.* **316**, Chap. 35, 2000 (this volume).
[12] I. Perrault, J.-M. Rozet, P. Calvas et al., *Nature Genet.* **14**, 461 (1996).
[13] D. G. Birch, R.-B. Yang, and D. L. Garbers, *Invest. Ophthalmol. Vis. Sci.* **39**, S643 (1998).
[14] A. Mendez, M. E. Burns, I. Sokal et al., *Invest. Ophthalmol. Vis. Sci.* **40** (1999).

a diagnostic restriction enzyme, and/or direct sequencing or sequencing after cloning. The following is a set of methods serving these purposes.

Isolation of Genomic DNA from Mouse Tail Biopsies

Required Materials

Clean, sterile scalpel blades
Microcentrifuge tubes, 1.5 ml
Tail buffer [50 mM Tris (pH 7.5), 50 mM EDTA (pH 8.0), 100 mM NaCl, 5 mM dithiothreitol (DTT), 0.5 mM spermidine, sodium dodecyl sulfate (SDS; 2%, w/v)]
Proteinase K, 20 mg/ml in water
Potassium acetate, 3 M/5 M
2-Propanol
Ethanol (70%, v/v)
Tris–EDTA (TE, 1×)
Water bath, 55°

To prepare 100 ml of tail buffer, combine

Tris (pH 7.5), 1.0 M	5 ml
EDTA (pH 8.0), 0.5 M	10 ml
NaCl, 5.0 M	2 ml
DTT, 1.0 M	0.5 ml
Spermidine, 1.0 M	50 μl
SDS	2 g

Bring the volume to 100 ml with sterile, deionized water.

Procedure

Using a scalpel blade dipped in ethanol and flamed, cut 1–2 cm of distal tail and place it in a microcentrifuge tube. Add 0.6 ml of tail buffer and 20 μl of proteinase K to each tube. Incubate overnight at 55° with gentle shaking.

Vortex each sample briefly to ensure that tails are digested. Pellet cellular debris by spinning the tubes at 15,000 rpm for 2 min at 4°. Transfer each supernatant to a clean 1.5-ml microcentrifuge tube, add 200 μl of potassium acetate solution, and vortex vigorously for 20 sec. Chill the samples on ice for 5 min to precipitate the proteins. Centrifuge for 4 min at 15,000 rpm at 4°. Decant the supernatant into a clean 1.5-ml microcentrifuge tube containing 0.6 ml of 2-propanol. Precipitate the DNA by inverting each tube until a white mass forms. Centrifuge at 15,000 rpm for 1 min at 4°. Pour off the supernatant, add 0.6 ml of 70% (v/v) ethanol, and wash the DNA pellet by inverting the tube several times. Centrifuge at 15,000 rpm for 1 min at 4°. Carefully decant the ethanol and air dry the DNA pellet for 15–20 min. Add 100 μl of TE and dissolve the pellet overnight at room temperature. The yield of nucleic acid approaches 1 μg/μl as estimated by agarose gel electrophoresis. Refrigerate at 4° for near-term analysis, or freeze at −20° for long-term storage. Although this DNA preparation is suitable for PCR amplification on appropriate dilution (1:25), trace amounts of salt and detergent may require reextraction with phenol–chloroform if, for example, the DNA is to be used for Southern blots and/or restriction digests.

Genomic Polymerase Chain Reaction for Genotyping

An excellent, practical overview of genotyping requirements and techniques used in support of transgenic colonies is available.[15] A stringent protocol for amplifications of genomic fragments takes into consideration the melting points of the oligonucleotide primers used, the length of the fragment to be amplified, other parameters such as $MgCl_2$ and pH sensitivity of given primer pairs, and peculiarities of the PCR cycler. The following are two sample protocols used in our laboratory for amplification of genomic DNA.

Example 1: Amplification of Diagnostic Exon 7 of Mouse PDE_β Subunit Gene

A nonsense ochre mutation (T → A transversion in codon 347 of the PDE_β gene) is responsible for the *rd* mouse phenotype.[6] The nonsense mutation introduces a new *Dde*I site that can be used in subsequent enzymatic digestions to verify its presence or absence. The pimer pair amplifies a 296-bp fragment that carries two *Dde*I sites in the *rd* PDE_β subunit allele and none in the normal allele. The amplification requires 5–20 pmol each of primer 1 (sense, 5'-CAT CCC ACC TGA GCT CAC AGA AAG) and primer 2 (antisense, 5'-GCC TAC AAC AGA GGA GCT TCT AGC), 25–75 ng of genomic DNA, and standard buffer conditions. Cycling is done 35 times at 92° for 30 sec, 55° for 30 sec, and 72° for 30 sec in a Perkin-Elmer, (Norwalk, CT) GeneAmp 960 cycler (or equivalent). A single fragment of the indicated size can be detected in agarose gels. The fragment is subsequently digested with *Dde*I as described below.

[15] B. Elder, *Lab. Anim.* 20 (1999).

Example 2: Amplification of Exon 1 of Mouse Rhodopsin Gene for Genotyping

A transgenic mouse line with three mutations in exon 1 and a restriction fragment length polymorphism (RFLP) deleting an intrinsic *Nco*I site was generated.[16] The diagnostic assay consists of amplification of exon 1 of the mouse rhodopsin gene, followed by *Nco*I digestion. Amplification with primers W75/W11 produces a 1.3-kb fragment, which must first be verified by electrophoresis on a 1% (w/v) agarose gel. After digestion of the fragment with *Nco*I (typical reaction given below), wild-type mice reveal three fragments of 689, 431, and 197 bp. Transgenic (heterozygous) mice have, in addition, an 886-bp fragment (689 + 197 bp) owing to the deletion of one *Nco*I site in the transgene.

Restriction Digest of Amplified Fragments

Restriction enzymes are supplied with 10× buffers for which activity is optimal. The conditions (pH, salt) optimal for *Taq* polymerase activity may, or may not, be compatible with those of the restriction enzyme. To minimize potential incompatibility, an aliquot of the amplification reaction is diluted in a restriction digest of larger volume (20 μl), e.g., to 12 μl of distilled H$_2$O, one adds 2 μl of 10× buffer, 1 μl of enzyme and 5 μl of amplification reaction, mixing after each addition. After incubation according to the enzyme manufacturer recommendation (e.g., 1 hr at 37°), the products of digestion are separated electrophoretically by loading 10 μl of the restriction digest per lane of a 1–2% (w/v) agarose gel. Small fragment sizes (<500 bp) are resolved more readily in gels of 2% (w/v) agarose concentration.

Sequencing of Amplified Fragments

If diagnostic restriction sites are not present, the DNA fragment must be sequenced directly to assess for mutations. The most common procedures sequence a DNA fragment directly after amplification, or after cloning. Most sequencing today is done with automatic sequencers and chemistry based on linear PCR (one primer only), primer extension, and chain terminators.

Direct Sequencing

By far the most used method is based on the Perkin-Elmer/Applied Biosystems large-scale sequencers available through core facilities. On a

---

[16] M. I. Naash, J. G. Hollyfield, M. R. Al-Ubaidi, and W. Baehr, *Proc. Natl. Acad. Sci. U.S.A.* **90,** 5499 (1993).

smaller scale, we found the Perkin-Elmer ABI 310 capillary sequencer to be the most reliable and useful. The ABI 310 works without gels, separates DNA fragments by capillary electrophoresis, and provides sequences up to 400-bp fragments in less than 2 hr. For all sequencing, the DNA template should be highly purified, using for example, a Qiagen (Chatsworth, CA) DNA preparation kit or a cesium chloride gradient.

ABI 310 Protocol: Two hundred to 500 ng of plasmid DNA template (or 30–90 ng of amplified fragment) is needed for each reaction. The primer extension chemistry uses labeled dideoxy terminators, each labeled with a different fluorescent dye emitting light at distinct wavelengths. Consequently, each reaction can be done in a single tube. Unincorporated terminators, however, will create a background during the electrophoresis, and must be removed by precipitation or column chromatography before loading the sample onto the capillary. Perkin-Elmer/Applied Biosystems markets a kit for this chemistry called the BigDye Terminator Cycle Sequencing Ready Reaction kit.

To set up a typical reaction, combine

H_2O	15.0 μl
DNA template	1.0 μl
Primer (3.2 pmol)	1.0 μl
Ready Reaction Mix	8.0 μl
	20.0 μl

Procedure. A layer of oil is not needed over the reactions if a thermal cycler with a heated top is used. A basic cycling program for a Perkin-Elmer 9700 thermal cycler consists of a denaturing step (94° for 2 min), and 25 cycles of 94° for 10 sec, 50° for 5 sec, and 60° for 4 min.

After cycling is complete, the reaction is passed through a column [Centri-Sep column (Princeton Separations, Adelphia, NJ) or Sephadex G-50 column] to remove excess unincorporated nucleotides. Desiccate the flow-through in a SpeedVac concentrator and add 25 μl of template suppression reagent (supplied with the kit); then load the sample into the sequencer for electrophoresis through the capillary.

Plasmid Sequencing after Cloning of Genomic Fragments

For large fragments (>3 kb), cloning followed by sequencing most often yields the best results. The fragment of interest can be cloned into PCR vectors such as PCR2.1 (InVitrogen, San Diego, CA), and the resulting plasmid sequenced with universal primers and automatic sequencers. The following example describes a cycle-sequencing method using a Li-Cor DNA4000 automatic sequencer (Li-Cor Technologies, Lincoln, NE). This system uses infrared-tagged primers, requires four lanes per reaction, and employs long polyacrylamide-based gels for separation of the DNA fragments. The Li-Cor sequences reliably more than 1 kb in one direction once parameters have been optimized.

Li-Cor DNA4000 Protocol. Two hundred to 1000 ng of template is needed for each reaction (depending on plasmid size). Relatively inexpensive prelabeled standard primers as well as custom labeled primers may be obtained directly from Li-Cor Technologies. Because the primer is prelabeled in this chemistry, termination reactions must be done (and loaded onto the gel) separately. Epicentre Technologies (Madison, WI) markets a kit for this chemistry called SequiTherm Excell II.

A typical reaction includes

H_2O	7.8 μl
DNA template (0.2 pmol)	1.0 μl
Labeled Primer (2 pmol)	1.0 μl
Excel II reaction buffer	7.2 μl
SequiTherm Excel II enzyme	1.0 μl
	18.0 μl

G, A, T, and C termination reactions are prepared by adding 2 μl of termination mix (supplied with the kit) to PCR tubes labeled accordingly. Transfer 4-μl aliquots into the four termination reaction tubes. A basic linear cycling program for a Perkin-Elmer 9700 thermal cycler consists of the following steps: (1) denature—94° for 5 min; (2) cycle (25 cycles)—94° for 30 sec, 50° for 15 sec, and 70° for 1 min.

After cycling is complete, add 4 μl of loading buffer (supplied with kit) to each termination reaction, and load 1.5 μl of each onto the gel to electrophorese. Six hundred to 1200 bases of sequence can be obtained per reaction. Data acquisition is fully automatic.

[35] Avian Models of Inherited Retinal Disease

By SUSAN L. SEMPLE-ROWLAND and NANCY R. LEE

Introduction

There has been a dramatic increase in the number of genes that have been linked to inherited retinal disease (see RetNet at *http://www.sph.uth.tmc.edu/Retent*), many of which encode proteins that directly affect the function of photoreceptor cells. The most devastating human retinal dystrophies affect cone cells, the photoreceptor cells responsible for color perception and high visual acuity in humans. Avian species, and in particular the domestic chicken, are excellent model systems for studies of cone cell function and their response to retinal disease. Unlike the retinas of most mammalian species, the retinas of chickens are cone dominant. Approximately 90% of the photoreceptors in central retina are cone cells.[1] The value of the chicken as a model system for studies of cone cell function is enhanced by the fact that the development, morphology, and electrophysiology of the chicken retina is relatively well characterized. In this chapter we briefly summarize the development and distinguishing characteristics of the domestic chicken retina. We then provide a detailed description of the retinal microdissection technique that we used in our analyses of cyclic nucleotide levels within the individual cell layers of the retinal degeneration (*rd*) chicken retina. We conclude our chapter with a brief discussion of two chicken models of inherited retinal disease and the advantages that avian models bring to research focused on the development of therapeutic treatments for cone photoreceptor disease.

Avian Retina

The organization and pattern of development of the chicken retina are similar to those observed in other vertebrate species. The process of neurogenesis proceeds along both central-to-peripheral and dorsal-to-ventral gradients and is complete within the first 11 days of embryonic development. Precursors of retinal ganglion, horizontal, and cone cells are the first to leave the mitotic cycle (embryonic days 3–7; E3–E7), the ganglion cell precursors generally appearing before any of the other cell types. The birthdates of the bipolar and Müller cells follow close behind (E4–E11).[2,3] Migration and differentiation of the various retinal cells begin shortly after the cells have left the mitotic cycle and continue over the course of development of the embryo. A distinct ganglion cell layer is first evident around E7.[4] Differentiation of the photoreceptor cells begins around E7, their inner and outer segments forming around E10 and E15, respectively.[5,6] Adult-like electroretinograms are recordable from E17 retina,[7] and by E19, the retina has completed the process of differentiation.[4,6] The mature morphological and functional characteristics of the retinas of

[1] V. B. Morris and C. D. Shorey, *J. Comp. Neurol.* **129**, 313 (1967).
[2] A. J. Kahn, *Dev. Biol.* **38**, 30 (1974).
[3] S. G. Spence and J. A. Robson, *Neuroscience* **32**, 801 (1989).
[4] A. J. Coulombre, *Am. J. Anat.* **96**, 153 (1955).
[5] K. Meller and W. Tetzlaff, *Cell Tissue Res.* **170**, 145 (1976).
[6] M. D. Olson, *Anat. Rec.* **193**, 423 (1979).
[7] G. Rager, *J. Comp. Neurol.* **188**, 225 (1979).

newly hatched chickens contrast with the relatively immature retinas of newborn mammals, which continue to undergo differentiation after birth.

The structure of the chicken retina is similar to that found in many vertebrate species. All cellular and fiber layers present in mammalian retina are present in chicken retina (Fig. 1A). Within the chicken retina, cone photoreceptor cells outnumber rod cells by six to one in the central retina

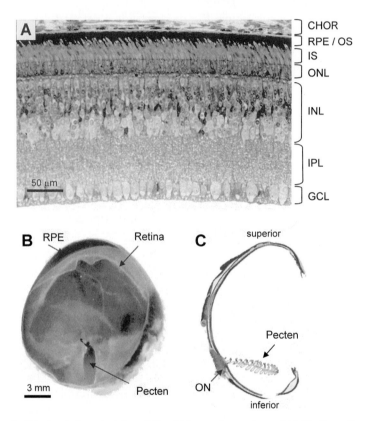

FIG. 1. Structural characteristics of the chicken retina and eye. (A) Plastic section of 1-day-old chicken retina stained with toluidine blue. All cell and fiber layers found in mammalian retina are present. The oil droplets present in the cone photoreceptor inner segments appear as white spheres adjacent to the pigment epithelium. (B) View inside chicken eye after removal of the anterior segment. The pecten, a highly vascular structure, projects into the inferior portion of the posterior chamber. (C) Cross section of an eyecup, showing the comblike appearance of the pecten and its relationship to the optic nerve. CHOR, Choroid; RPE, retinal pigment epithelium; OS, outer segment; IS, photoreceptor inner segment; ONL, outer nuclear layer; INL, inner nuclear layer; IPL, inner plexiform layer; GCL, ganglion cell layer; ON, optic nerve.

and by three to one in the peripheral retina.[1,8] The cone-dominant character of the avian retina contrasts with the retinas of mice, rats, and dogs, in which cones comprise less than 5% of the photoreceptor population. The abundance of cone cells in chicken retina has made it a useful model system for studies of cone cell-specific proteins, as well as for examining the function of these cells under normal and disease conditions.

The chicken retina, while exhibiting many attributes common to mammalian retina, also possesses distinguishing features. First, cone cells in the chicken retina contain colored oil droplets that are located at the apical ends of the inner segments. The two types of single cones contain red and yellow oil droplets and the primary member of the double cone possesses an orange oil droplet. Cone cells also possess four different types of visual pigments–red (iodopsin), green, blue, and violet—in addition to the rod visual pigment, rhodopsin. The absorption spectra of these pigments, in combination with the presence of the colored oil droplets, gives rise to tetrachromatic color vision in these animals.[9,10] Second, the avian neural retina is avascular. Two nutrient sources are present in the avian eye: the choroid and the pecten oculi. The choriocapillaris layer of the choroid, as in other vertebrate retinas, is situated behind the basal surface of the retinal pigment epithelium and provides nutrients to the retinal pigment epithelium and photoreceptor cells. The cells in the inner retina receive nutrients from the pecten, a vascular structure that consists of a pleated sheet of freely anastomosing capillaries that is continuous with the choroid and protrudes into the vitreous body at the base of the optic nerve head (Fig. 1B, and C).[11,12]

There are several advantages in using chickens as model systems for vision research. First, the chicken eye is large, a feature that simplifies surgical manipulations of the eye. In E7 embryos, the eyes are approximately 0.5 cm in diameter and by hatching the diameter of the eye has increased to 1 cm. Second, embryonic chicken retina can easily be accessed and manipulated *in ovo*. Use of chicken models simplifies studies designed to address the temporal relationship between disease progression and the efficacy of potential therapeutic treatment regimens because treatments can be delivered to the embryo at any stage during its development. Third, it is possible to maintain embryonic chicken retinal cells in culture. Primary

[8] V. B. Morris, *J. Comp. Neurol.* **140**, 3589 (1970).

[9] T. Okano, D. Kojima, Y. Fukada, Y. Shichida, and T. Yoshizawa, *Proc. Natl. Acad. Sci. U.S.A.* **89**, 5932 (1992).

[10] T. Okano, Y. Fukada, and T. Yoshizawa, *Comp. Biochem. Physiol. B. Biochem. Mol. Biol.* **112**, 405 (1995).

[11] M. Uehara, S. Oomori, H. Kitagawa, and T. Ueshima, Nippon Juigaku Zasshi **52**, 503 (1990).

[12] S. Liebner, H. Gerhardt, and H. Wolburg, *Brain Res. Dev. Brain Res.* **100**, 205 (1997).

cultures of embryonic chicken retinal cells have proved useful in the analyses of the cellular and molecular signals that control differentiation and survival of photoreceptor cells.[13-16] For example, chicken retinal cultures have been successfully used in studies to characterize the *cis*-acting response elements and *trans*-acting factors that regulate the expression of genes in photoreceptor cells.[17-19] The preponderance of cone photoreceptors in these cultures allows specific investigations into the mechanisms that control the level and specificity of gene expression in these cells, cells that play a critical role in human vision.

Laminar Microdissection of Chicken Retina

Analyses and understanding of retinal biochemistry have been facilitated by the layered arrangement of cell types in the retina. The following is a detailed description of a retinal microdissection technique that we developed for analyses of chicken retina. Our method is based on a microdissection procedure that was described by O. H. Lowry and colleagues for the isolation and biochemical analyses of retinal cell layers from quick-frozen eyes.[20] We have used this microdissection method to analyze cyclic nucleotide levels in the individual cell layers of the normal and the retinal degeneration (*rd*) chicken retina.[21,22] Variations of this technique have been employed in studies of the laminar distribution of second messengers, enzymes, metabolites, and neurotransmitters in vertebrate retina.[23-25] This technique, when coupled to the reverse transcription-polymerase chain

[13] R. Adler, J. D. Lindsey, and C. L. Elsner, *J. Cell Biol.* **99,** 1173 (1984).
[14] L. E. Politi and R. Adler, *Invest. Ophthalmol. Vis. Sci.* **27,** 656 (1986).
[15] A. Repka and R. Adler, *Dev. Biol.* **153,** 242 (1992).
[16] T. Saga, D. Scheurer, and R. Adler, *Invest. Ophthalmol. Vis. Sci.* **37,** 561 (1996).
[17] M. Werner, S. Madreperla, P. Lieberman, and R. Adler, *J. Neurosci. Res.* **25,** 50 (1990).
[18] R. Kumar, S. Chen, D. Scheurer, Q. L. Wang, E. Duh, C. H. Sung, A. Rehemtulla, A. Swaroop, R. Adler, and D. J. Zack, *J. Biol. Chem.* **271,** 29612 (1996).
[19] J. H. Boatright, D. E. Borst, J. W. Peoples, J. Bruno, C. L. Edwards, J. S. Si, and J. M. Nickerson, *Mol. Vis.* **3,** 15 (1997). <http://www.emory.edu/molvis/v3/boatright>
[20] O. H. Lowry and J. V. Passonneau, "A Flexible System of Enzymatic Analysis." Academic Press, New York, (1972).
[21] N. R. Lee, Ph.D. Dissertation. University of Florida, Gainesville, Florida, (1991).
[22] S. L. Semple-Rowland, N. R. Lee, J. P. Van Hooser, K. Palczewski, and W. Baehr, *Proc. Natl. Acad. Sci. U.S.A.* **95,** 1271 (1998).
[23] H. T. Orr, O. H. Lowry, A. I. Cohen, and J. A. Ferrendelli, *Proc. Natl. Acad. Sci. U.S.A.* **73,** 4442 (1976).
[24] G. W. De Vries, A. I. Cohen, O. H. Lowry, and J. A. Ferrendelli, *Exp. Eye. Res.* **29,** 315 (1979).
[25] S. J. Berger, G. W. De Vries, J. G. Carter, D. W. Schultz, P. N. Passonneau, O. H. Lowry, and J. A. Ferrendelli, *J. Biol. Chem.* **255,** 3128 (1980).

reaction (RT-PCR), could also be applied to studies of gene transcripts in the various cell layers of the retina.

Preparation of Freeze-Dried Retinal Sections for Microdissection

One- to 3-day-old chickens are sacrificed by decapitation and the heads are immediately submerged into liquid nitrogen. Immersion of the whole head in liquid nitrogen promotes rapid freezing of the posterior pole of the eye, the temperature of which reaches $-10°$ within 14.0 ± 2.9 sec after immersion in the nitrogen. The rate of cooling of the central retina in the posterior pole of the eye is $3.6 \pm 0.7°/\text{sec}$ ($n = 7$) as measured with a thermistor probe assembly. After freezing, the heads are transferred to a $-25°$ cryostat and bisected along the sagittal plane. Each half of the head is mounted eyelid side down onto a prechilled dissection chuck, using cryostat tissue-embedding medium. Tissue covering the posterior pole of the eye is removed with a chilled scalpel blade.

Frozen sections (7 μm thickness) containing paracentral and central retina are ideal for microdissection. Before beginning sectioning, the blade should be positioned tangential to the posterior pole of the eye so that the widths of the individual retinal cell layers are as large as possible in each section. Optimization of the layer widths at this point greatly facilitates hand dissection of the layers once the sections are freeze-dried. The thickness of the sections is critical for successful microdissection. Sections that are too thin (5 μm) crumble easily during the dissection procedure. Sections that are too thick (10 μm) do not allow clear visualization of the individual retinal layers. In addition, because the relative positions of the boundaries between the cell layers on the two sides of a thicker tissue section are not identical, microdissection of the thicker sections increases the chance of cross-contamination of the retinal layers. Once cut, the tissue sections are gently placed in a prechilled drying assembly section holder (Ace Glass, Vineland, NJ). The section holder is an aluminum slide containing two rows of holes that is secured between two glass slides by adjustable screws. Once filled with frozen retinal sections, the section holder is placed in a glass-drying assembly vacuum tube that has been prechilled. The entire drying unit must be kept cold throughout the drying process to prevent thawing of the tissue sections. We installed a vacuum port in a cryostat that allowed the assembly to be kept within the chamber of the cryostat during the drying process. Tissue sections are dried at $-40°$ overnight under a vacuum of less than 0.1 mmHg. After drying, the drying assembly is brought to room temperature before turning off the vacuum pump. Once freeze-dried, the sections are kept under vacuum within the drying assembly and the assembly is stored at $-70°$ until microdissection.

Microdissection of Retinal Layers

Dissection of the individual retinal layers is performed with the aid of a dissecting microscope ($\times 40$ magnification). The microscope is housed within a humidity- and draft-controlled glove box. Maintaining humidity levels within the range of 40–55% is critical for successful handling and dissection of freeze-dried tissue. At humidity levels above 55%, the sections become sticky as a result of absorbing moisture from the air. At humidity levels below 40%, excessive static electricity creates unwanted section movement and adhesion. To minimize moisture changes in the sections, it is recommended that only a few of the freeze-dried sections be transferred to the glove box at any one time.

To aid in visualization and identification of the various retinal layers during the dissection procedure, tissue paper can be placed under the Plexiglas dissection platform. A few intact, frozen sections can be stained and used as guides for locating the retinal layers. Figure 2A shows an unstained freeze-dried retinal section prior to dissection. The oil droplets located within the apical ends of the cone inner segments form an orange-colored layer. The photoreceptor outer segments are located above the oil droplet layer and are deeply embedded in the retinal pigment epithelium.

Each of the retinal layers is dissected freehand using hair points, hair loops, and 1- to 2-mm razor blade shards that are attached to wooden applicator sticks. Hair points are prepared by gluing individual eyelashes or paint brush hairs to the tapered end of a wooden applicator stick. Hair loops are similarly prepared using small loops of fine hair. These tools are used to manipulate and hold sections during dissection. The razor blade shards are prepared by placing double-edged tungsten razor blades into liquid nitrogen until brittle and then snipping off small pieces of the blade with pliers. Knives that are flexible can be prepared by attaching the razor blade shard to a small piece of plastic hair brush bristle and then gluing the other end of the bristle onto the applicator stick. Flexible knives were found to work best.

During dissection, the retinal pigment epithelium layer has a tendency to separate from the other retinal layers. This results in the detachment of some of the photoreceptor outer segments from their respective inner segments. Thus, in order to carry out biochemical analyses on photoreceptor outer segments, it is necessary to collect both pigment epithelium and oil droplet-containing layers with their associated outer segments. The dissected layers of a representative freeze-dried chicken retina section are shown in Fig. 2B. After dissection, the individual pieces of each of the retinal layers are transferred into small plastic wells of a microtiter plate, using a set of hair points. The wells should be small enough to be seated

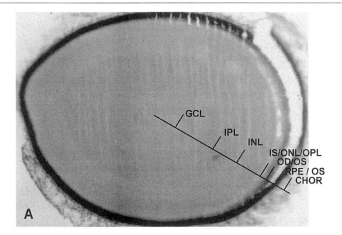
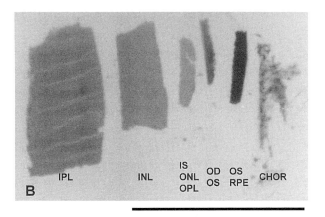

FIG. 2. Microdissection of freeze-dried chicken retina sections. (A) Unstained, freeze-dried section containing central chicken retina as viewed under a dissection microscope at (×40 original magnification). The retinal cell layers have been labeled and can be distinguished by their colors and shadings. The oil droplets that are found in the inner segments of the cone photoreceptor cells appear orange in unstained sections. (B) Fragments of the various retinal layers that are produced after microdissection of a freeze-dried retinal section. The length of the black bar shown below the retinal layer fragments is 1 mm. CHOR, Choroid; RPE, retinal pigment epithelium; OS, outer segment; OD, oil droplet; IS, photoreceptor inner segment; ONL, outer nuclear layer; OPL, outer plexiform layer; INL, inner nuclear layer; IPL, inner plexiform layer; GCL, ganglion cell layer.

into the aluminum section holder. Sections are picked up with a hair point that has been statically charged by rubbing it against tissue paper. A second, uncharged hair point is used to release the section from the charged hair point. Once transferred to the plastic wells, the plastic wells containing the sections are sealed in the drying assembly section holder and stored under vacuum at $-70°$ in the drying assembly vacuum tube until needed for analyses. In our studies of the normal and *rd* chicken retina, retinal layers obtained from 10 to 30 sections (1 eye) provided sufficient material for analyses of cAMP and cGMP content in the individual layers of the retina.[21,22]

Discussion

The chicken, which has long been a favorite model system for studies of development, also possesses many attributes that make it an excellent model system for studies of inherited retinal diseases that affect cone photoreceptor function and survival. Our analyses of the *rd* (retinal degeneration) chicken model of inherited retinal disease exemplify the value of chicken models of inherited retinal disease (for review, see Semple-Rowland[26]). The *rd* chicken, which was originally described by Cheng *et al.*,[27] is blind at hatching despite the presence of a normal, fully developed retina.[28,29] Degenerative changes first appear in the outer segments of the rod and cone photoreceptor cells 7–10 days after the birds hatch. Both scotopic and photopic electroretinograms are either absent or severely attenuated in 1-day-old animals.[28,30] Molecular and biochemical analyses of the retinas of these animals have revealed that the blind phenotype exhibited by these animals is due to a null mutation in the gene encoding photoreceptor guanylate cyclase 1 (GC1).[22] The absence of GC1 in both the rod and cone photoreceptor cells leads to a 80–90% reduction in levels of cGMP within these cells, a level too low to support phototransduction. In humans, GC1 frameshift and missense mutations have been shown to cause Leber congenital amaurosis type 1 (LCA1), a severe autosomal recessive disease that causes blindness at birth.[31] The striking similarities exhibited between the

[26] S. L. Semple-Rowland, *in* "Retinal Degenerative Diseases and Experimental Therapy" (J. G. Hollyfield, R. E. Anderson, and M. M. LaVail, eds.). Plenum, New York, (1999).
[27] K. M. Cheng, R. N. Shoffner, K. N. Gelatt, G. G. Gum, J. S. Otis, and J. J. Bitgood, *Poultry Sci.* **59**, 2179 (1980).
[28] R. J. Ulshafer, C. B. Allen, W. W. Dawson, and E. D. Wolf, *Exp. Eye Res.* **39**, 125 (1984).
[29] R. J. Ulshafer and C. B. Allen, *Exp. Eye Res.* **40**, 865 (1985).
[30] W. W. Dawson, R. J. Ulshafer, R. Parmer, and N. R. Lee, *Clin. Vis. Sci.* **5**, 285 (1990).
[31] I. Perrault, J. M. Rozet, P. Calvas, S. Gerber, A. Camuzat, H. Dollfus, S. Chatelin, E. Souied, I. Ghazi, C. Leowski, M. Bonnemaison, D. Le Paslier, J. Frezal, J. L. Dufier, S. Pittler, A. Munnich, and J. Kaplan, *Nature Genet.* **14**, 461 (1996).

phenotypes exhibited by the *rd* chicken and patients diagnosed with LCA1 support continued use of the *rd* chicken in studies directed toward understanding issues critical for successful treatment of cone photoreceptor disease.

Several strains of chickens have been described that possess genetic mutations that adversely affect ocular physiology (for review see Somes *et al.*[32]). Of these, only a few strains exhibit retinal degeneration. In addition to the *rd* chicken, there is a strain called the partial retinal dysplasia and subsequent degeneration (*rdd*) chicken. The *rdd* chicken carries an autosomal recessive mutation that produces abnormalities in both the retinal pigment epithelium and the neural retina that eventually lead to blindness in sexually mature birds.[33] In the *rdd* chicken, abnormalities in the pigment epithelium and neural retina are first detected at the light microscopic level in E9 embryos. These changes include the formation of small holes or gaps in the pigment epithelium cell layer within the temporal and nasal quadrants of the fundus and distortion of the orderly arrangement of the neural retinal layers in the vicinity of these holes. By E18, there is an increase in the number and size of the holes in the pigment epithelium and in the extent of the photoreceptor and outer plexiform layer buckling, which expands to include all quadrants of the neural retina. Few photoreceptor inner and outer segments are present in the E18 retina and loss of cells from both the inner nuclear and ganglion cell layers is evident. Between 1 and 5 weeks of age, cell loss from the photoreceptor and inner nuclear layers proceeds at a rapid rate. Despite the presence of retinal degeneration, behavioral signs of defective vision are not apparent until the animals reach 5–6 weeks of age. In blind, 9-month-old adult *rdd* chickens, occasional degenerated photoreceptor cells can still be identified. The identity of the genetic defect responsible for this disease has not yet been determined.

The *rd* and *rdd* chicken models described above represent mutants that spontaneously appeared in established strains of domestic fowl. Until relatively recently, many of the animals carrying genetic mutations that affect the visual system were discovered by chance. Advances in transgenic technology have allowed scientists to create animals that carry mutations identical to those that have been linked to human disease. Development of transgenic bird technology has been slowed by the unique problems created by the reproductive system of the laying hen. Many of the tech-

---

[32] R. G. Somes, Jr., K. M. Cheng, D. E. Bernon, and R. D. Crawford, *in* "Developments in Animal and Veterinary Sciences," Vol. 22: "Poultry Breeding and Genetics" (R. D. Crawford, ed.), p. 273. Elsevier, New York, 1990.

[33] C. J. Randall, M. A. Wilson, B. J. Pollock, R. M. Clayton, A. S. Ross, J. B. Bard, and I. McLachlan, *Exp. Eye. Res.* **37**, 337 (1983).

niques used in mammalian transgenesis require implantation of manipulated eggs into surrogate females, a procedure that is not feasible in birds. Despite this difficulty, technical approaches suitable for the creation of transgenic chickens have been developed (for reviews see Sang[34] and Etches and Verrinder Gibbins[35]). With increased worldwide commitment to mapping of the chicken genome,[36] interest in improving these methods for use in agriculture as well as in studies of basic biological processes is likely to increase. Current information about the chicken genome mapping project can be viewed on the internet at the ChickGBASE site maintained by the Roslin Institute in Edinburgh, United Kingdom (*http://www.ri.bbsrc.ac.uk/*). Generation and analyses of transgenic chickens that possess mutations in genes specifically expressed in cone photoreceptor could dramatically improve our understanding of cone cell function and the parameters that are critical for successful rescue of these cells in diseased retinas.

Acknowledgments

This work was supported by Grant EY11388 from the National Institutes of Health.

[34] H. Sang, *Trends Biotechnol.* **12,** 415 (1994).
[35] R. J. Etches and A. M. Verrinder Gibbins, *Methods Mol. Biol.* **62,** 433 (1997).
[36] D. W. Burt, N. Bumstead, J. J. Bitgood, F. A. Ponce de Leon, and L. B. Crittenden, *Trends Genet.* **11,** 190 (1995).

[36] Biochemical Analysis of Phototransduction and Visual Cycle in Zebrafish Larvae

By MICHAEL R. TAYLOR, HEATHER A. VAN EPPS, MATTHEW J. KENNEDY, JOHN C. SAARI, JAMES B. HURLEY, and SUSAN E. BROCKERHOFF

Introduction

The zebrafish (*Danio rerio*) has received considerable attention as a model organism for the study of vertebrate development.[1,2] More recently,

[1] P. Haffter, M. Granato, M. Brand, M. C. Mullins, M. Hammerschmidt, D. A. Kane, J. Odenthal, F. J. van Eeden, Y. J. Jiang, C. P. Heisenberg, R. N. Kelsh, M. Furutani-Seiki, E. Vogelsang, D. Beuchle, U. Schach, C. Fabian, and C. Nusslein-Volhard, *Development* **123,** 1 (1996).
[2] W. Driever, L. Solnica-Krezel, A. F. Schier, S. C. Neuhauss, J. Malicki, D. L. Stemple, D. Y. Stainier, F. Zwartkruis, S. Abdelilah, Z. Rangini, J. Belak, and C. Boggs, *Development* **123,** 37 (1996).

zebrafish have also been recognized as an excellent model organism for the study of retinal function.[3] The zebrafish visual system has been well characterized and is similar to that of other vertebrates.[4] The retina contains rods and four types of cones, making it possible to study both scotopic and photopic visual responses.[5,6] Zebrafish larvae develop rapidly, allowing behavioral analysis of visual function by 4 days postfertilization.[7,8] Zebrafish also provide many advantages for genetic analysis. For instance, the large number of progeny that can be generated from a single cross greatly simplifies and accelerates genetic analysis when compared with other vertebrate systems. Finally, molecular and genetic methods are available that will expedite the cloning of zebrafish genes.[9]

Zebrafish with recessive mutations that cause defective visual responses have been isolated by a behavioral screening strategy.[3,10,11] These mutants have been characterized by examining retinal histology and electroretinogram (ERG) responses.[3] In this way, the visual defects in several mutants were localized to the outer layers of the retina. The *no optokinetic response b* (nrb^{a13}) mutant, for example, exhibits normal retinal morphology at 5 days postfertilization, but has an abnormal ERG that is consistent with a light adaptation defect in cone photoreceptors.[3]

To further characterize mutants with vision defects, we adapted biochemical assays to analyze the phototransduction cascade and the visual cycle in zebrafish. These assays must be performed on larvae because mutants with visual defects do not survive to adulthood. In the *nrb* mutant, for example, the swim bladder develops normally and the larvae respond vigorously to touch stimuli. However, *nrb* larvae do not forage for food and die emaciated at 10–12 days postfertilization. In general, vision mutants probably die from starvation since normal larvae appear to rely on visual responses in order to feed (our unpublished observation, 1998). Alternatively, some vision mutants may also have additional unidentified defects that cause lethality later in development. Biochemical characterization of larvae with vision defects must therefore take place around 5 or

---

[3] S. E. Brockerhoff, J. E. Dowling, and J. B. Hurley, *Vision Res.* **38,** 1335 (1998).
[4] E. A. Schmitt and J. E. Dowling, *J. Comp. Neurol.* **344,** 532 (1994).
[5] J. Robinson, E. A. Schmitt, F. I. Harosi, R. J. Reece, and J. E. Dowling, *Proc. Natl. Acad. Sci. U.S.A.* **90,** 6009 (1993).
[6] I. J. Kljavin, *J. Comp. Neurol.* **260,** 461 (1987).
[7] D. T. Clark, Doctoral Dissertation. University of Oregon Press, Eugene, Oregon, 1981.
[8] S. S. Easter, Jr., and G. N. Nicola, *Dev. Biol.* **180,** 646 (1996).
[9] H. W. Detrich, M. Westerfield, and L. I. Zon (eds.), *Methods Cell Biol.* **60** (1999).
[10] S. E. Brockerhoff, J. B. Hurley, U. Janssen-Bienhold, C. F. Neuhauss, W. Driever, and J. E. Dowling, *Proc. Natl. Acad. Sci. U.S.A.* **92,** 10545 (1995).
[11] S. E. Brockerhoff, J. B. Hurley, G. A. Niemi, and J. E. Dowling, *J. Neurosci.* **20,** 1 (1997).

6 days postfertilization, which is after the visual system of wild-type larvae becomes functional but before the larvae die from starvation.

The methods described in this chapter include assays for transducin activation, phosphodiesterase (PDE) activity, guanylyl cyclase (GC) activity, rhodopsin and cone opsin phosphorylation, and the analysis of retinoids in the visual cycle. For each assay, the general principles are discussed, the necessary materials listed, and the procedure outlined. Results from representative experiments are also presented and discussed. These methods provide valuable tools for characterizing the phenotypes of an increasing number of zebrafish vision mutants. In addition, candidate genes for these mutants may be identified by these methods, facilitating the cloning of the genes responsible for the defects.

Materials and Methods

General Materials

$[^{35}S]GTP\gamma S$ (1000–1500 Ci/mmol), $[\alpha\text{-}^{32}P]GTP$ (800 Ci/mmol), $[\gamma\text{-}^{32}P]ATP$ (3000 Ci/mmol), all-*trans*-$[^{3}H]$ retinol (55–85 Ci/mmol): New England Nuclear (Boston, MA)

3′,5′-[8-^{3}H]cGMP (5–25 Ci/mmol): Amersham (Arlington Heights, IL)

ATP, cGMP, GDP, GTP, GTPγS, NADPH: Sigma (St. Louis, MO)

Rod outer segment (ROS) buffer (1×): 20 mM HEPES (pH 7.4), 120 mM KCl, 5 mM MgCl$_2$, 1 mM dithiothreitol (DTT), 100 μM phenylmethylsulfonyl fluoride (PMSF)

Dissecting scope

Darkroom

Night-vision goggles and infrared illumination source

Calibrated light source (Oriel, Stratford, CT) with shutter driver/timer (Vincent Associates, Rochester, NY)

Liquid scintillation counter

Zebrafish Stocks and Culture Conditions

AB strain zebrafish originally obtained from the University of Oregon[10] are maintained as an inbred stock in the University of Washington Zebrafish Facility. The ENU-induced mutant *partial optokinetic response b* (*pob^{a1}*) has been previously isolated,[10] and is maintained in the facility as described.[12] Adult fish and larvae are maintained at 28.5° in reverse osmosis-distilled water reconstituted for fish compatibility by addition of salts and vitamins.[12]

---

[12] M. Westerfield, "The Zebrafish Book." University of Oregon Press, Eugene, Oregon, 1995.

Preparation of Larval Eye Homogenates

One of the most critical steps in performing most of the assays described below is the proper preparation of the eye homogenates. Whole eyes are used in these assays instead of photoreceptor preparations owing to the technical difficulty and the excessive number of eyes that would be necessary to isolate outer segments (larval eyes are ~250 μm in diameter at 5 days postfertilization). In principle, signaling components from various retinal cell types could interact when brought together on homogenization. Our results suggest that this is not a significant problem with the assays described here.

The homogenization procedure outlined below is used in the transducin, PDE, GC, and phosphorylation assays. The retinoid analysis, however, uses a modified version of this procedure that is described below. In general, 50 to 100 eyes are collected per tube for each assay, which is sufficient for 10 to 100 reactions depending on the assay being performed. Once homogenized, the total protein concentration can be estimated by a bicinchoninic acid (BCA) protein assay (Pierce, Rockford, IL). Protein concentration has been estimated by this method to be approximately 1.5 μg/larval eye, using bovine serum albumin as a standard (data not shown).

Materials

ROS buffer (1×)
Disposable homogenizer and fitted 1.5-ml centrifuge tubes: Kontes (Vineland, NJ)
Dissecting scope
Darkroom
Night vision goggles and infrared illumination source
Intravenous syringe needles (21–22 gauge)

Procedure

1. Dark adapt zebrafish larvae (5–6 days postfertilization) for at least 3 hr.

2. Transfer the larvae into cold 1× ROS buffer (4°) for 5–10 min. The cold buffer anesthetizes and immobilizes the larvae.

3. Dissect the eyes into cold 1× ROS buffer, using two intravenous syringe needles. This can be accomplished by holding down the larva with one needle and removing the eye with the other needle, using a slicing motion.

4. Collect the dissected eyes with a pipet and transfer the eyes into a 1.5-ml homogenizer-fitted centrifuge tube (50–100 eyes per tube).

5. Pellet the eyes at 3,000g for 5 sec at room temperature.

6. Remove the buffer and resuspend the eyes in fresh 1× ROS buffer at a suitable concentration (0.2–2 eyes/μl).

7. Homogenize the eyes on ice with a disposable homogenizer by hand or with an electric drill, using short pulses to avoid heating the sample.

8. The homogenate can be used immediately or aliquoted and frozen at $-70°$ for later use.

(All steps should be performed under infrared illumination, using night vision goggles.)

Biochemical Assays

Transducin Activation

Two different [^{35}S]GTPγS-binding assays have been developed to measure light-stimulated transducin activity. The first assay described below uses an *in situ* approach to assess photoreceptor-specific activation, while the second assay uses an *in vitro* approach based on the homogenization procedure described above. Both assays use GTPγS, a nonhydrolyzable analog of GTP, as the substrate. On absorption of a photon, photoexcited rhodopsin and cone opsins catalyze the exchange of GTP for bound GDP on transducin α subunits. By using [^{35}S]GTPγS in the reaction buffer, transducin activation can be examined by measuring the light-dependent incorporation of radioactivity, using filter-binding assays.

[^{35}S]GTPγS in Situ Binding Assay

Principle. The *in situ* approach described here is modified from Yarfitz et al.[13] The advantage of the *in situ* assay is that it potentially maintains the segregation of signaling components from different cell types that could interact when brought together on homogenization. For example, transducin activation in cone subtypes may be selectively measured by this assay. In addition, it requires collecting only a few wild-type and mutant larvae for each assay, as opposed to dissecting and homogenizing hundreds of wild-type and mutant eyes for the *in vitro* assays. This assay has other potential applications in addition to the procedure described below. For example, larvae can be examined for light-dependent defects by growing the larvae in complete darkness and then subjecting them to this assay. The disadvantage of the *in situ* assay is that it is difficult to perform. While wearing night vision goggles under infrared illumination, the larvae must

---

[13] S. L. Yarfitz, J. L. Running-Deer, G. Froelick, N. J. Colley, and J. B. Hurley, *J. Biol. Chem.* **269**, 30340 (1994).

be precisely aligned in O.C.T. and then sectioned directly through the eyes in a cryostat.

A diagram of this procedure is shown in Fig. 1A. Larvae are dark adapted for several hours, frozen in O.C.T. blocks, sectioned, bound to nitrocellulose filters, and stored at $-70°$. The reactions are performed directly on the sections that are bound to the nitrocellulose filters. The sections are thawed, preincubated with GDP and NADPH to reduce background, covered with [^{35}S]GTPγS reaction buffer, and subjected to darkness or a flash of light. After a brief incubation period, the filters are washed several times with 1× ROS buffer to remove unincorporated [^{35}S]GTPγS. Incorporated [^{35}S]GTPγS binds irreversibly to the transducin α subunits on the nitrocellulose filters. The filters are then exposed to X-ray film for autoradiography and the sections are excised for liquid scintillation counting.

Materials

ROS buffer (1×)
Preincubation buffer: 1× ROS buffer, 100 μM GDP, 2 mM NADPH
[^{35}S]GTPγS reaction buffer: 1× ROS buffer, 100 μM GDP, 2 mM NADPH, 100 nM GTPγS, [^{35}S]GTPγS (~1 μCi/reaction)
O.C.T. compound and vinyl specimen molds, 15 × 15 × 5 mm: Miles (Elkhart, IN)
Cryostat
Nitrocellulose membrane filters (25 mm), 45-μm pore size
Six-well, flat-bottom tissue culture plates
Filter support connected to a vacuum pump
XAR film: Kodak (Rochester, NY)
Dissecting scope
Darkroom
Night vision goggles and infrared illumination source
Calibrated light source with shutter driver/timer
Liquid scintillation counter

Procedure

1. Dark adapt zebrafish larvae (5–6 days postfertilization) for at least 3 hours.

2. Place the petri dish of larvae on ice for 5–10 min to anesthetize and immobilize the larvae (immobile larvae are much easier to align in O.C.T.).

3. Transfer the larvae (5–20 larvae per mold) into O.C.T.-filled vinyl molds, being careful not to introduce too much larvae water. Submerge the larvae in the O.C.T. and align the heads along one edge of the mold,

using fine-tip forceps. Freeze the blocks by dipping in a dry ice–ethanol bath and store at −70° in darkness.

4. With a cryostat, cut 10-μm sections directly through the eyes. Apply the sections to nitrocellulose membrane filters, using a sable brush. Melt the section to the filter by briefly touching the backside of the filter with a fingertip. Store the filters individually in six-well, flat-bottom tissue culture plates wrapped in aluminum foil at −70°.

5. To perform the assay, place individual filters on a filter support, apply 50 μl of preincubation buffer to the sections, incubate at room temperature (23–25°) for 1 min, and filter by applying vacuum.

6. Immediately apply 50 μl of [^{35}S]GTPγS reaction buffer and flash with light. Allow the reaction to proceed in the dark at room temperature. Stop the reaction by applying vacuum to remove reaction buffer.

7. Wash the filters under vacuum five times (5 ml each) with 1× ROS, remove the filters from the filter support, and dry the filters at room temperature for 1 hr.

8. Expose the filters directly to X-ray film overnight and develop.

9. Cut out the sections, immerse in scintillation fluid, and quantify the amount of incorporated radioactivity, using a liquid scintillation counter.

(Steps 2–8 should be performed under infrared illumination while wearing night vision goggles.)

Results. The *in situ* binding assay was used to examine GTPγS incorporation in wild-type and *pob* mutant larvae (Fig. 1). The *pob* mutant serves as a good control for red cone-specific incorporation because red cones selectively degenerate at 4 days postfertilization in this mutant.[3] Five wild-type and 6 *pob* larvae were subjected to the procedure described above, using a flash of 640-nm red light (∼300 μW/cm^2) or white light (∼11.5 mW/cm^2) for 1 sec followed by a 1-min incubation at room temperature in the dark. In response to red light, wild-type larvae incorporated radioactivity, whereas the *pob* larvae incorporated only background levels (Fig. 1B). Both wild-type and *pob* larvae incorporated radioactivity in response to white light, although *pob* incorporated less, most likely due to the absence of red cones (Fig. 1B).

These results were quantified by liquid scintillation counting (Fig. 1C). The total incorporation for five wild-type larvae was approximately 1100 cpm in the white light reaction and 150 cpm in the dark reaction. The values used for the 6 *pob* larvae in Fig. 1C were normalized to the five wild-type larvae by multiplying by 0.83. The background incorporation (dark reaction) was subtracted from both the red and white light reactions. Wild-type larvae incorporated nearly one-third of the counts in red light compared with white light. *pob* larvae incorporate only background levels

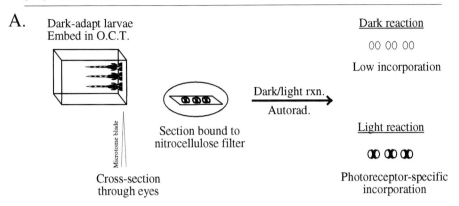

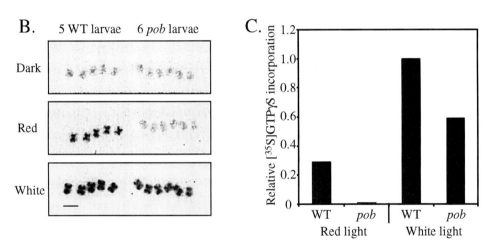

FIG. 1. [^{35}S]GTPγS *in situ* binding assay. (A) Diagram of the procedure as described in text. (B) Autoradiographs of sections containing 5 wild-type (WT) and 6 *pob* mutant larvae after dark, red light and white light reactions. Bar: 1 mm. (C) Quantification of incorporated radioactivity from the sections. After autoradiography, the sections were excised and counted on a liquid scintillation counter. The values from a single experiment were normalized to the wild-type larvae in white light and the background incorporation was subtracted from both the red and white light reactions.

of radioactivity in red light, but a significant amount in white light (about 60% of wild-type). The incorporation differences between wild-type and *pob* larvae appear to be due exclusively to the absence of red cones.

[^{35}S]GTPγS in Vitro Binding Assay

Principle. This assay is based on a method reported by Fung et al.[14] It is based on the same basic principles as the *in situ* binding assay described above except that it uses eye homogenates instead of eye sections. The reactions are performed directly in 1.5-ml centrifuge tubes before applying the homogenate to nitrocellulose filters. This assay is much easier to perform than the *in situ* binding assay; however, it requires a significantly larger number of larvae per assay (50–100 eyes per assay). In addition, a potential problem is that signaling components from different cell types may interact when brought together on homogenization.

Materials

GTPγS reaction buffer (2×): 1× ROS buffer, 200 μM GDP, 4 mM NADPH, 200 nM GTPγS, [^{35}S]GTPγS (~1 μCi/reaction)
Nitrocellulose membrane filters
Filter support connected to a vacuum pump
Darkroom
Night vision goggles and infrared illumination source
Calibrated light source with shutter driver/timer
Liquid scintillation counter

Procedure

1. Homogenize dark-adapted eyes as described above.
2. Add an equal volume of 2× GTPγS reaction buffer and mix thoroughly for 30 sec.
3. Initiate the reaction with a flash of light and allow the reaction to proceed in the dark at room temperature.
4. Stop the reaction by adding 1 ml of 1× ROS buffer, filter the homogenate through a nitrocellulose membrane filter by applying vacuum, and wash five times (5 ml each) with 1× ROS buffer.
5. Dry the filters and count the bound radioactivity with a liquid scintillation counter.

(Steps 1–4 should be performed under infrared illumination using night vision goggles.)

---

[14] B. K. Fung, J. B. Hurley, and L. Stryer, *Proc. Natl. Acad. Sci. U.S.A.* **78,** 152 (1981).

Results. The results from a [^{35}S]GTPγS *in vitro* binding assay are shown in Fig. 2. Fifty dark-adapted wild-type larval eyes were collected, homogenized, and diluted to 0.5 eyes/μl in 100 μl of 1× ROS buffer as described above. The homogenate was aliquoted into 20 1.5-ml centrifuge tubes (5 μl/tube) and frozen at −70°. Individual tubes were thawed just prior to the addition of 5 μl of 2× reaction buffer (~1 μCi). The reactions were mixed thoroughly by pipetting for 30 sec and exposed to a 1.5-sec flash of 640-nm red light (~250 μW/cm^2) or white light (~10 mW/cm^2). After the flash, the reactions were allowed to proceed for 1 or 4 min in darkness.

The data used for Fig. 2 were normalized from three independent assays performed in duplicate after subtracting the background incorporation (dark reactions). The average and standard deviation was 700 ± 76 pmol of GTPγS bound per microgram of protein in the white light reactions at 4 min, 161 ± 54 pmol of GTPγS bound per microgram protein in the dark reactions at 4 min, and 85 ± 24 pmol of GTPγS bound per microgram protein in the dark reactions at 1 min. GTPγS incorporation had reached an apparent maximum value by 1 min in red light, whereas incorporation had not reached maximum by 1 min in white light. The results from this

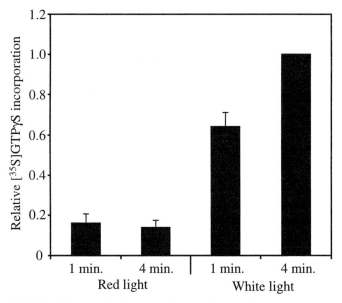

FIG. 2. [^{35}S]GTPγS *in vitro* binding assay. Transducin activation was analyzed with dark-adapted eye homogenates from wild-type zebrafish larvae. One- and 4-min reactions (2.5 eyes/10-μl reaction) were initiated with a flash of red or white light as described in text. The data were normalized to the 4-min white light reactions from three independent reactions that were performed in duplicate. The error bars are the standard deviations of the means.

assay are comparable to the results of the $[^{35}S]GTP\gamma S$ *in situ* binding assay (Fig. 1C), where the incorporation after 1 min in red light is approximately 30% of the value obtained after 1 min in white light (Fig. 2).

Phosphodiesterase Assay

Principle. Phosphodiesterase (PDE) activation is analyzed by using a 3′,5′-[8-^{3}H]cGMP-based assay developed for bovine rod outer segment and mouse retinal homogenates.[15,16] Briefly, 3′,5′-[8-^{3}H]cGMP is added to dark-adapted eye homogenates and exposed to various light intensities at several time points. Reactions are boiled to inactivate PDE, treated with snake venom to convert the PDE-catalyzed product 5′-[8-^{3}H]GMP to [8-^{3}H]guanosine, and run through DEAE-Sephadex columns (3′,5′-[8-^{3}H]cGMP binds to the column, whereas [8-^{3}H]guanosine does not). The eluate from the column is counted on a liquid scintillation counter. Ideally, 5′-[^{14}C]GMP would be used as an internal control to monitor recovery. However, owing to the limited availability of this compound, it was not used in these assays.

Materials

PDE preincubation buffer (4×): 1× ROS buffer, 400 mM ATP, 400 mM GTP, 8 mM NADPH
PDE reaction buffer (4×): 1× ROS buffer, 800 μM cGMP, 3′,5′-[8-^{3}H]cGMP (~250,000 cpm/assay)
Snake venom (*Ophiophagus hannah*), 1 mg/ml in distilled H$_2$O: Sigma
Heating block or water bath, 100°
DEAE-Sephadex, A-50: Pharmacia (Piscataway, NJ)
Column regeneration buffer: 20 mM Tris (pH 8.0), 0.5 M NaCl
Darkroom
Night vision goggles and infrared illumination source
Calibrated light source with shutter driver/timer
Liquid scintillation counter

Procedure

1. Prepare DEAE-Sephadex according to manufacturer instructions and pour several columns, each containing 0.5–1.0 ml of resin in distilled H$_2$O.
2. Homogenize dark-adapted zebrafish larval eyes as described above.

---

[15] J. B. Hurley, and L. Stryer, *J. Biol. Chem.* **257,** 11094 (1982).
[16] C. J. Raport, J. Lem, C. Makino, C. K. Chen, C. L. Fitch, A. Hobson, D. Baylor, M. I. Simon, and J. B. Hurley, *Invest. Ophthalmol. Vis. Sci.* **35,** 2932 (1994).

3. Add the appropriate volume of 4× preincubation buffer, mix thoroughly, and incubate for 2.5 min at room temperature.

4. Add the appropriate volume of 4× reaction buffer and mix thoroughly.

5. Initiate the reaction with a flash of light and allow the reaction to proceed at room temperature.

6. Stop the reaction by heating to 100° for 5 min, then place the tubes on ice.

7. Add an equal volume of snake venom (1 mg/ml) to the sample and incubate for 20 min at 30°.

8. Load the sample onto DEAE-Sephadex columns, elute with 3–4 ml of distilled H_2O into scintillation vials, add a sufficient volume of scintillation fluid, and count radioactivity with a liquid scintillation counter.

9. Regenerate the columns by washing three times (5 ml each) with column regeneration buffer followed by washing three times (5 ml each) with distilled H_2O. Columns can be stored in distilled H_2O at 4° for several months.

(Steps 2-6 should be performed under infrared illumination using night vision goggles.)

Results. The light sensitivity and the time course of activation were determined by the PDE assay (Fig. 3). To determine the light sensitivity, 50 dark-adapted larval eyes were collected and homogenized to 1 eye/μl in 50 μl of 1× ROS buffer, aliquoted into six 1.5-ml tubes (6 μl/tube), and stored on ice until needed. The homogenate was preincubated with 3 μl of 4× PDE preincubation buffer for 2.5 min at room temperature. Three microliters of 4× PDE reaction buffer (~100,000 cpm/μl) was added to the homogenate and mixed thoroughly for 30 sec. The reactions were initiated with a flash of white light at varying intensities (a flash intensity of −2 is equivalent to a 1-sec flash of ~10 mW/cm^2). The reactions were stopped after a 5-min incubation at room temperature by transferring two 5-μl aliquots into 1.5-ml tubes and heating to 100° for 5 min.

The graph shown in Fig. 3A uses data from one experiment, with each point representing the average from duplicate reactions. The values were normalized to the brightest flash intensity after subtracting the background activity. Approximately 35% of the total [8-^{3}H]cGMP was hydrolyzed in the brightest light reactions and about 11% was hydrolyzed in the dark reactions.

To determine the time course of PDE activation, 50 dark-adapted larval eyes were collected and homogenized to 1 eye/μl in 50 μl of 1× ROS buffer and used immediately. The homogenate was preincubated with 25 μl of 4× PDE preincubation buffer for 2.5 min at room temperature.

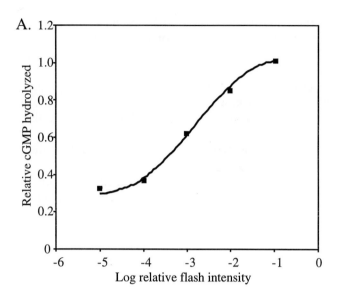

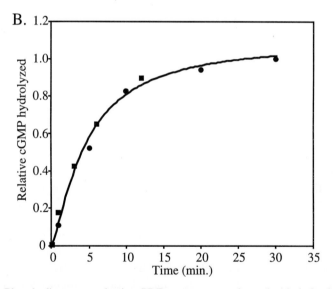

FIG. 3. Phosphodiesterase activation. PDE assays were performed with dark-adapted eye homogenates from wild-type zebrafish larvae. (A) Five-minute reactions were initiated with white light flashes of varying intensities as described in text. The values are normalized to the brightest flash intensity and represent the average of duplicate reactions after subtracting the background activity. (B) Reactions were initiated with a flash of light and aliquots were removed at several time points. The values were derived from two different experiments (solid circles and solid squares), where each time point represents the average of duplicate reactions after subtracting the background activity.

Twenty-five microliters of 4× PDE reaction buffer (~100,000 cpm/μl) was added to the homogenate and mixed thoroughly for 30 sec. At time zero, two 5-μl aliquots were transferred to 1.5-ml tubes and heated to 100° for 5 min to stop the reaction. The remainder (90 μl) was divided equally into two 1.5-ml tubes for the dark reactions and light reactions. The light reactions were initiated with a 1-sec flash of white light (~10 mW/cm^2) equivalent to the flash intensity of -2 in Fig. 3A. Two 5-μl aliquots were transferred to 1.5-ml tubes and heated to 100° for 5 min to stop the reactions at several time points for both the dark and light reactions.

The data used to generate Fig. 3B were from two different experiments (solid circles and solid squares), with each time point representing the average of duplicate reactions. The dark reaction values were subtracted from the light reaction values at the same time points and normalized to the 30-min time point. At 30 min, about 60% of the total 3′,5′-[8-^{3}H]cGMP was hydrolyzed in the light reactions and about 25% was hydrolyzed in the dark reactions. The dark reactions typically accounted for 10 to 15% of the ^{3}H counts per minute at time points less than 12 min.

Guanylyl Cyclase Assay

Principle. Calcium-sensitive guanylyl cyclase (GC) activity is measured by a modified method from Raport *et al.*[16] and Dizhoor *et al.*[17] Briefly, [α-^{32}P]GTP is added to dark-adapted eye homogenates in the presence of high and low Ca^{2+} concentrations and the reactions are allowed to proceed in darkness. All reactions are kept in darkness during this assay to avoid PDE activation. PDE inhibitors such as dipyridamol or zaprinast are used to reduce hydrolysis of the product of the cyclase reaction, cGMP. 3′,5′-[8-^{3}H]cGMP is also added to the reactions in order to monitor recovery of cGMP. The reactions are terminated by boiling, centrifuged to remove cellular debri, and developed on thin-layer chromatography (TLC) plates in 0.2 *M* LiCl to separate cGMP from GTP (cGMP migrates on the TLC plates, whereas GTP does not). Spots containing cGMP are visualized under UV light, excised, dried, and counted on a liquid scintillation counter. Initial amounts of ^{32}P and ^{3}H are determined from samples in which the homogenate was inactivated by boiling before incubation.

Materials

GC buffer (4×) 200 m*M* morpholinepropanesulfonic acid (MOPS, pH 7.4), 400 m*M* KCl, 32 m*M* NaCl, 40 m*M* MgCl$_2$, 2 m*M* ATP, 20

---

[17] A. M. Dizhoor, D. G. Lowe, E. V. Olshevskaya, R. P. Laura, and J. B. Hurley, *Neuron* **12**, 1345 (1994).

mM 2-mercaptoethanol, 20 μM dipyridamol, 40 μM CaCl$_2$ or 4 mM EGTA

Nucleotide mix (5×): 5 mM GTP, 25 mM 3′,5′-cGMP, [α-^{32}P]GTP (1 μCi/reaction), 3′,5′-[8-^{3}H]cGMP (0.1 μCi/reaction)

Polygram polyethyleneimine (PEI)–cellulose-coated thin-layer chromatography plates: EM Science (Gibbstown, NJ)

LiCl, 0.2 M

Heating block or water bath, 100°

Darkroom

Night vision goggles and infrared illumination source

Calibrated light source with shutter driver/timer

Liquid scintillation counter

Procedure

1. Homogenize dark-adapted zebrafish larval eyes as described above.
2. Add the appropriate volume of 4× GC buffer and mix thoroughly.
3. Add the appropriate volume of 5× nucleotide mix to initiate the reaction and incubate for 10 min at 30° in the dark.
4. Stop the reaction by heating to 100° for 2 min, and then place the tubes on ice.
5. Centrifuge the samples at 10,000g for 5 min at 4°.
6. Load 6 μl of the supernatant onto a TLC plate and develop in 0.2 M LiCl.
7. Visualize with UV light, excise and dry the spots containing cGMP, add scintillation fluid, and count the radioactivity in a liquid scintillation counter.

(Steps 1–4 should be performed under infrared illumination, using night vision goggles.)

Results. Ca^{2+}-sensitive guanylate cyclase activity was analyzed from zebrafish eye homogenates. Fifty to 100 dark-adapted larval eyes were collected and homogenized in 1× ROS buffer to a concentration of 1 eye/μl and used immediately. Five microliters of homogenate was added to 6.25 μl of 4× GC reaction buffer containing 40 μM CaCl$_2$ or 4 mM EGTA, and the volume was adjusted to 20 μl with distilled H$_2$O. The reactions were initiated with 5 μl of 5× nucleotide mix and allowed to proceed in the dark for 10 min. The values were adjusted for each assay by accounting for background GC and PDE activity from boiled homogenates. The data from several independent assays indicated that guanylyl cyclase activity was at least 3.5-fold higher in low Ca^{2+} (1 mM EGTA) than in high Ca^{2+} (10 μM CaCl$_2$) (data not shown). The fold stimulation varied significantly between assays, with a range of 3.5 to 30 in low Ca^{2+} (data not shown).

Rhodopsin and Cone Opsin Phosphorylation

Principle. The light-dependent incorporation of phosphate from $[\gamma\text{-}^{32}P]ATP$ is used to monitor phosphorylation of rhodopsin and cone opsins.[18] Briefly, $[\gamma\text{-}^{32}P]ATP$ is added to dark-adapted eye homogenates and exposed to red or white light. The reaction is terminated with the addition of a 6 M urea buffer that inactivates the kinase and phosphatase activities. Urea-treated bovine ROS (or the equivalent) is added as a carrier to help pellet the membranes that contain phosphorylated rhodopsin and cone opsins. Prior to centrifugation, the urea concentration is diluted with 1× ROS buffer to help pellet the membranes. After centrifugation, the supernatant containing unincorporated $[\gamma\text{-}^{32}P]ATP$ and soluble proteins is discarded, and the pellet is resuspended in 1× sodium dodecyl sulfate–polyacrylamide gel electrophoresis (SDS–PAGE) gel loading buffer. The samples are analyzed by SDS–PAGE, followed by autoradiography. To avoid forming aggregates of phosphorylated rhodopsin and cone opsins, do not boil or heat the samples prior to SDS–PAGE.

Materials

ATP reaction buffer (2×): 1× ROS, 40 μM ATP, $[\gamma\text{-}^{32}P]ATP$ (~5 μCi/reaction)
Stop buffer: 20 mM Tris (pH 7.5), 6 M urea, 5 mM EDTA
Bovine rod outer segments (or equivalent) for use as a carrier during centrifugation of membranes
SDS–PAGE gel loading buffer (reducing), 1×
SDS–polyacrylamide gel, 10% (w/v)
Darkroom
Night vision goggles and infrared illumination source
Calibrated light source with shutter driver/timer
Liquid scintillation counter

Procedure

1. Homogenize zebrafish larval eyes as described above.
2. Add an appropriate volume of 2× ATP reaction buffer and mix thoroughly for 30 sec.
3. Expose the sample to darkness, red light, or white light and incubate the reaction at room temperature.

---

[18] J. B. Hurley, M. Spencer, and G. A. Niemi, *Vision Res.* **38,** 1341 (1998).

4. Stop the reaction by adding an equal volume of stop buffer.

5. Add an appropriate quantity of bovine ROS (or equivalent) pretreated with stop buffer.

6. Dilute the sample with 1× ROS buffer to reduce the urea concentration below 50 mM.

7. Centrifuge the sample at 14,000g for 30 min at 4°.

8. Discard the supernatant and resuspend the pellet in 1× SDS–PAGE gel loading buffer. (Do not boil the sample!)

9. Run the samples on a 10% (w/v) SDS–polyacrylamide gel, dry the gel, and perform autoradiography overnight.

(Steps 1–4 should be performed under infrared illumination, using night vision goggles.)

Results. The red and white light-stimulated phosphorylation of rhodopsin and cone opsins was examined (Fig. 4). Fifty dark-adapted larval eyes were collected, homogenized at 0.2 eyes/μl in 1× ROS buffer, and stored as 5-μl aliquots in 1.5-ml centrifuge tubes at $-70°$. Individual tubes were thawed immediately prior to the addition of 5 μl of 2× ATP reaction buffer ($\sim$5 μCi). The samples were mixed thoroughly by pipetting for 30 sec and exposed to darkness, continuous 640-nm red light ($\sim$300 μW/cm^2), or white light ($\sim$11.5 mW/cm^2) for 1 min. After terminating the reactions with 10 μl of stop buffer, 10 μl of bovine ROS ($\sim$10 μM rhodopsin) pretreated with stop buffer and 1 ml of 1× ROS buffer were added. Samples were centrifuged at 14,000g for 30 min at 4°, the supernatant was discarded, and the pellet was resuspended in 10 μl of 1× SDS–PAGE gel loading buffer. Finally, the samples were subjected to 10% (w/v) SDS–PAGE followed by autoradiography for 16 hr.

The results from this experiment are shown in Fig. 4. Phosphorylated rhodopsin and cone opsin bands are clearly visible between 34 and 40 kDa in the red and white light reactions. No light-dependent phosphorylation was observed in the dark reaction. Preliminary experiments indicate that the prominent band in the red light-stimulated lane is significantly reduced in the *pob* mutant, suggesting that a significant fraction of the protein in this band is phosphorylated red opsin (data not shown).

Analysis of Retinoids

Principle. The distribution of retinoids is analyzed by a method modified from Saari *et al.*[19] and in [19] in this volume.[20] In this assay, the eyes are dissected under infrared illumination and immediately frozen in 1.5-ml

---

[19] J. C. Saari, G. G. Garwin, J. P. Van Hooser, and K. Palczewski, *Vision Res.* **38,** 1325 (1998).
[20] G. G. Garwin and J. C. Saari, *Methods Enzymol.* **316,** Chap 19, 2000 (this volume).

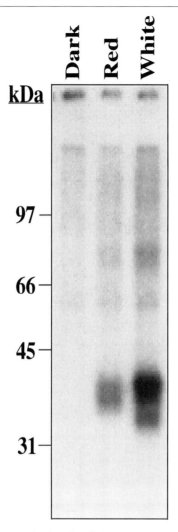

FIG. 4. Rhodopsin and cone opsin phosphorylation. Phosphorylation assays were performed with dark-adapted eye homogenates from wild-type zebrafish larvae. Samples were exposed to darkness, red light (640 nm), or white light for 1 min in the presence of [γ-^{32}P]ATP as described in text. The samples were subjected to 10% (w/v) SDS–PAGE followed by autoradiography.

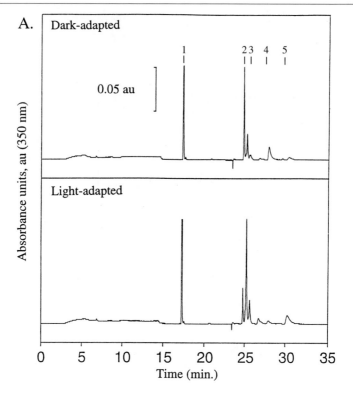
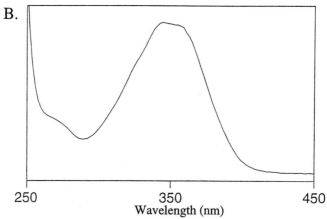

tubes. Prior to analysis, the tissue was thawed in the presence of 50% (v/v) ethanol, hydroxylamine (NH_2OH), and SDS to ensure complete extraction of retinals from the visual pigments. all-*trans*-[^3H]Retinol is added to allow the extraction yields to be determined for each reaction. The eyes are homogenized and incubated at room temperature to derivatize the retinals. The resulting retinal oximes are extracted in hexane and resolved on a silica HPLC column employing a hexane mobile phase and using ethyl acetate as a solvent polarity modifier.

Materials

NH_2OH buffer: 0.1 M MOPS (pH 6.5), 0.2% (w/v) SDS, 10 mM NH_2OH, and 0.2 pmol of all-*trans*-[^3H] retinol (10,000 dpm/reaction)
Duall glass homogenizer
Ethanol, hexane, ethyl acetate
HPLC and a silica HPLC column (Supelcosil LC-Si, 150 × 4.6 mm, 3-μm particle size): Supelco (Bellefonte, PA)
Darkroom
Night vision goggles and infrared illumination source
Liquid scintillation counter

Procedure

1. Dissect eyes from dark- or light-adapted zebrafish and immediately freeze in 1.5-ml plastic tubes immersed in dry ice–ethanol.
2. Transfer the eyes into a Duall glass homogenizer containing ethanol on dry ice just prior to analysis.
3. Add an equal volume of NH_2OH buffer, homogenize the sample, and incubate at room temperature for 30 min.
4. Add 4 volumes of hexane and mix well.
5. Centrifuge (in the homogenizer tube) at 900g for 2 min at 4°.
6. Remove the upper phase, add an additional 4 volumes of hexane, and repeat the extraction.

FIG. 5. Analysis of retinoids. Retinoids from 10 dark-adapted and 10 light-adapted adult zebrafish eyes were derivatized, extracted, and analyzed by HPLC as described in text. (A) Typical chromatographic traces of retinoids extracted from dark-adapted (top trace) and light-adapted (bottom trace) fish. The identity of the resolved components (determined using known retinoid standards) are as follows: 1, retinyl esters (truncated peak); 2, *syn*-11-*cis*-retinal oxime; 3, *syn*-all-*trans*-retinal oxime; 4, *anti*-11-*cis*-retinal oxime; and 5, *anti*-all-*trans*-retinal oxime. (B) The absorbance spectrum was measured as 11-*cis*-retinal oxime eluted from the column, indicating that the chromphore used by zebrafish is retinal$_1$.

7. Dry the extract under flowing argon, dissolve the residue in an appropriate volume of hexane, and remove an aliquot for liquid scintillation counting.
8. Equilibrate the silica HPLC column with hexane.
9. Resolve the retinal oximes at a flow rate of 1 ml/min in hexane–ethyl acetate at the following ratios: 100:0, 0–10 min; 99.5:0.5, 10–20 min; 90:10, 20–25 min; and 100:0, 25–35 min. Monitor the absorbance of the effluent at 350 nm.

(Steps 1–3 should be performed under infrared illumination, using night vision goggles; steps 3–9 should be performed under dim red light.)

Results. This assay was used to examine the distribution of 11-*cis*-retinal and all-*trans*-retinal in dark- and light-adapted zebrafish. Ten eyes were dissected from adult zebrafish that had been in complete darkness overnight or under bright room light for at least 30 min. The dark-adapted samples were not exposed to light at any time during the experiment. Once dissected, the eyes were immediately transferred into 1.5 ml of ethanol on dry ice or stored frozen until analyzed. An equal volume of NH_2OH buffer was added and then the eyes were homogenized in a Duall glass homogenizer. After a 30-min incubation at room temperature, the derivatized retinals were extracted two times with 12 ml of hexane. After drying under flowing argon, the residue was dissolved in 0.5 ml of hexane and a 50-μl aliquot was removed for liquid scintillation counting. The polar retinoids were resolved with a silica HPLC column at a flow rate of 1 ml/min.

Chromatographic traces of the retinoid separations are shown in Fig. 5A. 11-*cis*-Retinal (detected as the *syn-* and *anti*-retinal oximes) was the primary retinal observed in the dark-adapted samples (top trace); and all-*trans*-retinal was the primary retinal observed in the light-adapted samples (bottom trace). Retinyl esters were present in both dark- and light-adapted samples. Approximately 200 larval eyes were also subjected to the procedure described above. Although the absorbance peaks of the retinoids were only ~10% of those produced by the adults, the ratio of 11-*cis*-retinal to all-*trans*-retinal from dark- and light-adapted larvae was comparable to that generated by the adults (data not shown).

The absorbance spectrum of *syn*-11-*cis*-retinal oxime from adult zebrafish was measured as it eluted from the column to determine which retinal is used as the visual pigment chromophore. Two retinals, derived from either vitamin A_1 (retinal$_1$) or vitamin A_2 (retinal$_2$), are known to be used to form visual pigments. Typically, animals that live in fresh water such as fish and some amphibians use retinal$_2$, whereas marine fish and land animals use retinal$_1$.[21] The two retinals can be distinguished by their absorbance spectra (retinal$_1$ oxime absorbs maximally at 350 nm; retinal$_2$ oxime absorbs maximally at 375 nm).[22] As shown in Fig. 5B, the maximum absorbance peak of *syn*-11-*cis*-retinal oxime was located at ~350 nm, indicating that the chromophore used by zebrafish is retinal$_1$. Similar results were obtained with zebrafish larvae, suggesting that they also use retinal$_1$ (data not shown).

Discussion

This chapter demonstrates the feasibility of using biochemical methods to analyze phototransduction and the visual cycle in zebrafish. By identifying and characterizing mutants with defects in these pathways, candidate genes may be identified that will facilitate the cloning of the genes responsible for the vision defects. Novel proteins that are associated with these pathways may also be uncovered in this analysis. The genes for these new proteins will then be cloned, using the molecular and genetic techniques that are available for zebrafish. The ability to combine genetic approaches with biochemical analysis makes zebrafish indispensable for studying molecular mechanisms of vertebrate vision.

Acknowledgments

We are especially thankful to Glenda Froelick for sectioning the larvae in the [^{35}S]GTPγS *in situ* binding assay, to members of the Hurley laboratory for sharing their expertise, and to Gregory Garwin for assistance with retinoid analysis. This work was supported by PHS National Research Grant T32 GM07270 (M.R.T.), by the Howard Hughes Medical Institute (J.B.H.), and by National Institute of Health Grants EY06641 (J.B.H.) and EY12373 (S.E.B.).

---

[21] J. E. Dowling, "The Retina: An Approachable Part of the Brain." Harvard University Press, Cambridge, Massachusetts, 1987.
[22] R. Hubbard, P. K. Brown, and D. Bounds, *Methods Enzymol.* **XVIIIC,** 615 (1971).

[37] Genetic Models to Study Guanylyl Cyclase Function

By SUSAN W. ROBINSON and DAVID L. GARBERS

Introduction

The process of vision is initiated by light activation of the visual pigment and a subsequent decrease in cGMP levels, leading to closure of cGMP-gated ion channels and a decrease in intracellular calcium (reviewed in Refs. 1 and 2). It has been suggested that subsequent to this decrease in calcium concentration, an activation of retinal guanylyl cyclases by guanylyl cyclase-activating proteins (GCAP1 and GCAP2)[3,4] occurs, leading to, at least in part, a restoration of the cGMP-gated ion channel to its open or "dark" state. Two retina-specific guanylyl cyclases have been identified in mammals. In the human they are designated retGC1 and retGC2[5,6]; in the rat the homologs are referred to as GC-E and GC-F.[7] These cyclases are members of the family of membrane guanylyl cyclase receptors. The members of this family all contain an extracellular, putative ligand-binding domain, a single membrane-spanning segment, and intracellular protein kinase-like and cyclase catalytic domains.[8] With respect to the putative ligand-binding domain, these cyclases remain orphan receptors.

In the mouse, the gene for GC-E maps to chromosome 11 and the GC-F gene is localized on the X chromosome.[9] To date, no visual diseases have been mapped to Xq22, the region of the X chromosome where GC-F is located in humans. However, several retinal degenerations have been mapped near 17p13.1, the GC-E locus in humans. Missense and frameshift mutations in GC-E have been found in families with Leber

---

[1] S. Yarfitz and J. B. Hurley, *J. Biol. Chem.* **269,** 14329 (1994).
[2] K.-W. Yau, *Invest. Ophthalmol. Vis. Sci.* **35,** 9 (1994).
[3] K. Palczewski, I. Subbaraya, W. A. Gorczyca, B. S. Helekar, C. C. Ruiz, H. Ohguro, J. Huang, X. Zhao, J. W. Crabb, R. S. Johnson, K. A. Walsh, M. P. Gray-Keller, P. B. Detwiler, and W. Baehr, *Neuron* **13,** 395 (1994).
[4] A. M. Dizhoor, E. V. Olshevskaya, W. J. Henzel, S. C. Wong, J. T. Stults, I. Ankoudinova, and J. B. Hurley, *J. Biol. Chem.* **270,** 25200 (1995).
[5] A. W. Shyjan, F. J. de Sauvage, N. A. Gillett, D. V. Goeddel, and D. G. Lowe, *Neuron* **9,** 727 (1992).
[6] D. G. Lowe, A. M. Dizhoor, K. Liu, Q. Gu, M. Spencer, R. Laura, L. Lu, and J. B. Hurley, *Proc. Natl. Acad. Sci. U.S.A.* **92,** 5535 (1995).
[7] R.-B. Yang, D. C. Foster, D. L. Garbers, and H.-J. Fülle, *Proc. Natl. Acad. Sci. U.S.A.* **92,** 602 (1995).
[8] B. J. Wedel and D. L. Garbers, *Trends Endocrinol. Metab.* **9,** 213 (1998).
[9] R.-B. Yang, H.-J. Fülle, and D. L. Garbers, *Genomics* **31,** 367 (1996).

congenital amaurosis,[10] an autosomal recessive disease characterized by blindness at birth, a normal fundus, and extinguished or markedly reduced electroretinogram (ERG).[11] In addition, two mutations in GC-E have been identified in families with CORD6, a form of autosomal dominant cone–rod dystrophy.[12,13]

A classic approach to define the importance of a signaling pathway is to specifically block the pathway. However, specific inhibitors have not been available for most receptor guanylyl cyclases. In addition, in the retina the situation is further complicated by the presence of two closely related proteins that are expressed not only in the same tissue, but also within the same cell.[14] Thus, gene disruption appears to represent a particularly powerful method by which to define the functions of these guanylyl cyclases in vision and to delineate specific roles each cyclase might play. An advantage of this approach is the ability to identify possible developmental roles for the molecule, as well as defining its function *in vivo*. In addition, given the apparent involvement of GC-E, and perhaps also GC-F, in inherited retinal dystrophies these animals may represent useful models for the study and possible treatment of these diseases.

Methods

Generation of GC-E-Deficient Mice

The mouse GC-E gene has been cloned from a 129SV/J genomic library.[9] A neomycin resistance cassette has been introduced in exon 5. This disrupts the protein in the transmembrane domain and prevents expression of the catalytic domain, which is required for cGMP synthesis. The targeting vector has been introduced into mouse 129SV/J embryonic stem cells (SM1, from R. Hammer, University of Texas Southwestern Medical Center, Dallas, TX) and GC-E-deficient mice have been generated by standard techniques.[15] A similar targeting vector for the disruption of the GC-F gene has also been constructed (S. W. R. and D. L. G., unpublished data, 1999).

---

[10] I. Perrault, J.-M. Rozet, P. Calvas, S. Gerber, A. Camuzat, H. Dollfus, S. Châtelin, E. Souied, I. Ghazi, C. Leowski, M. Bonnemaison, D. Le Paslier, J. Frézal, J.-L. Dufier, S. Pittler, A. Munnich, and J. Kaplan, *Nature Genet.* **14**, 461 (1996).

[11] K. Mizuno, Y. Takei, M. L. Sears, W. S. Peterson, R. E. Carr, and L. M. Jampol, *Am. J. Ophthalmol.* **83**, 32 (1977).

[12] R. E. Kelsell, K. Gregory-Evans, A. M. Payne, I. Perrault, J. Kaplan, R.-B. Yang, D. L. Garbers, A. C. Bird, A. Y. Moore, and D. M. Hunt, *Hum. Mol. Genet.* **7**, 1179 (1998).

[13] I. Perrault, J.-M. Rozet, S. Gerber, R. E. Kelsell, E. Souied, A. Cabot, D. M. Hunt, A. Munnich, and J. Kaplan, *Am. J. Hum. Genet.* **63**, 651 (1998).

[14] R.-B. Yang and D. L. Garbers, *J. Biol. Chem.* **272**, 13738 (1997).

[15] R. Ramirez-Solis, A. C. Davis, and A. Bradley, *Methods Enzymol.* **225**, 855 (1993).

Histological Analysis

Mice are sacrificed by carbon dioxide asphyxiation and the eyes are collected. Eyes are typically fixed overnight in 10% (v/v) formalin at 4° and embedded in paraffin. Sections (3 μm thick) are cut through the optic nerve head. Sections are stained with hematoxylin and eosin to identify cell structure. Cone photoreceptors are identified on the basis of their nuclear morphology. Cone cell bodies characteristically have an oval shape with several clumps of chromatin and a large amount of lightly staining euchromatin, while rod nuclei are smaller and round with a single clump of chromatin and little euchromatin.[16]

Electroretinogram Recordings

Full-field corneal ERGs are recorded in a Ganzfeld dome from mice that are dark adapted for at least 12 hr. All manipulations are carried out under dim red light. Mice are anesthetized with ketamine (200 mg/kg) and xylazine (10 mg/kg) and pupils are dilated with cyclopentolate hydrochloride drops. A gold wire coil is placed on the cornea and referenced to a similar wire in the mouth; a needle electrode in the tail serves as ground. Signals are amplified 10,000-fold with a Tektronix (Beaverton, OR) AM502 differential amplifier (3 dB down at 2 and 10,000 Hz), digitized (sampling rate, 1.25–5 kHz), and averaged on a personal computer. Short-wavelength flashes [Wratten (Eastman Kodak, Rochester, NY) 47A: λ_{max} 470 nm, half-bandwidth 55 nm] from -3.0 to -1.0 log scotopic troland-seconds (scot td · sec) in 0.3 log unit steps are produced by a Grass (Grass Instrument Company, Quincy, MA) photostimulator. High-intensity short-wavelength flashes (Wratten W47B: λ_{max} 449 nm, half-bandwidth 47 nm) from 1.0 to 3.4 log scot td · sec in 0.3 log unit steps are produced by a Novatron (Dallas, TX) flash unit. Double flashes are produced by two Novatron flash units, with a 1.5 log scot td · sec test flash followed by a 3.4 log scot td · sec probe flash.

Single-Cell Electrophysiology

For recording from isolated rod photoreceptors, mice are dark adapted overnight and sacrificed under dim red light. All subsequent manipulations are carried out under infrared light. The retina is isolated as described by Sung et al.[17] Briefly, the retina is isolated in chilled oxygenated Leibovitz L-15 medium (GIBCO-BRL, Gaithersburg, MD) and placed on a glass capillary array (10-μm-diameter capillaries; Galileo Electro-Optics, San

---

[16] L. D. Carter-Dawson and M. M. LaVail, *J. Comp. Neurol.* **188,** 245 (1979).
[17] C.-H. Sung, C. Makino, D. Baylor, and J. Nathans, *J. Neurosci.* **14,** 5818 (1994).

Jose, CA) with the photoreceptors up. When needed, a piece of retina is chopped under L-15 medium containing DNase I (8 µg/ml; Sigma, St. Louis, MO) and a suspension of fragments is transferred to the recording chamber. The outer segment of a single rod is drawn into a suction electrode connected to a current-to-voltage converter. The electrode is filled with a solution containing 134.5 mM Na$^+$, 3.6 mM K$^+$, 2.4 mM Mg^{2+}, 1.2 mM Ca^{2+}, 136.3 mM Cl$^-$, 3 mM succinate, 3 mM L-glutamate, 10 mM glucose, 10 mM HEPES, 0.02 mM EDTA, Basal Media Eagle (BME) amino acid, and BME vitamin supplement. Membrane current is filtered with a low-pass eight-pole Bessel filter at 30 Hz and digitized. Unpolarized 8-msec flashes at 500 nm (10-nm bandwidth) are used for stimulation.

Discussion

Mice that contain a disrupted GC-E gene were obtained and the lack of GC-E expression in the retina was confirmed by Western blotting.[18] The development of the outer segment layer during the first 3 weeks postnatal appeared normal in GC-E null animals (data not shown). The overall retinal structure of these animals appears normal up to at least 18 months of age. In mice at 4 weeks of age, there is no difference in the number of cone photoreceptor nuclei between wild-type and knockout animals.[18] However, at 5 weeks of age, the number of identifiable cones is sharply reduced and at subsequent ages few cone cell nuclei are present (Fig. 1). These data suggest that GC-E is not required for the development of cones, as similar numbers are present in null and wild-type mice at 4 weeks of age. However, the presence of GC-E appears to be required for the survival of these cells. This raises the possibility that cGMP is an important survival factor for photoreceptor cells, and that the survival of rods in GC-E null mice is due to the presence of GC-F in these cells.[14]

ERGs were recorded from mice at various ages to determine how the absence of GC-E affects retinal function. The ERG is a reflection of the electrical activity of the retina and has two main components: the *a*-wave and the *b*-wave. The *a*-wave reflects the response of the photoreceptor cells to light and the *b*-wave is generated in the cells of the inner nuclear layer.[19] In the GC-E null mice, the ERG is markedly reduced as early as 4 weeks of age. Figure 2A shows representative ERGs from 2-month-old mice. Both the *a*- and *b*-waves are reduced in the GC-E null animal; however, the

---

[18] R.-B. Yang, S. W. Robinson, W.-H. Xiong, K.-W. Yau, D. G. Birch, and D. L. Garbers, *J. Neuroscience* **19,** 5889 (1999).

[19] E. L. Berson, in "Adler's Physiology of the Eye" (W. M. Hart, ed.), p. 641. Mosby-Year Book, St. Louis, Missouri, 1992.

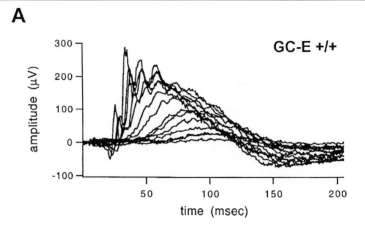

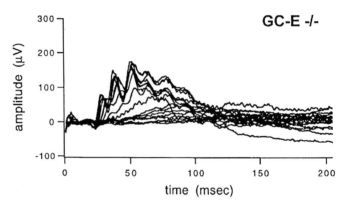

FIG. 2. ERGs from 2-month-old wild-type (GC-E +/+) and null (GC-E −/−) mice. (A) Responses to short-wavelength stimuli from −3 to 1 log scot td · sec in 0.3 log unit steps. The *a*-wave is the initial small negative response; the positive *b*-wave is the main component. (B) ERGs showing the *b*-wave response from a 2-month-old GC-E +/+ animal and from 1-month-, 4-month-, and 6-month-old GC-E −/− mice. *Left:* Rod responses to a −2 log scot td · sec flash. The response is reduced in 1-month-old null mice; however, rod responses continue to be detectable in older knockout mice and remain relatively stable. *Right:* Responses to a 1.44 log photopic td · sec flash in the presence of a rod-saturating (40 cd/m^2) background. This presumably represents only the cone response. The response is severely reduced in GC-E −/− animals at 1 month and is undetectable at later ages.

maximal amplitude of the rod response did not decrease significantly with age in null animals (Fig. 2B). In addition, the response to white flashes in the presence of a rod-saturating background, which presumably reflects cone activity, was barely detectable at 4 weeks in GC-E −/− mice and was

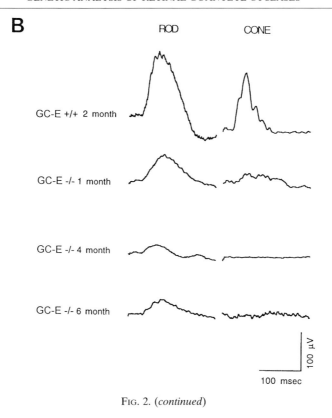

FIG. 2. (*continued*)

undetectable at later ages (Fig. 2). Double-flash experiments indicated that the recovery of the photoresponse from a test flash was substantially faster in GC-E knockout mice (205 ± 16 msec) than in wild-type animals (372 ± 32 msec). The ERG data are in agreement with the histological data, in that they reflect the disappearance of the cone photoreceptors and are consistent with the maintenance of the overall retinal structure in the knockout mice.

The effect of the absence of GC-E on rod function was examined by recording responses from isolated rods. The response to a saturating flash was comparable in wild-type and null animals. This is in contrast to the sharp reduction in the *a*-wave observed by ERG. Although the flash response rose with normal kinetics, the time-to-peak was increased, resulting in a higher sensitivity in GC-E knockout rods. In addition, the recovery of the flash response was faster in null than in wild-type animals, consistent with our observations in the double-flash ERG experiments. Thus, some discrepancies between ERG and single-cell recordings remain to be resolved in the GC-E null mice.

The use of gene targeting to study the function of retinal guanylyl cyclases has both confirmed previous studies of the role of this molecule in vision, and has provided important new insights. As expected, GC-E plays an important role in the retinal response to light. In GC-E –/– animals, the ERG, which reflects the activity of the photoreceptor and inner retinal neurons, is reduced. The remaining responsiveness to light may be due to the expression of the second retinal guanylyl cyclase, GC-F, in rods. It will be possible to answer this question when mice lacking GC-F are available.

Surprisingly, rather than a slowed recovery from a light flash, which might be expected in cells that are impaired in their ability to resynthesize cGMP, both ERG and single-rod recordings indicate that GC-E-deficient photoreceptors in fact have a speeded recovery process. This result suggests that rods in the GC-E null animals have undergone compensatory changes that have altered the kinetics of the photoresponse. Further studies will be important to pinpoint these changes and assess their significance.

The identification of a form of cone–rod dystrophy (CORD6) associated with mutations in GC-E[12,13] further confirms the importance of GC-E for cone survival. The lack of rod cell death in the mouse model may be a reflection of the small number of cones in the mouse retina. Their loss may not cause as severe a disruption of retinal structure as the loss of cones in humans or other species with a more cone-dominated retina. Despite this difference, the GC-E null mice represent a unique animal model for testing treatments of cone dystrophies, including gene therapy approaches.

Acknowledgments

We thank our collaborators in the analysis of the GC-E knockout mice: Dr. David G. Birch, electroretinograms; and Drs. Wei-hong Xiong and King-Wai Yau, rod single-cell electrophysiology.

[38] Analysis of Visual Cycle in Normal and Transgenic Mice

By J. Preston Van Hooser, Gregory G. Garwin, and John C. Saari

Introduction

Phototransduction and the visual cycle play complementary roles in mammalian vision. Phototransduction is initiated by the photoisomerization of 11-*cis*-retinal bound to opsin and ultimately results in a change in the release of neurotransmitter by photoreceptor cells. The visual cycle restores the product of photoisomerization, all-*trans*-retinal, to the 11-*cis* configuration and allows the regeneration of bleached visual pigments.[1,2] The biochemical mechanism of phototransduction has been extensively studied, and as a result, the process serves as a paradigm for understanding other G-protein-coupled receptors. In contrast, molecular understanding of the reactions of the visual cycle remains in a rudimentary state. Several factors are responsible for this: first, the enzymes involved are all membrane associated and in considerably lower abundance than those of phototransduction. Second, several of the reactions involve novel chemistries that have not been completely resolved. Third, partial reactions of the visual cycle occur in at least two different cell types: photoreceptor outer segments and retinal pigment epithelial (RPE) cells. Fourth, retinoids are virtually water insoluble, yet must diffuse through several membranes and aqueous compartments during their traverse of the visual cycle. Figure 1 depicts the current understanding of the reactions of the visual cycle in rod photoreceptors.[3–6] Both photoreceptor outer segments and RPE contain retinoid-binding components whose functions have not been defined, for example, RPE-65,[7–9]

---

[1] J. E. Dowling, *Nature* (*London*) **188,** 114 (1960).
[2] G. Wald, *Science* **162,** 230 (1968).
[3] D. Bok, *J. Cell Sci. Suppl.* **17,** 189 (1993).
[4] R. K. Crouch, G. J. Chader, B. Wiggert, and D. R. Pepperberg, *Photochem. Photobiol.* **64,** 613 (1996).
[5] R. R. Rando, *Chem. Biol.* **3,** 255 (1996).
[6] J. C. Saari, in "The Retinoids: Biology, Chemistry and Medicine," 2nd Ed. (M. B. Sporn, A. B. Roberts, and D. S. Goodman, eds.), p. 531. Raven Press, New York, 1994.
[7] C. P. Hamel, E. Tsilou, B. A. Pfeffer, J. J. Hooks, B. Detrick, and T. M. Redmond, *J. Biol. Chem.* **268,** 15751 (1993).
[8] C. O. Bavik, U. Eriksson, R. A. Allen, and P. A. Peterson, *J. Biol. Chem.* **266,** 14978 (1991).
[9] A. Nicoletti, D. J. Wong, K. Kawase, L. H. Gibson, T. L. Yang-Feng, J. E. Richards, and D. A. Thompson, *Hum. Mol. Genet.* **4,** 641 (1995).

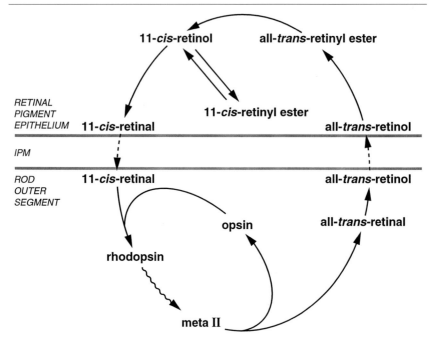

FIG. 1. Reactions of the visual cycle in rod photoreceptors. The double line denotes the extracellular compartment (IPM, interphotoreceptor matrix) separating the retinal pigment epithelium and the photoreceptor outer segment. Detailed discussions of the visual cycle are provided in several reviews.[3-6]

peropsin,[10] RGR (RPE-retinal G-protein-coupled receptor),[11] and ABCR (ATP-binding transporter.)[12] One or more of these components is likely to participate in the visual cycle in ways that will require alteration of current models.

We have established biochemical protocols for determining visual cycle function in normal mice and in mice bearing targeted disruptions of specific genes.[13,14] The levels of rhodopsin and of visual cycle retinoids are measured in dark-adapted animals and in animals recovering from a flash or from

[10] H. Sun, D. J. Gilbert, N. G. Copeland, N. A. Jenkins, and J. Nathans, *Proc. Natl. Acad. Sci. U.S.A.* **94,** 9893 (1997).
[11] D. Shen, M. Jiang, W. Hao, L. Tao, M. Salazar, and H. K. W. Fong, *Biochemistry* **33,** 13117 (1994).
[12] M. Illing, L. L. Molday, and R. S. Molday, *J. Biol. Chem.* **272,** 10303 (1997).
[13] K. Palczewski, J. P. Van Hooser, G. G. Garwin, J. Chen, G. I. Liou, and J. C. Saari, *Biochemistry* **38,** 12012 (1999).
[14] J. C. Saari, G. G. Garwin, J. P. Van Hooser, and K. Palczewski, *Vision Res.* **38,** 1325 (1998).

constant illumination. The flash subjects the resting visual cycle to a sudden pulse of substrate (all-*trans*-retinal), whereas constant illumination establishes a steady cycling state from which recovery to the dark-adapted state can be measured. In each protocol, the illumination conditions have been chosen to bleach approximately 40% of the visual pigment, thus avoiding the complete saturation of the cycle with all-*trans*-retinal, which occurs with prolonged, total bleaching. Additional information can be obtained by applying these bleaching parameters to excised, whole mouse eyes in which the visual cycle is blocked after bleaching or retinyl ester formation, depending on the type of illumination.[13]

Methods

General

All procedures are performed under dim, red illumination provided by a Kodak (Rochester, NY) darkroom lamp equipped with a Kodak No. 1 safelight filter. Mice are killed by cervical dislocation as approved by the American Veterinary Medical Association Panel on Euthanasia.[15] Animals are dark adapted for 3–6 hr, or overnight, before experimental protocols are carried out.

Housing of Animals

Mice bearing mutations that affect photoreceptor viability are born and raised in a 24-hr, dark environment. The animals are used for experiments at ~45 days of age, at which time adult levels of rhodopsin are achieved.[16]

Dissection of Eyes for Retinoid Analysis

Immediately after cervical dislocation, the left and right eyes are enucleated with a pair of curved, 140-mm Mayo dissecting scissors. The eyes are quickly rinsed with sterile H_2O and transferred to a 8 × 10 cm glass dissecting plate. A small incision is made at the anterior pole, using a sterile No. 11 surgical blade. The lens is gently removed through the incision by placing a small amount of pressure on the posterior part of the eye, using a pair of fine dissecting forceps. The remaining part of the eye is immediately transferred to a sterile, 1.5-ml polyethylene centrifuge tube immersed in a

---

[15] E. J. Andrews, B. T. Bennett, J. D. Clark, K. A. Houpt, P. J. Pascoe, G. W. Robinson, and J. R. Boyce, *J. Am. Vet. Med. Assoc.* **202**, 229 (1993).

[16] L. Carter-Dawson, R. A. Alvarez, S.-L. Fong, G. I. Liou, H.G. Sperling, and C. D. B. Bridges, *Dev. Biol.* **116**, 431 (1986).

dry ice–ethanol bath and stored at $-80°$ until analyzed. The process requires ~30 sec per eye.

Extraction of Retinoids and High-Performance Liquid Chromatography Analysis

These procedures are described in detail in [19] this volume.[17] Four mouse eyes are used for each high-performance liquid chromatography (HPLC) analysis, although two eyes produce a reliable signal.

Rhodopsin Determination

Mouse eyes (four per analysis) are dissected and rinsed, and the lenses are removed as described above. The eyes are then cut into several small pieces, which are collected in a sterile, 1.5-ml polyethylene centrifuge tube immersed in a dry ice–ethanol bath and stored at $-80°$ until analyzed. For rhodopsin analysis, 0.9 ml of 20 mM Bis–Tris propane {1,3-bis[tris(hydroxymethyl)amino]propane, or BTP, pH 7.5} containing 10 mM dodecyl-β-maltoside and 5 mM NH_2OH (freshly neutralized $NH_2OH \cdot HCl$) is added, and the sample is disrupted with 10 passes of a Dounce tissue homogenizer and shaken for 5 min at room temperature [Eppendorf (Hamburg, Germany) mixer]. The sample is then centrifuged at 14,000 rpm for 5 min (Eppendorf centrifuge) at room temperature. The supernatant is reserved and the pellet extracted with the preceding detergent a second time. The combined supernatants are centrifuged at 50,000 rpm for 10 min (Optima TLX centrifuge, TLA-45 rotor; Beckman Instruments, Palo Alto, CA), and absorption spectra are obtained before and after complete bleaching with a 60-W incandescent bulb. The concentration of rhodopsin is determined by the decrease in absorption at 500 nm and the molar extinction of rhodopsin ($\varepsilon = 42,000$ liter/mol/cm).

Flash Protocol: Living Mice

Hand-held, dark-adapted mice are individually subjected to a single flash from a photographic flash unit (SunPak Thyristor 433D; SunPak Division, Hackensack, NJ) set at maximum power (2.5-msec flash duration; distance from eyes, 3 cm). Animals are sacrificed for retinoid analysis or rhodopsin determination before the flash (dark adapted); immediately after the flash ($t = 0$); and 15, 30, 45, and 60 min after the flash.

---

[17] G. G. Garwin and J. C. Saari, *Methods Enzymol.* **316,** Chap. 19, 2000 (this volume).

Flash Protocol: Excised Eyes

Dark-adapted mouse eyes (two eyes per flash, on a 10 × 30 cm glass plate) are enucleated, rinsed, and subjected to a single flash from a photographic flash unit (see description above). Eyes are dissected before the flash (dark adapted); immediately after the flash ($t = 0$); and 15, 30, 45, and 60 min after the flash.

Constant Illumination Protocol: Living Mice

Mice are placed in a cage 60 cm beneath two 60-W fluorescent bulbs (54 ft-cd) for 45 min. After this period of illumination, which is sufficient to establish a steady state with ~40% of the visual pigment bleached, the animals are returned to a dark environment to allow regeneration of visual pigment. Animals are sacrificed after 0, 15, 30, 45, or 60 min in the dark, and the eyes are used for retinoid analysis or rhodopsin determination.

Constant Illumination Protocol: Excised Eyes

Eyes are enucleated from dark-adapted mice immediately after cervical dislocation as described above and placed in a six-well, flat-bottom culture dish (four eyes per well) filled with 1 ml of 10 mM HEPES, pH 7.5, containing 120 mM NaCl, 3.5 mM KCl, 10 mM glucose, 0.2 mM CaCl$_2$, 0.2 mM MgCl$_2$, and 1 mM EDTA. The culture dishes are obtained from Falcon (Becton Dickinson, Franklin Lakes, NJ). After positioning the eyes in a cornea-up attitude, the eyes are illuminated with a 60-W incandescent bulb, (30-cm distance; 54 ft-cd) at room temperature. Eyes are dissected under red illumination after 0, 15, 30, 45, and 60 min of illumination and processed as described above for either retinoid analysis or rhodopsin determination.

Preparation of Photoreceptor Membranes

Mice are sacrificed by cervical dislocation and the eyes are dissected and rinsed as described above. The muscles, optic nerve, and lens are removed and the retinas are dissected, collected in 1.5-ml polyethylene centrifuge tubes containing 0.5 ml 67 mM potassium phosphate (pH 7.0) and 5% (w/v) sucrose, and stored frozen at −80°. Frozen retinas were thawed after the addition of 1.5 ml of the preceding buffer (10 retinas/ ml), transferred to a 3-ml glass/glass homogenizer, and disrupted. The homogenate is transferred to a 1.5-ml centrifuge tube and agitated for 15 min at room temperature. The homogenate is layered on top of a discontinuous gradient of sucrose[18] and centrifuged at 13,000 rpm for 150

---

[18] D. S. Papermaster, *Methods Enzymol.* **81H**, 48 (1982).

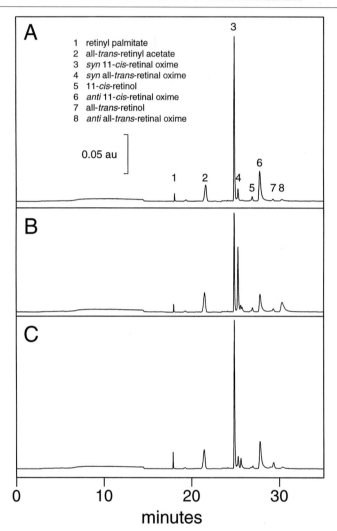

FIG. 2. HPLC analysis of retinoids in the mouse retina. (A) Retinas from dark-adapted mice. The major components are 11-*cis*-retinal oximes (*syn* and *anti* isomers). (B) Retina from flash-bleached mice. The major components are 11-*cis*- and all-*trans*-retinal oximes (*syn* and *anti* isomers). (C) Retinas from flash-bleached mice after 60 min in the dark. The major components are 11-*cis*-retinal oximes. The absorption was monitored at 350 nm, which emphasizes the retinal oximes relative to the retinols and retinyl esters.

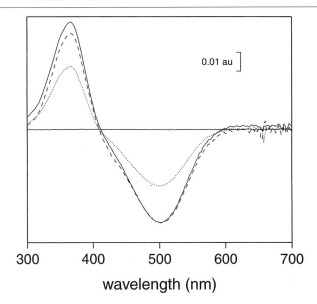

FIG. 3. Rhodopsin analysis. Detergent extracts of mouse eyes were prepared in the dark and spectra were obtained before and after bleaching. Difference curves were obtained by subtracting the nonbleached spectrum from the bleached spectrum. Solid line: extract prepared from dark-adapted mouse eyes. Dotted line: extract prepared from flash-bleached mouse eyes. Dashed line: extract prepared from bleached mouse eyes after 60 min in the dark. The dark-adapted rhodopsin content of these mice was 549 pmol/eye.

min at 4° in a swinging bucket rotor (Beckman JS 13.1 rotor). Photoreceptor membranes at the interface of the 1.11- and 1.15-g/ml sucrose layers are removed with a Pasteur pipette, collected in a 25-ml graduated cylinder, diluted 1:1 (v:v) with 67 mM potassium phosphate, pH 7.0, and centrifuged in 1.5-ml centrifuge tubes (Beckman Microfuge tubes) at 40,000 rpm for 15 min at 4° (Beckman Optima TLX centrifuge, TLA-45 rotor). The combined supernatant from the membrane suspension is decanted, and the membranes are stored at −80° until use. Membranes are subjected to sodium dodecyl sulfate–polyacrylamide gel electrophoresis (SDS–PAGE)[19] and analyzed by conventional staining procedures or by Western analysis with specific antibodies.

Discussion

The generation of transgenic mice bearing targeted disruptions of specific genes has become relatively routine. In ideal cases, this technique

[19] U. K. Laemmli, *Nature (London)* **227,** 680 (1970).

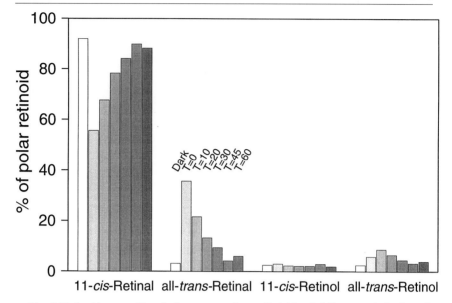

FIG. 4. Retinoid composition during recovery from a flash bleach. Mice were dark adapted, subjected to a flash that bleached ~40% of their visual pigment, and maintained in the dark. Samples for retinoid analyses were obtained before the flash (dark-adapted), and 0, 10, 20, 30, 45, and 60 min after the flash. The results are presented as a percentage of polar retinoids. No major changes in retinyl esters were observed in normal mice subjected to this experimental bleaching paradigm. [Modified from reference 13, with permission of the American Chemical Society.]

allows remarkable insight into the biological function of specific gene products. We and others[13,14,20] have analyzed visual cycle activity in mice with targeted disruptions of genes hypothesized to be involved in the visual cycle. To date, this technique has been applied only to retinoid-binding proteins and has considerably altered our understanding of their roles in the visual cycle. It is likely that targeted disruption of genes encoding enzymes involved in the visual cycle will prove to be equally instructive.

The methods described here for rhodopsin and retinoid analysis by HPLC are sufficiently sensitive to provide a reliable analysis with four mouse retinas (Figs. 2 and 3). Flash illumination of dark-adapted mice resulted in a decrease in 11-*cis*-retinal and rhodopsin and a concomitant increase in all-*trans*-retinal. During recovery in the dark, the rates of in-

[20] T. M. Redmond, S. Yu, E. Lee, D. Bok, D. Hamasaki, N. Chen, P. Goletz, J.-X. Ma, R. K. Crouch, and K. Pfeifer, *Nature Genet.* **20**, 344 (1998).

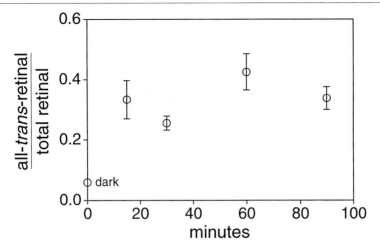

FIG. 5. Establishment of a steady state level of bleached visual pigment during constant illumination. Dark-adapted mice were subjected to constant illumination as described in the text. At the times indicated, eyes were processed for retinoid analysis. The results are presented as the ratio of all-*trans*-retinal to total retinal, a parameter proportional to the fractional bleach of rhodopsin. Means and standard deviations are shown.

crease in 11-*cis*-retinal and rhodopsin and of decrease in all-*trans*-retinal were approximately equal, and no other retinoid accumulated in major amounts[13,14] (Fig. 4). These results suggest that reduction of all-*trans*-retinal is slow and that all subsequent steps of the visual cycle, including transport, isomerization, and combination with opsin, are rapid in comparison. Constant illumination of mice established a steady state level of bleached visual pigment (Fig. 5) in which all-*trans*-retinal was the only retinoid that accumulated.[14] Recovery in the dark from a steady state bleach was more rapid than recovery from a flash ($t_{1/2}$ of 5 and 17 min, respectively[14,21]) for reasons that are not clear. Blocks in the mouse visual cycle were readily detected using these experimental paradigms. For instance, with excised mouse eyes, the visual cycle appeared to be blocked after all-*trans*-retinal following a flash, perhaps because of insufficient NADPH production by the pentose phosphate pathway. Continual bleaching of excised mouse eyes appeared to block the cycle after all-*trans*-retinyl ester (Fig. 1), perhaps because of ATP depletion.[13]

[21] H. Ohguro, J. P. Van Hooser, A. H. Milam, and K. Palczewski, *J. Biol. Chem.* **270**, 14259 (1995).

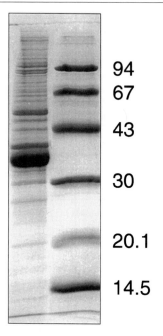

FIG. 6. Analysis of mouse photoreceptor outer segment membranes by SDS–PAGE. Left lane, membranes prepared from dark-adapted mice. The major band is opsin. Right lane, molecular weight ($\times 10^{-3}$) ladder.

Considerable variation in the absolute levels of retinoids in mice of different ages and strains was encountered in our studies, and bleaching regimens introduced further poorly controlled variables. However, we found that the data were reproducible and consistent when reported as percentages of either total or polar retinoids[13] (see Fig. 4). The average rhodopsin content in wild-type mice from our studies was 590 ± 50 pmol/eye, a value similar to that reported by others.[16] The retinyl palmitate content of the normal animals in our studies was ~25% of the total retinoids, or ~150 pmol/eye, similar to reported values.[22]

SDS–PAGE and immunostaining should be carried out with all presumptive knockouts to confirm a genotype obtained by Southern blotting and/or polymerase chain reaction analysis. In addition, the method could be used to determine whether alterations in the expression of other genes have occurred in response to the loss of expression of the gene in question.

[22] S. B. Smith, T. Duncan, G. Kutty, R. Krishnan Kutty, and B. Wiggert, *Biochem. J.* **300**, 63 (1994).

Photoreceptor membranes derived from five mouse retinas provided sufficient material for analysis by Coomassie blue staining (Fig. 6). Staining with silver reagent or with specific antibodies is generally definitive with membranes from two mouse retinas, depending on the level of the protein in question.

Acknowledgments

This research was supported by grants from The National Eye Institute (EY01730, 02317, and 09339) and by an unrestricted grant to the Department of Ophthalmology from Research to Prevent Blindness, Inc. (RPB). J.C.S. is a Senior Scientific Investigator of RPB. The authors thank Krzysztof Palczewski for helpful discussions and support during the development of these methods.

Section VII

Inherited Retinal Disease: From the Defective Gene to Its Function and Repair

[39] In Situ Hybridization Studies of Retinal Neurons

By LINDA K. BARTHEL and PAMELA A. RAYMOND

Introduction

In situ hybridization has become a standard technique widely used in the study of gene expression in the retinas of a variety of species, including both vertebrates and invertebrates.[1-7] With the development of colorimetric techniques that employ RNA probes, nonisotopic *in situ* hybridization provides a high-resolution method for detection and visualization of mRNA at the level of single cells.[8] Nonisotopic detection techniques have been modified and refined by many investigators, leading to marked improvements in sensitivity without the hazards involved with handling radioactive material.[9,10] Additional benefits of digoxigenin (DIG)-labeled or fluorescein (FL)-labeled RNA probes include improved subcellular resolution, increased probe stability, and shorter processing time.[9-11] These RNA probes provide another advantage in that they can also be used in combination with other histochemical techniques such as immunohistochemistry and fluorescent dyes used for axonal tract tracing.[12-18]

[1] N. Hirsch and W. A. Harris, *J. Neurobiol.* **32**, 45 (1997).
[2] D. Kojima, A. Terakita, T. Ishikawa, Y. Tsukahara, A. Maeda, and Y. Shichida, *J. Biol. Chem.* **37**, 22979 (1997).
[3] T. Furukawa, E. M. Morrow, and C. L. Cepko, *Cell* **91**, 531 (1997).
[4] T. Belecky-Adams, S. Tomarev, H.-S. Li, L. Ploder, R. R. McInnes, O. Sundin, and R. Adler, *Invest. Ophthalmol. Vis. Sci.* **38**, 1293 (1997).
[5] S. Glardon, L. Z. Holland, W. J. Gehring, and N. D. Holland, *Development* **125**, 2701 (1998).
[6] S. A. Spencer, P. A. Powell, D. T. Miller, and R. L. Cagan, *Development* **125**, 4777 (1998).
[7] K. Bumsted and A. Hendrickson, *J. Comp. Neurol.* **403**, 501 (1999).
[8] H.-J. Höltke and C. Kessler, *Nucleic Acids Res.* **18**, 5843 (1990).
[9] R. H. Singer, J. B. Lawrence, and C. Villnave, *BioTechniques* **4**, 230 (1986).
[10] A. Panoskaltsis-Mortari and R. P. Bucy, *BioTechniques* **18**, 300 (1995).
[11] J. M. Polak and J. O. D. McGee (eds.), "*In Situ* Hybridization: Principles and Practice," p. 247. Oxford University Press, Oxford, 1990.
[12] R. I. Dorsky, D. H. Rapaport, and W. A. Harris, *Neuron* **14**, 487 (1995).
[13] R. I. Dorsky, W. S. Chang, D. H. Rapaport, and W. A. Harris, *Nature (London)* **385**, 67 (1997).
[14] G. S. Mastick, N. M. Davis, G. L. Andrews, and S. S. Easter, Jr., *Development* **124**, 1985 (1997).
[15] D. L. Stenkamp, L. K. Barthel, and P. A. Raymond, *J. Comp. Neurol.* **382**, 272 (1997).
[16] S. A. Sullivan, L. K. Barthel, B. L. Largent, and P. A. Raymond, *Dev. Genet.* **20**, 208 (1997).
[17] M. Perron, S. Kanekar, M. L. Vetter, and W. A. Harris, *Dev. Biol.* **199**, 185 (1998).
[18] Q. Liu, J. A. Marrs, and P. A. Raymond, *J. Comp. Neurol.* **410**, 290 (1999).

This chapter describes methods that employ DIG–RNA and FL–RNA probes for *in situ* hybridization on cryosections and on whole mounts of retinal tissue.

General Methods

All materials and labware used in the steps prior to hybridization are designated "For RNA Use Only," and care is taken to prevent RNase contamination. Disposable gloves are worn throughout all prehybridization procedures. Plastic labware is thoroughly washed, rinsed in a solution of 0.1% (v/v) diethyl pyrocarbonate (DEPC), and autoclaved; glassware is baked at 210° overnight. Solutions are made with deionized water, then thoroughly mixed with 0.1% (v/v) DEPC by stirring overnight, and autoclaved the next day. (Some laboratories use ultrapure, deionized water that is generated immediately before use.) Any solutions requiring Tris buffer are mixed in DEPC-treated, autoclaved, deionized water (DEPC–H_2O), since solutions containing Tris cannot be autoclaved. Reagents used in steps following hybridization are made in deionized water without pretreatment for RNase. All reagents are stored at 4° unless otherwise noted.

Probe Synthesis

DIG–RNA probes are the standard method for visualizing expression of a single gene. The DIG–RNA probes are made according to the protocol described by Boehringer Mannheim Biochemicals (BMB, Indianapolis, IN).[19] To generate transcripts complementary to the mRNA, the plasmid containing the cDNA of interest is linearized at a restriction enzyme site located 5' of the cDNA insert, and RNA transcription is initiated at an RNA promoter that is 3' of the insert. This is the antisense RNA probe. To ensure fidelity of transcription, restriction enzymes that leave 5' overhangs or blunt ends should be avoided. To generate the sense probe, which serves as a negative control, the plasmid is linearized at a restriction site 3' of the cDNA insert, and transcription is initiated from an RNA promoter located 5' of the insert. A minimum of 1 μg of cDNA template (excluding the plasmid DNA) is needed for adequate probe synthesis. The linearized template is purified by phenol–chloroform extraction and ethanol precipitation. RNase-free, siliconized, microcentrifuge tubes (Life Science Products, Denver, CO), are used for the transcription and storage of the RNA probes. The transcription reagents (BMB DIG–RNA labeling kit)

[19] Boehringer GmbH, "Nonradioactive *in Situ* Hybridization Application Manual," 2nd Ed. Boehringer GmbH, Mannheim, Germany, 1996.

are added to the microcentrifuge tube at room temperature in the following order: 1 μg of purified DNA template; 2 μl of 10× NTP labeling mixture (10 mM ATP, 10 mM CTP, 10 mM GTP, 6.5 mM UTP, 3.5 mM DIG–UTP; in Tris–HCl, pH 7.5); 2 μl of 10× transcription buffer [400 mM Tris–HCl (pH 8.0), 60 mM MgCl$_2$, 100 mM dithioerythritol, 20 mM spermidine, and 100 mM NaCl; 1 unit of RNase inhibitor is optional]; DEPC–H$_2$O to bring the final volume to 20 μl; and RNA polymerase (SP6, T7, or T3) to achieve a final activity of 2 units/μl. To increase the yield of RNA, the reaction components can be scaled up while keeping the amount of the template at 1 μg. The reagents are mixed gently, briefly centrifuged, and incubated at 37° for 2 hr. After the transcription reaction the template can be digested with 2 μl of DNase I, RNase free (BMB), for 15 min at 37°, but because the amount of probe generated greatly exceeds the amount of DNA template, digestion with DNase I is not usually necessary. The reaction is terminated by adding 2 μl of EDTA (200 mM in DEPC–H$_2$O).

We have found that limited hydrolysis of probes to reduce their length is generally not necessary for hybridization either on cryosections or with whole mounts, even when the probes are generated from full-length cDNA templates up to 2 kb. The DIG–RNA probes are precipitated at $-90°$ for 30 min to overnight with 0.1 volume of 4 M LiCl and 2.5–3.0 volumes of prechilled 100% ethanol. The microcentrifuge tubes containing the probes are removed from the freezer and centrifuged at 13,000g for 15 min at 4°. The ethanol–LiCl is removed, and the pellet is washed with 100 μl of 70% (v/v) ethanol and centrifuged at 13,000g for 5 min at 4°. The ethanol wash is removed, and the pellet is air dried and resuspended in 50 μl of DEPC–H$_2$O. The probe is then divided into 2-μg aliquots in RNase-free, siliconized, microcentrifuge tubes and stored at $-90°$.

In Situ Hybridization on Cryosections

This protocol has been used extensively by our laboratory and others in studies of gene expression in the retinas of fish, *Xenopus,* and mouse.[20–29]

[20] L. K. Barthel and P. A. Raymond, *J. Neurosci. Methods* **50,** 145 (1993).
[21] P. A. Raymond, L. K. Barthel, M. E. Rounsifer, S. A. Sullivan, and J. K. Knight, *Neuron* **10,** 1161 (1993).
[22] P. A. Raymond, L. K. Barthel, and G. A. Curran, *J. Comp. Neurol.* **359,** 537 (1995).
[23] O. Hisatomi, T. Satoh, L. K. Barthel, D. L. Stenkamp, P. A. Raymond, and F. Tokunaga, *Vision Res.* **36,** 933 (1996).
[24] M. A. Passini, E. M. Levine, A. K. Canger, P. A. Raymond, and N. Schechter, *J. Comp. Neurol.* **388,** 495 (1997).
[25] O. Hisatomi, T. Satoh, and F. Tokanaga, *Vision Res.* **37,** 3089 (1997).
[26] M. A. Passini, P. A. Raymond, and N. Schechter, *Brain Res. Dev. Brain Res.* **109,** 129 (1998).
[27] M. A. Passini, A. L. Kurtzman, A. K. Canger, W. S. Asch, G. A. Wray, P. A. Raymond, and N. Schechter, *Dev. Genet.* **23,** 128 (1998).

The technique is compatible with both immunohistochemistry and other fluorescent markers used for double-labeling strategies.[12,13,15,18]

Tissue Fixation. The objective of fixation is to maintain the morphological integrity of the tissue, but one of the undesirable consequences of fixation is cross-linking of proteins, resulting in decreased probe penetration into the cell. It is therefore necessary to optimize the fixation of the tissue to preserve morphology, but at the same time allow access of the DIG–RNA probe to the cellular mRNA.[9,10,30,31] The technique described here preserves retinal tissue morphology, allows cryosections to be cut at ≥ 3 μm thickness, and provides adequate penetration of the DIG–RNA probes.[20,32] This protocol has been optimized for fish retinas but it works equally well on mouse retina (D. Gamm, L. Barthel, M. Uhler, and P. Raymond, unpublished observations, 1998).

Adult eyes are enucleated, radial cuts are made in the cornea, and the lens is removed to provide adequate penetration of the reagents. The vitreous can be removed enzymatically by treatment with hyaluronidase (200 U/ml) and collagenase (350 U/ml) in phosphate-buffered saline [PBS: $NaH_2PO_4 \cdot H_2O$ (0.26 g/liter), Na_2PO_4 (1.15 g/liter), NaCl (8.76 g/liter), KCl (0.187 g/liter); pH 7.4] for 5 min at room temperature. The eyecups are then fixed for 0.5 hr at room temperature in 4% (w/v) paraformaldehyde in 100 mM phosphate buffer [$NaH_2PO_4 \cdot H_2O$ (2.6 g/liter), Na_2HPO_4 (11.5 g/liter); pH 7.4] with 5% (w/v) sucrose. The eyecups are then removed from the fixative, bisected, and placed back into the fixative for an additional 0.5 hr. Fish embryos at ≤ 50 hr postfertilization (hpf) are fixed intact with 4% (w/v) paraformaldehyde in 100 mM phosphate buffer and 5% (w/v) sucrose, at room temperature for 1 hr.

After fixation the tissue is rinsed three times (10 min each) at room temperature in 100 mM phosphate buffer was 5% (w/v) sucrose. After the buffer rinse, the tissue is cryoprotected overnight at 4° in 100 mM phosphate buffer with 20% (w/v) sucrose, then embedded in a 1:2 mixture of Tissue Tek O.C.T. embedding compound (Sakura Finetek USA, Torrance, CA) and 20% (w/v) sucrose in 100 mM phosphate buffer, and then frozen in isopentane cooled with dry ice or liquid nitrogen.[32] For storage, the

---

[28] A. Escayg, J. M. Jones, J. A. Kearney, P. F. Hitchcock, and M. H. Meisler, *Genomics* **50**, 14 (1998).

[29] D. M. Starace and B. E. Knox, *Exp. Eye Res.* **67**, 209 (1998).

[30] S. Urieli-Shoval, R. L. Meek, R. H. Hanson, M. Ferguson, D. Gordon, and E. P. Benditt, *J. Histochem. Cytochem.* **40**, 1879 (1992).

[31] F. Uehara, N. Ohba, Y. Nakashima, T. Yanagita, M. Ozawa, and T. Muramatsu, *J. Histochem. Cytochem.* **41**, 947 (1993).

[32] L. K. Barthel and P. A. Raymond, *J. Histochem. Cytochem.* **38**, 1383 (1990).

tissue blocks are wrapped tightly with plastic wrap and then aluminum foil and stored at $-90°$. Cryosections are cut at $-20°$ and collected on poly-L-lysine-coated slides at room temperature. To ensure adhesion of the sections during subsequent processing, the slides are placed in a vacuum chamber with desiccant for 2 hr to overnight at room temperature, and stored with desiccant at $-90°$. Slides are removed from the freezer, allowed to warm to room temperature, and air dried just prior to use.

Prehybridization. After air drying the tissue is rehydrated in an ethanol series (95, 70, and 50%, v/v) to 2× SSC (made from 20× stock: 3 M NaCl, 300 mM sodium citrate) for 1 min each. Digestion with proteinase K (0.01 mg/ml) is performed at 37° in 100 mM Tris base, pH 8.0, with 50 mM EDTA. The optimal time of digestion (2–5 min) depends on several variables including the degree of fixation of the tissue, thickness of the sections, and the specific mRNA of interest. The appropriate time must be determined empirically. After digestion, the slides are rinsed briefly in DEPC–H_2O and incubated at room temperature in 100 mM triethanolamine (TEA), pH 8.0, for 3 min, followed by a 10-min rinse in 100 mM TEA with 0.25% (v/v) acetic anhydride. The TEA and acetic anhydride solution must be made immediately before use. The tissue is then dehydrated in an ascending ethanol series to 100% and allowed to dry for 1 hr at room temperature.

Hybridization. Prior to use the microcentrifuge tube containing 2 μg of the DIG–RNA probe is boiled for 10 min and placed immediately on ice. A TEN buffer stock solution is made by combining 5 ml of 1 M Tris–HCl (pH 7.5), 1 ml of 500 mM EDTA, and 30 ml of 5 M NaCl. The following reagents are added to the tube in order: 36 μl of TEN buffer stock solution, 250 μl of 100% (v/v) formamide, 100 μl of dextran sulfate [50% (w/v) in DEPC-treated water], 50 μl of 10% (v/v) blocking reagent (BMB), and DEPC–H_2O to bring to a final volume of 500 μl. The hybridization solution is pipetted onto an RNase-free, siliconized coverslip. For example, we use a 22 × 60 mm plastic, disposable coverslip from Grace Bio-Labs (Sunriver, OR) and pipette 75 μl of hybridization solution across the bottom edge of the coverslip, and then gently invert the slide onto the coverslip, trying to avoid trapping any air bubbles. Incubation is done in a humid chamber overnight at 56°; a hybridization oven is convenient for this purpose.

Posthybridization. (*Note: Do not allow the slides to dry between any of the rinsing steps.*) After hybridization the slides are removed from the oven, and the coverslips are gently pulled off. The slides are washed thoroughly in 2× SSC at room temperature for 0.5 hr, then in 50% (v/v) formamide in 2× SSC, at 65°, for 0.5 hr, followed by two 10-min washes in 2× SSC at 37°. Unbound probe is digested at 37° for 0.5 hr with RNase A (20 μg/ml) in 10 mM Tris (pH 7.5), 1 mM EDTA, and 500 mM NaCl. The RNase is inactivated by rinsing the slides for 0.5 hr in the same buffer without RNase A at 65°. The RNase step is optional; it effectively reduces the background, but some decrease in signal intensity may occur.

Immunohistochemical Detection of Digoxigenin–RNA Probe. To reduce nonspecific binding of the antibodies, the slides are incubated for 2–3 hr at room temperature in 2× SSC with 0.5% (v/v) Triton X-100 and 1% (v/v) blocking reagent (BMB). The alkaline phosphatase-conjugated anti-DIG Fab fragment antibodies (BMB) are diluted 1 : 1000–1 : 5000 in maleate buffer (100 mM maleic acid, 150 mM NaCl; pH 7.5) with 1% (v/v) blocking reagent and 0.3% (v/v) Triton X-100. The slides are incubated with the antibody for 3 hr to overnight at room temperature in a humid chamber.

After antibody incubation, the slides are washed twice in maleate buffer for 10 min each, followed by a 10-min wash in BMB Genius buffer 3 [100 mM Tris (pH 9.5), 100 mM NaCl, 50 mM MgCl$_2$]. Enzymatic detection of the alkaline phosphatase is accomplished by incubating the slides in the chromagen substrate containing 4-nitro blue tetrazolium chloride (NBT) and 5-bromo-4-chloro-3-indolyl phosphate (BCIP). This is the most commonly used substrate, and it can be purchased in the form of ready-to-use tablets (BMB). Incubation of the color reaction is carried out in the dark, with the slides in a humid chamber at room temperature. The length of time required to detect an adequate signal can vary from 4 hr to overnight. CoverWells from Grace Bio-Labs are useful for these incubations. They hold a volume of 200 μl and form a water-tight seal on the slides even when wet. When using the CoverWells for the enzymatic reaction, the slide is inverted so that the CoverWell faces down to prevent reaction product from precipitating on the tissue. After the color reaction is complete the slides are washed in 10 mM Tris-HCl (pH 7.5), with 1 mM EDTA and 150 mM NaCl, and coverslipped under 100% (v/v) glycerol. The signal is a deep blue/purple color (Fig. 1).

Alternative detection systems for alkaline phosphatase include Fast Red (BMB); Sigma *Fast* Fast Red (Sigma, St. Louis, MO); Vector Red, Vector Blue, and Vector Black (Vector Laboratories, Burlingame, CA); and enzyme-labeled fluorescence (ELF; Molecular Probes, Eugene, OR). Vector Red and ELF produce a reaction product that is visible under fluorescent illumination.[13,17,33]

Double-Label in Situ Hybridization on Cryosections

The simultaneous localization of two different mRNAs in the same tissue can provide additional information about gene expression patterns, and this can be accomplished with nonisotopic *in situ* hybridization techniques. RNA probes can be labeled with either DIG, fluorescein (FL), or

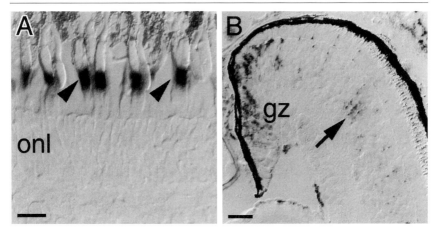

FIG. 1. Nonisotopic *in situ* hybridization with NBT/BCIP histochemistry on radial retinal cryosections. (A) Adult goldfish retina hybridized with a DIG–RNA probe to goldfish green cone opsin. Note the subcellular signal localized within the myoid region of the green cones (the accessory members of the double cone pairs; arrowheads). The clumps of black at the top of (A) are melanin in the pigmented retinal epithelium. (B) Retina of larval goldfish (6 days postfertilization) hybridized with a DIG–RNA probe to a goldfish neurogenic gene, *G-Notch3*. The signal is localized to regions associated with proliferating, neuroepithelial cells, including cells in and adjacent to the circumferential germinal zone (gz), and cells scattered throughout the inner retina (arrow). Bar: 10 μm (A) and 20 μm (B).

biotin, followed by enzymatic detection with antibodies conjugated to alkaline phosphatase (AP) or horseradish peroxidase (HRP), combined with substrates that produce contrasting colors.[33-37] We have used DIG–RNA probes in combination with FL–RNA probes, and antibodies conjugated to HRP and AP, respectively, to simultaneously detect two different mRNAs expressed by photoreceptors.[15,38] We visualize the HRP-conjugated antibody with the substrate 3,3′-diaminobenzidine tetrahydrochloride (DAB) in the presence of hydrogen peroxide, and the AP-conjugated antibody with Vector Blue (Vector Laboratories) substrate (Fig. 2A and B).

For optimal detection of the RNA probes when the signals are in

---

[33] T. Jowett, "Tissue in Situ Hybridization: Methods in Animal Development" (EMBO Practical Course). John Wiley & Sons, New York, 1997.
[34] J. W. O'Neill and E. Bier, *BioTechniques* **17,** 870 (1994).
[35] U. Strähle, P. Blader, J. Adam, and P. W. Ingham, *Trends Genet.* **10,** 75 (1994).
[36] T. Jowett and L. Lettice, *Trends Genet.* **10,** 73 (1994).
[37] T. Jowett and Y. L. Yan, *Trends Genet.* **12,** 387 (1996).
[38] D. L. Stenkamp, L. L. Cunningham, P. A. Raymond, and F. Gonzalez-Fernandez, *Mol. Vision* **4,** 26 (1998).

different cells, the immunocytochemical reactions should be done sequentially. The FL–RNA probe should be visualized first because FL-UTP is less stable than DIG–UTP. The AP- and HRP-conjugated antibodies are applied simultaneously; this is essential if it is not known whether different cells express the two genes of interest or whether they are expressed by the same cell. The horseradish peroxidase–DAB enzymatic detection is done first, followed by the alkaline phosphatase–Vector Blue color reaction. The brown precipitation product generated from DAB and the Vector Blue product provide good color contrast and signal resolution (Fig. 2C and D). If both signals originate in the same cell, the combination of DAB and Vector Blue reaction products is black (Fig. 2E).

Probe Synthesis, Prehybridization, Hybridization, and Posthybridization. These steps are essentially unchanged from the preceding description except as follows. FL–NTP 10× stock (BMB) is used to generate FL–RNA probes. The DIG–RNA and FL–RNA are combined in the same tube, mixed with the hybridization solution, and hybridized to the tissue simultaneously.

Immunohistochemical Detection of Digoxigenin– and Fluorescein–RNA Probes in Same Cell

After posthybridization, the slides are washed and blocked as described above. The HRP-conjugated, anti-DIG antibody (BMB) and the AP-conjugated anti-FL antibody (BMB) are diluted 1:50 and combined in the same tube. Slides are incubated with the antibodies overnight at room temperature and then washed as described above. Prior to detection of the HRP-conjugated antibodies, the slides are first rinsed in 100 mM Tris buffer [Tris-HCl (14.04 g/liter), Tris base (1.34 g/liter); pH 7.2] for 10 min. It is important that the water used in HRP reactions be free of organic impurities. Equal volumes of 0.02% (v/v) hydrogen peroxide and 0.1% (w/v) DAB (1 mg/ml) in 100 mM Tris buffer are mixed together, and the slides are incubated in this solution with occasional gentle agitation in the dark, at room temperature, for 10–15 min. After the color reaction is complete, the slides are rinsed in 100 mM Tris buffer for 15 min.

Detection of the AP-conjugated antibody with Vector Blue is done according to the manufacturer protocol (Vector Laboratories; alkaline phosphatase substrate kit III). The slides are rinsed in 100 mM Tris-HCl, pH 8.2, for 5–10 min, then incubated in the Vector Blue reagent in the dark without agitation. The color reaction can be quite strong after 0.5 hr, but the best and most consistent results are typically found with 4-hr incubations. After the color reaction, the slides are washed in distilled water for 5 min, dipped briefly in 100% ethanol, and coverslipped under

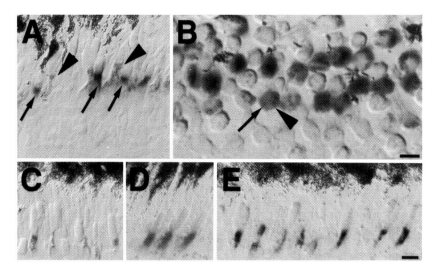

FIG. 2. Double-label *in situ* hybridization using DIG–RNA and FL–RNA probes. (A) Radial cryosection of adult goldfish retina hybridized with a DIG–RNA probe to goldfish red opsin and an FL–RNA probe to goldfish blue opsin. The DIG–RNA probe is revealed with anti-DIG, HRP-conjugated antibodies and detected by DAB histochemistry. The brown DAB reaction product is localized to the myoids of red cones (the principal members of the double cone pairs; arrowheads). The FL–RNA probe is localized with anti-FL, AP-conjugated antibodies detected by Vector Blue histochemistry. The blue reaction product is localized to the short single blue cone myoids (arrows). Melanin in the retinal pigmented epithelium can be seen at the top. (B) Tangential cryosection of adult goldfish retina at the level of the photoreceptor myoids, hybridized with a DIG–RNA green cone opsin probe, and an FL–RNA red cone opsin probe. Vector Blue reveals the green cone opsin probe (arrow) and DAB reveals the red cone opsin probe (arrowhead); the red and green opsins are expressed by members of double cone pairs. (C–E) *In situ* hybridization signals in the same cell. Two unique probes were generated against nonoverlapping regions of the goldfish green cone opsin cDNA. (C) The first was a DIG–RNA probe, detected with HRP-conjugated antibodies and DAB histochemistry. The brown reaction product is localized to the myoids of the green cones. (D) The second was an FL–RNA probe detected with AP-conjugated antibodies and Vector Blue histochemistry. The blue reaction product is also localized to the myoids of the green cones. (E) DIG–RNA and FL–RNA probes were hybridized and detected simultaneously with DAB and Vector Blue as described. The combination of DAB and Vector Blue in the same cell produces a dark, nearly black signal. Bars: 10 μm (A, B); 10 μm (C–E).

100% glycerol. The DAB-labeled cells will be brown, the Vector Blue cells will be bright blue, and double-labeled cells will be black (Fig. 2).

In Situ Hybridization on Whole Mounts

The introduction of nonisotopic methods for *in situ* hybridization provided, for the first time, a method to localize gene expression in whole tissues or embryos. This "whole mount" technique was first applied to *Drosophila* embryos using DNA probes.[39] Applications to vertebrate embryos have typically used RNA probes.[40] Several laboratories have published protocols for whole mount *in situ* hybridization, and we have adapted these for both whole adult retinas and embryonic tissue.[33,36,37,41,42]

Tissue Fixation. Preparation of retinal whole mounts requires that the retina be gently dissected from the eyecup prior to fixation. In fish, this is done by first dark adapting the retina, and then detaching it from the retinal pigmented epithelium by introducing a stream of saline into the subretinal space, then severing the optic nerve and flushing the retina away from the eyecup. The vitreous is enzymatically removed from the isolated retina with hyaluronidase and collagenase as described above. To aid in flattening the retina several radial cuts are made to a depth of one-third to one-fourth of the retinal radius. Embryos are fixed intact (within the chorion in the case of fish). Both adult retinas and embryos are fixed overnight in 4% (w/v) paraformaldehyde in 100 mM phosphate buffer, pH 7.4, with 5% (w/v) sucrose, at 4°. After fixation the tissue is rinsed twice in phosphate buffer, pH 7.4, with 5% (w/v) sucrose, at room temperature, 5 min each. At this point the fish embryos are removed from the chorion. The whole embryos or retinas are rinsed briefly with 70% (v/v) methanol, dehydrated in 100% methanol for 5 min, and then stored at $-20°$ in fresh 100% methanol.

Prehybridization and Hybridization. Prior to prehybridization the tissue is removed from the freezer, rinsed at room temperature in 100% methanol, and then rehydrated in a methanol–PBST [PBS with 0.1% (v/v) Tween 20] series at ratios of 3:1, 1:1, and 1:3, for 5 min each. The rehydration is completed with four rinses of 100% PBST, 5 min each. To enhance penetration of the probe, the tissue is digested with proteinase K (0.01 mg/ml in PBST) at room temperature; the length of time must be determined empirically for the tissue and the probe. For whole adult fish retinas,

---

[39] D. Tautz and C. Pfeifle, *Chromosoma* **98**, 81 (1989).
[40] R. M. Harland, *Methods Cell Biol.* **36**, 685 (1991).
[41] S. Schulte-Merker, R. K. Ho, B. G. Herrmann, and C. Nüsslein-Volhard, *Development* **116**, 1021 (1992).
[42] A. W. Püschel, P. Gruss, and M. Westerfield, *Development* **114**, 643 (1992).

15 min is adequate. Length of digestion required for embryos depends on the age. For example, zebrafish at 0–20 hr postfertilization (hpf) do not require treatment, but embryos at 20–30 hpf require 10 min, embryos at 30–48 hpf require 15 min, and embryos older than 50 hpf require 20 min. After digestion the tissue is rinsed briefly in PBST, washed for 5 min in fresh PBST, and fixed again in 4% (w/v) paraformaldehyde in 100 mM phosphate buffer for 20 min at room temperature. The fixative is rinsed from the tissue with PBST (five changes every 5 min).

The tissue is transferred to a fresh, siliconized, RNase-free microcentrifuge tube (1.5 ml). Prehybridization is performed in the hybridization solution as described above [100 mM Tris (pH 7.5), 1 mM EDTA, 300 mM NaCl, 50% (v/v) formamide, 10% (w/v) dextran sulfate, 1% (v/v) blocking reagent] without probe for a minimum of 2 hr at 56° with gentle agitation. The DIG–RNA probe is prepared in the hybridization solution as described above. Once the prehybridization is complete, the solution is removed and replaced with the hybridization solution containing the probe. Hybridization is carried out at 56°, overnight with gentle agitation, in a hybridization oven.

Posthybridization and Immunohistochemical Detection of Digoxigenin–RNA Probe

After overnight incubation the hybridization solution is removed, and the tissue is briefly rinsed at room temperature with 2× SSC. The tissue is then exposed to a series of high-stringency washes at 56°: 1 hr in 2× SSC with 50% (v/v) formamide, twice for 10 min each in 2× SSCT [2× SSC with 0.01% (v/v) Tween 20], then twice for 0.5 hr each in 0.2× SSCT. The tissue is then rinsed at room temperature in a 3:1, 1:1, 1:3 series of 0.2× SSC–PBST [PBS with 0.01% (v/v) Tween 20] for 10 min each, and transferred to a fresh microcentrifuge tube. Adequate rinsing is critical, so the rinse volume is at least 1 ml for each step.

The tissue is then incubated in the immunohistochemical blocking reagent [5% (v/v) goat serum, bovine serum albumin (BSA, 2 mg/ml), 1% (v/v) dimethyl sulfoxide (DMSO); in PBST] at room temperature for 2 hr. The blocking reagent is removed, and the anti-DIG, alkaline phosphatase-conjugated Fab antibody is diluted 1:5000 in fresh immunohistochemical blocking reagent, and added to the tissue. The tissue is incubated with the antibody overnight at 4°. The antibody is thoroughly rinsed out of the tissue with PBST, eight changes of 1 ml each, every 15 min. Prior to enzymatic detection the tissue is rinsed three times, 10 min each with BMB Genius buffer 3. The tissue is transferred to a 24-well tissue culture plate so that the progress of the color reaction can be easily monitored with a dissecting microscope. The NBT/BCIP substrate for the alkaline phosphatase antibody is prepared as described above. The color reaction is done in the dark, without agitation, and is complete within 4 hr to overnight. When the reaction is complete, as determined by examination with a dissection microscope, the tissue is rinsed in BMB Genius buffer 3 and then in PBST, 10 min each at room temperature. To minimize diffusion of the NBT/BCIP reaction product, the tissue is fixed again in 4% (w/v) paraformaldehyde in 100 mM phosphate buffer, overnight at 4° or 20 min at room temperature. Two final rinses are performed, 5 min each in PBST. The NBT/BCIP reaction product is a deep blue/purple color that is easily visualized when the tissue is mounted under 100% glycerol (Fig. 3).

Conclusions

Nonisotopic *in situ* hybridization has become a powerful method for the analysis of gene expression in the vertebrate retina. The methods described here have a wide range of applications that include simultaneous detection of multiple mRNA on the same section and within the same cell, comprehensive analysis of patterns of gene expression in the intact retina, and the ability to combine detection of mRNA expression with other fluorescent or immunohistochemical markers, including bromodeoxyuridine

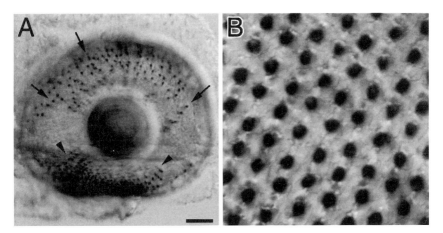

FIG. 3. Nonisotopic *in situ* hybridization with NBT/BCIP histochemistry on whole mounts. (A) Goldfish rod opsin DIG–RNA probe hybridized to a zebrafish embryo at 3.5 days postfertilization. The signal is localized to rod photoreceptors, which are concentrated on the ventral side of the retina (arrowheads), and arrayed in radial rows throughout much of the rest of the retina (arrows). (B) Goldfish blue opsin DIG–RNA probe hybridized to an adult zebrafish retinal whole mount. The signal is localized to the blue cone myoids, which form a regular array. Bar: 50 μm (A); 10 μm (B).

immunocytochemistry to detect proliferating cells[12,15,16]; the lipophilic fluorescent dye DiI (1,1'-dioctadecyl-3,3,3',3', tetramethylindocarbocyanine), for labeling of fiber tracts[14,18]; and the reporter gene encoding green fluorescent protein (GFP), to mark transfected cells.[13] The ease of use, extended stability in storage, and excellent subcellular resolution of the signal are additional benefits of nonisotopic RNA probes. The protocols described in this chapter have been optimized for fish retinas and embryos, but they have been used successfully on mouse and *Xenopus* retinas,[28,29] demonstrating their wide applicability.

[40] Cloning and Characterization of Retinal Transcription Factors, Using Target Site-Based Methodology

By SHIMING CHEN and DONALD J. ZACK

Introduction

The generation of the intricately structured vertebrate retina from individual pluripotential progenitor cells is a carefully choreographed process.[1] One of the important goals in current studies of retinal development is to understand the molecular mechanisms that regulate the complex patterns of gene expression that underlie this process.[2] Because most eukaryotic genes are regulated predominantly at the level of transcriptional initiation, retinal development, at least in part, can be viewed as a consequence of the selective turning on and off of specific genes. There has been progress in defining some of the *cis*-acting DNA elements that regulate the expression of retina-specific genes. More recently, there has also been progress in identifying and cloning the genes for some of the *trans*-acting factors that bind to these regulatory elements.[3–10] These studies are important not

---

[1] R. Adler, *Invest. Ophthalmol. Vis. Sci.* **34,** 1677 (1993).
[2] R. Kumar and D. J. Zack, in "Molecular Genetics of Ocular Disease" (J. Wiggs, ed.), p. 139. Wiley-Liss, New York, 1995.
[3] C. Freund, D. J. Horsford, and R. R. McInnes, *Hum. Mol. Genet.* **5,** 1471 (1996).
[4] A. Swaroop, J. Z. Xu, H. Pawar, A. Jackson, C. Skolnick, and N. Agarwal, *Proc. Natl. Acad. Sci. U.S.A.* **89,** 266 (1992).
[5] A. Rehemtulla, R. Warwar, R. Kumar, X. Ji, D. J. Zack, and A. Swaroop, *Proc. Natl. Acad. Sci. U.S.A.* **93,** 191 (1996).
[6] R. Kumar, S. Chen, D. Scheurer, Q. L. Wang, E. Duh, C. H. Sung, A. Rehemtulla, A. Swaroop, R. Adler, and D. J. Zack. *J. Biol. Chem.* **271,** 29612 (1996).
[7] T. Furukawa, E. M. Morrow, and C. L. Cepko, *Cell* **91,** 531 (1997).

only in terms of retinal development and function, but also have implications for understanding disease pathogenesis, as demonstrated by the finding that transcription factor mutations can cause a number of photoreceptor degenerations.[7,8,11-18]

A variety of approaches have been used to clone and characterize the genes that encode the transcription factors responsible for retinal gene expression. In this chapter, we discuss some of the approaches that are based on transcription factor DNA-binding activity. More specifically, we review the theory and provide methodological details for (1) the cloning of retinal transcription factors using the yeast one-hybrid assay, and (2) the subsequent biochemical characterization of the resulting cloned transcription factors using electrophoretic mobility shift (EMSA) and DNase I footprint assays.

Cloning Retinal Transcription Factors using Yeast One-Hybrid Assay

The yeast one-hybrid assay is a yeast genetics-based approach for cloning cDNAs that encode proteins that bind to specific DNA sequences. It is a modification of the yeast two-hybrid assay, which identifies proteins that interact with a given "bait" protein.[19] Both assays are based on the

---

[8] C. L. Freund, C. Y. Gregory-Evans, T. Furukawa, M. Papaioannou, J. Looser, L. Ploder, J. Bellingham, D. Ng, J. A. Herbrick, A. Duncan, S. W. Scherer, L. C. Tsui, A. Loutradis-Anagnostou, S. G. Jacobson, C. L. Cepko, S. S. Bhattacharya, and R. R. McInnes, *Cell* **91,** 543 (1997).

[9] S. Chen, Q. L. Wang, Z. Nie, H. Sun, G. Lennon, N. G. Copeland, D. J. Gilbert, N. A. Jenkins, and D. J. Zack, *Neuron* **19,** 1017 (1997).

[10] J. A. Martinez and C. J. Barnstable, *Biochem. Biophys. Res. Commun.* **250,** 175 (1998).

[11] C. L. Freund, Q. L. Wang, S. Chen, B. L. Muskat, C. D. Wiles, V. C. Sheffield, S. G. Jacobson, R. R. McInnes, D. J. Zack, and E. M. Stone, *Nature Genet.* **18,** 311 (1998).

[12] P. K. Swain, S. Chen, Q. L. Wang, L. M. Affatigato, C. L. Coats, K. D. Brady, G. A. Fishman, S. G. Jacobson, A. Swaroop, E. Stone, P. A. Sieving, and D. J. Zack, *Neuron* **19,** 1329 (1997).

[13] M. M. Sohocki, L. S. Sullivan, H. A. Mintz-Hittner, D. Birch, J. R. Heckenlively, C. L. Freund, R. R. McInnes, and S. P. Daiger, *Am. J. Hum. Genet.* **63,** 1307 (1998).

[14] A. Swaroop, Q.-L. Wang, W. Wu, J. Cook, C. Coats, S. Xu, S. Chen, D. Zack, and P. Sieving, *Hum. Mol. Genet.* **8,** 299 (1999).

[15] D. Bessant, A. M. Payne, K. P. Mitton, Q.-L. Wang, P. K. Swain, C. Plant, A. C. Bird, D. J. Zack, A. Swaroop, and S. S. Bhattacharya, *Nature Genet.* **21,** 355 (1999).

[16] S. G. Jacobson, A. V. Cideciyan, Y. Huang, D. B. Hanna, C. L. Freund, L. M. Affatigato, R. E. Carr, D. J. Zack, E. M. Stone, and R. R. McInnes, *Invest. Ophthalmol. Vis. Sci.* **39,** 2417 (1998).

[17] E. M. Morrow, T. Furukawa, and C. L. Cepko, *Trends Cell Biol.* **8,** 353 (1998).

[18] K. Gregory-Evans and S. S. Bhattacharya, *Trends Genet.* **14,** 103 (1998).

[19] S. Fields and O. Song, *Nature (London)* **340,** 245 (1989).

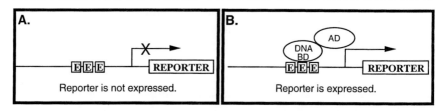

FIG. 1. Schematic diagram of the yeast one-hybrid assay. (A) BD–AD fusion protein is not present and, because the minimal promoter is not sufficient for transcription, the reporter is not expressed. (B) BD–AD fusion protein that binds to the E element is present; an AD is therefore positioned appropriately near the basal transcription complex, transcription is turned on, and the reporter is expressed.

observation that many eukaryotic transcriptional activators contain independent DNA-binding (BD) and *trans*-activating domains (AD) that can be physically separated and yet remain functional. In the two-hybrid assay, when two proteins expressed as BD and AD fusion proteins are physically brought together by protein–protein interaction, the resulting complex is localized correctly and *trans*-activates expression of a reporter gene. In the one-hybrid assay, instead of a protein bait, the bait is the DNA sequence element (designated as E) to which a putative transcription factor binds. The DNA element E, often in the form of a multimer so as to increase sensitivity, is placed upstream of a reporter gene driven by a minimal promoter. In the basal state, the minimal promoter is inactive so the reporter gene is not transcribed. However, if a DNA-binding protein (expressed as an AD fusion) binds to the E sequence, then the reporter gene is turned on (Fig. 1). An advantage of the one-hybrid system, compared with classic biochemical approaches, is that because it is genetically based, it is relatively simple, not labor intensive, and does not require protein purification. Among the many tissue-specific DNA-binding proteins that have been cloned by this approach are OLF-1,[20] REST,[21] CRX,[9] and ERT.[22]

Despite its power, the one-hybrid assay also has its limitations. One problem is that the success of the screen is clearly dependent on the quality of the cDNA–AD fusion library. A second problem is that if the desired transcription factor only binds DNA as a heterodimer, then it usually cannot be cloned by this approach because the transformed yeast generally expresses only one fusion cDNA. In addition, if the yeast strain being used

[20] M. M. Wang and R. R. Reed, *Nature* (*London*) **364,** 121 (1993).
[21] J. A. Chong, J. Tapia-Ramirez, S. Kim, J. J. Toledo-Aral, Y. Zheng, M. C. Boutros, Y. M. Altshuller, M. A. Frohman, S. D. Kraner, and G. Mandel, *Cell* **80,** 949 (1995).
[22] S. Choi, Y. Yi, Y. Kim, M. Kato, J. Chang, H. Chung, K. Hahm, H. Yang, H. Rhee, Y. Bang, and S. Kim, *J. Biol. Chem.* **273,** 110 (1998).

expresses an endogenous transcription factor that binds to the E site, this may preclude use of the screen because of autoactivation, i.e., the yeast grow in the selective medium even in the absence of added cDNA.

A kit containing most of the reagents required to perform a yeast one-hybrid screen is available from Clontech (Palo Alto, CA) (Matchmaker One-Hybrid System). (Regardless of whether one uses their system, a useful manual is downloadable from their website at *http://www.clontech.com/*. The information in the manual nicely complements the protocols that are provided in this chapter.) Using this system, we have successfully cloned several retinal transcription factors that bind to the bovine rhodopsin promoter region, including Nrl[6] and CRX.[9] On the basis of this experience, we discuss below several key points and hints for using this system.

Construction of Bait Plasmids

Choosing Appropriate Target DNA Sequences. Careful definition of the target sequence of a putative transcription factor is important for a successful screen. Use of too small a "bait" can lead to failure to clone the desired DNA-binding protein because of the absence of base pairs that are required for efficient binding. On the other hand, use of too large a "bait" sequence can lead to autoactivation of the reporter gene owing to binding by an endogenous yeast *trans*-activator, high background activation due to increased chance of nonspecific interaction with a library encoded protein, or the cloning of a real (but different from intended) transcription factor owing to the inadvertant inclusion of the binding site for another factor. The use of several complementary assays to define the target site can be helpful. These assays include EMSA and/or DNase I footprint analysis for protein–DNA interactions, and promoter analysis (using transient cell transfection, *in vitro* transcription, and/or transgenic animal approaches), for determining the functional importance of the putative DNA target site. Deletion and mutation studies, as well as the database searches, are often helpful.

Cloning Target DNA Sequence, in Tandem Copies, into Reporter Vectors. Once a target site is clearly defined, the next step is construction of the bait plasmids. The Matchmaker One-Hybrid System offers three reporter vectors (pHISi, pHISi-1, and pLacZi) for making these constructs. We suggest that the target DNA sequence be cloned into all three vectors, because the background level can vary significantly and unpredictably between these bait constructs when they are integrated into the yeast genome. We initially used ligation of small DNA oligomers to generate binding site multimers; but with the decreasing cost and increasing efficiency of large-size oligomer synthesis, we now prefer the use of long synthetic oligonucleotide pairs that contain tandem copies (two to five) of the desired target sequence with a restriction site at each end. Here is a generalized example of such an oligonucleotide pair:

```
                    EcoRI                                    SalI
Top strand:      5' AATTCXX(target site sequence)nXX  GTCGA 3'
Bottom strand:      3' G(complement to the top strand)CAGCTGCGC 5'
                                                              MluI
```

The strategy for this design is that the annealed oligomer, without gel purification of the double-stranded component, can be directly cloned into the *Eco*RI and *Mlu*I sites of the pHISi and pHISi-1 vectors. After *Sal*I digestion, assuming there is not an internal *Sal*I site, the same oligomer can also be cloned into the *Eco*RI/*Sal*I site of pLacZi. Other restriction sites in the polylinker regions of the three vectors can also be used. However, try to avoid using *Xho*I because its site will be used later for linearizing the pHISi and pHISi-1 bait constructs for integration of the plasmid into the yeast genome.

In the preceding sequence, n represents the number of tandem copies of the target sequence. Multiple copies can enhance the sensitivity of a screen. However, too many copies often lead to a high background. We have used three to five copies successfully in our one-hybrid screens. In general, it is best to start with three copies. n can also be limited by the length of the target sequence. Try to limit the length of one copy to 30 bp, because oligonucleotides longer than 120 nucleotides are difficult to obtain in high quality and quantity. Purification of these oligonucleotides, either by polyacrylamide gel electrophoresis or high-performance liquid chromatography (HPLC), is generally recommended.

Annealing of oligomer pairs can be achieved by mixing equal molar ratios of the two complementary oligonucleotides in 1× *Sal*I restriction digestion buffer, heating in a heat block at 80° for 5 min, and gradually cooling to room temperature. The annealed oligomer can be directly used for insertion to the *HIS3*-based bait vectors. A *Sal*I digestion is required for cloning the oligomer into the pLacZi vector.

Construction and Amplification of Retinal cDNA Library

A retinal cDNA library suitable for one-hybrid screening can be constructed using one- or two-hybrid cDNA library construction kits available from both Clontech and Stratagene (La Jolla, CA). About 5 μg of a poly(A)$^+$ RNA preparation is desirable as the starting material, but one can use less. The Clontech kit uses random primers and a lock-docking oligo(dT)$_{25}$d(A/C/G) primer for construction of bidirectional cDNA libraries. This may increase the ratio of full-length clones versus 3' ends represented in the library. In contrast, the Stratagene kit allows construction of unidirectional cDNA libraries, which theoretically doubles the chances of obtaining functional fusion proteins. Both kits provide adaptors for cloning the cDNA into appropriate one- or two-hybrid AD vectors (e.g., pGAD10, pGAD424, and pACT2 provided by Clontech). To achieve maximal library representation (complexity), transformation of the recombinant library plasmids into *Escherichia coli* should be performed by electroporation. A good library should contain more than 1×10^6 independent clones. Electrocompetent DH5α cells can be made according to Clontech protocol PT1113-1, which can be found in the Clontech Web site (*http://www.clontech.com*).

Although it is theoretically desirable to use a primary library for screening, because it will be the most representative, for practical reasons it is often necessary to amplify the primary library so as to have sufficient DNA available for multiple screens. We prefer to use solid rather than liquid medium for amplification because it limits the skewing caused by excessive growth of rapidly dividing plasmids. It is desirable to initiate the amplification with more than 2×10^6 clones, although the number somewhat depends on the size of the primary library. This can be accomplished by plating 80–100 Luria–Bertani (LB) agar plates (150 mm; with 50 μg of ampicillin per milliliter) with 20,000–25,000 colonies per plate. High-quality electrocompetent DH5α cells should be used. A preliminary electroporation should be carried out to maximize efficiency and determine how much DNA will be required to obtain more than 2×10^6 transformants. This should be done with actual library DNA because the transformation efficiency of such DNA is often lower than that of individual supercoiled vector. A suggested protocol is as follows.

1. Prepare 80–100 LB + ampicillin agar plates (150 mm), four 500-ml volumes (in 2-liter flasks) of LB + ampicillin broth, and electrocompetent DH5α cells. The transformation efficiency of the cells should be greater than 1×10^9 colony-forming units (cfu)/μg of supercoiled vector DNA.
2. Transform the competent cells with the desired amount (<1 μg) of library DNA by electroporation. Add the calculated volume of LB liquid medium (e.g., 10 ml for 100 plates) and incubate the transformation mixture for 1 hr at 37° with shaking. Plate the appropriate volume of transformation mixture on each of the LB + ampicillin agar plates. Incubate the plates overnight at 30° so as to minimize uneven growth of the colonies.
3. Scrape the colonies into 2 liters of LB + ampicillin broth. Incubate at 37° for 2 hr with shaking.

4. Save a small aliquot of the bacterial suspension and replate on an 85-mm LB + ampicillin agar plate for analysis of library quality. Also, if desirable, save an aliquot of the bacterial suspension for future use by addition of glycerol to 20% (v/v) and store at $-80°$. For the rest of the bacterial suspension, carry out a protocol for large-scale preparation of plasmid DNA. For example, the Qiagen (Chatsworth, CA) Maxi preparation protocol can be used. With this method, we have used 12 Maxi columns to obtain 6 mg of library DNA from 2 liters of bacterial suspension.
5. Assess the quality of the amplified cDNA library:
    a. Determine the percentage of recombinant clones and the average size of the inserts: a good library should have >80% recombinants with an average insert size of 1–2 kb. We typically test 20–30 randomly selected clones, either by colony polymerase chain reaction (PCR), using vector primer pairs that flank the cloning sites, or by restriction enzyme digestion of DNA minipreparations, using enzymes that cut at sites flanking the inserts. The PCR approach takes less time, but the minipreparation approach is more reliable, especially if large inserts are involved.
    b. Determine how representative the library is by testing for the presence of known genes: This can be performed by applying PCR to the library DNA with PCR primer pairs corresponding to the genes known to be expressed in the retina. It is best to use primers corresponding to genes whose expression level varies over a wide range.

Use of Dual-Reporter Selection to Minimize False Positives

One of the limitations of the one-hybrid approach is the not infrequent occurrence of false-positive clones. To reduce the frequency of such false positives, most one-hybrid systems take advantage of dual reporters, usually *HIS3* and *lacZ*. Because the minimal promoter of the *HIS3* bait (*HIS3*) is different from that of the *lacZ* bait (*CYC1*), nonspecific interactions often do not activate expression of both reporter genes. Thus, the first genetic selection can be performed with the *HIS3* readout, and the second selection can then be performed by measuring *lacZ* expression with a β-galactosidase assay. Using this dual-selection assay, we have found that about half of the *HIS3* positives from a library screen are *lacZ* negative, thus presumably representing false positives.

There are two ways to perform the dual selection. One is to use a single dual-reporter strain as described in the Clontech instruction manual; the other is to perform a diploid analysis. These two methods are described below.

Use of Single Dual-Reporter Strain

In both approaches the first step is to integrate all three bait construct vectors (pHISi, pHISi-1, and pLacZi) into the YM4271 genome. The resulting new reporter strains should be tested for background reporter expression. If the background expression is low for pLacZi, and at least one of the pHIS strains, a dual-reporter strain (e.g., pHISi/pLacZi or pHISi-1/pLacZi) can be created. The advantage of using a single dual-reporter strain over diploid analysis is simplicity, because the second test can be directly performed on the same clones pulled out by the *HIS3* selection without any additional steps. However, a disadvantage is that with dual-reporter strains it is more difficult to identify false positives derived from nonspecific binding to the target site (E). In addition, this approach does not provide any information about binding site specificity because it is not possible to directly test binding to heterologous control sequences.

Use of Separate Reporter Strains and Diploid Analysis

As an alternative to the single-strain dual-selection approach, we have used a modified diploid dual-reporter assay. In this assay, the two bait constructs are integrated into two different yeast strains of opposite mating type. For example, the *HIS3* bait is integrated into the genome of *MATa* strain YM4271, while the pLacZi bait is integrated into the genome of *MATα* strain BY380 (*MATα ura3-52 his3Δ200 leu2Δ1;* kindly provided by J. D. Boeke, Johns Hopkins University School of Medicine, Baltimore, MD). In this case, a library screen is first performed with a YM4271 derivative, using the *HIS3* readout. The selected positives are then examined for *lacZ* expression using the same target site as bait, as well as other related and unrelated bait sequences. This secondary screen is carried out in diploids derived from mating the YM4271 (*MATa*) clone carrying the selected one-hybrid plasmid encoding the putative DNA-binding protein and the BY381 strain (*MATα*) containing the pLacZi bait. Because the same positive clone can be analyzed for *lacZ* expression mediated by many different target sites, this approach allows us not only to exclude false positives, but also to examine the binding site specificity for a putative DNA-binding protein without doing *in vitro* protein–DNA-binding assays (see below). For example, by using this assay, we have found that CRX can bind to three sites within the rhodopsin promoter, BAT-1, Ret 1, and Ret 4, although CRX was originally cloned by using the Ret 4 site as a bait. This one-hybrid finding was confirmed by *in vitro* footprinting assays.[23]

---

[23] S. Chen and D. J. Zack, *J. Biol. Chem.* **271,** 28549 (1996).

Materials

Diploid selection plates: Only diploids can grow on these plates. SD/-Leu-Ura-His-Lys is recommended for selecting diploids of YM4271 clone X, using the BY380/pLacZi bait strain

Replica plating apparatus (Clontech) and velveteen squares (Clontech)

Procedures

1. Follow the manufacturer instructions to create YM4271/*HIS3* reporter strains and BY380/*lacZ* reporter strains containing various target sites and control sequences. Make sure that the background *lacZ* expression is low for the target sites to be analyzed. A positive control, p53BLUE, provided in the kit should also be integrated into the BY380 genome.

2. Perform a library screen, using the YM4271 bait strain on SD/-Leu-His/3AT plates.

3. Pick up growers from the preceding library screen. Restreak the colonies onto duplicate SD/-Leu-His/3AT plates, which will serve as master plates.

4. Mating: On a yeast extract–peptone–dextrose (YEPD) plate, make 30–40 patches of BY380/*lacZ* reporter yeast cells from a freshly prepared master plate. Immediately patch each of the library colonies on a BY380 patch for mating and label the double patches appropriately. Remember to leave a single BY380 patch and a YM4271 patch as haploid controls. Incubate the mating plate at 30° overnight.

5. Diploid selection: Replicate the mating patches to a diploid selection plate, using a replicating apparatus. Incubate at 30° for 3 days.

6. Perform a filter β-galactosidase assay on.diploid growers. A diploid of BY380/p53BLUE × YM4271/pGAD53m can serve as a positive control.

7. Further test the dual-positive clones for binding site specificity. Use the same diploid analysis described above to analyze cloned proteins for binding to various related and unrelated target sites presented by various BY380/*lacZ* bait strains. If the clone is positive when mated with many unrelated *lacZ* baits, it most likely represents a false positive.

Characterization of Positive Clones

Once a clone encoding a putative transcription factor has been obtained, the next step is to characterize it. An outline of some of the many available approaches is as follows.

1. DNA from positive clones should be recovered and the resulting plasmid should be retransformed into the original bait strain to ensure that activation of both reporter genes is reproducible.

2. Sequence the fusion cDNA and perform bioinformatic analysis (e.g., *http://www2.ncbi.nlm.nih.gov/*). The finding of a known DNA-binding motif in a predicted open reading frame is, of course, particularly encouraging.

3. Obtain and sequence a full-length cDNA. Possibly obtain orthologs from other species to assess conserved regions.

4. Obtain a genomic clone, determine the genomic structure of the gene and map its human chromosomal position. The map location may provide a hint as to the possible involvement of the gene in a human disease. Mapping the murine gene may provide insight into its possible connection with a known mouse mutation.

5. Determine expression pattern in adult and during development. Northern blot analysis and *in situ* hybridization are two commonly used approaches to obtain this information. It is important to demonstrate that the expression of the cloned protein coincides with the expression of its target gene(s).

6. Characterize DNA-binding activity. *In vitro* protein–DNA-binding assays are important to confirm the expected specific interaction between the cloned protein and the putative target DNA element (for a detailed description, see below).

7. Determine the biological significance of the protein–DNA interaction. It is important to assess whether the transcription factor–DNA interaction mediates a positive or negative regulatory role, or both, depending on the circumstances. One approach is to use transient transfection of either primary retinal cells[6] or established cell lines.[23] In such an assay, cotransfection is carried out with a transcription factor expression vector together with a reporter gene driven by a promoter containing the putative DNA target site(s). Transient transfection analysis can also be used to examine the effect of disease-causing mutations on transcription factor activity.[12,14,15] (It is also common to include an internal control to adjust for differences in transfection efficiency.) As an alternative, transgenic mice can be used to assess the effect of ectopic expression of the transcription factor, or the gene can be "knocked out" in mice to determine the consequences of deletion of the transcription factor.

8. Determine the subcellular localization of the cloned protein. A nuclear localization is important for a transcription factor. However, some transcription factors are normally in the cytoplasm and are transported to the nucleus only on "induction." To study subcellular localization of a protein, immunohistochemical analysis of the endogenous protein with a specific antibody is clearly the preferred approach. However, if an appro-

priate antibody is not available, it is possible to transiently transfect cells with an epitope-tagged construct and use an antibody against the tag.

Characterization of Cloned DNA-Binding Proteins: *In Vitro* Protein–DNA-Binding Assays

In vitro protein–DNA-binding assays can be used to demonstrate and characterize specific interaction between a cloned DNA-binding protein and its putative target DNA sequence. Two of the commonly used methods are electrophoretic mobility shift assays (EMSAs) and DNase I footprint analysis, and detailed methods have been published previously.[24,25] Below we provide detailed protocols for the modified versions of these assays that we have used for characterizing retinal DNA-binding proteins.

Analysis of Protein–DNA Interactions by Electrophoretic Mobility Shift Assay

The EMSA provides a simple, rapid, and sensitive *in vitro* assay to test if a protein encoded by a library cDNA binds to the expected target sequence, and also a quick way to determine binding sequence specificity. In this assay, a radiolabeled small DNA fragment or oligomer is incubated with a putative DNA-binding protein, and then separated by electrophoresis through a native polyacrylamide gel. If the protein binds to the labeled DNA probe, the protein–DNA complex migrates slower than the DNA probe alone. One of the major advantages of this assay over the DNase I footprint analysis (see below) is that it is simpler and does not require a large amount of protein. The protein preparation containing the putative DNA-binding protein can consist of (1) a crude whole cell or nuclear extract derived from tissue or cultured cells, (2) *in vitro*-translated protein derived from a candidate cDNA, (3) crude protein extracts made from cells transfected with a protein expression vector, or (4) a purified or partially purified recombinant protein. The high sensitivity of the EMSA makes it possible to detect protein–DNA complex in the nanogram range. Binding sequence specificity can also be studied by using mutant oligomer probes.[23]

Material

Mobility shift buffer (MSB, 5×): 125 mM HEPES (pH 7.6), 300 mM KCl, 25% (v/v) glycerol, 5 mM dithiothreitol (DTT), 5 mM EDTA

---

[24] F. Ausubel, R. Brent, R. Kingston, D. Moore, J. Seidman, S. JA and S. K. John Wiley & Sons, New York, 1992.

[25] M. Brenowitz, D. F. Senear, M. A. Shea, and G. K. Ackers, *Methods Enzymol.* **130**, 132 (1986).

Tris–borate buffer (TBE, 5×): Per liter, add 54 g of Tris base, 27.5 g of boric acid, and 20 ml of 0.5 M EDTA (pH 8.0)

Nondenaturing 5% (w/v) polyacrylamide gel mix: 5% (w/v) acrylamide–bisacrylamide (29:1, v/v) in 0.5× TBE buffer

Nonspecific carrier DNA, e.g., poly(dI–dC), at concentrations of 0.01, 0.1, and 1 µg/µl: Obtain poly(dI–dC) as a powder from Pharmacia (Piscataway, NJ), and dissolve it in TE buffer at appropriate concentrations

$MgCl_2$ (100 mM): This may be required for some proteins to bind to their targets. Final concentration should generally be 4–6 mM, depending on the individual protein

Polyacrylamide slab gel apparatus: The Bio-Rad (Hercules, CA) Protean II xi Cell is one of several suitable units

Oligonucleotides: Both top and bottom strands containing the desired binding site for the protein of interest. Gel purification of oligonucleotides is recommended. However, we have used unpurified oligomers successfully. Double-stranded oligomers can be made by mixing an equal molar ratio of an oligonucleotide pair in 1× restriction enzyme buffer [medium salt concentration, e.g., New England BioLabs (Beverly, CA) buffer 2], heating it to 90° for 3 min, and slowly cooling it to room temperature

[γ-^{32}P]ATP (3000–6000 Ci/mmol) for kinase reaction or [α-^{32}P]dNTP (3000–6000 Ci/mmol) for fill-in reaction with Klenow

Procedures

Labeling DNA Oligomers with ^{32}P. Oligomers can be ^{32}P labeled by either a kinase reaction (for blunt ends) or a fill-in reaction (for sticky ends). The latter procedure is recommended, because only the double-stranded oligomer will be labeled by a fill-in reaction. The specific [α-^{32}P]dNTP used should be chosen on the basis of the nucleotide(s) that are needed in the fill-in reaction. An example of a fill-in reaction with [α-^{32}P]dCTP is described below.

1. Mix well in a 1.5-ml microtube the following reagents:

Oligomer (1 µM or ~10 ng/µl)	1 µl
Klenow buffer (or NEB buffer 2), 10×	1 µl
dNTPs minus dCTP (3.3 mM)	1 µl
[α-^{32}P]dCTP (3000 Ci/mmol)	1 µl
Klenow enzyme	1 µl
H_2O	5 µl
Total:	10 µl

2. Incubate the mixture at room temperature for 15 min.

3. Add 2 μl of dNTP mix (1.25 mM for each nucleotide) and incubate at room temperature for an additional 2 min.

4. Add 120 μl of 0.3 M NaCl. Extract the DNA with phenol–CH$_3$Cl (1:1, v/v) and CH$_3$Cl. Precipitate the DNA with a 2× volume of 100% ethanol. Store at −20° for 20 min and spin the DNA at 14,000 rpm in a microcentrifuge at 4°.

5. Resuspend the pellet in 100 μl of TE. Measure the radioactivity by transferring a 1-μl aliquot to an aqueous scintillation cocktail and counting. Make the appropriate dilution such that approximately 20,000 cpm/μl will be used in each EMSA reaction.

Note: In general, removing unincorporated isotope from the labeled oligomer is not necessary. However, labeled single-stranded oligonucleotides by a kinase reaction with [γ-^{32}P]ATP could cause problems if proteins in the extract can recognize single-stranded DNA.

EMSA Reactions. To perform the EMSA reactions, follow these four steps.

1. For each reaction, mix the following contents in a microfuge tube on ice:

Probe (20,000 cpm/μl)	1 μl
Poly(dI–dC) (0.01–1.0 μg/μl)	1 μl
MSB (5×)	2 μl
MgCl$_2$ (optional)	0.4–0.6 μl (final concentration, 4–6 mM)
Protein (5–20 μg of crude extract)	X μl
H$_2$O	X μl
Total:	10 μl

Notes: (1) Always include a no-protein control, and add the protein last to the reaction; (2) to obtain an optimal result, the amount of a protein extract and poly(dI–dC) added to the reaction will need to be titered. This can be achieved by using either a fixed protein concentration with variable poly(dI–dC) concentration, or vice versa; and (3) to decrease experimental variables, we normally make a reaction mix containing nonvariable components, such as 5× MSB, MgCl$_2$, and a probe, then aliquot the mix to each reaction tube before addition of the variable components.

2. Incubate on ice (can be at room temperature) for 20 to 30 min.

3. Add 1 μl of DNA loading dye. Load samples on a 5% (w/v) polyacrylamide gel and run in 0.5× TBE at 110 V for 2 hr.

4. Dry the gel and expose to X-ray film with a screen at −80° for 12–24 hr.

Determination of Binding Site Specificity by Using Mutant DNA Probes. A simple way to assess binding site specificity is to perform EMSA mutation analysis. This can be performed with radiolabeled oligomers harboring a series of serial nucleotide mutations (e.g., mutate 3 bp each time). As an alternative, a cold competition EMSA with various molar ratios of cold versus hot oligomers (wild-type and mutant) can be performed.

Protein Identification by Supershift Assay. If an antibody against the cloned protein is available, or the protein of interest is epitope tagged, a supershift assay can be performed to identify the protein in the protein–DNA complex. In this assay, first allow the protein–DNA-binding reaction to reach equilibrium, then add the antibody to the reaction. If the antibody recognizes a specific protein in the protein–DNA complex, it will increase the size of the complex and thereby further retard its migration, leading to a "supershift."

Analysis of Protein–DNA Interactions by in Vitro DNase I Footprinting

Deoxyribonuclease I (DNase I) footprinting is an assay that localizes protein-binding sites on DNA. The assay is based on the fact that a bound protein can protect the phosphodiester backbone of DNA from DNase I hydrolysis. In this assay, a DNA fragment (100–400 bp) that is labeled at one end with ^{32}P is incubated with a protein, and then partially digested with DNase I. The resulting products are separated on a denaturing DNA sequencing gel along with a sequence ladder. The binding sites (protected regions) are visualized by autoradiography (see Fig. 2). Advantages of DNase I footprinting over EMSA is that the actual binding site is "visualized," multiple binding sites on the DNA can be identified, and the relative binding affinities of the different sites can be studied. This is particularly useful when a protein of interest requires appropriate sequences flanking its recognition site for optimal DNA binding, or there is cooperative binding involving separated sites. A disadvantage, however, is that in general the amount of protein needed for carrying out footprint analysis is more than that needed for an EMSA, and it is a more difficult and tedious assay to perform. In addition, purified or partially purified protein may be needed for a "clean" result. Crude protein extracts can often be used successfully, but usually require multiple titration experiments with different concentrations of extract and nonspecific carrier DNA in order to distinguish nonspecific from specific DNA binding.

A detailed procedure for DNase I footprint analysis is described below. The four main parts of the assay are covered: preparation of single end-labeled DNA fragments, generation of sequence ladders, protein–DNA-binding reactions and DNase I digestion, and gel electrophoresis and autoradiography.

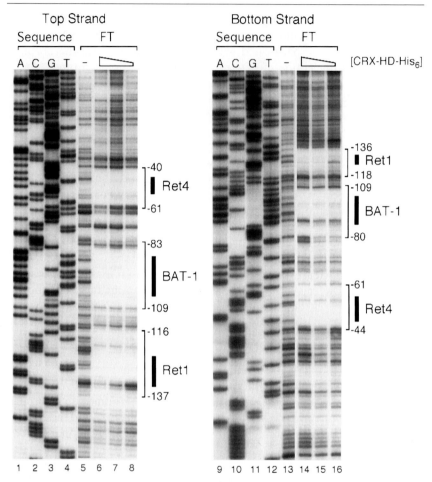

FIG. 2. Footprint analysis of the bovine rhodopsin promoter region with purified recombinant CrxHD–His$_6$ fusion peptide. A PCR fragment spanning −225 to +70 bp of the bovine rhodopsin promoter region was end labeled with ^{32}P at either the −225 bp end (top strand, lanes 5–8) or the +70 bp end (bottom strand, lanes 13–16). The end-labeled fragments were incubated with 1.2, 0.4, and 0.1 μg of purified CrxHD-His$_6$ fusion peptide (lanes 6–8 and 14–16, respectively). Lanes 5 and 13 did not receive any fusion peptide. The DNase I dilutions were 1:1000 for lanes 5 and 13, 1:500 for lanes 8 and 16, and 1:250 for lanes 6, 7, 13, and 14. Sequence ladders for the top (lanes 1–4) and bottom strand (lanes 9–12) are shown on the left side of the footprint lanes (FT). Protected regions are indicated by brackets, and the nucleotide positions for each protected region are indicated. The location of the Ret1, BAT-1, and Ret4 sites are indicated by solid bars.

Reagent and Solutions

Purified/partially purified proteins or crude protein extracts in an appropriate buffer for the protein being studied: We generally start with a buffer containing 20 mM HEPES (pH 7.6), 60 mM KCl, 10% (v/v) glycerol, 0.1 mM EDTA, and 1 mM DTT. For a purified protein, depending on the protein, as little as 10 ng of the protein may be sufficient. However, for a crude protein extract, as much as 300 µg of extract may be required

Plasmid DNA containing the promoter region of interest with protein-binding sites

5' and 3' primers for PCR amplification of the DNA fragment of interest

T4 polynucleotide kinase and 10× kinase buffer

[γ-^{32}P]ATP (3000 Ci/mmol)

Cycle sequencing kit from Amersham (Arlington Heights, IL)

Gel apparatus for DNA sequencing gel and agarose gel

MgCl$_2$ (10 mM) and CaCl$_2$ (5 mM): Store at room temperature

Polyvinyl alcohol (Sigma, St. Louis, MO): Molecular weight, 10,000; 10% (w/v) in H$_2$O. Store at 4°

DNase I (10 U/µl; Boehringer Mannheim, Indianapolis, IN): Store at −20°

Proteinase K, 12.5 mg/ml in TE buffer: Store in aliquots at −20°

Assay buffer* Z_{100} (100 mM KCl), Z_{140} (140 mM KCl) and Z_{160} (160 mM KCl)

Z_{100}	Z_{140}	Z_{160}
25 mM HEPES·K, pH 7.6	25 mM HEPES·K, pH 7.6	25 mM HEPES·K, pH 7.6
10 µM ZnSO$_4$	10 µM ZnSO$_4$	10 µM ZnSO$_4$
0.1% (v/v) Nonidet P-40	0.1% (v/v) Nonidet P-40	0.1% (v/v) Nonidet P-40
100 mM KCl	100 mM KCl	100 mM KCl
1 mM DTT	1 mM DTT	1 mM DTT

DNase I stop solution
EDTA, pH 8.0 (20 mM)
Sodium dodecyl sulfate (SDS; 1%, w/v)
NaCl (0.2 M)

---

* Different Z buffers are used to obtain a final concentration of 40–55 mM KCl for each protein–DNA-binding reaction. Make a stock solution for each component and mix all the components except DTT. Store at 4°. Add DTT just prior to use.

Glycogen (250 µg/ml): Dilute from a 20-mg/ml stock (Boehringer Mannheim)

Proteinase K (0.25 mg/ml)

Mix all the components except proteinase K and store at room temperature. Just before use, add a 1/50 volume of proteinase K (12.5 mg/ml)

Carrier DNA [poly(dI–dC), 1 mg/ml]:

0.1 to 5 µg may be needed for each reaction if a crude protein extract is used. The optimal amount of carrier DNA needed for removing interference from nonspecific DNA-binding proteins can be determined by using the same amount of crude protein but a variable amount of the carrier DNA (e.g., 0.1, 0.5, 2, and 5 µg). It is usually not necessary to add the carrier DNA when using purified protein

Procedures

Preparation of Single ^{32}P End-Labeled DNA Fragments. Two methods are commonly used to generate single end-labeled DNA fragments. One is based on restriction enzyme digestion followed by a kinase or fill-in reaction.[24] The other is a PCR-based method. Although both methods work well in our hands, the latter is often simpler and does not depend on the presence of restrictions sites at the proper locations, or the subcloning of DNA fragments. Below is an example of the PCR-based method.

1. Labeling 5′ and 3′ primers: PCR primer pairs corresponding to at least 25–50 bp upstream/downstream of the expected binding sites are recommended. The size of the PCR fragment can be 100–500 bp.

   a. For each reaction, add

Primer (20 µM)	1 µl
Kinase buffer (10×)	2 µl
[γ-^{32}P]ATP (3000 Ci/mmol)	7 µl
T4 polynucleotide kinase	1 µl
H$_2$O	9 µl
Total:	20 µl

   b. Incubate at 37° for 30 min. Add 80 µl of 0.3 M NaCl to the reaction. Extract once with phenol–CH$_3$Cl (1:1, v/v) and precipitate the "hot" primer with a 2.5× volume of ethanol. Store at −80° for 20 min and spin at 4° for 20 min. Air dry the pellet and dissolve it in 25 µl of H$_2$O.

2. PCRs with hot/cold primer pairs:
    a. Prior to a hot PCR, carry out a small-scale cold PCR to determine the optimal PCR conditions (e.g., annealing and extension temperatures) for a given template and primer pair.
    b. Carry out "hot" PCRs, using the defined optimal condition. Use hot 5' primer and cold 3' primer for labeling the top strand and vice versa for the bottom strand. For example, mix the following components in a 0.2 µl thin-wall PCR tube:

Component	Top strand	Bottom strand
Template DNA (1 ng/µl)	2 µl	2 µl
5' Primer	25 µl ^{32}P labeled	1 µl unlabeled (20 µM)
3' Primer	1 µl unlabeled (20 µM)	25 µl ^{32}P labeled
PCR buffer (10×)	10 µl	10 µl
dNTPs (1.25 mM)	10 µl	10 µl
Taq polymerase	1 µl	1 µl
H$_2$O	51 µl	51 µl
	Total: 100 µl	Total: 100 µl

Cycle on a PCR machine:
  Step 1: 94°, 4 min
  Step 2: 94°, 1 min
  Step 3: 55°, 1 min
  Step 4: 72°, 1 min
  Step 5: Repeat steps 2–4, 29 times
  Step 6: 94°, 1 min
  Step 7: 55°, 2 min
  Step 8: 72°, 5 min
  Step 9: 4°, hold

    c. Phenol–CH$_3$Cl extract once and ethanol precipitate the DNA. Resuspend the DNA in 20–30 µl of TE. Add DNA loading dye for gel purification.
3. Gel purification of the PCR products: Purify the labeled DNA by agarose gel electrophoresis, followed by any conventional gel purification method. As an example, we describe a method using low melting agarose gel followed by a hot phenol extraction procedure.
    a. Run the labeled DNA samples in a 1–2% low melting point agarose gel (GIBCO-BRL, Gaithersburg, MD), in TAE buffer.
    b. Visualize the band of interest by either ethidium bromide staining or autoradiography with a X-ray film for 5–10 min.

c. Cut the radiolabeled band from the gel and place the gel piece in a 1.5-ml microcentrifuge tube. Spin for 1–2 min in a microcentrifuge at 14,000 rpm, room temperature, to measure the gel volume (100–200 μl). Add a 1/10 volume of 3 M NaCl and an equal volume of 0.3 M NaCl.

d. Hot phenol extraction: Heat the samples to 65° for 10 min to melt the gel and add an equal volume of hot phenol (buffer saturated pH > 7.0; preheated to 65°). Vortex for 1 min and immediately spin at 14,000 rpm for 1–2 min at room temperature. Carefully remove the top aqueous layer to another tube and extract the DNA with phenol–CH$_3$Cl (1:1, v/v) an additional two times.

e. Precipitate the DNA with 2 volumes of ethanol and dissolve the DNA in 100 μl of TE buffer (pH 8.0). Store at −20°.

4. Measuring radioactivity of DNA: Transfer a 1-μl aliquot of the labeled DNA to an aqueous scintillation cocktail and count. Usually 10,000–15,000 cpm/lane is needed.

Generation of Sequence Ladders for Both Strands. Running sequence ladders adjacent to the footprint lanes on the same gel is necessary for determining sequences of the protected regions. Sequencing reactions can be carried out on the same template DNA by using a standard manual sequencing kit (e.g., Amersham Thermo Sequenase Cycle Sequencing Kit) with the 5' and 3' hot primers.

Protein–DNA-Binding Reactions and DNase I Digestion. Steps 5 and 6 cover the binding of proteins to DNA, and DNase I digestion.

5. To bind proteins to DNA:

a. Calculate the total number (n) of reactions. The number depends on the experimental data points planned. For example, the following reactions should be considered: (1) serial DNase I dilutions (e.g., 1:4000 to 1:500 for the enzyme at 10 U/μl) without any protein, (2) serial protein dilutions (e.g., start with 10 ng to 1 μg, a wide range for a purified protein), (3) at least two DNase I dilutions for each protein concentration point, and (4) control without DNase I and protein. For crude protein fractions, consider also a titration of carrier DNA with a fixed amount of a protein extract. Keep in mind that four extra lanes are needed to run the sequence ladder for each labeled probe.

b. Prepare $(n + 1) \times 25$ μl of reaction mixture. Each 25-μl reaction mix contains the following:

^{32}P-labeled DNA probe (10,000–15,000 cpm)	1 μl
Polyvinyl alcohol (10%, w/v)	10 μl

Carrier DNA (optional for purified protein)	1–5 μl
MgCl$_2$ (100 mM): final concentration is 4–6 mM, depending on the protein	2–3 μl
H$_2$O	X μl
Total:	25 μl

   c. Vortex to mix well and aliquot 25 μl to prechilled 1.5-ml tubes on ice.
   d. Prepare serial protein dilutions that cover the range of concentrations to be analyzed.
   e. To each assay tube, add a total of 25 μl of the following components:

	Amount of protein	
Component	0–12 μl	13–20 μl
Buffer Z140	X μl	—
Buffer Z100	25 − 2X μl	—
Buffer Z160	—	25 − X μl
Protein	X μl	X μl

     Gently mix and incubate on ice for 20–30 min until and equilibration is reached.

6. DNase I digestion: Prepare serial DNase I dilutions in ice-cold water. Usually for DNase I from Boehringer Mannheim (10 U/μl), a 1:4000 or 1:2000 dilution is effective for reactions without proteins, while 1:1000 and 1:500 dilutions work well for reactions with proteins. However, the optimal dilutions could be different for various proteins. Here is an example of a DNase I dilution.

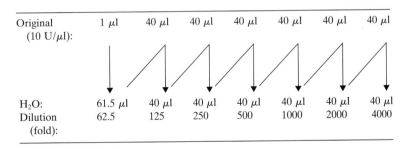

Perform DNase I digestion on a single sample as follows:
   a. Remove a single assay tube and let stand at room temperature for 1 min.

b. Add 50 μl of 10 mM MgCl$_2$, 5 mM CaCl. Mix by flicking. Let stand at room temperature for 1 min.
c. Add 2 μl of diluted DNase I, and quickly mix by flicking. Incubate at room temperature for exactly 1 min.
d. Add 100 μl of the DNase I stop solution, mix well by gentle vortexing, and let stand at room temperature for >5 min.
e. Repeat steps a–d for each sample, including a control sample containing no DNase I. When all the samples are processed, extract the samples with 200 μl of phenol–CH$_3$Cl (1:1, v/v).
f. Precipitate the DNA with 600 μl of ethanol. Store at $-80°$ for 20 min and spin at 4° for 20 min. Carefully remove the supernatant.
g. Wash the pellet with 800 μl of ice-cold 75% (v/v) ethanol.
h. Dry the pellet in a Speedvac evaporator for 10–15 min.

Gel Electrophoresis and Autoradiography: Steps 7–10 conclude this assay.

7. Resuspend the DNA in 10 μl of formamide loading buffer [40% (v/v) formamide, 10 mM Tris (pH 8.0), 1 mM EDTA, 0.1% (w/v) xylene cyanol, 0.1% (w/v) bromphenol blue]. Store at $-20°$.
8. Heat the samples at 90° for 5 min and immediately transfer to an ice bath.
9. Load 5 μl on a prerun sequencing gel (6–8%) and run for 1–2 hr until the predicted binding sites migrate into the middle of the gel.
10. Dry the gel, and expose it to X-ray film with an intensifying screen for 24–48 hr at $-80°$.

[41] In Vivo Assessment of Photoreceptor Function in Human Diseases Caused by Photoreceptor-Specific Gene Mutations

By ARTUR V. CIDECIYAN

Introduction

Many human hereditary retinal degenerations are now known to be caused by mutations in genes that encode proteins exclusively expressed in photoreceptors.[1-4] Some of these proteins, such as rhodopsin, are directly involved in signaling light, and visual dysfunction would be expected from mutant forms of the protein. However, specific functions of other proteins, such as peripherin/retinal degeneration slow (RDS), are currently not fully elucidated and the type of visual dysfunction resulting from mutations is more difficult to predict. In either case, a thorough understanding of human photoreceptor degenerations requires a targeted assessment of the function of the primary site of disease, the photoreceptor.

Currently, electroretinogram (ERG) photoresponses constitute the most direct noninvasive measure of rod and cone photoreceptor function in isolation. This chapter outlines the details of this new methodology and its application to patients with retinal degenerations caused by mutations in two photoreceptor-specific genes: *rhodopsin* and *peripherin/RDS*. Other methodological details, molecular biology results, and clinical information can be found in previous publications.[5-12]

---

[1] T. P. Dryja and T. Li, *Hum. Mol. Genet.* **4,** 1739 (1995).
[2] A. Gal, E. Apfelstedt-Sylla, A. R. Janecke, and E. Zrenner, *Prog. Retin. Eye Res.* **16,** 51 (1997).
[3] D. B. Farber and M. Danciger, *Curr. Opin. Neurobiol.* **7,** 666 (1997).
[4] R. S. Molday, *Invest. Ophthalmol. Vis. Sci.* **39,** 2493 (1998).
[5] A. V. Cideciyan and S. G. Jacobson, *Invest. Ophthalmol. Vis. Sci.* **34,** 3253 (1993).
[6] A. V. Cideciyan and S. G. Jacobson, *Vision Res.* **36,** 2609 (1996).
[7] S. G. Jacobson, C. M. Kemp, A. V. Cideciyan, J. P. Macke, C.-H. Sung, and J. Nathans, *Invest. Ophthalmol. Vis. Sci.* **35,** 2521 (1994).
[8] S. G. Jacobson, A. V. Cideciyan, C. M. Kemp, V. C. Sheffield, and E. M. Stone, *Invest. Ophthalmol. Vis. Sci.* **37,** 1662 (1996).
[9] S. G. Jacobson, A. V. Cideciyan, A. M. Maguire, J. Bennett, V. C. Sheffield, and E. M. Stone, *Exp. Eye Res.* **63,** 603 (1996).
[10] C. M. Kemp, S. G. Jacobson, A. V. Cideciyan, A. E. Kimura, V. C. Sheffield, and E. M. Stone, *Invest. Ophthalmol. Vis. Sci.* **35,** 3154 (1994).
[11] A. H. Milam, Z.-Y. Li, A. V. Cideciyan, and S. G. Jacobson, *Invest. Ophthalmol. Vis. Sci.* **37,** 753 (1996).

ERG Recording Technique

The ERG is a light-evoked field potential generated across the retina and is recorded with an electrode placed on the cornea.[13,14] ERGs have been used in the diagnosis of retinopathies for many decades as a noninvasive and objective means of evaluating retinal function.[15,16] Much of our understanding of the origins of this complex potential has come from animal experiments.[17,18] Careful evaluation of rare human retinopathies has also provided valuable pieces to solve the ERG puzzle,[19,20] especially in light of our molecular understanding of the underlying dysfunction.[21]

Full-field ERGs are performed with unipolar Burian–Allen contact lens electrodes, a custom-built ganzfeld, and a computer-based recording system. The contact lens electrode is referenced to an indifferent forehead electrode that is protected from light exposure; the ground electrode is located at the ear lobe. The signals are amplified (band-pass, 0.5–1000 Hz; four pole) and digitized with an 8-bit analog-to-digital converter over a 1- or 2-mV peak-to-peak dynamic range at a 2.5-kHz sampling rate for a duration of 200 msec. The stimulus is produced by a helical xenon flash tube coupled from above to the ganzfeld with an intervening infrared blocking filter and two movable filter trays carrying gelatin colored and neutral density filter combinations. The flash tube is driven by a power supply delivering 400 J per flash (unattenuated white ganzfeld luminance of 2300 cd · sec · m^{-2}, flash duration of ~1 msec) with a maximum repetition rate of 0.3 Hz. Flash-to-flash energy variability is less than 5% with a semirandom interflash interval.

ERG Photoresponses

ERGs evoked by blue (Wratten 47A; Eastman Kodak, Rochester, NY) flashes in a dark-adapted normal subject are shown in Fig. 1. The waveforms

---

[12] A. V. Cideciyan, D. C. Hood, Y. Huang, E. Banin, Z.-Y. Li, E. M. Stone, A. H. Milam, and S. G. Jacobson, *Proc. Natl. Acad. Sci. U.S.A.* **95,** 7103 (1998).
[13] D. C. Hood and D. G. Birch, *IEEE Eng. Med. Biol. Mag.* **14,** 59 (1995).
[14] E. N. Pugh, Jr., B. Falsini, and A. L. Lyubarsky, in "Photostasis and Related Phenomena" (T. P. Williams and A. B. Thistle, Eds.), pp. 93–128. Plenum Press, New York, 1998.
[15] G. Karpe, *Acta Ophthalmol.* **24**(Suppl), 1 (1945).
[16] J. R. Heckenlively and G. B. Arden, "Principles and Practice of Clinical Electrophysiology of Vision." Mosby-Year Book, St. Louis, Missouri, 1991.
[17] R. Granit, *J. Physiol.* **77,** 207 (1933).
[18] R. H. Steinberg, L. J. Frishman, and P. A. Sieving, in "Progress in Retinal Research" (N. Osborne and G. Chader, eds.), pp. 121–160. Pergamon, New York, 1991.
[19] D. C. Hood, A. V. Cideciyan, A. J. Roman, and S. G. Jacobson, *Vision Res.* **35,** 1473 (1995).
[20] D. C. Hood, A. V. Cideciyan, D. A. Halevy, and S. G. Jacobson, *Vision Res.* **36,** 889 (1996).
[21] A. V. Cideciyan, X. Zhao, L. Nielsen, S. C. Khani, S. G. Jacobson, and K. Palczewski, *Proc. Natl. Acad. Sci. U.S.A.* **95,** 328 (1998).

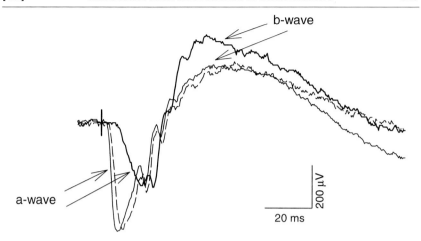

FIG. 1. Electroretinograms (ERGs) evoked by three different blue flash stimuli (thick, dashed, and thin traces, 2.3, 3.9, and 4.6 log scot-td · sec, respectively) in a dark-adapted normal subject. Each waveform shows the potential recorded between cornea and forehead as a function of time; cornea-positive direction is upward, vertical line marks timing of the stimulus.

have two prominent peaks, *a*-wave and *b*-wave. It has long been accepted that the *a*-wave reflects photoreceptor activity and the *b*-wave is produced by cells of the inner nuclear layer.[17] Evidence of a direct relationship between the ERG *a*-wave and the underlying photoreceptor hyperpolarization has been obtained with the use of high-energy stimuli and a computational model previously fitted to responses from single mammalian rods.[22] Better understanding of the properties of ERG *a*-wave families clearly showed that *a*-waves evoked by moderate energy stimuli (∼2 log scot-td · sec = ∼860 photoisomerizations per rod[23]), as commonly used in clinical settings, are insufficient for the quantitation of the underlying photoreceptor component. High-energy stimuli are needed to estimate the saturated amplitude, which in turn allows the estimation of activation kinetics.[24] In the current work, the term "ERG photoresponse" is used to refer to ERGs evoked by stimuli that are known to saturate *a*-wave amplitude (>4 log scot-td · sec for normal adult human rods). Figure 1 demonstrates (compare dashed and thin traces) the saturation of the *a*-wave where increasing the stimulus energy by 0.7 log units accelerates the latency of the waveform but causes no change in the amplitude. Discussion of ERG *b*-waves, which

[22] D. C. Hood and D. G. Birch, *Vis. Neurosci.* **5**, 379 (1990).
[23] M. E. Breton, A. W. Schueller, T. D. Lamb, and E. N. Pugh, *Invest. Ophthalmol. Vis. Sci.* **35**, 295 (1994).
[24] D. C. Hood and D. G. Birch, *Doc. Ophthalmol.* **92**, 253 (1997).

have been the mainstay of traditional ERG analysis, is neglected in this chapter for the sake of brevity; evidence suggests that they directly originate from bipolar cells[25] and attempts at physiologically based modeling are ongoing.[26,27]

Quantitative Modeling of Rod Activation

The first physiologically based quantitative model of phototransduction was developed to account for the hypothesized phototransduction activation steps in vertebrate rod photoreceptors.[28] Under several simplifying assumptions, this model predicts the photoreceptor dark current to fall as a delayed Gaussian function of time when presented with a short light stimulus. With appropriate scaling, this model was shown to describe leading edges of ERG photoresponse families.[5,23,29,30] The "delayed Gaussian" model is defined as

$$R(I, t) = R_{max} \left[1 - \exp\left(-\frac{1}{2} I\sigma(t - t_{eff})^2 \right) \right] \quad \text{for } t \gg t_{eff} \quad (1)$$

where $R(I, t)$ is the leading edge of the corneally measured rod-isolated potential in microvolts; I is the retinal illuminance of the stimulus in scot-td · sec; t is the time after stimulus onset in seconds; R_{max} is the maximum amplitude in microvolts; σ is the sensitivity in scot-td^{-1} · sec^{-3}; and t_{eff} is a brief time delay in seconds.

Figure 2A shows the leading edge of rod-isolated ERG photoresponses in a normal human subject. The delayed Gaussian model of rod phototransduction has been fit as an ensemble to the photoresponses. A careful examination reveals that the delayed Gaussian model fits the data well until the saturation of the a-wave (i.e., ~4 log scot-td · sec). The responses evoked by supersaturating stimulus energies differ from the model (arrowhead in Fig. 2A). An alternative model of phototransduction was developed[6] by reconsidering some of the simplifying assumptions made during the derivation of the delayed Gaussian model. Specifically, the alternative model considers the earliest stages of biochemical reactions as a cascade of three dominant first-order reactions instead of a simple time delay used in the delayed Gaussian model; furthermore, a low-pass filter is added

---

[25] X. Xu and C. J. Karwoski, *J. Neurophysiol.* **72**, 96 (1994).
[26] J. G. Robson and L. J. Frishman, *J. Opt. Soc. Am. A* **13**, 613 (1996).
[27] D. C. Hood and D. G. Birch, *J. Opt. Soc. Am. A* **13**, 623 (1996).
[28] T. D. Lamb and E. N. Pugh, Jr., *J. Physiol.* **449**, 719 (1992).
[29] E. N. Pugh, Jr., and T. D. Lamb, *Biochim. Biophys. Acta* **1141**, 111 (1993).
[30] D. C. Hood and D. G. Birch, *Vision Res.* **33**, 1605 (1993).

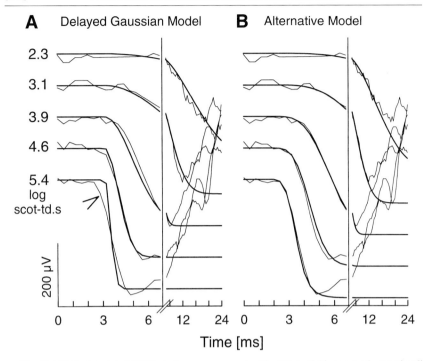

FIG. 2. (A) Rod-isolated ERG photoresponses evoked by 2.3- to 5.4 log scot-td · sec stimuli in a normal subject (thin traces) and the delayed Gaussian model of rod phototransduction (thick traces) fit to the leading edge of the data as an ensemble (R_{max} = 345 μV, σ = 1.54 log scot-td^{-1} · sec^{-3}, t_{eff} = 3.1 msec). Traces have been shifted vertically for clarity of presentation. Arrowhead points to misfit of model to data. (B) Same data as in (A), fit with the alternative model of phototransduction (R_{max} = 375 μV, σ = 1.73 log scot-td^{-1} · sec^{-3}, δ = 1.5 msec; constants τ and τ_m were 0.85 and 0.5 msec, respectively). Note change of scale on the time axis at 7 msec.

owing to the expected capacitive current of the outer segment (OS) membrane. The alternative model is defined as

$$R(I,t) = R_{max}\left(1 - \exp\left\{-\frac{1}{2}I\sigma\left[(\varepsilon^2 - 6\varepsilon\tau + 12\tau^2) - e^{-\varepsilon/\tau}(\varepsilon^2 + 6\varepsilon\tau + 12\tau^2)\right]\right\}\right) \star e^{-\varepsilon/\tau_m} \quad (2)$$

where $R(I, t)$ is the leading edge of the corneally measured rod-isolated potential in microvolts; I is the retinal illuminance of the stimulus in scot-td · sec; t is the time after stimulus onset in seconds; R_{max} is the maximum amplitude in microvolts; σ is the sensitivity in scot-td^{-1}·sec^{-3}; $\varepsilon = t - \delta$, time in seconds delayed by δ; τ is time constant of a cascade of three first-order decays presumed to correspond to three dominant photoactivation reactions; τ_m is the capacitive time constant of the photoreceptor membrane; and the star (★) represents the convolution operation. The details of Eq. (2) have been published.[6] Note that delayed Gaussian and alternative models have the same form when ε is much larger than the largest of δ, τ, or τ_m.

The fit of the alternative model to the same data (Fig. 2B) shows superior fidelity over the whole stimulus energy range. Specifically, the model describes families of waveforms that attain a constant slope and shortening latency at higher stimulus energies, which is the observed behavior of ERG photoresponses.[31]

Quantitative Modeling of Cone Activation

Cone phototransduction, like that of rods, is believed to be based on a cGMP phosphodiesterase cascade.[32] Cone ERG a-waves can be isolated by the use of rod-desensitizing backgrounds and they show saturation of amplitude when sufficiently high stimulus energies are used.[6,33–35] Although the relationship between these cone ERG photoresponses and underlying cone photocurrents is not as well established as that for the rods, there is evidence in the primate retina that early components evoked by high-energy stimuli are derived from cone photoreceptors.[36] Appropriately, cone photoresponse families isolated with rod-desensitizing backgrounds have been fit with variants of rod phototransduction models.[6,33–35] Until recently,[12] the rod and cone contributions to the dark-adapted ERG photoresponses had not been evaluated simultaneously. On the basis of the success of the alternative model in describing both rod- and cone-isolated photoresponses independently, it was hypothesized that a single model consisting of rod and cone components may describe the complete dark-adapted ERG photoresponse.[12] The combined rod + cone alternative model is defined as

$$R_{sc+ph}(I_{sc}, I_{ph}, t) = R_{sc}(I_{sc}, t) + R_{ph}(I_{ph}, t) \tag{3}$$

[31] M. E. Breton and D. P. Montzka, *Doc. Ophthalmol.* **79**, 337 (1992).
[32] T. D. Lamb, in "Color Vision: From Genes to Perception" (K. R. Gegenfurtner and L. T. Sharpe, eds.). pp. 89–101. Cambridge University Press, New York, 1999.
[33] D. C. Hood and D. G. Birch, *Vis. Neurosci.* **10**, 857 (1993).
[34] D. C. Hood and D. G. Birch, *Vision Res.* **35**, 2801 (1995).
[35] N. P. Smith and T. D. Lamb, *Vision Res.* **37**, 2943 (1997).
[36] R. A. Bush and P. A. Sieving, *Invest. Ophthalmol. Vis. Sci.* **35**, 635 (1994).

where $R_{sc+ph}(I_{sc}, I_{ph}, t)$ is the leading edge of the corneally measured potential, in microvolts, consisting of scotopic (sc) and photopic (ph) components; I_{sc} and I_{ph} are the retinal illuminance of the stimulus in scot-td · sec and phot-td · sec units, respectively; t is the time after stimulus onset in seconds; and R_{sc} and R_{ph} are the scotopic and photopic subcomponents of the photoresponses, respectively, each of which is defined similarly to $R(I, t)$ in Eq. (2).

Figure 3A shows the initial 25 msec of ERG photoresponses recorded in a dark-adapted normal subject with blue and red stimuli. Overlaid on the photoresponses is the combined rod + cone alternative model fit to the leading edges of all 10 photoresponses simultaneously. Dashed lines show the rod component in the blue responses and the cone component

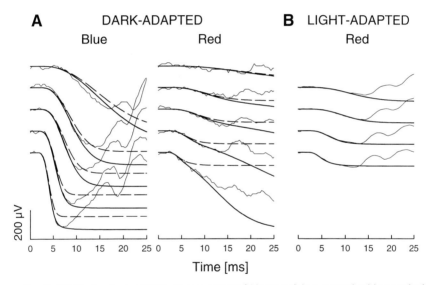

FIG. 3. (A) Dark-adapted ERG photoresponses (thin traces) in a normal subject evoked by blue (Wratten 47A; bottom to top, 4.6, 3.9, 3.3, 3.1, and 2.3 log scot-td · sec, respectively) and red (Wratten 26; 3.6, 3.0, 2.4, 2.1, and 1.4 log phot-td · sec) stimuli; blue and red pairs are photopically matched. The leading edges of all responses are fit as an ensemble with the combined rod + cone alternative model (thick traces); dashed lines depict the rod component of the model ($R_{max} = 444$ μV, $\sigma = 1.66$ log scot-td^{-1} · sec^{-3}) on blue responses and the cone component ($R_{max} = 91$ μV, $\sigma = 2.4$ log phot-td^{-1} · sec^{-3}) on red responses. (B) ERG photoresponses (thin traces) in the same normal subject as in (A) recorded on a 3.1 log phot-td white background with red (Wratten 26; bottom to top, 3.6, 3.0, 2.4, and 2.1 log phot-td · sec) stimuli. Each trace is the average of 12–16 individual responses presented at 0.3 Hz. The cone version of the alternative model of phototransduction (thick traces) is fit to the leading edges of the responses as an ensemble ($R_{max} = 94$ μV, $\sigma = 2.2$ log phot-td^{-1} · sec^{-3}). Vertical calibration is the same for (A) and (B), and is shown at lower left.

in the red responses. The rigor in estimating the underlying rod and cone phototransduction parameters is thus significantly increased by requiring a single set of physiologically based model parameters to describe data covering several log units of dynamic range and two extremes of the visible wavelength spectrum.

Figure 3B shows a cone-isolated ERG photoresponse series recorded on a white rod-desensitizing background in the same subject as in Fig. 3A. When the cone-only version of the alternative model is fit to this family of responses a reduction in sensitivity with light adaptation is observed; maximum amplitude does not change significantly. Analysis of the results from a group of normal subjects ($n = 14$) showed the dark-adapted cone sensitivity reduced from 2.58 ± 0.20 to 2.28 ± 0.14 log phot-td^{-1} · sec^{-3} on light adaptation to a 3.2 log phot-td white background; maximum amplitudes were unchanged (81 ± 12 μV dark adapted versus 83 ± 8 μV light adapted).

Assuming cone photoresponses represent the sum of underlying saturated cone photocurrents, these results allow speculation on the adaptation properties of normal human cone photoreceptors. The ~0.3 log unit reduction in sensitivity with light adaptation is similar to the ~0.2 log unit reduction reported in single-cell recordings of primate L or M cones.[37] The current results are also consistent with the speculative reason provided for higher cone photoresponse sensitivities reported on a rod-desensitizing blue background when compared with a white background.[35] In the first study of adaptive properties of human cone photoresponses, the reduction in sensitivity (calculated differently than in the current work) was somewhat smaller.[33] Dark-adapted cone photoresponses isolated by a two-flash technique have been compared with light-adapted cone photoresponses but sensitivity values were not provided.[38]

Phototransduction Activation: Mutations in Rhodopsin Gene

Retinitis pigmentosa (RP) refers to a group of hereditary retinal degenerations that affect photoreceptors. Approximately 30% of the cases of autosomal dominant (ad) RP are now believed to be caused by mutations in the *rhodopsin* gene.[2] Since the first report[39] nearly 100 different *rhodopsin* mutations have been associated with disease.[2] The rod cell-specific expres-

---

[37] J. L. Schnapf, B. J. Nunn, M. Meister, and D. A. Baylor, *J. Physiol.* **427,** 681 (1990).

[38] W. A. Verdon, M. E. Schneck, and G. Haegerstrom-Portnoy, *Invest. Ophthalmol. Vis. Sci.* **39,** S975 (1998).

[39] T. P. Dryja, T. L. McGee, E. Reichel, L. B. Hahn, G. S. Cowley, D. W. Yandell, M. A. Sandberg, and E. L. Berson, *Nature (London)* **343,** 364 (1990).

sion of this protein superficially explains the impaired night blindness commonly reported by patients with RP, but extensive laboratory and clinical investigations to date have not resolved either the mechanisms for the observed death of not only rod but also cone photoreceptors, or the differences between dying and surviving rods containing the same gene defect.[40] The biochemical dysfunction that presumably precedes cell death is also not fully understood.

We have studied a population of 63 patients representing 18 *rhodopsin* gene mutations by applying noninvasive tests of rod and cone function and correlative retinal histopathology with the aim of better understanding early retinal dysfunction.[12] We found two patterns of rod disease expression in the heterozygous patients with *rhodopsin* mutations, and there was an intrafamilial consistency of pattern. Class A families (mutations G89D, R135G, R135L, R135W, E181K, V345L, and P347L) had early onset of severely abnormal rod photoreceptor-mediated function. Class B families had one or more mutation-positive members with little or no night vision symptoms; rod b-waves and rod thresholds were relatively preserved. Some class B families, termed B1 (T17M, P23H, T58R, V87D, G106R, and D190G), had retinal regions with severe disease and other regions that were normal or far less affected. Other families, termed B2 (G51A, Q64ter, T193M, Q312ter, Q344ter), showed no regional retinal predilection for disease. Patients at advanced stages of both classes were not distinguishable by retinal function testing.[12]

Topographical maps of rod sensitivity loss (RSL) and cone sensitivity loss (CSL) by psychophysics provided evidence of a temporal sequence of rod and cone disease. Locus-by-locus examination of these maps suggested a common theme despite interindividual and intraretinal differences in severity of expression: detectable rod disease precedes that of cones. Specifically, no significant change in CSL was found for RSL of up to 1.5 log units. Larger RSL was associated with significant CSL and, on average, cones lost 0.5 log units of function for each log unit loss of rod function.[12]

Retina-wide ERGs support and extend the psychophysical findings. Figure 4A and B shows dark-adapted ERG photoresponses recorded with two stimulation colors in representative adRP patients with class B1 (D190G) and B2 (Q344ter) mutations. Overlaid on the photoresponse families are the combined rod + cone alternative model phototransduction fit to the leading edges. In addition, the rod-only component of photoresponses (recorded with blue stimuli) and cone-only component of photoresponses (recorded with red stimuli) are shown. The cone components are normal in each patient, whereas there are 51 and 33% reductions in rod

[40] E. L. Berson, *Proc. Natl. Acad. Sci. U.S.A.* **93,** 4526 (1996).

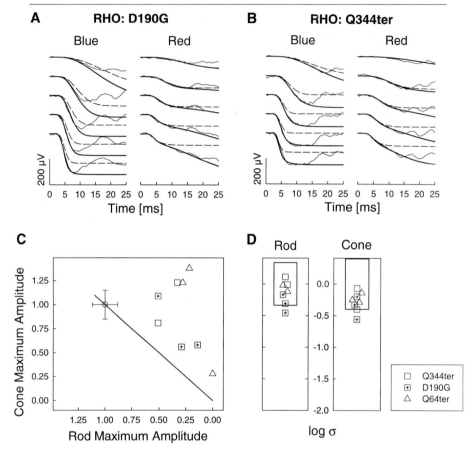

FIG. 4. (A and B) Dark-adapted ERG photoresponses in patients with D190G and Q344ter mutations in the *rhodopsin* gene. See caption to Fig. 3A for description. (C) Plot of cone-versus-rod maximum amplitude as a fraction of respective mean normal values (open circle and error bars represent mean ± SD). Hypothetical loci of equal rod and cone reduction lie on the diagonal line shown. (D) Sensitivity (log σ) of rod and cone transduction activation is shown as a logarithm of the fraction of mean normal. Boxes define normal limits (mean ± 2 SD).

maximal amplitude in the two patients, respectively; the sensitivities of rod phototransduction are borderline abnormal and normal for D190G and Q344ter, respectively.

Figure 4C summarizes the rod and cone phototransduction results in eight class B patients. Rod maximal amplitude was abnormally reduced to varying degrees in all patients but cone maximal amplitude could be normal or abnormally reduced. Relative reduction of rod amplitude was always greater than that of cone amplitude. The sensitivity of photoresponses was normal for most patients, suggesting that rod phototransduction activation may proceed normally in the presence of the mutant allele and the attendant biochemical defects. Psychophysical and electrophysiological results produced by patients with class B mutations, taken together, suggest a temporal dependence of cone disease on the extent of rod pathology.[12] Cone cells appear to retain normal function until a certain percentage of rods is sufficiently affected by disease. The decrease in cone function, once initiated, occurs at a slower rate than in rods.

Phototransduction Activation: Mutations in *Peripherin/RDS* Gene

Peripherin/RDS and ROM-1 are subunits of a photoreceptor-specific integral disk membrane protein complex that is believed to establish and maintain the unique rim region of rod and cone OS.[4,41,42] This complex may also be involved in OS membrane fusion processes during disk membrane formation and disk shedding.[43] Originally, a *peripherin/RDS* mutation was implicated in causing a naturally occurring murine model of retinal degeneration called *rds*.[44] More recently, many different heterozygous mutations in *peripherin/RDS* have been associated with human autosomal dominant retinal degenerations.[41,45–47] One of the early and confusing aspects about *peripherin/RDS*-associated retinal degeneration was the observation of considerable inter- and intrafamilial variability in phenotype, implying that there was no relationship between genotype and phenotype.[46] Most of these studies emphasized the ophthalmoscopic differences in phenotype. We approached the phenotype by examining rod and cone photoreceptor-specific retinal function in a group of patients with 10 different genotypes.[8–10]

Figure 5 shows a schematic drawing of peripherin/RDS and the altered amino acids as a result of the mutations we studied. There were seven point mutations (G86R, K153R, D157N, G167D, R172W, P210S, and R220Q); two mutations involving the deletion/insertion of one or more base pairs (bp), resulting in premature termination codons (I32ins and K193del); and

[41] R. S. Molday, *Prog. Retin. Eye Res.* **13,** 271 (1994).
[42] O. L. Moritz and R. S. Molday, *Invest. Ophthalmol. Vis. Sci.* **37,** 352 (1996).
[43] K. Boesze-Battaglia, O. P. Lamba, A. A. Napoli, S. Sinha, and Y. Guo, *Biochemistry* **37,** 9477 (1998).
[44] G. H. Travis, M. B. Brennan, P. E. Danielson, C. A. Kozak, and J. G. Sutcliffe, *Nature (London)* **338,** 70 (1989).
[45] A. C. Bird, *Am. J. Ophthalmol.* **119,** 543 (1995).
[46] S. Kohl, I. Giddings, D. Besch, E. Apfelstedt-Sylla, E. Zrenner, and B. Wissinger, *Acta Anat.* **162,** 75 (1998).
[47] R. G. Weleber, *Digit. J. Ophthalmol.* Vol. 5 (1999).

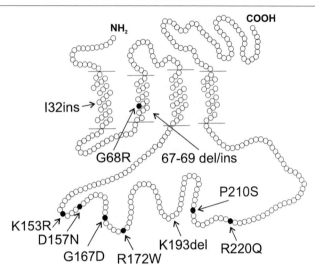

FIG. 5. Schematic of the peripherin/RDS protein, showing the loci of mutations. Note that I32ins and K193del mutations cause premature termination codons predicted to truncate the protein by 303 and 131 amino acids, respectively; the 67–69 del/ins mutation would be predicted to disrupt the second transmembrane domain.

one mutation consisting of an 8-bp replacement of 5 bp spanning codons 67–69 (67–69 del/ins); the latter three mutations are presumed to be functional *null* alleles.[8]

Dark-adapted ERG photoresponses were recorded in 23 patients. Figure 6A shows the results in a patient with the P210S mutation as a representative of 9 of 23 patients with normal rod and cone photoresponses. The patient with the G68R mutation in Fig. 6B represents 11 of 23 patients with abnormal rod and cone photoresponses. Results of all patients are summarized in Fig. 6C. Twenty patients show either normal photoresponses or a proportional reduction of rod and cone photoresponse maximum amplitudes compared with mean normal. The three exceptions are the patients with K153R and R172W mutations. K153R patients show a tendency toward greater rod abnormality compared with cones and the R172W patient shows a greater cone abnormality compared with rods. This tendency existed both in terms of maximum amplitudes (Fig. 6C) and activation kinetics (Fig. 6D).

On the basis of clinical examination, heterozygotes with different *peripherin/RDS* gene mutations showed variability of clinically determined phenotype and are associated with diagnoses of pattern dystrophy, macular degeneration, and widespread retinal degeneration. The results of ERG photoresponses allow three photoreceptor-specific phenotypic groupings

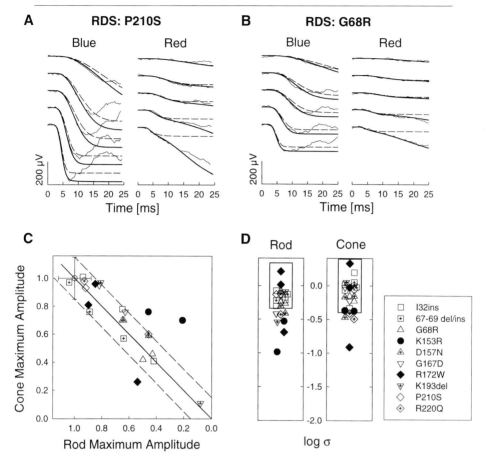

FIG. 6. (A and B) Dark-adapted ERG photoresponses in patients with P210S and G68R mutations in the *peripherin/RDS* gene. See the caption to Fig. 3A for description. (C) Plot of cone-versus-rod maximum amplitude, normalized by respective mean normal values (open circle and error bars = mean ± SD). Hypothetical loci of equal rod and cone reduction lie on the diagonal solid line; dashed lines define the uncertainty around the equal loss line. (D) Sensitivity (log σ) of rod and cone transduction activation is shown as a logarithm of the fraction of mean normal. Boxes define normal limits (mean ± 2 SD).

with intrafamilial consistency of functional phenotype: equal rod and cone, rod worse than cone, or cone worse than rod dysfunction. At the earliest stages of the disease, the majority of mutations showed evidence of normal rod and cone function. As the disease progressed, different patterns of dysfunction emerged. Results in more severely affected heterozygous patients appeared to reflect a progression of the patterns found in more mildly affected heterozygous patients.

Equal Rod and Cone Disease

In 7 of 10 mutations (G86R, D157N, G167D, R220Q, I32ins, 67–69 del/ins, and K193del), there is evidence of retina-wide involvement of equal rod and cone disease. This group of patients includes all putative *null* mutants, which would be expected to show a haploinsufficiency phenotype. Comparison can be made with the *rds*/+ mouse phenotype, which is also believed to exemplify the haploinsufficiency phenotype for the *peripherin/RDS* gene.[48] Histology in *rds*/+ mice shows equal relative reduction of rod and M-cone nuclei during the postnatal 3 months,[49] consistent with the functional results in patients with putative *null* alleles. Interspecies comparison of retinal functional results is complicated by the use of cone ERG *b*-waves (presumed to have inner retinal origins) in the mouse[49] and cone ERG photoresponses (presumed to have photoreceptor origins) in patients. Some of the point mutations that also show a phenotype with equal rod and cone involvement suggest the functional importance of these codons to be equal in rods and cones.

Rod Worse than Cone Disease

Patients with the *RDS:*K153R mutation show a disease pattern with greater rod than cone involvement, similar to those with *rhodopsin* mutations. Both patients tested show a reduction in rod photoresponse maximum amplitude, which would be consistent with a reduction in total OS membrane area caused by decreased OS length or receptor loss.[50,51] Both patients also show abnormal rod but normal cone phototransduction activation kinetics. It can be speculated that the observed abnormalities in rod activation are secondary to structural alterations of the ROS membrane structure caused by the mutant protein. It is also possible that concentration profiles of transduction proteins within the photoreceptor change secondary to the abnormalities caused by the mutant protein.[52]

[48] J. Ma, J. C. Norton, A. C. Allen, J. B. Burns, K. W. Hasel, J. L. Burns, J. G. Sutcliffe, and G. H. Travis, *Genomics* **28,** 212 (1995).
[49] T. Cheng, N. S. Peachey, S. Li, Y. Goto, Y. Cao, and M. I. Naash, *J. Neurosci.* **17,** 8118 (1997).
[50] D. C. Hood and D. G. Birch, *Invest. Ophthalmol. Vis. Sci.* **35,** 2948 (1994).
[51] M. A. Reiser, T. P. Williams, and E. N. Pugh, Jr., *Invest. Ophthalmol. Vis. Sci.* **37,** 221 (1996).
[52] J. Lem, N. V. Krasnoperova, P. D. Calvert, B. Kosaras, D. A. Cameron, M. Nicolo, C. L. Makino, and R. L. Sidman, *Proc. Natl. Acad. Sci. U.S.A.* **96,** 736 (1999).

Cone Worse than Rod Disease

Patients with R172W mutations sampled in this work either had normal photoresponses or showed a pattern of greater cone than rod involvement. Codon 172 appears to be the most frequently reported *peripherin/RDS* mutation.[9,53-60] The clinical disease usually presents between the late teens and early thirties as macular abnormalities, and slowly progresses to involve more of the posterior pole. It is difficult to compare the ERG photoresponse results of the current work with traditional ERG *b*-wave analyses reported in the literature. The photoresponse results allow the speculation that codon 172 is more important for the function of the peripherin/RDS-ROM-1 complex in cones than in rods. This speculation is consistent with *in vitro* work that found no difference in heterotetramer formation between wild-type and R172W peripherin/RDS and rod-specific wild-type ROM-1.[61]

Conclusion

ERG photoresponse recordings constitute a simple, direct, and *in vivo* method to quantify rod and cone dysfunction for better understanding of the initial functional insult in human photoreceptor degenerations. Naturally occurring or genetically engineered models of the human disease are being used increasingly to elucidate disease mechanisms. ERG photoresponses performed similarly in humans and animals[62] may provide a common ground for interspecies comparison of dysfunction and allow the outcomes of potential forms of treatment applied to animal models to be extrapolated to human patients.

[53] J. J. Wroblewski, J. A. Wells, A. Eckstein, F. Fitzke, C. Jubb, T. J. Keen, C. Inglehearn, S. Bhattacharya, G. B. Arden, M. Jay, and A. C. Bird, *Ophthalmology* **101**, 12 (1994).
[54] M. Nakazawa, Y. Wada, and M. Tamai, *Retina* **15**, 518 (1995).
[55] C. Reig, A. Serra, E. Gean, M. Vidal, J. Arumi, M. D. de la Calzada, and M. Carballo, *Ophthal. Genet.* **16**, 39 (1995).
[56] B. Piguet, E. Heon, F. L. Munier, P. A. Grounauer, G. Niemeyer, N. Butler, D. F. Schorderet, V. C. Sheffield, and E. M. Stone, *Ophthal. Genet.* **17**, 175 (1996).
[57] S. M. Downes, A. M. Payne, D. A. R. Bessant, F. W. Fitzke, G. E. Holder, S. S. Bhattacharya, and A. C. Bird, *Invest. Ophthalmol. Vis. Sci.* **38**, S798 (1997).
[58] S. M. Downes, G. E. Holder, A. M. Payne, D. A. R. Bessant, F. W. Fitzke, S. S. Bhattacharya, and A. C. Bird, in "International Society for Clinical Electrophysiology of Vision," p. 20. ATD Press, Hradec Kralove, Czech Republic, 1998.
[59] A. M. Payne, S. M. Downes, D. A. R. Bessant, A. C. Bird, and S. S. Bhattacharya, *Am. J. Hum. Genet.* **62**, 192 (1998).
[60] U. Ekstrom, S. Andreasson, V. Ponjavic, M. Abrahamson, O. Sandgren, P. Nilsson-Ehle, and B. Ehinger, *Ophthal. Genet.* **19**, 149 (1998).
[61] A. F. X. Goldberg and R. S. Molday, *Proc. Natl. Acad. Sci. U.S.A.* **93**, 13726 (1996).
[62] A. V. Cideciyan, T. S. Aleman, J. Bennett, and E. Banin, in "Vision Science and its Applications," pp. 60–63. Optical Society of America, Washington, D.C., 1999.

Acknowledgments

This work was supported in part by EY-05627, the Foundation Fighting Blindness (Hunt Valley, MD), and the Whitaker Foundation (Rosslyn, VA). The author is grateful for collaborations with Drs. S. G. Jacobson, D. C. Hood, T. D. Lamb, E. N. Pugh, Jr., C. M. Kemp, E. M. Stone, V. C. Sheffield, J. Nathans, J. Bennett, and A. H. Milam.

[42] Spectral Sensitivities of Human Cone Visual Pigments Determined *in Vivo* and *in Vitro*

By ANDREW STOCKMAN, LINDSAY T. SHARPE, SHANNATH MERBS, and JEREMY NATHANS

Introduction

Human color vision is trichromatic. It depends on three cone photoreceptors, each of which responds univariantly to absorbed quanta with different spectral sensitivity. The three types are now conventionally referred to as S (short-wavelength sensitive), M (middle-wavelength sensitive), and L (long-wavelength sensitive), according to the relative spectral positions of their peak sensitivities (see Fig. 1A). In the older literature and in some genetics literature, they are more often referred to as blue, green, and red. The overlapping spectral sensitivities of the three cone types (see Fig. 1A) are determined by the molecular properties of the photopigment that each contains. By comparing their different quantal catches, the brain obtains information about the spectral composition of the light arriving at the photoreceptors. Each human cone pigment is encoded by a separate gene; those encoding the M and L cone pigments are arranged in a head-to-tail tandem array on the X chromosome.[1]

That trichromacy is a property of the eye rather than of the physics of light was first formally postulated in 1802 by Thomas Young.[2] In 1860, James Clerk Maxwell described an instrument for producing and mixing monochromatic lights in defined proportions, and with this instrument Maxwell made the first careful, quantitative measurements of color matching and trichromacy.[3] However, the color-matching data of normal trichromats, obtained under standard viewing conditions, cannot uniquely define the

---

[1] J. Nathans, D. Thomas, and D. S. Hogness, *Science* **232,** 193 (1986).
[2] T. Young, *Phil. Trans. R. Soc.* **92,** 20 (1802).
[3] J. C. Maxwell, *Phil. Trans. R. Soc.* **150,** 57 (1860).

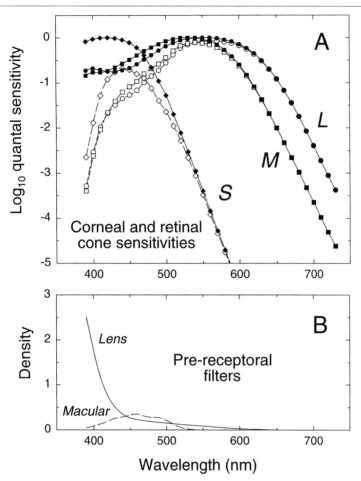

FIG. 1. Cone spectral sensitivities. (A) The spectral sensitivities of the L (circles) M (squares), and S (diamonds) cones measured at the cornea (open symbols, dashed lines) adjusted to the retinal level (filled symbols, solid lines) by removing the filtering effects of the macular and lens pigments. The sensitivities are linear transformations of color-matching functions guided by the spectral sensitivities of dichromats and S cone monochromats. The L cone spectral sensitivity takes into account diversity in the normal population. It is a weighted mixture of the two major polymorphic pigment variants L(S180) and L(A180) (see text) according to the ratio 63 to 37%. (B) Estimates of the optical density spectra of the macular (dashed line) and lens (solid line) pigments. [Adapted from Ref. 7.]

three fundamental sensitivity curves (with the exception of parts of the S curve[4–6]).

In the twentieth century, a number of strategies have been applied to determine the three human cone spectral sensitivities (for a review, see Ref. 7). One approach uses psychophysical techniques that isolate single cone sensitivities *in vivo* by exploiting the selective desensitization caused by either steady[8–11] or transient[12,13] chromatic adaptation. With these techniques it is possible to isolate psychophysically the L and M cones of normal subjects throughout the spectrum[5,13] (see dotted triangles, Fig. 7), and the S cones from short wavelengths to about 540 nm[6,14] (see diamonds, Fig. 7). The cone spectral sensitivities can also be defined by using the constraints imposed by color matching or spectral sensitivity data obtained from three types of congenital, partially colorblind individuals, called dichromats, who lack one of the three cone types. Of these, those lacking L cones (protanopes) and M cones (deuteranopes) are common but those lacking S cones (tritanopes) are rare.[15] König and Dieterici used this approach in 1893 to derive a set of cone sensitivity curves that are substantially correct in their shapes and locations along the wavelength axis,[16] and this work has since been refined by many researchers.[7,17–20] Relevant data may also be obtained from the much rarer congenital monochromats (e.g., S or blue cone monochromats) who lack two of the three cone types.[6] Over the past several decades, the techniques of fundus reflectometry,[21] microspectrophotome-

---

[4] M. M. Bongard and M. S. Smirnov, *Doklady Akad. S.S.S.R.* **102**, 111 (1954).
[5] A. Stockman, D. I. A. MacLeod, and N. E. Johnson, *J. Opt. Soc. Am.* **10**, 2491 (1993).
[6] A. Stockman, L. T. Sharpe, and C. C. Fach, *Vision Res.* **39**, 2901 (1999).
[7] A. Stockman and L. T. Sharpe, in "Color Vision: From Genes to Perception" (K. Gegenfurtner and L. T. Sharpe, eds.). Cambridge University Press, Cambridge, 1999.
[8] W. S. Stiles, *Proc. R. Soc. London B* **127**, 64 (1939).
[9] H. I. De Vries, *Physica* **14**, 367 (1948).
[10] W. S. Stiles, *Science* **145**, 1016 (1964).
[11] G. Wald, *Science* **145**, 1007 (1964).
[12] P. E. King-Smith and J. R. Webb, *Vision Res.* **14**, 421 (1974).
[13] A. Stockman, D. I. A. MacLeod, and J. A. Vivien, *J. Opt. Soc. Am.* **10**, 2471 (1993).
[14] W. S. Stiles, *Coloq. Probl. Opt. Vis. (UIPAP, Madrid)* **1**, 65 (1953).
[15] L. T. Sharpe, A. Stockman, H. Jagle, and J. Nathans, in "Color Vision: From Genes to Perception" (K. Gegenfurtner and L. T. Sharpe, eds.). Cambridge University Press, Cambridge, 1999.
[16] A. König and C. Dieterici, *Z. Psychol. Physiol. Sinnesorg* **4**, 241 (1893).
[17] F. H. G. Pitt, "Medical Research Council Special Report Series No. 200." His Majesty's Stationery Office, London, 1935.
[18] S. Hecht, *Doc. Ophthal.* **3**, 289 (1949).
[19] W. A. H. Rushton, D. S. Powell, and K. D. White, *Vision Res.* **13**, 1993 (1973).
[20] V. C. Smith and J. Pokorny, *Vision Res.* **15**, 161 (1975).
[21] W. A. H. Rushton, *J. Physiol.* **176**, 24 (1965).

try,[22] single-cell electrophysiology,[23] and electroretinography[24] have also been applied to the study of human cone pigment spectral sensitivities. The most recently developed approach to this problem is the *in vitro* study of recombinant cone pigments produced in tissue culture cells.[25–28]

Each of these techniques has strengths and weaknesses. Work based on the perceptions of dichromats and monochromats assumes that their color vision is a "reduced" form of normal color vision[3,29]; that is, that their surviving cones have the same spectral sensitivities as their counterparts in color-normal trichromats. However, it is now known that not all dichromats with alterations in the M or L cones conform to the reduction hypothesis, either because they have hybrid visual pigments or because they have multiple photopigment genes (Fig. 2[30,31]). Only M/L (i.e., X chromosome-linked or red–green) dichromats with a single, normal visual pigment gene or with multiple genes that produce identical visual pigments conform completely to the reduction hypothesis. This genetic complexity calls into question the conclusions of previous studies in which the genotypes of the M/L dichromats were unknown.

Desensitization techniques used to separate cone responses psychophysically or in the electroretinogram are limited by the requirement for a minimum separation between the relevant spectral sensitivity curves. As a result, the isolation of cones containing L–M or M–L hybrid pigments—present in the approximately 6% of Caucasian males with X-linked anomalous trichromacy (a phenotype in which trichromatic color vision is present but reduced in discriminatory power[15])—from the accompanying normal M or L cones has been difficult or impossible to achieve psychophysically owing to the similarities between the spectral sensitivities of these pigments.[32] Microspectrophotometry and especially fundus reflectometry have been limited by a low signal-to-noise (S/N) ratio, and microspectrophotometry and single-cell electrophysiology are limited by the requirement for fresh

---

[22] H. J. A. Dartnall, J. K. Bowmaker, and J. D. Mollon, *Proc. R. Soc. (London) Ser. B* **220,** 115 (1983).
[23] J. L. Schnapf, T. W. Kraft, and D. A. Baylor, *Nature (London)* **325,** 439 (1987).
[24] M. Neitz, J. Neitz, and G. H. Jacobs, *Vision Res.* **35,** 2095 (1995).
[25] D. D. Oprian, A. B. Asenjo, N. Lee, and S. L. Pelletier, *Biochemistry* **30,** 11367 (1991).
[26] S. L. Merbs and J. Nathans, *Nature (London)* **356,** 431 (1992).
[27] S. L. Merbs and J. Nathans, *Science* **258,** 464 (1992).
[28] A. B. Asenjo, J. Rim, and D. D. Oprian, *Neuron* **12,** 1131 (1994).
[29] A. König and C. Dieterici, *Sitz. Akad. Wiss. (Berlin)* 805 (1886).
[30] J. Nathans, T. P. Piantanida, R. L. Eddy, T. B. Shows, and D. S. Hogness, *Science* **232,** 302 (1986).
[31] S. S. Deeb, D. T. Lindsey, Y. Hbiya, E. Sanocki, J. Winderickx, D. Y. Teller, and A. G. Motulsky, *Am. J. Hum. Genet.* **51,** 687 (1992).
[32] T. P. Piantanida and H. G. Sperling, *Vision Res.* **13,** 2033 (1973).

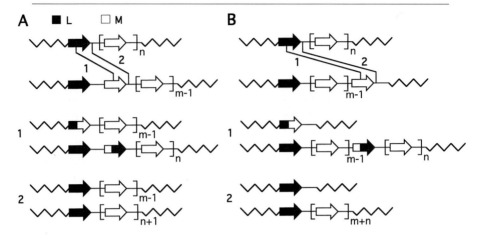

FIG. 2. Unequal recombination within the tandem array of L and M pigment genes responsible for the common anomalies of color vision. Each gene is represented by an arrow: the base corresponds to the 5' end and the tip to the 3' end. Filled arrows, L pigment genes; open arrows, M pigment genes. Unique flanking DNA is represented by zig-zag lines, and homologous intergenic DNA by straight lines. The total number of M pigment genes per array is indicated by m and n. For each recombination event, the reciprocal products are shown. (A) Unequal homologous recombination between two wild-type gene arrays, each containing one L pigment gene and a variable number of M pigment genes. In this example, recombination occurs within the first M pigment gene repeat unit. Intragenic and intergenic recombination events are indicated by 1 and 2, respectively. (B) The special case in which unequal homologous recombination occurs between the most 5' gene in one visual pigment gene array (an L pigment gene) and the most 3' gene in a second visual pigment gene array (an M pigment gene) thereby producing a single gene dichromat genotype. These single-gene recombination products would arise in (A) when $m = 1$. An intragenic recombination event (crossover 1) produces an array with a single 5' L–3' M hybrid gene resulting in a dichromatic (reduced) phenotype: either a classic protanopia (i.e., missing the L pigment, but retaining the normal M pigment) or an anomalous protanopia (i.e., missing the L pigment, but possessing a shifted M pigment). An intergenic recombination event (crossover 2) produces an array with a single L pigment gene also resulting in a dichromatic phenotype: a classic deuteranopia (i.e., missing the M pigment, but retaining the normal L pigment). When a 5' L–3' M hybrid gene is paired with a normal M pigment gene, an anomalous trichromatic phenotype, protanomalous trichromacy or protanomaly, results (i.e., a shifted L pigment is paired with a normal M pigment). In contrast, when a downstream 5' M–3' L hybrid gene is paired with a normal L pigment gene (see crossover 1 in panel (A), lower recombination product), deuteranomalous trichromacy or deuteranomaly results (i.e., a normal L pigment is paired with a shifted M pigment).

human retinal tissue. A strength of psychophysical, electrophysiological, and electroretinographic methods is that the signal amplification produced by the photoreceptor permits accurate measurements over a range of at least 4–5 log units in visual pigment sensitivity. Electroretinography appears to hold promise for future investigations as it is noninvasive and relatively rapid, and unlike classic psychophysical testing it does not require a high level of cooperation and sustained attention from the subject.

Sensitivity measurements made by microspectrophotometry and single-photoreceptor electrophysiology are made transversely through the photoreceptor outer segment, and measurements of recombinant visual pigment absorbance in solution are made with the pigment molecules oriented randomly. In the living human eye, absorbance occurs axially along the outer segment, so that sensitivity is affected by waveguiding[33] and self-screening.[34] Measurements of recombinant pigment absorbance are further handicapped by being accurate to only within approximately 1 log unit of the peak sensitivity, thus encompassing only a limited range of wavelengths near the peak sensitivity (λ_{max}). Recombinant pigments also differ from their *in situ* counterparts with respect to posttranslational modifications, local lipid environment, and the effects of detergent solubilization, which is known to produce blue shifts of several nanometers in some pigments.[35] However, this approach has the virtue that any visual pigment sequence can be created by site-directed mutagenesis, the recombinant pigment is studied free of other visual pigments, and the experiments do not require recruiting and screening of human subjects.

Some forms of sensitivity measurement, such as psychophysics and electroretinography, are made relative to light entering the eye at the cornea, whereas other forms of measurement, such as microspectrophotometry, electrophysiology, and visual pigment absorbance, are made relative to light at the isolated photoreceptor or photopigment. Consequently, before they can be compared, sensitivity curves must be adjusted to account for prereceptoral absorption. Figure 1B shows the changes in cone spectral sensitivity caused by the lens and macular pigment, two pigments that lie between the cornea and the photoreceptors, and that absorb mainly short-wavelength light. Notice that, owing to these prereceptoral filters the λ_{max} values measured at the cornea are substantially longer in wavelength than those measured at the photoreceptor, particularly in the case of the S cones, the λ_{max} of which shifts by more than 20 nm. Another consideration is that the effective photopigment optical density is higher in *in vivo* measurements, because light travels axially along the outer segment, than in *in vitro* measurement, in which light is passed transversely through the outer segment (such as in microspectrophotometry). For comparisons to be made, adjustments in photopigment optical density must be applied; but unlike the adjustments for the lens and macular pigment, such adjustments do not

[33] J. M. Enoch, *J. Opt. Soc. Am.* **51,** 1122 (1961).
[34] G. S. Brindley, *J. Physiol.* **122,** 332 (1953).
[35] G. Wald and P. K. Brown, *Science* **127,** 222 (1958).

TABLE I
Absorbance Spectrum Peaks (λ_{max} + SD) of Human Normal and Hybrid Cone Pigments

	In vitro				In vivo	
Genotype	Recombinant pigments (Refs. 26 and 27)	Recombinant pigments (Ref. 28)	Suction electrode (Refs. 62 and 63)	Microspectro-photometry (Ref. 22)	Psycho-physics (Ref. 43)	Electro-retinography (Ref. 24)
S	426.3 ± 1.0	424.0[a]	—	419.0 ± 3.6[b]	418.9 ± 1.5[b,c]	—
M(A180)/ L1M2(A180)	529.7 ± 2.0	532 ± 1.0	531[d]	530.8 ± 3.5[b]	527.8 ± 1.1	530
L2M3(A180)	529.5 ± 2.6	532 ± 1.0	—	—	528.5 ± 0.7	530
L3M4(S180)	533.3 ± 1.0	534 ± 1.0	—	—	531.5 ± 0.8	—
L4M5(A180)	531.6 ± 1.8	—	—	—	535.4	—
L4M5(S180)	536.0 ± 1.4	538 ± 1.0	—	—	534.2	537
M2L3(A180)	549.6 ± 0.9	—	—	—	—	—
M2L3(S180)	553.0 ± 1.4	559 ± 1.0	—	—	—	—
M3L4	548.8 ± 1.3	555 ± 1.0	—	—	—	—
M4L5	544.8 ± 1.8	551 ± 1.0	—	—	—	—
L(A180)	552.4 ± 1.1	556 ± 1.0	559	—	557.9 ± 0.4	558
L(M2, A180)	—	—	—	—	556.9	—
L(M2, S180)	—	—	—	—	558.5	—
L(S180)	556.7 ± 2.1	563 ± 1.0	564	—	560.3 ± 0.3	563

[a] Value from Ref. 25.
[b] Gene sequence not determined.
[c] Value from Ref. 6.
[d] Value from Ref. 63.

affect the λ_{max} value. When adjusted to the same level, the sensitivity curves obtained by the various methods are in relatively good agreement, especially near λ_{max}, but some differences remain (Table I).[7]

One potential source of variability associated with measurements in the living eye is uncertainty regarding the lens and macular density corrections that should be made, because both vary considerably between individuals. Another source of variability within and between studies derives from person-to-person differences in cone pigment spectral sensitivities. First inferred,[36] and then later fully established[37] psychophysically, the most prominent of these differences derives from single-nucleotide polymorphisms that create variant M or L pigments in which spectral sensitivity may be shifted by several nanometers. The most common polymorphic

[36] J. Neitz and G. H. Jacobs, *Nature* (London) **323,** 623 (1986).
[37] J. Winderickx, D. T. Lindsey, E. Sanocki, D. Y. Teller, A. R. Motulsky, and S. S. Deeb, *Nature* (London) **356,** 431 (1992).

variation occurs at codon 180 in the L pigment gene where site-directed mutagenesis experiments suggest that the presence of an alanine or a serine results in a shift to shorter or longer wavelengths, respectively, of approximately 4 nm[26] or 2–7 nm.[28] A second complication arises from variation in the number of M and L pigment genes between X chromosomes. In general, each X chromosome array has only a single L pigment gene, whereas the number of M pigment genes varies from one to at least five (Fig. 2).[1,38–41] The presence of more than one M pigment gene, or in the case of deuteranomalous trichromats (subjects with an altered M cone sensitivity) more than one 5' M–3' L hybrid gene and/or M pigment gene, complicates the correlation of genotype and phenotype because evidence indicates that only a subset of the M pigment genes is expressed.[40,42] There is currently no method for determining from the genotype which M or 5' M–3' L hybrid pigment genes are expressed in those individuals who carry multiple copies of these genes in their array. A partial solution to this genetic complexity can be achieved by studying male dichromats whose X chromosomes carry only a single visual pigment gene, an arrangement observed in approximately 1% of human X chromosomes (Fig. 2).[43] This simplified arrangement allows a straightforward correlation to be made between spectral sensitivity and visual pigment sequence, and it eliminates problems associated with dichromats who carry multiple genes that may differ subtly in spectral sensitivity.

In this chapter, we summarize the current status of the spectral sensitivity curves that underlie normal and anomalous human color vision, with an emphasis on *in vivo* psychophysical measurements in genetically well-characterized subjects and *in vitro* measurements with recombinant cone pigments.

Absorption Spectra of Recombinant Cone Pigments

The methods and results outlined in this section are from the work of Merbs and Nathans.[26,27,44,45]

[38] M. Drummond-Borg, S. S. Deeb, and A. G. Motulsky, *Proc. Natl. Acad. Sci. U.S.A.* **86,** 983 (1989).
[39] J. P. Macke and J. Nathans, *Invest. Ophthalmol. Vis. Sci.* **38,** 1040 (1997).
[40] T. Yamaguchi, A. G. Motulsky, and S. S. Deeb, *Hum. Mol. Genet.* **6,** 981 (1997).
[41] S. Wolf, L. T. Sharpe, H. J. Schmidt, H. Knau, S. Weitz, P. Kioschis, A. Poustka, E. Zrenner, P. Lichter, and B. Wissinger, *Invest. Ophthalmol. Vis. Sci.* **40,** 1585 (1999).
[42] J. Winderickx, L. Battisti, A. R. Motulsky, and S. S. Deeb, *Proc. Natl. Acad. Sci. U.S.A.* **89,** 9710 (1992).
[43] L. T. Sharpe, A. Stockman, H. Jägle, H. Knau, G. Klausen, A. Reitner, and J. Nathans, *J. Neurosci.* **18,** 10053 (1998).
[44] S. L. Merbs and J. Nathans, *Photochem. Photobiol.* **56,** 869 (1992).
[45] S. L. Merbs and J. Nathans, *Photochem. Photobiol.* **58,** 706 (1993).

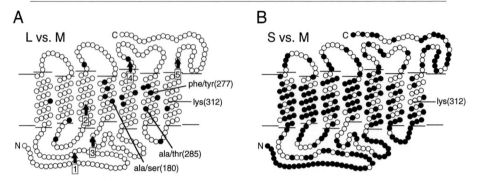

FIG. 3. Topographical model and pairwise comparison of human cone pigments showing amino acid identities (open circles) and differences (filled circles).[1] The seven α-helical segments are shown embedded within the membrane (horizontal lines). N and C denote the amino and carboxy termini, respectively, with the C terminus on the cytoplasmic side of the membrane. (A) L pigment versus M pigment. (B) S pigment versus M pigment. Ala/Ser(180) refers to the common L pigment polymorphism. Two other amino acid differences at codon positions 285 (Ala/Thr) and 309 (Phe/Tyr), which are relevant to the differential spectral tuning of the M and L pigments, are indicated, as well as the location of lysine at codon position 312, the site of covalent attachment of the 11-*cis*-retinal chromophore. The five intron positions in the L and M genes are indicated by numbered vertical arrows.

Cone Pigment Expression Constructs

Cone pigment expression vectors were constructed by inserting human cDNA clones hs37, hs2, and hs7, encoding, respectively, the S, M and L(A180) pigments (Fig. 3),[1] into the mammalian expression plasmid pCIS, which uses the cytomegalovirus (CMV) promotor and enhancer. Standard oligonucleotide-directed mutagenesis procedures were used to prepare single-amino acid substitutions. cDNAs encoding M–L hybrids were prepared by digesting either the M or L pigment cDNAs to varying extents with exonuclease III, followed by digestion with S1 nuclease. The resulting cDNA fragment was then used to prime synthesis on a single-stranded template containing L or M pigment cDNA, respectively. The partial heteroduplex products of this reaction were transformed into an *Escherichia coli* strain defective in mismatch repair and appropriate hybrids were identified by oligonucleotide hybridization and single-track sequencing. Prior to transfection, the entire insert was sequenced. To increase translation efficiency, the 5' untranslated region of the L and M pigment cDNAs was replaced with the last 10 base pairs (bp) of the bovine rhodopsin 5' untranslated region, a sequence known to give high levels of opsin expression with the pCIS vector.[46]

[46] J. Nathans, C. J. Weitz, N. Agarwal, I. Nir, and D. S. Papermaster, *Vision Res.* **29,** 907 (1989).

Production and Reconstitution of Recombinant Cone Pigments

Cone pigments were expressed in human embryonic kidney cell line 293S (ATCC CRL 1573) after transient transfection. In a typical transient transfection, twenty to forty 10-cm plates of 293S cells were transfected with 100 to 200 μg of the pCIS expression plasmid and 10 to 20 μg of pRSV-TAg [a simian virus 40 (SV40) T-antigen expression plasmid] by the calcium phosphate method. Sixty hours after transient transfection, the cells were collected by washing the plates with ice-cold phosphate-buffered saline (PBS) containing 5 mM ethylenediaminetetraacetic acid (EDTA). Cells were pelleted at 4° by centrifugation at 1000g for 10 min. Cell pellets were washed once with 25 ml of ice-cold PBS and then homogenized in 20 ml of ice-cold buffer A [50 mM N-(2-hydroxyethyl)piperazine-N'-(2-ethanesulfonic acid) (HEPES, pH 6.5), 140 mM NaCl, 3 mM MgCl$_2$, and 2 mM EDTA] containing 250 mM sucrose, aprotinin and leupeptin (10 μg/ml each), 0.2 mM phenylmethylsulfonyl fluoride (PMSF), and 1 mM dithiothreitol (DTT) for 45 sec with a Polytron (Brinkmann, Westbury, NY) homogenizer at a setting of 5.5. The homogenate was layered onto 15 ml of 1.5 M sucrose in buffer A, and centrifuged at 4° in a swinging bucket rotor (SW28) at 105,000g for 30 min. Cell membranes were collected from the interface in a volume of 6 to 9 ml and additional DTT was added to increase the concentration by 1 mM. All further manipulations were performed at room temperature either under dim red light or in the dark. Cone pigment reconstitution was accomplished by incubation of the purified cell membranes for 30 min to 2 hr with a 20-fold molar excess of 11-*cis*-retinal added in 1–5 μl of ethanol. More than 95% of the free 11-*cis*-retinal was then removed by diluting the membranes in buffer A containing 4% (w/v) bovine serum albumin and pelleting the membranes in a swinging-bucket rotor (SW28) at 105,000g for 30 min at 4°. The membrane pellet was rinsed with buffer A and resuspended in buffer A containing 2% (w/v) 3-(3-cholamidopropyl)dimethylammonio-1-propane sulfonate (CHAPS). To remove insoluble material, the membrane–detergent mixture was centrifuged at either 10,000g for 5 min at 4° in a microcentrifuge, or at 86,000g for 10 min at 4° in a table-top ultracentrifuge.

Ultraviolet–Visible Absorption Spectroscopy

Absorption spectra were recorded using a Kontron Instruments (Milan, Italy) Uvikon 860 equipped with a water-jacketed cuvette holder. Before photobleaching, four absorption spectra were measured and averaged. The sample was photobleached with light from a 150-W fiber optic light that had been passed through an appropriate filter to maximize pigment bleaching and minimize isomerization of residual retinal (S pigment, 1 min through a 420-nm short wavelength-cutoff filter; M pigment, 2 min through a 580-nm narrow-bandpass filter; and L pigment, 1 min through a 580-nm narrow-bandpass filter). After photobleaching, four absorption spectra were measured and averaged, and the difference absorption spectrum was calculated by subtracting the averaged postbleach curve from the averaged prebleach curve. For each S pigment difference spectrum, a 293S cell pellet was processed in parallel, and the resulting control membrane difference spectrum was subtracted from the S pigment difference spectrum to correct for the change in retinal absorbance that occurs when the >420-nm bleaching light is used. Some difference spectra, especially those of samples centrifuged at 10,000g in the last step, showed a downward sloping background at shorter wavelengths owing to an increase in light scattering as the spectra were collected. From those curves requiring background correction, a difference curve of 293S membranes (without added 11-*cis*-retinal), showing the absorbance change due to light scattering, was appropriately scaled and subtracted to equalize the absorbance values at 300 and 700 nm. An absorption spectrum of all-*trans*-retinal in 2% CHAPS, buffer A, was scaled and added to each S pigment difference curve to correct for the effect of the released all-*trans*-retinal (Fig. 4B).

The most significant sources of experimental variability in determining photobleaching difference spectra are baseline drift due to light scattering and distortions due to released all-*trans*-retinal or other retinal-based photoproducts. As seen in Fig. 4B, released all-*trans*-retinal significantly distorts the uncorrected S pigment photobleaching difference spectrum. The L, M, and L–M hybrid pigment spectra are not affected because retinal absorption is negligible at wavelengths above 500 nm. There is also some variability among experiments involving L, M, and L–M hybrid pigments in the 440-nm region of the photobleaching difference spectra, which most likely arises from photochemical events involving Schiff bases of 11-*cis*-retinal. Long-lived photoproducts appear not to accumulate to significant levels as determined by the minimal differences between spectra obtain during a 5-min period after photobleaching. Figure 4 shows the sum of multiple recombinant cone pigment spectra, giving a weighted-average curve for each pigment [six curves for the S pigment, two curves for the M pigment, seven curves for the L(A180) pigment, and seven curves for the L(S180) pigment].

Absorption Maxima Determined in Vitro

For each photobleaching difference absorption spectrum, the wavelength of maximal absorption (λ_{max}) was determined by calculating the best fitting fifth-order polynomial to a 100-nm segment of the spectrum centered at the approximate peak sensitivity (Fig. 5). Table I lists the absorption

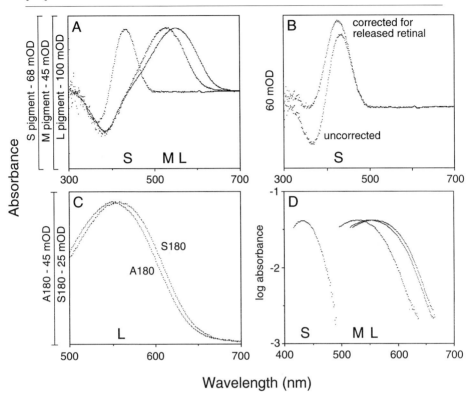

FIG. 4. Superimposed photobleaching difference absorption spectra of recombinant human cone pigments. (A) S, M, and L(A180) pigments. (B) S pigment, uncorrected and corrected for released all-*trans*-retinal. (C) L pigments, containing either alanine or serine at position 180. (D) S, M, L(A180), and L(S180) pigments (from left to right) plotted on a log scale. Only those regions of the spectra that are greater than 80% of the absorbance maximum on the short-wavelength side and greater that 5% of the absorbance maximum on the long-wavelength side are included. mOD, Optical density units $\times 10^{-3}$. [Adapted from Ref. 45.]

maxima of the normal human cone pigments and the 5' M–3' L and 5' L–3' M hybrid pigments that are commonly encountered in the human population. The hybrid pigments arise from recombination events within introns and therefore produce hybrids in which exons 1 to X ($X = 2, 3, 4$, or 5) derive from either an M or L pigment and exons $X + 1$ to 6 derive from either an L or M pigment, respectively (Fig. 2). As exons 1 and 6 are identical between M and L pigments, the crossover events that produce hybrid pigments are confined to introns 2, 3, and 4. Each hybrid pigment is referred to by an abbreviation that reflects the origin of its exons and, if exon 3 is derived from an L pigment, the identity of the polymorphic

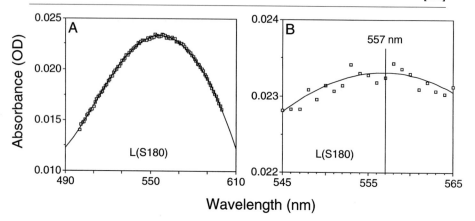

FIG. 5. The fifth-order polynomial calculated as the best fit to the 100-nm region of the spectrum centered about the approximate peak sensitivity of one absorption spectrum of the L(S180) cone pigment. (A) The best fitting polynomial superimposed on the raw data from 500 to 600 nm. (B) Expanded x axis and peak sensitivity determination (557 nm) from the local maximum of the fifth-order polynomial. OD, Optical density units. [Adapted from Ref. 45.]

residue (alanine or serine) at position 180 in the third exon. For example, L4M5(A180) is a hybrid pigment encoded by a gene in which exons 1–4 are derived from an L pigment gene, exons 5 and 6 are derived from an M pigment gene, and position 180 is occupied by alanine. L pigment genes are designated L(A180) or L(S180) to indicate the presence of alanine or serine, respectively, at position 180. Table I lists the absorption maxima determined by two research groups for recombinant pigments *in vitro*.[26-28] Although the values reported by Asenjo *et al.*[28] are systematically 4 ± 2 nm greater than those reported by Merbs and Nathans,[26,27] the two sets of data are in close agreement with respect to absorption differences between pigments. The systematic differences could arise from differences in the lipids, detergents, or buffers used by the different laboratories.

Psychophysical Determination of Cone Spectral Sensitivities

The methods and results outlined in this section are from the work of Sharpe and co-workers.[6,43]

M, L, and 5′ L–3′ M Hybrid Cone Sensitivities: Ascertainment of Subjects

In vivo estimates of the M, L, and 5′ L–3′ M hybrid pigment sensitivities at the cornea can be obtained most simply by studying male dichromats whose X chromosomes carry only a single visual pigment gene. Males with severe color vision deficiencies were recruited and screened by anomaloscopy, using the Rayleigh match.[47] Virtually all such subjects have defects in the M or L cones; S cone defects are far less common and are easily distinguished from M and L cone defects in preliminary screening tests. Prospective subjects had to behave as dichromats in the Rayleigh test; that is, they had to be able to match a spectral yellow light to a juxtaposed mixture of spectral red and green lights by adjusting the intensity of the yellow, regardless of the red-to-green ratio. This implies that quantal absorptions in a single photopigment are responsible for the matches. The choice of the wavelengths and intensities of the primary lights as well as the small field size (2–2.6° diameter) largely preclude absorptions in the S cones or rods from influencing the matches. Of 94 dichromat males identified by anomaloscope testing, 41 were found to carry a single L or L–M hybrid gene by whole genome Southern blot hybridization, and for these subjects the sequences of exons 2–5, which differ between L and M pigments, were determined by direct sequencing of polymerase chain reaction (PCR) products generated with flanking intron primers.[43]

Each single-gene dichromat made repeated matches (3 to 5 times) in random order for 17 different red-to-green mixture ratios, by adjusting only the intensity of the yellow primary light. Their individual matching range slopes (i.e., the slopes of regression lines fitted to their yellow intensity settings for the 17 red–green mixtures) and intercepts (i.e., the yellow intensity required to match the red primary alone) were then determined by a least-squares criterion. From the slope of the regression line, the subjects were categorized as protanopes (missing the L pigment) or deuteranopes (missing the M pigment).[15,48,49]

Flicker Photometry: Methodology

Foveal spectral sensitivities were determined in 37 single-gene dichromats by heterochromatic flicker photometry. A reference light of 560 nm was alternated at a rate of 16 or 25 Hz with a superimposed test light, the wavelength of which was varied in 5-nm steps from 400 to 700 nm. Subjects found the radiance of the test light that eliminated or "nulled" the perception of flicker produced by the alternation of the two lights. To saturate the rods and to desensitize the S cones, and thus prevent both from contributing to spectral sensitivity, the flickering stimuli were superimposed on a

---

[47] L. Rayleigh (J. W. Strutt), *Nature (London)* **25,** 64 (1881).
[48] J. Pokorny, V. C. Smith, G. Verriest, and A. J. L. G. Pinckers, "Congenital and Acquired Color Vision Defects." Grune & Stratton, New York, 1979.
[49] G. Wyszecki and W. S. Stiles, "Color Science." John Wiley & Sons, New York, 1982.

large, violet (430 nm) background with an intensity of 11.0 log quanta sec^{-1} deg^{-2}, which is a strong S cone stimulus and more than 1 log unit more radiant than the rod saturating level.

Because the S cones are desensitized by the background (and in any case make little or no contribution to flicker photometry[50,51]) and the rods are saturated, the null should occur when the test and reference lights produce the same levels of activation in the remaining single class of L, de facto M, or L–M hybrid cone in each single-gene dichromat. The radiance of the test light required to null the reference light as a function of wavelength is therefore an estimate of the spectral sensitivity of the single longer wavelength cone type of each subject.

Flicker Photometry: Apparatus

A Maxwellian-view optical system produced the flickering test stimuli and the steady adapting field, all of which originated from a xenon arc lamp.[43] Two channels provide the 2° in visual diameter flickering test and reference lights. The wavelength of the reference light was always set to 560 nm, while that of the test light was varied from 400 to 700 nm in 5-nm steps. A third channel provided the 18°-diameter, 430-nm adapting fields. The images of the xenon arc were less than 1.5 mm in diameter at the plane of the observer's pupil (i.e., smaller than the smallest pupil diameter, so that changes in pupil size have no effect). Circular field stops placed in collimated portions of each beam defined the test and adapting fields as seen by the observer. Mechanical shutters driven by a computer-controlled square-wave generator were positioned in each channel near focal points of the xenon arc to produce the square-wave flicker seen by the subjects. Fine control over the luminance of the stimuli was achieved by variable, 2.0 log unit linear (Spindler & Hoyer, Göttingen, Germany) or 4.0 log unit circular (Rolyn Optics, Covina, CA) neutral density wedges positioned at image points of the xenon arc lamp, and by insertion of fixed neutral density filters in parallel portions of the beams. The position of the observer's head was maintained by a rigidly mounted dental wax impression.

The radiant fluxes of the test and adapting fields were measured at the plane of the observer's pupil with a calibrated radiometer (model 80X optometer; United Detector Technology, Baltimore, MD) or with a Pin-10 diode connected to a picoammeter (model 486; Keithley, Cleveland, OH). The fixed and variable neutral density filters were calibrated *in situ* for all test and field wavelengths.

[50] A. Eisner and D. I. A. MacLeod, *J. Opt. Soc. Am.* **70,** 121 (1980).
[51] A. Stockman, D. I. A. MacLeod, and D. D. DePriest, *Vision Res.* **31,** 189 (1991).

L, M, and 5' L–3' M Hybrid Pigment Spectral Sensitivity Measurements

Corneal spectral sensitivity measurements were confined to the central 2° of the fovea. At the start of the spectral sensitivity experiment, the subject adjusted the intensity of the 560-nm reference flickering light until satisfied that the flicker was just at threshold. After five settings had been made, the mean threshold setting was calculated and the reference light was set 0.2 log unit above this value. The test light was then added to the reference light in counterphase. The subject adjusted the intensity of the flickering test light until the flicker perception disappeared or was minimized. This procedure was repeated five times at each wavelength. After each setting, the intensity of the flickering test light was randomly reset to a higher or lower intensity, so that the subject must readjust the intensity to find the best setting. The target wavelength was randomly varied in 5-nm steps from 400 to 700 nm. From two to six complete runs were carried out by each subject. Thus, each data point represents between 10 and 30 threshold settings.

Analysis of Flicker Photometry Data

Methods for elimination of clearly discrepant data are described in Sharpe *et al.*[43] In that study the cumulative rejection rate was about 6%. The λ_{max} of the L, M, or L–M hybrid spectral sensitivity at the retina of each subject was estimated by fitting a photopigment template to their flicker photometry data corrected to the retinal level. The photopigment template was derived from the M and L cone spectral sensitivities of Stockman *et al.*,[5] which are based on color-matching data and spectral sensitivity measurements made in dichromats and normal trichromats under conditions of selective desensitization (dotted triangles, Fig. 7). First, the M and L cone spectral sensitivities[5] were individually corrected to the retinal level by removing the effects of macular and lens pigmentation [Eq. (1), below]. Next, they were corrected to photopigment optical density (or absorbance) spectra by adjusting them to infinitely dilute photopigment concentrations [Eq. (2), below].

Calculating Photopigment Spectra from Corneal Spectral Sensitivities and Vice Versa

The calculation of photopigment optical density spectra from corneal spectral sensitivities is, in principle, straightforward, provided that the appropriate values of (1) D_{peak}, the peak optical density of the photopigment, (2) k_{lens}, the scaling constant by which the lens density spectrum [$d_{lens}(\lambda)$, Fig. 1B, solid line] should be multiplied, and (3) k_{mac}, the scaling constant by which the macular density spectrum [$d_{mac}(\lambda)$, Fig. 1B, dashed line] should be multiplied are known. Starting with the quantal cone spectral sensitivity [$S(\lambda)$], the effects of the lens pigment [$K_{lens}d_{lens}(\lambda)$] and the macular pigment [$k_{mac}d_{mac}(\lambda)$] are first removed, by restoring the sensitivity losses that they cause:

$$\log[S_r(\lambda)] = \log[S(\lambda)] + k_{lens}d_{lens}(\lambda) + k_{mac}d_{mac}(\lambda) \tag{1}$$

The functions $d_{lens}(\lambda)$ and $d_{mac}(\lambda)$ are the optical density spectra of the lens and macular pigment depicted in Fig. 1B.[7] They are scaled to the densities that are appropriate for a 2° viewing field for an average observer (a peak macular density at 460 nm of 0.35, and a lens density at, e.g., 400 nm of 1.765; Stockman and Sharpe[51a]). The values k_{mac} and k_{lens} are therefore 1 for the mean 2° spectral sensitivities, but should be adjusted for individual observers or small groups of observers, who are likely to have different lens and macular densities. Because macular pigment density decreases with retinal eccentricity, k_{mac} must also be adjusted for other viewing fields. $S_r(\lambda)$ is the cone spectral sensitivity at the retina, which by convention means in the absence of macular pigment absorption.

To calculate the photopigment optical density of the L cones scaled to unity peak [$S_{OD}(\lambda)$], from $S_r(\lambda)$:

$$S_{OD}(\lambda) = \frac{-\log_{10}[1 - S_r(\lambda)]}{D_{peak}} \tag{2}$$

D_{peak}, the peak optical density, was assumed to be 0.5, 0.5, and 0.4 for the L, M, and S cones, respectively. [$S_r(\lambda)$ should be scaled before applying Eq. (2), so that $S_{OD}(\lambda)$ peaks at one.] The optical densities of the cone photopigments are known to diminish with retinal eccentricity; so these values correspond only to the central 2° of the viewing field. Moreover, these calculations from corneal spectral sensitivities to retinal photopigment optical densities ignore changes in spectral sensitivity that may result from the structure of the photoreceptor or other ocular structures and pigments (unless they are incorporated in the lens or macular pigment density spectra).

The calculation of relative quantal corneal spectral sensitivities from photopigment or absorbance spectra is also straightforward, again if the appropriate values (D_{peak}, k_{lens}, and k_{mac}) are known. First, the spectral

---

[51a] A. Stockman and L. T. Sharpe, *Vision Res.*, in press (1999).

sensitivity at the retina, $S_r(\lambda)$, is calculated from the normalized photopigment optical density spectrum, $S_{OD}(\lambda)$, by the inversion of Eq. (2)[52]:

$$S_r(\lambda) = 1 - 10^{-D_{peak}S_{OD}(\lambda)} \tag{3}$$

Then, the filtering effects of the lens and macular pigments are added back:

$$\log[S(\lambda)] = \log[S_r(\lambda)] - k_{lens}d_{lens}(\lambda) - k_{mac}d_{mac}(\lambda) \tag{4}$$

Scales

A simple polynomial function was devised to describe the logarithm of the L, M, and 5′ L–3′ M hybrid photopigment spectra, after the L cone spectrum had been shifted horizontally along a $\log_{10}(\lambda)$ scale to align it with the M cone spectrum. In deriving this template, and analyzing the spectral sensitivity data, it was assumed that the family of L, M and 5′ L–3′ M hybrid photopigment spectra are invariant in shape when plotted as a function of $\log_{10}(\lambda)$.[53–55] This simplification provides a straightforward means of analyzing the spectral sensitivity data, because the λ_{max} of each photopigment can then be estimated from a simple shift of the polynomial curve.[56]

Attempts have been made previously to simplify cone photopigment spectra by finding an abscissa that produces spectra of a fixed spectral shape, whatever the photopigment λ_{max}. An early proposal was by Dartnall,[57] who described a "nomogram" or fixed template shape for photopigment spectra plotted as a function of wavenumber ($1/\lambda$, in units of cm^{-1}). Another proposal was that the spectra are shape invariant when plotted as a function of $\log_{10}$ frequency or wavenumber $[\log_{10}(1/\lambda)]$,[53,54] which is equivalent to $\log_{10}$ wavelength $[\log_{10}(\lambda)]$ or normalized frequency (λ_{max}/λ). For this scale, Lamb has proposed a template.[55] Barlow has also proposed an abscissa of the fourth root of wavelength ($\lambda^{1/4}$).[58]

Fitting of the retinal photopigment template to the corneal data was carried out by an iterative procedure that simultaneously (1) found the best-fitting shift of the template along the log wavelength scale, (2) adjusted

---

[52] A Knowles and H. J. A. Dartnall, "The Eye," Vol. 2B: "The Photobiology of Vision." Academic Press, London, 1977.
[53] R. J. W. Mansfield, *in* "The Visual System" (A. Fein and J. S. Levine, eds.), p. 89. Alan R. Liss, New York, 1985.
[54] E. F. MacNichol, *Vision Res.* **26,** 1543 (1986).
[55] T. D. Lamb, *Vision Res.* **35,** 3083 (1995).
[56] D. A. Baylor, B. J. Nunn, and J. L. Schnapf, *J. Physiol.* **390,** 145 (1987).
[57] H. J. A. Dartnall, *Br. Med. Bull.* **9,** 24 (1953).
[58] H. B. Barlow, *Vision Res.* **22,** 635 (1982).

the template to a peak photopigment optical density of 0.5, and (3) added back the effects of the best-fitting lens and macular pigment optical densities.[43] In one analysis the model was fitted at all measured wavelengths, and in a second analysis we carried out the fit only for measurements made at wavelengths ≥ 520 nm. Restricting the fit to ≥ 520 nm simplified the fitting procedure, because at those wavelengths macular pigment plays little role, and the lens is relatively transparent (having an average optical density of only 0.10 log unit at 520 nm that declines with wavelength). The ≥ 520-nm fit served, in part, as a control for the full-spectrum fit, and in particular for the reliance on best-fitting macular and lens densities, which could, in principle, distort the λ_{max} estimates. Given that small differences are expected between the two estimates, because the lens density assumed for the partial fit is the population mean density rather than the optimized individual density, the agreement between the two is extremely good.[43]

Systematic errors in both fits would be expected if the various L, M, or L–M hybrid pigments are not shape invariant when plotted against $\log_{10}(\lambda)$ or if the peak photopigment optical density varies with genotype. Such errors would cause small shifts in the λ_{max} estimates between genotypes, but would have little effect on estimates within genotypes. Individual differences in photopigment optical density within a genotype would increase the variability of the λ_{max} estimates within that group.

Figure 6 shows representative examples of the heterochromatic flicker data (symbols) for nine different genotypes, five of which produce protanopia (Fig. 6A) and four of which produce deuteranopia (Fig. 6B). Cone photopigment λ_{max} values obtained from subjects with identical visual pigment amino acid sequences show up to an $\sim$3-nm variation from subject to subject, presumably owing to a combination of inexact (or no) corrections for variation in preretinal absorption, variation in photopigment optical density, optical effects within the photoreceptor, and measurement error. This variation implies that spectral sensitivities must be averaged over multiple subjects with the same genotype to obtain accurate values for a given pigment.

Average values for each genotype, varying in the number of subjects from 1 to 19, are given in Table I (the complete data set for all single-gene dichromats can be seen in Sharpe et al.[43]). Note that to allow comparisons with the in vitro estimates, the λ_{max} values are for the psychophysical spectral sensitivities adjusted to the retinal level (see above). One limitation of single-gene dichromats in determining such estimations is that they do not carry 5' M–3' L hybrid genes, with the result that 5' M–3' L pigments can be studied psychophysically only in deuteranomalous trichromats.

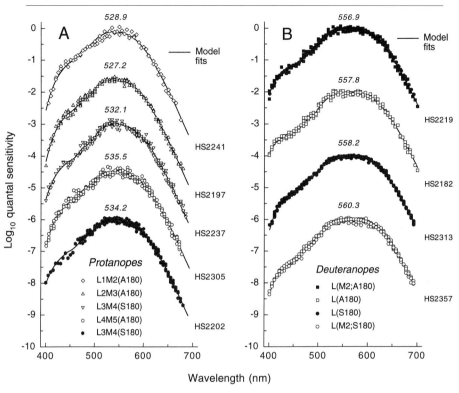

FIG. 6. Representative individual cone spectral sensitivity data (symbols) for nine genotypes found in single-gene dichromats; together with the fits (continuous lines) of the visual pigment template.[43] (A) Genotypes producing protanopia: L1M2(A180), L2M3(A180), L3M4(S180), L4M5(A180), and L4M5(S180). (B) Genotypes producing deuteranopia: L(M2; A180), L(A180), L(M2; S180), and L(S180). The λ_{max} of the fitted photopigment template is given above each curve. Data are from several single runs. (Note that the λ_{max} values are values for individual observers; therefore they do not necessarily correspond to the values averaged over different observers with the same genotype, as given in Table I.) Further details can be found in Ref. 43, in which the same subject code is used, including details of the gene sequences for each subject.

Psychophysical Determinations of S Cone Spectral Sensitivity

S cone spectral sensitivity was measured in five normal trichromats, using selective chromatic adaptation to suppress the M and L cones, and in three S (or blue) cone monochromats. The S cone monochromats are known to lack M and L cone function on genetic grounds.[59] One of the three observers has two X chromosome photopigment genes but an upstream deletion in the region that controls their expression. The two other observers have a single X chromosome photopigment gene with a point mutation that results in a cysteine-to-arginine substitution at position 203 in the opsin.

The same experimental apparatus that was used for the M and L cone measurements (see above) was also used for the S cone measurements, with some important modifications. For measurements in normal subjects, a single 2°-diameter target field was presented in the center of an intense 16°-diameter, yellow (580 nm) background field of 12.10 log quanta sec^{-1} deg^{-2} (5.93 log photopic td or 5.47 log scotopic td), which was chosen to selectively adapt the M and L cones and saturate the rods, but have comparatively little direct effect on the S cones. The target was square-wave flickered at 1 Hz. A much lower temporal frequency was used for the S cone measurements, because S cone pathways are typically sluggish.[22,60] For measurements in the S cone monochromats, the target was presented on an orange (620 nm) background of 11.24 log quanta sec^{-1} deg^{-2} (4.68 log photopic td or 3.36 log scotopic td), which was chosen to saturate the rods. The subject's task was to set the threshold for detecting the flicker as a function of target wavelength. Five settings at each target wavelength were made on each of four runs.

On long-wavelength backgrounds, S cone isolation could be achieved in normal subjects from short wavelengths (the lower limit of the measurements was 390 nm) to ~540 nm, and in the S cone monochromats throughout the spectrum. From 390 to 540 nm the data for the normal subjects and the S cone monochromats agree well, except that—consistent with their extrafoveal fixation—the monochromats have lower macular pigment and S cone photopigment optical densities. The mean centrally measured S cone thresholds, after individually adjusting the S cone monochromat data to normal density values, are shown in Fig. 7 (diamonds). The mean is based on normal and monochromat data up to 540 nm and on monochromat data alone at wavelengths longer than 540 nm.[6]

Absorption Maxima Determined Psychophysically

Figure 7 shows a compilation of the mean psychophysically determined corneal spectral sensitivities for S, M, L(A180), and L(S180) pigment-

---

[59] J. Nathans, I. H. Maumenee, E. Zrenner, B. Sadowski, L. T. Sharpe, R. A. Lewis, E. Hansen, T. Rosenberg, M. Schwartz, J. R. Heckenlively, E. Traboulsi, R. Klingaman, N. T. Bech-Hansen, G. R. LaRoche, R. A. Pagon, W. H. Murphey, and R. G. Weleber, *Am. J. Hum. Genet.* **53**, 987 (1993).

[60] G. S. Brindley, J. J. Du Croz, and W. A. H. Rushton, *J. Physiol.* **183**, 497 (1966).

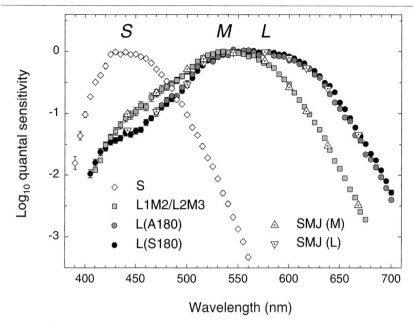

FIG. 7. Comparison of mean spectral sensitivity data: L cone data for 15 L(S180) subjects (black circles), 5 L(A180) subjects (gray circles), and M cone data for 9 protanopes (gray squares)[43]; S cone data for five normal trichromats and three S (or blue) cone monochromats (white diamonds)[6]; L cone data (white dotted inverted triangles) for 12 normal subjects and 4 deuteranopes and M cone data (white dotted triangles) for 9 normal subjects and 2 protanopes measured using transient adaptation.[5]

containing cones.[6,43] The M cone data were obtained from genotyped single- and multiple-gene dichromats carrying either an L1M2 hybrid gene, which encodes a de facto M pigment, an L2M3 hybrid gene, or either of these genes and one or more normal M pigment genes. The absorbance maximum of the L2M3 hybrid pigment does not differ significantly from that of the M pigment as determined psychophysically and by analysis of recombinant pigments (Table I). Therefore data from subjects carrying an L2M3 pigment can be combined with data from subjects carrying a L1M2 pigment, thereby improving the signal-to-noise ratio. Table I lists the λ_{max} of the S pigment determined psychophysically from normal trichromats, using selective desensitization, and from S cone monochromats.

Comparison of Spectral Sensitivities Determined by Different Methods

The spectral sensitivities of the L, M, and some 5' L'–3' M hybrid pigments have also been studied in five dichromats, using electroretinography[24] (Table I). The spectral sensitivity functions were measured twice for each subject and were corrected for preretinal absorption by the lens.[49] The λ_{max} of each function was then determined by translation of a standard visual pigment absorption curve on a log wavenumber axis.[61]

Suction electrode (current) recordings from single human cones have also provided spectral sensitivities for the L(A180) and L(S180) pigments[62] and for the M pigment.[63] In these measurements, spectral sensitivity was estimated by adjusting the intensity of light for as many as 20 wavelengths to produce a criterion response of about 25% of the maximum photocurrent. An eighth-order polynomial[56] was used to provide the estimates of λ_{max} along a log wavenumber axis (see Table I).

For comparison, miscrospectrophotometric estimates (transverse absorbance measurements) of the S cone and M cone λ_{max} values[22] are given in Table I. The values are based on averages of 11 (S cone) and 49 (M cone) spectra records obtained from cone outer segments excised from seven human eyes. The genotype was not determined. The absorbance spectra were constructed from the raw data by plotting them on a $\lambda^{1/4}$ scale. Sixty-nine records were also obtained from L cone phooreceptors, but because their mean λ_{max} value depends on an unknown mixture of L(S180) and L(A180) pigments, it is not tabulated.

As can be seen in Table I, the psychophysical, recombinant pigment, suction electrode, and ERG data are in rough agreement with respect to the λ_{max} values of the normal human cone pigments: M(A180), L(A180), and L(S180). However, as noted in the introduction, the recombinant pigments and those obtained from suction electrode recordings differ from their *in vivo* counterparts with respect to waveguiding, self-screening, and local chemical environment. For instance, compared with the spectroscopy of recombinant pigments and suction electrode recordings, spectra obtained from psychophysics are slightly steeper at longer wavelengths, presumably owing to waveguiding effects. This can result in a slightly shorter λ_{max} estimate.

The agreement between spectral sensitivities estimated *in vivo* and *in vitro* is also good for the L–M hybrid pigments. The data indicate that all of the hybrids derived from L and M pigments have spectral sensitivities between those of the two parental pigments, and that the spectral sensitivity of each hybrid depends on the position of the crossover and on the identity of other polymorphic amino acids, principally alanine or serine at position

---

[61] G. H. Jacobs, J. Neitz, and K. Krogh, *J. Opt. Soc. Am.* **13,** 641 (1996).
[62] T. W. Kraft, J. Neitz, and M. Neitz, *Vision Res.* **38,** 3663 (1998).
[63] T. W. Kraft, personal communication (1998).

180. For each exon, the set of amino acids normally associated with the L or M pigments produce, respectively, spectral shifts to longer or shorter wavelengths, thus producing a monotonic relationship between the λ_{max} and the fraction of the hybrid pigment derived from the parental L or M pigment.

The primary determinants of the spectral shift are located in exon 5, as seen by the clustering of the λ_{max} values of 5' L–3' M or 5' M–3' L hybrids within 10 nm of those of the normal M or L pigments, respectively (see Table I). A comparison of the spectral sensitivities of the L(A180) and L(S180) pigments with those of the L4M5(A180) and L4M5(S180) pigments indicates that L/M sequence differences in exon 5 result in a spectral shift of approximately 25 nm. Site-directed mutagenesis experiments have shown that threonine/alanine and tyrosine/phenylalanine differences at codons 285 and 309, respectively (see Fig. 3), account for essentially all of the effects of exon 5.[28,45] These data are consistent with inferences based on a comparison of primate visual pigment gene sequences and cone spectral sensitivity curves.[64–66]

Considerable effort has been made to determine the effect of the serine/alanine polymorphism at position 180.[24,26,28,37,67,68] All of the reported measurements have shown that serine-containing pigments are red shifted with respect to alanine-containing pigments, but the magnitude of the shift has been controversial. On the basis of anomaloscope matches, Sanocki et al. estimated that the substitution of alanine by serine in the L pigment or within 5' L–3' M hybrid pigments results in a red shift of λ_{max} by 4.3, 3.5, and 2.6–2.7 nm for deuteranopes, protanopes, and normal trichromats, respectively.[67,68] Neitz and Jacobs estimated a red shift of approximately 3 nm from Rayleigh match data obtained from 60 normal trichromats.[69] Red-shift estimates of 5–7 nm were obtained from five dichromats by electroretinography, but this interpretation was complicated by additional amino acid differences between the pigments.[24] In vitro measurements of recombinant pigments by Merbs and Nathans showed a red shift of 4.3–4.4 nm in the L pigment and in 5' L–3' M and 5' M–3' L hybrid pigments.[26,27] In contrast, Asenjo et al. found a range of red shifts depending on the parental pigment: a 2-nm shift in a 5' L–3' M hybrid and in the M pigment,

---

[64] M. Neitz, J. Neitz, and G. H. Jacobs, Science **252**, 971 (1991).
[65] R. E. Ibbotson, D. M. Hunt, J. K. Bowmaker, and J. D. Mollon, Proc. R. Soc. (London) B **247**, 145 (1992).
[66] A. J. Williams, D. M. Hunt, J. K. Bowmaker, and J. D. Mollon, EMBO J. **11**, 2039 (1992).
[67] E. Sanocki, D. T. Lindsey, J. Winderickx, D. Y. Teller, S. S. Deeb, and A. G. Motulsky, Vision Res. **33**, 2139 (1993).
[68] E. Sanocki, S. K. Shevell, and J. Winderickx, Vision Res. **34**, 377 (1994).
[69] J. Neitz and G. H. Jacobs, Vision Res. **30**, 621 (1990).

a 4-nm shift in a 5′ M–3′ L hybrid, and a 7-nm shift in the L pigment.[28] The spectral sensitivity curves obtained from single-gene dichromats show a mean separation of 2.5–2.9 nm, depending on which of two methods is used to calculate the spectral sensitivity curve, and whether individual or mean data are used.[43] These results are consistent with other psychophysical measures of the variability of the L cone λ_{max} in the normal population.[70,71]

Implications for Variant Color Vision and Derivation of Human Cone Fundamentals

The absorption differences that distinguish closely related L–M hybrid pigments and A180 and S180 variants most likely account for the observation that anomalous trichromats differ greatly in the location and range of their Rayleigh matching points.[15,48,72] These data support a model of anomalous trichromacy in which any one of many M-like or L-like anomalous pigments can be paired with one of the polymorphic versions of the more similar normal pigment.[27,73] As the spectral sensitivities of the normal or the anomalous pigments shift, the midpoint of the Rayleigh match will shift, and as the separation between the spectral sensitivities of the normal and anomalous pigments increases or decreases, the better or poorer will be the subject's chromatic discrimination.

The existence of polymorphisms among normal M and L pigment genes, most especially the A180/S180 polymorphism, means that a single set of cone fundamentals (i.e., the basic cone spectral sensitivities of trichromatic color vision) will accurately describe the color vision of only a subset of normal trichromats, and that in the construction of an average set of fundamentals it is important that the weighting of polymorphic types within the test population match that in the general population. Thus, the *in vivo* determination of the cone fundamentals requires an analysis of the spectral sensitivity curves for subjects whose visual pigment gene sequences reveal which of the various possible pigments they possess.[7]

In conclusion, by building on advances in molecular biology and exploiting high-precision *in vivo* and *in vitro* techniques, significant progress has been made toward the goal of fully cataloging the rich diversity of cone photopigments that underlie normal and anomalous human color vision.

---

[70] M. A. Webster, *J. Opt. Soc. Am.* **9,** 1419 (1992).
[71] M. A. Webster and D. I. A. MacLeod, *J. Opt. Soc. Am.* **5,** 1722 (1988).
[72] J. D. Mollon, *in* "Color Vision Deficiencies XIII" (C. R. Cavonius, ed.), p. 3. Kluwer Academic, Dordecht, 1997.
[73] J. Neitz, M. Neitz, and P. Kainz, *Science* **274,** 801 (1996).

[43] Molecular Analysis of Human Red/Green Visual Pigment Gene Locus: Relationship to Color Vision

By SAMIR S. DEEB, TAKAAKI HAYASHI, JORIS WINDERICKX, and TOMOHIKO YAMAGUCHI

Overview

Advances in the molecular biology of the human cone visual pigments and the genes that encode them have led to elucidation of the molecular basis of interindividual differences in red–green color vision. The retina has three classes of cone photoreceptors containing blue (short wave)-, green (middle wave)-, or red (long wave)-sensitive photopigments. Retinal cones are used for vision in bright light and for color vision. Normal color vision is trichromatic and is subserved by these three cone photoreceptors. The photopigments have characteristic absorption maxima with wide regions of overlap. Perception of color results from comparison between the output from the three cone photoreceptors. The visual pigments belong to the heptahelical transmembrane receptor family that includes the olfactory receptors. The red and green photopigments are similar in amino acid sequence, differing in 15 positions. Differences at three residues, at positions 180, 277 and 285, account for 25 nm of the 30-nm difference in absorption maxima (λ_{max}) between the absorption maxima of these two pigments.[1,2]

The genes encoding the red and green pigments are arranged in head-to-tail tandem arrays (Fig. 1A) on the X chromosome (Xq28). The arrays are composed of a single red pigment gene (six exons) 5' of one or more green pigment genes (six exons).[1,3] Approximately 25% of white males have a single green pigment gene, 50% have two, while the rest have three (or more) green pigment genes.[4–6] The high degree of homology between the red and green pigment genes (including introns and intergenic sequences) has predisposed the locus to unequal homologous recombination/gene conversion events. These illegitimate events cause a change in the number of green pigment genes (including their total elimination) and the formation

---

[1] J. Nathans, D. Thomas, and D. S. Hogness, *Science* **232,** 193 (1986).
[2] S. L. Merbs and J. Nathans, *Nature (London)* **356,** 433 (1992).
[3] D. Vollrath, J. Nathans, and R. W. Davis, *Science* **240,** 1669 (1988).
[4] A. L. Jorgensen, S. S. Deeb, and A. G. Motulsky, *Proc. Natl. Acad. Sci. U.S.A.* **87,** 6512 (1990).
[5] M. Drummond-Borg, S. S. Deeb, and A. G. Motulsky, *Proc. Natl. Acad. Sci. U.S.A.* **86,** 983 (1989).
[6] J. P. Macke and J. Nathans, *Invest. Ophthalmol. Vis. Sci.* **38,** 1040 (1997).

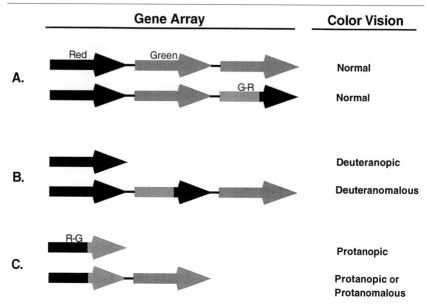

Fig. 1. Gene arrays of males with normal, deutan, and protan color vision defects. G–R and R–G denote 5′ green–red 3′ and 5′ red–3′ green hybrid genes, respectively. Diagram is not to scale.

of red–green hybrid genes.[7] The presence of the red and green pigment genes on the X chromosome and their propensity for unequal recombination account for the relatively high frequency of red–green color vision defects among males (~8% in the United States and northern Europe).

The deletion of green pigment genes leaves a single red pigment gene that is characteristically associated with deuteranopia and, therefore, dichromatic color vision. Exon 5 plays a major role in spectral tuning because it contains the two residues that account for the majority (22 nm) of the spectral difference between the red and green pigments. Therefore, illegitimate recombination that result in the exchange of exon 5 between the red and green pigment genes is sufficient to produce hybrid pigments with large spectral shifts that may have significant effects on color vision (Fig. 2). 5′ Green–red 3′ hybrid (G–R) genes, which encode redlike photopigments, with or without additional green pigment genes, usually are associated with deuteranomaly—a milder type of color vision defect. Surprisingly, G–R hybrid genes similar to those of deuteranomaly are also found in 4–8% of

---

[7] J. Nathans, T. P. Piantanida, R. L. Eddy, T. B. Shows, and D. S. Hogness, *Science* **232,** 203 (1986).

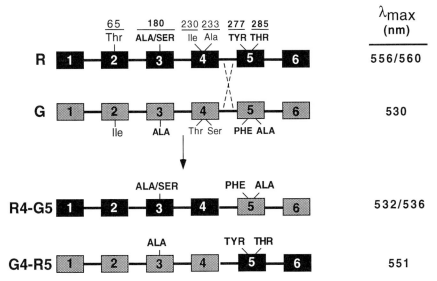

FIG. 2. Unequal homologous recombination (crossover) between the red and green pigment genes to form hybrid genes. Black and gray rectangles represent red (R) and green (G) exons, respectively. Dashed lines represent the point of recombination or crossover. R4–G5 and G4–R5 genes represent hybrids resulting from the crossover in intron 4. The seven amino acid residues (numbers given at top) that differentiate the red from the green pigments are indicated, with those that contribute the majority of the difference in absorption maxima (λ_{max}) between these pigments shown in bold face. Position 180 is polymorphic (alanine or serine) in the general population.

males with normal color vision.[8,9] The hybrid genes in deuteranomals are expressed instead of the normal green pigment genes but are not expressed in individuals with normal color vision. The position of the gene in the array is important for its expression. There is good evidence that only the first two genes of a visual pigment array are expressed in the retina and, therefore, participate in determining the color vision phenotype. Thus in deuteranomals, the hybrid gene occupies the second position whereas in normal individuals it occupies the third or more distal positions (Fig. 3). 5′ Red–green 3′ hybrid genes [which encode a greenlike photopigment (Fig. 2)] are always associated with protan abnormalities (Fig. 1C). Those who have only a single hybrid gene are always protanopic and are therefore dichromats. Those who have additional normal green pigment genes are either protanopic or protanomalous (i.e., a milder defect) depending on

[8] S. S. Deeb, D. T. Lindsey, Y. Hibiya, E. Sanocki, J. Windericks, D. Y. Teller, and A. G. Motulsky, *Am. J. Hum. Genet.* **51,** 687 (1992).
[9] T. Yamaguchi, A. G. Motulsky, and S. S. Deeb, *Hum. Mol. Genet.* **6,** 981 (1997).

A. Expression of the red pigment gene to form red cones.

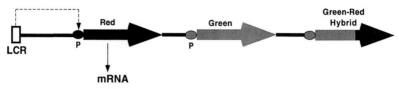

B. Expression of the green pigment gene to form green cones (Normal color vision).

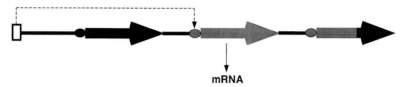

C. Expression of the G-R hybrid gene to form anomalous cones (Deuteranomaly).

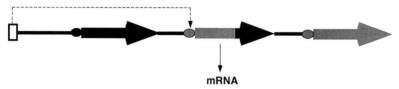

FIG. 3. Model illustrating the role of the position of the green–red hybrid gene in its expression and in determining the color vision phenotype. Red-sensitive cones are postulated to result from the exclusive expression of the red pigment gene as a result of stable coupling of the locus control region (LCR) to the red pigment gene promoter (P). Green-sensitive cones are formed if the LCR preferentially and permanently couples to the proximal green pigment gene promoter and turns on its expression. Genes that occupy a third or more of the distal positions are activated with low probability. Normal color vision results if the green–red hybrid gene occupies the distal position in the array, and deuteranomalous color results if the green–red hybrid gene occupies the second position in the array.

the spectral separation between the hybrid and normal pigments. The greater the separation the milder the defect.

A rare cause of red–green color vision deficiency (occurring in 1 of 64 color-defective males) is a point mutation of a critical cysteine residue of the green pigment (Cys203Arg).[10]

Slight variation in color discrimination capacity was observed among males with normal color vision. This variation was shown to be largely due to a common polymorphism (Ser180Ala) in the red pigment (Fig. 2). The absorption maxima of the photopigment with serine at position 180 is ~4 nm longer than that with alanine. Males who have the serine variant have

---

[10] J. Winderickx, E. Sanocki, D. T. Lindsey, D. Y. Teller, A. G. Motulsky, and S. S. Deeb, *Nature Genet.* **1,** 251 (1992).

more sensitivity to red light than those who have alanine at position 180, as shown by color matching.[11]

This chapter focuses on determining the gross structure, coding sequences, and expression of the red and green pigment genes.

Determining Number and Ratio of X-Linked Red–Green Photopigment Genes

Previously, the number and ratio of the red to green pigment gene sequences in the X-linked arrays of males were determined by conventional agarose gel electrophoresis and Southern blot analysis.[7,8] These and other studies indicated the presence of a single red pigment gene 5' of one or more (average of two among white individuals) green pigment genes per X chromosome.[6,12,13] The total number of genes (but not the ratio of green to red) was also determined by pulsed-field gel electrophoresis (PFGE) of *Not*I-digested genomic DNA followed by Southern blot analysis, because the entire array is flanked by two unique *Not*I sites.[3]

We developed a rapid and reliable polymerase chain reaction (PCR)-based method for determining the number and ratios of these genes in the array.[9] Competitive PCR, using primer pairs that completely match the sequence of both red and green pigment gene promoters and exons 2, 4, and 5 (Fig. 4A) was used for quantitative amplification. Exons 1 and 6 of the red and green pigment genes are identical in sequence and, owing to common polymorphisms, exon 3 cannot be assigned to either the red or green pigment genes.[11,14] A method for determining the genotype at position 180 is described below. Therefore, only the promoter and exons 2, 4, and 5 are informative in this strategy. The amplified segments derived from the red and green pigment genes were subsequently resolved and quantified by single-strand conformation polymorphism (SSCP). The ratios of fragments derived from the green and red pigment genes allowed us to determine the number and exonic composition of these genes in the arrays of male subjects. Inequality in this ratio between the promoter and exonic sequences indicates the presence of red–green hybrid genes, and defines their points of fusion. For example, for an array composed of one red and two green pigment genes, the ratio between green and red segments derived from the promoters and any of the exons would be 2:1. Assume that one of the

---

[11] J. Winderickx, D. T. Lindsey, E. Sanocki, D. Y. Teller, A. G. Motulsky, and S. S. Deeb, *Nature (London)* **356**, 431 (1992).

[12] M. Drummond-Borg, S. Deeb, and A. G. Motulsky, *Proc. Natl. Acad. Sci. U.S.A.* **86**, 983 (1989).

[13] R. Feil, P. Aubourg, R. Heiling, and J. L. Mandel, *Genomics* **6**, 367 (1990).

[14] J. Winderickx, L. Battisti, Y. Hibiya, A. G. Motulsky, and S. S. Deeb, *Hum. Mol. Genet.* **2**, 1413 (1993).

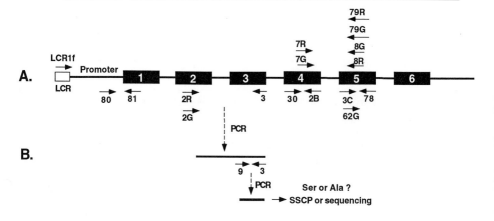

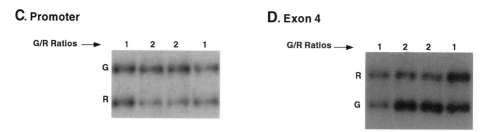

FIG. 4. (A) Position of primers used in amplification reactions. LCR, locus control region. Rectangles with numbers inside represent exons. Arrows denote the direction of primers. Primer identification numbers designated by R or G indicate that they are specific for red or green gene sequences, respectively. The sequence of primers is given in Table I. (B) Strategy for determining whether alanine or serine occupies residue 180 of the red and green pigments. A first round of amplification is performed using either the red- or green-specific primer 2R or 2G, respectively and the nonspecific primer 3. A second round of PCR is performed using primers 9 and 3 and the product is either sequenced or subjected to SSCP analysis. (C) An autoradiograph of an SSCP gel showing separation and ratios of green (G) to red (R) PCR fragments derived from the promoters of four males. (D) An autoradiograph of an SSCP gel showing separation and ratios of green to red PCR fragments derived from exon 4.

green pigment genes is replaced by a G4–R5 hybrid in which the point of fusion is in intron 4. The green-to-red exon 5 ratio in this array would now be 1:2. If the point of fusion were in intron 3, then the green-to-red ratios of exons 4 and 5 would be 1:2. In addition, hybrid genes in which the fusion has occurred in intron 4 can be detected simply by PCR amplification, using green- and red-specific primers in exons 4 and 5, respectively (see below).

TABLE I
SEQUENCE OF PRIMERS

Primer number	Position and sequence (5' → 3')[a]		Location
LCR1F	(−3636)	CAC CCT TCT GCA AGA GTG TGG G	LCR[b]
80	(−190)	*CCAGCAAATCCCTCTGAGCCG	Promoter
81	(41)	GGCTATGGAAAGCCCTGTCCC	Promoter
55	(286)	*AGAAGCTGCGCCACCCGCT	Exon 2
24B	(449)	ACACAGGGAGACGGTGTAGC	Exon 2
2R	(368)	CAGCATTGTGAACCAGGTCTC	Exon 2
2G	(368)	CAGCGTTGTGAACCAGGTCTAT	Exon 2
30	(621)	*TACTGGCCCCACGGCCTGAAG	Exon 4
2B	(785)	CGCTCGGATGGCCAGCCACAC	Exon 4
7G	(730)	ACCCCACTCAGCATCATCGT	Exon 4
7R	(730)	ATCCCACTCGCTATCATCAT	Exon 4
3B	(1025)	CTG CCG GTT CAT AAA GAC ATA G	Exon 5
3C	(786)	GTGGCAAAGCAGCAGAAAGAG	Exon 5
78	(922)	*TTGGCAGCAGCAAAGCATGCG	Exon 5
8G	(878)	GAAGCAGAATGCCAGGACC	Exon 5
62G	(855)	TGATGGTCCTGGCATTCTGCT	Exon 5
8R	(878)	GACGCAGTACGCAAAGAT	Exon 5
79G	(884)	CCAGCAGAAGCAGAATGCCAGGAC	Exon 5
79R	(884)	CCAGCAGACGCAGTACGCAAAGATC	Exon 5
157		TAG CCC CCA TGT CGG TCT CAG AGA ACC TTC TC	3' of locus

[a] Asterisks indicate primers that were 5' end-radiolabeled for SSCP analysis.
[b] LCR, Locus control region.

Polymerase Chain Reaction Amplification

Promoter. DNA fragments derived from the red and green pigment gene promoters (from position −190 to +41 with respect to the transcription start site) are competitively amplified by PCR, using genomic DNA as template and primers (80 and 81 in Table I) that completely match both red and green gene sequences (Table I and Fig. 4A). Primer 80 is radiolabeled with ^{32}P at its 5' end, which results in tagging of the sense strands of the red and green PCR products. The labeling reaction for primers contains, in a total volume of 15 μl, 15 pmol of deoxyoligonucleotide, 90 μCi of [^{32}P] ATP [New England Nuclear (Boston, MA); specific activity of 3000 Ci/mmol], 15 units of bacteriophage T4 polynucleotide kinase, and kinase buffer (Bethesda Research Laboratories, Gaithersburg, MD). The reaction mixture is incubated at 37° for 45 min and then at 70° for 15 min. The labeled oligonucleotide is purified on a Sephadex G-25 spin column (Pharmacia Piscataway, NJ) according to the manufacturer protocol. Typically, the oligonucleotide primer is labeled to a specific activity of 1×10^6

cpm/pmol, diluted with cold oligonucleotide to a specific activity of 5×10^4 cpm/pmol, and the concentration adjusted to 5 μM for use in PCR amplification.

The amplification reaction contains, in a total volume of 10 μl, 100 ng of genomic DNA, 0.5 unit of *Taq* polymerase, amplification buffer (Perkin-Elmer, Norwalk, CT), dNTPs 200 μM (each), a 0.5 μM concentration of each primer (one of which is labeled), and 10% (v/v) glycerol. An initial denaturation at 96° for 5 min is followed by 26 cycles of amplification at 94° for 30 sec and extension at 64° for 1 min and a final 4 min of extension.

Single-Strand Conformation Polymorphism Analysis

The PCR-amplified products are diluted 25 to 50 times in 10 mM EDTA–0.1% (w/v) sodium dodecyl sulfate (SDS). Equal volumes of this dilution and formamide–dye mix are denatured at 100° and then placed on ice. One to μl is applied to a 6% (w/v) nondenaturing polyacrylamide gel and electrophoresis is carried out for 5–6 hr at a gel temperature of 35°. Temperature-sensitive color strips are used to monitor the gel temperature throughout the run. If the gel electrophoresis apparatus does not have a temperature control unit, then the temperature can be controlled with a fan. The radioactivity in the two bands on the gel derived from the red and green gene promoters is determined by phosphorimage analysis. SSCP analysis can be performed on up to approximately 30 samples on one gel. A representative autoradiograph of an SSCP gel of the red and green promoter fragments of 12 males is shown in Fig. 4C.

The frequency distribution of the total number of genes per array (mean of 2) is similar to that reported previously for unselected white males, using Southern blot analysis and pulsed-field electrophoresis. PCR/SSCP analysis of 20 samples of DNA that we had previously studied by Southern blot analysis shows that these two methods give similar values for green-to-red pigment gene ratios of up to 4.

As already mentioned, in addition to normal pigment genes, arrays may also contain a variety of hybrids between red and green pigment gene sequences. These can be detected by analysis of ratios of green to red exons by the same methodology. The hybrid pigment genes are referred to here by the abbreviations originally used by Merbs and Nathans[15] to indicate the origin of various exons. For example, a G4–R5 gene (found among deuteranomalous and normal color vision subjects) is a hybrid in which exons 1 to 4 are derived from the green pigment gene and exons 5 and 6 are derived from the red pigment gene (point of fusion in intron 4) (Figs. 1 and 2). An

---

[15] S. L. Merbs and J. Nathans, *Science* **258**, 464 (1992).

R4–G5 hybrid, found among subjects with protan color vision, is a hybrid in which exons 1–4 are derived from the red pigment genes and exons 5 and 6 from the green pigment genes (Figs. 1 and 2). A G2, 3–R4–G5 is a hybrid in which only exon 4 of a red pigment gene is substituted with that of a green pigment gene. The designation G2, 3 is used because exon 3 of the red pigment gene is indistinguishable from that of a green pigment gene owing to the existence of several shared polymorphisms.[14]

Polymerase Chain Reaction/Single-Strand Conformation Polymorphism Analysis of Exons 2, 4, and 5. The primer pairs, including the member that is 5' radiolabeled indicated with an asterisk, are given in Table I (and shown in Fig. 4A). The PCR and SSCP conditions for the exons are the same as for the promoter except for the following: no glycerol is included in the PCRs; the SSCP gel temperatures for exons 2, 4, and 5 are 32, 28, and 36°, respectively; the exon 2 PCR products are run on a 5% (w/v) polyacrylamide gel. A representative autoradiograph of an SSCP gel for PCR fragments derived from exon 4 of the red and green pigment genes is shown in Fig. 4D.

Another group has used PCR followed by enzyme digestion of the products to distinguish between PCR products derived from the green and red pigment genes.[16] However, this method often gives a significant overestimate of the ratios of green to red sequences. In addition to being more accurate, the PCR/SSCP method has the added advantage of revealing the presence of all polymorphisms and rare sequence variants.[14]

We have also developed a rapid method for detection of G4–R5 and R4–G5 hybrid genes in arrays using gene-specific primers.[8] For example, primer pairs 7G and 8R in exons 4 and 5, respectively, amplify a fragment across intron 4 only if a G4–R5 hybrid gene is present in the array (Fig. 4A). The sequence of the red- and green-specific primers are given in Table I. The PCR conditions have been described.[8]

Determining Genotype at Amino Acid Residue 180 of Red
  Pigment Gene

As mentioned above, the Ser180Ala polymorphism (60% serine, 40% alanine) plays a role in both normal and defective color vision. The pigment with serine has a longer (by 4 nm) λ_{max} than that with alanine at position 180. This polymorphism also exists in the green pigment gene, but at a much lower frequency (85% alanine, 15% serine). Therefore, it would be important to determine the genotype at this position in order to be able to completely infer the λ_{max} of a pigment from its amino acid sequence.

Exon 3 of the red and green pigment genes contains polymorphisms at

---

[16] M. Neitz and J. Neitz, *Science* **267**, 1013 (1995).

positions 151, 153, 155, 171, 174, 178, and 180.[14] The 3' half of this exon contains the polymorphism at position 180. The strategy involves, first, amplification of either a red or green gene fragment encompassing exon 3, purifying this fragment, and subsequently using it as template in a second round of PCR to amplify the 3' half of exon 3[11] (Fig. 4B).

Experimental Protocol

The red (2R)- or green (2G)-specific primers in exon 2 together with the nonspecific primer (3) in exon 3 are first used to amplify the respective gene-specific fragments (Fig. 4B). The sequences of the primers are given Table I. Thirty cycles of amplification (15 sec at 94°, 1 min at 64°, and 3 min at 72°) are performed. The amplified fragment is purified by electrophoresis on a 1% (w/v) low-melting point agarose gel (Seaplaque GTG; FMC, Philadelphia, PA). An aliquot (1–2 μl) of the melted gel slices containing fragments is then used in a second round of amplification (30 cycles of 15 sec at 94° and 1 min at 64°) to amplify the 3' part of exon 3 with primers 9 and 3 (Table I and Fig. 4B). The resultant fragments are either sequenced directly or subjected to SSCP analysis on a 5% (w/v) polyacrylamide gel for 5 hr at 32°. In the latter case, the PCR products are radiolabeled by the inclusion of [^{32}P]dCTP in the amplification reaction as described.[17] The fragments encoding serine or alanine at position 180 are well distinguished on an SSCP gel.[11] SSCP is the method of choice when analyzing a large number of samples.

Analysis of First Gene in Array and Rapid Detection of 5' Red–Green 3' Hybrid Genes among Females

The first gene in the array together with the promoter and locus control region (LCR) can be amplified in a 15.8-kb segment, using a forward primer within the LCR and a reverse primer in exon 5 (Fig. 4A). The reverse primer may be one that matches red (79R), green (79G), or both (78) sequences in exon 5. In individuals with normal or deutan color vision, the first gene in the array is always red. However, in protan subjects the first gene in the array is a 5' red–green 3' (R–G) hybrid with various points of fusion (Fig. 5A), all of which include the exchange of exon 5. Thus, using gene-specific exon 5 primers, it may be easily determined whether a red or R–G hybrid gene (Fig. 5B) occupies the first position in the array. This represents a rapid diagnostic test for protan color vision defects in males

[17] M. Orita, H. Iwahana, H. Kanazawa, K. Hayashi, and T. Sekiya, *Proc. Natl. Acad. Sci. U.S.A.* **86**, 2766 (1989).

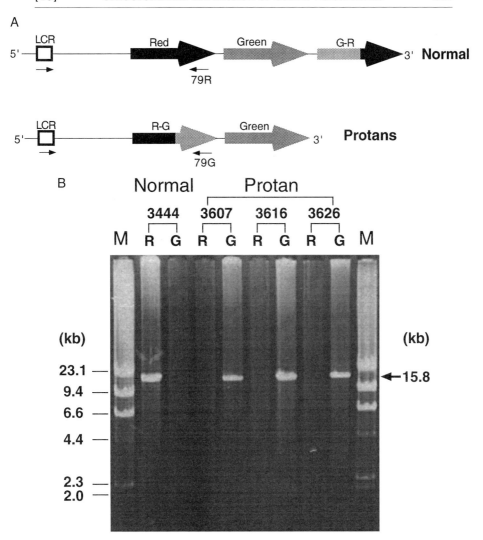

FIG. 5. Analysis of the first (5') gene in arrays of subjects with normal and protan color vision. (A) Diagrammatic representation of the gene arrays of subjects with normal (subject 3444) and protanomalous (subjects 3607, 3616, and 3626) color vision.[22] Primers used in amplification are indicated with arrows. Primer sequences are given in Table I. (B) Agarose gel electrophoresis of the PCR products (stained with ethidium bromide) from the normal and protan subjects shown in (A). R and G denote the red or green specificity of primer 79 in exon 5. Lanes labeled with M contain a molecular size marker (bacteriophage DNA digested with HindIII).

and for detection of female carriers of arrays that cause protanomaly in males. A different version of this method, which involves a similar strategy but did not include amplification of the region between the LCR and the proximal promoter of the red pigment gene, has been used to detect female carriers of protan visual pigment gene arrays.[18]

The amplified DNA segment may also be used as a template in direct sequencing of any desired region, with the appropriate internal primers. Identifying the amino acid residues at positions that are critical for spectral tuning would allow precise inference of the absorption maxima of the encoded photopigments. This would be particularly useful in determining whether serine or alanine occupies the polymorphic position 180.

Experimental Protocol

The following primers are used (see Table I and Fig. 5A): Forward primer LCR1F is complementary to the LCR region and reverse primer 79R (red pigment gene specific) is complementary to exon 5 of the red pigment gene. Reverse primer 79G (green gene-specific primer) is complementary to exon 5 of the green pigment gene (Table I and Fig. 4A). The PCR mixture contains, in a total volume of 50 μl, 0.5–1.3 μg of total genomic DNA, a 0.2 μM concentration of each primer, a 400 μM concentration of each dNTP, 1× LA PCR buffer II (Mg^{2+} plus) and 2.5 units of TaKaRa (Shiba, Japan) LA *Taq* polymerase. An initial step of denaturation at 94° for 1 min is followed by 14 cycles of 98° for 10 sec and 70° for 15 min; 16 cycles of 98° for 10 sec and 70° for 15 min plus 15 sec; and 1 cycle of 72° for 10 min. The PCR products are electrophoresed on a 0.6% agarose gel and visualized by ethidium bromide. Amplification products (15,787 bp) using either red or green exon 5-specific primers on DNA templates from the color normal subject and the three protans, whose gene arrays are diagrammed in Fig. 5A. The red exon 5-specific primer gives a PCR product only from the normal color vision subject. Conversely, the green-specific exon 5 primer gives a product from protan subjects only.

Characterization of Gene Occupying 3′ Terminal Position in Array

The gene arrays of most deuteranomalous males are composed of one normal red, one G–R hybrid, and one or more normal green pigment genes (Fig. 1B). However, G–R hybrid genes similar to those of deuteranomaly are also found in 4–8% of males with normal color vision. To explain this paradox, we suggested that only the first two genes in a visual pigment array are expressed, and that deuteranomaly results only if the G–R hybrid

---

[18] P. M. Kainz, M. Neitz, and J. Neitz, *Vision Res.* **38**, 3365 (1998).

gene occupies the second position in an array. To test this hypothesis, we ascertained the identity of the visual pigment gene occupying the 3'-terminal or third position among males carrying one normal red, one normal green, and one G–R hybrid gene.[19] The order of the genes was then correlated with the color vision status and with expression of the pigment genes into mRNA in the retina (see below).

The sequence of the entire red–green array and flanking regions has been determined (GenBank accession numbers Z68193, Z46936, and Z49258), and the 3' terminus of the visual pigment gene array has been defined[20] (Fig. 6a). Therefore, it became possible to design PCR primers to amplify a DNA segment (27,398 bp in length) extending from the 5' end of exon 5 of the 3'-terminal pigment gene (red or green) to just beyond the 3' terminus of the array (Fig. 6b) and determine its identity.[19]

Experimental Protocol

Long-Range Polymerase Chain Reaction Amplification. Human genomic DNA is extracted from either peripheral white blood cells or from retina, choroid, and sclera of postmortem eyes by either the salting out method[21] or using the Puregene kit (Gentra Systems, Minneapolis, MN). Long-range PCR (predicted product size of 27.4 kb) using forward primer 3C, which is complementary to the 5' region of exon 5 of both the red and green pigment genes, and reverse primer 157, which is complementary to a sequence in exon 1 of the *TEX28* gene located just 3' of the 3'-terminal repeat unit of the locus (Table I, Fig. 6b). The PCR mixture contains, in a total volume of 50 μl, 0.5 μg of total genomic DNA, a 0.2 μM concentration of each primer, a 400 μM concentration of each dNTP, 1× LA PCR buffer II (Mg^{2+} plus), and 2.5 units of LA *Taq* polymerase (TaKaRa). An initial denaturation step at 94° for 1 min is followed by 14 cycles of 98° for 10 sec and 68° for 20 min; 16 cycles of 98° for 10 sec and 68° for 20 min plus 15 sec; and 1 cycle of 72° for 10 min. The PCR products are electrophoresed on a 0.8% (w/v) agarose gel and visualized by ethidium bromide or are subjected to pulsed-field gel electrophoresis (PFGE) and Southern blot analysis as described in the next section.

Pulsed-Field Gel Electrophoresis and Southern Blot Analysis. The long-range PCR products are subjected to pulsed-field gel electrophoresis on a 1.5% (w/v) PFGE agarose gel (Boehringer Mannheim, Indianapolis, IN), using a CHEF-DRII apparatus (Bio-Rad, Hercules, CA) in 0.5× Tris–borate–EDTA (TBE) buffer for 24 hr at 200 V, 14° with pulse time from

---

[19] T. Hayashi, A. G. Motulsky, and S. S. Deeb, *Nature Genet.* **22,** 90 (1999).
[20] M. C. Hanna, J. T. Platts, and E. F. Kirkness, *Genomics* **43,** 384 (1997).
[21] S. A. Miller, D. D. Dykes, and H. F. Polesky, *Nucl. Acids Res.* **16,** 1215 (1988).

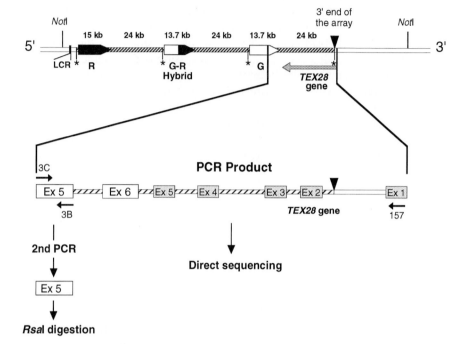

FIG. 6. Strategy for amplification of the 3'-terminal repeat unit of the red–green pigment gene array. (a) Map of the visual pigment gene array of a deuteranomalous individual who has one normal red, one 5' green–red 3' hybrid, and one normal green gene. Asterisks (*) delineate the three repeats of this family, which consists of a red (R) pigment gene, a green–red hybrid gene (G–R hybrid), and a normal green pigment gene (G). Hatched segments denote intergenic DNA that contains a gene termed *TEX28*. A complete *TEX28* gene is located at the end of the visual pigment gene array (dotted arrow). Exon 1 of *TEX28* is located 700 bp 3' of the end of the visual pigment gene locus. Truncated (lacking exon 1) and, therefore, nonfunctional copies of the *TEX28* gene are also found within each of the two intergenic DNA (not shown). LCR, Locus control region. The two *Not*I sites that flank the array are used to determine the number of genes in the array by pulsed-field electrophoresis. (b) Enlarged segment of part of the terminal (third position) repeat that is amplified by long-range PCR from total human genomic DNA with primers 3C and 157. This PCR product is used as template either for direct sequencing of exon 5 of the pigment gene, or to amplify exon 5 (with primers 3C and 3B) followed by digestion with *Rsa*I, which specifically cleaves exon 5 of the red pigment gene.[5]

1.2 to 1.6 sec. Cosmid 42A 1-1, which contains the entire human red pigment gene and flanking sequences,[22] is linearized with NotI and used as a marker (approximately 42 kb). Figure 7a shows an ethidium bromide-stained PFGE gel of PCR products (27.4 kb) from two males. To verify that the observed bands are derived from the visual pigment gene locus, the gel is capillary blotted onto a nylon membrane (Hybond-XL; Amersham, Cleveland, OH) by alkaline transfer. A red pigment gene cDNA probe [HS 7, which detects both red and green pigment gene sequences; kindly provided by J. Nathans (Johns Hopkins University, Baltimore, MD) and has been described[7]] is labeled with ^{32}P (3000 Ci/mmol) with the random primed DNA labeling kit (Boehringer Mannheim) to a specific activity of approximately 4×10^8 cpm/µg and used as a probe (6×10^5 cpm/ml of hybridization solution). The blot is visualized by autoradiography. The 27.4-kb band hybridizes to the red–green cDNA probe (Fig. 7b).

Identification of exon 5 of the 3'-terminal pigment gene amplified in this manner is achieved by either direct sequencing of the PCR product or by a second round of PCR to amplify exon 5 and subjecting the secondary product to digestion by *Rsa*I as described in the following section.

Sequencing and Restriction Enzyme Analysis. The long-range PCR products are purified by electrophoresis on a 0.8% (w/v) low-melting point agarose gel (GIBCO-BRL, Gaithersburg, MD) and used directly as templates to sequence (Thermosequenase kit; Amersham) exon 5 by using primer 3C, which matches both red and green sequences. In another test, the extensively diluted long-range PCR product is used as a template to amplify exon 5 in its entirely, using primers 3C and 3B, which match both red and green sequences (Table I, Fig. 6b). After 1 min of denaturation, amplification (30 cycles of 94°C for 15 sec), annealing (64° for 30 sec), and extension (72° for 30 sec) are performed. These cycles are followed by an extension step of 1 min at 72°. The amplified exon 5 products are digested with *Rsa*I (which cleaves only red exon 5) and electrophoresed on a 3% (w/v) agarose gel.

Determination of Ratio of Red to Green mRNA Transcripts in Retinas

Knowing the pattern of expression of the red and green pigment genes in the retina is an important intermediate step in attempting to relate the genotype to the color vision phenotype. As already discussed, regulation of gene expression at this locus has important and interesting facets. First, despite the proximity and high degree of sequence identity between the red and green pigment genes, only one is expressed in a given photoreceptor

[22] R. Feil, P. Aubourg, R. Heilig, and J. L. Mandel, *Genomics* **6,** 367 (1990).

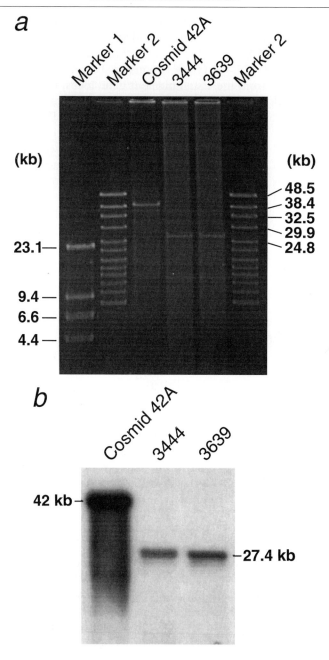

cell, a requirement for the evolution of trichromatic color vision among primates. The mechanism by which this segregated expression occurs is unknown. The expression of either the red or green pigment genes in photoreceptors may be a stochastic process, favoring the red pigment gene, perhaps because it is closest to the LCR. Second, not all the genes of the red–green array seem to be expressed in the retina. Knowing which of the genes are expressed will help resolve the paradox of why males who have arrays composed of a normal red, a normal green, and a G–R hybrid gene may have either normal or deuteranomalous color vision. We hypothesized that only the first two genes in the array are expressed and that deuteranomaly would result only if the G–R hybrid, but not the normal green pigment gene, is expressed in photoreceptors. To test this hypothesis, we determined the pattern of expression of genes in male postmortem retinas and related it to the genotype.

Experimental Protocol

Determination of Ratio of Red to Green Pigment mRNAs in Whole Retinas by Polymerase Chain Reaction/Single-Strand Conformation Polymorphism. This protocol is performed to determine the ratio between retinal mRNAs encoding the red and green pigment genes. This ratio most likely corresponds to the ratio of red to green photoreceptors in the retina. Although this protocol is described for whole human retinas, it is also applicable to analysis of small sections (1–2 mm in diameter) of retinas.

The general procedure involves competitive amplification of exon 4 derived from red and green mRNA sequences followed by SSCP analysis and quantification of the separated fragments derived from these two genes. We have chosen exon 4 for this competitive PCR-based assay because it contains the least number of nucleotide differences (3 of 165) between red and green sequences, so that differences in amplification rate based solely on sequence differences would be minimized. Furthermore, the amplified DNA strands derived from red and green exon 4 sequences can easily be separated by conformational differences on SSCP.

Total cellular RNA is prepared from whole retinas of eye donors, and aliquots (approximately 1 μg) are reverse transcribed into cDNA (first

---

FIG. 7. Amplification of the 3′ terminal repeat unit of the red–green gene array. (a) Photograph of PFGE gel (stained with ethidium bromide) of the long-range PCR products from subjects 3444 and 3639. Molecular size markers: cosmid 42A, Marker 1, and Marker 2 (8.3- to 48.5-kb ladder; Bio-Rad). (b) Autoradiogram of a Southern blot of the pulsed-field gel shown in (A). The band corresponding to cosmid 42A and the 27.4-kb amplified fragment hybridized with the red–green pigment cDNA probe.

strand synthesis kit; Ambion, Austin, TX) in a 20-μl reaction. Aliquots (0.5–1.0 μl) of the cDNA solution are then used as templates to amplify exon 4 and subject the products to SSCP analysis as described.[9]

Red and green exon 4 sequences are amplified competitively in 25 cycles, using primers 30 and 2B (Fig. 4A, Table I), with primer 30 radiolabeled at its 5' end with ^{32}P as described above under Polymerase Chain Reaction Amplification. The experimental protocol is identical to that used to amplify exon 4, using genomic DNA as template as described above under Determining Number and Ratio of X-Linked Red–Green Photopigment Genes. An autoradiograph of an SSCP gel showing ratios of red to green exon 4 mRNAs in retinas of various subjects is shown in Fig. 8. The red-to-green mRNA ratio varied from 1 to 10, with a mode of 4.

Templates containing known ratios of red to green sequences are used to construct a standard curve. The known mixtures are prepared by mixing known concentrations of the exon 4 fragments amplified from genomic DNA of individuals who have only red or green exons 4 in their genomic DNA. The fragments are purified by agarose gel electrophoresis and their concentration determined. Various proportions of fragments are mixed and a 100-fold excess of excess yeast DNA is added. There is excellent correlation between the input and observed ratios of red to green exon 4 sequences.[9]

Expression of Green and G–R Hybrid Genes in Arrays. To test our hypothesis that not all genes in an array are expressed in the human retina, we earlier determined if all normal green pigment genes of an array are expressed. We took advantage of a common silent polymorphism in exon 5 (A or C at the third position of codon 283) of the green pigment gene. Approximately 75% of males with more than one green pigment gene have both the A and C forms of the green pigment gene, which can be

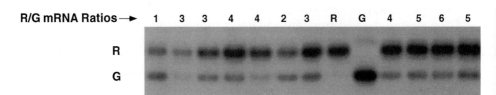

FIG. 8. Ratio of red to green pigment mRNAs in whole retinas of male eye specimens. Total RNA was extracted from retinas of human male donors and analyzed for relative abundance of red to green exon 4-encoded sequences. This autoradiograph of an SSCP gel shows intensities of bands derived from red (R) and green (G) exon 4 mRNA sequences. Lanes marked R and G on top are markers for the red and green pigment bands. Phosphorimage analysis was used to determine band intensity. Ratios of red to green band intensities were rounded to the nearest whole numbers.

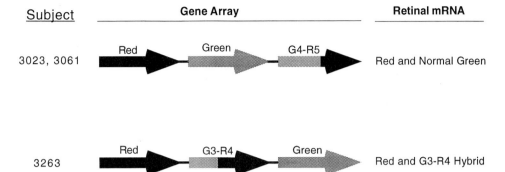

FIG. 9. Expression of hybrid versus normal green pigment genes in male retinas. Total retinal RNA from three postmortem eye donors, whose arrays were composed of a normal red, a normal green, and a green–red hybrid gene, was used as template in competitive RT/PCR/SSCP analysis to determine the ratio of mRNA derived from red and green exons 2, 4, and 5.

distinguished by PCR/SSCP analysis. The strategy we adopted was to select retinas from donors who had both forms of the green exon 5 and determine if both are represented in retinal mRNA.[23] The results clearly showed that, in the retina of any one male, only one of the green pigment mRNA forms was represented.

The next question that we addressed was related to expression of G–R hybrid genes, which is relevant to the paradox of why males who have a normal red, a normal green, and a G–R hybrid gene have either normal or deuteranomalous color vision (see preceding discussion). We found 3 of 51 unselected male eye donors to have gene arrays composed of one normal red, one normal green, and one G–R hybrid gene.[9] Two of these subjects (3023 and 3061) had a G4R5 hybrid gene and the third (3263) had a G3R4 hybrid gene (Fig. 9). We examined retinal mRNA from these subjects for the presence of green and G-R hybrid transcripts according to the following protocol:

Experimental Protocol

Detection of mRNA Transcripts Containing Green-Red Hybrid Sequences with Fusion Points in Intron 4 (G4–R5 Hybrids). Competitive PCR amplification (after reverse transcription of total retinal mRNA; see above) is performed with primers 7G (specific for green exon 4) and 78 (identical

[23] J. Winderickx, L. Battisti, A. G. Motulsky, and S. S. Deeb, *Proc. Natl. Acad. Sci. U.S.A.* **89**, 9710 (1992).

in sequence to both red and green exon 5) (Table I and Fig. 4A). These primers will amplify both G4–R5 hybrid and normal green pigment cDNA sequences with equal efficiency. Primer 78 (Table I) is end labeled with ^{32}P in order to radiolabel a single strand from each sequence and simplify the SSCP pattern. The amplification conditions are the same as for amplification of exon 4 described above. The amplified fragments (derived both from green–red hybrid and normal green pigment genes) are resolved on an SSCP gel for 5 hr at a gel temperature of 33°. The radioactivity in the bands corresponding to hybrid and normal DNA segments is quantified by phosphorimage analysis.

The results show that subjects 3063 and 3061 both express the red and normal green genes, but not the G4R5 hybrid gene, that they carried on their X chromosomes.[9] We have subsequently showed (see above) that the hybrid genes of these individuals occupy the third position in their arrays (Fig. 9).

Analysis of the mRNA of individual 3263 is performed by competitive PCR amplification/SSCP analysis of exons 2, 4, and 5, using primers that are common to both red and green cDNA sequences. The primers and conditions for amplification and SSCP analysis are identical to those given above for analysis of genomic DNA. The results of this analysis indicate that subject 3263 expresses only red exons 4 and 5 and only green exon 2, indicating that he expresses the red and G3–R4 hybrid genes but not the normal green pigment gene (Fig. 9), and presumably had deuteranomalous color vision. In this subject the normal green pigment gene occupies the third position in his array as determined by long-range PCR (see above).

These results are consistent with our hypothesis that only the first two genes in the array are expressed into retinal mRNA.

Acknowledgments

This research was supported by National Institutes of Health Grant EY 08395.

[44] Expression and Characterization of Peripherin/ rds–rom-1 Complexes and Mutants Implicated in Retinal Degenerative Diseases

By ANDREW F. X. GOLDBERG and ROBERT S. MOLDAY

Introduction

A large number of genes implicated in a variety of retinal dystrophies have been identified and mutations in these genes that cause specific disease phenotypes have been cataloged. In most cases, the mechanisms by which mutations in these genes alter protein structure and function and ultimately lead to photoreceptor degeneration and visual loss remain to be determined at a molecular level. These considerations prompted us to investigate the application of heterologous protein expression for the molecular analysis of disease-causing mutations. This chapter describes the expression of peripherin/rds and rom-1 in COS-1 cells and methods that have been used to examine the effect of disease-causing mutations on peripherin/rds–rom-1 structure and subunit assembly.

Inherited defects in peripherin/rds result in heterogeneity of human retinal degenerative diseases. Clinical manifestations of diseases linked to mutations in the human *RDS* gene (which encodes the peripherin/rds polypeptide) include autosomal dominant retinitis pigmentosa, digenic retinitis pigmentosa, rod–cone dystrophies, fundus flavimaculatus, adult vitelliform dystrophy, butterfly pigment dystrophy, cone–rod dystrophy, and generalized macular degeneration.[1–5] The involvement of peripherin/rds in both central and peripheral dystrophies suggests that it is a crucial element in both rod and cone photoreceptor cell biology. To date, several mutations in the human *ROM1* gene have been linked to a novel digenic form of autosomal dominant retinitis pigmentosa (adRP). In this case, the disease phenotype is expressed in individuals who coinherit the *rom-1* mutation

---

[1] R. G. Weleber, R. E. Carr, W. H. Murphey, V. C. Sheffield, and E. M. Stone, *Arch. Ophthalmol.* **111,** 1531 (1993).

[2] E. Apfelstedt-Sylla, M. Theischen, K. Ruther, H. Wedemann, A. Gal, and E. Zrenner, *Br. J. Ophthalmol.* **79,** 28 (1995).

[3] J. Wells, J. Wroblewski, J. Keen, C. Inglehearn, C. Jubb, A. Eckstein, M. Jay, G. Arden, S. Bhattacharya, F. Fitzke, and A. Bird, *Nature Genet.* **3,** 213 (1993).

[4] B. S. Shastry, *Am. J. Med. Genet.* **52,** 467 (1994).

[5] T. P. Dryja, L. B. Hahn, K. Kajiwara, and E. L. Berson, *Invest. Ophthalmol. Vis. Sci.* **38,** 1972 (1997).

with a *peripherin/rds* mutation, but not in family members who inherit only the *rom-1* mutation.[5]

In addition to the diversity of diseases described for humans, a defect in the *RDS* gene that prevents expression of peripherin/rds is the cause of the retinal degeneration slow (*rds*) phenotype in mice.[6-8] Early morphological studies have shown that mice homozygous for the *rds* defect lack photoreceptor outer segments and mice heterozygous for this defect produce highly disorganized outer segments containing whorls of disk membranes.[9,10] Over a period of many months, the photoreceptor cells gradually die by the process of apoptosis.[11,12] These studies firmly establish the importance of the peripherin/rds polypeptide in vertebrate outer segment morphogenesis and structure. More recently, a transgenic mouse model for peripherin/rds-based human disease has also been reported to undergo retinal degeneration.[13]

To date, 11 orthologs of peripherin/rds have been cloned from a variety of mammals, amphibia, and birds.[6,14-19] In addition, three mammalian orthologs of rom-1 (a peripherin/rds homolog) have been identified.[20-22] The peripherin/rds orthologs show a high degree of amino acid conservation (~50–90% similarity) among themselves, as do the rom-1 orthologs (~85%

[6] G. H. Travis, M. B. Brennan, P. E. Danielson, C. Kozak, and J. G. Sutcliffe, *Nature (London)* **338,** 70 (1989).
[7] G. H. Travis, K. R. Groshan, M. Lloyd, and D. Bok, *Neuron* **9,** 113 (1992).
[8] G. L. Connell, R. Bascom, L. L. Molday, D. Reid, R. R. McInnes, and R. S. Molday, *Proc. Natl. Acad. Sci. U.S.A.* **88,** 723 (1991).
[9] S. Sanyal and H. G. Jansen, *Neurosci. Lett.* **21,** 23 (1981).
[10] R. K. Hawkins, H. G. Jansen, and S. Sanyal, *Exp. Eye Res.* **41,** 701 (1985).
[11] N. Agarwal, I. Nir, and D. S. Papermaster, *J. Neurosci.* **10,** 3257 (1990).
[12] C. Portera-Cailliau, C.-H. Sung, J. Nathans, and R. Adler, *Proc. Natl. Acad. Sci. U.S.A.* **91,** 974 (1994).
[13] W. Kedzierski, M. Lloyd, D. G. Birch, D. Bok, and G. H. Travis, *Invest. Ophthalmol. Vis. Sci.* **38,** 498 (1997).
[14] G. J. Connell and R. S. Molday, *Biochemistry* **29,** 4691 (1990).
[15] G. H. Travis, L. Christerson, P. E. Danielson, I. Klisak, R. S. Sparkes, L. B. Hahn, T. P. Dryja, and J. G. Sutcliffe, *Genomics* **10,** 733 (1991).
[16] M. B. Gorin, S. Synder, A. To, K. Narfstrom, and R. Curtis, *Mamm. Genome* **4,** 544 (1993).
[17] W. N. Moghrabi, W. Kedzierski, and G. H. Travis, *Exp. Eye Res.* **61,** 641 (1995).
[18] J. Weng, T. Belecky-Adams, R. Adler, and G. H. Travis, *Invest. Ophthalmol. Vis. Sci.* **39,** 440 (1998).
[19] W. Kedzierski, W. N. Moghrabi, A. C. Allen, M. M. Jablonski-Stiemke, S. M. Azarian, D. Bok, and G. H. Travis, *J. Cell Sci.* **109,** 2551 (1996).
[20] R. A. Bascom, S. Manara, L. Collins, R. S. Molday, V. I. Kalnins, and R. R. McInnes, *Neuron* **8,** 1171 (1992).
[21] R. A. Bascom, K. Schappert, and R. R. McInnes, *Hum. Mol. Genet.* **2,** 385 (1993).
[22] O. L. Moritz and R. S. Molday, *Invest. Ophthalmol. Vis. Sci.* **37,** 352 (1996).

similarity)—a lesser degree of conservation is seen between the two groups (~30%). Although peripherin/rds and rom-1 were initially suggested to comprise a novel family of proteins, they appear instead to be distant relatives of the larger tetraspanin or transmembrane-4 superfamily (TM4SF).[23] Tetraspanins have been implicated in a diversity of cellular processes, including membrane morphogenesis; however, their molecular function is currently unknown.[24]

Peripherin/rds and rom-1 are polypeptide subunits that form an integral membrane protein complex in vertebrate photoreceptor outer segments. These complexes require detergent to be solubilized from outer segment membranes and are extracted as heterotetramers under mildly reducing conditions.[25] Preliminary studies suggest that some heterotetramers may disulfide bond to form higher order structures.[26] Peripherin/rds is posttranslationally modified by N-linked glycosylation and inter- and intramolecular disulfide bonds, whereas rom-1 appears to possess disulfide but not carbohydrate modification.[14,22,27] The membrane-bound complex shows a striking subcellular localization, being restricted exclusively to the rim regions of both rod and cone photoreceptor outer segment disks.[22,27,28] This pattern of localization was the first indication that peripherin/rds might play a role in outer segment disk structure and morphogenesis. More recent studies suggest that an amphipathic helix in the C-terminal domain of peripherin/rds possesses fusogenic activity and may function in membrane fusion events required for outer segment formation and renewal.[29]

We have developed the COS-1 cell culture system for the expression and characterization of peripherin/rds and rom-1.[30] Heterologous expression and site-specific mutagenesis is combined with biochemical, immunochemical, and biophysical analysis to probe various aspects of protein structure and function. This approach offers the advantages of high throughput and economy, and allows relatively rapid screening of many mutations. The techniques we have adapted are general and may be applied to other proteins of interest involved in photoreceptor cell degenerations.

---

[23] M. D. Wright and M. G. Tomlinson, *Immunol. Today* **15**, 588 (1991).
[24] H. T. Maecker, S. C. Todd, and S. Levy, *FASEB J.* **11**, 428 (1997).
[25] A. F. X. Goldberg and R. S. Molday, R. S. *Biochemistry* **35**, 6144 (1996).
[26] C. R. Loewen and R. S. Molday, *Invest. Ophthalmol. Vis. Sci.* **39**(4), S963 (1998).
[27] R. S. Molday, D. Hicks, and L. L. Molday, L. L, *Invest. Ophthalmol. Vis. Sci.* **28**, 50 (1987).
[28] K. Arikawa, L. L. Molday, R. S. Molday, and D. S. Williams, *J. Cell Biol.* **116**, 659 (1992).
[29] K. Boesze-Battaglia, O. P. Lambda, A. A. Napoli, S. Sinha, and Y. Guo, *Biochemistry* **37**, 9477 (1998).
[30] A. F. X. Goldberg, O. L. Moritz, and R. S. Molday, *Biochemistry* **34**, 14213 (1995).

Materials and Methods

Growth and Maintenance of COS-1 Cells

Reagents

Growth medium: Dulbecco's modified Eagle's medium, containing L-glutamine and D-glucose (4.5 g/liter each), sodium pyruvate and pyridoxine hydrochloride (110 mg/liter each), streptomycin (50 μg/ml), penicillin G (50 U/ml), and 10% (v/v) fetal calf serum (FCS)

Freezing medium: Growth medium (as above) containing 10% (v/v) dimethyl sulfoxide (DMSO) and 30% (v/v) fetal calf serum

Phosphate-buffered saline (PBS)

Trypsin solution: 0.2% (w/v) in PBS; store filter-sterilized aliquots at $-20°$

Procedures. The COS-1 cell line was established from African green monkey kidney cells (CV-1) cells by transformation with an origin-defective mutant of simian virus 40 (SV40).[31] COS-1 contains an integrated copy of the complete early region genes of SV40 and produces the SV40 T-antigen, allowing plasmid DNA containing the SV40 origin of replication to replicate autonomously. This cell line also offers the advantages of ready availability, hardiness, and ease of culture. We typically culture cells in 10 ml of growth medium at 37°, 5% CO_2, 100% humidity in 100-mm dishes. All operations are performed using sterile techniques in a laminar flow hood.

Cell stocks may be obtained for a nominal fee from the American Type Cell Collection (ATCC, Rockville, MD) and should generally be thawed and propagated as soon as possible after receipt. If required, cells may be stored frozen in a liquid nitrogen vapor phase. Each aliquot of frozen cells should be thawed rapidly in a 37° water bath, diluted with a 10× volume of growth medium, plated into a 100-mm tissue culture dish, and grown at 5% CO_2, 100% humidity, 37°. The cells should begin adhering to plates within several hours. It is good practice to change the culture media approximately 24 hr after thawing to remove residual cryoprotectant.

General maintenance of cells requires passaging approximately twice weekly. For a typical division of cells on a 100-mm plate, the medium is aspirated, cells are washed twice with 3 ml of PBS, and then 2 ml of trypsin solution is applied. Plates are incubated at 37° until cells begin to detach (typically 2–10 min). When floating cells are visible, growth medium is added (8 ml), and a pipette is used to flush any remaining adherent cells off the dish. Resuspended cells are plated into five new dishes (2 ml each), and 8 ml of fresh growth medium is added to each.

---

[31] Y. Gluzman, *Cell* **23,** 175 (1981).

Cells can typically be passaged for many generations without ill effect, although stocks of frozen cells are handy, since contamination or other interruptions in passaging do occasionally occur. A frozen stock is made by collecting cells (via trypsinolysis) from a 50–70% confluent 100-mm plate, and pelleting in a clinical centrifuge. The cell pellet should be resuspended carefully in 1 ml of cold freezing medium, transferred to a cryogenic vial, wrapped in several layers of paper towels, and placed in a $-70°$ freezer overnight. The frozen cells should be transferred to liquid nitrogen storage the next day, and should be viable for several years.

Preparation of Wild-Type and Mutant Expression Vectors

Reagents

Mammalian expression vector: pcDNAI/Amp or equivalent (InVitrogen, Carlsbad, CA)
Synthetic oligonucleotides: Desalted (appropriate commercial supplier)
Vent polymerase (New England BioLabs, Beverly, MA)
10× Vent polymerase buffer (New England BioLabs)
Bluescript II cloning vector: pBS II KS(+) (Stratagene, La Jolla, CA)
dNTP mix (Boehringer Mannheim, Indianapolis, IN)
$MgSO_4$, 100 mM

Procedures. We use the mammalian expression vector pcDNAI/Amp (InVitrogen) for the transient transfection of adherent COS-1 cells in culture. This vector contains a multiple cloning site (MCS) cassette, T7 and SP6 priming sites for sequencing, cytomegalovirus (CMV) promoter and enhancer, splice segment and polyadenylation signal, SV40 origin of replication, ColE1-like plasmid origin, and an ampicillin (Amp) resistance gene. An updated version of pcDNAI/Amp, pcDNA3, includes a neomycin resistance gene for the selection of stably transformed mammalian lines and is also suitable for transient transfections. Full-length coding sequences (including upstream Kozak sequences) of peripherin/rds and rom-1 were cloned into the MCS of pcDNAI/Amp to generate pcPER and pcROM, respectively.[22,30]

Site-specific mutagenesis is performed by an adaptation of the approach of Nelson and Long.[32] This polymerase chain reaction (PCR)-based method is versatile, relatively inexpensive, and requires a minimum of DNA sequencing; however, it does require several steps (including subcloning)

---

[32] R. M. Nelson and G. L. Long, *Anal. Biochem.* **180,** 147 (1989).

TABLE I
SYNTHETIC OLIGONUCLOTIDES USED FOR SITE-SPECIFIC MUTAGENESIS

Primer	Type	Sequence
A(L185P)	Mutagenic	5'-CCGCTATCCGGATTTTTCC
B	Flanking	5'-GGAGTACTAGTAACCCTGGCCCCAGTCACGACGTTGTAA
C	Flanking	5'-CAGGAAACAGCTATGACCAT
D	Flanking	5'-GGAGTACTAGTAACCCTGGC

to complete. We describe here the procedures used to generate the disease-linked L185P peripherin/rds mutant involved in digenic RP.[33,34]

The desired target sequence (coding for amino acids 64–243) is isolated from a bovine peripherin/rds cDNA as a 422-bp double-stranded DNA (dsDNA) fragment, using unique *Bgl*II and *Sac*II restriction sites. It is cloned into *Sac*II/*Eco*RI-digested pBS II KS(+) cloning vector, using a synthetic *Eco*RI–Bg/II adapter in an orientation similar to the β-galactosidase gene (pBgScKS). Three flanking oligonucleotides designed for the PCR amplification of the target region are described in Table I.

Mutagenic primer A (sense strand), flanking primer B (antisense strand), and pBgScKS are used in PCR I to amplify a fragment that extends from the mutation site to the flanking primer B, which lies beyond one end of the MCS. In PCR II, amplified product from PCR I is heat denatured and reannealed to pBgScKS, and mutation-containing product is extended beyond the site of flanking primer C. The extended product serves as the mutagenic template for the final PCR III. In this reaction, flanking primer C is used in conjunction with primer D (which binds to a unique region within primer B) to specifically amplify the mutagenized 652-bp target sequence. The unique *Bgl*II/*Sac*II restriction sites are utilized to subclone the mutagenized fragment back into the wild-type (WT) peripherin/rds cDNA background. The mutagenized target is sequenced in its entirety to confirm desired, and rule out spurious, mutations.

PCR is performed with Vent polymerase (New England BioLabs), as it offers a good compromise between fidelity and economy. Other thermostable polymerases may be substituted with slight adjustments of solution conditions and cycling parameters. Each PCR is carried out with 1 unit of Vent polymerase (New England BioLabs) and the manufacturer-supplied buffer in a 100-μl volume sealed with mineral oil. In addition, PCR I

---

[33] K. Kajiwara, E. L. Berson, and T. P. Dryja, *Science* **264**, 1604 (1994).
[34] A. F. X. Goldberg and R. S. Molday, *Proc. Natl. Acad. Sci. U.S.A.* **93**, 13726 (1996).

reactions contains 1.6 mM dNTPs, 0.4 μM primer A, 0.4 μM primer B, ~1 nmol of pBgScKS, and 0–5 mM MgSO$_4$ as required for optimization. PCR I is denatured at 94° for 5 min and then 25 amplification cycles are performed as follows: denaturation at 94° for 15 sec, annealing at $T_m - 5°$ for 20 sec, and extension at 72° for 30 sec per 500 bp. The product is assayed by agarose gel electrophoresis; a robust band of the expected size (~215 bp) should be obtained before proceeding to subsequent steps.

PCR II and PCR III are performed successively in a single tube that contains 1.6 mM dNTPs, 2–5 μl of PCR I, ~1 nmol of pBgScKS, buffer, and polymerase as described above. PCR II/III is denatured at 94° for 5 min, and 27 amplification cycles are performed as follows: denaturation at 94° for 15 sec, annealing at 50° for 20 sec, and extension at 72° for 30 sec per 500 bp. The thermocycler is paused after the first two cycles, primers C and D are added to 0.4 μM each, and then cycling is resumed. Reactions are assayed by agarose electrophoresis after the remaining 25 thermocycles have been completed. The mutagenized PCR product is subcloned into the WT pcPER background, using unique *Bgl*II/*Sac*II restriction sites and verified by DNA sequencing (using T7 or SP6 priming sites) prior to use.

Wild-type and mutagenized expression plasmids used for transfection are grown in a suitable strain of *Escherichia coli* (such as DH5-α) and purified with a silica-based chromatography kit (Qiagen, Valencia, CA). Regardless of the particular method chosen for purification, care should be taken to avoid introducing nicks into the DNA, as highest transfection efficiencies are achieved with covalently closed plasmids. Approximately 5–15 μg of vector DNA is required to transfect each 60-mm dish of cells by a calcium phosphate procedure.

Transient Transfection of COS-1 Cells and Immunofluorescence Assay

Transfection Reagents

CaCl$_2$ (1 M): Store filter-sterilized aliquots at $-20°$

BBS (2×): 50 mM N,N-bis(2-hydroxyethyl)-2-aminoethanesulfonic acid (BES), 280 mM NaCl, 1.5 mM Na$_2$HPO$_4$, pH 6.95. Store filter-sterilized aliquots at $-20°$

PBS

Immunofluorescence Reagents

Blocking solution: PBS containing 4% (v/v) goat serum (Jackson Immunoresearch, West Grove, PA)

Primary antibody solutions: 1:10 dilution of per2B6 hybridoma cell supernatant in PBS; 1:10 dilution of rom 1D5 hybridoma cell supernatant in PBS

Secondary antibody solution: 1:200 dilution of FluoroLink Cy3-labeled goat anti-mouse IgG (H+L) (Amersham-Pharmacia Biotech, Piscataway, NJ) in PBS containing 0.4% (v/v) goat serum
PBS

Procedures. Several transient transfection methods are commonly used for cultured cells and increasing numbers of commercial reagents are also becoming available. We examined three methods for transient cotransfections of COS-1 cells with peripherin/rds and rom-1: electroporation, DEAE-dextran, and calcium phosphate. Although each method offers the advantage of economy over commercially available reagents, we find that only the calcium phosphate protocol produces significant levels of cotransfected cells. The method of Chen and Okayama[35] requires neither expensive reagents, nor complex procedures, but does necessitate overnight growth in a 3% CO_2 incubator. An experimental titration of the total mass and relative ratio of plasmid DNA(s) is required for optimal results. Expression levels do not always show a direct or simple relationship to the mass of transfected plasmids, and experimental conditions need to be determined empirically.

One day prior to transfection, 2×10^5 cells are plated in 4.5 ml of growth medium in a 60-mm dish and are grown overnight in 5% CO_2, 100% humidity, at 37°. If transfection efficiency is to be assayed, several sterile glass coverslips are laid into the dishes prior to the addition of cells. On the day of transfection, 12 μg of purified plasmid DNA (i.e., 4 μg of pcPER and 8 μg of pcROM) is diluted to 124 μl with doubly distilled H_2O, then mixed with 41 μl of 1 M $CaCl_2$ in a sterile microcentrifuge tube. The calcium–DNA mixture is added to 250 μl of 2× BBS with gentle vortexing. After a 5- to 30-min incubation at room temperature, the solution is added dropwise with swirling to the growth medium of a single 60-mm dish of COS-1 cells. Plates should be incubated at 3% CO_2, 37° overnight. On the next day, the medium is aspirated and replaced with fresh growth medium. The cells are typically assayed 48–72 hr posttransfection. Time points may be taken to determine the optimal time for cell harvesting.

Transfection efficiency is assessed by estimating the percentage of expressing cells as measured by immunofluorescent labeling. Cells grown on glass coverslips are simultaneously fixed and permeabilized in a 4° methanol bath for 20 min, then covered with blocking solution for 30 min at room temperature. The blocking solution is aspirated, and coverslips are rinsed once with PBS and then floated (cell side down) onto a 500-μl drop of primary antibody solution resting on a piece of Parafilm. Primary antibody is allowed to bind for 1 hr at room temperature, after which it is aspirated,

---

[35] C. Chen and H. Okayama, *Mol. Cell Biol.* **7,** 2745 (1987).

and the coverslip is rinsed serially in three beakers of PBS. Coverslips are floated onto secondary antibody solution (as described above), and allowed to bind for 30 min at room temperature. The secondary solution is aspirated, and the coverslips are again washed serially in three beakers of PBS. Coverslips are mounted onto slides and viewed with an epifluorescence microscope using the appropriate filter set (Cy3 emission is maximal at 570 nm). Routine controls include labeling of mock-transfected cells and labeling of cells with secondary antibody only. Transfection efficiencies can vary from 5 to 30% for singly transfected cells, and depend strongly on the particular protein being expressed. In contrast to rhodopsin, peripherin/rds and rom-1 are localized primarily within internal membranes (Fig. 1).

Analysis of Wild-Type and Mutant Peripherin/rds-rom-1 Complexes

Reagents

Cell lysis buffer: 1× PBS, 1.0% (v/v) Triton X-100, 1 mM dithiothreitol (DTT), pH 7.4

Immunoaffinity matrix: per 2B6–Sepharose 4B (~2 mg of antibody/ml of matrix)

Wash and sedimentation (WS) buffer: 1× PBS, 0.1% (v/v) Triton X-100, 1 mM DTT, pH 7.4.

Elution buffer: WS buffer containing synthetic peptide DAGQA-PAAG (0.1 mg/ml)

Density gradient solutions: 20, 15, 10, and 5% sucrose (w/w) in WS buffer

Phosphate-buffered saline containing 0.1% (v/v) Tween 20 (PBST)

Sodium dodecyl sulfate (SDS) sample buffer (3×): 60 mM Tris(hydroxymethyl)aminomethane hydrochloride (pH 6.8), 6% (w/v) SDS, 24% (v/v) glycerol, 0.01% (w/v) bromphenol blue [±7.5% (v/v) 2-mercaptoethanol (2-ME)]

Primary antibody solutions: per2B6 hybridoma cell supernatant diluted 1:25 in PBST; rom1D5 hybridoma cell supernatant diluted 1:50 in PBST

Secondary antibody solution: Sheep anti-mouse immunoglobulin linked to horseradish peroxidase (Amersham-Pharmacia Biotech) diluted 1:5000 in PBST containing nonfat dry milk (1 mg/ml)

Stripping buffer: 60 mM Tris(hydroxymethyl)aminomethane hydrochloride (pH 6.7), 2% (w/v) SDS, 100 mM 2-mercaptoethanol (2-ME)

ECL kit: Chemiluminescence Western development system (Amersham-Pharmacia Biotech)

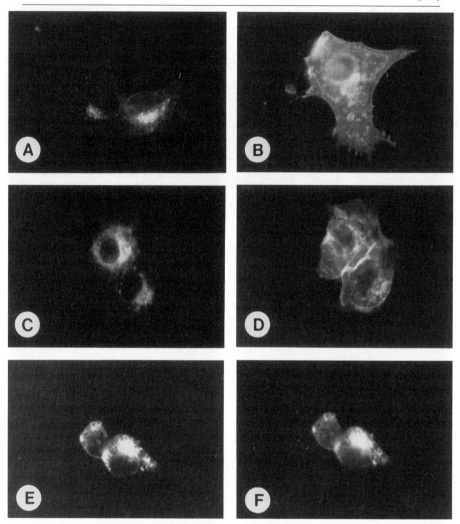

Fig. 1. Immunofluorescent localization of expressed photoreceptor proteins: peripherin/rds, rom-1, and rhodopsin. COS-1 cells individually transfected with either pcPER or pcRHO and labeled with anti-peripherin/rds or anti-rhodopsin antibodies illustrate contrasting localization patterns for peripherin/rds (A) and rhodopsin (B). Double-labeling of a single pair of cells cotransfected with both pcPER and pcRHO indicates that coexpression does not affect the localization of either peripherin/rds (C) or rhodopsin (D). Double-labeling of cells cotransfected with pcPER and pcROM demonstrates that rom-1 is localized to internal cell membranes (F) and does not affect peripherin/rds localization (E). All immunofluorescent labelings were dependent on transfection with the appropriate plasmid(s) and could be specifically inhibited by the addition of peptides corresponding to primary antibody epitopes (not shown) [Reproduced from A. F. X. Goldberg, O. L. Moritz, and R. S. Molday, *Biochemistry* **34,** 14213 (1995).]

Procedures. Cells are harvested at ~48 hr posttransfection; plates are washed twice with PBS, excess buffer is aspirated, and then the plates are chilled for approximately 10 min on ice. Cold lysis buffer (300 μl) is added, and the plates are swirled to distribute the buffer. The plates are maintained on ice for an additional 10 min. Cells and debris are then scraped into the buffer with a rubber policeman and collected into a microcentifuge tube. The postnuclear supernatant obtained after a 10-min centrifugation (14,000g) at 4° in a microcentrifuge is collected and maintained on ice until use.

Biosynthesis, protein folding, glycosylation, and dimerization of individually expressed peripherin/rds and rom-1 are assayed by Western blot analysis; representative data are presented in Fig. 2. Samples (20 μl of cell lysate plus 10 μl of 3× loading buffer) are subjected to discontinuous SDS–PAGE on 10% (w/v) polyacrylamide minigels (Bio-Rad, Hercules, CA). Gels are electroblotted onto Immobilon-P membranes (Millipore, Bedford, MA) in a Towbin buffer containing 10% (v/v) methanol for ~1 hr at 100 V. Membranes are blocked with PBST containing nonfat dry milk (1 mg/ml) for 30 min at room temperature, and then incubated with primary antibody solution for 1 hr at room temperature. The membranes are washed with three 10-min washes of PBST, then incubated with secondary antibody solution for 30 min at room temperature. After three 10-min washes with

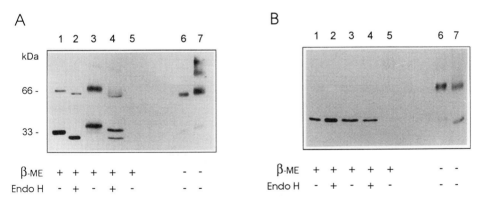

FIG. 2. Western blot analysis of individually expressed peripherin/rds and rom-1 proteins in COS-1 cells. Detergent extracts of COS-1 cells transiently transfected with either pcPER (A; lanes 3, 4, and 7), pcROM (B; lanes 3, 4, and 7) or vector alone (A and B; lanes 5) were electrophoresed under reducing or nonreducing (±2-mercaptoethanol) denaturing conditions, and then Western blotted with monoclonal antibodies specific for peripherin/rds (per2B6; A) or rom-1 (rom1D5; B). Mobility shifts induced by Endo H indicate high-mannose or hybrid carbohydrate modifications. A detergent extract from ROS membranes is shown for comparison in lanes 1, 2, and 6 (A and B). [Reproduced from A. F. X. Goldberg, O. L. Moritz, and R. S. Molday, *Biochemistry* **34**, 14213 (1995).]

PBST, the blots are developed with an enhanced chemiluminescence (ECL) kit as described by the manufacturer (Amersham-Pharmacia Biotech).

Samples are run under reducing conditions by the inclusion of 2.5% (v/v) 2-ME in the denaturing sample buffer. Wild-type peripherin/rds expressed in COS-1 cells migrates at approximately 35 kDa. Multiple bands are often observed and are thought to indicate multiple glycosylated forms. Rom-1 migrates at a similar size, but as a sharper band; it is not glycosylated.

Cell lysates are also run under nonreducing conditions to assay whether peripherin/rds mutants retain the ability to form disulfide-linked dimers, akin to WT. For these studies, the cell lysis buffer does not contain DTT. The samples are prepared in SDS sample buffer in the absence of reducing agent (2-ME), but otherwise treated similarly. A blank lane or two should be left between samples of differing reducing potentials to avoid potential diffusion of 2-ME into neighboring sample lanes.

To assay expressed proteins for N-linked glycosylation, cell lysates are digested with Endoglycosidase H_f (Endo H; New England BioLabs) for 16 hr at room temperature (200 units/μg total protein). Inclusion of a protease inhibitor cocktail is recommended. A mobility shift to lower molecular weight on glycosidase treatment indicates N-linked carbohydrate modification.

Immunoprecipitation is used to assay the formation of peripherin/rds–rom-1 complexes and to prepare samples for velocity sedimentation. An anti-peripherin/rds monoclonal antibody (per2B6) covalently coupled to Sepharose 4B is used to purify solubilized peripherin/rds from cell lysates. The combined lysates of four 60-mm dishes of cotransfected COS-1 cells are combined with 75 μl of per2B6–Sepharose beads and gently shaken for 2 hr at 4°. The beads are gently pelleted in a clinical centrifuge, then washed three times with WS buffer (10 ml each). Peripherin/rds-containing complexes are eluted for 1.5 hr at room temperature in 125 μl of elution buffer, which contains 0.1 mg of the synthetic peptide DAGQAPAAG (the per2B6 epitope), per milliliter. The immunopurified peripherin/rds is used both for coprecipitation and velocity sedimentation analyses.

Immunopurified peripherin/rds is assayed for associated rom-1 by Western blot analysis, using the anti-rom-1 monoclonal antibody rom1D5. Samples from (1) rom-1-transfected, (2) peripherin/rds-transfected, and (3) cotransfected cells are prepared as described, and 2-μl aliquots of each immunopurified peripherin/rds sample are run side by side on a 10% (w/v) SDS–polyacrylamide gel under reducing conditions. The gel is transferred to Immobilon-P as described, and is developed with rom1D5. A peripherin/rds-dependent copurification of rom-1 demonstrates peripherin/rds–rom-1 complex formation.

Velocity sedimentation analysis is performed in detergent solution on continuous sucrose gradients as first described by Goldberg et al.[30] Extensive use is made of a Beckman (Fullerton, CA) Optima TLX tabletop ultracentrifuge and its companion TLS-55 swinging bucket rotor. Continuous density gradients are prepared by sequentially layering 0.5 ml each of 20, 15, 10, and 5% (w/w) sucrose solutions into TLS-55 Ultraclear centrifuge tubes with a micropipette. The step gradients are allowed to become continuous by diffusion at room temperature for 1 hr. The prepared gradients are then chilled for at least 30 min on ice prior to loading the samples.

Sedimentation coefficients are estimated on the basis of distance traveled in the gradients: measured from point of sample application to final position at fractionation (peak fraction). The sedimentation coefficient ($S_{T,m}$) at the half-distance of travel (r_{avg}) at temperature T in solvent m is given by Eq. (1)[36]:

$$S_{T,m} = [(r - r_0)/t]/\omega^2 r_{avg} \tag{1}$$

where r_0 is the initial distance of the species of interest from the center of rotation, r is the distance at some later time t, and ω is the angular velocity. This sedimentation coefficient ($S_{20,w}$) can be corrected to standard conditions (water at 20°) by the following relationship:

$$S_{20,w} = S_{T,m}(\eta_{T,m}/\eta_{20,w})[(1 - \nu\rho_{20,w})/(1 - \nu\rho_{T,m})] \tag{2}$$

using density (ρ) and estimated viscosity (η) values at one-half the radial distance traveled.[37] Partial specific volume (ν) is taken as 0.83 cm^3/g, determined previously for the Triton X-100-solubilized peripherin/rds-rom-1 complex from bovine rod outer segments (ROS),[25] and r_0 is taken at 47.5 mm, from the dimensions of the TLS-55 rotor. Although these calculations can be performed manually, we prefer to use a QBasic computer program ("Svedberg's best friend") to facilitate multiple computations and avoid arithmetic errors; it is available on request.

Immunopurified samples (90–100 µl, described above) are gently layered onto gradients and centrifuged overnight at 50,000 rpm, 16 hr, 4°. Gradual acceleration and deceleration programs are employed to avoid disturbing the gradients. The next morning, tubes are carefully removed from their buckets, and gradients are fractionated dropwise into collection tubes (5 drops/tube) by careful puncturing of the tube bottoms with a 26-gauge syringe needle. This typically results in about 15 fractions per gradient. The sucrose concentration in each is measured by refractometry or gravitometry.

---

[36] S. Clarke and M. D. Smigel, *Methods Enzymol.* **172**, 696 (1989).
[37] ISCO, "ISCO Tables," 9th Ed. ISCO, Lincoln, Nebraska, 1987.

The gradient samples (20 µl each) are prepared in reducing loading buffer and run on 10% (w/v) polyacrylamide gels, using a 20-well comb. The gels are electroblotted onto Immobilon-P membranes (as described above), and are developed with rom1D5 antibody. Blots are stripped (30 min at 50°, submerged in stripping buffer), washed three times with PBS, reblocked for 30 min in PBST, and finally reprobed with per2B6 antibody. Laser densitometry is used to convert data into a graphical form. Figure 3A shows a comparison of the heterologously expressed peripherin/rds–rom-1 complex with its counterpart solubilized from bovine ROS membranes— known to be heterotetrameric.[25] Individually expressed peripherin/rds and rom-1 also sediment (with or without prior immunopurification) in a similar stoichiometric form (Fig. 3B).

This type of analysis is easily extended to analyze disease-linked mutations in peripherin/rds and rom-1 (Fig. 4). For example, when expressed in COS-1 cells, both WT and the L185P mutant peripherin/rds migrate as single major species, although their differing positions (WT, 6.9s; L185P, 4.7s) reveal a defect in L185P subunit assembly. Coexpression with rom-1 indicates that L185P peripherin/rds is recruited into a complex (6.6 s) that is indistinguishable from the WT peripherin/rds–rom-1 complex (6.4 s), and provides a molecular rationale for the digenic pattern of inheritance.[33,34]

Summary

Nearly 40 disease-linked mutations have been reported for peripherin/rds to date; heterologous expression in tissue culture cells offers a valuable means of efficiently characterizing the biochemical properties of the various mutants. Peripherin/rds is proposed to act as an essential structural element in outer segment disk morphogenesis, and a present transgenic mice offer the sole tractable system in which recombinant peripherin/rds may be examined functionally *in situ*. Because the generation and characterization of transgenic animals are both expensive and time consuming, heterologous expression in cultured cells offers an important and complementary means of addressing protein structure and function.

The immunopurification and detection of the peripherin/rds–rom-1 complex are performed using specific immunochemical reagents, monoclonal and polyclonal antibodies, that are not commonly available. Several laboratories have developed antibodies to peripherin/rds and rom-1 in rabbits and mice, using a variety of immunogens: purified ROS membranes, purified *E. coli* fusion proteins, and synthetic peptides coupled to proteins.[7,20,22,27] The C-terminal regions appear to be most highly antigenic,

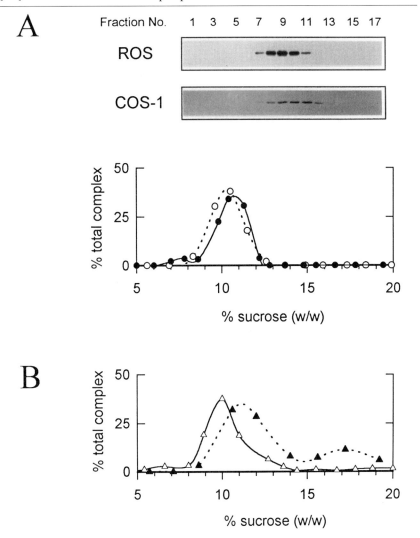

FIG. 3. Sedimentation velocity analysis of complexes formed by peripherin/rds and rom-1 in bovine ROS membranes and COS-1 cells. Peripherin/rds-rom-1 complexes from Triton X-100-solubilized bovine ROS membranes (○) and per2B6-immunopurified complexes from cotransfected COS-1 cells (●) shows similar sedimentation velocities in 5–20% (w/w) sucrose gradients (A). Chemiluminescent Western blots using anti-peripherin antibody per2B6 (ROS) and anti-rom-1 antibody rom1D5 (COS-1) are shown, as are corresponding plots of laser densitometric scans. Plots of individually COS-1 expressed peripherin/rds (△) and rom-1 (▲) proteins were derived in an analogous manner and are shown in (B). [Reproduced from A. F. X. Goldberg, O. L. Moritz, and R. S. Molday, *Biochemistry* **34,** 14213 (1995).]

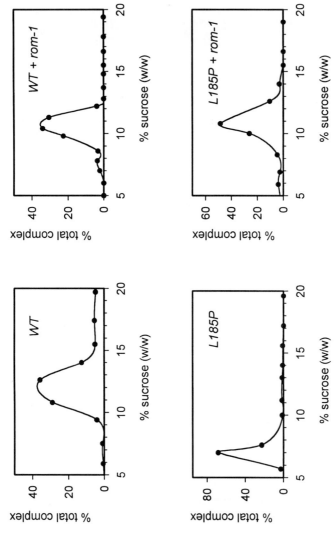

FIG. 4. Sedimentation velocity analysis of complexes formed in transfected COS-1 cells by WT (*top*) and L185P mutant (*bottom*) peripherin/rds with (*right*) and without (*left*) rom-1. Complexes immunopurified from transfected COS-1 cells were sedimented in 5–20% (w/w) sucrose gradients, fractionated, and identified by Western blotting. L185P mutant shows a conditional defect in subunit assembly; coexpression with rom-1 allows assembly of a WT stoichiometric form.

although antibodies have been generated to other regions as well.[38] Regardless of their source, antibodies must be thoroughly characterized; specificity is often a function of solution conditions and must be determined empirically.

The approach as described here has provided explanations for several instances of peripherin/rds-associated disease, including digenic RP linked to as L185P mutation, and adRP associated with C118/119del and C214S mutations.[34,39] In addition, the R172W mutation, linked to macular dystrophy[40] and preferential loss in cone function,[41] is shown to behave normally with respect to biosynthesis and subunit assembly[34]; it likely involves a more subtle functional defect that remains to be described. Finally, the methodology reported here has suggested the existence of a novel (homotetrameric) form of peripherin/rds in individuals lacking rom-1[34]; this hypothesis has been confirmed in rom-1 knockout mice.[42] The information obtained thus far demonstrates the utility of using heterologously expressed peripherin/rds and rom-1 to investigate the consequences of disease-linked mutations in these polypeptides. Heterologous cell expression coupled with transgenic mouse methodologies should continue to provide a more detailed understanding of molecular mechanisms underlying inherited retinal degenerative diseases.

Acknowledgments

The authors thank Beatrice Tam, Orson Moritz, Andrea Dose, and Laurie Molday for technical advice and discussion. This work was support by grants from the National Eye Institute (EY02422), the RP Foundation of Canada, and the Medical Research Council.

---

[38] R. Y. Kim, M. A. Bedolli, and J. Goodarzi, *Ophthalmic Genet.* **19,** 165 (1998).
[39] A. F. X. Goldberg and R. S. Molday, *Biochemistry* **37,** 680 (1998).
[40] J. A. Wells, J. J. Wroblewski, T. J. Keen, C. Inglehearn, C. Jubb, A. Eckstein, M. Jay, G. Arden, S. Ghattachary, F. Fitzke, and A. Bird, *Nature Genet.* **3,** 213 (1993).
[41] S. G. Jacobson, A. V. Cideciyan, A. M. Maguire, J. Bennett, V. C. Sheffield, and E. M. Stone, *Exp. Eye Res.* **63,** 603 (1996).
[42] A. F. X. Goldberg, G. A. Clarke, L. Molday, R. McInnes, and R. S. Molday, unpublished data (1998).

[45] Isolation of Retinal Proteins That Interact with Retinitis Pigmentosa GTPase Regulator by Interaction Trap Screen in Yeast

By RONALD ROEPMAN, DIANA SCHICK, and PAULO A. FERREIRA

Introduction

Retinitis pigmentosa (RP) is a clinically and genetically heterogeneous group of retinal disorders, leading to blindness. More than 15 different genetic loci have been reported and 10 disease genes have been identified.[1] Positional cloning led to the cloning of the gene underlying X-linked retinitis pigmentosa type 3 (RP3), retinitis pigmentosa GTPase regulator (*RPGR*). The N-terminal half of the RPGR protein shows homology with RCC1, the guanidine-nucleotide exchange factor (GEF) of the Ras-like GTPase, Ran.[2,3] *RPGR* is mutated in 20% of patients with X-linked RP.[4] All RP-associated missense mutations reported so far are located in the RCC1-homologous domain, which is therefore thought to be the key functional domain of RPGR in the retina. The retina-specific phenotype of RP3 contrasts with the ubiquitous expression of RPGR.[2,3] To elucidate the molecular mechanism underlying this severe form of blindness, we have made use of the yeast two-hybrid system to isolate proteins that interact with RPGR in the retina. The methods we used are described in this chapter.

The yeast two-hybrid system is a genetic method that initially was developed to study protein–protein interactions *in vivo*.[5] The system is based on the ability of some eukaryotic transcription factors to act in a modular fashion and their ability to activate adjacent reporter genes, usually *lacZ* and a nutritional marker (Fig. 1). The level of expression of the reporter gene(s) is a measure for the interaction of the two proteins. This

---

[1] Retnet Internet site at URL: *http://www.sph.uth.tmc.edu/Retnet/*.

[2] R. Roepman, G. van Duynhoven, T. Rosenberg, A. J. L. G. Pinckers, E. M. Bleeker-Wagemakers, A. A. B. Bergen, J. Post, A. Beck, R. Reinhardt, H.-H. Ropers, F. P. M. Cremers, and W. Berger, *Hum. Mol. Genet.* **5**, 1035 (1996).

[3] A. Meindl, K. Dry, K. Herrmann, F. Manson, A. Ciccodicola, A. Edgar, M. R. S. Carvalho, H. Achatz, H. Hellebrand, A. Lennon, C. Migliaccio, K. Porter, E. Zrenner, A. Bird, M. Jay, B. Lorenz, B. Wittwer, M. D'Urso, T. Meitinger, and A. Wright, *Nature Genet.* **13**, 35 (1996).

[4] M. Buraczynska, W. Wu, R. Fujita, K. Buraczynska, E. Phelps, S. Andreasson, J. Bennett, D. G. Birch, G. A. Fishman, D. R. Hoffman, G. Inana, S. G. Jacobson, M. A. Musarella, P. A. Sieving, and A. Swaroop, *Am. J. Hum. Genet.* **61**, 1287 (1997).

[5] S. Fields and O. Song, *Nature (London)* **340**, 245 (1989).

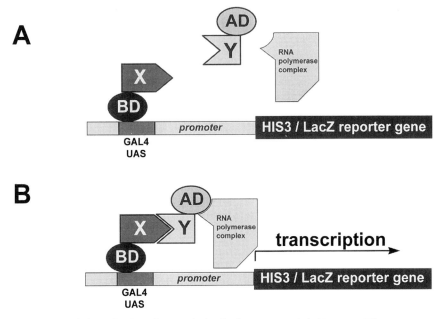

FIG. 1. Modular activation of transcription in the yeast two-hybrid system. The upstream activating sequence (UAS) and reporter genes are integrated into the yeast chromosome. (A) The GAL4 BD hybrid protein (BD and the bait protein X) binds to the *GAL4* UAS present upstream of the *lacZ* (colorimetric marker) and *HIS3* (auxotrophic marker) reporter genes. The GAL4 AD hybrid protein (AD and target protein Y) binds transcription factors of the RNA polymerase complex in the nucleus, but does not localize to the *GAL4* UAS. (B) If the bait (X) and target (Y) proteins interact, the GAL4 AD and the GAL4 BD are brought close to each other and act together with the bound transcription factors to initiate transcription of *lacZ* and *HIS3* reporter genes.

feature of the system has been used with success for a rapid screening of encoded proteins from a library of cDNAs, fused to different kinds of transcription activation domains, for *in vivo* protein–protein interactions with a target or "bait" protein, fused to the DNA-binding domain of the cognate transcription factor (Fig. 1).[6–9] We have made use of the *GAL4*-based HybriZAP yeast two-hybrid system (Stratagene, La Jolla, CA) (Fig. 1).

[6] C. T. Chien, P. L. Bartel, R. Sternglanz, and S. Fields, *Proc. Natl. Acad. Sci. U.S.A.* **88,** 9578 (1991).
[7] T. Durfee, K. Becherer, P. L. Chen, S. H. Yeh, Y. Yang, A. E. Kilburn, W. H. Lee, and S. J. Elledge, *Genes Dev.* **7,** 555 (1993).
[8] A. S. Zervos, J. Gyuris, and R. Brent, *Cell* **72,** 223 (1993).
[9] T. Munder and P. Furst, *Mol. Cell. Biol.* **12,** 2091 (1992).

There are some advantages of using the yeast two-hybrid system for isolating possible substrates for RPGR, above other biochemical methods. First, it is an *in vivo* interaction assay in a eukaryotic cell, in which better physiological conditions are present than in *in vitro* interaction assays. Second, the system is more sensitive than some *in vitro* assays such as coimmunoprecipitation or protein affinity chromatography, and has proved to detect relatively weak and, more importantly, transient interactions (with a high k_{off}) that are often observed in large protein complexes.[9–12] The third advantage is that the end product of this assay is a cDNA clone of the interacting protein, which shortcuts downstream analysis of the isolated gene drastically. The fourth advantage is that interactions might be detected if the proteins require posttranslational modification for proper interaction, the machinery of which is present in the eukaryotic yeast. The fifth advantage is that the system often allows quick identification of the substrate-binding domain in the target protein product. The yeast two-hybrid system, however, also presents some disadvantages. For example, protein–protein interactions will not be identified if they are blocked by one or more of the fusion partners or if the proteins are not properly targeted to the nucleus owing to specific sequence signals or posttranslational modifications. Also, the "bait" fusion protein might cause intrinsic transcriptional activation of the reporter gene. This is especially true when this reflects the true function of a protein in transcription, but it might also be caused by stretches of acidic residues that can act as activators when fused to the DNA-binding domain. However, in many cases random proteins or protein domains can cause activation.[13] Therefore, the "bait" proteins that are used should always be tested for intrinsic activation of transcription of the reporter gene before they are used in a library screen. Finally, because there is not a direct functional basis for the interactions, they should always be verified by, often much less sensitive, biochemical or immunohistochemical techniques. In this respect, it is important to use cDNA libraries for two-hybrid screening that are derived from the tissue in which the bait protein is known to be expressed, and pathology of the involved disease is manifested.

We have chosen to screen retinal cDNA libraries from bovine because of the availability of fresh bovine eyes. This allowed us to isolate intact RNA from retina and prepare cDNA libraries of high quality. The cDNA libraries contained more than 3 million independent recombinants. A highly

---

[10] L. Van Aelst, M. Barr, S. Marcus, A. Polverino, and M. Wigler, *Proc. Natl. Acad. Sci. U.S.A.* **90,** 6213 (1993).
[11] M. Yang, Z. Wu, and S. Fields, *Nucleic Acids Res.* **23,** 1152 (1995).
[12] B. Li and S. Fields, *FASEB J.* **7,** 957 (1993).
[13] J. Ma and M. Ptashne, *Cell* **51,** 113 (1987).

representative library is an important prerequisite for experiments in which there is no *a priori* knowledge of the level of expression of the genes that encode the putative interacting proteins. In addition, only one-third of the products in a unidirectional [oligo(dT) primed] library and one-sixth of the products in a bidirectional (randomly primed) library are translated in the correct reading frame. As a "bait," we used only the RCC1-homologous domain of RPGR, denoted N-RPGR, which does not contain the isoprenylation site. The isoprenylation of the cysteine in the CAAX box of RPGR may interfere with nuclear translocation, as shown before for other CAAX-modified proteins.[10]

Procedures

Descriptions of general molecular techniques (cloning methods, DNA/ RNA isolations, preparation of growth media for bacteria and yeast) can be found elsewhere.[14,15] The cloning vectors that we use are from Stratagene and all restriction endonucleases and DNA-modifying enzymes are from Life Technologies (Gaithersburg, MD) unless noted otherwise.

Construction of Retinitis Pigmentosa GTPase Regulator Bait Plasmids

The RPGR bait plasmids are constructed by inserting the coding sequence of RPGR in-frame with the *GAL4* DNA-binding domain into pBD-T2C, a vector that is derived from the HybriZap pBD-GAL4 Cam phagemid vector (Stratagene), but that contains a modified multiple cloning site (gift from D. de Bruijn), as shown in Fig. 2. N-RPGR is obtained by performing a polymerase chain reaction (PCR) on cDNA clone R36, as previously reported.[2] The 5' sequence of *RPGR* (exons 1–13, a cDNA fragment of 1249 bp) is amplified using the sense primer 1412 (gtg ata gga tcc atg agg gag ccg gaa gag c) and antisense primer 1394 (tct tcg ccg cat acg tgc t). The reaction is carried out in a total volume of 50 μl and in the presence of 20 mM Tris-HCl (pH 8.75), 10 mM KCl, 10 mM $(NH_4)_2SO_4$, 2 mM $MgSO_4$, 0.1% (v/v) Triton X-100, a 0.5 mM concentration of each dNTP, 10 ng of bovine serum albumin (New England BioLabs, Beverly, MA), 0.1% (v/v) glycerol, 100 ng of template DNA, and 1.25 units of *Pfu* DNA polymerase (Stratagene). Amplification is performed on an ABI 9600 thermal cycler (Perkin-Elmer, Norwalk, CT) for 15 cycles, each 30 sec at 95°, 45 sec at

---

[14] J. Sambrook, E. F. Fritsch, and T. Maniatis, "Molecular Cloning: A Laboratory Manual." Cold Spring Harbor Laboratory Press, Cold Spring Harbor, New York, 1989.

[15] F. M. Ausubel, R. Brent, R. E. Kingston, D. D. Moore, J. G. Seidman, J. A. Smith, and K. Struhl (eds.), "Current Protocols in Molecular Biology." John Wiley & Sons, New York, 1995.

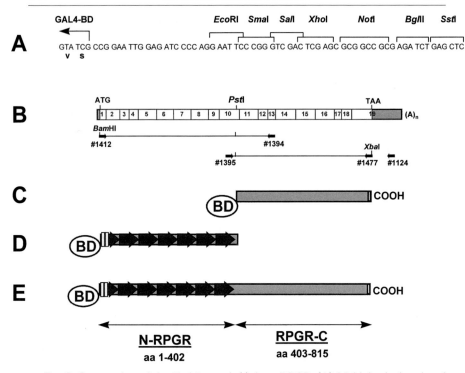

FIG. 2. Construction of the "bait" protein(s) from *RPGR*. (A) Multiple cloning site of pBD-T2C. (B) Construction of the inserts by PCR, using the indicated oligonucleotides as primers. (C) pBD/N-RPGR fusion product. (D) pBD/RPGR-C fusion product. (E) pBD/RPGR fusion product.

58°, and 3 min at 72°. Oligonucleotide 1412 contains an additional (unique) 5' *Bam*HI site and *RPGR* contains a unique *Pst*I site at position 1204, which is present in this amplified fragment. These sites are used to isolate N-RPGR (codons 1–401, exons 1–10 of full-length *RPGR*).

The 3' sequence of RPGR is amplified by reverse transcriptase (RT)-PCR on poly(A)$^+$ mRNA that is isolated from HeLa cells as described elsewhere.[2,16] We use the antisense primer 1124 (aca ata cac ttg gtg act gtg a) to perform the first strand synthesis, and the sense primer 1395 (tct gcc gta tag cag ttt aac) and antisense primer 1477 (att cgg tct aga tta tag tat tgt aca gga ttt) for PCR amplification of the 3' RPGR cDNA fragment of 1388 bp. The reaction is carried out exactly as described above for the 5' RPGR fragment. Oligonucleotide 1477 contains an additional (unique) 3'

---

[16] H. van Bokhoven, J. A. van den Hurk, L. Bogerd, C. Philippe, S. Gilgenkrantz, P. de Jong, H. H. Ropers, and F. P. Cremers, *Hum. Mol. Genet.* **3**, 1041 (1994).

*Xba*I site and the *Pst*I site at position 1204 is also present in this amplified fragment. These sites are used to isolate RPGR-C (codons 402–815, exons 10–19 of full-length RPGR).

The amplified fragments are isolated from a 1% (w/v) agarose gel and purified with a QIAquick gel extraction kit (Qiagen, Chatsworth, CA). The 5′ RPGR cDNA fragment (2 μg) is digested with 10 units each of *Bam*HI and *Pst*I restriction endonuclease (4 hr at 37°). The 1.2-kb N-RPGR fragment is isolated from a 1% (w/v) agarose gel, purified by QIAquick gel extraction, and ligated (16 hr at 16°) into the *Bam*HI- and *Pst*I-digested plasmid pBluescript, using T4 DNA ligase. RPGR-C is constructed in a similar manner using the *Xba*I and *Pst*I restriction sites. The ligation products are transformed into competent XL1-Blue cells (Stratagene) and the plasmids are isolated by minipreparation, using a QIAgen spin plasmid kit. The correct sequence of the cDNA fragments is confirmed by sequencing on an ABI 370A automatic DNA sequencer, using a *Taq* DyeDeoxy terminator cycle sequencing kit (Applied Biosystems, Foster City, CA). Full-length RPGR (codons 1–815, exons 1–19) is constructed by fusing the N-RPGR and RPGR-C fragments at the *Pst*I site. The inserts with the correct sequence are isolated from pBluescript, using the previously mentioned restriction sites, and gel purified, and the sticky ends are filled in by Klenow DNA polymerase. These blunt-ended fragments are ligated in-frame with the GAL4-binding domain into the *Sma*I-digested modified HybriZap vector pBD-T2C. The plasmid constructs are electroporated into XL1-Blue cells and screened for the correct orientation of the insert by digestion with *Eco*RI. The correct reading frame at the 5′ end of the selected "bait" clones is confirmed by sequence analysis. Large-scale isolations of the relevant plasmids are prepared with a QIAgen Maxiprep kit.

Preparation of Target Plasmids from cDNA Library

The custom cDNA libraries from female bovine retinas are prepared in the HybriZAP vector, which will accommodate DNA inserts up to 10 kb in length. A HybriZAP λ library can be converted to a pAD-GAL4 phagemid library by *in vivo* mass excision, using the same excision mechanism as has been widely used in the LambdaZAP vectors.[17,18] The bacterial host strains are from Stratagene. The strain XLOLR, genotype Δ(*mcrA*)183 Δ (*mcrCB-hsdSMR-mrr*)173 *endA1 thi-1 gyrA96 relA1 lac* [F′ *proAB lac-I*^{q}*Z*Δ*M15* Tn*10* (Tet^r)] Su$^-$ (nonsuppressing) λ^R (λ resistant), is used for plating excised phagemids, using the interference-resistant helper phage ExAssist (Stratagene) to efficiently excise the pAD-GAL4 phagemid vector

---

[17] J. M. Short, J. M. Fernandez, J. A. Sorge, and W. D. Huse, *Nucleic Acids Res.* **16,** 7583 (1988).
[18] J. M. Short and J. A. Sorge, *Methods Enzymol.* **216,** 495 (1992).

from the HybriZAP vector while preventing problems with helper phage coinfection. XLOLR cells are also resistant to λ infection, thereby ensuring that the library is not lysed by residual λ phage. The strain XL1-Blue MRF', genotype Δ(mcrA)183 Δ(mcrCB-hsdSMR-mrr) 173 endA1 supE44 thi-1 gyrA96 relA1 lac [F' proAB lacI^{q}ZΔM15 Tn10 (Tet^r)], is used for all other manipulations.

Oligo(dT)-primed cDNA libraries are constructed into the EcoRI and XhoI cloning sites and randomly primed cDNA libraries are constructed into the EcoRI cloning site. These sites are also used to control the inserts of the phagemids after the excision reaction. From 12 clones picked at random from the libraries, restriction analysis with EcoRI and XhoI yielded an average insert size of 1.9 kb (0.85–3.4 kb) for the oligo(dT)-primed library and of 1.4 kb (0.6–2 kb) for the randomly primed library. cDNA libraries are size fractionated into two pools on a Sepharose sizing column. The first collected fraction contains the largest cDNAs. The number of independent recombinants of the first and second fractions are, respectively, 4.5×10^6 and 4.2×10^6 pfu for the oligo(dT)-primed library, and 2.2×10^6 and 3.0×10^6 pfu for the randomly primed library. Primary libraries are subdivided into pools representing each about 1 million independent recombinants, and each of these is then amplified. Amplified libraries from the first collected fractions are used in our screen.

Yeast Host Strain

The yeast reporter host strain we use is Saccharomyces cerevisiae YRG-2 (Stratagene). This strain is derived from strain HF7c and has the following genotype: Matα ura3-52 his3-200 ade2-101 lys2-801 trp1-901 leu2-3 112 gal4-542 gal80-538 LYS2::UAS$_{GAL1}$-TATA$_{GAL1}$-HIS3, URA3::$_{UASGAL4\ 17mers(\times 3)}$-TATA$_{CYC1}$-lacZ.[19] This strain contains two reporter genes, lacZ and HIS3, to detect protein–protein interactions. Three copies of the GAL4 17-mer consensus sequence (GAL4 DNA-binding sites) and the TATA box of the iso-1-cytochrome c promoter (p_{CYC1}) are fused to the lacZ reporter gene and regulate its expression. The expression of the HIS3 reporter gene is regulated by fusion of the upstream activating sequence (UAS$_{GAL1}$) of the GAL1 promoter (p_{GAL1}), which contains four GAL4 DNA-binding sites, and its TATA box. The strain also carries mutations that prevent the expression of the endogenous GAL4 gene and a mutation in GAL80, whose product inhibits function of the GAL4 gene product. For selection of yeast clones that have been transformed with the AD and BD plasmids, it carries the auxotrophic markers leucine (leu2, to select for the AD plasmid) and

---

[19] H. E. Feilotter, G. J. Hannon, C. J. Ruddell, and D. Beach, Nucleic Acids Res. **22,** 1502 (1994).

tryptophan (*trp1*, to select for the BD plasmid). The auxotrophic marker histidine (*his3*) allows selection for activation of transcription of the *HIS3* reporter gene.

Prescreen Testing of Bait Plasmid pBD/N-RPGR

The amount of intrinsic transcriptional activation by the bait plasmids pBD/N-RPGR, pBD/RPGR-C, and pBD/RPGR, is determined by prescreen testing of the activation of the two reporter genes by the bait protein alone. For this purpose, 50 μl of frozen competent yeast cells is thawed for 5 min on ice. The cells are mixed gently and transferred into a 12-ml polypropylene tube. While the tubes are kept on ice and the cells are mixed gently, the following components are subsequently added: 0.5 μl of transformation reagent ATR-1 (Stratagene), 0.5 μl of ATR-2, and 1 μg of pBD/N-RPGR bait plasmid DNA. This mixture is incubated for 30 min at 30° (mixed gently every 5 min) followed by a heat shock for 5 min at 42°. To this mixture, 0.45 ml of SD (synthetic dextrose minimal media) dropout medium without Trp (at 30°) is added to allow selection for the bait plasmid. The cells are allowed to recuperate for 3 hr (about two divisions) at 30° with shaking at 300 rpm. The cells are plated on plates containing SD-agar −Trp and SD-agar −Trp −His. The transformants, which grow well on media lacking Trp because of the *trp1* gene in the pBD plasmid, will also grow on media lacking His if the bait protein causes intrinsic activation of the *his3* reporter gene. This is not the case with any of the three RPGR bait plasmids: in our experiments no transformants appeared on the SD −Trp −His plates, while the −Trp plate was covered with transformants. Two clones from these plates were picked, using a toothpick, and streaked onto a fresh plate containing SD-agar −Trp, which was incubated for 3 days at 30°. The activity of the second (*lacZ*) reporter gene is determined with these clones in a filter lift assay as described below. In addition, no intrinsic transcriptional activation can be determined in any of the yeast transformants containing one of the three RPGR bait constructs. Therefore, any of the bait constructs is suitable as a bait in the two-hybrid system.

Two-Hybrid cDNA Library Screening

YRG-2 yeast competent cells (Stratagene) are cotransformed with bait pBD/N-RPGR and prey (target) pAD/cDNA library plasmids as follows: 1 ml of frozen competent yeast cells is thawed for 5 min on ice. The cells are mixed gently and transferred into a 50-ml conical polypropylene tube. While the tubes are kept on ice and the cells are mixed gently, the following components are subsequently added: 10 μl of transformation reagent

ATR-1 (Stratagene), 10 μl of ATR-2, 10 μg of bait plasmid DNA, and 10 μg of library plasmid DNA. This mixture is incubated for 30 min at 30° (mixed gently every 5 min) and after that, a heat shock is given for 5 min at 42°. To this mixture is added 9 ml of SD dropout medium (at 30°), which contains all necessary amino acids, including His but not Leu and Trp, to allow selection for both bait and target plasmids. The cells are allowed to recuperate for 3 hr (about two divisions) at 30° with shaking at 300 rpm.

To determine the transformation efficiency, 10 μl of cotransformation reaction mixture is plated on 9-cm SD-agar plates (−Leu −Trp) that are incubated at 30° for 3 days. On average we achieved 5×10^5 cotransformants per transformation. We screened a total of 4×10^6 clones of the oligo(dT)-primed bovine retina cDNA library and a total of 2×10^6 clones of the randomly primed bovine retina cDNA library. Thus, the same 1 million independent recombinant clones were screened four and two times, respectively, to ensure a statistical representation of every translated protein in the fraction (375,000 and 175,000 proteins in the respective libraries) and saturation of the screen. The remainder of the cotransformation reaction mixture is centrifuged (2500g for 5 min at 25°) and then resuspended in 400 μl of SD dropout-medium without Leu, Trp, and His to allow selection for protein–protein interactions. Each transformation sample is then plated on two 15-cm dropout plates, each containing 50 ml of SD-agar (−Leu −Trp −His). It is important to do this gently with only a few streaks, because the cells are quite fragile at this point. The plates are incubated at 30° for up to 7 days. All colonies that can be distinguished from the background growth of the yeast cells are transferred to a fresh dropout plate (15 cm, SD-agar −Leu −Trp −His) in a gridded pattern and this plate is wrapped with Parafilm to prevent excessive drying. The background growth of the yeast cells is due to the leaky expression of the *HIS3* reporter gene. The first colonies usually start to show up after 3 days, but for weaker interactions, and therefore a lower expression of the *HIS3* reporter gene, it might take longer to generate enough histidine to allow the yeast cells to grow. From this screen, we have isolated 36 putatively interacting clones from the oligo(dT)-primed library and 34 from the randomly primed library. The streaked clones are grown for another 3 days until the transcription activation of the second reporter gene, *lacZ,* is detected by a β-galactosidase assay.

β-Galactosidase Filter Lift Assay

The production of β-galactosidase by the activation of the *lacZ* reporter gene is detected by a filter lift assay. In this assay, the colorless 5-bromo-4-chloro-3-indolyl-β-D-galactopyranoside (X-Gal) is used as a substrate by β-galactosidase, which turns the X-Gal into a blue product. To transfer the gridded transformants from the first selection to a piece of qualitative filter paper (12.5-cm Whatman No. 1, grade A; Fisher Scientific, Pittsburgh, PA), this paper is placed on the surface of the plate that contains the streaked clones. Any air bubbles are carefully removed and the filter paper is pressed firmly onto the surface of the plate to ensure contact of all clones with the filter paper for 3 min. The filter paper is carefully lifted, using forceps, and placed colony side up in liquid nitrogen for 15 sec. It is then placed on a new filter paper to thaw for 2 min and, subsequently, placed colony side up in a 15-cm petri dish onto a filter paper soaked in 4.5 ml of Z-buffer (60 mM $Na_2HPO_4 \cdot 7H_2O$, 40 mM $NaH_2PO_4 \cdot H_2O$, 10 mM KCl, 1 mM $MgSO_4 \cdot 7H_2O$, pH 7.0) containing 2-mercaptoethanol (2.7 ml/liter) and X-Gal stock solution [16.7 ml/liter; 20 mg of X-gal per milliliter of N,N-dimethylformamide (DMF)]. The petri dishes are wrapped with Parafilm and incubated at 30°. They are monitored for the production of blue color indicating β-galactosidase activity. The amount of blue color produced is compared with the production of blue color by the proper positive and negative controls that are included to validate the assay. A yeast transformant containing pGAL4 (grown on SD −Leu) is used as a positive control for activation of the reporter genes by the full-length GAL4. It produces an intense blue color after 1 hr of incubation. The p53 control plasmid (yeast transformants grown on SD −Trp) expresses the binding domain of GAL4 and amino acids 72–390 of murine p53 as a hybrid protein.[20] The pSV40 control plasmid (grown on SD −Leu) expresses the activation domain of GAL4 and amino acids 84–708 of the SV40 large-T antigen as a hybrid protein.[6,12] These portions of p53 and pSV40 are known to interact *in vivo,* and therefore yeast cotransformants with these plasmids (grown on SD −Leu −Trp −His) are used as positive controls for true interactions in the system.[21] They produce a blue color after incubation for 3 hr, although much less intense than the GAL4 control.

The pLamin C control plasmid (yeast transformants grown on SD −Leu) expresses the binding domain of GAL4 and amino acids 67–230 of human lamin C as a hybrid protein.[21] The expressed protein from this plasmid has been shown not to interact with p53 and therefore yeast cotransformants with these plasmids (grown on SD −Leu −Trp, with or without His) are used as negative controls.[21] Only after incubation for 20 hr may a weak blue color be determined, probably caused by nonspecific interactions.

From the 70 gridded library transformants in our experiment, 9 clones showed a blue color after 3 hr of incubation, 16 clones after 8 hr, and most

---

[20] K. Iwabuchi, B. Li, P. Bartel, and S. Fields, *Oncogene* **8,** 1693 (1993).
[21] P. Bartel, C. T. Chien, R. Sternglanz, and S. Fields, *Biotechniques* **14,** 920 (1993).

of the other ones did not turn blue after 20 hr of incubation (or as weakly as the negative control). Therefore, this screening yielded 25 putatively interacting clones [17 from the oligo(dT)-primed library and 8 from the randomly primed library], which were selected for verification of the interaction.

Verification of Interaction

Although the yeast two-hybrid method is a powerful and sensitive method to search for interacting protein substrates, a pitfall might be the isolation of false positives. Numerous false positives have been described, some of which are reported more often by different investigators, performing entirely unrelated screens.[22] Because in yeast, different unrelated multicopy plasmids can be present in the same cell, the target plasmids are isolated from yeast and used for a series of control (co-)transformations to rule out any nonspecific interactions.[21,23]

The plasmid DNA is isolated from yeast by the following quick procedure.[24] Five milliliters of SD-medium −Leu −Trp −His in a 50-ml propylene conical tube is inoculated with a His$^+$ LacZ$^+$ colony. This culture is incubated for 4 days until the medium was saturated. One milliliter of culture is transferred to a freeze-vial containing 1 ml of 40% (v/v) glycerol. This sample is put at −80° for long-term storage. The remaining cells are pelleted in a 1.5-ml microcentrifuge tube at 14,000g for 10 sec at 4°. The supernatant is decanted, 0.2 ml of yeast lysis solution [2% (v/v) Triton X-100, 1% (w/v) sodium dodecyl sulfate (SDS), 100 mM NaCl, 10 mM Tris-HCl (pH 8.0), 1 mM EDTA] is added, and the yeast cells are resuspended by vortexing. To this mixture, 0.2 ml of phenol–chloroform–isoamyl alcohol (25:24:1, v/v/v) is added together with 0.3 g of acid-washed glass beads (Sigma, St. Louis, MO). This suspension is vortexed for 2 min and subsequently centrifuged at 14,000g for 5 min at room temperature. The top aqueous phase containing the DNA is transferred to a new microcentrifuge tube and the DNA is precipitated with a one-tenth volume of 3M sodium acetate (pH 5.2) and 2 volumes of ethanol. This suspension is centrifuged at 14,000g for 10 min at 4°, the supernatant is decanted, and the DNA pellet is washed with 70% (v/v) ethanol and dried under vacuum. The DNA pellet is redissolved in 50 μl of Tris–EDTA (TE) buffer and 10 μl of this solution is used to transform electrocompetent XL1-Blue MRF' cells by

---

[22] Golemis laboratory Internet site at URL: http://www.fccc.edu/research/labs/golemis/intro.html.
[23] J. W. Harper, G. R. Adami, N. Wei, K. Keyomarsi, and S. J. Elledge, *Cell* **75**, 805 (1993).
[24] C. S. Hoffman and F. Winston, *Gene* **57**, 267 (1987).

electroporation, as has been described by others.[25] Selection for the target plasmid is performed by plating on LB–ampicillin agar plates, which are incubated overnight at 37°. Colonies are transferred from the plates to 5 ml of LB–ampicillin medium, the cultures are incubated overnight at 37° with shaking, and plasmid DNA is isolated with a QIAgen spin plasmid miniprep kit.

The nucleic acid sequence of these target inserts is determined by sequencing on an ABI 370A automatic DNA sequencer (with a *Taq* Dye-Deoxy terminator cycle sequencing kit; Applied Biosystems) from both ends, using the forward primer GAL5-AD and the reverse primer GAL3-AD. Among the 25 putatively positive clones from different screens, 5 clones from the oligo(dT)-primed library have been found to contain an identical insert: A13, A65, A81, A92, and B37. The open reading frames of this insert is in-frame with the activation domain of GAL4. These clones also produce a strong blue color after 3 hr in the *lacZ* filter lift assay. Further analysis of the sequence of the remaining clones has revealed one clone, C80, whose sequence largely overlaps with these five identical clones. This clone was identified in the screen of the randomly primed library. Compared with clone A13, C80 contains 120 bp of additional sequence at the 5' end. The isolation of the same statistical number of clones per million reassures us that the interaction is not fortuitous.

To further rule out any nonspecific interactions and define the specificity for the RCC1-homologous domain in N-RPGR, plasmids pAD/A13 and pAD/C80 have been cotransformed with the following HybriZap bait plasmids: pBD/N-RPGR, pBD/RPGR-C, pBD/RPGR, pBD without insert, p53, and pLamin C. The same small-scale transformation procedure is used as described above in prescreen testing of the bait plasmids. The results of these control transformations are shown in Table I. Yeast cells transformed with pAD/A13 and pAD/C80 are able to grow well on SD-agar −Leu −Trp when cotransformed with any of the pBD bait plasmids, thus selecting for the presence of both bait and target pladmids. When these cotransformants are transferred to dropout plates containing SD-agar −Leu −Trp −His, only cotransformants with pBD/N-RPGR or pBD/RPGR (and not with RPGR-C) are able to grow, as well as the original positive A13 and C80 clones and the p53/pSV40 positive control. When a filter lift β-galactosidase assay is performed on both plates, only these clones produce a blue product, as is shown in Fig. 3. This production, however, is much lower with the interaction of (full-length) RPGR and the protein encoded by A13, than with the interaction of N-RPGR and A13. As described above, this may

---

[25] D. C. Chang, B. M. Chassy, J. A. Saunders, and A. E. Sowers (eds.), "Guide to Electroporation and Electrofusion." Academic Press, San Diego, California, 1992.

TABLE I
INTERACTIONS OF YEAST CELLS COTRANSFORMED WITH INDICATED HybriZap PLASMID CONSTRUCTS[a]

pBD-	D-	−Leu	−Trp	−Leu −Trp	HIS3 expression (−Leu −Trp −His)	lacZ expression (β-Galactosidase)
p53	−	−	+[b]	−	−	−
pLamin C	−	−	+[b]	−	−	−
pBD	−	−	+[b]	−	−	−
−	pGAL4	+	−	−	−	+
−	pLamin C	+	−	−	−	−
N-RPGR	−	−	+[b]	−	−	−
RPGR	−	−	+[b]	−	−	−
RPGR-C	−	−	+[b]	−	−	−
p53	pSV40	+	+	+	+	+
pLamin C	pSV40	+	+	+	−	−
pBD	pSV40	+	+	+	−	−
p53	A13	+	+	+	−	−
pLamin C	A13	+	+	+	−	−
pBD	A13	+	+	+	−	−
N-RPGR	A13	+	+	+	+	+
RPGR	A13	+	+	+	+	+
RPGR-C	A13	+	+	+	−	−

[a] Assayed for growth on the indicated SD-agar dropout plates and assayed for production of a blue substrate in the *lacZ* filter lift assay. The results show the absence of intrinsic activation of the reporter genes by the bait plasmids (no growth on SD-agar −Trp −His, therefore no *HIS3* expression, and no β-galactosidase activity of clones growing on SD-agar −Trp, therefore no *lacZ* expression). They also show the activation of both reporter genes by interaction of pAD/A13 with pBD/N-RPGR and pBD/RPGR, but not with pBD/RPGR-C or any of the control plasmids.

be caused by isoprenylation of the C terminus of RPGR by the yeast posttranslational machinery and/or less efficient nuclear translocation of RPGR compared with N-RPGR.

These results confirm the true, specific interaction of the A13 and C80 protein products with (N-)RPGR in yeast, which are therefore referred to as (part of an) RPGR-binding protein (RPGR-BP).

Quantification of Interaction

RPGR interactions with its substrates are quantified by a liquid β-galactosidase assay, which can be regarded as a measure of the binding affinity of the interactions.[12] For this liquid assay, the luminescent β-galac-

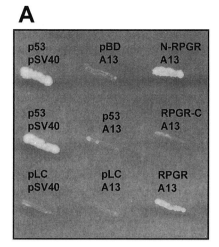 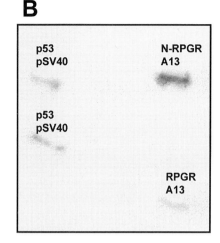

HIS3 activation　　　　　　　　　　*lacZ* activation

Fig. 3. Specific activation of the reporter genes. (A) Cotransformants of pAD/A13 and pBD/N-RPGR or pBD/RPGR are able to grow on SD-agar −Leu −Trp −His, caused by activation of the *HIS3* reporter gene by interaction of the protein products. (B) Cotransformants of pAD/A13 and pBD/N-RPGR or pBD/RPGR produce a blue product in the filter lift assay of the *lacZ* reporter gene activation. Clearly less β-galactosidase activity is visible with pBD/RPGR.

tosidase detection kit II (Clontech, Palo Alto, CA) is used. This kit uses a reaction buffer that contains the Galacton-Star substrate and Sapphire-II enhancer. Cleavage of the galactoside moiety from Galacton-Star by β-galactosidase yields a dioxethane anion, which further degrades with the concurrent production of light.[26] The Sapphire-II enhancer amplifies the chemiluminescent light signal. Light emission is used as a quantitative measure of β-galactosidase activity.

For this assay, yeast colonies are transferred into 50-ml conical polypropylene tubes containing 5 ml of appropriate SD dropout medium and grown for 36 hr at 30° with shaking (250 rpm). SD dropout medium is used to maintain selective pressure on interacting cognate plasmids. The OD_{600} of each culture is determined (usually between 0.5 and 1.0) and ≤8 ml of YPAD (yeast extract, peptone, adenine sulfate, dextrose) medium (at 30°) is inoculated with ≥2 ml of the yeast culture, to an OD_{600} of approximately 0.2. This culture is grown for an additional 5 hr with shaking to an OD_{600} of 0.4–0.6. The tube is vortexed, the OD_{600} is determined, and 1.5-ml aliquots of yeast cultures are transferred to microcentrifuge tubes. Samples are centrifuged at

[26] I. Bronstein, B. Edwards, and J. C. Voyta, *J. Chemiluminesc. Bioluminesc.* **4,** 99 (1989).

12,000g for 30 sec at 25°, and cell pellets are resuspended in 1.5 ml of Z-buffer (60 mM Na$_2$HPO$_4 \cdot$ 7H$_2$O, 40 mM NaH$_2$PO$_4 \cdot$ H$_2$O, 10 mM KCl, 1 mM MgSO$_4 \cdot$ 7H$_2$O, pH 7.0) to wash the cells. These are then pelleted at 12,000g for 30 sec at 25° and resuspended in 300 μl of fresh Z-buffer. Cell suspensions are vortexed and 100-μl aliquots are transferred to new tubes, which are placed in liquid nitrogen for 1 min and then thawed at 37° for 1 min. This freeze–thaw cycle is repeated twice and the lysed cell extracts are centrifuged at 14,000g for 5 min at 4°. Cleared supernatants are transferred to fresh tubes (on ice) and 30 μl is transferred into 5-ml tubes (Sarstedt, Newton, NC) at room temperature. Chemiluminescence reactions are started by the addition of 200 μl of Galacton-Star/Sapphire II reaction buffer (Clontech) followed by gentle mixing. After 60 min of incubation, sample light emissions are recorded at 5-sec integrals in an Auto Lumat LB 953 luminometer (Berthold, Bad Wilbad, Germany).

From these measurements, β-galactosidase activity is calculated in average relative light units (RLU)/OD$_{600}$ after normalizing for the amount of cells. The average is taken from three readings and three clones that contain the same bait and target plasmids. The results are shown in Fig. 4. The β-galactosidase activities of cotransformants N-RPGR/A13 and N-RPGR/C80 are 3826 and 1908 RLU/OD$_{600}$, respectively. The β-galactosidase activities of the positive controls, pGAL4 and p53/pSV40, are 8995 and 843

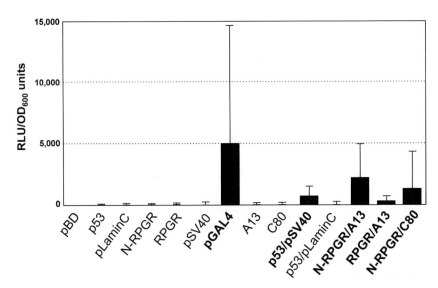

FIG. 4. Quantification of two-hybrid interactions by a liquid β-galactosidase assay. The average activity is shown in RLU/OD$_{600}$ for the different cotransformants, confirming quantitatively the results of the qualitative filter lift assay of *lacZ* reporter gene activation.

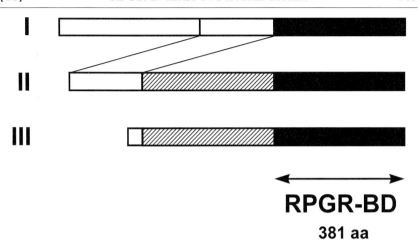

FIG. 5. Diagram of the different variants of RPGR-binding protein (RPGR-BP). The bovine variants (I and II) are highly homologous to the human homolog (III). The shared RPGR-binding domain (RPGR-BD) contains 381 amino acids.

RLU/OD_{600}, respectively. The β-galactosidase activity of full-length RPGR/A13 is lower, 339 RLU/OD_{600}, and RPGR-C/A13 shows no β-galactosidase activity.

Retina-Specific Expression of Retinitis Pigmentosa GTPase Regulator-Binding Protein and Isolation of Spliced Variants

The 1.3-kb insert of clone pAD/A13 is used as a probe on a multiple-tissue Northern blot, containing 3 μg of poly(A)$^+$ mRNA from bovine retina, brain, retinal pigment epithelium (RPE), kidney, spleen, liver, and skeletal muscle to determine the expression pattern, by methods described elsewhere.[27] This reveals two signals, one of 4.5 kb and one of 7.5 kb, that are present only in the retina.[28] One smaller signal of 0.8 kb is present in all tissues tested. To isolate the full-length bovine clones and its human counterparts, A13 is used as a probe to screen random and oligo(dT)-primed HybriZap and lambda-ZAP cDNA libraries from bovine retina, and an oligo(dT)-primed HybriZap cDNA library from human retina, using standard methods described elsewhere.[27] As shown in Fig. 5, sequence analysis of the bovine cDNA clones reveals different spliced isoforms of the 4.5-kb transcript. They all share a 381-amino acid RPGR-interacting domain. The human counterparts are largely overlapping with these clones

[27] P. A. Ferreira, R. D. Shortridge, and W. L. Pak, *Proc. Natl. Acad. Sci. U.S.A.* **90,** 6042 (1993).
[28] R. Roepman, N. Bernoud-Hubac, D. Schick, A. Maugeri, W. Berger, H.-H. Ropers, F. Cremers, and P. Ferreira, submitted (2000).

and highly homologous. Overall, they are 75 and 84% identical at the amino acid and nucleic acid level, respectively. Northern blot analysis of human retina tissue, using a human RPGR-BP cDNA clone as a probe, reveals mainly a 4.5-kb signal. One of these human clones contains an insert that is in-frame with the activation domain. This clone is analyzed in the two-hybrid system for interaction with N-RPGR, as described above for A13 and C80, and interaction is confirmed.

Conclusion

We have used a GAL4-based yeast two-hybrid screen to isolate different bovine substrates that specifically interact with the N-terminal RCC1-homologous half of RPGR and with full-length RPGR, but not with the C-terminal half of RPGR. They represent clones that are derived from the same cognate gene, which encodes several alternatively spliced transcripts, in bovine as well as in humans. All clones share an RPGR-interacting domain and are specifically expressed in the retina. Because no striking homologies have been identified with any known gene or protein, the function of this novel RPGR-binding protein (RPGR-BP) remains unknown. In summary, the methods described here will provide clues to the molecular pathogenesis of RP3 and possibly lead to the identification of novel candidate genes for chorioretinal degenerations. These methods may be extended to the identification of other, as yet unknown, target substrates for other gene products that are involved in retinal and neurodysplasia.

Acknowledgments

We thank F. P. M. Cremers for critical reading of the manuscript and H.-H. Ropers and W. Berger for helpful discussions and encouragement. This work was supported in part by the Foundation Fighting Blindness, the Ter Meulen Fund, the Royal Netherlands Academy of Arts and Sciences (R.R.), NIH Grant EY11993 the Karl Kirchgessner Foundation, and Fight for Sight, Inc. (P.A.F.).

[46] Genetic Analysis of RPE65: From Human Disease to Mouse Model

By T. MICHAEL REDMOND and CHRISTIAN P. HAMEL

Introduction

RPE65 is a highly expressed, conserved retinal pigment epithelium (RPE)-specific[1,2] protein preferentially associated with the RPE microsomal membrane fraction, although it is not an integral membrane protein.[2,3] Comparison of bovine,[3] human,[4] rat,[5] mouse,[6] dog,[7] and salamander[8] RPE65s shows a rather high degree of conservation of this protein and Northern analysis detects RPE65 mRNA only in RPE.[3] RPE65 expression is lost by cultured RPE cells within 2 weeks of explantation,[2] and cultured RPE cells (7 weeks in primary culture) containing RPE65 mRNA did not express protein,[3] implicating a mechanism of post-transcriptional regulation.[3,9] In neonatal rats, morphologically well-differentiated RPE cells do not express RPE65 at birth but it is detectable on postnatal day 4 (P4),[2] that is, 1–2 days before the photoreceptors develop their outer segments (OS). Interestingly, RPE65 mRNA accumulates slowly from embryonic day 18 to a steady level by P12.[5] This chronology coincides with the initial appearance of rhodopsin,[10] 11-*cis*-retinoids,[10] and isomerase[11] activity in the mouse and rat. The localization, RPE specificity, conservation, timing of

---

[1] J. J. Hooks, B. Detrick, C. Percopo, C. Hamel, and R. Siraganian, *Invest. Ophthalmol. Vis. Sci.* **30,** 2106, (1989).

[2] C. P. Hamel, E. Tsilou, E. Harris, B. A. Pfeffer, J. J. Hooks, B. Detrick, and T. M. Redmond, *J. Neurosci. Res.* **34,** 414 (1993).

[3] C. P. Hamel, E. Tsilou, B. A. Pfeffer, J. J. Hooks, B. Detrick, and T. M. Redmond, *J. Biol. Chem.* **268,** 15751 (1993).

[4] A. Nicoletti, D. J. Wong, K. Kawase, L. H. Gibson, T. L. Yang-Feng, J. E. Richards, and D. A. Thompson, *Hum. Mol. Genet.* **4,** 641 (1995).

[5] G. Manes, R. Leducq, J. Kucharczak, A. Pages, C. F. Schmitt-Bernard, and C. P. Hamel, *FEBS Lett.* **423,** 133 (1998).

[6] T. M. Redmond, unpublished (1999).

[7] G. D. Aguirre, V. Baldwin, S. Pearce-Kelling, K. Narfstrom, K. Ray, and G. M. Acland, *Mol. Vis.* **4,** 23 (1998). ⟨http://www.molvis.org/molvis/v4/p23⟩

[8] J.-X. Ma, L. Xu, D. K. Othersen, T. M. Redmond, and R. K. Crouch, *Biochim. Biophys. Acta* **1443,** 225 (1998).

[9] S.-Y. Liu, and T. M. Redmond, *Arch. Biochem. Biophys.* **357,** 37 (1998).

[10] L. Carter-Dawson, R. A. Alvarez, S. L. Fong, G. I. Liou, H. G. Sperling, and C. D. Bridges, *Dev. Biol.* **116,** 431 (1986).

[11] C. D. B. Bridges, *Vision Res.* **29,** 1711 (1989).

expression, and unique nature of RPE65 thus suggest a conserved function specific and unique to the RPE.

What might this function be? As a protein highly and specifically expressed in the RPE, RPE65 has been strongly suspected to play a role in the vitamin A metabolism of the eye.[2,12,13] Evidence for this is provided by its biochemical association with the retinol-binding protein (RBP)[12,14] and 11-*cis*-retinol dehydrogenase.[15] As a result, RPE65 (also known as p63[12,14]) has been proposed to be the RPE basolateral RBP receptor.[12] However, association with the 11-*cis*-retinol dehydrogenase suggests that RPE65 is part of the visual cycle pathway, the process by which the 11-*cis*-retinal chromophore of visual pigments is photoisomerized to the all-*trans* isomer, and then regenerated in the dark.[16,17] While the localization and properties of RPE65 are not consistent with a role as the RPE RBP receptor, they do correlate with a role in the visual cycle. If RPE65 is part of the visual cycle, what is its role? Most of the components of the visual cycle, with the exception of the isomerase,[18–20] have been identified at the molecular level. Because of the lability of the suspected isomerase enzyme complex and its resistance to biochemical purification, the molecular identity of the protein(s) involved is still unknown.[19,20] Consequently, our understanding of visual cycle retinoid metabolism, and particularly 11-*cis*-retinoid production, remains incomplete. Application of the tools of modern molecular genetics may have provided an opening in understanding this process. Mutations in *RPE65* have been discovered in Leber congenital amaurosis[21,22] and autosomal recessive childhood-onset severe retinal dystrophy (arCSRD)[23] (conditions characterized by blindness at birth or early child-

---

[12] C.-O. Båvik, F. Lévy, U. Hellman, C. Wernstedt, and U. Eriksson, *J. Biol. Chem.* **268**, 20540 (1993).
[13] A. F. Wright, *Nature Genet.* **17**, 132 (1997).
[14] C.-O. Båvik, C. Busch, and U. Eriksson, *J. Biol. Chem.* **267**, 23035 (1992).
[15] A. Simon, U. Hellman, C. Wernstedt, and U. Eriksson, *J. Biol. Chem.* **270**, 1107 (1995).
[16] G. Wald, *Science* **162**, 230 (1968).
[17] J. C. Saari, in "The Retinoids: Biology, Chemistry and Medicine," 2nd Ed. (M. B. Sporn, A. B. Roberts, and D. S. Goodman, eds.), p. 351. Raven, New York, 1994.
[18] P. S. Bernstein, W. C. Law, and R. R. Rando, *J. Biol. Chem.* **262**, 16848 (1987).
[19] A. Winston and R. R. Rando, *Biochemistry* **37**, 2044 (1998).
[20] R. J. Barry, F. J. Canada, and R. R. Rando, *J. Biol. Chem.* **264**, 9231 (1989).
[21] F. Marlhens, C. Bareil, J.-M. Griffoin, E. Zrenner, P. Amalric, C. Eliaou, S.-Y. Liu, E. Harris, T. M. Redmond, B. Arnaud, M. Claustres, and C. P. Hamel, *Nature Genet.* **17**, 139 (1997).
[22] H. Morimura, G. A. Fishman, S. A. Grover, A. B. Fulton, E. L. Berson, and T. P. Dryja, *Proc. Natl. Acad. Sci. U.S.A* **95**, 3088 (1998).
[23] S. Gu, D. A. Thompson, C. R. S. Srikumari, B. Lorenz, U. Finckh, A. Nicoletti, K. R. Murthy, M. Rathmann, G. Kumaramanickavel, M. J. Denton, and A. Gal, *Nature Genet.* **17**, 194 (1997).

hood) that point to the critical role of RPE65 in vision. The complete night blindness that characterizes these human diseases as well as the congenital stationary night blindness (csnb) in the Briard dog (linked to a 4-bp deletion in the canine *RPE65* gene[7]) are consistent with disruption of the visual cycle. In addition, the phenotype of *Rpe65* knockout mice shows that RPE65 is directly involved in all-*trans*-to-11-*cis* isomerization, because *Rpe65*-deficient RPE does not produce 11-*cis*-retinoids although the immediate precursor, all-*trans*-retinyl ester, overaccumulates.[24]

In this chapter we describe the methods and reagents we have used for the characterization of mutations in human RPE65 and for the genetic analysis of knockout mice and how such studies are helping in our understanding of the function of RPE65.

Procedures

Genetic Analysis of Human RPE65 Gene

Patients. While a tissue-specific gene such as *RPE65* was an obvious candidate for genetic disease involving the RPE, what diseases to screen were not. This was due to uncertainty regarding the expected phenotype together with the wide importance of the RPE to the outer retina. In addition, at the time when *RPE65* was first mapped to 1p31[25] this region was not associated with a known retinal dystrophy. Despite this uncertainty, and assuming a key role for RPE65 in retinoid metabolism, we anticipated that a genetic loss of function would cause a total or near total absence of vision. We therefore screened an initial series of 12 patients (3 from France and 9 from Germany) affected with congenital blindness, i.e., Leber congenital amaurosis (LCA). Criteria for the diagnosis of this condition include poor vision from birth (less than 1/20 of visual acuity), relatively few lesions in the fundus, unrecordable photopic and scotopic electroretinograms (ERGs), and absence of extraocular signs. Later, we screened a second series of 184 unrelated patients with various retinal dystrophies including retinitis pigmentosa (RP; 65% of cases) as well as cone–rod and macular dystrophies (Table I).

Sequencing of RPE65. By screening with a bovine cDNA probe,[3] a clone containing exon 7 was isolated from a human genomic library in bacteriophage λ. Working from these data, exon sequences and exon/intron

---

[24] T. M. Redmond, S. Yu, E. Lee, D. Bok, D. Hamasaki, N. Chen, P. Goletz, J.-X. Ma, R. K. Crouch, and K. Pfeiffer, *Nature Genet.* **20,** 341 (1998).

[25] C. P. Hamel, N. A. Jenkins, D. J. Gilbert, N. G. Copeland, and T. M. Redmond, *Genomics* **20,** 509 (1994).

TABLE I
PHENOTYPES OF PATIENTS SCREENED IN *RPE65*[a]

Diagnostic	Mode of inheritance	Number of unrelated patients
Retinitis pigmentosa	S	72
	AD	25
	AR	22
Cone–rod dystrophies	S, AD, and AR	9
Cone dystrophies	S and AR	2
Stargardt's disease	AR	8
Usher's syndromes	AR	19
Best's disease	AD	2
Others	Various	25
	Total:	184

[a] Simplex cases (S) and autosomal dominant (AD) or recessive (AR) inheritances are indicated.

structure were characterized by direct sequencing of a longer P1 clone containing the entire sequence of *RPE65*. It was found to contain 14 coding exons spanning 20 kb. These data were used to generate pairs of intronic primers for each exon.

Mutation Screening. Polymerase chain reaction single-strand conformation analysis (PCR SSCA) is used because of its ease of use and capacity for multiple parallel analyses.[26] If an aberrant migration pattern is observed, follow PCR SSCA by direct sequencing of the exon amplimer.

Labeled Polymerase Chain Reaction Single-Chain Conformation Analysis. Synthesize intronic primers that amplify fragments from 207 to 281 bp in length (Table II). Obtain blood samples (with the informed consent of patients) and extract genomic DNA by standard salting-out procedures.[27] Amplify each of the 14 exons of the *RPE65* gene by combining 8 pmol of the relevant pair of forward and reverse intronic primers, 100 ng of genomic DNA, 0.8 U of *Taq* DNA polymerase (Perkin-Elmer, Norwalk, CT), and 8 nCi of [^{32}P]dCTP (Amersham, Arlington Heights, IL) in a 20-μl volume containing 10 mM Tris-HCl (pH 8.3), 50 mM KCl, 1.5 mM MgCl$_2$, a 70 μM concentration of each dNTP, and 0.001% (w/v) gelatin. After a denaturation step (94° for 5 min), carry out amplification for 30 (exons 3, 11, and 12) or 35 (exons 1, 2, 4–10, 13, and 14) cycles at 94° for 1 min, 56 or 60° for 1 min (Table II), and 72° for 2 min. Mix aliquots (5 μl) of labeled PCR products with 7 μl of stop solution [95% formamide, 10 mM NaOH, 0.05%

[26] M. Orita, H. Iwahana, H. Kanazawa, K. Hayashi, and T. Sekiya, *Proc. Natl. Acad. Sci. U.S.A.* **86,** 2766 (1989).
[27] S. A. Miller, D. D. Dykes, and H. F. Polesky, *Nucleic Acids Res.* **16,** 1215 (1988).

TABLE II
PRIMER SEQUENCES OF *RPE65* FOR PCR AMPLIFICATION OF INDIVIDUAL EXONS[a]

	Primers and conditions for PCR amplification of individual exons		Length of fragment amplified (bp)	Amplification temperature[b] (°C)
Exon	Forward primer (5' → 3')	Reverse primer (5' → 3')		
1	CAACTTCTGTTCCCCTCCCTCAG	CTCTTCAGGAGCCCTTGAAATAGC	251	60
2	TCTATCTCTGCGGACTTTGAGCAT	ATAGGAAGCCAGAGAAGAGAGACT	207	60
3	CCCAAGGCAGGGATAAGAAGCAAT	AAGCTAGGCCCTACTTTGAGGAGG	248	60
4	ACATGGGCTGTACGGATTGCTCCT	GAGAGAAAAAGGGCTAATATAAAA	225	56
5	ATGGCTTGAAAATTACTGGACTGA	TGGTGAATTAATTTTAAGTTCCAA	250	56
6	AGGTATAAATGTATCTTCCTTCTCT	TTATCTTTCTCACAATACAGTAAC	280	56
7	CTGTTCTAAATGCTTTGTATTAAA	TAGGCCAAGCAATCTTTAAACTT	217	56
8	TGTGGCTTGAGAATCAGCCCTTTC	GTGTACATTATTAAACACATCTTC	250	56
9	GTACACTTTTTCCTTTTTAAATG	ACCCGTAATTTCCAGGAACAATG	250	56
10	AGAATCATCTCTCTAAAATTATTT	CTGAGAGATGAAACATTCTGGT	249	56
11	GAATTCTTTCCTGCTCACTGAGGT	GAGCACATGCTTAGGAAAACTCTT	244	60
12	ACTTCACACGGGAGTGAACAAATG	GTCAATATGCTTTACTTGACTAGC	240	60
13	GCTAGTCAAGTAAAGCATATTGAC	CATACAGAGCTGCAGTAAGAAGAG	224	60
14	TGACACTCAATCTATAGCTTCGGC	GATTGCAGACCTGAAGCTGATTTT	281	60

[a] Adapted from Marlhens et al.[21] (copyright holder, Nature America, Inc.).
[b] See text for details.

(w/v) bromphenol blue, 0.05% (w/v) xylene cyanol] and denature for 2 min at 92°. Load samples (2.8 µl) on a Hydrolink MDE gel (FMC Bioproducts, Philadelphia, PA) and electrophorese overnight at room temperature in 0.6× Tris–borate–EDTA (TBE). Dry the gels and autoradiograph at room temperature for 6–12 hr.

Unlabeled Polymerase Chain Reaction Single-Chain Conformation Analysis. In our hands, this technique gives comparable results to the labeled PCR SSCA but requires more reagents. Amplify each exon by combining 20 pmol of forward and reverse intronic primers, 200 ng of genomic DNA, 1 U of *Taq* DNA polymerase (Eurogentec, Seraing, Belgium) in a 50-µl volume containing 75 mM Tris-HCl (pH 9.0), 20 mM $(NH_4)_2SO_4$, 2.0 mM $MgCl_2$, a 200 µM concentration of each dNTP, and 0.01% (v/v) Tween 20. PCR conditions are the same as for labeled SSCA except that both groups of exons are amplified for 35 or 40 cycles (instead of 30 and 35 cycles, respectively). Mix aliquots (12 µl) of each reaction with 4 µl of stop solution, and denature for 2 min at 95°. Load 7 µl of the mix on a Hydrolink MDE gel and electrophorese overnight at room temperature in 0.6× TBE. Because SSCA detects only 70–90% of mutations, an alternative method is necessary to increase the probability of finding mutations. Therefore, run some samples on an 5% (w/v) total acrylamide/2% (w/v) bisacrylamide (49:1)/5% (v/v) glycerol gel, and electrophorese overnight at room temperature at 3 W in 0.6× TBE.[28] Silver stain both gels (Promega, Madison, WI). To do this, fix the gel, adherent to one of the glass plates, for 20 min in 10% (v/v) acetic acid, wash three times for 2 min in ultrapure water, stain for 30 min in an $AgNO_3$ solution (1 g/liter) containing 0.05% (v/v) formaldehyde, rinse for 10 sec in water, and develop at 11° in a solution containing anhydrous Na_2CO_3 (30 g/liter), 0.05% (v/v) formaldehyde, and sodium thiosulfate (2 mg/liter). Stop development of the gel by adding 10% (v/v) acetic acid. Finally, air dry the gel and read directly.

Restriction Digest Analysis. The R234X mutation creates an *Alw*NI restriction site. To detect this, mix 15 µl of PCR products from an unlabeled SSCA amplification of exon 7 with 10 units of *Alw*NI (New England BioLabs, Beverly, MA) and digest overnight at 37°. After addition of 5 µl of stop solution [10 mM Tris-HCl (pH 7.8), 25 mM EDTA, 0.1% (w/v) bromphenol blue, 80% (v/v) glycerol], separate the products on a 3% (w/v) ethidium bromide–agarose gel.

Direct Sequencing. Amplify each exon as for labeled PCR SSCA except that 20 pmol of each primer and 300 ng of genomic DNA are added to the reaction mix in a 50-µl volume without radioactive dCTP. Then separately

---

[28] G. Berx, F. Nollet, K. Strumane, and F. van Roy, *Hum. Mutat.* **9**, 567 (1997).

amplify each strand of DNA by asymmetric PCR with 80 pmol of one primer, 3 μl of the previous amplification, and 1 U of *Taq* DNA polymerase (Perkin-Elmer) in a 100-μl volume. Filter purify the products on QIAquick (Qiagen, Hilden, Germany), concentrate to 10–20 μl, and sequence 7 μl by the dideoxynucleotide chain-termination method, using Sequenase (U.S. Biochemical, Cleveland, OH) and [^{35}S] dATP and following the manufacturer recommendations.

Analysis of Pathogenic Mutations in Human RPE65 Gene

Survey of Mutations in RPE65. In our series of 196 cases (including 12 LCA) we have found 2 unrelated patients carrying mutations in *RPE65*, one with LCA, the other with RP. Both are compound heterozygotes. A third patient was found to be heterozygous for a single mutation. The LCA patient has a paternal 700C→T substitution (exon 7) in a CpG site, resulting in the nonsense codon R234X, and a maternal 1067delA (exon 10), resulting in a frameshift and a premature termination at codon 373, 47 base pairs downstream from the deletion.[21] While 1067delA induced a bandshift on SSCA, R234X was undetectable under both SSCA conditions (see procedures). The heterozygous carriers of either one of the two mutations were asymptomatic. The possibility that they carry an SSCA-undetected mutation in the other *RPE65* allele with no visual impairment was excluded by sequencing all 14 *RPE65* exons in both parents. Both mutated *RPE65* genes, if translated, would result in severely truncated proteins with C-terminal deletions of 33.2 and 56.3% for 1067delA and R234X, respectively. As such, they probably do not encode functional proteins and therefore are likely to be null alleles.

The RP patient carried a paternal 65T→C substitution (exon 2) resulting in a missense codon (L22P), and a double maternal 201G→T–202C→T substitution (exon 3) resulting in another missense codon (H68Y).[29] Mutations in exon 3 were detectable in both SSCA conditions while that in exon 2 was visible only on MDE gels.

In another unrelated RP patient we found a 1047insTGG that results in the insertion of an additional tryptophan at position 351, following tryptophan at position 350. This was detectable under both SSCA conditions (see procedures). This mutation was not observed in the other 183 unrelated patients. We have not found another mutation by sequencing the 13 other exons. Therefore, while the mutated allele could encode a defective protein, it is uncertain if this mutation could be involved in the disease.

To date, a total of 22 different mutations have been reported (Table

---

[29] F. Marlhens, J.-M. Griffoin, C. Bareil, B. Arnaud, M. Claustres, and C. P. Hamel, *Eur. J. Hum. Genet.* **6,** 527 (1998).

III) in 35 patients from 17 families or sporadic cases.[21–23,29] Among these 17 unrelated cases, 10 reside in the United States or Canada (ethnic origin not indicated), 3 in Germany, 2 in India, 1 in France, and 1 in Italy. Except for one case (family RP188[23]), in which the presumed mutation in the other allele was not reported, all cases displayed autosomal recessive inheritance, with nine cases being compound heterozygotes and eight cases homozy-

TABLE III
MUTATIONS FOUND IN *RPE65*[a]

Number	Exon	Mutation	Protein	Allele	Diagnostic	Case	Ref.
1	1	2T→C	Met1Thr	Ho	LCA	Sp	22
2	1	11+5G→A	Sd?	Ho/He	CSRD	1 Fam, 1 Sp	23
1	2	65T→C	Leu22Pro	He	RP	Sp	29
1	2	90insT	Fs	He	CSRD	Sp	23
1	3	118G→A	Gly40Ser	He	LCA	Fam	22
1	3	202C→T	His68Tyr	He	RP	Sp	29
3	4	271C→T	Arg91Trp	Ho/He	LCA/RP	1 Fam, 2 Sp	22
1	4	304G→A	Glu102Lys	He	LCA	Sp	22
1	5	370C→T	Arg124X	He	LCA	Sp	22
1	5	394G→A	Ala132Thr	Ho	RP	Fam	22
1	6	544C→G	His182Tyr	He	LCA	Fam	22
1	7	644−2A→T	Sa?	Ho	LCA	ND	22
1	7	700C→T	Arg234X	He	LCA	Fam	21
1	8	777del8	Fs	He[b]	CSRD	Fam	23
1	8	858+1G→T	Sd?	Ho	CSRD	Fam	23
1	9	961insA	Fs	He	LCA	Sp	22
1	9	1022T→C	Leu341Ser	He	RP	Fam	22
1	9	1047insTGG	350insTrp	He[c]	RP	Sp	29
1	10	1067delA	Fs	He	LCA	Fam	21
1	10	1087C→A	Pro363Thr	Ho	CSRD	Fam	23
1	11	1210insCTGG	Fs	He	RP	Fam	22
1	13	1355T→G	Val452Gly	He	RP	Sp	22
1	13	1418T→A	Val473Asp	Ho	LCA	Sp	22

[a] Mutation description follows standard nomenclature, using the *RPE65* cDNA sequence as a reference (GenBank U118991) with the ATG translation initiation codon counted as +1. All mutations were analyzed at the DNA level only. Indicated are the number of times a mutation was identified (unrelated cases), the mutated exon, the mutation at the DNA level, the consequence of the mutation at the protein level (X, stop codon; Fs, frameshift; Sd, splice donor site; Sa, splice acceptor site; ?, not yet confirmed at the RNA level), reference reporting mutation (as in text), if the mutation was present at the homozygous (Ho) or heterozygous (He) state, the clinical phenotype (LCA, Leber's congenital amaurosis; RP, retinitis pigmentosa; CSRD, childhood-onset severe retinal dystrophy), and if the mutation was found in a familial (Fam) or sporadic (Sp) case.
[b] Mutation in the other allele not reported.
[c] Uncertain pathogenicity with mutation in the other allele not found.

gotes. This indicates that the pathogenic mechanism is due to a loss of function or impairment of the RPE65 protein. Variability in the coding sequence of *RPE65* seems relatively low because only one polymorphism is frequently found in the population (i.e., 1056G/A, with nucleotide A found in 14–15% individuals), all other polymorphisms being present in only one individual.[22,28] It has been estimated that *RPE65* mutations could represent about 16% of all cases of LCA.[22] Given that LCA accounts for 5% of the retinal dystrophies, which themselves affect 1 individual (over age 3) in 4000,[30] 1 of 400,000 persons would be affected by LCA due to mutations in *RPE65*. This would represent 150 to 200 persons in each of the larger European countries and 600 to 800 persons in the United States and Canada. In addition, *RPE65* could also cause 2% of autosomal recessive retinitis pigmentosa,[22] accounting for an equivalent number of affected patients.

Clinical Phenotypes due to Mutations in RPE65. Mutations in *RPE65* cause a retinal dystrophy whose hallmarks include a severe visual impairment starting in early infancy and primarily affecting rod function, although a prominent macular dystrophy occurs rapidly, with the presence of many whitish retinal spots in the fundus. While rod and cone ERG responses are undetectable at an early stage, the fundus shows relatively mild signs of degeneration at the beginning of the disease (scarcity of pigmentary deposits, moderate attenuation of retinal vessels), suggesting that impairment of vision may be in part due to retinal dysfunction rather than to complete destruction of photoreceptors. The retinal spots are suggestive of lesions at the level of the RPE. This phenotype belongs to the autosomal recessive childhood-onset severe retinal dystrophy (arCSRD) group and the disease caused by *RPE65* mutations has been described as such.[23] However, variations in the severity of the disease encompass a spectrum from severe cases reported as LCA[21,22] to milder phenotypes described as RP.[23,28] As a result, LCAs due to mutations in *RPE65* have been designated LCA type II (OMIM # 204100), with LCA type I (OMIM # 204000) being due to mutations in *RETGC1* encoding retina-specific guanylate cyclase 1.[31] LCA I is essentially a nonevolving disease causing greatly impaired vision from birth while LCA II patients often have a measurable visual acuity in early infancy but undergo a rapid aggravation toward total blindness in the second or third decade. At the other end of the spectrum are

---

[30] J. Kaplan, D. Bonneau, J. Frézal, A. Munnich, and J.-L. Dufier, (1990), *Hum. Genet.* **85,** 635 (1990).

[31] I. Perrault, J.-M. Rozet, P. Calvas, S. Gerber, A. Camuzat, H. Dollfus, S. Châtelin, E. Souied, I. Ghazi, C. Leowski, M. Bonnemaison, D. Le Paslier, J. Frézal, J.-L. Dufier, S. Pittler, A. Munnich, and K. Kaplan, *Nature Genet.* **14,** 461 (1996).

the RP patients with normal day vision during childhood but often with night blindness reflecting primary rod impairment.[32] As the disease progresses over several decades, patients lose their peripheral visual field, and later central vision, eventually leading to blindness. In comparing the described phenotypes with the predicted protein mutations (Table III) we see that in RP most of the mutations found are amino acid substitutions, whereas in the case of LCA II, both translation-terminating mutations (nonsense and frameshifts) and amino acid substitutions are present. This suggests that the level of impairment of the RPE65 protein, either residual function (in amino acid substitution mutations) or abolished function (in null mutations), controls the rate of progression of the disease.

Generation of Rpe65-Disrupted Mice and Genotype Analysis

Derivation of Targeted ES Cell Line. A P1 clone containing the entire mouse *Rpe65* gene was isolated from a 129SV mouse genomic library (Genome Systems, St. Louis, MO). From this, a 7.7-kb *Eco*RI fragment containing exons 1, 2, and 3 and 2.8 kb of 5' flanking region[6] was identified by Southern blot hybridization with a 5' region bovine RPE65 cDNA probe[3] and subcloned into Bluescript II SK(−). This subcloned fragment was sequenced by automated fluorescent dideoxy sequencing (Applied Biosystems, division of Perkin-Elmer, Foster City, CA). The locations of all restriction sites were identified by the GeneWorks version 2.4 sequence analysis program (Oxford Molecular, Campbell, CA) and, on the basis of these data (Fig. 1a, i), a subcloning strategy was devised for the generation of a targeting vector. A 1.7-kb *Eco*RI/*Afl*II fragment from the 5' flanking region was made by *Eco*RI and *Afl*II. This was then subcloned into pLPG9 and linearized as an *Eco*RI–*Bam*HI fragment. A downstream 2.2-kb fragment from intron c was made by *Ecl*136II and *Eco*RI double digestion of the Bluescript II SK(−) clone containing the 7.7-kb *Eco*RI fragment, followed by gel purification of the 2.2-kb band. This was inserted into pLPG9 and linearized as an *Xho*I–*Not*I fragment. These linearized segments were inserted into the targeting vector X-pPNT[33] to generate pPNT-*Rpe65* (Fig. 1a, ii). All junctions were verified by sequencing. pPNT-*Rpe65* was linearized with *Not*I and electroporated into 129SV strain-derived R1 ES cells.[34] ES clones with a targeted disruption were selected by the positive–negative

---

[32] J. R. Heckenlively, S. L. Yoser, L. H. Friedman, and J. J. Oversier, *Am. J. Ophthalmol.* **105**, 504 (1988).

[33] V. L. Tybulewicz, C. E. Crawford, P. K. Jackson, R. T. Bronson, and R. C. Mulligan, *Cell* **65**, 1153, (1991).

[34] A. Nagy, J. Rossant, R. Nagy, W. Abramow-Newerly, and J. C. Roder, *Proc. Natl. Acad. Sci. U.S.A.* **90**, 8424 (1993).

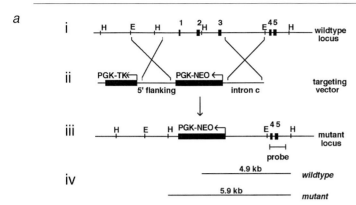

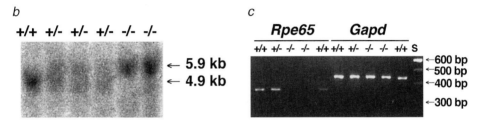

FIG. 1. Targeted disruption of the *Rpe65* locus. (a) Strategy. (i) 5' end of the wild-type *Rpe65* gene; (ii) targeting vector; (iii) targeted locus; (iv) sizes of *Hin*dIII fragments from wild-type and mutant loci, respectively, hybridized to probe indicated in (iii). H, *Hin*dIII sites; E, *Eco*RI restriction sites. (b) Southern analysis of genomic DNA from progeny of a heterozygous cross, digested with *Hin*dIII, using a flanking probe showing wild-type (4.9-kb) and/or mutant (5.9-kb) *Hin*dIII restriction fragments [see (a), iv]. (c) RT-PCR analysis of *Rpe65* mRNA expression in progeny of a heterozygous cross. The left set of lanes presents results of *Rpe65*-specific RT-PCR on cDNA reverse transcribed from total RNA of whole eyes from $Rpe65^{+/-}$, $Rpe65^{-/-}$, and $Rpe65^{+/+}$ mice, while the right set are results of *Gapd*-specific RT-PCR on the same cDNAs. The *Rpe65* mRNA-specific 366-bp band is absent from $Rpe65^{-/-}$, while the GAPDH-specific 452-bp band is present. [Adapted from Redmond *et al.*[24]; copyright holder, Nature America, Inc.]

G418–ganciclovir selection method[35] and isolated. Homologous recombination with this targeting vector resulted in the disruption of the *Rpe65* locus (Fig. 1a, iii), replacing a 5' region containing 1.1 kb of 5' flanking region; exons 1, 2, and 3; intervening introns; and 0.5 kb of intron c with the PGK-*neo* gene. A *Hin*dIII site in intron b was lost, providing a facile assay for the disruption of the allele. Targeted clones were identified by Southern blot analysis after *Hin*dIII digestion of genomic DNA, using a flanking

[35] S. L. Mansour, K. R. Thomas, and M. R. Capecchi, *Nature* (*London*) **329**, 348 (1988).

probe 3' to the replacement (see Fig. 1a, iv). A DNA fragment comprising exon 4, intron d, exon 5, and part of intron e was amplified by PCR and used as template for a random-primed labeled[36] probe. Mouse genomic DNA, isolated from ES cells, mouse spleen, or mouse tail biopsies was digested with *Hin*dIII, resolved by agarose electrophoresis, and blotted onto nylon membranes. The blot was hybridized with the flanking probe. This probe hybridizes to a 4.9-kb *Hin*dIII fragment in wild-type *Hin*dIII-digested ES cell or mouse genomic DNA while the disrupted *Rpe65* allele yields a 5.9-kb *Hin*dIII fragment. The mutant ES cell lines contained both the wild-type 4.9-kb fragment and the mutant 5.9-kb fragment.

Generation of Rpe65-Deficient Mouse Line. Chimeric mice were generated from one of these ES cell clones. Disrupted 129SV R1 ES cells were microinjected into blastocysts obtained from female C57BL/6J mice. The blastocysts were implanted into pseudopregnant CD1 foster mothers and allowed to develop to term. Because the 129SV mouse has an agouti coat color, chimeric mice were identified by the extent of the agouti coat color on the black C57BL/6 background. Chimeric animals of the highest or total agouti coloration were preferentially selected for breeding. Because colonization of the gonadal lineage by disrupted ES cells is the measure of success of this experiment, germline transmission was assessed by mating 2-month-old male chimeric mice with C57BL/6 females. Germline transmission of the mutation was verified by Southern blot analysis of tail DNA of the progeny of these crosses, using the probe described above (Fig. 1b). Wild-type ($Rpe65^{+/+}$) mice show only the wild-type 4.9-kb *Hin*dIII fragment, $Rpe65^{-/-}$ mice only the 5.9-kb *Hin*dIII fragment and heterozygous ($Rpe65^{+/-}$) mice have single copy of each. Heterozygous F_1 progeny ($Rpe65^{+/-}$) were bred and the resultant F_2 progeny were scored for genotype. A normal Mendelian averaged ratio of $Rpe65^{+/+}$: $Rpe65^{+/-}$: $Rpe65^{-/-}$ of 1:2:1 was observed, indicating no effect of the disrupted allele on embryonic development. Normal litter sizes were also seen in $Rpe65^{-/-} \times Rpe65^{-/-}$ crosses, indicating no effect of the disruption on normal development of the gonads or maturation of gametes.

Rpe65-Specific Polymerase Chain Reaction. In addition, genotype can be assessed by *Rpe65*-specific PCR using three primers to amplify from the wild-type and/or the mutant allele. This is much more convenient than Southern blotting for rapid genotyping of progeny. Take tail snip biopsies from weanling progeny of matings by methods approved by the institutional Animal Care and Use Committee. Purify genomic DNA from the tail snip biopsies, using a sodium dodecyl sulfate (SDS)–salt extraction method and redissolve in Tris–EDTA (TE) to approximately 100 ng/μl. Synthesize the

---

[36] A. P. Feinberg, and B. Vogelstein, *Anal. Bichem.* **132,** 6 (1983).

three specific oligonucleotide primers: oligo(A) (5' GGG AAC TTC CTG ACT AGG GGA GG-3') is a reverse primer for the PGK-*neo* gene (reversed in the mutant allele); oligo(B) (5'-GAT GTG GGC CAG GGC TCT TTG AAG-3') and oligo(C) (5'-CCC AAT AGT CTA GTA ATC ACA GAT G-3') are forward and reverse primers from exon 3 and intron C of the *Rpe65* gene, respectively. We use Pharmacia Biotech (Piscataway, NJ) Ready-to-Go PCR beads routinely. Reconstitute these with 25 μl of a reaction mix containing 2 μl of 25 mM MgCl$_2$, the three primers at a concentration of 1 μM each, 0.25 μl of Stratagene Perfect Match PCR enhancer (Stratagene, La Jolla, CA), and 100 ng of DNA in ultrapure sterile water. (*Note:* The Stratagene Perfect Match PCR enhancer is essential for optimal amplification of these fragments. It is possible that other PCR additives/adjuncts, such as dimethyl sulfoxide, are equally effective, although this has not been tested.) Perform an amplification program consisting of 94° hot start; 94° denaturation for 5 min; 40 cycles of 94° denaturation for 1 min, 60° annealing for 30 sec, and 72° extension for 30 sec; followed by 72° terminal extension for 5 min on a Perkin-Elmer GeneAmp PCR System 2400, 9700, or 9600 cycler. Conditions may need to be altered slightly for use on thermal cyclers produced by other manufacturers. Analyze aliquots of these reactions on 3% Nusieve 3:1 agarose gels (FMC Bioproducts) run in 1× Tris–borate–EDTA. This reaction generates a 546-bp wild-type, and/or a 459-bp mutant, allele fragment, respectively, from mouse genomic DNA (Fig. 2). When deemed necessary, the presence or absence of the *neo* gene, either in $Rpe65^{+/-}$ or $Rpe65^{-/-}$ animals, can be assayed for by *neo*-specific PCR, using primers recommended by The Jackson Laboratory Induced Mutant Resource. These are as follows: oIMR013 (5'-CTT GGG TGG AGA GGC TAT TC-3'; T_m 58.7°) and oIMR014 (5'-

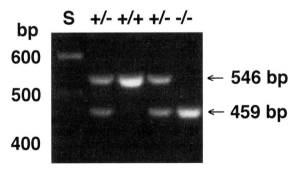

FIG. 2. *Rpe65*-specific 3-primer PCR genotype analysis of $Rpe65^{+/+}$, $Rpe65^{+/-}$, and $Rpe65^{-/-}$ mice showing the wild-type and mutant allele-specific amplimers of 546 and 459 bp, respectively. S, Molecular size markers.

AGG TGA GAT GAC AGG AGA TC-3'; T_m 53.9°). Use these in a reaction consisting of 100 ng of genomic DNA, 1× reaction buffer, 1.5 mM $MgCl_2$ (final), a 0.5 μM concentration of each primer (final), and 0.25–0.5 units of *Taq* DNA polymerase (Perkin-Elmer). Perform an amplification program consisting of 94° hot start; 94° denaturation for 5 min; 40 cycles of 94° denaturation for 1 min, 50° annealing for 30 sec, and 72° extension for 30 sec; followed by 72° terminal extension for 5 min on a Perkin-Elmer GeneAmp PCR System 2400, 9700, or 9600 cycler. Together, the oIMR013 and oIMR014 primers amplify a 280-bp product from the *neo* insert.

Analysis of RPE65 mRNA Expression by Rpe65-Deficient Mice

Rpe65 mRNA Reverse Transcriptase-Polymerase Chain Reaction Assay. The targeted disruption of the *Rpe65* locus described above results in the loss of 5' flanking sequence containing putative promoter elements critical for the expression of the mouse *Rpe65* gene[37] and homologous to those required for expression of the human *RPE65* gene.[38] In addition, the first three exons of the gene are lost. Transcription of the *Rpe65* gene, therefore, is not likely under these conditions. This is confirmed by RT-PCR of a region of the mRNA corresponding to exons 4–6, downstream of the disruption. Isolate total RNA from whole mouse eyes, using an RNeasy kit (Qiagen). Reverse transcribe the entire RNA sample into cDNA, using a Retroscript First-Strand synthesis kit (Ambion, Austin, TX). Denature the isolated RNA at 80° in the presence of kit-supplied first-strand random decamer oligonucleotide primers and dNTPs in a total volume of 20 μl. After cooling, add 10× RT-PCR buffer to a final concentration of 1×, along with 10 units of placental RNase inhibitor and 100 units of Moloney murine leukemia virus (Mo-MuLV) reverse transcriptase. Incubate the reaction at 42° for 1 hr, followed by heat inactivation of the Mo-MuLV reverse transcriptase at 92° for 10 min. PCR amplify the cDNA, using the mouse *Rpe65* exon 4 forward primer 5'-ATG ATC GAG AAG AGG ATT GTC-3' and exon 6 reverse primer 5'-CTG CTT TCA GTG GAG GGA TC-3'. The primers specific for the ubiquitously expressed mouse glyceraldehyde 3-phosphate dehydrogenase (GAPDH) gene mRNA transcript are as follows: forward (starting at nucleotide 586), 5'-ACC ACA GTC CAT GCC ATC AC-3; and reverse (starting at nucleotide 1037), 5'-TCC ACC ACC CTG TTG CTG TA-3'. Primers are used at 0.5 μM each in the presence of SuperTaq *Taq* polymerase (Ambion). The cycler conditions used are as follows: hot start at 94°, followed by 35 cycles (94° for 30 sec,

---

[37] A. Soto Prior, S. Y. Liu, A. A. Henningsgaard, S. Yu, and T. M. Redmond, submitted (1999).
[38] A. Nicoletti, K. Kawase, D. A. Thompson, *Invest. Ophthalmol. Vis. Sci.* **39**, 637 (1998).

55° for 30 sec, and 72° for 30 sec) in a Perkin-Elmer GeneAmp PCR System 2400 or 9700 cycler. The product sizes are 366 and 452 bp for the *Rpe65* and *Gapd* genes, respectively (Fig. 1c). No *Rpe65*-specific product is amplified in *Rpe65*-deficient mice although the ubiquitous *Gapd* product is reliably amplified. Even in the unlikely event that any mRNA were transcribed from the disrupted allele, it is unlikely that it could be translated into a product with any functional activity.

Expression of Mouse RPE65 Protein

The lack of *Rpe65* gene transcription seen by RT-PCR would be expected to preclude any translation of RPE65 protein in *Rpe65*-deficient mouse. This can be verified by Western blot analysis of ocular tissues. Because these mice are pigmented, detection of RPE65 on histological sections by immunocytochemical or immunofluorescence techniques is not convenient.

Antibodies to RPE65. Efforts to produce specific monoclonal antibodies against RPE cell or membrane fractions have invariably resulted in antibodies to RPE65. Several among these have been more extensively characterized,[1,2,14,39] although the epitopes recognized by any of these antibodies have not been identified. However, there are likely to be several, because certain monoclonal antibodies such as RET-PE10[37] and RPE9[1,2] recognize a wide variety of mammalian RPE65s, including rodent RPE65s, whereas others do not recognize rodent RPE65s. Furthermore, RET-PE10[39] also recognizes amphibian (toad) RPE65 quite strongly, whereas RPE9 only weakly recognizes frog RPE65.[2]

To supplement these monoclonal antibodies, one of the authors (TMR) has made antibodies to multiple antigenic peptides (MAPs) in rabbit. MAPs are synthesized in series on an octameric polyalanine core.[40] These MAP antibodies were raised against four different regions of the bovine/human protein. One of these peptides is completely conserved between mammalian and amphibian RPE65s. In general, these antibodies exhibit a wide cross-reactivity. Only one, however, that against human/bovine amino acid (aa) 150–164 peptide NFITKVNPETLETIK, reacts strongly and specifically with mouse RPE65. The corresponding mouse/rat aa 150–164 peptide is NFITKINPETLETIK, with substitution of an isoleucine at aa 155 for the valine in both human and bovine RPE65s. Another two antibodies, against the bovine peptides aa 293–307 (DKKRKKYINNKYRTS) and aa 354–367 (KKNARKAPQPEVRR), do not recognize mouse RPE65 protein effec-

---

[39] J. M. Neill, S. C. Thornquist, M. C. Raymond, J. T. Thompson, and C. J. Barnstable, *Invest. Ophthalmol. Vis. Sci.* **34,** 453 (1993).
[40] J. P. Tam, *Proc. Natl. Acad. Sci. U.S.A.* **85,** 5409 (1988).

tively. Unfortunately, the fourth antibody, generated against the completely conserved peptide (aa 339–354, NYLYLANLRENWEEVK), is limited in its usefulness for murine studies because of a strong and specific cross-reactivity against murine (although not bovine) serum albumin (a major component of eye extracts). The specificity of the immunoreactivity has been verified by sequence comparison of serum albumins with RPE65. Murine serum albumin contains the epitope NLREN found in the peptide, whereas the corresponding bovine region is SLRET.

Bovine Retinal Pigment Epithelium Membranes. Preparations containing bovine RPE65 are used as positive controls for immunoblotting. Bovine RPE cells are prepared from fresh slaughterhouse eyes as previously described.[2,41] Total RPE extract is obtained by homogenizing fresh bovine RPE cells in buffer A (150 mM NaCl, 10 mM Tris-HCl, pH 7.4), in some cases containing a detergent as indicated. In addition, bovine RPE microsomal membranes are prepared by differential centrifugation. Brush RPE cells from each eyecup into 1 ml of extraction buffer B [0.32 M sucrose in 0.1 M phosphate buffer, (pH 7.4) plus complete protease inhibitor cocktail (Boehringer Mannheim, Indianapolis, IN; one tablet per 50 ml of extraction buffer)], pool (50–70 eyes) and homogenize (10–12 up-and-down strokes in a Dounce glass homogenizer), and centrifuge at 30,000g for 30 min at 4° to sediment unbroken cells, nuclei, mitochondria, lysosomes, and melanin granules. Centrifuge the supernatant at 105,000g for 1 hr at 4° to sediment the microsomal membrane fraction. Resuspend the membrane pellet in 100 mM phosphate buffer (pH 7.4) plus complete protease inhibitor cocktail and store at −80°. To solubilize RPE65 from this membrane preparation, add CHAPS to a final concentration of 0.3% (w/v) and incubate the mixture at 4° for 1 hr, then dilute with 3 volumes of 100 mM phosphate buffer (pH 7.4) plus complete protease inhibitor cocktail, followed by centrifugation at 105,000g for 30 min at 4°.

Mouse Eye Tissues. Enucleate eyes from euthanized mice and dissect. Working under a stereo microscope, trim an eye of ophthalmic muscles, optic nerve, and other adnexa. Then grip the eye lightly using No. 5 watchmaker's forceps (Dumont, Switzerland) and pierce at the limbus, to penetrate just inside the retina, with the point of a new No. 11 scalpel blade (Bard-Parker, Rutherford, NJ). Starting with this aperture, cut the eye along the limbus with a pair of 0.3-mm straight Vannas scissors (Roboz Surgical, Rockville, MD). Remove and discard the anterior segment and lens. Using the No. 5 forceps and with cutting at the optic nerve head, remove the retina and reserve for the analysis of rhodopsin. Homogenize each pair of RPE/choroid/sclera in 200 μl of phosphate-buffered saline–

---

[41] E. Tsilou, C. P. Hamel, S. Yu, and T. M. Redmond, *Arch. Biochem. Biophys.* **346**, 21 (1997).

0.3% (w/v) CHAPS–complete protease inhibitor cocktail (Boehringer Mannheim; one tablet per 50 ml of extraction buffer) in a No. 20 Duall glass homogenizer (Kontes, Vineland, NJ). Clear these homogenates of unbroken cells, nuclei, and pigment granules by centrifugation at 14,000g for 15 min at 4°. This cleared supernatant can be used for immunoblot analysis. Combine aliquots of each sample with 1 volume of 2× Laemmli buffer, boil for 3 min, and cool on ice. Separate these samples on 12% (w/v) SDS–polyacrylamide gels, together with positive control samples of bovine RPE membrane (see below) and molecular weight standard proteins, and electroblot to nitrocellulose membranes. These blots can be incubated in the presence of the antibodies described above. For routine use, use the anti-NFITKVNPETLETIK MAP antibody at a dilution of 1:5000 in 3% (w/v) bovine serum albumin (BSA) in PBS–0.05% (v/v) Tween 20 in an overnight incubation at room temperature. After three 10-min washes in PBS–0.05% (v/v) Tween 20, use an alkaline phosphatase-conjugated goat anti-rabbit IgG, at a dilution of 1:3000 in 3% (w/v) BSA in PBS–0.05% (v/v) Tween 20 as the secondary antibody for an incubation period of 1 hr. Then wash the blot with another round of three 10-min washes in PBS–0.05% (v/v) Tween 20, followed by incubation in substrate solution until the reaction end point is deemed to be reached. The results of a typical immunoblot experiment analyzing RPE65 protein expression in $Rpe65^{+/+}$, $Rpe65^{+/-}$, and $Rpe65^{-/-}$ mice are shown in Fig. 3. It can be

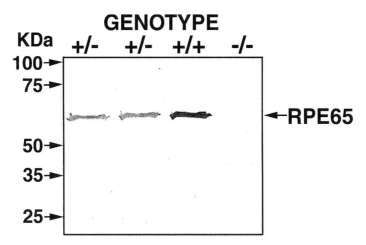

FIG. 3. Immunoblot analysis of RPE65 protein expression in ocular tissue of $Rpe65^{+/+}$, $Rpe65^{+/-}$, and $Rpe65^{-/-}$ mice. The primary antibody used was the rabbit anti-NFITKVNPETLETIK RPE65 MAP at a dilution of 1:5000. Alkaline phosphatase-conjugated goat anti-rabbit IgG was used as the secondary antibody.

seen that the RPE65 immunoreactivity in $Rpe65^{+/-}$ mice is roughly half that seen in the $Rpe65^{+/+}$ mice, and that the $Rpe65^{-/-}$ mice do not express RPE65 protein at all.

Use of Rpe65-Deficient Mice to Elucidate Function of RPE65

The phenotype of *Rpe65*-deficient mice has been described in detail elsewhere.[24] The most salient features of the phenotype revolve around the disruption in RPE retinoid metabolism observed in these mice. The RPE of *Rpe65*-deficient mice overaccumulate all-*trans*-retinyl esters as determined by normal-phase high-performance liquid chromatography (HPLC) of total retinoids extracted from RPE.[24]

High-Performance Liquid Chromatography Analysis of Retinoids. Extract retinyl esters with methanol–hexane from dark-adapted whole mouse eyes or dissected mouse eye tissue homogenized in 0.1 M phosphate buffer, pH 7.0.[10] Separate these retinyl esters on a Lichrosorb SI-60 normal-phase HPLC column (Alltech Associates, Deerfield, IL) in an Applied Biosystems or Waters (Milford, MA) HPLC system, using hexane–dioxane[42] or hexane–methyl tert-butyl ether[10] mobile phase elution with detection at 320 nm. Experimental data can be compared with 11-*cis*-retinal and 11-*cis*-retinol, all-*trans*-retinol and all-*trans*-retinyl palmitate (Sigma, St. Louis, MO) standards run on the same column. This overaccumulation is associated with the presence of lipid droplet inclusions in the RPE.[24] In addition, 11-*cis*-retinoids are not produced by the RPE, as indicated by lack of detection of 11-*cis*-retinol by saponification followed by HPLC.[24] Perform saponification by incubating the dried ester HPLC fractions with 1 ml of 0.06 N ethanolic KOH at 55° for 30 min in the dark.[43] A minor amount of isomerization (<3%) occurs with this method. Analyze hexane extracts by HPLC as described above, using 11.2% (v/v) ethyl acetate–2% (v/v) dioxane–1.4% (v/v) 1-octanol in hexane as mobile phase.[44]

Lack of 11-*cis*-retinoids was also seen, most importantly, by the complete absence of rhodopsin holoprotein in the retinas of *Rpe65*-deficient mice.[24]

Rhodopsin Analysis. For extraction of rhodopsin carry out all procedures under red light. Remove and dissect retinas from $Rpe65^{+/+}$, $Rpe65^{+/-}$, and $Rpe65^{-/-}$ mice and store at −70° until required. Homogenize retinas in 0.5 ml of 10 mM Tris-HCl, 1.0 mM EDTA buffer, pH 7.5 containing 1.0 mM phenylmethylsulfonyl fluoride (PMSF) and 10 μg of DNase, by a series of syringe triturations using progressively smaller needle sizes

---

[42] C. D. B. Bridges and R. A. Alvarez, *Methods Enzymol.* **81**, 463 (1982).
[43] H. Stecher, M. H. Gelb, J. C. Saari, and K. Palczewski, *J. Biol. Chem.* **274**, 8577 (1999).
[44] G. M. Landers and J. A. Olson, *J. Chromatogr.* **438**, 383 (1988).

(18, 20, and then 26 gauge). Rinse the syringes and tubes with an additional 0.5 ml of buffer and pool these with the homogenate. Centrifuge at 88,200g at r_{max} for 15 min at 4°. For regeneration experiments, first resuspend the pellet in PBS and add a fivefold excess of 11-*cis*-retinal (~4 nM), incubate for 1 hr at 4° on a rotator, centrifuge as described above, and resuspend the recovered pellet in 5% (v/v) BSA in PBS and incubate for 30 min on a rotator at 4°. Centrifuge the regenerated sample as described above to recover the pellet. Otherwise, after a PBS wash, resuspend the pellet in 100 μl of 1% (w/v) dodecyl maltoside in PBS and solubilize overnight at 4° on a rotator. Remove unsolubilized material by centrifugation at 109,000g at r_{max} for 15 min at 4°. Analyze the supernatant by spectroscopy on a SpectroPette microspectrophotometer (World Precision Instruments, Sarasota, FL). Sum a total of five spectra for each sample (both dark and bleached). For the bleached spectrum, limit light exposure to 30 sec, because of the extremely small sample volume (60 nl/mm light path length). Data can be exported into Kaleidagraph (Synergy Software, Reading, PA) (or other suitable analysis software package) for difference spectra calculations. The amount of rhodopsin present in *Rpe65*$^{+/-}$ mice is about half that seen in wild-type animals, suggesting a dose-dependent effect of RPE65 expression on rhodopsin level.[24] Despite this lack of rhodopsin, opsin apoprotein is expressed, as seen by immunoblotting with anti-rhodopsin monoclonal antibody 1D4,[45] and immunoelectron microscopy demonstrates that it is appropriately expressed in the rod outer segments.[24]

As might be expected from lack of rhodopsin, the functional rod electrophysiological response as measured by electroretinography (ERG) is abolished in *Rpe65*-deficient mice,[24] although those of *Rpe65*$^{+/-}$ mice are similar to the *Rpe65*$^{+/+}$ mice, suggesting a wide range of tolerance for rhodopsin level.[24] Intriguingly, cone ERG is relatively unaffected.[24] This suggests that cone pigment regeneration may occur independently of the RPE. The histology and ultrastructure of the RPE and retina of the *Rpe65*-deficient mice show minimal changes compared with the total loss of rod ERG activity[24] but appear slightly disorganized compared with wild-type OS. From these findings, we can conclude that the biochemical feature underlying the *Rpe65*-deficient phenotype is blockade of the all-*trans*- to 11-*cis*-retinoid isomerization, because we see accumulation of the putative substrate of the isomerase/isomerohydrolase and lack of formation of the 11-*cis*-retinoid product. Thus, it is clear that RPE65 is directly involved as an important component of this hallmark reaction of the RPE.

---

[45] R. S. Molday and D. MacKenzie, *Biochemistry* **22**, 653 (1983).

Concluding Remarks

RPE65 mutations resulting in amino acid substitutions can cause early- or late-onset disease depending on the residue changed, while null mutations invariably cause severe early-onset changes in the vision of both humans and mice. The extent of these changes varies, perhaps in a species-specific manner. The detailed physiological and biochemical phenotype of the *Rpe65*-deficient mouse is consistent with disruption of the RPE visual cycle proximate to isomerization. The clinical features of the human *RPE65* mutations (and Briard dog) are consistent with such a model. In view of these parallels, the *Rpe65*-deficient mouse provides a valuable model to better understand the role of RPE65 and the mechanism of chromophore regeneration in the visual cycle. In addition it will be useful as a model to test potential therapeutic modalities for *RPE65*-related retinal dystrophies in humans. The mouse phenotype described above provides potential methods to test for restored function of RPE65 activity.

Acknowledgments

We thank our coauthors of our published works discussed in this chapter. We also thank Dr. Sue Gentleman for critical reading of the manuscript.

[47] Construction of Encapsidated (gutted) Adenovirus Minichromosomes and Their Application to Rescue of Photoreceptor Degeneration

By RAJENDRA KUMAR-SINGH, CLYDE K. YAMASHITA, KEN TRAN, and DEBORA B. FARBER

Introduction

We described the development of a novel class of adenovirus (Ad) vectors termed encapsidated adenovirus minichromosomes (EAMs),[1] now also referred to in the literature as gutless,[2] high-capacity,[3] helper-depen-

---

[1] R. Kumar-Singh and J. S. Chamberlain, *Hum. Mol. Genet.* **5**, 913 (1996).
[2] N. Whittle, *Trends Genet.* **14**, 136 (1998).
[3] G. Schiedner, N. Morral, R. J. Parks, Y. Wu, S. C. Koopmans, C. Langston, F. L. Graham, A. L. Beaudet, and S. Kochanek, *Nature Genet.* **18**, 180 (1998).

dent,[4–6] and miniadenoviral[7] vectors. In our initial study we used EAMs to transfer a 14-kb dystrophin cDNA to myoblasts *in vitro*.[1] More recently, we have demonstrated the use of EAMs to rescue photoreceptor degeneration *in vivo*.[8]

The *rd* mouse is an animal model for retinitis pigmentosa in humans. Homozygous *rd* mice show the first signs of retinal abnormalities by the beginning of the second week of life, and they undergo complete degeneration of the photoreceptors during the following 10–12 days. The disease is caused by a mutation in the gene encoding for the β subunit of the cyclic GMP phosphodiesterase (βPDE).[9] The use of EAMs to deliver a βPDE cDNA to *rd* mice allows rescue of the rod photoreceptor cells up to 12 weeks postinjection.[8] These results demonstrate an advantage in the use of EAMs in gene delivery to photoreceptors over first-generation Ad vectors.[10]

The methods described in this chapter include all the protocols required to construct and test EAMs with any transgene of interest. We provide a brief discussion of the main features of the different types of EAMs and of the modified protocols we have developed for analysis of EAM-mediated expression of transgenes in the retina.

The construction of EAMs depends on two seminal observations regarding the nature of the origin of replication and encapsidation signals of Ad. Previously it was shown that the Ad inverted terminal repeats (ITRs) are sufficient to act as origins of replication when embedded in a circular plasmid and that they allowed replication of DNA between two ITRs in the presence of Ad DNA replication proteins. These replicated strands of DNA were initially referred to as minichromosomes.[11] The minimal *cis*-acting sequences of ad required to encapsidate viral DNA were identified by extensive mutational analysis of the left end of the adenoviral genome.[12,13] In our original work we had shown that these replicated minichromosomes could be encapsidated in the presence of helper virus, and there-

---

[4] K. Mitani, F. L. Graham, C. T. Caskey, and S. Kochanek, *Proc. Natl. Acad. Sci. U.S.A.* **92,** 3854 (1995).

[5] R. J. Parks and F. L. Graham, *J. Virol.* **71,** 3293 (1997).

[6] R. Alemany, Y. Dai, Y. C. Lou, E. Sethi, E. Prokopenko, S. F. Josephs, and W. W. Zhang, *J. Virol. Methods* **68,** 147 (1997).

[7] A. Lieber, C. Y. He, and M. A. Kay, *Nature Biotechnol.* **15,** 1383 (1997).

[8] R. Kumar-Singh and D. B. Farber, *Hum. Mol. Genet.* **7,** 1893 (1998).

[9] C. Bowes, T. Li, M. Danciger, L. C. Baxter, M. L. Applebury, and D. B. Farber, *Nature (London)* **347,** 677 (1990).

[10] J. Bennett, T. Tanabe, D. Sun, Y. Zeng, H. Kjeldbye, P. Gouras, and A. M. Maguire, *Nature Med.* **2,** 649 (1996).

[11] R. T. Hay, N. D. Stow, and I. M. McDougall, *J. Mol. Biol.* **175,** 493 (1984).

[12] M. Grable and P. Hearing, *J. Virol.* **64,** 2047 (1990).

[13] M. Grable and P. Hearing, *J. Virol.* **66,** 723 (1992).

fore we had referred to these novel class of vectors as encapsidated adenovirus minichromosomes (EAMs).[1,8]

We (and others) have developed EAMs to address the issue of the strong host immune response observed during clinical trials using first-generation Ad vectors.[14] These are vectors that have the adenoviral region E1 deleted and replaced with the transgene of interest.[15] Also included in the first-generation vectors are usually those Ad vectors with deletions in region E3. First-generation Ad vectors are propagated in human embryonic kidney 293 cells (HEK 293), which provide the E1 proteins in *trans*.[15] It was expected that deletion of region E1 in first-generation vectors would prevent the virus from replicating and synthesizing viral proteins outside 293 cells. However, because of E1-like activity in human cells (or simply because high multiplicities of infection are used), low levels of DNA replication and protein synthesis take place in cells infected with first-generation Ad vectors. Peptide fragments from Ad proteins are displayed on the surface of infected cells via MHC molecules, labeling these cells for T cell-mediated clearance. Retention of MHC or viral peptides in the endoplasmic reticulum (such as by Ad region E3) or prevention of transport of MHC–peptide complex to the cell surface results in the activation of natural killer cells that constantly patrol and eliminate cells deficient in surface MHC.[16]

To reduce the number of viral proteins being expressed in infected cells, second-generation viral vectors have been constructed, usually with deletions in genes coding for the viral replication machinery,[17,18] further crippling these viruses *in vivo*. However, it is unclear if such viruses result in prolonged expression when compared with first-generation Ad vectors.[19]

In addition, all of these vectors suffer from a limited capacity for cloning a foreign gene (8–9 kb). Although this capacity is sufficient to accommodate most cDNAs, it is insufficient to carry the large upstream regulatory elements that are required to allow the transgene to partake in normal cellular feedback mechanisms involved in regulation of gene expression.

EAMs are composed of the minimal region of Ad required for DNA replication, i.e., the ITRs, and the minimal *cis*-acting signals required for

---

[14] J. M. Wilson, J. F. Engelhardt, M. Grossman, R. H. Simon, and Y. Yang, *Hum. Gene Ther.* **5,** 501 (1994).
[15] F. L. Graham and L. Prevec, *Mol. Biotechnol.* **3,** 207 (1995).
[16] S. Paabo, L. Severinsson, M. Andersson, I. Martens, T. Nilsson, and P. A. Peterson, *Adv. Cancer Res.* **52,** 151 (1989).
[17] Y. Yang, F. A. Nunes, K. Berencsi, E. Gonczol, J. F. Engelhardt, and J. M. Wilson, *Nature Genet.* **7,** 362 (1994).
[18] H. Zhou, W. O'Neal, N. Morral, and A. L. Beaudet, *J. Virol.* **70,** 7030 (1996).
[19] B. Fang, H. Wang, G. Gordon, D. A. Bellinger, M. S. Read, K. M. Brinkhous, S. L. Woo, and R. C. Eisensmith, *Gene Ther.* **3,** 217 (1996).

encapsidation, Ad nucleotides 194–358 (Fig. 1A). For propagation, EAMs depend on the availability of all the proteins required to replicate and encapsidate DNA, most of which are provided by a first-generation recombinant Ad that has been modified to reduce its ability to propagate or package compared with wild-type Ads. Because there is also a requirement for E1 proteins, and the genes coding for them have been deleted in first-

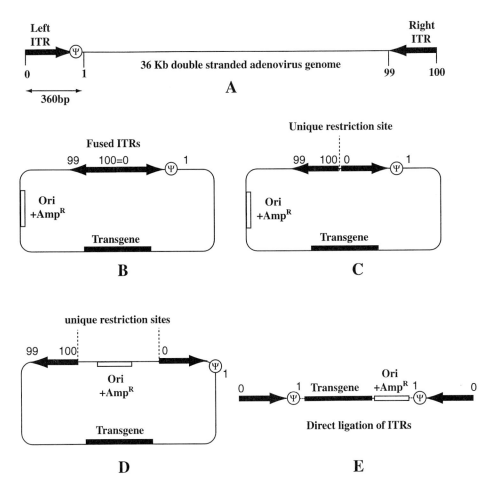

FIG. 1. Different starting points for the construction of EAMs. Each construct contains two adenovirus inverted terminal repeats (ITRs) and a packaging signal (Ψ) derived from the left and right end of the adenovirus genome (A). The ITRs can be cloned in various orientations and positions with respect to the transgene, including fused (B), fused with a restriction site (C), a spacer element between the ITRs (D), or direct ligation of ITRs onto the predigested plasmid (E).

generation Ads, EAMs are usually propagated in HEK 293 cells or cells derived from these, depending on the modifications in the helper viral genome. EAMs are initially constructed as plasmids in bacteria by standard recombinant DNA protocols and linearized and packaged in eukaryotic cells (HEK 293), using slightly modified virology protocols.

General Considerations for Construction of Encapsidated Adenovirus Minichromosomes

We, as well as other investigators, have used a variety of different plasmid structures to generate EAMs. All have a common intermediate step that results in the ITRs flanking the transgene and positioned at the ends of the EAM genome. In addition, a packaging signal is placed at one or both ends of the EAM genome, adjacent to the ITR. The most commonly used plasmid structures that serve as starting points for the construction of EAMs are shown in Fig. 1. Each has some advantages and disadvantages.

Construct B in Fig. 1 is of the type we have developed and used to deliver several different cDNAs including dystrophin,[1] βPDE, and *lacZ*.[8] This construct is distinguished from the others in that the 5' ends of the ITRs are fused. This fused origin of replication was originally cloned from pFG140, a plasmid propagated in bacteria and capable of generating infectious virus on transfection into 293 cells.[20] The main advantage this type of construct has is that it is ready for use after plasmid isolation, unlike others (Fig. 1C–E) that require more manipulations such as digestion and/ or ligation steps and further purification. The disadvantage of construct B is that it is not as efficient at being converted from the circular to the linear form as the other vectors. Constructs of type C (Fig. 1) have partly overcome this limiting step by introducing a unique restriction enzyme site at the junction of the ITRs, obviating the need for conversion from the circular to the linear form in 293 cells. This conversion is not optimal, however, as linearization after digestion does not lead to the 5' CATCATCAAT 3' sequence always found at the extreme ends of the Ad (serotype 5) genome. In the wild-type virus, an 80-kDa precursor to the terminal protein (TP) is covalently linked to the 5' terminal deoxycytosine of each DNA strand. One other disadvantage of this structure is the juxtaposition of two ITRs, one or both of which are sometimes deleted during amplification in bacteria. In the case of construct B, the fused ITRs are derived from the plasmid pFG140, which has been selected for stability in bacteria. Some of the DNA at the point of fusion between the ITRs has been deleted in pFG140. This deletion does not prevent formation of viable virions.[20] The problems

---

[20] F. L. Graham, J. Rudy, and P. Brinkley, *EMBO J.* **8,** 2077 (1989).

of recombination between juxtaposed ITRs have been partly addressed in constructs of type D (Fig. 1), which has a spacer between the ITRs, making it less likely that the ITRs will be deleted in bacteria; also, after digestion, the eukaryotic sequences linked to the ITRs and the Ad packaging signals can be purified before transfection. It is obvious that the final EAM construct will no longer contain any prokaryotic sequences. Finally, other constructs have been described[21] in which the ITRs are directly ligated into the predigested plasmid containing the transgene (E in Fig. 1), but we find this vector to be rather tedious to prepare and inefficient. We recommend construction of vectors of type B or D. The main components required to construct these plasmids can be obtained from several sources. The fused ITRs and packaging sequences (Ψ) can be cloned from pFG140 (Microbix, Toronto, Canada). ITRs (for C, D, and E) can be cloned from or directly amplified by polymerase chain reaction (PCR) from Ad DNA (protocol for Ad DNA preparation is presented below).

General Considerations for Construction of Helper Virus

Helper viruses are of three main varieties. They are usually based on first- or second-generation Ad backbones. Because almost all are deleted in regions E1, they need to be propagated in 293 cells. The Ad genome is divided into 100 map units (mu), each 360 bp in length. First-generation Ads are constructed by homologous recombination (Fig. 2A) between a plasmid containing the gene of interest flanked by Ad 0 to 1 mu and Ad 9.2–16 mu, and a plasmid containing the adenoviral backbone. Examples of such plasmids include pΔE1sp1A (the shuttle vector) and pBHG11 or pJM17 (the viral backbone). Homologous recombination results in the transfer of the transgene from the shuttle vector to the Ad backbone (Fig. 2B). These plasmids can be purchased from Microbix. These plasmids have the advantage that they can be propagated in bacteria before transfection. However, in our hands they are not as efficient at producing recombinant Ads as transfections with linearized shuttle vectors (available from Quantum Biotechnologies, Montreal, Canada) and Ad DNA prepared directly from virions, a more reliable but highly time-consuming process during the amplification and purification steps (protocol below). Although not essentially necessary, inclusion of a marker gene such as human placental alkaline phosphatase (hpAP) makes it easier to monitor relative populations of helper virus and EAMs during propagation.

Our initial experiments involved the use of an E1/E3-deleted helper virus.[1] Subsequent to these studies, an improved helper virus has been

---

[21] S. Kochanek, P. R. Clemens, K. Mitani, H. H. Chen, S. Chan, and C. T. Caskey, *Proc. Natl. Acad. Sci. U.S.A.* **93,** 5731 (1996).

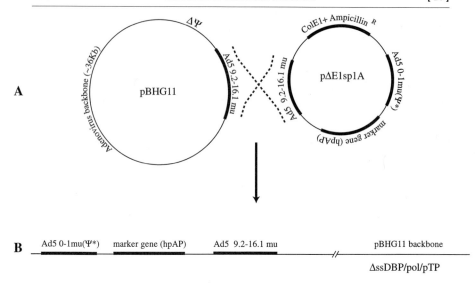

FIG. 2. Construction of first-generation Ad vectors by homologous recombination between a shuttle plasmid containing a transgene and a plasmid containing the Ad genome (A). Recombination between these two plasmids results in the transfer of the transgene into the viral backbone, which may be a first-generation E1/E3 or second-generation Ad with deletions in the single-stranded DNA-binding protein (ssDBP), the DNA polymerase (Pol), or the preterminal protein (pTP) in addition to the E1/E3 deletion (B). Modification of the left end of the shuttle plasmid by the addition of *loxP* sites flanking the packaging sequence (Ψ) leads to recombinant Ads where the (Ψ) can be deleted by the action of Cre recombinase (C).

described by Parks and colleagues.[22] Their helper virus contains a packaging signal flanked by *loxP* sites from phage λ. The action of Cre recombinase produced in 293 cells on the *loxP* sites results in the excision of the sequences between the *loxP* sites, including the encapsidation signal (Fig. 2C). This results in decreased packaging efficiency of the helper virus and better ratios of EAM to helper virus during serial propagation.[22] We are presently combining this scheme with Ad backbones with deletions in some early

[22] R. J. Parks, L. Chen, M. Anton, U. Sankar, M. A. Rudnicki, and F. L. Graham, *Proc. Natl. Acad. Sci. U.S.A.* **93,** 13565 (1996).

regions including the Ad single-stranded DNA-binding protein, the DNA polymerase, or the preterminal protein (pTP).[23] Helper viruses with mutations (deletions) in the packaging signal have also been previously utilized to propagate EAMs.[21] We have found that expression of pTP enhances the efficiency of conversion of circular EAMs to linear forms approximately twofold (our unpublished results, 1998). We determined this initially by observing that pFG140 results in a larger number of plaques when contransfected with an expression cassette for pTP. This observation held true for cotransfections of EAMs (containing fused ITRs) with pTP expression cassettes. Furthermore, we have also found that the use of helper viral DNA with the TP attached enhances the production of EAMs approximately 50-fold. We have also established cell lines that constitutively express the Ad pTP and these can be used to enhance the production of EAMs (our unpublished results, 1998).

Construction of Encapsidated Adenovirus Minichromosomes (pAd5βPDE) in Bacteria and of Helper Virus

Cloning is carried out by standard recombinant DNA protocols. For construction of the EAM designated pAd5βPDE (Fig. 3), the murine βPDE cDNA was obtained as a 2.75 EcoRI fragment from pBB1.[24] This cDNA was cloned downstream of a 350-bp human βPDE promoter.[25] We have previously shown that when a tissue-specific promoter is placed in the antisense orientation with respect to E1 enhancer sequences (which overlap with Ad packaging elements), tissue-specific/temporal regulation of expression can be retained.[1] Because of the intrinsic strength of viral promoters it is usually difficult to retain tissue specificity in the context of viral genomes. The inverted Ad origin of replication and five encapsidation signals were cloned from pFG140.[20] A plasmid (pAd5βN) that was similar in all respects to pAd5βPDE except that it did not contain a βPDE cDNA was used as a negative control.

Detailed protocols for the construction of a first-generation helper viruses have been published previously.[15] Essentially, it is necessary to clone a cassette for the marker gene, e.g., green fluorescent protein (GFP), into a shuttle vector such as pΔE1sp1A and cotransfect with either a plasmid containing the adenoviral genome (such as pBHG11 or pJM17) or linearized viral DNA. Construction of clones containing *loxP* sites flanking the encap-

---

[23] M. A. Hauser, A. Amalfitano, R. Kumar-Singh, S. D. Hauschka, and J. S. Chamberlain, *Neuromusc. Disord.* **7,** 277 (1997).
[24] N. I. Piriev, C. Yamashita, G. Samuel, and D. B. Farber, *Proc. Natl. Acad. Sci. U.S.A.* **90,** 9340 (1993).
[25] A. Di Polo, C. B. Rickman, and D. B. Farber, *Invest. Ophthalmol. Vis. Sci.* **37,** 551 (1996).

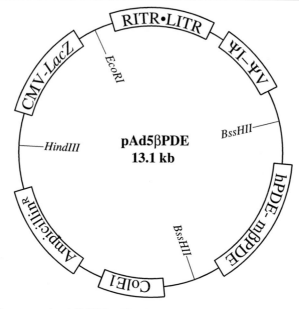

FIG. 3. Structure of pAd5βPDE. The fused left and right inverted terminal repeats (LITR · RITR) of adenovirus function as origins of replication in HEK 293 cells. The bacterial origin of DNA replication (ColE1) and ampicillin selection (ampicillin^R) for propagation of the circular EAM in bacteria are encoded by pBSIIKS (Stratagene). After replication, polar encapsidation of linear EAM DNA into preformed capsids is performed by the five packaging signals (ΨI–ΨV). The β subunit of *Mus musculus* cGMP phosphodiesterase (mβPDE) and the *Escherichia coli* β-galactosidase (*lacZ*) cDNAs are regulated by a human βPDE (hPDE) and cytomegalovirus (CMV) enhancer/promoters, respectively.

sidation signals have also been described elsewhere.[22] A protocol for the preparation of viral DNA for transfection is presented below. The preparation of viral DNA with retention of the TP has been described elsewhere.[26,27]

Summary of Protocol for Packaging Transgenes into Encapsidated Adenovirus Minichromosomes

The steps involved in the conversion of plasmids to linear EAMs are presented in Fig. 4. In summary, the circular plasmid (EAM) containing the transgene (with a marker gene) is cotransfected with helper virus DNA into 293 cells. Proteins produced by the helper viral DNA linearize and replicate the EAM in addition to their own (helper virus) DNA. Capsid

[26] A. J. Robinson, H. B. Younghusband, and A. J. Bellett, *Virology* **56**, 54 (1973).
[27] R. Pronk, M. H. Stuiver, and P. C. van der Vliet, *Chromosoma* **102**, S39 (1992).

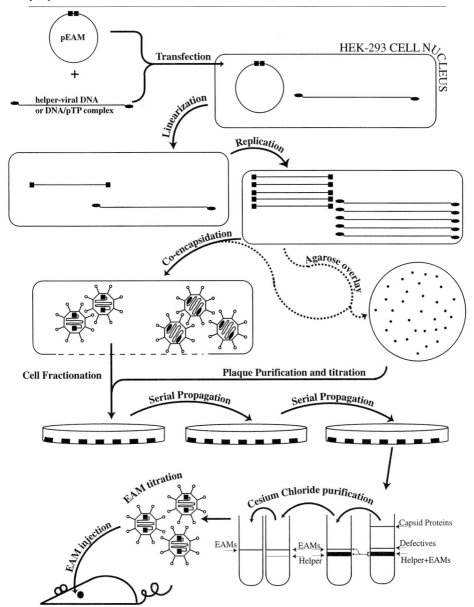

FIG. 4. Summary of the process involved in the packaging of circular EAMs into virions. The circular plasmid (pEAM) is cotransfected with helper viral DNA into HEK 293 cells, where through the production of viral replication proteins by the helper virus, pEAM is linearized and amplified. Individual plaques can be picked or an entire plate can be harvested for serial propagation. After amplification, the viral lysate is loaded on CsCl gradients and purified EAMs are injected into the subretinal space of *rd/rd* mice.

proteins are produced and both types of viral genomes (helper and EAMs) are packaged into preformed capsids. At this stage one can either pick plaques and identify those that contain the largest ratios of EAMs to helper virus or take a whole plate lysate and use this for serial propagation. After amplification by serial propagation, the virions are purified on the basis of buoyancy differences in CsCl gradients, titrated, and injected into animals.

Transfection of Encapsidated Adenovirus Minichromosome and Helper DNA

Prepare circular EAM plasmid by the endotoxin-free plasmid preparation method (Qiagen, Hilden, Germany). High-quality DNA is essential for good transfections. Prepare helper virus DNA as indicated below. Plate 8×10^5 cells/60-mm dish the day before transfection. This should result in approximately 1×10^6 cells/60-mm dish the following day (or about 80% confluency). Depending on their passage, different clones of 293 cells grow at slightly different rates, and so it is necessary to calculate the cell plating that will result in about 80% confluency in 24 hr. We have used low-passage 293 cells (Microbix) or 293 cells obtained from the American Type Culture Collection (ATCC, Rockville, MD) and in our hands the ATCC 293 cells are easier to transfect (and less expensive). We routinely use 1,2-dioleoyl-3-trimethylammoniumpropane [DOTAP; Boehringer Mannheim Biochemicals (BMB), Indianapolis, IN] or the slightly less efficient but much more economical calcium phosphate transfection procedures. We usually transfect six to eight plates per experiment, as not all plates will contain EAMs.

DOTAP Transfection for 60-mm Dish

A total of 10 μg of circular transgene-containing plasmid (with ITRs, etc.) and 0.6 μg of linear helper viral DNA is suspended in 70 μl of 20 mM HEPES buffer (pH 7.4) and incubated with 30 μl of DOTAP (BMB) in a polystyrene tube for 20 min at room temperature. The mixture is resuspended in 2 ml of Dulbecco's modified Eagle's medium (DMEM) with 2% (v/v) fetal calf serum and added to a 60-mm plate of 293 cells from which that medium has been removed. Approximately 5 hr later the transfection medium is replaced with DMEM and 10% (v/v) FCS. Observe daily for cytopathic effect, which usually appears within 7–10 days.

Calcium Phosphate Transfection for 60-mm Dish

Prepare 10 ml of $2 \times$ HEPES-buffered saline (HBS): 0.25 M NaCl–50 mM HEPES (pH 7.1)–1.5 mM Na$_2$PO$_4$(H$_2$O)$_7$, pH 7.0. Adjust the pH with NaOH, filter the HBS through a 0.2-μm pore size filter, and store at 4°.

For transfection, add 25 µl of 2 M $CaCl_2$ to the DNA mixture (as described above) and suspend it in deionized H_2O up to 200 µl. Add this mixture dropwise (and slowly) to 200 µl of 2 × HBS. Position a glass Pasteur pipette at the bottom of the transfection mixture, and push approximately 20 bubbles slowly through it. The solution should turn slightly cloudy, indicating the formation of a precipitate. Incubate for 20 min at room temperature and add dropwise to the cells, distributing across as much of the plate as possible. The medium will momentarily turn transparent. Return the cells to the CO_2 incubator.

Posttransfection

After transfection, we have used two methods to amplify EAMs and helper virus by serial propagation. We either allow the cytopathic effect to appear within 7–12 days and harvest the viral lysate, or we overlay (see protocol below) the transfected plate 24 hr posttransfection and pick plaques. These plaques are then tested by infection of 293 cells for identification of those that contain the highest ratio of EAM to helper virus (β-galactosidase/alkaline phosphatase staining).

Agarose Overlay and Plaque Formation

Prepare 5% (w/v) agarose (Seaplaque; FMC Bioproducts, Philadelphia, PA) in phosphate-buffered saline (PBS), autoclave, and store at 4° in 10-ml aliquots in 50-ml conical tubes.

Before overlaying, melt 10 ml of 5% (w/v) agarose in a microwave oven (avoid boiling) and cool it to 44–45° in a water bath. Add 30 ml of DMEM [with 3–5% (v/v) fetal bovine serum (FBS)] that has been prewarmed to 37° and mix gently. Use the mixture immediately or return it to the 45° waterbath. The mixture should be used within the next few minutes or it will begin to solidify. Remove medium (transfection or infection) from the plate and wash once with PBS prewarmed to 37°. Gently tilt the tissue culture plate and pour approximately 4–6 ml of the 1.25% (w/v) agarose mixture into the side of the plate (60 mm), onto the cells. It is easy to dislodge the cells at this stage. A cell monolayer will often fold over at the edges if this step is done too rapidly. Allow the plate to stand in a biological safety cabinet at room temperature for at least 20 min, until the agarose solidifies. Return the plate to the CO_2 incubator at 37°. Approximately every 5 days, add 3 ml of overlay to the plate until plaques appear (1–3 weeks).

Picking Plaques Containing Encapsidated Adenovirus Minichromosomes and Helper Virus

Plaques can be identified by holding the plate up under a strong light source; they appear as small transparent dots within the 293 cell layer.

Alternatively, stain the plaques with an agarose mixture containing 0.014% (w/v) neutral red. Examination of the plaques under low-power magnification usually distinguishes plaques from cell aggregates. With a water-proof marker, circle plaques to be picked (the ones well separated from others). Take an autoclaved cotton-plugged glass Pasteur pipette and insert it into the agarose vertically downward toward the plaque. Rotate the pipette concentrically and in a circular motion to loosen the agarose plug. Gently pull the plug into the pipette, using a tissue culture pippetor, and eject the plug into a sterile microcentrifuge tube. Draw 1 ml of DMEM (no serum) into the same pipette to wash any virus remaining in the pipette and pool with the original plug. Plaques can be stored at 4° before performing three cycles of freeze–thawing to release virus.

Measurement of Encapsidated Adenovirus Minichromosome-to-Helper Virus Ratio by Alkaline Phosphatase and LacZ Staining

We normally measure the titer of EAMs in individual plaques by infecting 293 cells with either a portion of the plaque or total viral lysate followed by staining for alkaline phosphatase activity (for helper virus titer) or *lacZ* activity (EAM titer).

Remove the medium from confluent six-well plates of 293 cells and wash once with PBS prewarmed at 37°. It will be necessary to infect two wells per plaque or lysate. Resuspend 5% of the viral plaque (or lysate) in 500 μl of DMEM with 2% (v/v) FCS; deposit the mixture gently on the cells and spread the virus by gently rocking the plate three times (in a cross shape). Return the plate to the CO_2 incubator at 37° for 1 hr. Rock the plate gently at 20-min intervals to ensure even spread of the virus. We find that the outer perimeter of the plates usually receives higher doses of infection. Add growth medium [with 10% (v/v) FCS] to the plates and incubate for 34 hr.

For detection of alkaline phosphatase, rinse the plates twice with PBS and fix the cells for 10 min at room temperature in 0.5% (v/v) glutaraldehyde in PBS. Wash the plates twice for 10 min each with 0.05 mM $MgCl_2$ in PBS. Endogenous alkaline phosphatase activity is inactivated by incubating the plates at 65° for 1 hr in PBS prior to addition of the chromogenic substrate 5-bromo-4-chloro-3-indolyl phosphate (BCIP, 0.15 mg/ml) and nitroblue tetrazolium (0.3 mg/ml; Sigma, St. Louis, MO). Cells are incubated at 37° in darkness for 3–10 hr. For β-galactosidase assays, the cells are fixed and washed as described above prior to the addition of 5-bromo-4-chloro-3-indolyl-β-D-galactoside (X-Gal, 1.2 mg/ml in PBS) containing 10 mM each of $K_3Fe(CN)_6$ and $K_4Fe(CN)_6 \cdot 3H_2O$, 2 mM $MgCl_2$. Counting the number of blue cells leads to lacZ-forming units (lfu) as opposed to plaque-forming units (pfu). Only a fraction of all virions are capable of undergoing the viral life cycle and producing infectious progeny; EAMs cannot form any plaques without the presence of a helper virus. Thus, a measurement of the number of transducing particles (lfu) is an accurate method by which to titrate EAMs.

Serial Amplification

After plaque isolation or preparation of viral lysate from transfected cells (see below), the EAMs and helper virus are released from the 293 cells as indicated above. For further amplification, select the plaque with the highest ratio of EAMs to helper virus by titrating for each virus as indicated and use this plaque to reinfect a plate of 293 cells. Because of the smaller size of the EAM genome, the EAMs generally amplify at a faster rate than the helper virus. Repeat the infection on 293 cells for about three cycles on 60-mm plates and then use a portion (75%) of the viral lysate to serially infect a larger number of plates, e.g., two 150-mm plates, four 150-mm plates, and so on, until there are at least ten 150-mm plates to harvest at the end of the seventh or eighth cycle. It is important not to exceed this number of cycles, as we have found that at high multiplicities of infection (MOIs), there is a greater rate of recombination between EAMs and helper virus, resulting in the appearance of new conformations of viral genomes.

Preparation of Lysate after Amplification

Depending on the MOI, cells will round up and show cytopathic effects after 2 to 4 days. To obtain high virus yield, it is important to let this process proceed until the cells are just starting to come off the plate. If the medium turns yellow before this happens add more medium but do not remove any of the previous medium, as this will result in lower virus yields. Pipette cells by gentle washing into 50-ml tubes. Do not disrupt the cells too much, as this will release virus into the medium. Spin the cells at 2000 rpm in a clinical centrifuge for 5 min at 4° and remove supernatant. Resuspend the cells in 500 μl of 10 mM Tris-HCl, pH 8.0. The cells can be stored at this stage at $-70°$. The viral lysates are pooled and subjected to four freeze–thaw cycles in ethanol–dry ice and a water bath at 37° to release most of the virus. Spin the lysates at 2000 rpm in a clinical centrifuge for about 10 min at 4° and retain the supernatant. For higher viral yields, resuspend the cell pellet in 10 mM Tris-HCl, pH 8.0, and spin as described above. Pool the supernatant with the original and add an equal volume of 1,1,2-trichlorotrifluoroethane (Sigma) in a glass tube. Spin at 2000 rpm for 5 min at 4° and transfer the top phase to a new tube on ice. Repeat the extraction with the bottom phase and pool the aqueous fractions. Extract the pooled aqueous fractions with an equal volume of 1,1,2-trichlorotrifluoroethane and retain the supernatant for loading onto a CsCl gradient.

Purification of Encapsidated Adenovirus Minichromosomes on Step and Self-Forming CsCl Gradients

Prepare CsCl in 10 mM Tris, pH 8.0, at densities of 1.45 (heavy), 1.20 (light), and 1.34 (medium). The medium CsCl needs to be more accurately measured than the heavy or light CsCl.

Heavy CsCl: 42.23 g of CsCl plus 57.77 ml of 10 mM Tris, pH 8.0
Light CsCl: 22.4 g of CsCl plus 77.61 ml of 10 mM Tris, pH 8.0
Medium CsCl: 20.01 g resuspended to 42 ml with 10 mM Tris, pH 8.0

Extracted lysate is loaded sequentially onto two step and two self-forming CsCl relaxation gradients. For the first step gradient, layer 15 ml of light (ρ 1.20) CsCl over 15 ml of heavy (ρ 1.45) CsCl in an SW28 tube. Layer the viral lysate over the light CsCl (total volume, approximately 38 ml). Spin at 20,000 rpm (SW28 rotor) in an ultracentrifuge at 4° for 3 hr. The virus-containing band will be the thickest band and will run approximately between the junction of the heavy and light CsCl. Depending on the MOI, there will be a number of lighter bands above the main band (see Fig. 4), and all of them should have a blue hue. The band immediately above the thickest band (containing the helper virus and EAMs) usually contains defective virions and should be avoided during the isolation of the band, which is performed by using a 5-ml syringe with an 18-gauge needle. The CsCl tube is sensitive to mixing of the bands at this stage and should be handled gently. Anchor the tube with a three-pronged clamp. Insert the needle just below the thick viral band and angle the needle tip such that the bevel is facing up and in contact with the bottom of the virus-containing band. Pull approximately 3 ml of CsCl gently while avoiding any of the defectives. The viscosity of the bands sometimes makes it difficult to pull the band without disturbing the defectives, but a small amount of contamination at this stage does not affect the final preparation too much. We usually collect the band inside a biological safety cabinet. Be prepared to collect the CsCl, which continues to spill after retraction of the needle. The virus-containing band is reloaded onto a similar step CsCl gradient to remove any residual defective virions.

After isolation of the major band in the lower gradient, the virus is passed through a self-forming gradient of ρ 1.34 at 37,000 rpm for 24 hr in an SW41 rotor at 12°. These conditions result in resolution between a thick band containing helper virus and a thinner band containing the EAMs. The EAM-containing band is pulled from the CsCl tube and passed once more through CsCl of ρ 1.34, initially at 37,000 rpm for 24 hr (to remove any residual helper virus) followed by a relaxation of the gradient (this compacts the bands) to 10,000 rpm for 10 hr. Do not use a brake to stop the centrifuge. This may add about 2 hr to the run but is necessary for optimal conditions. The final viral isolate is pulled by an 18-gauge needle and stored at 4° until further purification. To remove the CsCl, the final viral isolate is passed through a 30-ml column of Sephadex G-50–PBS and 250- to 500-μl fractions are assayed by optical density (OD) readings and β-galactosidase/alkaline phosphatase activity measurements. The CsCl fractions containing the largest amount of virus tend to retain a blue hue. One OD_{260} unit corresponds to 10^{12} particles/ml. For long-term storage of virus, split the stock into smaller aliquots and resuspend in 10 mM Tris-HCl, 100 mM NaCl, 0.1% (w/v) bovine serum albumin (BSA) and 10% (v/v) glycerol. Store at -20 or $-70°$. Repeated freezing and thawing should be avoided in order to maintain the quality (infectious dose) of the virus. Glycerol must be removed prior to use of the virus *in vivo*.

Preparation of Helper Viral DNA for Transfection

After CsCl banding of helper virus and desalting, digest 250–500 μl with an equal volume of pronase [0.1 M Tris (pH 8.0), 2 mM EDTA, 1% (w/v) sodium dodecyl sulfate (SDS), and pronase (1 mg/ml)] for 4 hr in a water bath at 37°. Extract with an equal volume of phenol–chloroform and ethanol precipitate DNA overnight. Resuspend in TE prior to transfection. When preparing viral DNA for generation of a recombinant, it is necessary to digest the left end of the virus with an enzyme that has a single restriction site at the left end, e.g., *Cla*I for dl7001, a commonly used first-generation adenovirus. Alternatively, there are kits available for the construction of first-generation Ads (Quantum Biotechnologies).

Crude Preparation of Viral DNA without CsCl Purification

This protocol saves the time needed for CsCl purification of virus and yields DNA of a quality useful for Southern blot analysis. After viral infection, pellet the cells as described above. Freeze–thaw three times in dry ice–ethanol and a 37° water bath. Spin out cell debris as described above. Add 0.5% (w/v) SDS, 10 mM EDTA, and pronase (1 mg/ml) directly to the viral lysate and incubate at 37° for 4 hr. Phenol extract and ethanol precipitate overnight.

Detection of Wild-Type Virus

Only one recombination event between a first-generation virus and E1 sequences in 293 cells is required to obtain a replication-competent adenovirus (RCA). Before administration *in vivo,* all stocks of EAMs should be tested for contamination with RCA by infection of HeLa cells at an MOI of 10 (infection for 12 days) as described above.

Reverse Transcriptase-Polymerase Chain Reaction of Encapsidated Adenovirus Minichromosome-Mediated Transgene

To distinguish expression of the transgene from the native mutant βPDE gene in *rd/rd* homozygotes, we have used an reverse transcriptase (RT)-PCR approach that relies on the presence of a novel *Dde*I site at the location of the *rd* mutation. To avoid RNA degradation retinas should be frozen as soon as possible after enucleation. RNA is extracted using TRIzol (Gibco-BRL Gaithersburg, MD) according to the manufacturer instructions. The RT reaction is carried out in PCR buffer [dNTPs (1 mM each), 500 mM KCl, 200 mM Tris-HCl (pH 8.4), 25 mM MgCl$_2$] with RNase inhibitor (1 unit/µl; Perkin-Elmer, Norwalk, CT), 100 pmol of random hexamers (Amersham, Arlington Heights, IL), 1 µg of total RNA, nuclease-free BSA (1 mg/ml), and 100 units of murine leukemia virus (MuLV) reverse transcriptase in a volume of 20 µl. The reaction is incubated at 23° for 10 min and at 42° for 30 min. The RNA–cDNA hybrid is denatured and reverse transcriptase is inactivated at 95° for 10 min. The sample is quick-chilled on ice. For cDNA amplification, the RT reaction is brought up to 100 µl in 1× PCR buffer with 100 ng of each oligonucleotide primer (5' CCA CAA CTG TGA GAC ACG CAG 3' and 5' TAC GGT TCT CCA GCT TGT TCA TC 3') and 3 U of *Taq* polymerase. PCR cycles are performed in a Stratagene (La Jolla, CA) robocycler with an initial denaturation at 94° for 4 min followed by 40 cycles at 94, 55, and 72° for 1 min each. PCR products are extracted with phenol–chloroform (1:1, v/v) and digested with *Dde*I prior to fractionation on agarose gels. Digestion products of 217 and 433 bp are associated with the transgene, whereas products of 295, 217, and 138 bp are associated with the mutant gene. An example of a typical gel has been previously shown.[8]

Biochemical Analysis of β-Phosphodiesterase Expression

At the appropriate postinjection ages, mice are sacrificed, the eyes are enucleated, and the neural retinas are dissected. Each retina is then disrupted in glass homogenizers containing 120 µl of hypotonic buffer [10 mM Tris (pH 7.6), 1 mM EDTA, 2 mM EGTA, 1 mM dithiothreitol (DTT), leupeptin (1 µg/ml), and pepstatin A (1 µg/ml)]. Homogenates are centrifuged at 14,000g for 20 min at 4° and the supernatant is removed and used for immunoblots, measurements of cyclic GMP phosphodiesterase activity (protocol below), and protein determination by the method of Peterson.[28]

Immunoblots

Proteins (20 µg) are separated by SDS–PAGE utilizing a Tris–Tricine buffer system. Additional resolution in the 80-to 100-kDa range is obtained by modification of a double-inverted gradient acrylamide gel.[29] Specifically, we use three layers: an inverted gradient [6.5% (w/v) acrylamide, 0.8% (w/v) bis-acrylamide to 9.5% (w/v) acrylamide, 0.8% (w/v) bis-acrylamide], a separating gel [10.5% (w/v) acrylamide, 0.8% (w/v) bisacrylamide], and a stacking gel [4% (w/v) acrylamide, 0.8% (w/v) bisacrylamide]. Proteins are transferred to nitrocellulose and probed with a polyclonal antibody[30] to PDE (1 : 2000 dilution) and visualized by using the amplified (secondary biotinylated goat anti-rabbit and tertiary biotin-conjugated alkaline phosphatase) AP kit (Bio-Rad, Hercules, CA) according to the manufacturer instructions. An example of a typical gel has been previously shown.[8]

Cyclic GMP Phosphodiesterase Activity

Measurement of the rate of $3',5'$-cyclic guanosine monophosphate (cGMP) hydrolysis is based on the two-step radioisotopic tracer method originally published by Farber and Lolley.[31] Prepare a mixture of all reagents so that their final concentration in a 100-µl reaction volume is 40 mM Tris (pH 7.6), 5 mM MgCl$_2$, 1 mM DTT, 250 µM cGMP, and 200,000 cpm of [$8'$-^{3}H]cGMP. Start the assay by adding 80 µl of the reaction mixture to 20 µl of sample (usually between 2 and 10 µg of protein), followed by incubation at 37° for 15 min. In this step, cGMP phosphodiesterase hydrolyzes cGMP to $5'$-GMP. Stop the reaction by heating the mixture to 85° for 3 min and then cool it quickly on ice. In the second step of the procedure, $5'$-GMP is converted by alkaline phosphatase to guanosine and is separated from other nucleotides by a resin slurry. Add 0.4 unit of alkaline phosphatase (EC 3.1.3.1, calf intestine; Sigma) to the cooled tubes and incubate at 37° for 10 min. Stop the reaction by the addition of a 1.0-ml slurry (1 : 4 bed volume to total volume in deionized distilled water) of AG1-X2 resin (Bio-Rad, 200-400 mesh anion-exchange resin, Cl$^-$ form) and keep the tubes on ice for 20 min. Centrifuge at 3500g for 20 min at 4°. Remove 300

---

[28] G. L. Peterson, *Anal. Biochem.* **83,** 346 (1977).
[29] R. Zardoya, A. Diez, P. J. Mason, L. Luzzatto, A. Garrido-Pertierra, and J. M. Bautista, *BioTechniques* **16,** 270 (1994).
[30] B. K. Fung, J. H. Young, H. K. Yamane, and I. Griswold-Prenner, *Biochemistry* **29,** 2657 (1990).
[31] D. B. Farber and R. N. Lolley, *J. Cyclic. Nucleotide Res.* **2,** 139 (1976).

µl of supernatant, add scintillation cocktail, and count the radioactivity in a scintillation counter. Tissue blanks are prepared by heating 20 µl of sample for 3 min at 85° prior to the start of the assay.

Phosphodiesterase activity is calculated by dividing the counts per minute in the sample (after subtraction of the counts per minute in the tissue blank) by the specific activity of the cGMP used as substrate, the number of minutes in which the reaction took place, and the micrograms of protein in the sample. The total number of counts per minute added per tube is determined by counting 300 µl of reaction mixture (containing no tissue) and adding 1.0 ml of water instead of 1.0 ml of slurry. The estimated activity values are corrected for losses in counts per minute due to nonspecific adsorption of [8'-^{3}H]guanosine by the AG1-X2 resin, using a correction factor determined spectrophotometrically from the ratio of the OD_{252} of 20 µM guanosine in 1 ml of water to the OD_{252} of 20 µM guanosine incubated with 1 ml of resin slurry.

Trypsin Activation

The activity of rod photoreceptor cGMP phosphodiesterase is modulated by the inhibitory γ subunit of the enzyme (PDEγ). Trypsin has been shown to activate phosphodiesterase through proteolytic degradation of PDEγ.[32] Measurement of the effect of trypsin on cGMP phosphodiesterase activity is carried out by adding trypsin to the retinal extract (4–7 µg of retinal protein per unit of trypsin) and incubating for 1.5 min at room temperature. The digestion is terminated by the addition of soybean trypsin inhibitor (20 units/unit trypsin) and phenylmethylsulfonyl fluoride (PMSF) to a final concentration of 0.1 mM. The phosphodiesterase assay is then performed as described above, with the exception that the first step incubation is carried out for 1 min at 37°.

Conclusions

Developments in molecular biology have enabled us to elucidate some of the molecular mechanisms involved in a number of retinal degenerations. We are now in a position to rescue some of those phenotypes with the tools of gene therapy. Preliminary data from several laboratories including our own indicate that delivery of the normal gene to photoreceptors leads to a reduced rate of photoreceptor degeneration. There are many previously tested candidate vectors for gene delivery to photoreceptors, including

---

[32] N. Miki, J. M. Baraban, J. J. Keirns, J. J. Boyce, and M. W. Bitensky, *J. Biol. Chem.* **250**, 6320 (1975).

adeno-associated virus (AAV) and lentivirus (HIV). It is not yet clear whether randomly integrating vectors such as recombinant AAV or lentivirus will be safe enough for use in a clinical setting. The concerns of wild-type contamination with HIV vectors does not need to be stressed. The episomal nature and mild symptoms of human Ad infection make them favorable candidates for gene therapy, and many clinical trials using Ads are already in progress.

Acknowledgments

This study was supported by grants from the Foundation Fighting Blindness, the National Institutes of Health (EY08285 and EY02651), and the Chatlos Foundation. D.B.F. is recipient of a Senior Investigator Award from Research to Prevent Blindness. R.K.-S. is recipient of a Young Investigator Award from the Foundation Fighting Blindness.

[48] Production and Purification of Recombinant Adeno-associated Virus

By WILLIAM W. HAUSWIRTH, ALFRED S. LEWIN, SERGEI ZOLOTUKHIN, and NICHOLAS MUZYCZKA

Introduction

Recombinant adeno-associated virus (rAAV), because of its simplicity, ability to infect a wide variety of dividing and nondividing cells, and lack of human pathogenicity, has proved to be a useful vector for efficient and long-term gene transfer *in vivo*. A variety of tissues have been successfully transduced, including retina,[1,2] lung,[3] muscle,[4–7] brain,[8] spinal

---

[1] J. G. Flannery, S. Zolotukhin, M. I. Vaquero, M. M. LaVail, N. Muzyczka, and W. W. Hauswirth, *Proc. Natl. Acad. Sci. U.S.A.* **94**, 6916 (1997).
[2] A. S. Lewin, K. A. Drenser, W. W. Hauswirth, S. Nishikawa, D. Yasumura, J. G. Flannery, and M. M. LaVail, *Nature Med.* **4**, 967 (1998).
[3] T. R. Flotte, S. A. Afione, C. Conrad, S. A. McGrath, R. Solow, H. Oka, P. L. Zeitlin, W. B. Guggino, and B. J. Carter, *Proc. Natl. Acad. Sci. U.S.A.* **90**, 10613 (1993).
[4] P. D. Kessler, G. M. Podsakoff, X. Chen, S. A. McQuiston, P. C. Colosi, L. A. Matelis, G. J. Kurtzman, and B. J. Byrne, *Proc. Natl. Acad. Sci. U.S.A.* **93**, 14082 (1996).
[5] X. Xiao, J. Li, and R. J. Samulski, *J. Virol.* **70**, 8098 (1996).
[6] K. R. Clark, T. J. Sferra, and P. R. Johnson, *Hum. Gene Ther.* **8**, 659 (1997).
[7] K. J. Fisher, K. Jooss, J. Alston, Y. Yang, S. E. Haecker, K. High, R. Pathak, S. E. Raper, and J. M. Wilson, *Nature Med.* **3**, 306 (1997).
[8] R. L. Klein, E. M. Meyer, A. L. Peel, S. Zolotukhin, C. Meyers, N. Muzyczka and M. A. King, *Exp. Neurol.* **150**, 183 (1998).

cord,[9] and liver.[10] Recombinant AAV vectors consist of a simple capsid with a single-stranded DNA genome containing short viral inverted terminal repeats (ITRs) but no viral coding sequences.[11,12] Most often rAAVs have been generated by cotransfection of adenovirus (Ad)-infected 293 cells with an rAAV vector plasmid and a wild-type AAV (wtAAV) helper plasmid.[13] Improvements in AAV helper design[14] as well as construction of noninfectious mini-Ad helper plasmids[15–17] have eliminated the need for adenovirus infection and improved the yield of rAAV per transfected cell. For normal retinal therapy experiments in small animal models, this general packaging protocol produces sufficient recombinant virus for injection typically into 50–100 eyes. However, for larger animals with human-sized eyes, scalable methods of rAAV production that do not rely on DNA transfection have also been developed.[18–21] These methods, generally involving the construction of producer cell lines and helper virus infection, are suitable for the higher volume production that may be needed for testing in larger eyes.

The traditional rAAV purification protocol involved the stepwise precipitation of rAAV with ammonium sulfate, followed by two or three rounds of CsCl density gradient centrifugation. Each gradient required fractionation and identification of the virus-containing regions by dot-blot hybridization or by polymerase chain reaction (PCR) analysis. Not only did this require up to 2 weeks to complete, it also often resulted in poor recovery and poor-quality virus. Such a lengthy production protocol com-

[9] A. L. Peel, S. Zolotukhin, G. W. Schrimsher, N. Muzyczka, and P. J. Reier, *Gene Ther.* **4,** 16 (1997).
[10] R. O. Snyder, C. H. Miao, G. A. Patijn, S. K. Spratt, O. Danos, D. Nagy, A. M. Gown, B. Winther, L. Meuse, L. K. Cohen, A. R. Thompson, and M. A. Kay, *Nature Genet.* **16,** 270 (1997).
[11] P. L. Hermonat, M. A. Labow, R. Wright, K. I. Berns, and N. Muzyczka, *J. Virol.* **51,** 329 (1984).
[12] S. K. McLaughlin, P. Collis, P. L. Hermonat, and N. Muzyczka, *J. Virol.* **62,** 1963 (1988).
[13] P. L. Hermonat, and N. Muzyczka, *Proc. Natl. Acad. Sci. U.S.A.* **81,** 6466 (1984).
[14] J. Li, R. J. Samulski, and X. Xiao, *J. Virol.* **71,** 5236 (1997).
[15] D. Grimm, A. Kern, K. Rittner, and J. Kleinschmidt, *Hum. Gene Ther.* **9,** 2745 (1998).
[16] X. Xiao, J. Li, and R. J. Samulski, *J. Virol.* **72,** 2224 (1998).
[17] A. Salvetti, S. Oreve, G. Chadeuf, D. Favre, Y. Cherel, P. Champion-Arnaud, J. David-Ameline, and P. Moullier, *Hum. Gene Ther.* **9,** 695 (1998).
[18] J. A. Chiorini, C. M. Wendtner, E. Urcelay, B. Safer, M. Hallek, and R. M. Kotin, *Hum. Gene Ther.* **6,** 1531 (1995).
[19] J. E. Conway, S. Zolotukhin, N. Muzyczka, G. S. Hayward, and B. J. Byrne, *J. Virol.* **71,** 8780 (1997).
[20] N. Inoue and D. W. Russell, *J. Virol.* **72,** 7024 (1998).
[21] K. R. Clark, F. Voulgaropoulou, D. M. Fraley, and P. R. Johnson, *Hum. Gene Ther.* **6,** 1329 (1995).

bined with the growing demand for a wide variety of different rAAV stocks often taxed the capacities of vector production facilities. The need for a protocol that substantially reduces preparation time without sacrificing the quality and purity of the final product has been clear for some time. Such a protocol is described here. It is based on two improvements: (1) the observation that AAV binds to cell surface heparin sulfate proteoglycan,[22] and (2) a new bulk purification technique employing the nonionic gradient medium, iodixanol, which allows efficient binding of the virus to the affinity medium. This combination of techniques results in high recovery rates, improved viral infectivity, and rapid purification. The temporal order of steps for producing a pure rAAV includes (1) production of rAAV in crude lysates, (2) iodixanol density step gradient viral purification, (3) heparin affinity chromatography and virus concentration, and (4) characterization and quantification of the purified rAAV.

Production of Crude Cell Lysates Containing Recombinant Adeno-Associated Virus

The basic strategy for producing rAAV involves DNA transfection of a suitable human host cell in culture. Two plasmid DNAs are cotranfected, an AAV vector plasmid containing the AAV terminal repeat sequences (ITRs) and the promoter/gene of interest[1]; and a helper plasmid containing both the *rep* and *cap* AAV genes, required for packaging ITR flanked DNA, and the early genes of Ad, required for a productive AAV infection.[2] Human 293 cells have been and remain the primary host for rAAV production. Although alternatives exist, calcium phosphate-mediated cell transfection remains our standard technique for introducing plasmid DNAs into host cells. We employ the basic rAAV vector plasmid pTR-UF5[23] containing the humanized *gfp* gene under control of the cytomegalovirus (CMV) promoter as our usual starting vector. We also use the helper plasmid pDG, which contains both the AAV genes (*rep* and *cap*) and the adenovirus genes required for AAV propagation.[15] No replication-competent adenovirus has been detected with the use of this helper plasmid. To simplify the transfection protocol, the $CaPO_4$–DNA precipitate can be left in the medium for the entire incubation period of 48 hr without compromising cell viability. Transfection efficiency routinely reached 60% as judged by green fluorescent protein (GFP) fluorescence in host 293 cells. After harvesting the cells, virus is extracted by freezing and thawing the

---

[22] C. Summerford and R. J. Samulski, *J. Virol.* **72,** 1438 (1998).

[23] S. Zolotukhin, M. Potter, W. W. Hauswirth, J. Guy, and N. Muzyczka, *J. Virol.* **70,** 4646 (1996).

cells and the supernatant is then clarified by low-speed centrifugation. The use of sonication, microfluidizing, or detergent extraction (e.g., deoxycholate) does not appear to significantly increase the viral yields.

It is also possible to concentrate and purify rAAV from the cell culture medium instead of from material in the cell pellet. For such bulk purification from medium, virus from the cell supernatant can be precipitated with 50% (w/v) ammonium sulfate. However, this procedure is inefficient because, at the time of harvest (48 hr posttransfection), about 90% of the virus remains intracellular.

Host Cells. Low passage number (passage 29–40) 293 cells are propagated in Dulbecco's modified Eagle's medium (DMEM) with 10% (v/v) fetal bovine serum (FBS). After passage 40 we observe a reproducible loss in recombinant viral titer. Therefore host cells are discarded after this passage number.

Vector Plasmid. The construction of pTR-UF5 has been described[8] and is available on request. Standard cloning procedures are used to delete the CMV promoter and/or the *gfp* gene in pTR-UF5 and insert the gene and regulatory sequences of interest into pTR-UF5. Because the packaging limit of the AAV capsid is about 4.9 kb, all DNA between (and including) ITRs must remain below this limit. Maximal space in the vector can be obtained by also deleting the herpesvirus thymidine kinase (TK) promoter and *neo* cassette in pTR-UF5, using a single *Sal*I digestion and reclosure.

Containment. All stages of rAAV purification are carried out in a designated area physically separated from the rest of the facility. Virus-containing reagents are handled exclusively in biosafety cabinets. Only Quick-Seal tubes (Beckman, Palo Alto, CA) are used to purify recombinant virus.

Virus Production. To produce rAAV, a double cotransfection procedure at a 1:1 molar ratio is used to introduce the rAAV vector plasmid together with the helper plasmid pDG. Plasmid DNA for transfection is purified by a conventional alkaline lysis/CsCl gradient protocol. Care must be taken to avoid deleting ITR sequences during the growth of the vector plasmid in *Escherichia coli*. Because both flanking ITRs must be present in the vector plasmid DNA for rAAV packaging, it is important to confirm their presence prior to 293 cell transfection. We propagate vector plasmids in recombination-deficient Sure 2 cells (Stratagene, LaJolla, CA) for no more than 12 hr at 37°. To check the resultant vector plasmid DNA for intact ITRs, a small sample is digested with *Sma*I, which releases all DNA between the ITRs. ITR deletion will result in a loss of one of these flanking *Sma*I sites. Vector plasmid DNA containing less than about 10% deleted ITR (loss of a flanking *Sma*I site) will retain efficient packaging properties.

Protocol

1. Human 293 cells are split 1:2 the day prior to the experiment, so that, when transfected, the cells are about 75–80% confluent. Ten 15-cm plates are transfected as one batch.

2. To make the $CaPO_4$ precipitate, 180 μg of pDG is mixed with 180 μg of vector plasmid DNA in a total volume of 12.5 ml of 0.25 M $CaCl_2$.

3. Formation of the $CaPO_4$ precipitate is initiated at 37° by adding 12.5 ml of prewarmed 2× HBS (0.3 M NaCl, 10 mM KCl, 1.5 mM Na_2HPO_4, 10 mM dextrose, and 40 mM HEPES at pH 7.05) to the DNA–$CaCl_2$ solution.

4. The DNA is incubated for 1 min, then transferred into 200 ml of prewarmed DMEM–10% (v/v) FBS, stopping formation of the precipitate.

5. Medium is removed from the cells, 22 ml of the precipitated DNA–medium is immediately dispensed into each plate and cells are incubated at 37° for 48 hr. The $CaPO_4$ precipitate can remain on the cells during the whole incubation period without compromising cell viability.

6. Cells are harvested after 48 hr of incubation by centrifugation at 1140g for 10 min at room temperature, and the medium is discarded.

7. Cells are then lysed in 15 ml of 0.15 M NaCl, 50 mM Tris-HCl (pH 8.5) by three freeze–thaw cycles using alternating dry ice–ethanol and 37° baths.

8. Benzonase [Nycomed Pharma A/S (Roskilde, Denmark), pure grade] is then added to the mixture at 50 U/ml, final concentration, and the lysate incubated for 30 min at 37°.

9. The crude lysate is clarified by centrifugation at 3700g for 20 min at 4°. The virus-containing supernatant is considered the crude lysate that will require further purification by density gradient centrifugation.

Iodixanol Density Step Gradient Virus Purificaton

During initial attempts to design a more efficient purification method, it appeared that rAAV and cell proteins were forming aggregates in the crude cell lysate. When such complexes formed, the virus failed to display uniform biochemical properties, leading to poor viral recovery at all purification stages and making it difficult to develop a reproducible purification strategy. Furthermore, these apparently nonspecific interactions resulted in viral preparations contaminated with adventitious proteins, even after several rounds of CsCl gradient centrifugation. Therefore bulk purification of the crude lysate became the most crucial new step in rAAV purification.

To avoid aggregation, we investigated several alternative methods including the use of detergents and limited proteolytic digestion. Although these methods provided some improvement during subsequent purification, they resulted in preparations that were more heat labile and/or sensitive to DNase. Ammonium sulfate precipitation, which has been used previously for concentrating virus, did not provide significant purification. Furthermore, residual ammonium sulfate in the virus pellet interfered with subsequent ion-exchange chromatography. Dialysis at this stage in the purification also led to aggregation and precipitation of virus and proteins resulting in poor rAAV recovery. A combination of $(NH_4)_2SO_4$ precipitation and hydrophobic phenyl-Sepharose chromatography also failed to produce a purified virus without substantial loss of viral infectivity. Cellufine sulfate chromatography had been proposed as a way to purify and concentrate rAAV from crude lysates.[24] In our hands, however, cellufine sulfate fails to reproducibly bind rAAV in the crude lysates made as described. To circumvent these problems, we developed a new strategy of purification based on density gradient centrifugation using iodixanol.

Iodixanol is an iodinated density gradient medium originally produced as an X-ray contrast compound for clinical use. Unlike CsCl and sucrose solutions commonly used in gradient fractionation of macromolecules, iodixanol solutions can be made isosmotic at all densities. This property suggests iodixanol could be a useful reagent for viral purification and analysis steps. The nonionic and inert nature of iodixanol allows electrophoretic analysis and viral infectivity assays to be carried out directly on gradient fractions. Because the viscosity of iodixanol solutions is also lower than that of sucrose of the same density, it is possible to use iodixanol fractions directly in subsequent chromatographic purification steps without dialysis or dilution.

Because the apparent density of macromolecules in iodixanol solutions is different from that in CsCl, the banding density of purified rAAV-UF5 was determined empirically by banding an aliquot of UF5 virus in a continuos iodixanol gradient. The density of rAAV-UF5 was found to be 1.415 g/ml, equivalent to a 52% (w/v) solution of iodixanol. Opting to generate preformed step gradients in order to expedite this purification step, we adopted a 40% iodixanol (1.21 g/ml) step to accommodate rAAV/capsid protein complexes banding at slightly lower densities. This density is underlaid with a 60% step that acts as a cushion for any slightly denser rAAV particles. To locate the 40% iodixanol step and the 40%–60% density interface after centrifugation, we include phenol red dye (0.01 μg/ml) in the upper 25% and lower 60% density steps. The step gradient, consisting of 15, 25, 40, and 60% iodixanol, is run in a Beckman Ti70 rotor as described in the protocol below.

Our initial attempts to purify rAAV with discontinuous iodixanol gradi-

[24] K. Tamayose, Y. Hirai, and T. Shimada, *Hum. Gene Ther.* **7,** 507 (1996).

ents gave inconsistent results. As mentioned earlier, rAAV aggregates with proteins in the cell lysate. This apparently changes virion buoyant density and results in distribution of rAAV particles throughout the length of the gradient. To solve this problem, we modified the top (15%) iodixanol step so that it contained 1 M NaCl to disrupt ionic interactions between macromolecules. High salt was not added to the remaining steps in order to band the virus under isosmotic conditions, hence allowing direct purification through subsequent chromatographic stages.

A typical gradient, before and after the 1-hr centrifugation, is shown in Fig. 1A, left and right tubes, respectively. A plot of the refractive index at the end of the run is shown in Fig. 1B. rAAV is distributed through the 40% density step (Fig. 1A, shown by the vertical bar, and Fig. 1C). During a typical production run, the virus is recovered by inserting an 18-gauge needle connected to a 5-ml syringe about 2 mm below the 60-to-40% density junction (indicated by the arrow in Fig. 1A) and collecting a total of 4 ml. In Fig. 1, however, the gradient was collected by dripping from the bottom and collecting 1-ml fractions for detailed analysis. The bulk of the rAAV virus bands within the 40% density step (Fig. 1C, fractions 5–8). An opaque band that migrates at the 40-to-25% iodixanol junction consists mostly of cellular proteins (Fig. 1C, fractions 9 and 10) and contains less then 5% of the total rAAV. A small amount of rAAV also bands at the 40-to-60% junction (Fig. 1B and C, fraction 5). Approximately 75–80% of the rAAV in the crude lysate is recovered in the iodixanol fraction.

The nucleic acid–protein ratio in rAAV-UF5 is different from wtAAV because the size of the DNA packaged is 3.4 kb, or approximately 70% of the wild-type genome size. We have used the same protocol with no modifications to purify more than 50 different rAAV vectors, in which the size of the packaged genome ranged from 2.3 to 4.9 kb. Regardless of the size, there is no substantial difference in the banding pattern of rAAV on the iodixanol step gradient. Therefore, there is no need to modify the concentrations of iodixanol in the gradient steps to accommodate rAAV genomes in this size range.

To determine the resolving capacity of the iodixanol gradient, an experiment was performed in which viral lysates obtained from 5, 10, or 15 plates (15-cm diameter) were loaded onto separate gradients. This was equivalent to 1.6, 3.1, or 4.7×10^8 human 293 cells, respectively. After centrifugation, rAAV was recovered from each gradient as described above, and aliquots from each gradient subjected to sodium dodecyl sulfate–polyacrylamide gel electrophoresis (SDS–PAGE) (Fig. 2). The three viral capsid proteins VP1, VP2, and VP3 constituted the major species seen at all concentrations, suggesting that good separation of rAAV was achieved even at the highest concentration of lysate used. However, a significant increase in background contaminating protein was seen when the pooled lysate from 15 plates was

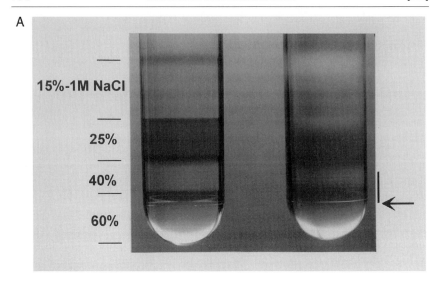
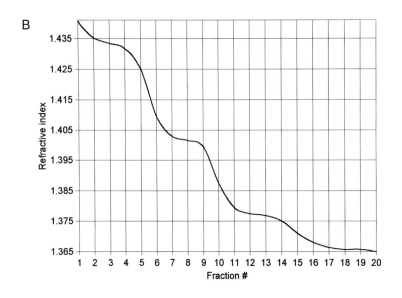

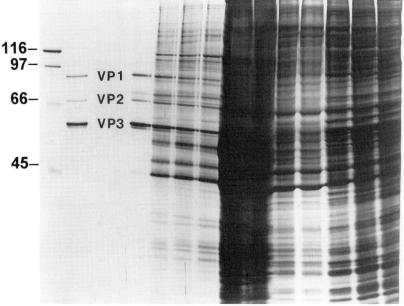

Fig. 1. Iodixanol step gradient for the purification of rAAV. (A) Preformed gradients shown before (left tube) and after (right tube) a 1-hr spin. The red or yellow tinge in the 60 and 25% steps indicates inclusion of phenol red in these steps. The positions of the density steps are shown on the left and the distribution of rAAV through the gradient is shown by the vertical bar. Typically, virus was collected by a needle puncture at the position on the side of the tube as indicated by the arrow. (B) A plot of the refractive index of 1-ml fractions collected from the bottom of the tube on the left in (A). (C) Silver-stained SDS–protein gel analysis of iodixanol fractions. rAAV was collected by dripping 1.0-ml fractions from the bottom of the gradient shown on the right in (A). Equivalent amounts of each fraction were then loaded onto a 12% (w/v) SDS–acrylamide gel and electrophoresed for 5 hr at 200 V. The numbers on the top of the gel correspond to the fraction numbers; only fractions 3–15 are shown. VP1, VP2, and VP3 indicate the position of viral capsid proteins. Lane (+) contains purified rAAV virus as a positive control. Lane M, protein standards whose molecular masses are shown on the left in kilodaltons.

used. Therefore, we routinely load rAAV lysate from only 10 plates per gradient. As a practical matter, this means that a 1-hr iodixanol gradient run in a Ti70 rotor would accommodate 3.1×10^9 cells from which approximately 10^{14} virus particles or about 10^{12} infectious units could be purified.

Protocol

1. A discontinuous step gradient is formed by underlayering and displacing the less dense cell lysate with iodixanol, 5,5′-[(2-hydroxy-1-3-propan-

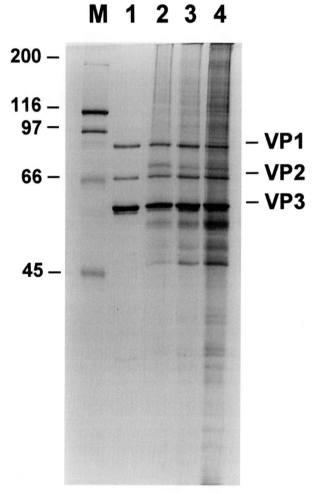

FIG. 2. Silver-stained SDS–protein gel analysis of the resolving capacity of an iodixanol step gradient. The virus lysate from 5, 10, or 15 dishes (lanes 2, 3, and 4, respectively) was purified on a single iodixanol gradient of the type shown in Fig. 1A. After the virus was collected, equivalent amounts of rAAV were loaded from each gradient onto an SDS–acrylamide gel as described in Fig. 1C. VP1, VP2, and VP3 indicate the positions of the viral capsid proteins. Lane 1 contains purified virus as a positive control. Lane M, standard proteins whose molecular weights are indicated on the left. See text for additional details.

ediyl)bis(acetylamino)]bis[N,N'-bis(2,3 dihydroxypropy)-2,4,6-triiodo-1,3-benzenecarboxamide], prepared using a 60% (w/v) sterile solution of Opti-Prep (Nycomed). Specifically, 15 ml of the clarified lysate is transferred into Quick-Seal Ultra-Clear 25 × 89 mm centrifuge tubes (Beckman), using a syringe equipped with a 1.27 × 89 mm spinal needle. Care is taken to avoid bubbles that could interfere with subsequent filling and sealing of the tube. A variable-speed peristaltic pump (model EP-1; Bio-Rad, Hercules, CA) is used to underlay in order: 9 ml of 15% (w/v) iodixanol and 1 M NaCl in PBS–MK buffer [1× phosphate-buffered saline (PBS), 1 mM MgCl$_2$, and 2.5 mM KCl]; 6 ml of 25% iodixanol in PBS–MK buffer containing phenol red [2.5 μl of a 0.5% (w/v) stock solution per milliliter of the iodixanol solution]; 5 ml of 40% iodixanol in PBS–MK buffer; and finally, 5 ml of 60% iodixanol in PBS–MK buffer containing phenol red (0.01 μg/ml).

2. The centrifuge tubes are then sealed and centrifuged in a type 70 Ti rotor (Beckman) at 350,000g for 1 hr at 18°.

3. Using the phenol red in the 25 and 60% iodixanol steps to identify density, 4 ml of the clear 40% step is aspirated after puncturing the tube on the side with a syringe equipped with an 18-gauge needle with the bevel facing upward.

Recombinant AAV in the 40% iodixanol fraction is further purified by the method described below.

Heparin Affinity Chromatography

Heparinized chromatographic supports have been successfully used for the purification of many heparin-binding macromolecules, including viruses such as CMV.[25] Heparin is the glycosaminoglycan moiety covalently bound to the protein core of proteoglycans found in mast cells. It is closely related to heparan sulfate (HS), the glycosaminoglycan (GAG) chain of the HS proteoglycan. The latter has been shown to be a cell surface receptor that mediates AAV cellular binding.[22] Covalent binding of heparin molecules to the matrix through its reducing end mimics the orientation of the naturally occurring GAGs.[26] To take advantage of the structural similarities between heparin and HS, we tested heparin affinity chromatography as a final purification step after iodixanol gradients.

Heparin is a heterogeneous carbohydrate molecule composed of long unbranched polysaccharides modified by sulfation and acetylation. The

---

[25] J. Neyts, R. Snoeck, J. Balzarini, J. D. Esko, A. Van Schepdael, and E. De Clercq, *Virology* **189**, 48 (1992).

[26] V. D. Nadcarni, A. Pervin, and R. J. Linahrdt, *Anal. Biochem.* **222**, 59 (1994).

degree of sulfation strongly correlates with its virus-binding capacity.[27] Therefore, we anticipated that heparinized matrices from different vendors might display different affinity toward rAAV. We therefore tested a number of heparin ligand-containing media, including three column chromatography media manufactured by Sigma (St. Louis, MO), heparin–agarose type I, heparin–agarose type II-S, heparin–agarose type III-S, as well as Affi-Gel heparin gel (Bio-Rad). Affi-Gel heparin gel and heparin–agarose type III-S columns bound less than 50% of the applied virus and, therefore, were not considered further. Either heparin–agarose type I or heparin–agarose type II-S prepacked 2.5-ml columns were efficient in both retaining and subsequently releasing rAAV. The type II-S column, however, is less selective, binding many cellular proteins along with the virus. In contrast, heparin–agarose type I columns were the best among those tested in terms of binding specificity and virus recovery, and is currently the virus-binding medium of choice.

Silver-stained SDS–acrylamide gel electrophoresis of rAAV-UF5 fractions at different stages of purification is shown in Fig. 3. Virus in iodixanol gradient fractions prepared from cells transfected with pTR-UF5/pDG (lane 2, Fig. 3) was directly applied to a heparin–agarose type I column (lane 4, Fig. 3) and eluted with 1 M NaCl as described below. The 1 M NaCl fraction contained 34.6% of the input rAAV (Table I), and was more than 95% pure, as judged by silver-stained SDS gel analysis. The heparin–agarose affinity fractionation yielded consistently higher purity virus than the traditional protocol that used ammonium sulfate precipitation and two rounds of CsCl gradient centrifugation (Fig. 3, lane 7). Iodixanol/heparin–agarose preparations also typically had the best particle-to-infectivity ratios.

Protocol

1. A prepacked 2.5-ml heparin–agarose type I column (Sigma) is equilibrated with 20 ml of PBS–MK (see above) under gravity. Alternatively, the columns can be placed inside 15-ml screw-cap conical tubes (Sarstedt) and spun at 200 rpm for 5 min in a type J6-HC centrifuge (Beckman). After each spin the flow-through is discarded and fresh buffer is added, and the washing repeated three more times.

2. The rAAV iodixanol fraction is then applied to the equilibrated column under gravity, and the column is washed with 10 ml of PBS–MK buffer either under gravity or by gentle batch spinning as described above.

3. Recombinant AAV is eluted with the same buffer containing 1 M NaCl under gravity or gentle centrifugation (100 rpm). After applying 10

[27] B. C. Herold, S. I. Gerber, T. Polonsky, B. J. Belval, P. Shaklee, and K. Holme, *Virology* **206**, 1108 (1995).

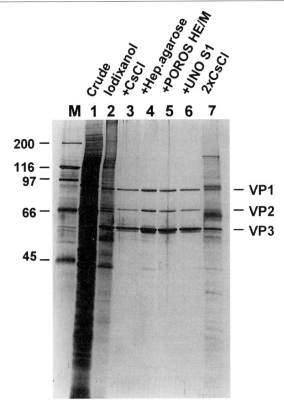

Fig. 3. Silver-stained SDS–acrylamide gel electrophoresis of rAAV-UF5 at various stages of purification. VP1, VP2, and VP3 indicate the position of the three AAV viral capsid proteins. Lanes 1–6 are various fractions obtained from virus preparations efirst employing an iodixanol gradient and then the indicated step from crude viral lysate as described. Lanes 5 and 6 are results from POROS HE/M and UNO S1 columns, respectively, and are not discussed specifically in text. Lane 7 contains a separate rAAV-UF5 preparation that was purified by the conventional method of ammonium sulfate precipitation followed by two CsCl gradients. A comparison of viral titers from each purification scheme is shown in Table I.

ml of the elution buffer, the first 2 ml of the eluant is discarded, and the virus collected in the subsequent 3.5 ml of elution buffer.

Concentration of Recombinant Adeno-Associated Virus

The heparin sulfate column-purified rAAV is then concentrated and desalted by centrifugation through BIOMAX 100 K filters (Millipore, Bedford, MA) according to the manufacturer instructions. The 1 M salt in the

TABLE I
COMPARISON OF IODIXANOL/HEPARIN AGAROSE AND $(NH)_4SO_4$/CsCl PURIFICATION[a]

Purification	Particles by QC-PCR ($\times 10^{-11}$)	Infectious units by ICA ($\times 10^{-9}$)	Particle-to-infectivity ratio
$NH_4SO_4/2\times$ CsCl	0.2	0.012	1667
Iodixanol/heparin–agarose	1.0	1.5	67

[a] Crude rAAV stock was made by cotransfection of 293 cells with pDG and pTR-UF5 and purified by either ammonium sulfate and two cesium chloride gradients or by iodixanol/heparin agarose fractionation. Note that the quoted physical and biological titers are assayed before viral concentration, a step that further increases titers at least 10-fold.

initial buffer is effectively exchanged by three cycles of concentration, dilution with lactated Ringer solution, and recentrifugation.

Characterization of Purified Recombinant Adeno-Associated Virus

An important index of virus quality is the ratio of physical particles to infectious particles in a given preparation. To characterize rAAV, we routinely titer both physical and infectious rAAV particles. We have found that a simple quantitative competitive (QC)-PCR assay (Fig. 4A) for physical particle titers is easier and more reliable than the conventional dot-blot assay. As the competitive standard a pTR-UF5 plasmid DNA containing a small internal deletion in the *neo* gene has been created. Infectious titers are determined by an infectious center assay (ICA). To avoid adventitious contamination of rAAV stocks with wtAAV, the use of wtAAV as a helper in the ICA has been eliminated. This was made possible by the use of the C12 cell line[21] containing integrated wtAAV *rep* and *cap* genes, for both the infectious center assay and the fluorescent cell assay. Adenovirus serotype 5 (Ad5), used to coinfect C12 along with rAAV for the ICA assay, is titered using the same C12 cell line in a serial dilution assay of cytopathic effect (CPE). The amount of Ad5 producing well-developed CPE at 48 hr postinfection on C12 cells is used to provide rAAV helper function. The physical particle titer and infectious titers typically differ by about a factor of 100 or less in an acceptable preparation (Table I).

Protocol

Quantitative-Competitive Polymerase Chain Reaction Assay for Recombinant Adeno-Associated Virus Physical Particles

1. The purified, concentrated viral stock is first treated with DNase I to digest any contaminating unpackaged DNA. Ten microliters of a purified

virus stock is incubated with 10 U of DNase I (Boehringer Mannheim, Indianapolis, IN) in a 100-μl reaction mixture, containing 50 mM Tris-HCl (pH 7.5), 10 mM MgCl$_2$ for 1 hr at 37°.

2. At the end of the reaction, 10 μl of 10× proteinase K buffer [10 mM Tris-HCl (pH 8.0), 10 mM EDTA, 1% (w/v) SDS final concentration] is added, followed by the addition of 1 μl of proteinase K (18.6 mg/ml; Boehringer Mannheim). The mixture is then incubated at 37° for 1 hr.

3. Viral DNA is purified by two phenol–chloroform extractions, followed by chloroform extraction and ethanol precipitation using 10 μg of glycogen as a carrier. The DNA pellet is then dissolved in 100 μl of water.

4. PCR mixtures contain 1 μl of the diluted viral DNA and twofold serial dilutions of the internal standard plasmid DNA. The most reliable range of standard DNA was found to be between 1 and 100 pg. Aliquots of each PCR are then analyzed by 2% (w/v) agarose gel electrophoresis until the two PCR products, one from the rAAV DNA template and one from the competitor template, are resolved (Fig. 4A). An image of the ethidium bromide-stained gel is then digitized using an ImageStore 7500 system (UVP, Upland, CA) and the densities of the target and competitor bands in each lane measured using the ZERO-Dscan Image Analysis System, version 1.0 (Scanalytics, Billerica, MA). The ratios are plotted as a function of the standard DNA concentration. At ratio of 1, the mass of viral DNA equals the mass of competitor DNA. This value is then used to determine the DNA concentration of the virus stock, thus allowing an estimate of the physical viral titer.

Infectious Center Assay

A modification of a previously published protocol[12] is used to measure the ability of the recombinant virus to infect C12 cells, unpackage, and replicate its DNA.

1. C12 cells are plated in DMEM in a 96-well dish at about 75% confluence. Serial dilutions of the rAAV to be titered are set up as follows: add 250 μl of medium to the first well and add 225 μl of medium to the adjacent wells. Add 2.5 μl of virus to be titered to the first well and serially dilute (10× steps) by transferring 25 μl per dilution to the adjacent well, being certain to change tips after each dilution. Add adenovirus to the cells at a multiplicity of infection (MOI) of 20. Leave a few wells as "adenovirus only" controls.

2. At 40 hr postinfection, set up a 12-port filter manifold built to hold 4-mm-diameter filters as follows: first, wet two pieces of Whatman (Clifton, NJ) paper with PBS and apply it to the entire manifold. Next, apply a nylon DNA transfer membrane (MagnaGraph; Micron Separations, Westboro, MA) wetted with PBS to the Whatman paper. Tighten the assembled

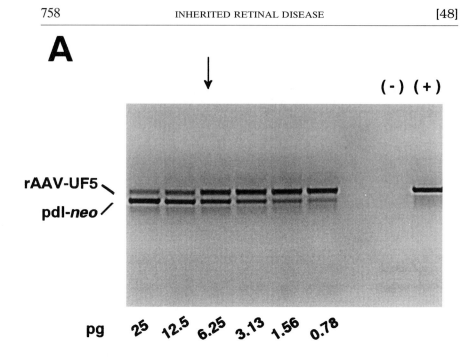

manifold and fill each well of the manifold with 5 ml of PBS. (Alternatively, individual glass-fritted disk 4-mm-diameter filter holders fitted into a vacuum manifold can be used.)

3. Detach cells from the 96-well dish by pipeting vigorously eight times and apply each infected cell sample to one well of the manifold. Wash the well with 200 µl of PBS and apply this also to the appropriate well. After transferring all of the cells to the filter manifold, allow 5 min to pass before gentle vacuum (about 1 cm of water) is applied.

4. Allow the nylon membrane to air dry for 5 min on Whatman 3MM paper. Denature the DNA on the filter for 5 min in 0.5 M NaOH, 1.5 M NaCl and then blot on Whatman paper. Neutralize the membrane for 5 min in 1.5 M NaCl, 0.5 M Tris-HCl, pH 7.8, and blot on Whatman paper. Rinse in 4× SSC (1× SSC is 0.15 M Nacl plus 0.015 M sodium citrate) for 30 sec and air dry for 10 min. Microwave the nylon membrane on a high setting for 4 min to fix the DNA. *Caution:* Be sure to have about 300 ml of water in a separate beaker in the microwave or the membrane may catch on fire.

5. Prehybridize the membrane in HM [7% (w/v) SDS, 0.25 M NaHPO$_4$ (pH 7.2), 1 mM EDTA (pH 8.0)] at 65° for at least 2 hr before adding the probe. The probe is a ^{32}P-labeled riboprobe against any portion of the DNA in the rAAV being titered. Add the probe in 100 ml of HM to the membrane in a glass hybridization cylinder and hybridize at 65° for 12 hr. Wash the membrane twice for 1 hr each in HM at 60°, dry, and expose the filter to standard X-ray film.

An example of filter autoradiographs is shown in Fig. 4B. To calculate the infectious titer of the rAAV preparation, count the number of positive cells on the filters (optimally 20–200 positive dots per filter) and correct for the dilution of the stock used on that filter.

FIG. 4. Titering of rAAV stocks by the QC-PCR assay (A) or infectious center assay (B). (A) Agarose gel electrophoresis of QC-PCR products (negative image). The top band is the product obtained from rAAV-UF5 viral DNA; the bottom band (pdl-*neo*) is the PCR product obtained from the standard competitor plasmid, which contains a small deletion in the *neo* gene sequence that is amplified. The numbers below each lane show the amount of input competitor DNA in each reaction. The viral DNA sample was diluted 1000-fold to achieve a template concentration that results in linear amplification of the input DNA. Symbols (−) and (+) indicate negative (no template) and positive (vector DNA without competitor) controls in the PCR. The arrow indicates the concentration at which the target template and the standard template were equivalent. (B) Autoradiographic image of nylon filters containing products of an infectious center assay, in which C12 cells had been infected with rAAV-UF5 and Ad5 and hybridized with ^{32}P-labeled GFP probe. Numbers at the top show the respective dilutions of the rAAV used to infect the C12 cells. The assay was typically linear in the range of 10–200 spots.

Discussion

The history of rAAV production is not without controversy. While some investigators report efficient rAAV-mediated transduction, others have found a strong dependence of transduction on helper virus adenovirus contaminants,[28] wtAAV contaminants,[12,29] or the growth state of the cells being transduced.[30] Crude rAAV preparations have also been considered a source of protein for artifactual transduction.[31] Some of the variability associated with rAAV transduction *in vivo* is undoubtedly due to intrinsic properties of the target cells. Some cells, for example, are relatively deficient in the high-affinity heparin proteoglycan receptor[22] and others may be incapable of efficiently synthesizing the transcriptionally active form of the rAAV genome.[28,32] However, a significant aspect of rAAV cell transduction variation undoubtedly relates to the methods used to purify rAAV and resultant contaminants present in the final preparation. In general, there has been a correlation between the success of rAAV vectors *in vivo* and the generation of high-titer, helper virus-free rAAV. Under optimal conditions, as few as 10–40 infectious particles of rAAV have been found to be sufficient to transduce one cell *in vivo*.[2,8,9]

The use of an iodixanol gradient rather than CsCl gradients as the initial rAAV purification from a crude lysate deserves examination because CsCl density centrifugation remains the standard in many laboratories. We encounter at least three problems during purification of rAAV using conventional CsCl centrifugation methods. First, rAAV is often nonspecifically bound to cellular protein and helper adenovirus. Such associations lead to an aggregation of rAAV, reducing yields and purity. Often the final rAAV stock contains cell remnants or serum proteins that may compromise subsequent interpretation of *in vivo* data by triggering immune responses. Second, the conventional purification method often produces viral preparations with particle-to-infectivity ratios greater than 1000, whereas ratios approaching 20:1 should be achievable[12] (Table I). Third, conventional purification takes up to 2 weeks to complete and often results in a substantially lower overall yield. In contrast, iodixanol has proved to be an excellent

---

[28] F. K. Ferrari, T. Samulski, T. Shenk, and R. J. Samulski, *J. Virol.* **70,** 3227 (1996).

[29] R. J. Samulski, L. S. Chang, and T. Shenk, *J. Virol.* **63,** 3822 (1989).

[30] D. W. Russell, A. D. Miller, and I. E. Alexander, *Proc. Natl. Acad. Sci. U.S.A.* **91,** 8915 (1994).

[31] I. E. Alexander, D. W. Russell, A. M. Spence, and A. D. Miller, *Hum. Gene Ther.* **7,** 841 (1996).

[32] K. J. Fisher, G. P. Gao, M. D. Weitzman, R. DeMatteo, J. F. Burda, and J. M. Wilson, *J. Virol.* **70,** 520 (1996).

bulk purification method that accomplishes three things: first, rAAV from the crude lysate is purified at least 100-fold. Even if helper adenovirus helper is present, this contamination is reduced by a factor of 100. Second, rAAV is purified and concentrated in a nonionic and relatively nonviscous medium that can be loaded onto virtually any kind of chromatographic matrix. Finally, for reasons that remain unclear, iodixanol prevents rAAV aggregation and the associated loss of virus accompanying most other bulk purification and column chromatography methods. Typically, 70–80% of the starting infectious units are recoverable after iodixanol gradient fractionation and, unlike other purification methods, this step is reproducible. In combination with final purification on heparin resins, iodixanol gradients therefore allow a fast, simple, and reproducible rAAV purification protocol, one that is amenable to scale-up and yields stocks of high titer and purity.

Acknowledgments

This work was supported by the National Institutes of Health, The Foundation Fighting Blindness, March of Dimes, Macular Vision Foundation/Ronald McDonald House and Research to Prevent Blindness Inc.

[49] Ribozymes in Treatment of Inherited Retinal Disease

By Lynn C. Shaw, Patrick O. Whalen, Kimberly A. Drenser, Weiming Yan, William W. Hauswirth, and Alfred S. Lewin

Introduction

The medical applications of ribozymes were recognized soon after RNA catalysis was discovered in the early 1980s.[1] RNA enzymes, or ribozymes, promote a variety of reactions involving RNA and DNA molecules including site-specific cleavage, ligation, polymerization, and phosphoryl exchange.[2] Naturally occurring ribozymes fall into three broad classes: (1) RNase P, (2) self-splicing introns, and (3) self-cleaving viral agents. RNase P is required for tRNA processing. Self-splicing introns include the group I and II introns of bacteria, mitochondria, and chloroplasts. Self-cleaving agents include hepatitis delta virus and components of plant viroids that

---

[1] T. R. Cech, *JAMA* **260**, 3030 (1988).
[2] T. R. Cech and B. L. Bass, *Annu. Rev. Biochem.* **55**, 599 (1986).

sever the RNA genome as part of a rolling-circle mode of replication. More recently, a process known as *in vitro* selection has permitted the generation of ribozymes of broader catalytic capacity and more varied sequence.[3,4]

Both RNase P[5,6] and group I ribozymes[7,8] have promise for gene therapy, but because of their small size and great specificity, the self-cleaving ribozymes have the greatest potential for medical applications. The ability of these ribozymes to cleave other RNA molecules at specific sites makes them useful as inhibitors of viral replication or of cell proliferation.[9-20] A phase I clinical trial is underway to test the safety of ribozyme therapy for human immunodeficiency virus type 1 (HIV-1) infection.[21]

The catalytic properties and greater specificity of ribozymes give them potential advantage over antisense RNA. In a head-to-head study comparing antisense RNA and ribozymes with identical target sequences, ribozymes inhibited viral replication 2- to 10-fold better than the comparable antisense molecules.[22] For the degradation of mRNA molecules whose protein products lead to retinal degeneration, the ability of ribozymes to

---

[3] D. P. Bartel and J. W. Szostak, *Science* **261,** 1411 (1993).
[4] T. Tuschl, P. A. Sharp, and D. P. Bartel, *EMBO J.* **17,** 2637 (1998).
[5] F. Liu, and S. Altman, *Genes Dev.* **9,** 471 (1995).
[6] C. Guerrier-Takada, R. Salavati, and S. Altman, *Proc. Natl. Acad. Sci. U.S.A.* **94,** 8468 (1997).
[7] B. A. Sullenger and T. R. Cech, *Nature (London)* **371,** 619 (1994).
[8] J. T. Jones and B. A. Sullenger, *Nature Biotechnol.* **15,** 902 (1997).
[9] J. J. Rossi, E. M. Cantin, J. A. Zaia, P. A. Ladne, J. Chen, D. A. Stephens, N. Sarver, and P. S. Chang, *Ann. N.Y. Acad. Sci.* **616,** 184 (1990).
[10] J. J. Rossi, D. Elkins, N. Taylor, J. Zaia, S. Sullivan, and J. O. Deshler, *Antisense Res. Dev.* **1,** 285 (1991).
[11] M. Koizumi, H. Kamiya, and E. Ohtsuka, *Gene* **117,** 179 (1992).
[12] J. O. Ojwang, A. Hampel, D. J. Looney, F. Wong-Staal, and J. Rappaport, *Proc. Natl. Acad. Sci. U.S.A.* **89,** 10802 (1992).
[13] M. Kashani-Sabet, T. Funato, T. Tone, L. Jiao, W. Wang, E. Yoshida, B. I. Kashfinn, T. Shitara, A. M. Wu, and J. G. Moreno, *Antisense Res. Dev.* **2,** 3 (1992).
[14] N. R. Taylor and J. J. Rossi, *Antisense Res. Dev.* **1,** 173 (1991).
[15] F. von Weizsacker, H. E. Blum, and J. R. Wands, *Biochem. Biophys. Res. Commun.* **189,** 743 (1992).
[16] M. Yu, J. Ojwang, O. Yamada, A. Hampel, J. Rapapport, D. Looney, and F. Wong-Staal, *Proc. Natl. Acad. Sci. U.S.A.* **90,** 6340 (1993).
[17] Z. Xing and J. L. Whitton, *J. Virol.* **67,** 1840 (1993).
[18] M. Yu, E. Poeschla, O. Yamada, P. Degrandis, M. C. Leavitt, M. Heusch, J. K. Yees, F. Wong-Staal, and A. Hampel, *Virology* **206,** 381 (1995).
[19] M. Kiehntopf, M. A. Brach, T. Licht, S. Petschauer, L. Karawajew, C. Kirschning, and F. Herrmann, *EMBO J.* **13,** 4645 (1994).
[20] E. Little and A. S. Lee, *J. Biol. Chem.* **270,** 9526 (1995).
[21] F. Wong-Staal, E. M. Poeschla, and D. J. Looney, *Hum. Gene Ther.* **9,** 2407 (1998).
[22] R. Hormes, M. Homann, I. Oelze, P. Marschall, M. Tabler, F. Eckstein, and G. Sczakiel, *Nucleic Acids Res.* **25,** 769 (1997).

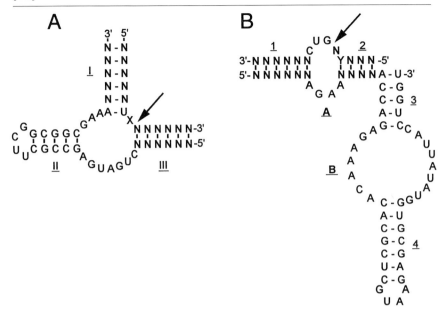

FIG. 1. Schematic secondary structures of the hammerhead (A) and the hairpin (B) ribozymes. Arrows indicate the cleavage sites in target RNAs. N can be any nucleotide, X is any nucleotide but guanosine, and Y indicates a pyrimidine nucleotide. The helices in the hammerhead ribozyme are numbered with roman numerals, and the helices in the hairpin are designated with arabic numerals. The large loops in the hairpin are designated A and B.

selectively cleave the mutant mRNA is their chief advantage over antisense oligonucleotides. While a single-base mismatch is sufficient to block the cleavage of wild-type mRNA by a ribozyme, a similar mismatch in an antisense molecule will have little impact: both wild-type and mutant messages will be bound and their translation into protein inhibited.

Two kinds of ribozyme have been widely employed: hairpins and hammerheads. Both were originally derived from the satellite RNA of tobacco ringspot virus, and each derives its name from its schematic secondary structure (Fig. 1). Both catalyze sequence-specific cleavage resulting in products with a 5'-hydroxyl and a 2',3'-cyclic phosphate. Hammerhead ribozymes have been used more commonly, because they impose few restrictions on the target site.[23] The basic strategy is first to select a region in the RNA of interest that contains the triplet GUC or, in general, NUX, where N stands for any nucleotide and X is any nucleotide except guanosine. Then it is necessary to create two stretches of antisense nucleotides 6 to 8

[23] K. R. Birikh, P. A. Heaton, and F. Eckstein, *Eur. J. Biochem.* **245**, 1 (1997).

nucleotides long that flank the 21-nucleotide sequence forming the catalytic hammerhead between them. The principle is the same for hairpin ribozymes, except that the catalytic core is larger (34 nucleotides) and this type of ribozyme requires more specificity in the target site.[24] Hairpins recognize the sequence NNNYNGUCNNNNNN, where N is any nucleotide and Y is a pyrimidine. *In vitro* selection and site-specific mutagenesis have been used to identify nucleotides essential for activity of both types of ribozyme and to broaden the target range for both hairpins and hammerheads.[25–32] Hairpin ribozymes are more stable and, consequently, function better than hammerheads at physiologic temperature and magnesium concentrations. Replacing a naturally occurring 3-base loop in the catalytic core with a tetraloop increases the stability of the hairpin ribozyme.[18,33] In addition, a U-to-C mutation at position 39 of the hairpin increases activity twofold compared with the original sequence.[33]

We have successfully tested both hammerheads and hairpins for gene therapy of rat models of autosomal dominant retinitis pigmentosa (ADRP).[34,35] The underlying hypothesis is that ribozymes can reduce the synthesis of mutated proteins whose accumulation eventually leads to apoptosis of rod photoreceptors. The issues involved in developing ribozyme therapy for dominant retinal disease include (1) the existence of a target sequence created by the disease-causing mutation or located near the relevant mutation, (2) ribozyme accessibility of that target sequence with respect to RNA secondary structure, and (3) delivery of ribozymes in a cell type-specific manner. Our strategy has been to express ribozymes in a cell type-specific manner, using the promoter from the rod opsin gene. The delivery system we have used, adeno-associated virus (AAV),

---

[24] D. J. Earnshaw and M. J. Gait, *Antisense Nucleic Acid Drug Dev.* **7**, 403 (1997).
[25] A. Berzal-Herranz, S. Joseph, and J. M. Burke, *Genes Dev.* **6**, 129 (1992).
[26] J. M. Burke and A. Berzalherranz, *FASEB J.* **7**, 106 (1993).
[27] S. Joseph and J. M. Burke, *J. Biol. Chem.* **268**, 24515 (1993).
[28] P. Anderson, J. Monforte, R. Tritz, S. Nesbitt, J. Hearst, and A. Hampel, *Nucleic Acids Res.* **22**, 1096 (1994).
[29] P. Degrandis, A. Hampel, S. Galasinski, J. Borneman, A. Siwkowski, and M. Altschuler, *PCR Methods Appl.* **4**, 139 (1994).
[30] R. Shippy, A. Siwskowski, and A. Hampel, *Biochemistry* **37**, 564 (1998).
[31] N. K. Vaish, P. A. Heaton, and F. Eckstein, *Biochemistry* **36**, 6495 (1997).
[32] N. K. Vaish, P. A. Heaton, O. Fedorova, and F. Eckstein, *Proc. Natl. Acad. Sci. U.S.A.* **95**, 2158 (1998).
[33] B. Sargueil, D. B. Pecchia, and J. M. Burke, *Biochemistry* **34**, 7739 (1995).
[34] K. A. Drenser, A. M. Timmers, W. W. Hauswirth, and A. S. Lewin, *Invest. Ophthalmol. Vis. Sci.* **39**, 681 (1998).
[35] A. S. Lewin, K. A. Drenser, W. W. Hauswirth, S. Nishikawa, D. Yasumura, J. G. Flannery, and M. M. LaVail, *Nature Med.* **4**, 967 (1998).

is discussed in [48] in this volume.[35a] An excellent handbook for ribozyme technology is available and contains detailed descriptions for the production and testing of several types of ribozyme.[36] We confine our discussion to methods tested on retinal targets.

Identification of Target Sites

While hammerhead ribozymes are flexible with respect to target sites, not all target triplets (NUX) are equally susceptible to cleavage.[37] In bimolecular reactions, the cleavage site triplet GUC is cleaved five times more efficiently than the next best sequence (CUC), which is cleaved at more than twice the rate of the triplet UUC. Cleavage of AUU is more than 100-fold less efficient than of GUC. Fortunately, the most easily cleaved triplets are relatively common in mRNA. For example, the coding region of the human rod opsin gene contains 22 UUC triplets, 17 GUC triplets, and 14 CUC triplets distributed throughout the sequence. Selection of a target site, then, involves identifying one of these triplets near the site of a sequence alteration leading to a dominant genetic mutation. As an example, we have targeted a hammerhead ribozyme toward the mRNA present in a P23H rhodopsin rat model of ADRP[34,38] (Fig. 2A).

Ribozymes designed to discriminate between mutant and wild-type sequences should have relatively short hybridizing arms, the ribozyme sequences that are complementary to the target sequence and that comprise helix I and helix III of the secondary structure. Hertel *et al.* have analyzed the specificity of the hammerhead ribozyme in a bimolecular reaction.[39] They concluded that a recognition sequences of up to 12 nucleotides (in helix I and helix III together) could be used to discriminate a 1-nucleotide mismatch in one of the hybridizing arms. However, ribozymes with longer arms can distinguish between mismatched and fully complementary targets if the mismatch affects the cleavage triplet or is adjacent to it.

Hairpin ribozymes require a GUC triplet in a more constrained context (see above) and the length of one of the hybridizing arms is fixed: while helix 1 can be of variable length, helix 2 is limited to 4 base pairs and is separated from helix 1 by an 8-nucleotide loop (loop A).[28,40] Hairpins can

---

[35a] W. W. Hauswirth, A. S. Lewin, S. Zolotukhin, and N. Muzyczka, *Methods Enzymol.* **316**, Chap. 48, 2000 (this volume).

[36] Anonymous, "Ribozyme Protocols." Humana Press, Totowa, New Jersey, 1997.

[37] T. Shimayama, S. Nishikawa, and K. Taira, *Biochemistry* **34**, 3649 (1995).

[38] M. I. Naash, J. G. Hollyfield, M. R. Al-Ubaidi, and W. Baehr, *Proc. Natl. Acad. Sci. U.S.A.* **90**, 5499 (1993).

[39] K. J. Hertel, D. Herschlag, and O. C. Uhlenbeck, *EMBO J.* **15**, 3751 (1996).

[40] A. Hampel, R. Tritz, M. Hicks, and P. Cruz, *Nucleic Acids Res.* **18**, 299 (1990).

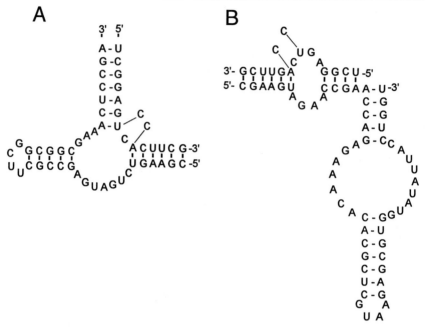

FIG. 2. Diagrams of the hammerhead (A) and the hairpin (B) ribozymes used to target the P23H opsin mRNA.[34] The positions of cytosine nucleotides present in wild-type rhodopsin are indicated.

cut mismatched targets with reduced cleavage efficiency, provided that the mismatches are distal to the cleavage site.[41] Therefore, in targeting a mutation in mRNA, it is desirable to keep helix 1 short to promote selective cleavage (Fig. 2B). While selection of the optimal target will result in increased ribozyme activity on oligonucleotide targets *in vitro,* Yu *et al.* used a combinatorial library of ribozymes to demonstrate that target accessibility determines which ribozymes are most active in cells.[42] Whether a possible target site is exposed or concealed (by RNA secondary structure) should be determined experimentally, because RNA-folding algorithms are not reliable for long sequences.[43,44]

[41] S. Joseph, A. Berzalherranz, B. M. Chowrira, S. E. Butcher, and J. M. Burke, *Gene Dev.* **7,** 130 (1993).
[42] Q. Yu, D. B. Pecchia, S. L. Kingsley, J. E. Heckman, and J. M. Burke, *J. Biol. Chem.* **273,** 23524 (1998).
[43] T. B. Campbell, C. K. McDonald, and M. Hagen, *Nucleic Acids Res.* **25,** 4985 (1997).
[44] K. R. Birikh, Y. A. Berlin, H. Soreq, and F. Eckstein, *RNA* **3,** 429 (1997).

Materials

Plasmids, Nucleotides, and Oligonucleotides

Transcription vectors such as pT7T3-19 (GIBCO-BRL, Gaithersburg, MD) and pBluescriptII KS+ (Stratagene, La Jolla, CA) contain promoter sequences for bacteriophage RNA polymerases such as T7 RNA polymerase. Unlabeled nucleoside triphosphates should be obtained from a high-quality vendor and purchased as a buffered stock solution (typically 100 mM). Labeled nucleotides for ribozyme assays can be purchased from Amersham (Arlington Heights, IL), ICN (Irvine, CA), or Du Pont/NEN (Boston, MA). DNA oligonucleotides encoding the ribozyme sequences with appropriate flanking restriction sites can be ordered from a variety of commercial sources. RNA oligonucleotides to serve as targets for ribozyme cleavage assays can be ordered from fewer companies; we have used Dharmacon (Boulder, CO). RNA oligonucleotides are protected on their 2′ residues, and must be deblocked according to the manufacturer recommendations prior to use.

Chromatography and Electrophoresis

Sephadex G-25 fine or G-50 fine resins are purchased directly from Pharmacia (Piscataway, NJ). Reagents for denaturing electrophoresis [acrylamide, methylene(bis)acrylamide, urea] should be highly purified and are available from a variety of companies. Silica gel thin-layer chromatography (TLC) plates for UV-shadowing should contain fluorescent (F_{254}) indicator. These plates are available from Fisher Scientific (Pittsburgh, PA) and other vendors.

Enzymes and Reagent Chemicals

T7 RNA polymerase, DNA polymerase, restriction endonucleases, and polynucleotide kinase are available from several manufacturers including GIBCO-BRL and New England BioLabs (Beverly, MA). Recombinant placental ribonuclease inhibitor can be obtained as RNasin from Promega (Madison, WI). Pure water is critical for RNA experiments. Deionized water should be purified by a filtration and ion-exchange chromatography system such as Barnstead NanoPure II. We steam-sterilize such water and use it directly without diethyl pyrocarbonate (DEPC) treatment. Should RNase contamination prove a problem, water can be treated for 1–15 hr with DEPC at 0.02% final concentration prior to autoclaving.

Methods

Cloning of Ribozymes and Targets

Coding sequences for ribozymes are generated by extension of two overlapping synthetic DNA oligonucleotides coding for the ribozyme sequence and flanked by restriction sites. The large fragment of DNA Pol I (Klenow; NEB) is used to fill out the DNA duplexes. Overlapping oligonucleotides are heated to 65° for 2 min and then annealed by slow cooling to room temperature over 30 min. The annealed oligonucleotides prime each other and are mutually extended by DNA polymerase in the presence of 5mM deoxynucleoside triphosphates and polymerase buffer [10 mM Tris-HCl (pH 7.5), 5 mM MgCl$_2$, 7.5 mM dithiothreitol] for 1 hr at 37°. The fully duplex fragments are then digested and ligated into the T7 RNA polymerase expression plasmid of choice. Ligated plasmids are used to transform *Escherichia coli* cells; clones are screened by colony hybridization and verified by sequencing. In addition to producing clones for active ribozymes this way, clones for inactive ribozymes may be prepared by mutating one or more of the nucleotides essential for catalysis by hammerhead or hairpin ribozymes.[25,45] Inactive ribozymes are useful in distinguishing physiological effects that depend on catalytic activity.

Clones for the production of target RNA can be prepared in a similar fashion by mutually primed synthesis, or can consist of partial or full-length cDNA clones. The latter are particularly useful in determining whether a ribozyme target site is accessible in the full-length mRNA.

In Vitro Transcription

Plasmid DNAs containing target sequences or ribozyme genes are linearized with a restriction enzyme that cuts downstream of the sequence to be transcribed. Endonucleases and contaminating ribonucleases are inactivated by extraction with phenol–chloroform–isoamyl alcohol (50:50:1, v/v/v) followed by ethanol precipitation. Transcripts are generated with T7 (or other) RNA polymerase and labeled by incorporation of [α-^{32}P]UTP (ICN, Costa Mesa, CA) using the protocol of Grodberg and Dunn.[46] A 100-μl reaction contains

Autoclaved distilled water	48 μl
Sodium phosphate buffer (pH 7.7), 100 mM	20 μl
Template DNA (0.4–0.8 μg) in 10 mM Tris-HCl–0.1 mM EDTA (pH 7.5)	1 μl
Dithiothreitol (DTT, 1 M)	4 μl
MgCl$_2$ (40 mM), 20 mM spermidine hydrochloride	20 μl
NTP mix (40 mM; 40 mM each of ATP, CTP, GTP, and UTP, freshly mixed from 100 mM stocks)	4 μl
RNasin (Promega)	1 μl
T7 RNA polymerase (5 units)	1 μl
[α-^{32}P]UTP or [α-^{32}P]CTP (3000 Ci/mmol; 10 μCi/μl)	1 μl

[45] D. E. Ruffner, G. D. Stormo, and O. C. Uhlenbeck, *Biochemistry* **29**, 10695 (1990).
[46] J. Grodberg and J. J. Dunn, *J. Bacteriol.* **170**, 1245 (1988).

For labeling of target molecules, a high specific activity is desirable, and the level of unlabeled UTP or CTP is reduced fourfold. (Reducing unlabeled nucleoside triphosphates further severely reduces the yield of full-length transcripts.) For labeling of ribozymes, the labeled nucleotide is diluted 10-fold, and the unlabeled nucleotide is used at 40 mM in the stock, because the radioactivity serves only as a tracer to allow quantitation of ribozyme. Reactions are routinely performed at 37° for 2 hr, although labeling will continue under these conditions for at least 6 hr.

After synthesis, transcription reactions should be brought to 0.5% (w/v) sodium dodecyl sulfate (SDS), extracted with phenol–chloroform–isoamyl alcohol (50:50:1,v/v/v), and passed through a 1-ml spin column of Sephadex G-50.[47] Hammerhead ribozymes or target transcripts are then ethanol precipitated, washed twice with 70% (v/v) ethanol, and resuspended in 10 mM Tris HCl–0.1 mM EDTA, pH 7.4. Ethanol precipitation may denature or lead to irreversible aggregation of hairpin ribozymes. If the concentration is kept low (<1 nM), heating and slow cooling should restore activity of denatured hairpins.

Labeling of RNA Oligonucleotides

Protecting groups are removed from the 2' positions of RNA oligonucleotides according to the manufacturer instructions, and oligonucleotides (typically 14-mers) are assembled with the labeling reagents as follows:

RNA oligonucleotide (2.5 nmol)	10 μl
[γ-^{32}P]ATP	1 μl
Kinase buffer [0.7 M Tris-HCl (pH 7.6), 0.1 M MgCl$_2$, 50 mM DTT]	2 μl
RNasin (diluted 1:10 from stock in 10 mM DTT)	1 μl
Polynucleotide kinase (10 units)	1 μl
Sterile water	5 μl

The mixture is incubated at 37° for 30 min and then is diluted to 100 μl with sterile water and extracted with phenol–chloroform–isoamyl alcohol (50:50:1, v/v/v) to inactivate the enzyme. Unincorporated nucleotides are removed by passing the aqueous phase over a Sephadex G-25 spin column. The sample can be used as is or can be ethanol precipitated, keeping the salt (ammonium acetate) concentration 200 mM or less. Addition of Dextran Blue as a carrier enhances recovery of small RNA molecules, without affecting their cleavage.

Gel Purification

Ribozymes and target RNA molecules produced by *in vitro* transcription may require purification by gel electrophoresis to assure homogeneity.

---

[47] H. S. Penefsky, *J. Biol. Chem.* **252,** 2891 (1977).

Ribozymes or larger targets (>30 nucleotides) can be purified on 8% (w/v) acrylamide, 8 M urea sequencing gels run in TBE buffer [89 mM Tris–borate (pH 8.3), 20 mM EDTA].[48] Labeled molecules are visualized by autodradiography, excised with a sterile scalpel, and eluted from the gel in 1 M ammonium acetate 50 mM Tris-HCl (pH 7.5), 20 mM EDTA, 0.5% (w/v) SDS at 37° for 1–4 hr. If gel purification is omitted, the transcript should be treated with RNase-free DNase I to remove DNA sequences that may anneal to the target or to the ribozyme.

RNA oligonucleotides require purification if the synthesis reaction or the deblocking step are incomplete. Purification should precede labeling so that the recovery of the purified nucleotide can be determined before labeling. One hundred nanomoles of deprotected oligonucleotide is suspended in sterile water, mixed with an equal volume of formamide loading dye [90% (v/v) formamide, TBE buffer, 0.4% (w/v) xylene cyanol, and 0.4% (w/v) bromphenol blue] and separated on a 20% (w/v) acylamide, 8 M urea gel run in TBE. After the bromphenol blue tracking dye has run two-thirds of the gel, electrophoresis is stopped and the gel is transferred to a 20 × 20 cm silica gel TLC plate containing F_{254} fluorescent indicator. The TLC plate should be covered with plastic wrap to prevent transfer of the powder to the gel. In a dark room, using a short-wavelength UV hand lamp, the RNA band can be detected as a dark band against the fluorescent background. A 14- or 15-nucleotide molecule will run just above the bromphenol blue dye. This band is excised and eluted as described above except that carrier is omitted. The ethanol precipitate of the oligonucleotide is dissolved in sterile water (100 μl), and the concentration is determined by absorbance at 260 nm.

Retinal RNA Extraction

Total RNA from the retina is conveniently recovered by extraction with Trizol (GIBCO-BRL) according to the manufacturer recommendations. Retinas can be snap frozen in liquid nitrogen prior to RNA extraction and stored at −70°. A single mouse or rat retina is resuspended in 1 ml of Trizol reagent and homogenized by repeated passage though an 18-gauge syringe needle. After a 5- to 15-min incubation at room temperature, the suspension is extracted with 0.2 ml of chloroform. After centrifugation to separate phases, RNA is precipitated from the aqueous phase. A typical preparation from rat results in 100 mg of total RNA.

Specific Activity

To determine the specific radioactivity of transcripts and oligonucleotides, small samples (1–5 μl) are spotted on glass fiber filters, dried, and

---

[48] A. M. Maxam and W. Gilbert, *Proc. Natl. Acad. Sci. U.S.A.* **74**, 560 (1977).

analyzed in a liquid scintillation counter with scintillation fluid. Counting efficiency is presumed to be 90%. For end-labeled oligonucleotides, the specific activity is determine by dividing the disintegrations per minute by the concentration of oligonucleotide determined by UV absorbance. This may be converted to microcuries per mole by dividing by 2.22×10^6 dpm/μCi. For transcribed molecules, the moles of cold UTP used in the transcription reaction (40 or 10 nmol) is divided by the radioactivity in microcuries of the [α-^{32}P]UTP added (10 or 1 μCi). (Radioactivity is determined by the concentration of radioactivity on the vial times the decay factor). This number is divided by the number of uridine residues in the transcript, and the result divided by 2.22×10^6 dpm/μCi. The dividend is multiplied by the disintegrations per minute in the sample to determine moles of transcript in the sample.

Ribozyme Cleavage Reactions

Specific radioactivity of each molecule is used to calculate the concentration of target and ribozyme molecules. Initial cleavage conditions consist of 50 nM substrate RNA and 10 nM ribozyme, 20 mM MgCl$_2$, 40 mM Tris-HCl, pH 7.5, and incubation at 37°. Prior to incubation with substrate, hammerhead ribozymes are renatured by incubation in 40 mM Tris-HCl (pH 7.4), 10 mM MgCl$_2$ at 37° for 15 min to several hours. (Short periods of renaturation are usually adequate, but some ribozymes need to incubate in magnesium buffer for hours to achieve full potency). Hairpin ribozymes are denatured at 85° for 2–3 min in 0.1 mM EDTA and renatured at room temperature for 5–15 min. Ribozyme and target are then mixed in digestion buffer and preincubated for several minutes, and cleavage is initiated by the addition of MgCl$_2$ to 20 mM. Reactions can be stopped by adding a one-tenth reaction volume of 0.5 M EDTA followed by an equal volume of formamide loading dye (see above). Experimental conditions such as time of incubation, magnesium concentration, and ribozyme concentration should be varied independently. The magnesium dependence of ribozymes can be lowered by addition of spermine (0.5 mM). Ribozymes requiring high magnesium concentrations or high ribozyme concentration for cleavage may be eliminated from further analysis at this stage.

Cleavage products for target RNAs produced by *in vitro* transcription are analyzed by electrophoresis on 8 M urea, 8 or 10% (w/v) acrylamide sequencing gels run in TBE. Digests of oligonucleotides are analyzed on 20% (w/v) gels. Gels should be fixed in 10% (v/v) acetic acid, 10% (v/v) methanol (for 10% gels) or in 40% methanol, 10% acetic acid, 3% glycerol for high-percentage gels. Gels may be imaged by autoradiography with X-ray film, but quantitation is performed with a radioanalytic phosphorescent screen scanner. We have used a PhosphorImager from Molecular Dynamics

(Sunnyvale, CA) and similar instruments are available from Fuji Medical Electronics (Tokyo, Japan) and from Bio-Rad (Hercules, CA). If such an instrument is not available, cleavage can be quantitated by counting the radioactivity in gel slices. Conversion of the PhosphorImager output to disintegrations per minute can be made by exposing a slot blot of serial dilutions of a sample of known radioactivity (usually the target molecule) to the same screen as the gel. The fraction cleaved is determined from ratio of radioactivity in the product to the sum of the radioactivity in the product and the original target:

$$\text{Fraction cleaved (FC)} = P/(P + T)$$

The reaction velocity is the fraction cleaved per minute.

Multiple Turnover Kinetic Analysis

Analyses to determine multiple turnover kinetic constants are typically carried out in 20 mM MgCl$_2$, 40 mM Tris-HCl, pH 7.5, at 37° for short intervals. The appropriate interval is determined by a time-course experiment under multiple turnover conditions (i.e., substrate excess). Initial rates should be measured when the amount of cleavage is linear with time and when no more than 10% of substrate has been converted to product. Rates are measured at several intervals (e.g., 5, 10, and 20 min) to ensure linearity. Samples are preincubated at 37° prior to initiation of cleavage and contain 1–10 nM ribozyme and increasing concentrations of substrate RNA, holding the ribozyme concentration constant. Substrate concentrations should greatly exceed ribozyme concentration, the lowest being in fivefold excess. Values for V_{max} and K_m are obtained by double-reciprocal plots of velocity versus substrate concentration (Lineweaver–Burke plots) or by plots of reaction velocity versus the ratio of velocity to substrate concentration (Eadie–Hofstee plots) (Fig. 3). The turnover number K_{cat} is determined by dividing V_{max} by the ribozyme concentration: $k_{cat} = V_{max}/[\text{Rz}]$. The values for k_{cat} and K_m cannot be determined accurately unless the saturating substrate level (which yields V_{max}) is approached. This level can be difficult to achieve with cloned targets if the K_m of the ribozyme is high. In such a case, synthetic oligonucleotides are recommended.[49]

Single Turnover Kinetics

The rate of the cleavage step of a ribozyme (k_{obs}), as distinct from the overall rate depending on association and dissociation steps, can be determined in ribozyme excess experiments. This is often a useful statistic

---

[49] M. B. DeYoung, A. Siwkowski, and A. Hampel, in "Ribozyme Protocols" (P. C. Turner, ed.), pp. 209–220. Humana Press, Totowa, New Jersey, 1997.

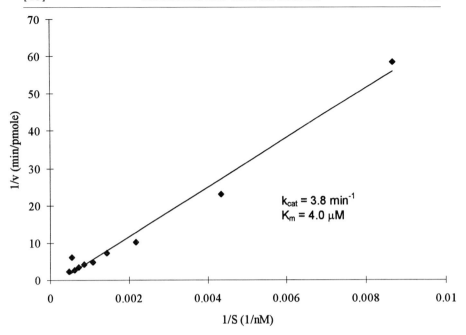

FIG. 3. Multiturnover kinetics of a P347S-specific hammerhead ribozyme. The target is a synthetic 14-nucleotide target corresponding to the P347S target present in transgenic pigs (F. Wong, personal communication). Kinetic constants based on this Lineweaver–Burke plot are shown.

in comparing a new ribozyme with others that have been used successfully in the past. The relative amounts of ribozyme and substrate must be experimentally determined, but a 20-to-1 ratio is a reasonable starting condition. (Rates should not vary with ribozyme concentration if an adequate excess has been achieved.) Ribozymes are renatured and mixed with substrate as described above, except that a single large reaction (e.g., 50 μl) is set up and samples (e.g., 4 μl) are withdrawn at various intervals. At least one sample should be taken after a long incubation to serve as an estimate of the fraction cleaved at $t = \infty$. Reactions are stopped with EDTA and loading buffer as described and analyzed on denaturing acylamide gels. The fraction cleaved at each time point should be determined using a PhosphorImager (or similar instrument). If the substrate is fully complexed with the ribozyme at the outset of the experiment, cleavage should follow an exponential curve (Fig. 4). The data fit an equation of the type

$$Fc_t = Fc_\infty - [\exp(-k_{obs}t)Fc_\Delta]$$

where Fc_t is the fraction cleaved at time t, Fc_∞ is the fraction cleaved after

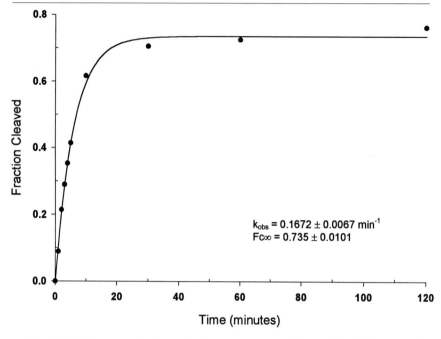

FIG. 4. Single turnover kinetics using the same target and hammerhead ribozyme as in Fig. 3. The first-order rate constant and fraction cleaved at $t = \infty$ are indicated.

a long interval (t_∞), k_{obs} is the first-order rate constant for the reaction, and Fc_Δ is the difference in fraction cleaved between t_0 and t_∞.

Reverse Transcription-Polymerase Chain Reaction

Retinal messenger RNA can be reverse transcribed using avian mycloblastosis virus (AMV) reverse transcriptase (Superscript; GIBCO-BRL) and oligo(dT) primer. To distinguish wild-type opsin mRNA from mutant transgene mRNA we have used a three-primer system for PCR amplification, in which the downstream primer anneals to both the mutant and wild-type cDNA and two separate upstream primers anneal to either the mutant or wild-type cDNA.[34] This reaction results in PCR products of different lengths from wild-type and mutant mRNA. To detect RT-PCR products, [α-^{32}P]dATP, included during the final PCR cycle. The fragments are analyzed by electrophoresis on 4 or 5% (w/v) nondenaturing polyacrylamide gels. Radioactivity of PCR fragments is quantitated by by radioanalytic scanning with a PhosphorImager. An additional primer set to amplify an invariant mRNA should be added to control for differences in sample preparation. We use β-actin as an internal standard.

As an alternative to distinguishing mutant retinal genes from wild-type using distinctive primers, restriction site differences between the mutant and wild-type cDNA can be employed if they exist. We have used this method to distinguish between RT-PCR products resulting from heterozygous *rd* mice.[50] To obtain a quantitative estimate of product based on restriction enzyme digestion, one must account for the fraction of heteroduplex molecules formed during the last stages of PCR. Apostolakos *et al.*[51] describe a method for making this correction, and it is described in the context of retinal RNA in the article by Yan *et al.*[50]

To test ribozyme cleavage of full-length opsin mRNA in the context of total retinal RNA, one simply precedes the RT-PCR with a ribozyme cleavage incubation or with a mock incubation. Cleavage assays with the total retinal RNAs typically contain 0.1 mg of total RNA extract and 20–50 nM ribozyme.

Discussion

For ribozyme gene therapy of dominant mutations to be applicable, expression from one allele must maintain the normal phenotype. Many loss-of-function mutations are dominant because of haploinsufficiency: too little protein is produced from the remaining wild-type allele to preserve function. Fortunately, for rhodopsin and other visual cycle proteins, one normal allele seems to be adequate to preserve vision.[52] It is theoretically possible, therefore, to limit the expression of the dominant mutant allele, permitting the wild-type gene to direct the synthesis of opsin.

The major limitation in applying ribozyme technology to eliminate the consequences of dominant mutations, such as those leading to autosomal dominant retinitis pigmentosa, is the distance between mutation and cleavage site: if the cleavage site is more than 3 or 4 residues away from the altered base, then selective degradation of the mutant mRNA may not be possible. Note that the base change need not be that leading to RP; a linked, silent mutation is a perfect target. Indeed, the ribozymes we have employed in the rat model of ADRP (Fig. 2) cleave following a silent mutation (C to U) in codon 22 that is linked to the C-to-A transversion in

---

[50] W. Yan, A. Lewin, and W. Hauswirth, *Invest. Ophthalmol. Vis. Sci.* **39**, 2529 (1998).

[51] M. J. Apostolakos, W. H. Schuermann, M. W. Frampton, M. J. Utell, and J. C. Willey, *Anal. Biochem.* **213**, 277 (1993).

[52] M. M. Humphries, D. Rancourt, G. J. Farrar, P. Kenna, M. Hazel, R. A. Bush, P. A. Sieving, D. M. Sheils, N. McNally, P. Creighton, A. Erven, A. Boros, K. Gulya, M. R. Capecchi, and P. Humphries, *Nature Genet.* **15**, 216 (1997).

codon 23. However, the target site problem has led other investigators[53,54] to propose an alternative ribozyme strategy: generating ribozymes that destroy both mutant and wild-type mRNA while, at the same time, delivering a novel functional allele lacking the ribozyme target site.

Detailed kinetic analysis is not absolutely required for the development of therapeutic ribozymes. The prediction of RNA structure is problematic, however, and ribozymes that look reasonable on paper may fold improperly in solution.[55,56] Therefore, it is advisable to test ribozyme turnover under standard conditions *in vitro*, before any "heavy lifting" (testing the ribozyme in animals) is attempted. In general, we have found the k_{cat} determined in 10 or 20 mM MgCl$_2$ to be a good predictor of the relative activity of ribozymes *in vivo*. Ribozymes must also be tested on long RNA substrates, transcripts of cDNA clones, or mRNA extracted from tissue, to be sure that the intended cleavage site is exposed and not part of a stable helical domain.

Finally, the rate of even the most active ribozyme pales in comparison to the turnover number of protein ribonucleases. Consequently, for application as therapeutic agents, ribozymes must be expressed at a high level. We have relied on the proximal rod opsin promoter[35,57] to drive the synthesis of ribozymes in rod photoreceptor cells. Other have used RNA Pol III promoters, because these are normally responsible for the synthesis of small nuclear RNA and transfer RNA,[58] but Pol II promoters are more certain to produce a transcript that trafficks to the cytoplasm with its intended mRNA target.[59] For mRNA targets, therefore, active, cell type-specific promoters are recommended.

Acknowledgments

This work was supported by grants from the National Institutes of Health, the March of Dimes Foundation, the Foundation Fighting Blindness, Macular Vision Foundation and Research to Prevent Blindness, Inc.

[53] S. Millington-Ward, B. O'Neill, G. Tuohy, N. Al-Jandal, A. S. Kiang, P. F. Kenna, A. Palfi, P. Hayden, F. Mansergh, A. Kennan, P. Humphries, and G. J. Farrar, *Hum. Mol. Genet.* **6,** 1415 (1997).
[54] J. Sullivan and K. Pietras, *Invest. Ophthalmol. Vis. Sci.* **39,** S722 (1998). [Abstract]
[55] H. A. Heus, O. C. Uhlenbeck, and A. Pardi, *Nucleic Acids Res.* **18,** 1103 (1990).
[56] M. J. Fedor and O. C. Uhlenbeck, *Proc. Natl. Acad. Sci. U.S.A.* **87,** 1668 (1990).
[57] J. G. Flannery, S. Zolotukhin, M. I. Vaquero, M. M. LaVail, N. Muzyczka, and W. W. Hauswirth, *Proc. Natl. Acad. Sci. U.S.A.* **94,** 6916 (1997).
[58] A. Lieber and M. Strauss, *Mol. Cell. Biol.* **15,** 540 (1995).
[59] E. Bertrand, D. Castanotto, C. Zhou *et al.*, *RNA* **3,** 75 (1997).

[50] Cross-Species Comparison of *in Vivo* Reporter Gene Expression after Recombinant Adeno-Associated Virus-Mediated Retinal Transduction

By JEAN BENNETT, VIBHA ANAND, GREGORY M. ACLAND, and ALBERT M. MAGUIRE

Introduction

Technologic advances have been made that allow somatic transfer of functional genes to target cells in the retina *in vivo*. Although the retinal cell gene transfer field is still embryonic, the promising results to date ensure that this technology will be applied in future studies to evaluate a variety of biochemical, cell biological, developmental, and physiological events involving wild-type and mutant retinal proteins. In addition, gene transfer techniques will likely be harnessed to provide therapy in retinal degenerative diseases such as retinitis pigmentosa, as proofs of principle of such approaches have been achieved in relevant animal models for these diseases.[1-5] While a number of vectors are presently available to deliver genes (and their respective proteins) to the retina, one of the most promising vectors at this time is recombinant adeno-associated virus (rAAV). The methods described here aim to provide the reader with information relevant to performing subretinal rAAV-mediated gene transfer in eyes of several different species of mammals. The transgene cassette used in the present study is a cDNA encoding green fluorescent protein (GFP) driven by a cytomegalovirus (CMV) promoter. Although any gene that fits into the rAAV vector could be used in such a study, the advantage of the *GFP* transgene is that the presence of its protein product can be evaluated noninvasively as a function of time after injection. A comparative study reveals a similar cellular specificity of transgene expression, stability of transgene expression, and immunological response to the rAAV across

---

[1] J. Bennett, T. Tanabe, D. Sun, Y. Zeng, H. Kjeldbye, P. Gouras, and A. Maguire, *Nature Med.* **2,** 649 (1996).

[2] A. S. Lewin, K. A. Drenser, W. W. Hauswirth, S. Nishikawa, D. Yasumura, J. G. Flannery, and M. M. LaVail, *Nature Med.* **4,** 967 (1998).

[3] W. M. Peterson, J. G. Flannery, W. W. Hauswirth, R. L. Klein, E. M. Meyer, N. Muzyczka, D. Yasumura, M. T. Matthes, and M. M. LaVail, *Invest. Ophthalmol. Vis. Sci.* **39,** S1117 (1998).

[4] R. Kumar-Singh and D. B. Farber, *Invest. Ophthalmol. Vis. Sci.* **39,** S1118 (1998).

[5] J. Bennett, Y. Zeng, R. Bajwa, L. Klatt, Y. Li, and A. M. Maguire, *Gene Ther.* **5,** 1156 (1998).

species. However, differences in delivery methods and time of onset of rAAV-mediated transgene expression are noted.

Methods

Virus Preparation

In preparing and evaluating purified recombinant virus, the investigator should assure that all manipulations are performed in accordance with institutional and national biosafety guidelines. The transgene cassette used in the comparative studies described here consists of the fluorescence-enhanced (EGFP) variant of the GFP-encoding cDNA (Clontech, Palo Alto, CA) under control of a CMV promoter. This is cloned into the plasmid psub201[6] in order to flank the transgene cassette by the rAAV inverted terminal repeats (ITRs).

The approach for generating rAAV-*GFP* is to package the recombinant DNA into AAV particles by complementation with a system expressing AAV *rep* and *cap* genes. This is performed by a method described originally by Samulski and colleagues: cotransfection with a *cis* plasmid containing the transgene cassette flanked by rAAV ITRs and a *trans* plasmid containing the *rep* and *cap* genes and infection of the cells with recombinant adenovirus at a low multiplicity of infection.[7–10] A number of improvements on this system have been reported, which generate high-titer purified virus in large scale. For example, plasmid vectors have been developed that contain all AAV and adenovirus functions required for amplification and packaging of AAV vector plasmids.[11,12] These can be co-transfected into cells, thereby avoiding potential contamination with helper virus. Another approach uses a cell line stably expressing the AAV genes *rep* and *cap* and an adenovirus–AAV (Ad/AAV) hybrid vector.[13,14] This method results in high-titer rAAV

[6] R. J. Samulski, L.-S. Chang, and T. Shenk, *J. Virol.* **61**, 3096 (1987).
[7] J. Bennett, D. Duan, J. Engelhardt, and A. M. Maguire, *Invest. Ophthalmol. Vis. Sci.* **38**, 2857 (1997).
[8] D. Duan, K. J. Fisher, J. F. Burda, and J. F. Engelhardt, *Virus Res.* **48**, 41 (1997).
[9] L. Dudus, V. Anand, S.-J. Chen, J. Wilson, K. J. Fisher, G. Acland, A. M. Maguire, and J. Bennett, *Vision Res.* **39**, 2545 (1999).
[10] R. J. Samulski, L.-S. Chang, and T. Shenk, *J. Virol.* **63**, 3822 (1989).
[11] D. Grimm, A. Kern, K. Rittner, and J. A. Kleinschmidt, *Hum. Gene Ther.* **9**, 2745 (1998).
[12] R. W. Herzog, E. Y. Yang, L. B. Couto, J. N. Hagstrom, D. Elwell, P. A. Fields, M. Burton, D. A. Bellinger, M. S. Read, K. M. Brinkhous, G. M. Podsakoff, T. C. Nichols, G. J. Kurtzmann, and K. A. High, *Nature Med.* **5**, 56 (1999).
[13] G.-P. Gao, L. Z. Faust, R. K. Engdahl, G. Qu, C. D. Nguyen, E. M. Krapf, T. J. Miller, W. Xiao, J. V. Hughes, and J. M. Wilson, *Hum. Gene Ther.* **9**, 2353 (1998).
[14] X. Ye, V. M. Rivera, P. Zoltick, F. Crasoli, Jr., M. A. Schnell, G.-P. Gao, J. V. Hughes, M. Gilman, and J. M. Wilson, *Science* **283**, 88 (1999).

free of replication-competent AAV. Other advantages of this system are that there is high production of vector per cell (resulting in titers typically of 5×10^{13} particles/ml) and that it is easy to scale up the procedure to obtain large quantities of the high-titer virus. Whatever the exact method used to generate the virus, quality control is essential. It is important to determine whether there are any contaminants (including replication-competent AAV and adenovirus). Contaminating adenovirus can be removed through heat denaturation and sedimentation through cesium. It is currently difficult to remove contaminating replication-competent AAV.

Contamination with replication-competent AAV can be assessed by Western blot or immunohistochemical analyses of Rep and Cap proteins.[9] Alternatively, 293 cells can be infected with the rAAV in the presence of adenovirus. The DNA harvested from the resulting lysate can be analyzed by Southern analysis, using an *rep* probe. Rep, reflecting the presence of wild-type AAV, should not be detected.

Contamination with adenovirus vectors can induce a strong cell-mediated immune response that can result in diminished levels of transgene expression and immune-mediated damage.[15] Purified rAAV-*GFP* can be tested for adenovirus contamination, using a variety of procedures including Southern analysis, Western analysis, polymerase chain reaction (PCR), and evaluation of cytopathic effects in infected cells in culture.[7]

Finally, it is important to verify the titer of the purified virus. This can be determined by infectious unit assay on 293 cells or another appropriate cell line and by slot-blot analysis of virion DNA particles.[9,16]

Intraocular Administration of Recombinant Virus

Any evaluation of the effect of rAAV on animals should be performed in accordance with the ARVO Statement for the Use of Animals in Ophthalmic and Vision Research and with federal, state, and local regulations. The minimal number of animals should be used to obtain statistically significant results. Aseptic technique should be used and all reagents and instruments that come in contact with the eye should be sterile. Animals should be appropriately anesthetized.

The surgical approach to subretinal injection depends mainly on the size of the eye. In animals with large eyes (such as the rabbit, dog, or primate) the relative volume of the vitreous space occupied by the crystalline lens is small. This allows for instruments to be introduced via an anterior

---

[15] Y. Yang, F. A. Nunes, K. Berensci, E. E. Furth, E. Gonczol, and W. J. M., *Proc. Natl. Acad. Sci. U.S.A.* **91**, 4407 (1994).

[16] K. J. Fisher, G.-P. Gao, M. D. Weitzman, R. DeMatteo, J. F. Burda, and J. M. Wilson, *J. Virol.* **70**, 520 (1996).

approach across the vitreous cavity (Fig. 1A and B). The area of retina to be injected is observed by direct visualization. A cannula tip penetrates the retina in the area to be injected and the injection is performed, causing a localized retinal detachment (Fig. 1B). The details are as follows: the

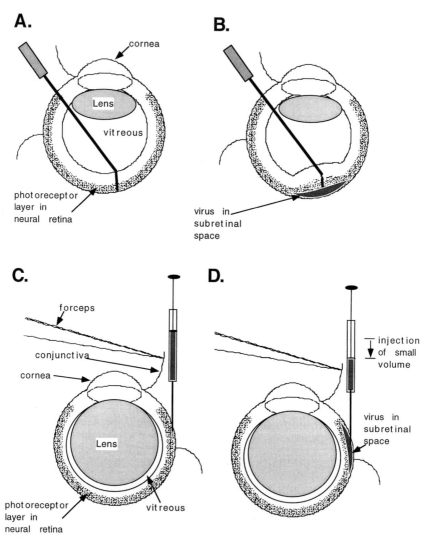

FIG. 1. Technique for subretinal injection is dependent on the size of the eye and percentage of vitreous cavity occupied by the lens. (A and B) Method for large eye (e.g., rabbit); (C and D) method for small eye (e.g., mouse).

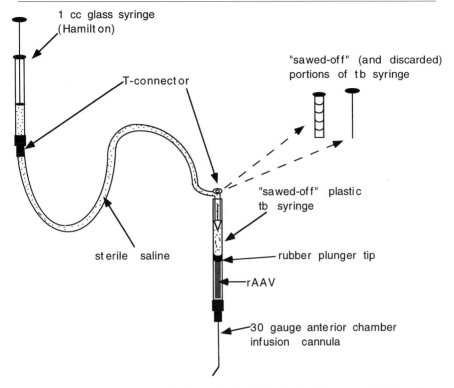

FIG. 2. Syringe set-up recommended for subretinal injection in a large eye. tb, Tuberculin.

pupils are dilated with 1% (w/v) Mydriacyl (tropicamide) and 2.5% (w/v) Mydfrin (phenylephrine hydrochloride) and a fornix-based conjunctival peritomy is performed in one quadrant to expose the underlying sclera. The external sclera is lightly treated by cautery at the incision site. An anterior chamber paracentesis is performed with a 30-gauge needle to soften the eye prior to injection (and to obtain fluid, if desired, for baseline intraocular antibody measurements). The injection apparatus is made by a modification of that described by Wallace and Vander.[17] The syringe/tubing assembly is not available commercially and must be assembled by the investigator. Briefly, a 30-gauge anterior chamber cannula (Storz) is connected to a syringe containing the material to be injected. This syringe is in turn connected through flexible extension tubing and a T connector (Abbott Hospitals, North Chicago, IL) with a second syringe containing saline (Fig. 2). The cannula is then inserted through a sclerotomy incision

[17] R. T. Wallace and J. F. Vander, *Retina* **13**, 260 (1993).

and a 27-gauge needle is used to incise the sclera approximately 2 to 3 mm posterior to the corneoscleral limbus. As shown in Fig. 2, the syringe/tubing assembly allows the surgeon flexibility in directing the needle to the target region. An assistant holds the saline-filled syringe. A 30-gauge blunt-tipped anterior chamber infusion cannula is introduced through the scleral incision into the vitreous. Direct visualization is achieved with coaxial illumination from an operating microscope as a plano convex fundus contact lens is held to the corneal surface. The tip of the 30-gauge cannula is gently pressed against the retina in a tangential orientation and the assistant briskly injects a defined amount of fluid from the saline-containing syringe. This, in turn, results in a controlled injection of purified rAAV-*GFP* into the subretinal space (Fig. 1A). Retinal vessels should be assessed to assure that they remain well perfused during and after the procedure. Ointment (preferably PredG; Allergan Pharmaceuticals, Irvine, CA) should be applied to the corneas to minimize drying of this tissue while the animal recovers from anesthesia.

In large animals, there is a risk of increased intraocular pressure if more than 100 μl of solution is injected into the subretinal space. A recommended dose is 3×10^{11} infectious units (IU); however, pilot experiments evaluating the effect of dose should be performed for each vector/species to be tested.

In smaller eyes, e.g., rodents, the vitreous space is largely occupied by the crystalline lens. Introduction of surgical instruments via an anterior approach would either necessitate removal of the lens or cause a cataract as the instruments penetrate the lens capsule. Therefore, a posterior approach is used, in which an incision is made across the posterior sclera and choroid and a cannula is placed in the subretinal space (Fig. 1C and D). The injection is not performed by direct visualization, as the pupil is rotated away from the surgeon in order to expose the posterior part of the globe. Accuracy of the injection should be assessed after the procedure, as accidental delivery to the vitreal cavity can lead to transduction of ganglion cells.[9]

Details of the surgical procedures recommended for mice and other small rodents have been described previously.[18] Briefly, pupils are dilated by the topical application of 1 drop of Mydriacyl [1.0% (w/v) tropicamide; Alcon, Humacao, Puerto Rico]. The anesthetized mouse is positioned on the base of a dissecting microscope with the eye to be injected under view ($\sim \times 15$ magnification). Vannas iridotomy scissors and jeweler's forceps are used to incise the conjunctiva near the equator of the globe. The conjunctiva and underlying episcleral tissue are opened circumferentially (i.e., conjunc-

---

[18] J. Bennett, Y. Zeng, A. R. Gupta, and A. M. Maguire, in "Application of Adenoviral Vectors: Analysis of Eye Development" (R. S. Tuan and C. W. Lo, eds.). Humana Press, Totowa, New Jersey, in press (1999).

tival peritomy) to expose the underlying sclera (Fig. 1C). The conjunctiva adjacent to the cornea is grasped with forceps, providing traction to rotate the globe and allowing optimal surgical exposure. The tip of a sterile 30-gauge needle is passed obliquely through the sclera, choroid, and retina into the vitreous avoiding the lens. A bead of clear vitreous will prolapse through this scleral incision and occasionally the translucent edge of the retina can be seen internally. At this point, the globe is gently retracted to expose the sclerotomy incision. The tip of a 33-gauge blunt needle mounted on a 10-μl Hamilton syringe is introduced into the incision oriented tangentially to the neural retina (Fig. 1C and D). Guidance is aided by use of jeweler's forceps in the opposite hand. The 33-gauge needle passes through the sclera and the choroid (an external transscleral transchoroidal approach) and then terminates in the subretinal space. The surgeon steadies the hand holding the injection syringe on a fixed surface (e.g., roll of towels) as the assistant delivers the injection (Fig. 1D). After injection, ointment (preferably PredG; Allergan Pharmaceuticals) should be applied to the corneas to minimize drying of this tissue while the animal recovers from anesthesia.

The volume of solution to be injected into the subretinal space of mice should not exceed 1 μl. It is difficult to measure the exact amount that remains in this space as there is usually some leakage through the needle entry site. Expression of *GFP* is difficult if not impossible to detect ophthalmoscopically unless at least 1×10^8 IU ($\sim 1 \times 10^9$ particles) has been delivered.

In Vivo Assessment of GFP Expression

Animals can be monitored for *GFP* expression by indirect ophthalmoscopy at weekly intervals after injection as described.[7] Illumination through conventional cobalt blue filters used in clinical evaluations (e.g., for fluorescein angiography) will excite the GFP chromophore and result in emission of green fluorescence. If the ophthalmoscope is not equipped with a blue filter, a Wratten 47B gelatin excitation filter (Kodak, Rochester, NY) can be taped to the ophthalmoscope light source. This filter transmits blue light at 450–490 nm, which corresponds to excitatory wavelengths for GFP.

An alternative to ophthalmoscopy for identifying fluorescing cells is based on a modification of an inexpensive method described by Clyne and Hume.[19] An excitation filter is attached to a fiber optic light source. The animal is placed on the stage of a dissecting microscope and the dilated pupil is viewed while illuminating with the exciting wavelength.

---

[19] J. D. Clyne and R. I. Hume, *BioTechniques* **23,** 1018 (1997).

GFP, if present, is detectable ophthalmoscopically in pigmented animals only. We have not successfully detected the GFP protein *in vivo* in albino animals (even when expression is apparent later on histological examination). For large animals, retinas should be viewed through a 20 or 28 diopter lens. A 78 or 90 diopter lens is recommended for use with rodents. Pupils must be dilated prior to examination (see above for details about dilating drops).

Fundus photographs can be taken with a Kowa camera (Keeler Instruments, Broomall, PA) equipped with a Wratten 47B gelatin excitation filter. This is technically difficult as it requires the photographer to simultaneously focus the lens on the area to be photographed and focus the camera—all done in dim illumination.

Ophthalmoscopic evaluation of retinas injected subretinally with rAAV reveals that there is a different time of onset of expression in different species of animals (Fig. 3). Expression is apparent much earlier in mice and rabbits than in primates. The reason for the different times of onset in different species is unknown but may relate to conditions in the cell allowing the rAAV-delivered transgene cassette to transform from its initial single-stranded conformation to a transcriptionally competent double-stranded form.[16] As indicated in Fig. 3, there is variability in exact time of onset from animal to animal. The most variability is observed in monkeys. It should also be noted that expression levels are quite faint initially. They

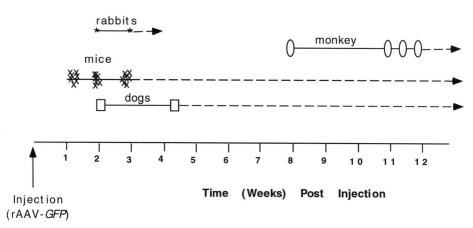

FIG. 3. Onset of expression in various animal species. Onset is defined as the time point at which green fluorescence is first observed ophthalmoscopically in animals receiving a subretinal injection of rAAV-*GFP*. The time of onset is indicated on the graph for individual animals. In monkeys, dogs, and mice, expression persists through the longest time point evaluated (indicated by dotted lines). Evaluation of long-term expression of *GFP* in rabbits is pending.

increase in intensity and ultimately plateau over a course of approximately 6 weeks. Expression persists for at least 2 years in mice, dogs, and primates, indicated in Fig. 3 by dashed lines with arrows. Long-term evaluations of expression in rabbits are pending.

Histological Analyses of GFP Transgene Expression

At necropsy, eyes are fixed with 4% (w/v) paraformaldehyde in phosphate-buffered saline as described.[18] The optimal method for evaluating GFP expression is to cryoprotect the tissue and prepare frozen sections. The GFP protein is remarkably stable, however, and can be appreciated under a variety of different fixation/processing conditions. The protein can be visualized after fixation with 10% (v/v) neutral buffered formalin and fixative containing glutaraldehyde. Glutaraldehyde is not recommended, however, as this fixative results in a high amount of background autofluorescence. GFP also persists through the dehydration and xylene incubation steps necessary for evaluating paraffin-embedded sections. Background fluorescence is high, however, after paraffin processing. To distinguish GFP-specific fluorescence from background fluorescence, sections should be evaluated not only after illumination with a fluorescein isothiocyanate (FITC) filter (which excites the GFP chromophore) but also with a filter that transmits light outside the wavelengths that excite GFP (e.g., rhodamine). This will define regions that may appear, under illumination with the FITC filter alone, to possess GFP but that are, in fact, autofluorescent. GFP will be apparent only after illumination with the FITC filter, not after illumination with the control (rhodamine) filter.

Accurate subretinal injection in mice, rabbits, dogs, and monkeys results in transduction of photoreceptors and variable numbers of retinal pigment epithelial (RPE) cells.[7-20a] Occasional cells of the inner retina can also be transduced. Shown in Fig. 4 are representative results 3 weeks after subretinal injection of rAAV-*GFP* in a mouse.

Evaluation of Humoral Immune Response

Exposure to wild-type or recombinant AAV can induce a strong antibody response.[9] Although this, in of itself, is not toxic, the presence of neutralizing antibody due to prior exposure to AAV might limit successful rAAV-mediated transduction events. The presence of rAAV-specific antibodies can also limit additional transduction events on readministration of

---

[20] C. Grant, S. Ponnazhagan, X.-S. Wang, A. Srivastava, and T. Li, *Curr. Eye Res.* **16,** 949 (1997).
[20a] J. Bennett, A. M. Maguire, A. V. Cideciyan, M. Schnell, E. Glover, V. Anand, T. S. Aleman, N. Chirmule, A. R. Gupta, Y. Huang, G. P. Gao, W. C. Nyberg, J. Tazelaar, J. Hughes, J. M. Wilson, and S. G. Jacobson, *PNAS* **96,** 9920 (1999).

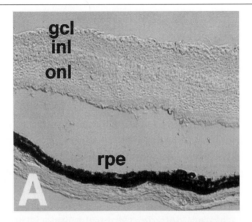
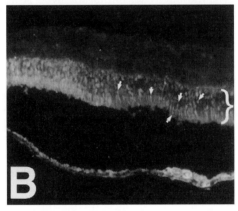

FIG. 4. Cell specificity and efficiency of rAAV after subretinal injection is similar across different species. Subretinal injection targets primarily photoreceptors and some RPE cells. Shown are *GFP*-expressing cells, 3 weeks after subretinal injection of 5×10^{10} IU of rAAV-*GFP* in a 1-month-old C57BL/6 mouse. (A) View of cryosection by light microscopy; (B) view of same sample shown in (A) after illumination through an FITC filter. Photoreceptors and RPE cells express *GFP*. This was confirmed through illumination with a rhodamine filter as described in text (not shown); retinal pigment epithelium, rpe; onl, outer nuclear layer; inl, inner nuclear layer; gcl, ganglion cell layer; arrows: individual *GFP*-expressing photoreceptors. Bracket in (B) indicates that GFP is found in all of the layers occupied by photoreceptors including the outer plexiform layer, outer nuclear layer, and inner and outer segment layers.

rAAV.[21,22] Therefore, in experiments involving readministration of rAAV, it may be desirable to identify and quantify rAAV-specific antibodies. This can be achieved by enzyme-linked immunosorbent assay (ELISA) of serum or intraocular fluid from experimental animals.

Serum from large animals can be collected by an ear vein puncture (rabbits) or by leg vein puncture (monkeys or dogs). Serum should be separated from the other blood components by centrifugation and stored at 80° until time of analysis. Serum samples can be obtained from mice by tail vein puncture. Collection is aided considerably by warming the tail under a heat lamp (for example, a 250-W infrared heat lamp), taking care that the animal does not become overheated. Collection of blood from the retro-orbital sinus is not recommended as that often causes damage to ocular structures.

While the small size of the rodent eye prohibits sampling of intraocular fluid, up to 100 μl of fluid from the anterior chamber can be obtained from the large eyes of the rabbit, dog, or monkey. Care must be taken when introducing the needle (30-gauge connected to a 1-ml syringe) through the limbus under direct visualization; avoid contact with the lens and corneal endothelium. An anterior chamber tap can result in mild inflammation lasting up to 2 weeks. Therefore, this procedure is not recommended prior to a fundus photography session.

To analyze samples for antibodies to viral capsid proteins, enhanced protein-binding ELISA plates are coated overnight at 4° with antigen (purified AAV). Alternatively, plates can be coated with purified AAV capsid (VP1, VP2, VP3) structural proteins. Plates are then washed, blocked, and incubated with diluted (1:100, 1:400, 1:1600) serum (or anterior chamber fluid). Saline is used as negative control for the mice. Positive controls can consist of serum from human patients previously found to have high levels of anti-AAV antibodies or from previously immunized experimental animals. Samples are incubated with a 1:1000 dilution of a peroxidase-conjugated goat anti-mouse IgG [Jackson ImmunoResearch (West Grove, PA); 100 μl/well] or alkaline phosphatase-conjugated sheep anti-mouse IgG [Sigma (St. Louis, MO); 100 μl/well] and washed. The Sigma fast p-nitrophenyl phosphate substrate system is used for the color reaction and the plates are read at an optical density of 405 nm. Results from ELISA analyses of samples obtained from mice, rabbits, dogs, and monkeys demonstrate that subretinal injection results in a small but significant increase in systemic antibodies to rAAV capsid protein (Fig. 5A).[9] Increase is primarily in the

---

[21] C. Halbert, T. Standaert, M. Aitken, I. E. Alexander, D. W. Russell, and A. D. Miller, *J. Virol.* **17**, 5932 (1997).

[22] X. Xiao, J. Li, and R. J. Samulski, *J. Virol.* **70**, 8098 (1996).

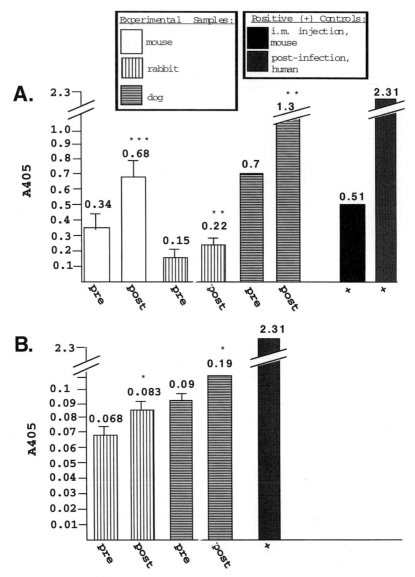

FIG. 5. ELISA for IgG to AAV in sera from mice, rabbits, and dogs receiving intraocular injection of rAAV-*GFP*. Cohorts of mice received a subretinal injection of 1×10^9 IU of rAAV-*GFP* and rabbits and dogs received a subretinal injection of 1.9×10^{10} IU of rAAV-*GFP*. (A) There is a small but significant increase in serum anti-rAAV capsid antibody (as measured by A_{405}) over control in all animals. Data from monkeys is pending (Bennett *et al.*[20a]). (B) There is an even smaller increase in intraocular fluid anti-rAAV capsid antibodies after subretinal injection [note The difference in scale of the *y* axes between (A) and (B)]. Positive controls include a mouse receiving the same amount of rAAV by intramuscular injection (as described by Fisher *et al.*[24]) and serum from a human with antibodies versus AAV. Significant differences compared with negative controls are indicated as follows: *p = 0.05; **p = 0.003; ***p < 0.0001.

Ig(G_{2b} + G_3) class of antibodies, thereby indicating that the immune response directed against AAV capsid proteins is a helper T cell type 2 (Th2) (or Th1-independent) response. As would be predicted given the Th2 response, there is a lack of immune cell infiltrate after subretinal injection of rAAV.[7,9,20,23] The rise in systemic antibodies after subretinal injection does not lead to a parallel increase in intraocular antibodies. Variable results are obtained with respect to intraocular antibodies in all species. (Ref. 9; and V. Anand and J. Bennett, unpublished data, 1998), but generally there is only a small increase in intraocular antibodies after subretinal injection (Fig. 5B).

Acknowledgments

This work was supported by a Research to Prevent Blindness Departmental Award, NEI RO1s EY10820 and EY12156 (J.B.), NEI Center Grant EY06855 (G.M.A.), a Center Grant from Foundation Fighting Blindness, Inc.; AHA F98223E (V.A.), the Pennsylvania Lions Sight Conservation and Eye Research Foundation (J.B.); the Paul and Vebanina Mackall Foundation Trust and the F. M. Kirby Foundation. The authors are grateful to Dr. James Wilson for studies involving monkeys (supported by NIH Grant P30 DK47757-05).

[23] J. Flannery, S. Zolotukin, M. Vaquero, M. LaVail, N. Muzyczka, and W. Hauswirth, *Proc. Natl. Acad. Sci. U.S.A.* **94,** 6916 (1997).

[24] K. J. Fisher, K. Jooss, J. Alston, Y. Yang, S. E. Haecker, K. High, R. Pathak, S. E. Raper, and J. M. Wilson, *Nature Med.* **3,** 306 (1997).

Author Index

Numbers in parentheses are footnote reference numbers and indicate that an author's work is referred to although the name is not cited in the text.

A

Abdelilah, S., 536
Abdulaev, N. G., 87, 98, 98(4), 99, 488
Aberijon, C., 346
Abrahamson, E. W., 70
Abrahamson, M., 625
Abramow-Newerly, W., 714
Achatz, H., 688
Acland, G. M., 519, 520, 705, 707(7), 777, 778, 779(9), 782(9), 785(9), 787(9), 789(9)
Adachi, A., 279, 280(2), 282(2), 283, 283(2), 284
Adam, J., 585
Adami, G. R., 698
Adams, A. J., 69
Adamus, G., 104, 107(12), 168
Adler, A. J., 235, 324
Adler, J., 453
Adler, R., 519, 530, 579, 590, 593(6), 599(6), 672
Aebersold, R. H., 493
Affatigato, L. M., 591
Afione, S. A., 743
Agarwal, N., 590, 634, 672
Agarwal, R. P., 87, 92
Aghazdeh, B., 464
Aguirre, G. D., 519, 520, 705, 707(7)
Aharon, A., 84, 85(61)
Aitken, M., 787
Akino, T., 452, 456, 457(6), 458, 462, 462(36), 463(36), 465, 466(4, 6), 467(1, 4, 6–8), 468(4, 6), 469(4, 6, 8), 475(8), 476, 476(4, 6), 478(4, 7), 481(1, 4)
Albert, A. D., 65, 66(6), 67, 68(6), 69(22), 71(6), 72, 75(6), 86
Aleman, T. S., 625
Alemany, R., 725
Alex, L. A., 22

Alexander, C. A., 519, 520
Alexander, I. E., 760, 787
Alizon, M., 82, 83(56)
Allen, A. C., 624, 672
Allen, C. B., 534
Allen, N. D., 655
Allen, R. A., 348, 565
Allen, R. G., 117
Alloway, G. P., 82, 83(56)
Alones, V., 282
Alonso, A., 82
Alpern, M., 519, 520
Alston, J., 743, 788(24), 789
Altman, S., 762
Altschuler, M., 764
Altshuller, Y. M., 592
Al-Ubaidi, M. R., 51, 519, 524, 765
Alvarez, R. A., 313, 316, 324, 327, 383, 388, 567, 574(16), 705, 722, 722(10)
Amalfitano, A., 731
Amalric, P., 706, 709(21), 711(21), 712(21), 713(21)
Amaya, E., 50, 58(14), 59(14), 61(14), 62(14), 294
Ames, A., 99
Ames, G. F., 492, 493(5)
Ames, J. B., 13, 34, 112(32), 113, 121, 122, 126, 127(15), 128, 129, 130(17), 131(7, 17)
Amundson, D., 104, 107(12)
Anand, V., 777, 778, 779(9), 782(9), 785(9), 787(9), 789, 789(9)
Anant, J. S., 453
Anderson, D. H., 65
Anderson, J. L., 204
Anderson, P., 764, 765(28)
Anderson, W. F., 127
Andersson, M., 726
Andreasson, S., 625, 688
Andrews, E. J., 567

Andrews, G. L., 579, 590(14)
Andrews, J. S., 316, 387
Angus, C. W., 120
Ankoudinova, I., 558
Anonymous, 764
Anotnny, B., 254
Antoch, M. P., 167
Anton, M., 730, 732(22)
Apfelstedt-Sylla, E., 611, 618(2), 621, 671
Apostakos, M. J., 775
Appel, R. D., 494, 505(21)
Applebury, M. L., 169, 283, 725
Arai, K., 285
Arai, N., 285
Araki, E., 519, 520
Arden, G. B., 612, 625, 671, 687
Arendt, A., 168
Argos, P., 372, 373(10)
Arikawa, K., 69, 673
Armes, L. F., 107
Armes, L. G., 107(21), 108
Arnaud, B., 706, 709(21), 711, 711(21), 712(21, 29), 713(21)
Arshavsky, V. Y., 22, 39(6), 167, 193
Artamonov, I. D., 283, 284, 488
Artemiev, N., 22, 39(6)
Arulanantham, P. R., 31
Arumi, M., 625
Asano, M., 105
Asano, T., 464, 465, 466(5), 467, 467(5, 9), 471(5), 473(5), 475, 476(5), 477, 478(5, 9), 479, 481(5, 9)
Asato, A. E., 429
Asch, W. S., 581
Asenjo, A. B., 629, 632(25, 28), 633(28), 638(28), 649(28)
Ashby, M. N., 436
Asson-Batres, M. A., 107, 116(19), 119(19), 363, 364(20), 365(18), 366(18, 20), 367(18), 372
Atrian, S., 372
Attwell, D., 196
Aubourg, P., 665
Ausubel, F. M., 600, 606(24), 691
Azarian, S. M., 519, 520, 672

B

Babu, Y. S., 126, 127(28)
Bachner, D., 519, 520

Bacon, D. J., 127
Bader, C. R., 188
Badwey, J. A., 464
Baehr, W., 103, 104, 121, 122, 122(2), 132, 331, 515, 519, 520, 521, 524, 524(6), 530, 558, 765
Baek, D. J., 425, 428(11, 12)
Baer, B., 411
Baizer, L., 104, 107, 107(12), 116(19), 117, 119(19), 120
Bajwa, R., 777
Baker, M. E., 378
Baldisseri, D. M., 112(33), 113
Baldwin, A. N., 34
Baldwin, J. M., 414
Baldwin, V., 519, 520, 705, 707(7)
Ball, L. E., 435
Balogh-Nair, V., 427, 428, 428(15), 430(15, 17), 431(17), 433(17)
Balzarini, J., 753
Bandarchi, J., 148, 157(16), 158(16), 161(16), 162(16), 163(16), 247
Bang, Y., 592
Banin, E., 611(12), 612, 616(12), 619(12), 621(12), 625
Baraban, J. M., 742
Barad, M., 99
Bard, J. B., 535
Bareil, C., 706, 709(21), 711, 711(21), 712(21, 29), 713(21)
Barenholtz, Y., 70, 85(38)
Barlow, H. B., 643
Barlow, R. B., 41, 49(5)
Barnett, K. C., 519, 520
Barnstable, C. J., 282, 590(10), 591, 719
Barr, M., 690, 691(10)
Barry, G. F., 380
Barry, R. J., 340, 370, 401, 706
Bartel, D. P., 762
Bartel, P. L., 120, 689, 697, 698(21)
Barteling, S. J., 109
Barthel, L. K., 579, 581, 582, 582(15), 585(15), 590(15, 16)
Bartlett, G. R., 67, 82(24)
Bascom, R. A., 672
Bass, B. L., 761
Basu, D. K., 146, 159
Batni, S., 50, 51(13), 52
Battisti, L., 633, 655, 659(14), 660(14), 669

AUTHOR INDEX

Bauer, K., 28
Bautista, J. M., 741
Båvik, C.-O., 348, 565, 706, 719(14)
Baxter, J. G., 315
Baxter, L. C., 725
Bayley, H., 270, 271
Baylor, D. A., 3, 123, 149, 167, 168(1, 2), 178, 178(1), 180(31), 184, 185, 187, 188, 189, 189(3), 190, 193, 195(7), 197, 199(7, 9, 21), 201(9, 21), 203, 214(19), 219(19), 222(19), 223, 226, 228, 233, 243, 247, 250(9), 253, 519, 520, 546, 549(16), 560, 618, 629, 643, 648(56)
Beach, D., 694
Beapalov, I. A., 488
Beatty, D. D., 383
Beaudet, A. L., 724, 726
Becherer, K., 689
Bech-Hansen, N. T., 646
Beck, A., 688, 691(2), 692(2)
Beckingham, K., 112
Beddington, R., 172
Bedolli, M. A., 687
Belak, J., 536
Belecky-Adams, T., 579, 672
Bellett, A. J., 732
Bellinger, D. A., 726, 778, 785(12)
Bellingham, J., 590(8), 591
Belval, B. J., 754
Benditt, E. P., 582
Bennett, B. T., 567
Bennett, J., 611, 621(9), 625, 625(9), 687, 688, 725, 777, 778, 779(7, 9), 782, 782(9), 783(7), 785, 785(7, 9, 18), 787(9), 788(20a), 789, 789(7, 9)
Bennett, N., 167, 425
Benovic, J. L., 23, 24(9), 25(9)
Bentz, J., 75, 76, 78, 80, 84(41), 86
Berencsi, K., 726, 779
Bergen, A. A. B., 688, 691(2), 692(2)
Berger, E. A., 66
Berger, S. J., 87, 530
Berger, T., 277
Berger, W., 519, 520, 688, 691(2), 692(2)
Berlin, Y. A., 766
Berman, E. R., 383
Bernon, D. E., 535
Berns, K. I., 744
Bernstein, P. S., 325, 401, 706
Berson, E. L., 180, 519, 520, 561, 618, 619, 671, 672(5), 676, 706, 712(22), 713(22), 722(22)
Bertrand, D., 188
Bertrand, E., 776
Bertsch, T. W., 50
Berx, G., 710, 713(28)
Berzal-Herranz, A., 764, 765, 768(25)
Besch, D., 621
Besharse, J., 50, 51(16), 52, 58(15), 59(15), 64(15), 65, 85(8), 296, 302(16)
Bessant, D., 591
Bessant, D. A. R., 132, 625
Beuchle, D., 536
Bhakdi, S., 270, 277
Bhattacharya, S. S., 132, 428, 515, 519, 520, 590(8), 591, 625
Bickel, M. H., 388
Biemann, K., 488, 489(13)
Bienvenue, A., 65
Bier, E., 585
Bigay, J., 27, 461, 467
Billeci, T., 510
Bingham, E. L., 519
Birch, D. G., 202, 203, 204, 205(22–24), 207(9), 208, 208(24), 210(5, 6, 9), 211(24), 212, 212(12, 24), 213(18), 214(22, 24), 215(24), 216(22–24), 218(9), 219(22, 23, 31), 221(24), 222(22–24), 223, 519, 520, 521, 561, 591, 612, 613, 614, 616, 618(33), 624, 672, 688
Birch, E. E., 208, 219(31)
Bird, A. C., 132, 559, 564(12), 591, 621, 625, 687, 688
Bird, D., 687
Birge, R. R., 425, 429
Birikh, K. R., 763, 766
Bishop, J. M., 347
Biswas, A., 398
Biswas, M. G., 365, 368(21), 371(21)
Bitensky, M. W., 742
Bitgood, J. J., 534, 536
Blackshaw, S., 279
Blader, P., 585
Blaner, W. S., 371, 385, 400
Bleasby, A. J., 508, 510(29)
Bleeker-Wagemakers, E. M., 688, 691(2), 692(2)
Blinger, B., 28
Blum, H. E., 762
Blum, W. P., 315

Blumenthal, R., 68, 69(28), 75, 84, 85(61)
Boatright, J. H, 530
Bodoia, R. D., 196
Boehm, M. F., 425, 428(11, 12)
Boehringer GmbH, 580
Boer, T. D., 66, 67(21), 71(21)
Boerman, M. H., 371, 372, 378(4)
Boesze-Battaglia, K., 65, 66, 66(6), 67, 68(6), 69(22), 70, 70(9, 12), 71(6, 9, 10, 12), 72, 73, 74(9, 10, 12), 75(6, 9–13), 76(9, 10), 85(9, 10, 13), 621, 673
Bogerd, L., 692
Boggs, C., 536
Boikov, S., 131, 132, 132(33)
Bojkova, N., 429
Bok, D., 325, 331, 359, 397, 400, 401, 402, 403(16), 406, 410(16), 467, 519, 520, 565, 566(3), 572, 672, 707, 715(24), 723(24)
Bongard, M. M., 628
Bongiorno, A., 383
Boni, L. T., 86
Bonneau, D., 713
Bonnemaison, M., 534, 559, 713
Bonner, L., 467
Bonting, S. L., 313, 315, 318(9), 344, 359
Bordoli, R. S., 107
Borhan, B., 429, 431
Borneman, J., 764
Boros, A., 182
Borst, D. E., 530
Bouchard, D., 24
Bounds, D., 235, 557
Bourne, H. R., 455
Bourret, R. B., 22
Boutros, M. C., 592
Bovee, P. H, 425
Bowes, C., 725
Bowmaker, J. K., 629, 632(22), 648(22), 649
Bownds, D., 313, 314(10), 315, 316(10), 317(10), 339, 422
Bownds, M. D., 3(12), 4, 13(12), 14, 22, 34, 39(6), 123
Boyartchuck, V. L., 436
Boyce, J. J., 742
Boyce, J. R., 567
Brach, M. A., 762
Bradford, M. M., 332
Bradley, A., 559
Bradley, M. E., 397, 402
Brady, K. D., 591

Brand, M., 536
Braun, K., 103
Brautigan, D. L., 137
Brazil, D., 455, 462, 464(44), 465
Bredberg, D. L., 331, 332, 358, 360, 363, 364(16), 365(16, 18), 366(18), 367(18), 370, 372, 388, 401
Brennan, M. B., 519, 520, 621, 672
Brent, R., 600, 606(24), 689, 691
Breton, M. E., 203, 214(10), 221(10), 613, 616
Bridge, M., 85
Bridges, C. D. B., 50, 51(5), 228, 313, 315, 316, 324, 327, 383, 388, 567, 574(16), 705, 722, 722(10)
Brigham-Burke, M., 29
Brindley, G. S., 631, 648
Brinkhous, K. M., 726, 778, 785(12)
Brinkley, P., 728, 731(20)
Brinster, R. L., 172
Brockerhoff, S. E., 536, 537, 538(10), 542(3)
Broder, C. C., 66
Brolley, D., 3, 13, 13(9), 14(9), 111, 121
Bronson, D., 121, 122
Bronson, J. D., 515
Bronson, R. T., 714
Bronstein, I., 701
Brothers, C. A., 24
Brothers, G. M., 24
Brousseau, R., 24
Brown, B., 24, 25(13), 40(13)
Brown, J., 41
Brown, M. F., 29
Brown, P. K., 235, 313, 314(10), 315, 316(10), 317(10), 339, 422, 476, 557, 631
Bruckert, F., 254
Brunger, A. T., 126
Brunner, M., 84
Bruno, J., 530
Bucy, R. P., 579, 582(10)
Buczylko, J., 19, 104, 107(11, 22), 108, 109, 167, 270, 359, 360, 363, 364(16), 365(16, 18), 366(18), 367(18), 372
Bugg, C. E., 126, 127(28)
Bumstead, N., 536
Bumsted, K., 579
Buraczynska, K., 688
Buraczynska, M., 688
Burda, J. F., 760, 778, 779, 785(8)
Burgers, A. C. J., 306
Burke, J. M., 764, 765, 766, 768(25)

AUTHOR INDEX

Burns, J. B., 624
Burns, J. L., 624
Burns, M. E., 167, 168(2), 193, 519, 520, 521
Burt, D. W., 536
Burton, M., 778, 785(12)
Burton, M. D., 120
Busch, C., 706, 719(14)
Bush, R. A., 182, 214, 616
Busman, M., 435
Butcher, S. E., 765
Butler, N., 625
Butrynski, J. E., 465
Buzdygon, B. E., 11
Byrne, B. J., 743, 744

C

Cabezon, E., 82
Cabot, A., 559, 564(13)
Cagan, R. L., 579
Cahill, G., 50
Caldwell, C., 427, 433(16)
Calvart, P. D., 3(12), 4, 13(12)
Calvas, P., 521, 534, 559, 713
Calvert, P. D., 34, 123, 181, 182(33), 193, 519, 520, 521, 624
Cama, H. R., 315
Cameron, D. A., 181, 182(33), 624
Campbell, T. B., 766
Camps, M., 455, 465
Camuzat, A., 534, 559, 713
Canada, F. J., 326, 331, 332(3), 340, 370, 401, 452, 457, 462, 465, 467(2), 706
Canger, A. K., 581
Canonica, S., 329
Cantin, E. M., 762
Cao, Y., 624
Capdevielle, J., 503
Capecchi, M. R., 182, 715
Carafoli, E., 510
Carballo, M., 625
Carey, S. C., 429
Carlson, F. D., 70, 85(38)
Caron, M. G., 452, 454(9)
Carr, R. E., 559, 591, 671
Carr, S. A., 107, 107(21), 108
Carriker, J. D., 427, 428(15), 430(15)
Carter, B. J., 743
Carter, J. G., 87, 530

Carter-Dawson, L., 560, 567, 574(16), 705, 722(10)
Carty, D. J., 459
Carvalho, M. R. S., 688
Carver, J. M., 82, 83(56)
Casey, P. J., 436, 437, 438(16), 455, 465, 466, 467
Caskey, C. T., 725, 729, 731(21)
Castanotto, D., 776
Catty, P., 254
Cawley, J. D., 315
Cech, T. R., 761, 762
Cepko, C. L., 519, 520, 579, 590, 590(8), 591, 591(7)
Cerione, R. A., 464
Cervetto, L., 133, 139, 145(4), 146, 147
Chabre, M., 27, 267, 269(10), 461, 467
Chader, G. J., 224, 235, 236, 237(20), 319, 324, 359, 565, 566(4)
Chadeuf, G., 744
Chai, X., 372, 378(4)
Chaiken, I. M., 31
Chakrabarti, I., 453
Chakravati, A., 383
Chamberlain, J. S., 724, 725(1), 726(1), 728(1), 729(1), 731, 731(1)
Chambers, J. P., 331, 386, 395(16)
Champer, R. J., 363, 364(20), 366(20)
Champion-Armaud, P., 744
Chan, S., 729, 731(21)
Chandraratna, R. A. S., 411, 429
Chang, D. C., 699
Chang, J. H., 519, 520, 592
Chang, L.-S., 760, 778, 785(10)
Chang, W. S., 579, 582(13), 584(13), 590(13)
Chapman, J. R., 494, 508(15)
Chassy, B. M., 699
Châtelin, S., 534, 559, 713
Chauhan, Y. S., 429
Chazin, W. J., 112(34), 113
Chechnak, A. J., 315
Cheley, S., 270
Chen, C., 678
Chen, C.-K., 13, 19, 23, 34, 34(8), 36(8), 122, 123, 167, 168(2), 264, 269(7), 546, 549(16)
Chen, H. H., 729, 731(21)
Chen, J., 3, 167, 168(1, 2), 169, 178(1), 184, 193, 313, 359, 519, 520, 566, 567(13), 572(13), 573(13), 574(13)
Chen, L., 730, 732(22)

Chen, N., 435, 572, 707, 715(24), 723(24)
Chen, P., 413
Chen, P. L., 689
Chen, S., 530, 590, 590(9), 591, 592(9), 593(6, 9), 597, 599(6, 23)
Chen, S.-J., 778, 779(9), 782(9), 785(9), 787(9), 789(9)
Chen, X., 743
Chen, Y., 332
Chenchik, A., 376
Cheng, H., 457, 462(33)
Cheng, K. M., 534, 535
Cheng, T., 624
Cherel, Y., 744
Cherenson, A. R., 347
Cheung, E., 330
Chien, C., 120
Chien, C. T., 689, 697, 698(21)
Chinchar, G. D., 301
Chinchilla, D., 87, 98(4)
Chiorini, J. A., 744
Choi, S., 592
Chong, J. A., 592
Chowrira, B. M., 765
Christensen, R. L., 322
Christerson, L., 672
Chung, H., 592
Chung, S., 24
Chun-Huey, C., 82, 83(56)
Churcher, C. M., 122
Ciardelli, T. L., 31
Ciccodicola, A., 688
Cideciyan, A. V., 203, 223, 591, 611, 611(12), 612, 614(5, 6), 616(6, 12), 619(12), 621(8–10, 12), 622(8), 625, 625(9), 687
Citovsky, V., 75
Clague, M. J., 68, 69(28)
Clapham, D. E., 103, 455
Clark, D. T., 537
Clark, J. D., 567
Clark, K. R., 743, 744
Clarke, G. A., 519, 520, 687
Clarke, S., 436, 438, 451, 452, 453, 467, 683
Claustres, M., 706, 709(21), 711, 711(21), 712(21, 29), 713(21)
Clayton, R. M., 535
Clemens, P. R., 729, 731(21)
Clements, P. J. M., 519, 520
Clore, G. M., 125, 126
Clyne, J. D., 783, 785(19)

Coats, C., 591
Coats, C. L., 591
Coen, A. I., 530
Coffino, P., 497
Cohen, A. I., 530
Cohen, G. B., 519, 520
Cohen, L. K., 744
Colley, N. J., 540
Collins, F. D., 315
Collins, F. S., 383
Collins, L., 672
Collis, P., 744, 757(12), 760(12), 761(12)
Colosi, P. C., 743
Colowick, S. P., 110
Combs, C., 464
Connell, G., 69, 71(29)
Connell, G. J., 672, 673(14)
Connell, G. L., 672
Connolly, M. L., 347
Conrad, C., 743
Contreras, C., 398
Conway, J. E., 744
Cook, J., 591
Cook, W. J., 126, 127(28)
Cooke, F. T., 455
Cooper, A., 425
Cooper, T. M., 429
Copeland, N. G., 566, 590(9), 591, 592(9), 593(9), 707
Copenhagen, D. R., 187
Cornel, E., 50
Cornwall, M. C., 3, 133(8), 134, 145(9), 148, 151(15), 153, 157(15, 16), 158(15, 16), 160(15), 161(16), 162(15, 16), 163(16), 197, 224, 233, 236, 237(20), 238(21), 242, 242(23), 243(23), 245(23, 24), 246(23), 247, 248(24), 249(23, 24), 251(42), 252, 252(3)
Corson, D. W., 236, 238(21), 258, 319
Costantini, F., 172
Coulombre, A. J., 527
Couto, L. B., 778, 785(12)
Covell, D. G., 86
Cowden, L. B., 435
Cowley, G. S., 180, 618
Cozzi, B., 294
Crabb, J. W., 103, 104, 105, 107, 107(11, 21, 22), 108, 109, 109(18), 116(19), 119(19), 120, 121, 122(2), 167, 332, 363, 364(20), 366(20), 482, 491, 492, 558

Craft, C. M., 24, 25(13), 40(13)
Crasoli, F., Jr., 778, 785(14)
Crawford, C. E., 714
Crawford, R. D., 535
Creighton, P., 182
Cremers, F. P., 688, 691(2), 692, 692(2)
Crescenzi, O., 112(34), 113
Crittenden, L. B., 536
Crognale, M. A., 661(22), 669
Crouch, R. K., 168, 224, 236, 237(20), 238(21), 242, 242(23), 245(23, 24), 246(23), 247, 248(24), 249(23, 24), 319, 359, 363, 365(18), 366(18), 367(18), 372, 425, 435, 483, 511, 565, 566(4), 572, 705, 707, 715(24), 723(24)
Cruz, P., 765
Cunningham, L. L., 585
Curran, G. A., 581
Curtis, R., 519, 520, 672

D

Dacey, D. M., 14
Daemen, F. J. M., 313, 315, 318(9), 344, 359, 419, 420(8, 9), 421(8)
Dai, Y., 725
Daiger, S. P., 591
Danciger, M., 611, 725
Dane, D. A., 536
Danielson, P. E., 519, 520, 621, 672
Daniolos, A., 292, 294(2)
Danos, O., 744
Dartnall, H. J. A., 153, 227, 228, 233(5), 629, 632(22), 643, 648(22)
Das, S. R., 385
Dascal, N., 41
Davenport, P. A., 455
Davey, M. P., 448
David-Ameline, J., 744
Davis, A. C., 559
Davis, L. G., 380
Davis, N. M., 579, 590(14)
Davis, R. W., 651, 655(3)
Dawson, W. W., 534
De Clercq, E., 753
Deeb, S. S., 629, 632, 633, 649, 649(37), 651, 653, 655, 655(8, 9), 659(8, 14), 660(14), 661(22), 664(5), 668(9), 669
de Graan, P. N. E., 293

Degrandis, P., 764
DeGrip, W. J., 86, 279, 291, 419, 420(8, 9), 421(8), 425, 428
Deigner, P. S., 326, 331, 332(3), 401
de Jong, P., 692
De Kruijf, B., 86
de la Calzada, M. D., 625
de Leeuw, T. L. M., 359
DeLuca, H. F., 313(12), 314
Demaille, J. G., 110
DeMatteo, R., 760, 779
Dencker, L., 371
Denny, M., 429
Denton, M. J., 706, 712(23), 713(23)
DePriest, D. D., 640
Deretic, D., 51, 184
Dergachev, A. E., 488
Derguini, F., 329, 429, 430(30)
de Sauvage, F. J., 558
Deshler, J. O., 762
Deterre, P., 254
Detrich, H. W., 537
Detrick, B., 565, 705, 706(2), 707(3), 716(3), 719(1, 2), 720(2)
Detwiler, P. B., 3, 103, 121, 122(2), 123, 133, 133(7), 134, 134(5), 137, 137(5, 10, 12), 139(12), 142(12), 144(5), 145(5), 147, 195, 196, 558
Deutman, A. F., 360, 368(15)
Devlin, K., 122
DeVries, G. W., 87, 530
De Vries, H. I., 628
Dewey, R. D., 294
Deyashiki, Y., 378
DeYoung, M. B., 772
Dharmasiri, K., 435
Diachenko, L., 376
Dickerson, C. D., 519, 520
Dienes, Z., 29
Dieplinger, H., 411
Dieterici, C., 628, 629
Dieterle, J. M., 315
Dietrich, A., 455, 465
Diez, A., 741
Ding, J., 464
Dingus, J., 25
Dionne, V. E., 41, 49(5)
DiPolo, A., 731
DiSepio, D., 411
Dizhoor, A. M., 3, 13, 13(9), 14, 14(9), 111,

121, 122, 123(10), 130, 130(10), 131, 132, 132(33), 549, 558
Dodd, R. L., 167, 193, 519, 520
Doi, T., 465, 466(5), 467(5), 471(5), 473(5), 475, 476(5), 477, 478(5), 479, 481(5)
Dollfus, H., 534, 559, 713
Donner, K., 187
Doolittle, R. F., 345, 378
Dorsky, R. I., 51, 579, 582(12, 13), 584(13), 590(12, 13)
Dowling, J. E., 313, 359, 383, 519, 520, 537, 538(10), 542(3), 557, 565
Downes, C. P., 455
Downes, S. M., 132, 625
Downing-Park, J., 247
Doyle, M. L., 31
Drager, U. C., 319
Dratz, E. A., 167, 318, 320(16), 365
Dratz, E. S., 482, 483(4)
Drayna, D., 411
Drenser, K. A., 743, 745(2), 760(2), 761, 764, 765(34), 766(34), 777
Drenthe, E. H. S., 86
Driessen, A. J. M., 82
Driessen, C. A. G. G., 359, 360, 368(15)
Driever, W., 536, 537, 538(10)
Drohat, A. C., 112(33), 113
Drummond-Borg, M., 633, 651, 655, 664(5)
Dry, K., 688
Dryja, T. P., 178, 180, 515, 519, 520, 611, 618, 671, 672, 672(5), 676, 706, 712(22), 713(22), 722(22)
Duan, D., 778, 779(7), 783(7), 785(7, 8), 789(7)
Du Croz, J. J., 648
Dudas, L., 778, 779(9), 782(9), 785(9), 787(9), 789(9)
Duecher, A., 411
Dufier, J.-L., 534, 559, 713
Duh, E., 530, 590, 593(6), 599(6)
Duke, R. C., 277
Dumke, C. L., 22, 39(6)
Dumont, J., 43
Duncan, A., 590(8), 591
Duncan, T., 574
Dunis, D., 50
Dunn, J. J., 768
Durfee, T., 689
D'Urso, M., 688
Duvic, M., 411

Düzgünes, N., 65, 66, 67(19), 75(17, 18), 80, 81(18), 84
Dykes, D. D., 708

E

Earnshaw, D. J., 763
Easter, S. S., Jr., 537, 579, 590(14)
Ebihara, S., 279, 280(2), 282(2), 283, 283(2), 284
Eckert, R. L., 411
Eckstein, A., 625, 671, 687
Eckstein, F., 267, 763, 764, 766
Eddedge, S. J., 689
Eddinger, C. C., 315
Eddy, R. L., 629, 652, 655(7)
Edebratt, F., 28
Edgar, A., 688
Edwards, B., 701
Edwards, C. L., 530
Ehinger, B., 625
Eisensmith, R. C., 726
Eisenstein, E., 87, 98(4)
Eisner, A., 640
Ekstrom, U., 625
Elder, B., 523
Eliaou, C., 706, 709(21), 711(21), 712(21), 713(21)
Elkins, D., 762
Elledge, S. J., 698
Ellens, H., 75, 76, 78, 84(41)
Ellis, H. M., 50
Elsner, C. L., 530
Elwell, D., 778, 785(12)
Emeis, D., 425
Eng, J. K., 448
Engdahl, R. K., 778, 785(13)
Engel, L. S., 277
Engelhardt, J. F., 726, 778, 779(7), 783(7), 785(7, 8), 789(7)
Enoch, J. M., 631
Erdjument-Bromage, H., 84
Erdman, R. A., 471
Ericcson, L. H., 122
Erickson, M. A., 123
Ericsson, L. H., 168, 483, 485
Eriksson, U., 331, 340(2), 344, 348, 350(10), 356(10), 357, 358(10), 359, 372, 565, 706, 719(14)

Ermolaeva, M. V., 464, 465
Ernst, O. P., 29
Erven, A., 182
Escayg, A., 581(28), 582, 590(28)
Escobar, A. L., 148
Esko, J. D., 753
Etches, R. J., 536
Evan, G. I., 347
Evanczuk, A. T., 254, 258(2)
Evanczuk, T., 14
Evans, B. L., 519, 520
Evans, C. D., 235
Every, D., 493

F

Faber, J., 50
Fabian, C., 536
Fabrizio, P., 120
Fach, C. C., 628, 632(6), 638(6), 646(6), 647(6)
Fager, R. S., 70
Fägerstam, L., 20
Fain, G. L., 3, 133(8), 134, 145(9), 146, 148, 149, 151(15), 157(15, 16), 158(15, 16), 160(15), 161(16), 162(15, 16), 163(16), 197, 224, 242, 242(23), 243(23), 245(23, 24), 246(23), 247, 248(24), 249(23, 24), 250, 251(42), 252, 252(3), 402
Fales, H. M., 317
Falk, J. D., 169
Falsini, B., 519, 520, 612
Fancy, D. A., 120
Fang, B., 726
Farber, D. B., 397, 402, 611, 724, 725, 726(8), 728(8), 731, 740(8), 741, 741(8)
Fariss, R. N., 122, 269, 270
Farnsworth, C. C., 436, 437, 438, 438(11, 16, 17), 444(17), 445, 446(24), 447(24), 451(11, 24), 452, 467
Farr, A. L., 500
Farrar, G. J., 182, 519, 520, 775
Faurobert, E., 461, 467
Faust, L. Z., 778, 785(13)
Favre, D., 744
Fawzi, A. B., 456
Fedor, M. J., 776
Fedorova, O., 764
Feeney, L., 383
Fei, Y., 519, 520

Feifle, C., 587
Feil, R., 655, 665
Feilotter, H. E., 694
Fein, A., 9, 153, 233
Feinberg, A. P., 719
Feldman, K., 111
Feng, D. F., 378
Feng, L., 519, 520
Fenner, M., 66, 75(13), 85(13)
Fenselau, C., 494
Ferguson, M., 582
Fernald, R. D., 283
Fernandez, J. M., 66, 148, 693
Ferrara, P., 503
Ferrari, F. K., 760
Ferreira, P. A., 688, 703
Ferrendelli, J. A., 87, 530
Ferrin, T., 127
Fielding, C. J., 411
Fielding, D., 411
Fields, P. A., 778, 785(12)
Fields, S., 120, 591, 688, 689, 690, 697, 698(21)
Findlay, J. B., 167
Firestein, S., 202
Fisher, K. J., 743, 760, 778, 779, 779(9), 782(9), 785(8, 9), 787(9), 788(24), 789, 789(9)
Fisher, R. J., 31
Fisher, S. K., 65
Fishman, G. A., 591, 688, 706, 712(22), 713(22), 722(22)
Fitch, C. L., 546, 549(16)
Fitzke, F., 625, 687
Fivash, M., 31
Flaherty, K. M., 13, 126, 127(27), 129(27), 130(27), 131(27)
Flanagan, T., 79
Flannery, J. G., 169, 402, 743, 745(2), 760(2), 776, 777, 789
Fletcher, J. E., 471
Fliesler, S. J., 70
Flood, M. T., 385
Florenes, V. A., 104
Florini, J. R., 93
Flotte, T. R., 743
Fodstad, O., 104
Folander, K., 43
Folch, J., 68, 70(26)
Fong, H. K. W., 413, 414, 416(4, 7), 417, 417(4), 418(3), 419(4), 420(3), 421(3), 422(4), 467, 566

Fong, S.-L., 316, 324, 383, 567, 574(16), 705, 722(10)
Foni, R. M., 82
Foord, S. M., 41
Forster, T. Z., 81
Foster, D. C., 558
Foster, R. G., 279
Fox, D. A., 167
Fox, L. E., 283
Fraley, D. M., 744
Frambach, D. A., 402
Frampton, M. W., 775
Franco, M., 461, 467
Francone, O. L., 411
Franklin, P. J., 429, 430(30)
Franson, W. K., 519, 520
Fraser, N. J., 41
Frederick, J., 515
Fresht, A., 389
Freund, C. L., 590, 590(8), 591
Frézal, J., 534, 559, 713
Friedman, L. H., 714
Frishman, L. J., 203, 204, 212, 212(13), 214(28), 612, 614
Fritsch, E. F., 94, 171, 408, 691
Froelick, G., 540
Frohman, M A., 592
Fujimura-Kamada, K., 436
Fujino, M., 437, 444
Fujioka, N., 428
Fujisawa, J., 285
Fujita, R., 688
Fujiwara, J. H., 87, 98(4)
Fukada, Y., 19, 278, 279, 280(2), 281, 282(2, 7), 283, 283(1, 2, 7), 284, 288, 288(1), 290(1), 430, 452, 456, 457(6), 458, 462, 462(36), 463(36), 464, 465, 466(4–6), 467, 467(1, 4–9), 468(4, 6), 469, 469(4, 6, 8), 471(5), 473(5), 475, 475(8), 476, 476(4–6), 477, 478(4, 5, 7, 9), 479, 481(1, 4, 5, 9), 511, 529
Fülle, H.-J., 558, 559(9)
Fulton, A. B., 706, 712(22), 713(22), 722(22)
Fulton, B. S., 326
Fulton, T. D., 219
Funato, T., 762
Fung, B. K., 438, 452, 453, 457, 467, 544, 741
Furst, P., 689, 690(9)
Furth, E. E., 779
Furukawa, K., 96

Furukawa, T., 579, 590, 590(8), 591, 591(7)
Furutani-Seiki, M., 536
Furuyoshi, S., 401
Futterman, S., 316, 319, 321(18), 339, 387

G

Gad, M. Z., 385
Gaidarov, I. O., 87, 98(4), 99
Gait, M. J., 763
Gaitan, A. E., 519
Gal, A., 519, 520, 611, 618(2), 671, 706, 712(23), 713(23)
Galasinki, S., 764
Galijatovic, A., 435
Gallen, E., 50, 51(5)
Gamby, C., 117
Gamm, D., 582
Gao, G.-P., 760, 778, 779, 785(13, 14)
Garbers, D. L., 519, 520, 521, 558, 559, 559(9), 561, 564(12, 14)
Garlick, N., 50
Garrett, D. S., 125
Garrido-Pertierra, A., 741
Garrison, J. C., 471
Garwin, G. G., 313, 359, 360, 362, 363, 364(16, 20), 365(16, 17), 366(17, 20), 382, 552, 565, 566, 567(13), 568, 572(13, 14), 573(13, 14), 574(13)
Gatlin, C. L., 448
Gautam, N., 464, 465
Gawinowicz, M. A., 425, 428(11)
Gaze, R. M., 50
Gean, E., 625
Gee, R. L., 103, 109, 110(25)
Gelatt, K. N., 534
Gelb, M. H., 331, 335(5), 436, 437, 438, 438(11, 16, 17), 440, 444(17), 445, 446(24), 447(24), 451(11, 24), 452, 467
Gelfand, C., 294
Gerber, S., 534, 559, 564(13), 713, 754
Gerhardt, H., 529
Gerhart, J., 58
Geromanos, S., 84
Gershon, P. D., 28
Geschwind, I. I., 294
Gesteland, R., 383
Geysen, H. M., 109
Ghattachary, S., 687

Ghazi, I., 534, 559, 713
Ghomashchi, F., 436, 437, 445, 446(24), 447(24), 451(24)
Ghosh, D., 372
Ghosn, C., 411
Gibbs, D., 70, 85(38)
Gibbs, J. B., 451
Gibson, F., 519, 520
Gibson, L. H., 565, 705
Giddings, I., 621
Gierschik, P., 455, 464, 465
Gilbert, B. A., 325, 331, 401, 403(16), 410(16), 457, 462, 462(33), 464
Gilbert, D. J., 566, 590(9), 591, 592(9), 593(9), 707
Gilbert, W., 770
Gilgenkrantz, S., 692
Gilles, A.-M., 90
Gillespie, J., 425
Gillett, N. A., 558
Gilliland, G. L., 87, 98, 98(4)
Gilman, A. G., 24, 455, 458, 465, 467
Gilman, M., 778, 785(14)
Gimelbrant, A. A., 3(11), 4
Ginty, D. D., 103
Gitler, C., 85
Glardon, S., 579
Glenn, J. S., 452
Glenney, J. R. J., 115
Glomset, J. A., 436, 437, 438, 438(11, 16, 17), 444(17), 445, 446(24), 447(24), 451(11, 24), 452, 467
Gluzman, Y., 674
Goeddel, D. V., 558
Goff, S. P., 193
Goldberg, A. F. X., 83, 625, 671, 673, 675(30), 676, 680, 681, 683(30), 684(25), 685, 687, 687(34)
Goldberg, N. D., 99
Goldflam, S., 332
Goldin, A. L., 43
Goletz, P., 572, 707, 715(24), 723(24)
Gonczol, E., 726, 779
Gonnet, G., 510
Gonzalez-Duarte, R., 372
Gonzalez-Fernandez, F., 324, 585
Goodarzi, J., 687
Goodman, D. S., 385, 400, 429
Goodman, M., 122
Goodwin, G. W., 437

Gorczyca, W. A., 103, 113, 114(35), 115, 116(35), 119(35), 121, 122(2)
Gordon, D., 582
Gordon, G., 726
Gordon, J. I., 13, 34, 122, 130(17), 131(17)
Gordon, J. W., 180, 410, 519, 520
Gordon, S. E., 137, 247
Gorin, M. B., 672
Gorodovikova, E. N., 3(11), 4, 19
Gorzyca, W. A., 558
Gosser, Y. Q., 464
Goto, K., 4
Goto, Y., 172, 178(28), 204, 216(27), 624
Gottschalk, A., 120
Gouaux, J. E., 270
Gour, P., 385
Gouras, P., 193, 519, 520, 521, 725, 777
Gown, A. M., 744
Goy, M. F., 453
Grable, M., 725
Graeff, R. M., 99
Graham, F. L., 724, 725, 726, 728, 730, 731(15, 20), 732(22)
Grainger, R. M., 296
Granato, M., 536
Granit, R., 202, 612
Grant, C., 785, 789(20)
Gray, D., 425
Gray, P., 196
Gray-Keller, M. P., 3, 103, 121, 122(2), 123, 133, 133(7), 134, 134(5), 137, 137(5), 144(5), 145(5), 147, 195, 558
Green, C., 50, 58(15), 59(15), 64(15)
Green, C. B., 291, 296, 302(16)
Green, R. S., 493
Gregg, R. G., 519, 520
Gregory, C. Y., 519, 520
Gregory-Evans, C. Y., 590(8), 591
Gregory-Evans, K., 515, 559, 564(12), 591
Griffin, P. R., 510
Griffoin, J.-M., 706, 709(21), 711, 711(21), 712(21, 29), 713(21)
Grimley, C., 510
Grimm, D., 744, 745(15), 778, 785(11)
Griswold-Prenner, I., 741
Grodberg, J., 768
Groenendijk, G. W. T., 313, 315, 318(9), 419, 420(8, 9), 421(8)
Groesbeck, M., 428
Gronenborn, A. M., 125, 126

Groshan, K. R., 672
Grossman, M., 726
Grounauer, P. A., 625
Grover, S. A., 706, 712(22), 713(22), 722(22)
Gruss, P., 587
Gryczan, C., 204, 216(27)
Grynkiewicz, G., 141, 148
Gu, Q., 558
Gu, S., 706, 712(23), 713(23)
Guerrier-Takada, C., 762
Guggino, W. B., 743
Guillemot, J. C., 503
Gulya, K., 182
Gum, G. G., 534
Guo, Y., 66, 71(10), 73, 74(10), 75(10), 76(10), 85(10), 621, 673
Gupta, A. R., 782, 785(18)
Gupta, B. D., 233
Gustafsson, A.-L., 348, 350(10), 356(10), 358(10)
Gutmann, C., 19
Gutowski, S., 467
Guy, J., 745
Gyuris, J., 689

H

Haecker, S. E., 743, 788(24), 789
Haegerstrom-Portnoy, G., 618
Haeseleer, F., 121, 122, 359, 365(7), 366(7), 368(7), 372, 373, 382
Haffter, P., 536
Haga, T., 19
Hageman, G., 65, 85(8)
Hagen, M., 766
Hagins, W. A., 203, 254
Hagiwara, H., 497, 498(22)
Hagstrom, J. N., 778, 785(12)
Hahm, K., 592
Hahn, L. B., 618, 671, 672, 672(5)
Haiech, J., 110
Hajnal, A., 411
Halbert, C., 787
Hale, I. L., 273
Halevy, D. A., 612
Haley, T. L., 104, 107, 107(12), 116(19), 119(19), 120
Hall, R. A., 24
Hall, S. W., 87, 88(2), 92(2), 167

Hallek, M., 744
Hamaguchi, S., 117
Hamalainen, M., 28
Hamanaka, T., 425
Hamasaki, D., 519, 520, 572, 707, 715(24), 723(24)
Hamburger, V., 50
Hamel, C. P., 565, 705, 706, 706(2), 707, 707(3), 709(21), 711, 711(21), 712(21, 29), 713(21), 716(3), 719(1, 2), 720, 720(2)
Hamm, H. E., 22, 39(6)
Hammerling, V. L., 400
Hammerschmidt, M., 536
Hampel, A., 762, 764, 765, 765(28), 772
Handel, M. A., 655
Handelman, G. J., 318, 320(16), 365
Hanna, D. B., 591
Hanna, M. C., 663
Hannon, G. J., 694
Hansen, E., 646
Hansen, R. M., 219
Hanson, R. H., 582
Hansson, A., 28
Hanzawa, Y., 429
Hanzel, W., 411
Hao, W., 413, 414, 416(4, 7), 417, 417(4), 418(3), 419(4), 420(3), 421(3), 422(4), 566
Hao, Y., 519, 520
Hapala, I., 270
Hara, A., 378
Hara, R., 315, 413, 420, 421(10), 422
Hara, T., 315, 413, 420, 421(10), 422
Hargrave, P. A., 76, 168, 482, 483, 483(4)
Harland, R. M., 296, 587
Hárosi, F. I., 232, 233(10), 537
Harper, J. W., 698
Harris, E., 705, 706, 706(2), 709(21), 711(21), 712(21), 713(21), 719(2), 720(2)
Harris, W. A., 50, 51, 579, 582(12, 13), 584(13, 17), 590(12, 13)
Harrison, E. H., 385, 389
Harvey, T. S., 13, 112(32), 113, 122, 127(15)
Hase, S., 92
Hasel, K., 82, 83(56), 624
Hashimoto, Y., 465, 466(5), 467, 467(5), 471(5), 473(5), 475, 476(5), 477, 478(5), 479, 481(5)
Haskell, J., 519, 520
Hauschka, S. D., 731
Hausen, P., 51

Hauser, M. A., 731
Hauswirth, W. W., 743, 745, 745(2), 760(2), 761, 764, 765(34), 766(34), 775, 776, 777, 789
Havel, C. M., 452
Hawkins, P. T., 455
Hawkins, R. K., 672
Hay, R. T., 725
Hayashi, F., 503
Hayashi, K., 660, 708
Hayashi, T., 651
Hayden, J. B., 448
Hayes, A., 180
Hayes, W. P., 291
Hays, L. G., 448
Hayward, G. S., 744
Hazard, E., 319
Hazel, M., 182
Hbiya, Y., 629
He, C. Y., 725
Hearing, P., 725
Hearst, J., 764, 765(28)
Heath, T., 76, 78(45)
Heaton, P. A., 763, 764
Heaton, R. J., 170, 171(24)
Hecht, S., 186, 628
Heckenlively, J. R., 591, 612, 646, 714
Heckman, J. E., 766
Hedrick, J., 54
Heilig, R., 665
Heisenberg, C. P., 536
Heizmann, C. W., 103, 104, 109, 110(25)
Helekar, B. S., 103, 121, 122(2), 558
Hellebrand, H., 688
Heller, J., 319, 321(18)
Hellman, U., 331, 340(2), 344, 359, 372, 706
Henderson, R., 425
Hendrickson, A., 579
Hendriks, W., 519, 520
Henningsgaard, A. A., 718
Henzel, W. J., 510, 558
Heokstra, D., 85
Heon, E., 625
Herbrick, J. A., 590(8), 591
Hermonat, P. L., 744, 757(12), 760(12), 761(12)
Herold, B. C., 754
Herrmann, B. G., 587
Herrmann, K., 688
Herschlag, D., 765

Hertel, K. J., 765
Herzberg, O., 126
Herzog, R. W., 778, 785(12)
Hetling, J. R., 204, 205(29), 214(29)
Heus, H. A., 776
Heyduk, T., 168
Heyman, R. A., 99
Heynen, H., 203
Heyse, S., 29
Hibiya, Y., 655, 659(14), 660(14)
Hicks, D., 673
Hicks, M., 765
Hidaka, H., 103, 105, 112(34), 113, 119
Higgins, J. B., 465
High, K. A., 743, 778, 785(12), 788(24), 789
Hildebrandt, J. D., 25
Hill, J. M., 277
Hirai, Y., 748
Hirata, H., 488
Hirs, C. H. W., 493
Hirsch, N., 579
Hirschberg, C. B., 346
Hirunagi, K., 283, 284
Hisatomi, O., 3, 11, 12, 12(10, 22), 13, 13(22), 14(22), 284, 581
Hitchcock, P. F., 581(28), 582, 590(28)
Ho, M.-T. P., 395
Ho, M. Y., 172
Hobaugh, M. R., 270
Hobson, A., 546, 549(16)
Hochstrasser, D. F., 494, 505(21)
Hodges, R. S., 170, 171(24)
Hodgkin, A. L., 146, 228, 239, 241(22), 242(22), 243, 245(22), 247(22)
Hoekstra, D., 65, 66, 67(21), 68, 71(21), 81, 82, 84
Hoffman, C. S., 698
Hoffman, D. R., 688
Hofmann, K. P., 29, 204, 205(23), 216(23), 219(23), 222(23), 359, 363, 365(18), 366(18), 367(18), 372, 425
Hogan, B., 172
Hogben, L., 292
Hogness, D. S., 626, 629, 633(1), 634(1), 651, 652, 655(7)
Hojrup, P., 508, 510, 510(29)
Holder, G. E., 132, 625
Holland, L. Z., 579
Hollyfield, J. G., 50, 51, 51(5, 7), 395, 519, 524, 765

Holme, K., 754
Holst, B., 347
Holt, C. E., 50
Höltke, H.-J., 579
Homann, M., 762
Honegger, A., 28
Hood, D. C., 202, 203, 204, 205(22–24), 207(9), 208(24), 210(5, 6, 9), 211(24), 212, 212(12, 24), 213(18), 214(22, 24), 215(24), 216(22–24), 218(9), 219(22, 23), 221(24), 222(22–24), 223, 611(12), 612, 613, 614, 616, 616(12), 618(33), 619(12), 621(12), 624
Hood, L. E., 493
Hooks, J. J., 565, 705, 706(2), 707(3), 716(3), 719(1, 2), 720(2)
Horio, T., 497, 498(22)
Hormes, R., 762
Horowitz, J. M., 294
Horsford, D. J., 590
Horvath, P., 282
Horwitz, J., 402
Hosoda, A., 429
Houghten, R. A., 347
Houpt, K. A., 567
Hovig, E., 104
Howald, W., 437, 438, 438(16), 452, 467
Hoy, P., 285
Hrycyna, C. A., 436
Hsu, S.-C., 99
Hsu, Y.-T., 104, 137, 247
Hu, D., 109, 110(25)
Hu, S., 429, 430(30)
Huang, C., 127
Huang, C. H., 388
Huang, J., 103, 121, 122(2), 359, 363, 364(20), 365(7), 366(7, 20), 368(7), 373, 558
Huang, P. C., 519
Huang, Y., 591, 611(12), 612, 616(12), 619(12), 621(12)
Hubacek, I., 401
Hubbard, R., 235, 313, 314(10), 315, 316(10), 317(10), 322, 324, 339, 359, 383, 422, 557
Hughes, J. V., 778, 785(13, 14)
Hughes, R. E., 122, 123(10), 130(10)
Hui, S. W., 86
Hume, R. I., 783, 785(19)
Humphries, M. M., 182, 519, 520, 775
Humphries, P., 182

Hunkapillar, 510
Hunt, D. F., 471
Hunt, D. M., 559, 564(12, 13), 649
Hunziker, W., 103
Hup, D. R. W., 293
Hurley, J. B., 3, 13, 13(9), 14(9), 19, 23, 34, 34(8), 36(8), 111, 121, 122, 123, 123(10), 130, 130(10), 131, 132(33), 167, 168, 168(2, 4), 264, 269(7), 457, 482, 511, 536, 537, 538(10), 540, 542(3), 544, 546, 549, 549(16), 551, 558
Huse, W. D., 94, 693
Huzoor-Akbar, 464

I

Ibbotson, R. E., 649
Iigo, M., 283
Ikenaka, T., 92
Ikura, M., 13, 34, 112(32), 113, 121, 122, 126, 127(15), 128, 129, 130(17), 131(7, 17)
Ilg, E. C., 104
Illing, M., 69, 566
Imanishi, Y., 80
Inana, G., 519, 520, 688
Ingham, P. W., 585
Inglehearn, C., 625, 671, 687
Inglese, J., 13, 34, 123, 452, 454(9), 455
Iñiguez-Lluhl, J. A., 455, 465
Inoue, N., 744
ISCO, 683
Ishibashi, Y., 444
Ishikawa, N., 96
Ishikawa, T., 105, 279, 282(6), 579
Ishima, R., 13, 34, 122, 130(17), 131(17)
Ishino, T., 13
Isler, O., 321, 339
Isobe, T., 467
Isogai, A., 437, 444
Isono, K., 503
Ito, H., 212
Ito, M., 428, 430
Iuvone, P. M., 50, 51(16)
Ivarsson, B., 20
Iwabuchi, K., 697
Iwahana, H., 660, 708
Iwakabe, H., 519, 520
Iwasa, T., 488

J

Jablonski-Stiemke, M. M., 672
Jackson, A., 590
Jackson, K. W., 501(30), 503, 508, 511(30)
Jackson, M., 70, 71(36)
Jackson, P. K., 714
Jackson, T. R., 455
Jacobs, G. H., 629, 632, 632(24), 648, 648(24), 649, 649(24)
Jacobson, S. G., 203, 223, 590(8), 591, 611, 611(12), 612, 614(5, 6), 616(6, 12), 619(12), 621(8–10, 12), 622(8), 625(9), 687, 688
Jager, S., 363, 365(18), 366(18), 367(18), 372
Jägle, H., 628, 629(15), 632(43), 633, 638(43), 639(15, 43), 640(43), 641(43), 644(43), 645(43), 647(43), 650(15)
James, M. N., 112, 126
James, P., 510
James, S. R., 455
Jamison, J. A., 214
Jampol, L. M., 559
Jan, L. Y., 455
Jan, Y. N., 455
Janecke, A. R., 611, 618(2)
Jang, G.-F., 359, 382, 436, 445, 446(24), 447(24), 451(24)
Jansen, H. G., 672
Jansen, P. A. A., 313, 315, 318(9)
Janssen, A. P. M., 360, 368(15)
Janssen, B. P. M., 359
Janssen, J. J. M., 359, 360, 368(15)
Janssen-Bienhold, U., 537, 538(10)
Jaouni, T. M., 317
Jarvis, L., 127
Jay, M., 625, 671, 687, 688
Jeffery, J., 372
Jelonek, M. T., 21
Jenkins, N. A., 566, 590(9), 591, 592(9), 593(9), 707
Jermyn, M. A., 69
Ji, M., 50
Ji, X., 590
Jiang, G., 279, 291
Jiang, H., 519, 520
Jiang, M., 413, 416(7), 417, 566
Jiang, S., 519, 520
Jiang, Y. J., 536
Jin, J., 236, 238(21)

Johnson, C. M., 107, 107(21), 108
Johnson, G. L., 456
Johnson, N. E., 628, 641(5), 647(5)
Johnson, P. R., 743, 744
Johnson, R. S., 103, 121, 122, 122(2), 168, 483, 485, 558
Johnson, W. C., 113, 114(35), 115, 116(35), 119(35)
Johnsson, B., 20
Johnston, S. A., 120
Jonas, D., 277
Jones, G. J., 153, 197, 224, 233, 236, 237(20)
Jones, J. M., 581(28), 582, 590(28)
Jones, J. T., 762
Jones, P., 84, 85(61)
Jönsson, U., 20
Jooss, K., 743, 788(24), 789
Jordan, E., 383
Jorgensen, A. L., 651
Jörnvall, H., 372
Joseph, S., 764, 765, 768(25)
Josephs, S. F., 725
Jowett, T., 584(33), 585, 587(33, 36, 37)
Jubb, C., 625, 671, 687
Jun, L., 70

K

Kahan, J., 321
Kahn, A. J., 527
Kahn, E. S., 503
Kaibuchi, K., 452
Kainosho, M., 126
Kainz, P. M., 650, 662
Kajiwara, K., 671, 672(5), 676
Kakuev, D. L., 87, 98, 98(4)
Kakuno, T., 497, 498(22)
Kalicharan, R. D., 82
Kalman, V. K., 471
Kalnins, V. I., 672
Kameya, S., 519, 520
Kameyama, K., 19
Kamiya, H., 762
Kanazawa, H., 660, 708
Kanekar, S., 579, 584(17)
Kao, J. P., 148, 159(18)
Kaplan, J., 534, 559, 564(12, 13), 713
Kaplan, K., 713

Kaplan, M. W., 14, 167, 482
Kapron, J., 332
Karaschuk, G. N., 87, 98(4)
Karlsson, A. F., 28
Karlsson, R., 20
Karpe, G., 612
Karwoski, C. J., 614
Kashani-Sabet, M., 762
Kasianowicz, J., 271
Kato, K., 467
Kato, M., 103, 592
Katz, A., 465
Katz, B. M., 236, 238(21)
Kawaguchi, T., 425
Kawamura, S., 3, 5, 7, 8, 9(1), 10, 11, 11(1, 18), 12, 12(10, 22), 13, 13(1, 20, 22), 14, 14(22), 15, 16, 16(14), 17, 18, 19(14, 27), 20, 20(27)
Kawase, K., 565, 705, 718
Kawata, M., 436, 438, 438(11), 451(11), 452
Kay, M. A., 725, 744
Kayada, S., 3, 12(10), 284
Kearney, J. A., 581(28), 582, 590(28)
Kedzierski, W., 204, 672
Keeler, C. E., 519, 520
Keen, J., 671
Keen, T. J., 625, 687
Kefalov, V. J., 224, 242
Kehoe, M., 270
Keirns, J. J., 742
Keller, P. M., 66, 67(20)
Kelsell, R. E., 559, 564(12, 13)
Kelsey, D., 79
Kelsh, R. N., 536
Kemp, C. M., 611, 621(8, 9), 622(8), 625(9)
Kenna, P., 182
Kennedy, M. J., 536, 537
Kenrick, K. G., 492
Kent, S. B., 493
Kerbel, R. S., 104
Kern, A., 744, 745(15), 778, 785(11)
Kessler, C., 579
Kessler, P. D., 743
Ketchem, R. R., 112(34), 113
Keyomarsi, K., 698
Khani, S. C., 223, 612
Khilko, S. N., 21, 28
Khorana, H. G., 41, 45(3), 46(3), 47(3), 49(4), 426, 428
Khorlin, A. Ya., 92

Kiehntopf, M., 762
Kielland-Brandt, M. C., 347
Kikkawa, S., 488
Kikuchi, A., 452
Kilburn, A. E., 689
Kim, D. R., 24
Kim, R. Y., 80, 81(50), 83(50), 687
Kim, S., 592
Kim, Y., 592
Kimura, A. E., 611, 621(10)
Kimura, N., 96, 98
Kimura, S., 80
Kincaid, R. L., 105
King, M. A., 743, 746(8), 760(8)
Kingsley, S. L., 766
King-Smith, P. E., 628
Kingston, R., 600, 606(24), 691
Kinumi, T., 503
Kioschis, P., 633
Kirkness, E. F., 663
Kirschner, M., 54
Kise, H., 117
Kisselev, O. G., 464, 465
Kitada, C., 437, 444
Kitagawa, H., 529
Kitamura, K., 10, 11(18)
Kito, Y., 425
Kjeldbye, H., 725, 777
Klappe, K., 66, 67(21), 68, 71(21), 85
Klatt, L., 777
Klausen, G., 632(43), 633, 638(43), 639(43), 640(43), 641(43), 644(43), 645(43), 647(43)
Klee, C., 110, 112
Kleemann, G. R., 448
Klein, R. L., 743, 746(8), 760(8)
Kleinschmidt, J. A., 50, 744, 745(15), 778, 785(11)
Klemenz, R., 411
Klenchin, V. A., 3(12), 4, 13(12), 34, 123
Kliger, Y., 84, 85(61)
Klingaman, R., 646
Klisak, I., 672
Kljavin, I. J., 537
Klose, J., 492
Knapp, D. R., 168, 435, 483, 511
Knau, H., 632(43), 633, 638(43), 639(43), 640(43), 641(43), 644(43), 645(43), 647(43)
Knight, J. K., 581

Knowles, A., 643
Knox, B. E., 41, 45(3), 46(3), 47(3), 49(4, 5), 50, 51(13), 52, 58(15), 59(15), 64(15), 279, 296, 297(14), 302(16), 428, 581(29), 582, 590(29)
Knudsen, C. G., 429
Ko, R. K., 587
Kobayashi, K., 122
Kobayashi, N., 429
Kobayashi, R., 105, 119
Kobayashi, Y., 13, 429
Koch, K.-W., 24, 34(10), 161
Koch, W. J., 452, 454(9)
Kochanek, S., 724, 725, 729, 731(21)
Kodadek, T., 120
Kodama, A., 430
Kohl, S., 621
Kohler, B. E., 322
Kohnken, R. E., 25
Kohr, W., 411
Koizumi, M., 762
Kojima, D., 279, 282(6), 288, 529, 579
Kokame, K., 458, 462(36), 463(36), 465, 466(4), 467, 467(4), 468(4), 469(4), 476(4), 478(4), 481(4)
Komori, N., 492, 493, 501(30), 503, 508, 510, 511(30)
Kondo, H., 4
Kong, F., 66, 69(9), 70(9), 71(9), 74(9), 75(9), 76(9), 85(9)
König, A., 628, 629
Konig, B., 425
Koopmans, S. C., 724
Korenbrot, J. I., 3, 133, 147, 283
Kornfeld, R., 346
Kornfeld, S., 346
Kornhauser, R., 464
Kort, E. N., 453
Kosaras, B., 181, 182(33), 519, 520, 624
Koshland, D. E., Jr., 453
Kotin, R. M., 744
Koutalos, Y., 199, 202(23), 247, 515
Kozak, C. A., 519, 520, 621, 672
Kozak, M., 406
Kozasa, T., 458
Kozawa, O., 467
Kraft, T. W., 203, 214(21), 216(21), 222(21), 629, 632(62, 63), 648
Kraner, S. D., 592
Krapf, E. M., 778, 785(13)

Krapivinsky, G. B., 455
Krasnoperova, N. V., 167, 181, 182(33), 519, 520, 624
Kraulis, P. J., 127
Krebber, A., 28
Krebber, C., 28
Kremmer, E., 28
Kretsinger, R. H., 122
Krishna, R. G., 494
Krishnan Kutty, R., 574
Krishnasatry, M., 271
Krogh, K., 648
Kroll, K. L., 50, 58(14), 59(14), 61(14), 62(14), 294
Krook, M., 372
Krozowski, Z., 372
Krumins, A. M., 24
Kucharczak, J., 705
Kuhlmann, E. D., 360, 368(15)
Kühn, H., 87, 88(2), 92(2), 167, 265, 266(8), 425
Kumar, R., 530, 590, 593(6), 599(6)
Kumar, S., 3, 13(9), 14(9), 111, 121, 122
Kumaramanickavel, G., 706, 712(23), 713(23)
Kumar-Singh, R., 724, 725, 725(1), 726(1, 8), 728(1, 8), 729(1), 731, 731(1), 740(8), 741(8)
Kunkel, T. A., 353
Kunsch, C., 467
Kunz, L., 277
Kunz, R., 429
Kuo, C.-H., 3, 4, 12(10)
Kurien, B. T., 503
Kuroda, S., 440
Kurtzman, A. L., 581
Kurtzmann, G. J., 743, 778, 785(12)
Kutty, G., 574
Kuwata, O., 11, 12, 12(22), 13(22), 14(22)
Kyte, J., 345

L

Labow, M. A., 744
Lacy, E., 172
Ladant, D., 122
Ladner, J. E., 87, 98, 98(4)
Laemmli, U. K., 339, 493, 571

Lagnado, L., 123, 133, 139, 145(4), 146, 147, 197
Lahiri, D., 51
Lai, R. K., 452, 465, 467(2)
Lai, Y.-L., 324
Lakey, J. H., 21
Lam, D., 436
Lamb, T. D., 149, 185, 190, 199, 199(9), 201(9, 22), 202(22), 203, 214(10), 216, 218, 219, 221(10), 226, 228, 242, 250, 250(9), 613, 614, 616, 618(35), 643
Lamba, O. P., 66, 69(9), 70(9), 71(9, 10), 73, 74(9, 10), 75(9, 10), 76(9, 10), 85(9, 10), 621, 673
Landers, G. M., 313(11), 314, 333, 722
Landers, R. A., 51
Lange, C., 24, 34(10)
Langlois, G., 264, 269(7)
Langridge, R., 127
Langston, C., 724
Largent, B. L., 579, 590(16)
LaRoche, G. R., 646
Larsen, S. H., 453
Larsson, A., 24
Lascu, I., 90, 96
Laura, R., 121, 549, 558
LaVail, M. M., 519, 520, 560, 743, 745(2), 760(2), 776, 777, 789
Law, W. C., 325, 326, 329, 331, 332(3), 401, 706
Lawler, A. M., 519, 520
Lawn, R., 411
Lawrence, J. B., 579, 582(9)
Lean, J., 411
Leavitt, J., 493
Lebioda, L., 109, 359, 365(7), 366(7), 368(7), 373
LeBot, N., 294
Leducq, R., 705
Lee, A. S., 762
Lee, E., 455, 519, 520, 572, 707, 715(24), 723(24)
Lee, L., 235
Lee, M. G., 41
Lee, N., 629, 632(25)
Lee, N. R., 519, 520, 521, 526, 530, 534
Lee, R. H., 467
Lee, S. C., 380
Lee, W. H., 689
Lees, M., 68, 70(26)

Lefkowitz, R. J., 13, 24, 34, 123, 452, 454(9), 455
Lehmann, M. N., 282
Lem, J., 167, 169, 181, 182(33), 517, 519, 520, 546, 549(16), 624
Lennon, A., 688
Lennon, G., 590(9), 591, 592(9), 593(9)
Leon, A., 519, 520
Leowski, C., 534, 559, 713
Le Paslier, D., 534, 559, 713
Lerner, A. B., 292, 294(2)
Lerner, M. R., 292, 294, 294(2), 307
Lerner, R. A., 347
Lerro, K. A., 425, 428(11, 12)
Lettice, L., 585, 587(36)
Levay, K., 23, 24(9), 25(9)
Levine, E. M., 581
Lévy, F., 706
Levy, S., 673
Lewin, A. S., 743, 745(2), 760(2), 761, 764, 765(34), 766(34), 775, 777
Lewis, G. K., 347
Lewis, M. S., 235
Lewis, R. A., 646
Li, B., 690, 697
Li, H.-S., 579
Li, J., 743, 744, 787
Li, L., 519, 520
Li, N., 121, 122, 132
Li, S., 172, 178(28), 624
Li, T., 178, 515, 519, 520, 611, 725, 785, 789(20)
Li, X. B., 159
Li, Y., 777
Li, Z.-Y., 519, 520, 611, 611(12), 612, 616(12), 619(12), 621(12)
Liberman, E. A., 66
Licht, T., 762
Lichter, P., 633
Lidholm, J., 28
Lieber, A., 725, 776
Lieberman, P., 530
Liebetrau, M., 277
Liebman, P. A., 11, 14, 167, 254, 258(2)
Liebner, S., 529
Lien, T. H., 425, 428(11, 12)
Lim, Y.-H., 325, 331, 401, 403(16), 410(16), 464
Lin, H. N., 388
Lind, T., 319
Linder, M. E., 455

Lindner, P., 28
Lindorfer, M. A., 471
Lindsey, D. T., 629, 632, 649, 649(37)
Lindsey, J. D., 530
Linhardt, R. J., 754
Liniai, M., 84
Link, A. J., 448
Lion, F., 344, 359
Liou, G. I., 313, 359, 519, 520, 566, 567, 567(13), 572(13), 573(13), 574(13, 16), 705, 722(10)
Liou, G. J., 324
Lipman, D. J., 374
Lisman, J. E., 49
Litman, B. J., 67, 69, 70, 71(36), 82(25), 85(38), 167
Little, E., 762
Litwin, V., 82, 83(56)
Liu, F., 762
Liu, K., 558
Liu, L., 445, 446(24), 447(24), 451(24)
Liu, Q., 579, 582(18), 590(18)
Liu, R. S. H., 429
Liu, S.-Y., 705, 706, 709(21), 711(21), 712(21), 713(21), 718
Liu, X. Z., 519, 520
Liu, Y. P., 324
Livera, M. A., 383
Lloyd, M., 672
Loew, L. M., 85
Loewen, C. R., 673
Löfås, S., 20
Logan, C., 120
Lok, J. M., 456
Lolley, R. N., 24, 25(13), 40(13), 741
Long, G. L., 675
Longstaff, C., 456
Looney, D. J., 762
Looser, J., 590(8), 591
Lorenz, B., 688
Lorrain, P., 258
Lou, Y. C., 725
Loutradis-Anagostou, A., 590(8), 591
Lowe, D. G., 121, 549, 558
Lowry, O. H., 87, 500, 530
Loyter, 75
Lu, D. J., 464
Lu, L., 558
Lubienski, M. J., 112(34), 113
Luckow, V. A., 380

Lugtenburg, J., 425, 428
Luhrmann, R., 120
Lukyanov, S., 376
Lundh, K., 20
Lutz, R. J., 451
Luzzato, L., 741
Lyubarsky, A. L., 204, 216(26), 250, 519, 520, 612

M

Ma, C., 519, 520
Ma, J., 624, 690
Ma, J.-X., 572, 705, 707, 715(24), 723(24)
Ma, Y.-T., 436, 464
MacDonald, P. N., 401
Macke, J. P., 611, 633, 651, 655(6)
MacKenzie, D., 723
MacLeod, D. I. A., 628, 640, 641(5), 647(5), 650
MacNichol, E. F., 236, 238(21), 643
MacNichol, E. F., Jr., 153, 227, 233, 233(5)
Maddon, P. J., 82, 83(56)
Maddox, J. F., 464
Madreperla, S., 530
Maecker, H. T., 673
Maeda, A., 279, 282(6), 579
Maelandsmo, G. M., 104
Magnusson, K., 28
Maguire, A. M., 611, 621(9), 625(9), 687, 725, 777, 778, 779(7, 9), 782, 782(9), 783(7), 785(7, 9, 18), 787(9), 789(7, 9)
Main, M. J., 41
Makino, C. L., 3, 167, 168(1), 178, 178(1), 180(31), 181, 182(33), 184, 193, 233, 517, 519, 520, 546, 549(16), 560, 624
Makous, W. L., 214, 221(37)
Malicki, J., 536
Malmqvist, M., 20
Malsbury, D. W., 386
Maltese, W. A., 436, 471
Manara, S., 672
Mandel, G., 592
Mandel, J. L., 665
Manes, G., 705
Mangion, J., 24
Mani, S. S., 50
Maniatis, T., 94, 171, 408, 691

Mann, M., 119, 120, 510
Manor, D., 464
Mansfield, R. J. W., 643
Manson, F., 688
Mansour, S. L., 715
Marcus, S., 690, 691(10)
Margolis, J., 492
Margulies, D. H., 21
Mariani, G., 397
Markin, V. A., 92
Marlhens, F., 706, 709(21), 711, 711(21), 712(21, 29), 713(21)
Marrell, B. G., 122
Marrs, J. A., 579, 582(18), 590(18)
Martens, I., 726
Martin, E., 277
Martin, K. J., 324
Martinez, J. A., 590(10), 591
Mason, P. J., 741
Masson, L., 24
Mastick, G. S., 579, 590(14)
Masu, M., 519, 520
Masuda, K., 467
Mata, J. R., 383, 387, 388, 390(20a)
Mata, N. L., 331, 383, 386, 387, 388, 390(20a), 391, 392(26), 394, 395(16), 396(26), 397, 398(26), 399(31), 400(27)
Matelis, L. A., 743
Mathies, R., 425
Mathiesz, K., 41, 49(5)
Matsuda, S., 11, 12, 12(22), 13, 13(22), 14(22)
Matsuda, T., 458, 462(36), 463(36), 464, 465, 466(4, 5), 467, 467(4, 5, 9), 468(4), 469(4), 471(5), 473(5), 475, 476(4, 5), 477, 478(4, 5, 9), 479, 481(4, 5, 9)
Matsudaira, P., 493
Matsumoto, B., 65, 273
Matsumoto, H., 492, 493, 497, 498(23), 501(30), 503, 508, 510, 511(30)
Matsuura, Y., 465, 466(5), 467(5), 471(5), 473(5), 475, 476(5), 477, 478(5), 479, 481(5)
Matthews, G., 187, 189(3), 243
Matthews, H. R., 3, 133(8), 134, 145(9), 146, 148, 149, 151(15), 157(15, 16), 158(15, 16), 160(15), 161(16), 162(15, 16), 163(16), 224, 242, 242(23), 245(23, 24), 246(23), 247, 248(24), 249(23, 24), 250, 251(42), 252, 252(3)
Maumenee, I. H., 646

Maune, J. F., 112
Maxam, A. M., 770
Maxwell, J. C., 626, 629(3)
Mayhew, E., 76, 78(45)
Mazza, A., 24
Mburu, P., 519, 520
McCaffery, P., 319
McCaffrey, J. M., 348, 350(10), 356(10), 358(10)
McCall, M. A., 519, 520
McCarthy, S. T., 3, 133, 133(8), 134, 145(6), 147, 163(12)
McClintock, T. S., 294, 307
McCloskey, J. A., 494, 508(14)
McCormack, A. L., 448
McCormick, A. M., 313(12), 314
McDonald, C. K., 766
McDougall, I. M., 725
McDowell, J. H., 76, 168, 483
McGeady, P., 440
McGee, J. O. D., 579
McGee, T. L., 618
McGrail, S. H., 464
McGrath, S. A., 743
McHenry, C. S., 24
McInnes, R. R., 519, 520, 579, 590, 590(8), 591, 672, 687
McKay, D. B., 13, 126, 127(27), 129(27), 130(27), 131(27)
McLachlan, I., 535
McLaren, M. J., 519, 520
McLatchie, L. M., 41
McLaughlin, S. K., 744, 757(12), 760(12), 761(12)
McNally, N., 182
McNaughton, P. A., 133, 139, 145(4), 146, 147, 219, 243
McQuinston, S. A., 743
Mead, D., 429
Meador, W. E., 112(31), 113
Means, A. R., 112(31), 113
Meek, R. L., 582
Meindl, A., 688
Meisler, M. H., 581(28), 582, 590(28)
Meister, M., 223, 455, 465, 618
Meitinger, T., 688
Melcher, K., 120
Meller, K., 527
Mellersh, C. S., 519, 520
Mellingsaeter, T., 104

Meloen, R. H., 109
Melton, D., 50
Mendez, A., 167, 519, 520, 521
Menini, A., 202
Merbs, S. L., 283, 626, 629, 632(26, 27), 633, 637(45), 638(26, 27, 45), 649(26, 27, 45), 650(27), 651, 658
Mertz, J. R., 371
Messersmith, L., 51
Messing, J., 353
Meuse, L., 744
Mey, J., 319
Meyer, E. M., 743, 746(8), 760(8)
Meyers, C., 743, 746(8), 760(8)
Miao, C. H., 744
Michaelis, S., 436
Michel-Villaz, M., 425
Migliaccio, C., 688
Miki, N., 4, 742
Mikuckis, G. M., 294
Milam, A. H., 14, 122, 168, 270, 363, 364(20), 366(20), 486, 487(12), 511, 519, 520, 573, 611, 611(12), 612, 616(12), 619(12), 621(12)
Miller, A. D., 760, 787
Miller, D. L., 3, 133, 147
Miller, D. T., 579
Miller, J. L., 167
Miller, S. A., 708
Miller, S. S., 397, 402
Miller, T. J., 778, 785(13)
Millington-Ward, S., 776
Milos, N. C., 302
Minta, A., 148, 159(18)
Mintz-Hittner, H. A., 591
Mircheff, A. K., 397, 402
Mirzayanova, M. N., 92
Misui, T., 425
Mitani, K., 729, 731(21)
Mitsui, Y., 378
Mitton, K. P., 591
Miyazaki, K., 497, 498(22)
Mizota, A., 519, 520
Mizuno, K., 467, 559
Mizuno, N., 519, 520
Mocikat, R., 28
Moghrabi, W. N., 672
Mohamadzadeh, M., 277
Molday, L. L., 69, 170, 171(24), 566, 672, 673, 687

Molday, R. S., 69, 71(29), 83, 99, 104, 137, 170, 171(24), 180, 247, 566, 611, 621, 621(4), 625, 671, 672, 673, 673(14, 22), 675(22, 30), 676, 680, 681, 683(30), 684(25), 685, 687, 687(34), 723
Mole, E. J., 41, 49(5)
Mollevanger, L., 86
Mollon, J. D., 629, 632(22), 648(22), 649, 650
Monck, J. R., 66, 148
Moncrief, N. D., 122
Monforte, J., 764, 765(28)
Montzka, D. P., 203, 616
Moody, S. A., 50, 51(13), 52
Moore, A. Y., 559, 564(12)
Moore, D. D., 600, 606(24), 691
Moqadam, F., 376
Moquin-Pattey, C., 425, 428(11, 12)
Morales, J. M., 455
Moreau, J. M., 277
Morimura, H., 706, 712(22), 713(22), 722(22)
Morishita, R., 467
Moritz, L., 83
Moritz, O. L., 69, 621, 672, 673, 673(22), 675(22, 30), 680, 681, 683(30), 685
Morral, N., 724, 726
Morris, V. B., 527, 529
Morrow, E. M., 590, 591, 591(7)
Morton, R. A., 315
Morton, T. A., 31
Motani, K., 725
Motulsky, A. G., 629, 633, 649, 651, 653, 655, 655(9), 659(14), 660(14), 661(22), 664(5), 668(9), 669
Motulsky, A. R., 632, 633, 649(37)
Moullier, P., 744
Mukai, S., 180
Mukanata, Y., 96
Mullen, R. J., 519, 520
Muller-Klieser, W., 277
Mulligan, R. C., 714
Mullins, M. C., 536
Mumby, S. M., 467
Munder, T., 689, 690(9)
Munier, F. L., 625
Munnich, A., 534, 559, 564(13), 713
Murakami, M., 3, 5, 7, 8, 9(1), 11(1), 13(1), 16(14), 19(14)
Muramatsu, T., 582
Murata, M., 464, 465, 467(9), 478(9), 481(9)
Murphey, W. H., 646, 671

Murphy, D., 172
Murphy, R. L., 250
Murray, A., 61
Murrow, E. M., 579
Musarella, M. A., 688
Muskat, B. L., 591
Muszaka, D. G., 31
Muzyczka, N., 743, 744, 745, 746(8), 757(12), 760(8, 9, 12), 761(12), 764, 776, 789

N

Naarendorp, F., 519
Naash, M. I., 172, 178(28), 204, 216(27), 519, 524, 624, 765
Nadcarni, V. D., 754
Nagashima, K., 82, 83(56)
Nagesh, R., 121
Nagpal, S., 411
Nagy, A., 714
Nagy, D., 744
Nagy, R., 714
Naka, M., 117
Nakagawa, M., 488
Nakamura, A., 283, 284, 290
Nakamura, Y., 519, 520
Nakanishi, K., 329, 425, 427, 428, 428(11, 12, 15), 429, 430(15, 17, 30), 431, 431(17), 433(16, 17)
Nakanishi, M., 378
Nakanishi, S., 519, 520
Nakashima, Y., 582
Nakatani, K., 3, 7, 146, 193, 199, 202(23), 203, 214(20), 222(20), 243, 247, 250
Nakayama, H., 467
Nakayama, T. A., 426
Nakazawa, M., 625
Napoli, A. A., 66, 71(10), 73, 74(10), 75(10), 76(10), 85(10), 621, 673
Napoli, A. A., Jr., 66, 75(13), 85(13)
Napoli, J. L., 313(12), 314, 318, 371, 372, 378(4)
Narfström, K., 519, 520, 672, 705, 707(7)
Nash, P. B., 277
Nasi, E., 41, 45(3), 46(3), 47(3), 49(4)
Nasonov, V. V., 92
Nathans, J., 178, 180(31), 283, 519, 560, 566, 611, 626, 628, 629, 629(15), 632(26, 27, 43), 633, 633(1), 634, 634(1), 637(45), 638(26, 27, 43, 45), 639(15, 43), 640(43), 641(43), 644(43), 645(43), 646, 647(43), 649(26, 27, 45), 650(15, 27), 651, 652, 655(3, 6, 7), 658, 672
Nawrocki, J. P., 76, 168, 483
Neer, E. J., 455, 456
Negishi, E. I., 428
Neill, J. M., 719
Neitz, J., 629, 632, 632(24, 62), 648, 648(24), 649, 649(24), 650, 659, 662
Neitz, M., 629, 632(24, 62), 648, 648(24), 649, 649(24), 650, 659, 662
Nelson, R. M., 675
Nenashev, V. A., 66
Nesbitt, S., 764, 765(28)
Neubauer, G., 120
Neubert, T. A., 122, 123
Neuhauss, C. F., 537, 538(10)
Neuhauss, S. C., 536
Newport, J., 54
Neyts, J., 753
Ng, D., 590(8), 591
Nguyen, C. D., 778, 785(13)
Nguyen, N. Y., 235
Nichols, T. C., 778, 785(12)
Nickerson, J. A., 96
Nickerson, J. M., 530
Nicola, G. N., 537
Nicoletti, A., 565, 705, 718
Nicolo, M., 181, 182(33), 624
Nie, Z., 590(9), 591, 592(9), 593(9)
Nieba, L., 28
Nieba-Axmann, S. E., 28
Nielsen, L., 223, 612
Niemeyer, G., 625
Niemi, G. A., 3, 13, 13(9), 14(9), 111, 121, 167, 168(2, 4), 511, 537, 551
Nieva, J. L., 82
Niewkoop, P. D., 50
Nikaido, K., 492, 493(5)
Niki, I., 103
Nikonov, S., 250
Nilges, M., 126
Nilsson, I. M., 347
Nilsson, P., 24
Nilsson, S. E., 519, 520
Nilsson, T., 726
Nilsson-Ehle, P., 625
Niman, H. L., 347

Nir, I., 634, 672
Nir, S., 85, 86
Nishikawa, S., 743, 745(2), 760(2), 765, 777
Nishizawa, Y., 501(30), 508, 511(30)
Niwa, M., 467
Nodes, B. R., 234
Noguchi, T., 122
Noll, G. N., 234
Nollet, F., 710, 713(28)
Nomanbhoy, T. K., 464
Nomura, A., 519, 520
Nomura, K., 98
Nonaka, T., 378
Northup, J. K., 456
Norton, J. C., 624
Noy, N., 358
Nunes, F. A., 726, 779
Nunn, B. J., 133, 146, 188, 203, 214(19), 219(19), 222(19), 223, 239, 241(22), 242(22), 243, 245(22), 247(22), 253, 618, 643, 648(56)
Nusinowitz, S., 204, 205(22), 208, 214(22), 216(22), 219(22, 31), 222(22)
Nussbaum, O., 66
Nüsslein-Volhard, C., 536, 587
Nygren, P., 24

O

Oatis, J. E., Jr., 168, 435, 483, 511
O'Brien, P. J., 453
O'Callaghan, R. J., 277
O'Day, W. T., 397, 402
Odenthal, J., 536
Oelze, I., 762
O'Farrell, P. H., 492, 497(3)
Ohashi, M., 503
Ohba, N., 582
Ohguro, H., 103, 107, 116(19), 119(19), 121, 122(2), 168, 270, 452, 456, 457(6), 462, 465, 466(6), 467(1, 6, 7, 8), 468(6), 469(6, 8), 475(8), 476(6), 478(7), 481(1), 482, 483, 485, 486, 487(12), 491, 511, 558, 573
Ohl, S., 633
Ohno, S., 467
Ohtsuka, E., 762
Oishi, T., 281, 282(7), 283(7), 284
Ojcius, D. M., 277
Ojwang, J. O., 762
Ok, H., 427, 433(16)
Oka, H., 743
Okami, T., 397
Okamoto, N., 519, 520
Okamura, W. H., 429
Okano, T., 278, 279, 280(2), 281, 282(2, 7), 283, 283(1, 2, 7), 284, 288, 288(1), 290(1), 467, 529
Okayama, H., 678
Olins, P. O., 380
Olshevskaya, E. V., 121, 122, 123(10), 130(10), 131, 132, 132(33), 549, 558
Olson, F., 76, 78(45)
Olson, J. A., 333, 722
Olson, M. D., 527
Olson, W. C., 82, 83(56)
Olsson, J. E., 178, 180, 519
Omer, C. A., 451
Omori, K., 397
O'Neal, W., 726
O'Neill, B., 776
O'Neill, J. W., 585
Ong, D. E., 401
Ong, O. C., 453, 467
Oomori, S., 529
Oprian, D. D., 283, 519, 520, 629, 632(25, 28), 633(28), 638(28), 649(28)
Oreve, S., 744
Orita, M., 660, 708
Orr, H. T., 530
O'Shannessy, D. J., 29, 31
Östlin, H., 20
Othersen, D. K., 705
Otis, J. S., 534
Otto-Bruc, A. E., 122, 254, 269, 270
Ottonello, S., 397
Ovchinnikov, Y. A., 488
Ovchinnikova, T. V., 99
Oversier, J. J., 714
Owczarcyk, Z., 428
Owen, W. G., 3, 133, 133(8), 134, 145(6), 147, 163(12)
Ozaki, K., 315, 420, 421(10)
Ozawa, M., 582

P

Paabo, S., 726
Pages, A., 705

Pagon, R. A., 646
Pak, W. L., 497, 498(23), 503, 703
Palczewski, K., 14, 19, 103, 104, 105, 107, 107(11, 22), 108, 109, 109(18), 113, 114(35), 115, 116(19, 35), 119(19, 35), 121, 122, 122(2), 123, 132, 167, 168, 195, 223, 264, 269(7), 270, 313, 330, 331, 335(5), 359, 360, 363, 364(16), 365(7, 16, 18), 366(7, 18), 367(18), 368(7), 372, 373, 382, 436, 447, 482, 486, 487(12), 511, 515, 519, 520, 521, 530, 558, 566, 572(14), 573, 573(14), 612
Palmer, M., 270, 277
Palmer, T., 391
Palmiter, R. D., 172
Palzaewski, K., 483, 485, 486, 491, 492, 552, 566, 567(13), 572(13), 573(13), 574(13)
Pandey, S., 413, 414
Pandolpho, L., 383
Panoskaltsis-Mortari, A., 579, 582(10)
Papac, D. I., 168, 483, 511
Papahadjopoulos, D., 66, 67(19), 76, 78(45)
Papaioannou, M., 590(8), 591
Papermaster, D. S., 184, 429, 484, 519, 520, 569, 634, 672
Pappin, D. J. C., 508, 510(29)
Pardi, A., 776
Pardue, M. T., 519, 520
ParHayes, W., 279
Parish, C. A., 451, 456, 457, 458, 458(31), 459, 459(38), 461(38, 39), 462, 462(31, 33), 463(31, 38, 39), 464(44), 465, 467
Parker, J. M. R., 170, 171(24)
Parks, R. E., 92
Parks, R. E., Jr., 87
Parks, R. J., 724, 725, 730, 732(22)
Parmentier, M., 120
Parmer, R., 534
Parshall, C. J., 519, 520
Partearroyo, M. A., 82
Pascoe, P. J., 567
Passini, M. A., 581
Passonneau, J. V, 530
Passonneau, P. N., 87, 530
Pathak, R., 743, 788(24), 789
Patijn, G. A., 744
Patrinos, A., 383
Pawar, H., 590
Pawlyk, B. S., 180, 519
Payne, A. M., 132, 559, 564(12), 591, 625

Payne, R., 133, 147
Peachey, N. S., 3, 167, 168(1), 172, 178(1, 28), 184, 193, 204, 216(27), 519, 520, 624
Pearce-Kelling, S., 519, 520, 705, 707(7)
Pearson, W. R., 374
Pecchia, D. B., 764, 766
Pechuer, E., 65
Peel, A. L., 743, 744, 746(8), 760(8, 9)
Pelletier, S. L., 629, 632(25)
Penefsky, H. S., 769
Penn, R. D., 203
Peoples, J. W., 530
Pepperberg, D. R., 202, 204, 205(22–24, 29), 208(24), 211(24), 212(24), 214(22, 24, 29), 215(24), 216(22–24, 27), 219(22, 23), 221(24), 222(22–24), 223, 224, 234, 359, 565, 566(4)
Percopo, C., 705, 719(1)
Perez-Sala, D., 452, 457, 462, 464, 465, 467(2)
Perlman, J. I., 234
Perrault, I., 521, 534, 559, 564(12, 13), 713
Perron, M., 579, 584(17)
Perry, R. J., 139, 146
Person, S., 66, 67(20)
Persson, A., 28
Persson, B., 20, 24, 372
Pervin, A., 754
Peterson, G. L., 741
Peterson, P. A., 348, 565, 726
Peterson, W. M., 777
Peterson, W. S., 559
Peterson-Jones, S. M., 519, 520
Petrella, E. C., 29
Pettenati, M., 121
Petters, R. M., 519, 520
Pfeffer, B. A., 402, 565, 705, 706(2), 707(3), 716(3), 719(2), 720(2)
Pfeiffer, K., 519, 520, 572, 707, 715(24), 723(24)
Pfister, C., 254
Phan, K. B., 467
Phelps, E., 688
Philipov, P. P., 3, 3(11), 4, 13(9), 14(9), 19, 111, 121, 122, 167
Philippe, C., 692
Philips, M. R., 453, 464(12)
Philp, A. R., 279
Piantanida, T. P., 629, 652, 655(7)
Piantedosi, R., 371
Picco, C., 202

Pierce, M. E., 50, 51(16), 283
Pietras, K., 776
Piguet, B., 625
Pillinger, M. H., 453, 464(12)
Pinckers, A. J. L. G., 639, 650(48), 688, 691(2), 692(2)
Pirenne, M., 186
Piriev, N. I., 731
Pitt, F. H. G., 628
Pittler, S. J., 519, 520, 521, 524(6), 534, 559, 713
Plaksin, D., 21
Plant, A. L., 29
Plant, C., 591
Platts, J. T., 663
Ploder, L., 579, 590(8), 591
Pluckthun, A., 28
Pochet, R., 120
Podsakoff, G. M., 743, 778, 785(12)
Poenie, M., 141, 148
Poeschla, E. M., 762, 764(18)
Pokorny, J., 628, 639, 650(48)
Polak, J. M., 579
Polans, A. S., 11, 14(21), 103, 104, 105, 107, 107(11, 12, 22), 108, 109, 109(18), 110(25), 113, 114(35), 115, 116(19, 35), 119(19, 35), 120, 123, 167, 331, 340(2), 482, 515
Polesky, H. F., 708
Politi, L. E., 530
Pollock, B. J., 535
Polonsky, T., 754
Polverino, A., 690, 691(10)
Ponce de Leon, F. A., 536
Ponjavic, V., 625
Ponnazhagan, S., 785, 789(20)
Popescu, G., 372
Popov, V. I., 87, 98(4)
Porter, K., 688
Portera-Cailliau, C., 519, 672
Porumb, T., 13, 34, 129
Post, J., 688, 691(2), 692(2)
Potenza, M. N., 292
Potter, J., 67, 69(23), 85(23)
Potter, M., 745
Poustka, A., 633
Powell, D. S., 628
Powell, P. A., 579
Powers, R., 125
Pownall, H. J., 395
Pozzan, T., 266

Presecan, E., 90, 96
Prevec, L., 726, 731(15)
Prokopenko, E., 725
Pronin, A. N., 23, 24(9), 25(9)
Pronk, R., 732
Provencio, I., 279, 291
Prytowsky, J. H., 385
Ptashne, M., 690
Pugh, E. N., 153, 199, 201(22), 202(22), 519, 520, 613
Pugh, E. N., Jr., 167, 203, 204, 214(10), 216(26), 218, 221(10), 242, 250, 519, 520, 612, 614, 624
Puleo Scheppke, B., 184
Pulvermüller, A., 359
Püschel, A. W., 587

Q

Qin, N., 122, 519, 520, 521
Qing, Z. X., 117
Qu, G., 778, 785(13)
Quadroni, M., 510
Quiocho, F. A., 112(31), 113

R

Rager, G., 527
Raggett, E. M., 21
Rahmatullah, M., 465
Rajandream, M. A., 122
Ramirez-Solis, R., 559
Ramsay, G., 347
Rancourt, D., 182, 519, 520, 775
Randal, R. J., 500
Randall, C. J., 535
Rando, R. R., 224, 324, 325, 326, 326(8), 327(8), 328(12), 329, 330, 331, 332(3), 340, 341(4), 359, 370, 401, 436, 451, 452, 456, 457, 458, 458(31), 459, 459(38), 461(38, 39), 462, 462(31, 33), 463(31, 38, 39), 464, 464(44), 465, 467, 467(2), 511, 565, 566(5), 706
Rangini, Z., 536
Rao, V. R., 519, 520
Rapaport, D., 51, 84, 85(61)

Rapaport, D. H., 579, 582(12, 13), 584(13), 590(13, 13)
Raper, S. E., 743, 788(24), 789
Raport, C. J., 519, 520, 546, 549(16)
Rappaport, J., 762
Ratto, G. M., 133, 147
Ray, K., 467, 519, 520, 705, 707(7)
Ray, S., 3, 13, 13(9), 14(9), 111, 121
Rayborn, M. E., 51
Rayleigh, L., 639
Raymond, M. C., 719
Raymond, P. A., 579, 581, 582, 582(15, 18), 585, 585(15), 590(15, 16, 18)
Read, M. S., 726, 778, 785(12)
Reddy, M. G., 212
Redmond, T. M., 235, 519, 520, 565, 572, 705, 706, 706(2), 707, 707(3), 709(21), 711(21), 712(21), 713(21), 715(24), 716(3, 6), 718, 719, 719(2), 720, 720(2), 723(24)
Reece, R. J., 537
Reed, R. R., 592
Reh, T. A., 50
Rehemtulla, A., 530, 590, 593(6), 599(6)
Reichel, E., 618
Reichert, J., 425
Reid, D., 672
Reier, P. J., 744, 760(9)
Reig, C., 625
Reik, W., 655
Reinhardt, R., 688, 691(2), 692(2)
Reiser, M. A., 624
Reitner, A., 632(43), 633, 638(43), 639(43), 640(43), 641(43), 644(43), 645(43), 647(43)
Repka, A., 530
Reuter, T., 187
Reuveny, E., 455
Rhee, H., 592
Riazance-Lawrence, J. H., 113, 114(35), 115, 116(35), 119(35)
Rich, K. A., 414
Richards, J. E., 519, 565, 705
Richardson, C. D., 472
Rickman, C. B., 731
Ridge, K. D., 87, 98, 98(4), 428
Riebesell, M., 51
Rieke, F., 186, 189, 195(7), 197, 199(7, 21), 201(21)
Rim, J., 629, 632(28), 633(28), 638(28), 649(28)

Rine, J., 436, 451
Ripps, H., 50, 51(5), 172, 178(28)
Rising, J. P., 294
Rispoli, G., 134, 137(12), 139(12), 142(12), 195
Rittner, K., 744, 745(15), 778, 785(11)
Rivera, V. M., 778, 785(14)
Roberts, A. B., 429
Roberts, G. D., 107
Roberts, J. D., 353
Robertson, S., 67, 69(23), 85(23)
Robeson, C. D., 315
Robey, F. A., 235
Robinson, A. J., 732
Robinson, D. W., 139, 146
Robinson, G. W., 567
Robinson, J., 537
Robinson, N., 411
Robinson, S. W., 558, 559, 561
Robishaw, J. D., 455, 465, 467, 471
Robison, B., 87
Robson, J. A., 527
Robson, J. G., 203, 204, 212, 212(13), 214(28), 614
Roden, L. D., 31
Roder, J. C., 714
Rodriguez, K. A., 383
Roepman, R., 688, 691(2), 692(2), 703
Roepstorff, P., 510
Rollag, M. D., 279, 291, 292, 294
Roman, A. J., 612
Romano, J. D., 436
Romer, J., 388
Romert, A., 344, 348, 350(10), 356(10), 357, 358(10), 371
Rönnberg, I., 20
Roof, D., 180
Roos, H., 20
Ropers, H.-H., 688, 691(2), 692, 692(2)
Rosen, M. K., 464
Rosenberg, I., 85
Rosenberg, T., 646, 688, 691(2), 692(2)
Rosenbrough, N. J., 500
Rosenfeld, J., 503
Rosenfeld, M. G., 453, 464(12)
Ross, A. C., 400
Ross, A. S., 535
Ross, P. D., 254
Rossant, J., 519, 520, 714
Rossi, J. J., 762
Rossman, M. G., 372, 373(10)

Rothblat, G. H., 395
Rothman, J. E., 84
Rothschild, K. J., 425
Rotmans, J. P., 344, 359
Rounsifer, M. E., 581
Rozet, J.-M., 521, 534, 559, 564(13), 713
Ruddell, C. J., 694
Rudnicka-Nawrot, M., 270
Rudnicki, M. A., 730, 732(22)
Rudy, J., 728, 731(20)
Ruffner, D. E., 768
Rugh, R., 50
Ruiz, A., 325, 331, 400, 401, 403(16), 406, 410(16)
Ruiz, C. C., 103, 121, 122(2), 558
Ruiz Silva, B. E., 429, 430(30)
Running-Deer, J. L., 540
Rushton, W. A. H., 628, 648
Russell, D. W., 365, 368(21), 371(21), 744, 760, 787
Rustandi, R. R., 112(33), 113
Ruther, K., 671
Ruun, A. W., 347
Ryder, A. M., 82, 83(56)

S

Saari, J. C., 107(21), 108, 313, 331, 332, 335(5), 339, 344, 358, 359, 360, 362, 363, 364(16, 20), 365(7, 16–18), 366(7, 17, 18, 20), 367(18), 368(7), 370, 373, 382, 388, 400, 401, 536, 537, 552, 565, 566, 566(6), 567(13), 568, 572(13, 14), 573(13, 14), 574(13), 706, 722
Saavedra, R. A., 493
Sadowski, B., 646
Safer, B., 744
Saga, T., 530
SainteMarie, J., 65
Saito, T., 456, 462, 465, 466(6), 467(6, 7), 468(6), 469(6), 476(6), 478(7)
Saitoh, S., 122
Sakagami, Y., 437, 444
Sakitt, B., 186
Salamon, Z., 29
Salavati, R., 762
Salazar, M., 416(7), 417, 566
Salinas, J. E., 398
Salvetti, A., 744

Sambrook, J., 94, 171, 408, 691
Sampath, A. P., 3, 133(8), 134, 145(9), 148, 149, 151(15), 157(15, 16), 158(15, 16), 160(15), 161(16), 162(15, 16), 163(16), 247
Samuel, G., 731
Samulski, R. J., 743, 744, 745, 754(22), 760, 760(22), 778, 785(10), 787
Samulski, T., 760
Sana, T., 31
Sanada, K., 19
Sandberg, M. A., 519, 618
Sandgren, O., 625
Sang, H., 536
Sanger, B., 50, 58(15), 59(15), 64(15)
Sanger, B. M., 296, 302(16)
Sankar, U., 730, 732(22)
Sanocki, E., 629, 632, 649, 649(37)
Sanyal, S., 672
Sapperstein, S. K., 436
Sargan, D. R., 519, 520
Sargueil, B., 764
Sasaki, T., 117
Sastry, L., 425, 428(11, 12)
Sastry, M., 112(34), 113
Sather, W. A., 134, 137(10–12), 139(11, 12), 142(12), 195
Sato, N., 13, 16, 17, 18, 19(27), 20(27)
Satoh, T., 581
Satpaev, D. K., 19, 20, 23, 24(9), 25(9), 34(8), 36(8)
Saunders, J. A., 699
Saxen, L., 50
Scalzetti, L., 50, 51(13)
Schach, U., 536
Schaefer, B. C., 376
Schaefer, J., 41, 49(5)
Schäfer, B. W., 104
Schafer, R., 411
Schafer, W. R., 451
Schaffer, W., 326
Schagger, H., 76
Schappert, K., 672
Schechter, N., 581
Scheele, G. A., 492
Scheer, A., 464
Scherer, S. W., 590(8), 591
Scherphof, G., 82
Schertler, G. F. X., 425
Scheurer, D., 530, 590, 593(6), 599(6)

Schick, D., 688
Schiedner, G., 724
Schieltz, D., 436
Schier, A. F., 536
Schlueter, C., 50, 58(15), 59(15), 64(15), 296, 302(16)
Schmidt, H.-J., 633
Schmidt, W. K., 436
Schmitt, E. A., 537
Schmitt, M. C., 401
Schmitt-Bernard, C. F., 705
Schnapf, J. L., 203, 214(19, 21), 216(21), 219(19), 222(19, 21), 223, 618, 629, 643, 648(56)
Schnapp, J. L., 253
Schneck, M. E., 618
Schneeweis, D. M., 203, 214(21), 216(21), 222(21)
Schnell, M. A., 778, 785(14)
Schnetkamp, P. P. M., 145, 146, 159
Schoch, C., 68, 69(28)
Schorderet, D. F., 625
Schrimsher, G. W., 744, 760(9)
Schroeder, D. R., 427, 433(16)
Schuck, P., 21, 31
Schueller, A. W., 203, 214(10), 221(10), 613
Schuermann, W. H., 775
Schulte-Merker, S., 587
Schultz, D. W., 87, 530
Schuster, S. C., 22
Schwartz, E. A., 188
Schwartz, M., 519, 520, 646
Schwieter, U., 339
Scott, K., 519
Scotti, A., 19, 23, 34(8), 36(8)
Seabra, M. C., 437
Sears, M. L., 559
Segal, N., 383
Seidman, J., 600, 606(24), 691
Seiki, M., 285
Seiple, W. H., 204, 216(27)
Sejnowski, P., 70
Sekiya, T., 660, 708
Seldenrijk, R., 293
Semple-Rowland, S. L., 519, 520, 521, 526, 530, 534
Sen, A., 86
Sen, R., 427, 428, 428(15), 430(15, 17), 431(17), 433(17)
Senin, I. I., 3(11), 4
Seraphin, B., 120
Serra, A., 625
Sethi, E., 725
Severinsson, L., 726
Seymour, A. B., 519, 520
Sferra, T. J., 743
Shai, Y., 84, 85(61)
Shaklee, P., 754
Shang, E., 371
Shapiro, M. J., 453
Sharp, P. A., 762
Sharpe, L. T., 626, 627(7), 628, 629(15), 632(6, 7, 43), 633, 638(6, 43), 639(15, 43), 640, 640(43), 641(43), 642, 644(43), 645(43), 646, 646(6), 647(6, 43), 650(7, 15)
Shastry, B. S., 671
Shaw, L. C., 761
Sheffield, V. C., 591, 611, 621(8–10), 622(8), 625, 625(9), 671, 687
Sheils, D. M., 182
Shelhamer, J. H., 120
Sheline, Y., 49
Shen, D., 414, 416(7), 417, 566
Shenk, T., 760, 778, 785(10)
Sherman, N. E., 471
Sheshberadaran, H., 283
Shevchenko, A., 119
Shevell, S. K., 649
Shi, Y.-Q., 401, 464
Shichi, H., 69
Shichida, Y., 279, 282(6), 288, 430, 529, 579
Shidou, H., 491, 492
Shigemoto, R., 519, 520
Shiku, H., 96
Shimada, N., 96, 98
Shimada, T., 748
Shimayama, T., 765
Shimizu, F., 19
Shimizu, K., 440
Shimonishi, Y., 452, 457(6), 458, 462, 462(36), 463(36), 464, 465, 466(4, 5), 467(1, 4, 5, 8, 9), 468(4), 469(4, 8), 471(5), 473(5), 475, 475(8), 476(4, 5), 477, 478(4, 5, 9), 479, 481(1, 4, 5, 9)
Shippy, R., 764
Shiyan, S. D., 92
Shlaer, S., 186
Shoffner, R. N., 534
Shorey, C. D., 527
Short, J. M., 94, 693

Shortridge, R. D., 703
Shows, T. B., 629, 652, 655(7)
Shustak, C., 270
Shyjan, A. W., 558
Si, J. S., 530
Siderovski, D. P., 24
Sidman, R. L., 181, 182(33), 519, 520, 624
Siebert, P. D., 376
Sierra, D. A., 398
Sieving, P. A., 182, 212, 214, 519, 591, 612, 616, 688
Sillivan, L. S., 591
Sillman, A. J., 212
Silver, R., 282
Simon, A., 344, 348, 350, 350(10), 356(10), 357, 358(10), 359, 371, 372, 706
Simon, M. I., 3, 19, 22, 23, 24, 25(12), 34(8), 36(8), 40(12), 167, 168(1, 2), 169, 178(1), 184, 193, 465, 519, 520, 546, 549(16)
Simon, R. H., 726
Simonds, W. F., 465
Sinclair, J. H., 301
Sinelnikova, V. V., 122
Sinensky, M., 451
Singer, R. H., 579, 582(9)
Singh, A. K., 427, 428, 430(17), 431(17), 433(16, 17)
Sinha, S., 66, 71(10), 73, 74(10), 75(10), 76(10), 85(10), 621
Sipila, I., 519, 520
Siraganian, R., 705, 719(1)
Sitaramayya, A., 167
Sive, H. L., 296
Siwskowski, A., 764, 772
Sjölander, S., 20
Skolnick, C., 590
Slepak, V. Z., 19, 20, 22, 23, 24, 24(9), 25(9, 12, 13), 34(8), 36(8), 39(6), 40(12, 13)
Slesinger, P. A., 455
Sloane-Stanley, M. G. A., 68, 70(26)
Slome, D., 292
Sloop, G. D., 277
Smigel, M. D., 683
Smirnov, M. S., 628
Smith, J. A., 691
Smith, J. E., 385
Smith, N. P., 616, 618(35)
Smith, S. B., 519, 520, 574
Smith, V. C., 628, 639, 650(48)

Smrcka, A. V., 455, 458, 459, 459(21, 38), 461(38, 39), 463(38, 39), 465, 467
Snipes, W., 66, 67(20)
Snoeck, R., 753
Snow, B. E., 24
Snyder, R. O., 744
Snyder, S., 672
Snyder, S. H., 279
Snyder, W. K., 178, 519
Sobacan, L., 22, 39(6)
Sohocki, M. M., 591
Sokal, I., 121, 132, 519, 520, 521
Sokoloff, L., 395
Solari, R., 41
Sollner, T., 84
Solnica-Krezel, L., 536
Solow, R., 743
Somes, R. G., Jr., 535
Sondek, J., 22, 39(6)
Song, L., 270
Song, O., 591, 688
Soni, B. G., 279
Soreq, H., 766
Sorge, J. A., 94, 693
Soto Prior, A., 718
Souied, E., 534, 559, 564(13), 713
Soulages, J. L., 29
Souto, M. L., 425, 429, 431
Sowers, A. E., 699
Sparkes, R. S., 672
Speicer, S., 510
Spence, A. M., 760
Spence, S. G., 527
Spencer, M., 3, 13(9), 14(9), 111, 121, 167, 168(2, 4), 511, 551, 558
Spencer, S. A., 579
Sperling, H. G., 567, 574(16), 629, 705, 722(10)
Spiegel, A. M., 465
Sporn, M. B., 429
Spratt, S. K., 744
Srivastava, A., 785, 789(20)
Staehlin, T., 410
Ståhlberg, R., 20
Stainier, D. Y., 536
Standaert, T., 787
Starace, D. M., 581(29), 582, 590(29)
Staud, R., 453, 464(12)
Stecher, H., 330, 331, 335(5)
Steel, G., 519, 520

Stefano, F. P., 66, 69(9), 70(9), 71(9), 74(9), 75(9, 13), 76(9), 85(9, 13)
Stegmann, T., 86
Steinberg, R. A., 497
Steinberg, R. H., 65, 212, 612
Stemple, D. L., 536
Stenberg, E., 20
Stenhag, K., 24
Stenkamp, D. L., 579, 581, 582(15), 585, 585(15), 590(15)
Stephens, L., 455
Stern, M. H., 315
Sternglanz, R., 120, 689, 697, 698(21)
Sternweis, P. C., 455, 459(21), 467
Stewart, T. P., 86
Stiemke, M., 51
Stiles, W. S., 628, 639, 648(49)
Stillwell, W., 395
Stock, J. B., 453, 464, 464(12)
Stockman, A., 626, 627(7), 628, 629(15), 632(6, 7, 43), 633, 638(6, 43), 639(15, 43), 640, 640(43), 641(5, 43), 642, 644(43), 645(43), 646(6), 647(5, 6, 43), 650(7, 15)
Stone, E. M., 591, 611, 611(12), 612, 616(12), 619(12), 621(8–10, 12), 622(8), 625, 625(9), 671, 687
Stormo, G. D., 768
Stow, N. D., 725
Strähle, U., 585
Strauss, M., 776
Straznicky, K., 50
Struhl, K., 691
Strumane, K., 710, 713(28)
Stryer, L., 3, 13, 13(9), 14(9), 34, 81, 111, 112(32), 113, 121, 122, 123, 126, 127(15, 27), 128(11), 129, 129(27), 130(17, 27), 131(7, 17, 27), 161, 267, 425, 453, 454(15), 457, 544, 546
Strynadka, N. C. J., 112
Stubbs, G. W., 339
Stuhmer, W., 44
Stuiver, M. H., 732
Stults, J. T., 510, 558
Subbaraya, I., 103, 121, 122, 122(2), 558
Suber, M. L., 519, 520, 521
Sudo, T., 103
Sugden, D., 291, 293
Sullenger, B. A., 762
Sullivan, J., 776
Sullivan, S. A., 579, 581, 590(16), 762

Sulston, J. E., 494
Summerford, C., 745, 754(22), 760(22)
Sun, D., 725, 777
Sun, H., 566, 590(9), 591, 592(9), 593(9)
Sun, P., 50
Sundin, O., 579
Sung, C.-H., 178, 180(31), 519, 530, 560, 590, 593(6), 599(6), 611, 672
Surgucheva, I., 132
Suslov, O. N., 99
Sutcliffe, J. G., 519, 520, 621, 624, 672
Suzuki, A., 437, 444
Suzuki, Y., 105
Swain, P. K., 591
Swanson, R., 41, 43, 45(3), 46(3), 47(3)
Swanson, R. V., 22
Swaroop, A., 530, 590, 591, 593(6), 599(6), 688
Swieter, U., 321
Swindells, M., 128
Szerencsei, R. T., 146, 159
Szoka, F. C., 76, 78, 78(45)
Szostak, J. W., 762
Szuts, E. Z., 9

T

Tagawa, Y., 519, 520
Taguchi, T., 429
Tagushi, T., 429
Taira, K., 765
Takada, T., 397
Takagi, Y., 503
Takahashi, J. S., 283
Takai, Y., 436, 438, 438(11), 440, 451(11), 452
Takamatsu, K., 10, 11(18)
Takamiya, K., 96
Takanaka, Y., 283, 284
Takao, T., 452, 457(6), 458, 462, 462(36), 463(36), 464, 465, 466(4), 467(1, 4, 8, 9), 468(4), 469(4, 8), 475(8), 476(4), 478(4, 9), 481(1, 4, 9)
Takebe, Y., 285
Takei, Y., 559
Takekuma, S., 425, 428(11, 12)
Takematso, S., 122
Takemoto, H., 92
Tako, T., 465, 466(5), 467(5), 471(5), 473(5), 475, 476(5), 477, 478(5), 479, 481(5)

Tam, A., 436
Tam, J. P., 719
Tamai, M., 625
Tamayose, K., 748
Tamura, S., 437, 444
Tamura, T., 203, 214(20), 222(20)
Tan, E. W., 457, 462
Tanabe, T., 725, 777
Tanaka, M., 69
Tanaka, N., 378
Tanaka, T., 13, 34, 112(32), 113, 117, 122, 126, 127(15), 129, 130(17), 131(7, 17)
Tanaka, Y., 428
Tang, M., 519, 520
Tang, W. J., 455
Tao, L., 414, 416(7), 417, 566
Tapia-Ramirez, J., 592
Tarabykin, S., 376
Tashiro, Y., 397
Tate, R. L., 254
Taussig, R., 455
Tautz, D., 587
Tawara, I., 117
Taylor, J. A., 270
Taylor, M. R., 536, 537
Taylor, N. R., 762
Taylor, R., 132
Taylor, W. R., 233
Teller, D. Y., 629, 632, 649, 649(37), 661(22), 669
Tempst, P., 84
Terakita, A., 279, 282(6), 315, 420, 421(10), 579
Tesoriere, L., 383
Tetzlaff, W., 527
Theischen, M., 671
Thomas, D., 626, 633(1), 634(1), 651
Thomas, D. D., 425
Thomas, K. R., 715
Thomas, M. M., 216
Thomason, P. A., 455
Thompson, A. R., 744
Thompson, D., 41, 45(3), 46(3), 47(3)
Thompson, D. A., 565, 705, 706, 712(23), 713(23), 718, 719
Thompson, N., 41
Thompson, P., 167
Thompson, T., 70, 85(38)
Thomson, A. J., 321, 322(23)
Thornquist, S. C., 719

Timasheff, S. N., 493
Timmers, A. M., 764, 765(34), 766(34)
To, A., 672
Tobin, S. L., 501(30), 503, 508, 511(30)
Todd, S. C., 673
Tokunaga, F., 3, 11, 12, 12(10, 22), 13, 13(22), 14(22), 284, 581
Toledo-Aral, J. J., 592
Tollin, G., 29
Tomarev, S., 579
Tomita, T., 212
Tomlinson, M. G., 673
Tone, T., 762
Tordova, M., 87, 98, 98(4)
Towbin, H., 410
Towler, E. M., 31
Traboulsi, E., 646
Tran, K., 724
Tranum-Jensen, J., 270
Travis, G. H., 204, 519, 520, 621, 624, 672
Trehan, A., 401, 429
Trippe, C., 184
Tritz, R., 764, 765, 765(28)
Tsang, S. H., 193, 519, 520, 521
Tsien, R. Y., 133, 141, 147, 148, 159(18), 266
Tsigelny, I., 378
Tsilou, E., 565, 705, 706(2), 707(3), 716(3), 719(2), 720, 720(2)
Tsin, A. T. C., 331, 383, 386, 387, 388, 390(20a), 394, 395(16), 397, 398, 399(31), 400(27)
Tsuda, M., 488, 491, 492
Tsuda, T., 488
Tsugane, S., 119
Tsui, L. C., 590(8), 591
Tsukahara, Y., 279, 282(6), 579
Tsukida, K., 430
Tuma, M. C., 294
Tuohy, G., 776
Tuschl, T., 762
Tuvendal, P., 371
Tybulewicz, V. L., 714

U

Ueda, H., 465, 466(5), 467(5), 471(5), 473(5), 475, 476(5), 477, 478(5), 479, 481(5)
Ueda, N., 455

Uehara, F., 582
Uehara, M., 529
Ueshima, T., 529
Uhlen, M., 24
Uhlenbeck, O. C., 765, 768, 776
Uhler, M., 582
Ulshafer, R. J., 534
Um, J. M., 429
Unson, C. G., 465
Urano, T., 96
Urbaniczky, C., 20
Urcelay, E., 744
Urieli-Shoval, S., 582
Uriksson, U., 371
Usuda, N., 119
Usukura, J., 493, 501(30), 508, 511(30)
Utell, M. J., 775
Uy, R., 494
Uyama, M., 397

V

Vail, W., 76, 78(45)
Vaish, N. K., 764
Valentini, P., 519, 520
Valeva, A., 270
Valle, D., 519, 520
Van Aelst, L., 690, 691(10)
van Amsterdam, L. J. P., 428
van Bokhoven, H., 519, 520, 692
Van den Berg, E. M., 425
van den Hurk, J. A., 519, 520, 692
van de Pol, D. J., 519, 520
Vander, J. F., 781, 785(17)
van der Steen, R., 428
van der Vliet, P. C., 732
van de Veerdonk, F. C. G., 293
van Duynhoven, G., 688, 691(2), 692(2)
Van Echteld, C. J. A., 86
van Eeden, F. J., 536
Van Epps, H. A., 536, 537
van Ginkel, P. R., 103, 109, 110(25)
Van Hooser, J. P., 168, 269, 270, 313, 359, 486, 487(12), 511, 519, 520, 521, 530, 552, 565, 566, 567(13), 572(13, 14), 573, 573(13, 14), 574(13)
Van Hooser, P., 359
Van Kuijk, F. J. G. M., 318, 320(16), 365

van Norren, D., 203
van Oordt, G. J., 306
van Oostrum, J., 428
van Roy, F., 710, 713(28)
Van Schepdael, A., 753
van Vugt, A. H. M., 359, 360, 368(15)
Vaquero, M. I., 743, 776, 789
Verdon, W. A., 618
Vergara, J. L., 148
Verkleij, A. J., 86
Vernos, I., 294
Verriest, G., 639, 650(48)
Verrinder Gibbins, A. M., 536
Veske, A., 519, 520
Vestling, C. S., 93
Vetter, M. L., 579, 584(17)
Vidal, M., 625
Vieira, J., 353
Villa, C., 425
Villazana, E. T., 383, 394, 398, 400(27)
Villnave, C., 579, 582(9)
Viswanathan, S., 204, 214(28)
Vivien, J. A., 628
Vogel, H., 29
Vogelsang, E., 536
Vogelstein, B., 719
Volker, C., 453, 464, 464(12)
Vollrath, D., 651, 655(3)
von Heijne, G., 347
Vonica, A., 90, 96
von Jagow, G., 76
von Weizsacker, F., 762
Vorm, O., 119
Voulgaropoulou, F., 744
Voyta, J. C., 701
Vuong, T. M., 253, 254, 264, 267, 269(7, 10)

W

Waage, M. C., 117
Wachtmeister, L., 203
Wada, A., 428
Wada, Y., 279, 280(2), 282(2), 283(2), 284, 625
Wald, G., 313, 315, 324, 359, 383, 476, 565, 628, 631
Walev, I., 270, 277
Walkeapaa, L. P., 429
Walker, B., 270, 271

Walker, T. M., 103, 109, 110(25)
Wallace, R. T., 781, 785(17)
Waller, S. J., 172
Walseth, T. F., 99
Walsh, J., 519, 520
Walsh, K. A., 3, 13(9), 14(9), 103, 111, 121, 122, 122(2), 168, 270, 483, 485, 558
Walsh, S. V., 122
Walters, L., 383
Wands, J. R., 762
Wang, A. Y., 429
Wang, G., 388
Wang, H., 726
Wang, J., 429, 430(30), 435
Wang, M. M., 592
Wang, Q.-L., 530, 590, 590(9), 591, 592(9), 593(6, 9), 599(6)
Wang, T., 519, 520
Wang, W., 464
Wang, X.-S., 785, 789(20)
Wang, Y., 29
Ware, J. A., 464
Warren, M. J., 132
Warwar, R., 590
Wassall, 395
Watanabe, C., 510
Watanabe, K., 96, 98
Watanabe, Y., 105, 119
Waterson, R., 494
Watson, J. A., 452
Watson, J. T., 510
Weaver, J. C., 278
Webb, J. R., 628
Weber, C., 112(34), 113
Weber, D. J., 112(33), 113
Weber, K., 115
Webster, M. A., 650
Wechter, R., 121
Wedel, B. J., 558
Wedemann, H., 671
Wei, N., 698
Wei, S., 371
Weisler, L., 315
Weiss, E. R., 519, 520
Weissmann, C., 326
Weissmann, G., 453, 464(12)
Weitz, C. J., 634
Weitz, S., 633
Weitzman, M. D., 760, 779
Weleber, R. G., 621, 646, 671

Well, W. W., 96
Wells, J. A., 625, 671, 687
Wells, K. D., 519, 520
Welte, W., 425
Wendtner, C. M., 744
Weng, J., 519, 520, 672
Werner, M., 530
Wernstedt, C., 331, 340(2), 359, 372, 706
Wessling-Resnick, M., 456
West, K., 332
Westerfield, M., 537, 538, 587
Whalen, P. O., 761
Wheeler, T., 360, 364(16), 365(16)
White, J. M., 65, 452
White, K. D., 628
Whiteheart, S. W., 84
Whitten, M. E., 436
Whittle, N., 724
Whitton, J. L., 762
Wickman, K. D., 455
Wiggert, B., 224, 235, 236, 237(20), 319, 324, 359, 565, 566(4), 574
Wigler, M., 690, 691(10)
Wikovsky, P., 50, 51(5)
Wilcox, M. D., 25
Wilden, U., 167
Wiles, C. D., 591
Wilkins, M. R., 494, 505(21)
Willey, J. C., 775
Williams, A. J., 649
Williams, D. S., 66, 69, 69(9), 70(9), 71(9), 74(9), 75(9), 76(9), 85(9), 673
Williams, K. L., 494, 505(21)
Williams, R. J. P., 103
Williams, T. P., 153, 233, 624
Wilm, M., 119
Wilschut, J., 66, 67(19, 21), 71(21), 82, 86
Wilson, H. C., 302
Wilson, I. A., 347
Wilson, J. M., 726, 743, 760, 778, 779, 779(9), 782(9), 785(9, 13, 14), 787(9), 788(24), 789, 789(9)
Wilson, M. A., 535
Winderickx, J., 629, 632, 633, 649, 649(37), 651, 654, 655, 659(14), 660(11, 14), 669
Winkens, H. J., 359, 360, 368(15)
Winston, A., 324, 325, 326, 326(8), 327(8), 328(12), 331, 341(4), 401, 403(16), 410(16), 706
Winston, F., 698

Winther, B., 744
Winther, J. R., 347
Winzor, D. J., 31
Wise, A., 41
Wissinger, B., 621, 633
Witkovsky, P., 50, 282
Witkowska, D., 104, 107, 107(12), 113, 114(35), 115, 116(19, 35), 119(19, 35)
Witter, R. Z., 277
Wittwer, B., 688
Wolburg, H., 529
Wold, F., 494
Wolda, S. L., 437, 452
Wolf, D., 54
Wolf, E. D., 534
Wolf, L. G., 456
Wolf, S., 633
Wolgemuth, D. J., 371
Womack, F. C., 110
Wong, D. J., 565, 705
Wong, F., 519, 520, 773
Wong, S. C., 510, 558
Wong-Staal, F., 762
Woo, S. L., 726
Woodford, B. J., 169
Woodfork, K. A., 471
Wray, G. A., 581
Wright, A. F., 688, 706
Wright, M. D., 673
Wright, R., 744
Wroblewski, J. J., 625, 671, 687
Wu, D., 24, 25(12), 40(12), 465
Wu, M., 58
Wu, T., 120
Wu, W., 591, 688
Wu, Y., 724
Wu, Z., 31, 690
Wyszecki, G., 639, 648(49)

X

Xiao, W., 778, 785(13)
Xiao, X., 743, 744, 787
Xie, H. Y., 438, 452, 453, 467
Xing, Z., 762
Xiong, W.-H., 561
Xu, J., 24, 25(12, 13), 40(12, 13), 167, 193, 519, 520

Xu, J. Z., 590
Xu, L., 705
Xu, S., 591
Xu, X., 614

Y

Yakhyaev, A. V., 87, 98(4)
Yamada, A., 429
Yamada, M., 11, 12, 12(22), 13(22), 14(22)
Yamada, O., 762, 764(18)
Yamada, T., 503
Yamagata, K., 4
Yamaguchi, K., 13
Yamaguchi, T., 651
Yamagushi, T., 653, 655(9), 668(9)
Yamamoto, A., 397
Yamamoto, T., 425, 428(11, 12)
Yamanaka, G., 267
Yamane, H. K., 438, 452, 467, 741
Yamashita, C. K., 519, 520, 521, 724, 731
Yamazaki, A., 122
Yan, W., 761, 775
Yan, Y. L., 585, 587(37)
Yanagita, T., 582
Yandell, D. W., 618
Yang, E. Y., 778, 785(12)
Yang, H., 592
Yang, M., 690
Yang, R.-B., 519, 520, 521, 558, 559, 559(9), 561, 564(12, 14)
Yang, Y., 689, 726, 743, 779, 788(24), 789
Yang-Feng, T. L., 565, 705
Yao, X. L., 120
Yap, K., 128
Yarfitz, S. L., 482, 540, 558
Yasumura, D., 743, 745(2), 760(2), 777
Yates, J. R. III, 436, 448, 494, 510
Yau, K.-W., 3, 7, 146, 149, 185, 187, 189(3), 190, 193, 199, 199(9), 201(9), 202(23), 203, 204, 214(20), 219, 222(20), 226, 243, 247, 250, 250(9), 515, 558, 561
Yau, T. D., 228
Ye, X., 778, 785(14)
Yeagle, P. L., 65, 66, 66(6), 68(6), 70, 70(12), 71(6, 12), 72, 74(12), 75(6, 12), 79, 86
Yeh, S. H., 689
Yi, Y., 592

Yokokura, H., 103
Yokota, K., 285
Yokota, M., 497, 498(22)
Yokoyama, K., 436, 437, 445, 446(24), 447(24), 451(24)
Yokoyama, S., 378
Yoser, S. L., 714
Yoshida, M., 285
Yoshida, N., 488, 491, 492
Yoshida, Y., 436, 438, 438(11), 451(11)
Yoshihara, K., 429
Yoshikami, S., 234, 254
Yoshikawa, T., 281, 282(7), 283(7)
Yoshizawa, T., 279, 283, 283(1), 284, 288, 288(1), 290(1), 430, 452, 456, 457(6), 458, 462, 462(36), 463(36), 464, 465, 466(4, 6), 467, 467(1, 4, 6–9), 468(4, 6), 469(4, 6, 8), 475(8), 476(4, 6), 478(4, 7, 9), 481(1, 4, 9), 529
Young, J. D., 277
Young, J. E., 70, 79
Young, J. H., 741
Young, R. W., 65
Young, T., 626
Younger, J. P., 3, 133, 133(8), 134, 145(6), 147, 163(12)
Younghusband, H. B., 732
Yu, M., 762, 764(18)
Yu, Q., 766
Yu, S., 519, 520, 572, 707, 715(24), 718, 723(24)

Z

Zack, D. J., 530, 590, 590(9), 591, 592(9), 593(6, 9), 597, 599(6, 23)
Zahn-Poe, X., 24, 25(13), 40(13)
Zaia, J. A., 762
Zakour, R. A., 353
Zang, F. L., 466
Zardoya, R., 741
Zech, L., 68, 69(28)
Zeitlin, P. L., 743
Zeng, Y., 725, 777, 782, 785(18)
Zervos, A. S., 689
Zhai, Y., 372, 378(4)
Zhang, F. L., 436
Zhang, H. Z., 425, 428(11, 12)
Zhang, J., 50
Zhang, K., 122
Zhang, Q., 519, 520
Zhang, W. W., 725
Zhang, Z., 283
Zhao, J., 80
Zhao, X., 103, 121, 122(2), 223, 270, 558, 612
Zheng, Y., 592
Zhou, B., 429
Zhou, C., 776
Zhou, H., 726
Zhu, X., 24, 25(13), 40(13)
Zhu, Y., 22, 39(6)
Zill, A., 294
Zimmerman, A. L., 137, 247
Zimmerman, W. F., 313, 383
Ziroli, N. E., 204, 216(27)
Zolotarev, A. S., 488
Zolotukhin, S., 743, 744, 745, 746(8), 760(8, 9), 764, 776, 789
Zoltick, P., 778, 785(14)
Zon, L. I., 537
Zozulya, S., 13, 34, 122, 123, 126, 127(27), 128(11), 129(27), 130(27), 131(27)
Zrenner, E., 611, 618(2), 621, 633, 646, 671, 688, 706, 709(21), 711(21), 712(21), 713(21)
Zwartkruis, F., 536

Subject Index

A

AAV, *see* Adeno-associated virus
Adeno-associated virus
  helper plasmid, 745
  infectious center assay, 756–757, 759–760
  purification of recombinant virus
    concentrating with filters, 755–756
    containment facility, 746
    cotransfection of viral and helper plasmids, 746–747
    heparin affinity chromatography, 753–756
    host cells, 746
    iodixanol density step gradient purification
      advantages over cesium chloride, 748, 760–761
      aggregation prevention, 747–748, 760
      density, 748
      efficiency and resolving capacity, 749, 751, 753
      gradient formation and centrifugation, 753
      iodixanol properties, 748
    lysis of cells, 747
    overview, 744–746, 760–761
    plasmid vector, 746
    technique effects on gene transfer efficiency, 760
  recombinant gene transfer applications, 743–744, 777
  retinal transduction
    animal species comparisons
      humoral immune response, 785, 787, 789
      time of onset for expression, 784–785
    green fluorescent protein reporter
      advantages, 777
      histological analysis, 785
      ophthalmoscopy, 783–784
    intraocular administration in animals, 779–783
    promoter, 777
    recombinant virus preparation, 778–779
    titering by quantitative polymerase chain reaction, 756–757
Adenovirus minichromosome, *see* Encapsidated adenovirus minichromosome
All-*trans* retinol dehydrogenase
  high-performance liquid chromatography assay, 365, 368, 370–371
  phase partion assay
    all-*trans* retinol tritiation, 362
    endogenous substrate assay, 363
    enzyme preparation, 363
    exogenous substrate assay, 363–364
    materials, 361
    membrane preparation, 363
    nicotinamide adenine dinucleotide tritiation, 361
    pH, 360
    principle, 360
  visual cycle role, 359
Argon ion laser, *see* Confocal microscopy
Arrestin, promoter analysis in transiently transfected *Xenopus* embryos, 57
Autosomal dominant retinitis pigmentosa, *see* Retinitis pigmentosa

B

BIACORE, *see* Surface plasmon resonance
Bis(*p*-nitrophenyl) phosphate, all-*trans*-retinyl ester hydrolase inhibition studies, 391–392
Bleaching, *see* Photobleaching recovery, photoreceptors

C

Calcium-binding proteins, purification and identification in eye

conformational changes on calcium binding, 112–114
flow dialysis analysis of calcium binding, 110–112
hydrophobic interaction chromatography, 105–106
protein–protein interactions
  coimmunoprecipitation, 117
  cross-linking, 117
  gel overlay assay, 115–117
  identification of target proteins, 117, 119–120
  mass spectrometry identification of binding proteins, 12
  yeast two-hybrid system, 119–120
reversed-phase high-performance liquid chromatography, 107
sequence analysis and protein identification, 107–109
uveal melanoma tissue extract preparation, 105
Calcium flux
  bleaching adaptation role, 249–252
  fluorescence indicators for photoreceptor studies, 146–147
  photoreceptor measurements using spot confocal microscopy with Fluo-3
    advantages, 148–149
    argon ion lasers, 149, 151–154
    calibration, 159–162
    detectors, 154–156
    dye loading, 157
    instrumentation, 149
    light-induced decline of calcium, 157–159
    photopigment cycle changes in calcium, 162–163
    photoreceptor preparation from tiger salamanders, 156–157
    rationale, 147–148
  phototransduction role, 133, 146
  rod outer segment, cytoplasmic measurements
    circulating current relationship to free cyclic GMP, 137, 139
    current sources, 137, 139
    Indo measurements
      calcium dependence of cation exchange rate, 145
      calibration, 141–142

    dextran-conjugated dye, 139–140
    light-evoked changes, 145–146
    optical setup for fluorescence measurements, 140–141
    snapshots of calcium, 142–145
    rod outer segment isolation, 134
    solution compositions, 134–135
    whole-cell patch-clamp recording, 135–137
Calmodulin
  phototransduction role, 103–104
  surface plasmon resonance
    GRK1 binding kinetics analysis, 30–31, 33
    immobilization for studies, 24–25
Calorimetry, see Microcalorimetry, time-resolved studies of phototransduction
Cellular retinaldehyde-binding protein, preparation of recombinant apoprotein from *Escherichia coli*, 332–333
cGMP, see Cyclic GMP
Chicken retina
  advantages as model system, 529–530, 535–536
  cone dominance, 527–529
  development, 527–528
  laminar microdissection
    apparatus and instruments, 532
    applications, 530–531
    freeze-dried retinal section preparation, 531
    section picking, 532, 534
  oil droplets, 529
  rd model, 534–535
  rdd model, 535
  structure, 528–529
Childhood-onset severe retinal dystrophy, retinal pigment epithelium p65 mutations, 711–713
Circular dichroism (CD), recoverin, calcium binding and conformational changes, 113–114
Color blindness, see Cone
Cone
  absorption of cone pigments
    peaks, 631–632
    polymorphisms and variant human color vision, 649–650
    recombinant pigments
      expression constructs, 634

human embryonic kidney cell transfection, 635
maximum wavelength determination, 636–638
mutation effects on maximum absorption wavelength, 649–650
photobleaching difference spectra acquisition, 635–636
reconstitution, 635
chicken retina, cone dominance, 527–529
deuteranomaly
green–red pigment hybrid characterization
long-range polymerase chain reaction amplification, 663
pulsed-field gel electrophoresis, 663, 665
sequencing and restriction enzyme analysis, 665
Southern blot analysis, 665
hybrid pigment genes, 652–653, 662–663
electroretinograms, activation modeling, 616–618
green and green–red hybrid pigment gene expression in arrays, polymerase chain reaction/single-strand conformation polymorphism, 668–670
green pigment
deletion, 652
point mutation, 654
opsin phosphorylation assay
incubation conditions and gel autoradiography, 551–552
materials, 551
principle, 551
white light versus red light response, 552
paired-flash electroretinography studies of photoreceptor response, subtraction of cone contribution, 209–211, 218
photobleaching, see Photobleaching recovery, photoreceptors
recombination of pigments in red–green color vision defects, 652
red–green hybrid gene, polymerase chain reaction amplification and detection in females, 660, 662

red/green pigment transcripts, ratio determination in retina
polymerase chain reaction/single-strand conformation polymorphism, 667–668
segregated expression, 665, 667
red pigment Ser180Ala polymorphism
color perception in males, slight variations, 653–654
single-strand conformational polymorphism analysis, 659–660
sequence homology between pigments, 651
spectral sensitivity determination
chromatic adaptation and selective desensitization, 628
electroretinography, 629–631, 647–648
fundus reflectometry, 628
microspectrophotometry, 628–629, 631, 648
pigment absorption effects, 631–632
psychosocial determination of M, L and hybrid cones
comparison with in vitro determinations, 648–649
flicker photometry, 639–640
male dichromat studies, 633, 639
maximum wavelength determination, 641
overview, 638–639, 641
spectra calculation from corneal spectral sensitivities, 641–644
psychosocial determination of S cones
flicker photometry, 646
maximum wavelength determination, 646–647
overview, 645–646
single-cell electrophysiology, 629, 631, 648
variability sources, 632–633
trichromacy, 626, 628, 651
X-linked red–green photopigment genes, determination of number and ratio
competitive polymerase chain reaction amplification, 658
data analysis, 655–656
exons 2, 4, and 5 analysis, 659
overview, 655
primers, 657–658

single-strand conformational polymorphism, 655, 658–659
pulsed-field gel electrophoresis, 655
Southern blot analysis, 655
Cone worse than rod disease, electroretinogram studies, 625
Confocal microscopy
  immunofluorescence localization in α-toxin permeabilized photoreceptors
    antibody preparation, 272–273
    bovine retina sectioning and staining, 273
    cell death analysis, 277
    data collection, 273, 275
    neurobiotin distribution following permeabilization, 277–278
    toxin distribution, 277
  photoreceptor calcium flux measurements using spot confocal microscopy with Fluo-3
    advantages, 148–149
    argon ion lasers, 149, 151–154
    calibration, 159–162
    detectors, 154–156
    dye loading, 157
    instrumentation, 149
    light-induced decline of calcium, 157–159
    photopigment cycle changes in calcium, 162–163
    photoreceptor preparation from tiger salamanders, 156–157
    rationale, 147–148
CRALBP, see Cellular retinaldehyde-binding protein
Cross-linking
  calcium-binding protein targets, 117
  S-modulin–rhodopsin kinase, 19–20
Cyclic GMP
  circulating current relationship to free cyclic GMP in isolated rod outer segments, 137, 139
  heat of hydrolysis, see Microcalorimetry, time-resolved studies of phototransduction
Cyclic GMP phosphodiesterase
  abundance in rod outer segment, 253
  calcium flux, bleaching adaptation role, 249–252
  calorimetric assay, see Microcalorimetry,

time-resolved studies of phototransduction
electrophysiology assay in photoreceptor
  bleaching adaptation assays, 243, 245, 249
  calculations, 242
  rapid solution changes, 239–241
γ subunit, surface plasmon resonance analysis of mutants, 39
genotyping of mouse retinitis pigmentosa models, 523
rescue in mouse rd model of retinitis pigmentosa using adenovirus minichromosome
  adenovirus minichromosome construction
    bacterial expression, 731
    helper virus construction, 731–732
    purification on cesium chloride gradients, 738
    transgene packaging, 732, 734
  agarose overlay and plaque formation, 735
  expression detection
    activity assay, 741–742
    reverse transcriptase–polymerase chain reaction, 740
    trypsin activation assay, 742
    Western blot analysis, 740–741
  helper virus DNA preparation for transfection, 739
  lysate preparation following amplification, 737–738
  minichromosome/helper virus ratio determination, 736–737
  overview, 725
  plaque picking, 735–736
  safety, 742–743
  serial propagation, 735, 737
  transfections
    calcium phosphate, 734–735
    lipofection, 734
  wild-type virus expression, 739–740
S-modulin regulation of activation, 3, 14
zebrafish larva eye, assay
  incubation conditions and product separation, 546–547
  light sensitivity determination, 547
  materials, 546

principle, 546
time course of activation, 547, 549

D

DNase I footprinting
  advantages and limitations, 600, 603
  principle, 600, 603
  retinal transcription factor characterization
    binding reaction, 608–609
    digestion, 609–610
    gel electrophoresis and autoradiography, 610
    materials, 605–606
    radiolabeled DNA preparation with polymerase chain reaction, 606–608
    sequence ladder genration for both strands, 608

E

EAM, *see* Encapsidated adenovirus minichromosome
Electrophoretic mobility shift assay
  principle, 600
  retinal transcription factor characterization
    binding reaction, 602
    gel electrophoresis and autoradiography, 602
    materials, 600–601
    mutant probes, determination of binding site specificity, 603
    radiolabeling of probe, 601–602
    supershift assay for protein identification, 603
Electroretinography
  applications, 203–204, 222
  cone activation modeling, 616–618
  cone worse than rod disease, 625
  equal rod and cone disease, 624
  GC-E knockout mouse recordings, 560–564
  paired-flash studies of photoreceptor response

amplitude–intensity relation, 219, 221–222
blink reflex, 209
b-wave component, 212
clinical applications, 222
cone contribution, subtraction, 209–211, 218
data acquisition, 207–208
desensitization of rod-mediated probe response, 214
postreceptor negative components, 212–214
principle, 204–205
probe flash unit, 208–209
prospects, 222–223
time course of derived response, 214, 216, 218–219
peripherin/rds mutation studies, 621–624
photoresponses, 612–614
recording technique, overview, 612
rhodopsin mutation studies, 618–621
rod activation modeling, 614–616
rod worse than cone disease, 624
spectral sensitivity determination of cones, 629–631, 647–648
waves, 202–203, 613
EMSA, *see* Electrophoretic mobility shift assay
Encapsidated adenovirus minichromosome
  cyclic GMP phosphodiesterase rescue in mouse *rd* model of retinitis pigmentosa
    adenovirus minichromosome construction
      bacterial expression, 731
      helper virus construction, 731–732
      purification on cesium chloride gradients, 738
      transgene packaging, 732, 734
    agarose overlay and plaque formation, 735
    expression detection
      activity assay, 741–742
      reverse transcriptase–polymerase chain reaction, 740
      trypsin activation assay, 742
      Western blot analysis, 740–741
    helper virus DNA preparation for transfection, 739

lysate preparation following amplification, 737–738
minichromosome/helper virus ratio determination, 736–737
overview, 725
plaque picking, 735–736
safety, 742–743
serial propagation, 735, 737
transfections
  calcium phosphate, 734–735
  lipofection, 734
wild-type virus expression, 739–740
gene therapy applications, 725
helper virus constructs, overview, 729–731
host immune response, 726
plasmid constructs, overview, 728–729
replication
  origin, 725
  requirements, 726–728
Equal rod and cone disease, electroretinogram studies, 624
ERG, *see* Electroretinography

F

Flicker photometry, *see* Cone
Fluo-3, photoreceptor clacium flux measurements using spot confocal microscopy
  advantages, 148–149
  argon ion lasers, 149, 151–154
  calibration, 159–162
  detectors, 154–156
  dye loading, 157
  instrumentation, 149
  light-induced decline of calcium, 157–159
  photopigment cycle changes in calcium, 162–163
  photoreceptor preparation from tiger salamanders, 156–157
  rationale, 147–148

G

GCAP1, *see* Guanylate cyclase-activating protein 1

GCAP2, *see* Guanylate cyclase-activating protein 2
Gel overlay, calcium-binding protein target identification, 115–117
Gene therapy, *see* Adeno-associated virus; Encapsidated adenovirus minichromosome; Ribozyme
GRK1
  adenosine nucleotide effects on protein–protein interactions, 34, 36
  autophosphorylation, 36
  binding kinetics analysis with recoverin or calmodulin using surface plasmon resonance, 30–31, 33
  fragment binding to recoverin, 39–40
  protein–protein interactions, 34
Guanylate cyclase-activating protein 1, calcium effects, 103
Guanylate cyclase-activating protein 2
  calcium dependency of guanylate cyclase activity modulation, 123
  homology with other guanylate cyclase-activating proteins, 132
  myristoylation, 123
  structure determination by nuclear magnetic resonance
    data acquisition and analysis, 125–126
    isotopic labeling and purification of recombinant protein, 124–125
    sample preparation, 125
    three-dimensional structures
      recoverin homology, 131–132
      unmyristoylated protein with calcium, 130–131
Guanylyl cyclase
  electrophysiology assay in photoreceptors
    bleaching adaptation assays, 245, 247, 249
    calculations, 242
    rapid solution changes, 239–241
  GC-E knockout mouse
    electroretinogram recordings, 560–564
    generation, 559
    histological analysis, 560–561
    single-cell electrophysiology, 560–561, 563–564
  gene loci, 558
  mutation in disease, 558–559
  types, 558
  zebrafish larva eye, assay

calcium dependence, 550–551
incubation conditions and product separation, 550
materials, 549–550
principle, 549

H

High-performance liquid chromatography
  enzyme assay
    all-*trans* retinol dehydrogenase, 365, 368, 370–371
    11-*cis*-retinol dehydrogenase, 365–366, 368, 370–371
  retinoid analysis in visual cycle, *see* Retinoids, high-performance liquid chromatography analysis
HPLC, *see* High-performance liquid chromatography

I

Indo-dextran, calcium flux measurements in rod outer segments
  calcium dependence of cation exchange rate, 145
  calibration, 141–142
  dextran-conjugated dye, 139–140
  light-evoked changes, 145–146
  optical setup for fluorescence measurements, 140–141
  snapshots of calcium, 142–145
Iodixanol gradient, *see* Adeno-associated virus
Isoelectric focusing, *see* Two-dimensional gel electrophoresis
Isomerhydrolase
  assay
    enzyme preparation, 326
    high-performance liquid chromatography of products, 327
    incubation conditions, 326–327
    materials, 325
    principle, 325
    tracer concentrations of substrate, 327–328
  11-*cis*-retinol

generation, 325
product inhibition, 328–330
simultaneous detection with lecithin:retinol acyltransferase and retinyl ester hydrolase activities
  all-*trans*-retinol preparation, 333
  cellular retinaldehyde-binding protein preparation and effects, 332–333, 340–341
  denaturing gel electrophoresis, 339
  high-performance liquid chromatography
    calculations, 339
    extraction of retinoids, 333–334
    retinyl ester hydrolysis, 337
    separation of retinoids, 334–335, 340
  incubation conditions, 333
  kinetic effects of phosphates, ATP, and alcohols, 341–343
  materials, 332
  stability of proteins, 339–340
  ultraviolet treatment of retinal pigment epithelium microsomes, 332, 340
visual cycle role, 325, 331

L

Leber's congenital amaurosis, retinal pigment epithelium p65 mutations, 711–714
Lecithin:retinol acyltransferase
  catalytic reaction, 401
  complementary DNA
    characterization, 406–407
    library screening, 405
  expression in human embryonic kidney cells
    subcloning into expression vector, 407
    transfection, 407–408
  inhibitors, sulfhydryl-directed, 401–402
  messenger RNA isolation, 402–403
  Northern blot analysis of tissue distribution, 408–410
  retinal pigment epithelial cell preparation for complementary DNA library preparation, 402
  reverse transcriptase–polymerase chain

reaction amplification of RNA, 403–405
sequence analysis
  nucleic acid, 412–413
  protein, 410–412
simultaneous detection with retinyl ester hydrolase and retinol isomerase activities
  all-*trans*-retinol preparation, 333
  cellular retinaldehyde-binding protein preparation and effects, 332–333, 340–341
  denaturing gel electrophoresis, 339
  high-performance liquid chromatography
    calculations, 339
    extraction of retinoids, 333–334
    retinyl ester hydrolysis, 337
    separation of retinoids, 334–335, 340
  incubation conditions, 333
  kinetic effects of phosphates, ATP, and alcohols, 341–343
  materials, 332
  stability of proteins, 339–340
  ultraviolet treatment of retinal pigment epithelium microsomes, 332, 340
visual cycle role, 325, 331, 401
Western blot analysis, 410
LRAT, *see* Lecithin:retinol acyltransferase

M

Mass spectrometry
  calcium-binding proteins
    binding protein identification, 120
    identification of calcium-binding proteins, 108–109
  peptide mass fingerprinting of retinal proteins separated by two-dimensional gel electrophoresis
    applications, 494–495, 510–511
    in-gel digestion
      protease selection, 502–503
      trypsin digestion, 503–504
    matrix-assisted laser desorption ionization–time of flight analysis
      criteria for identification using MS-Fit, 508–509

database searching, 505–506
equipment, 505
large fragment fingerprinting, 509–510
principle, 504–505
spectra acquisition, 505
prospects, 510–511
prenylation analysis
  gas chromatography–mass spectrometry analysis of cleaved prenyl groups, 444–445
  high-performance liquid chromatography–electrospray ionization mass spectrometry of peptides, 448–449, 451
rhodopsin phosphopeptide analysis, 485, 487–488, 492
Melanophore, *see* Melanopsin
Melanopsin
  function, 279
  gene cloning, 279–282
  G protein coupling, 294
  tissue distribution in amphibians, 291
  Xenopus melanophore
    advantages and prospects for system, 308–309
    isolation from embryos
      dissociation of tissue, 300
      Ficoll gradient purification, 300–301
      maintenance of cultures, 301–302
      materials, 299–300
      primary culture, 300
    microtiter plate reader absorbance assay of responses
      data acquisition, 308
      illumination artifacts, 306
      materials, 305–306
      sample preparation, 307–308
    phenotype switching in development, 293–294
    response to light and scoring, 291–293
    transgenic embryo preparation for melanopsin expression
      materials, 296–297
      nuclear transplantation, 298–299
      overview, 294–296
      restriction-enzyme-mediated integration, 298
      sperm and egg preparation, 297–298
      transgene incorporation analysis, 302

video microscopy assay of light response
image acquisition, 303–304
materials, 302–303
transgenic melanophore responses, 304–305
Membrane fusion, *see* Rod outer segment
Methylation, proteins
electrospray ionization mass spectrometry, analysis of peptide methyl esters
high-performance liquid chromatography–mass spectrometry, 448–449, 451
preparation of peptides, 445–447
rhodopsin kinase trypsinization, 447–448
examples of vision proteins, 452
functions, 452–453
transducin, *see* Transducin
Microcalorimetry, time-resolved studies of phototransduction
brass blocks in instrumentation, 256
cyclic GMP hydrolysis enthalpy, 253–254
cyclic GMP phosphodiesterase quantification, 265–266, 269
electronics, 260–261
GTP analog studies, 266–268
impulse response, time resolution, and calibration, 261–262, 264
polystyrene block in Faraday cage, 257
polyvinylidene difluoride film, electric dipole and principle of operation, 257–260
sample chamber, 254–256
sample preparation, 265–266
sensitivity, 268–269
S-Modulin
cyclic GMP phosphodiesterase, regulation of activation, 3, 14
discovery, 3
electrophysiological detection in truncated rod outer segment
internal perfusion and current measurements, 6–9
rod outer segment
isolation, 4
truncation, 5–6
suction electrode preparation, 4–5
localization in retina, 14
physical properties, 13

purification
anion-exchange chromatography, 11–12
frog retina soluble protein preparation, 9–10
hydrophobic affinity chromatography, 10–11
recombinant protein from *Escherichia coli*, 13
rhodopsin phosphorylation regulation analysis
phosphorous-32 radiolabeling and gel electrophoresis, 18–19
rhodopsin kinase cross-linking, 19–20
rod outer segment preparation from frog, 16–18
MS, *see* Mass spectrometry
Müller cell opsin, *see* Retinal pigment epithelium retinal G-protein-coupled receptor

N

NDP kinase, *see* Nucleoside diphosphate kinase
NMR, *see* Nuclear magnetic resonance
Northern blot
lecithin:retinol acyltransferase tissue distribution, 408–410
retinitis pigmentosa GTPase regulator, 703
short-chain dehydrogenase/reductase, 379–380
Nuclear magnetic resonance
guanylate cyclase-activating protein 2 structure determination
data acquisition and analysis, 125–126
isotopic labeling and purification of recombinant protein, 124–125
sample preparation, 125
three-dimensional structures
recoverin homology, 131–132
unmyristoylated protein with calcium, 130–131
recoverin structure determination
data acquisition and analysis, 125–126
isotopic labeling and purification of recombinant protein, 124–125

sample preparation, 125
three-dimensional structures
　calcium-induced conformational changes, 128–129
　myristoylated protein with and without calcium, 127
　unmyristoylated protein, 126–127
Nucleoside diphosphate kinase
　assay and kinetic parameters for bovine retina enzyme, 92–93, 99
　expression of bovine retina enzyme in *Escherichia coli*
　　cell growth, 95–96
　　cloning of isoforms, 93–94
　　physical properties, 96
　　purification, 96
　　vectors, 94–95
　glycosylation analysis, bovine retina enzyme, 90, 92, 98
　homology between species, 98–99
　isoform comparisons of bovine retina enzyme, 96–98
　phototransduction role, 87
　purification from bovine retina, 88–89
　quaternary structure, bovine retina enzyme, 89–90

O

Opsin, *see also specific opsins*
　phylogenetic relationships of retinal and extraretinal proteins, 279–280
　promoter analysis in transiently transfected *Xenopus* embryos, 57
　Xenopus oocyte expression
　　advantages, 41, 49
　　glycosylation forms, 46–47
　　immunoprecipitation of radiolabeled proteins, 45–46
　　Limulus opsin, 49
　　materials and solutions, 41–43
　　membrane preparation, 45
　　nucleic acid injection, 43–44
　　oocyte preparation, 43
　　reconstitution, 49
　　voltage clamp recording, 44–45

P

PCR, *see* Polymerase chain reaction
PDE, *see* Cyclic GMP phosphodiesterase
Peripherin/rds
　domains involved in membrane fusion, 75–76
　electroretinogram studies
　　cone worse than rod disease, 625
　　equal rod and cone disease, 624
　　mutation studies, 621–624
　　rod worse than cone disease, 624
　membrane fusion assays, *see also* Rod outer segment
　　aggregation–adhesion analysis, 84–85
　　bilayer destabilization analysis, 85–86
　　characteristics of fusion with target membranes, 74–75
　　fusion competency assay of canine kidney cells expressing mutant proteins
　　　advantages, 83
　　　cell culture, 80–81
　　　pitfalls, 83–84
　　　probe labeling of membranes, 82
　　　resonance energy transfer-based lipid mixing assay, 81–83
　　peptide fragment assay
　　　contents mixing assay, 78–79
　　　materials, 78
　　　peptide effects on fusion, 79–80
　　　pitfalls, 80
　　　quenching pair assay principle, 76
　　　vesicle preparation, 78
　　purification from bovine retina and recombination with vesicles, 69–71
　mouse mutant phenotype, 672
　mutations in retinal dystrophies, 671, 684
　posttranslational modification, 673
　sequence homology between species, 672–673
　transient transfection with rom-1 in COS-1 cells
　　analysis of protein complexes
　　　reagents, 679, 681
　　　retinal disease analysis, 684, 687
　　　velocity sedimentation analysis, 682–684
　　　Western blot analysis, 681–682
　　cell growth and maintenance, 674–675

expression vector construction for wild-type and mutant proteins, 675–677
immunofluorescence assay of transfection efficiency, 677–679
transfection, 677–678
Phospholipase A$_2$, all-*trans*-retinyl ester hydrolase inhibition studies, 392–393
Phosphorylation, rhodopsin
gel assay of radiolabeled protein, 271–272
inhibitors, *see* S-Modulin; Recoverin
kinase, *see* Rhodopsin kinase
kinetics, 168
mutant studies in transgenic mice
DNA preparation for microinjection, 171
founder screening, 173–174
gene dosage control with knockout crosses, 179–182, 184
overexpression and retinal degradation, 178–179
rationale, 168–169
reverse transcriptase–polymerase chain reaction for transgene quantification, 176
single-cell recordings from mutant-containing photoreceptors, 185
Southern blot analysis for sorting different lines from same founder, 174
trafficking of mutants, 184
transgene constructs, 169–171
transgenic mouse generation, 172–173
phototransduction role, 167
regulatory function, 482
site identification
bovine rhodopsin
Asp-N digestion, 484
mass spectrometry analysis of fragment phosphopeptides, 487–488
overview, 482–484
peptide separation and identification, 484–486
phosphorylated protein preparation, 484
subdigestion of C-terminal peptide, 486–487
octopus rhodopsin
overview, 488
peptide identification, 492

phosphorylated protein preparation, 488, 491
proteolysis with trypsin and thermolysin, 491–492
overview of sites, 167–168
stoichiometry, 167
zebrafish larva eye, assay
incubation conditions and gel autoradiography, 551–552
materials, 551
principle, 551
white light versus red light response, 552
Photoaffinity labeling, rhodopsin
challenges, 425–426
photolysis, 430–432
proteolysis and sequencing, 433–435
reconstitution with pigment analogs, 429–430
retinal analog synthesis, 426–429
Photobleaching recovery, photoreceptors
absorption spectra of pigments, 227–228
adaptation, definition, 224
bleaching estimation
absorption cross-sections, 228, 230
fraction of pigment bleached, 226–227
solution photosensitivity, 233–234
specific optical density of rod outer segment, 230–233
calcium role in bleaching adaptation, 249–252
cone electrophysiology, 236, 238–239
enzyme assays with electrophysiology
calculations, 242
cyclic GMP phosphodiesterase, 243, 245, 249
guanylyl cyclase, 245, 247, 249
rapid solution changes, 239–241
isolated rod electrophysiology, 236, 239
light flash apparatus, 226
photoreceptor preparation from tiger salamander, 225–226
rationale for study, 224–225
regeneration in retina, 224
11-*cis*-retinal solution preparation
ethanol solution, 235
exogenous application to isolated cells, 235–236
interphotoreceptor retinoid binding protein solution, 235

phospholipid vesicles, 234–235
Pinopsin
  absorption spectrum, 290
  function, 279
  gene cloning, 279–282
  immunohistochemical localization, 282–283
  overexpression in transfected mammalian cells
    expression plasmid construction, 284–285
    human embryonic kidney cell growth, 285
    overview, 283–284
    transfection, 287–288
  purification from transfected cells
    anion-exchange chromatography, 288
    cation-exchange chromatography, 289
    concentration determination, 290
    dialysis and centrifugation, 289
    extraction, 288
    nickel affinity chromatography of histidine-tagged protein, 289
PLA$_2$, see Phospholipase A$_2$
Polymerase chain reaction
  adeno-associated virus titering by quantitative polymerase chain reaction, 756–757
  deuteranomaly, green–red pigment hybrid characterization with long-range polymerase chain reaction amplification, 663
  genotyping of mouse retinitis pigmentosa models
    cyclic GMP phosphodiesterase gene amplification, 523
    overview, 523
    rhodopsin gene amplification, 524
  genotyping of transgenic mice, 716–718
  green and green–red hybrid pigment gene expression in arrays, polymerase chain reaction/single-strand conformation polymorphism, 668–670
  radiolabeled DNA preparation for DNase I footprinting, 606–608
  red–green hybrid gene, polymerase chain reaction amplification and detection in females, 660, 662
  red/green pigment transcripts, ratio determination in retina, 667–668
  retinal pigment epithelium p65, amplification and single-strand conformation analysis
    labeled DNA, 708, 710
    primers, 709
    unlabeled DNA, 710
  retinitis pigmentosa mouse models, genotyping
    DNA isolation from tail biopsy, 522–523
    overview, 521–522
    polymerase chain reaction amplification, 523–524
    restriction digestion of amplified fragments, 524
    sequencing, 524–526
  RNA amplification, see Reverse transcriptase–polymerase chain reaction
  X-linked red–green photopigment genes, determination of number and ratio with competitive polymerase chain reaction
    amplification, 658
    data analysis, 655–656
    exons 2, 4, and 5 analysis, 659
    overview, 655
    primers, 657–658
    single-strand conformational polymorphism, 655, 658–659
Polyvinylidene difluoride film, see Microcalorimetry, time-resolved studies of phototransduction
Prenylation
  electrospray ionization mass spectrometry, analysis of prenylated peptides and peptide methyl esters
    high-performance liquid chromatography–mass spectrometry, 448–449, 451
    preparation of peptides, 445–447
    rhodopsin kinase trypsinization, 447–448
  examples of vision proteins, 436, 452
  functions, 452–453
  methyl iodide cleavage, 437
  Raney nickel cleavage and analysis
    exogenous lipid removal by gel electrophoresis, 439–440
    extraction of proteins
      cells, 439–440

gels, 440–441
gas chromatography–mass spectrometry analysis of nonradiolabeled prenyl groups, 444–445
quantification of radiolabeled proteins, 440
radioprenylation of proteins, 438–439
Raney nickel cleavage of prenyl groups, 441–442
reversed-phase high-performance liquid chromatography analysis, 442–444
Trypanosoma brucei proteins, 439–444
sequence motifs, 436, 452
structure elucidation, overview, 437–438

R

Raney nickel, prenyl group cleavage, *see* Prenylation
Recoverin
  calcium binding
    conformational change studies
      circular dichroism, 113–114
      hydrophobic site exposure, 114
      intrinsic fluorescence, 113
    flow dialysis analysis, 110–112
  calcium effects on protein–protein interactions, 34
  discovery, 3
  GRK1 surface plasmon resonance
    binding kinetics analysis, 30–31, 33
    fragment binding analysis, 39–40
  homologs, 122
  immobilization for surface plasmon resonance studies, 23–24
  localization in retina, 14
  myristoyl switch
    mechanism of action, 122, 129
    rhodopsin deactivation inhibition, 123
  physical properties, 13
  purification
    anion-exchange chromatography, 11–12
    bovine retina soluble protein preparation, 10
    hydrophobic affinity chromatography, 10–11

recombinant protein from *Escherichia coli*, 13
rhodopsin phosphorylation, inhibition, 20
structure determination by nuclear magnetic resonance
  data acquisition and analysis, 125–126
  isotopic labeling and purification of recombinant protein, 124–125
  sample preparation, 125
  three-dimensional structures
    calcium-induced conformational changes, 128–129
    myristoylated protein with and without calcium, 127
    unmyristoylated protein, 126–127
Retina
  avian, *see* Chicken retina
  neurons, *in situ* hybridization
    cryosection hybridization for double-label detection, 584–587
    cryosection hybridization for single-label detection
      applications, 581–582
      fixation of tissue, 582–583
      hybridization, 583
      immunohistochemical detection, 584
      posthybridization, 583–584
      prehybridization, 583
    detection techniques, 579, 584–585, 589–590
    digoxigenin–RNA probe synthesis, 580–581
    ribonuclease contamination prevention, 580
    whole mounts
      fixation of tissue, 587
      hybridization, 588
      immunohistochemical detection, 588–589
      posthybridization, 588
      prehybridization, 587–588
    transcription factors
      cloning and identification, *see* DNase I footprinting; Electrophoretic mobility shift assay; Yeast one-hybrid assay
      overview, 590–591
  zebrafish, *see* Zebrafish retina
11-*cis*-Retinal
  challenges of retinal quantification, 313

high-performance liquid chromatography analysis, *see* Retinoids, high-performance liquid chromatography analysis
solution preparation for photoreceptor bleaching studies
  ethanol solution, 235
  exogenous application to isolated cells, 235–236
  interphotoreceptor retinoid binding protein solution, 235
  phospholipid vesicles, 234–235
Retinal degeneration slow, *see* Peripherin/rds
Retinaldehyde dehydrogenase, phase partion assay
  enzyme preparation, 363
  incubation conditions, 364–365
  materials, 361
  membrane preparation, 363
  nicotinamide adenine dinucleotide tritiation, 361
  pH, 360
  principle, 360
Retinal pigment epithelium
  microsomal enzyme preparations, *see* Isomerhydrolase; Lecithin:retinol acyltransferase; Retinal pigment epithelium retinal G-protein-coupled receptor; 11-*cis*-Retinol dehydrogenase; Retinyl ester hydrolase
  visual cycle role, 324, 331, 359, 400, 565–566
Retinal pigment epithelium p65
  function, 706, 723–724
  gene regulation, 705
  genotyping of mice with polymerase chain reaction, 716–718
  knockout mouse
    embryonic stem cell derivation, 714–716
    generation, 716
    phenotype, 722
    retinoids, high-performance liquid chromatography analysis, 722
    rhodopsin quantification and characterization, 722–723
  mutation in disease
    clinical phenotypes, 713–714
    patient selection for analysis, 707
  polymerase chain reaction/single-strand conformation analysis
    labeled DNA, 708, 710
    primers, 709
    unlabeled DNA, 710
    restriction digest analysis, 710
    sequencing, 707–708, 710–711
    types of mutations, 706–707, 711–713
  reverse transcriptase–polymerase chain reaction of gene expression in mice, 718–719
  sequence homology between species, 705
  Western blot analysis
    antibodies, 719–720
    bovine retinal pigment epithelium membranes, 720
    mouse eye, 720–722
Retinal pigment epithelium retinal G-protein-coupled receptor
  absorption spectroscopy, 417–418
  chromophore
    extraction by hydroxylamine derivatization, 418–420
    high-performance liquid chromatography analysis of oximes, 420–421
    identification, 414
    irradiated protein chromophore, 422
  homology
    rhodopsin similarity, 413
    species comparison, 414
  hydroxylamine reactivity, 422
  purification
    extraction, 416
    immunoaffinity chromatography, 416–417
    retinal epithelium microsome preparation, 414–415
Retinitis pigmentosa
  animal models
    avian models, 526–536
    cyclic GMP, genes affecting levels, 521
    cyclic GMP phosphodiesterase rescue in mouse *rd* model, *see* Encapsidated adenovirus minichromosome
    genotyping of mouse models
      DNA isolation from tail biopsy, 522–523
      overview, 521–522
      polymerase chain reaction amplification, 523–524

SUBJECT INDEX 841

  restriction digestion of amplified fragments, 524
  sequencing, 524–526
 laboratory-generated models, 517–519
 natural models, 515, 517
 overview, 515
autosomal dominant retinitis pigmentosa
 paired-flash electroretinography in humans, 222
 ribozyme targeting of mutant protein transcripts in rat, see Ribozyme
 rom-1 mutation, 671–672
 retinal pigment epithelium p65 mutations, 711–713
 rhodopsin mutations, 618–619
Retinitis pigmentosa GTPase regulator
 mutation in X-linked retinitis pigmentosa type 3, 688
 RCC1 homologous domain, 688
 splice variant isolation, 703–704
 tissue distribution analysis by Northern blot, 703
 yeast two-hybrid system, detection of interacting retinal proteins
  advantages and disadvantages, 689
  bait
   plasmid preparation, 691–693
   plasmid, prescreen testing, 695
   selection, 691
  complementary cDNA library characteristics, 690–691
  β-galactosidase filter lift assay, 696–698
  host strain, 694–695
  principle, 688–689
  quantification of interaction, 700–703
  target plasmid preparation, 693–694
  transformation, 695–696
  verification of interaction, 698–700
Retinoids, high-performance liquid chromatography analysis
 absorption properties, 314–315, 322, 324
 calculation of results, 319
 challenges of retina retinoid studies, 313
 equipment, 318
 esterification
  retinoic acid, 317
  retinol, 316
 extraction
  reaction mixtures, 314
  tissues, 317–318, 320–321

 fluorescence detection, 321–322
 identification of peaks, 321–322, 324
 normal phase chromatography, 318–319
 oxime formation, 317
 reductions
  retinal, 314, 316
  11-*cis*-retinal with tritiated sodium borohydride, 316
 retinal pigment epithelium retinal G-protein-coupled receptor chromophore
  extraction by hydroxylamine derivatization, 418–420
  high-performance liquid chromatography analysis of oximes, 420–421
  irradiated protein chromophore, 422
 retinol oxidation to retinal, 316
 reversed-phase chromatography, 318–320
 saponification, 316–317, 319
 sensitivity, 319
 simultaneous assay of retinoid-utilizing enzymes in retina
  calculations, 339
  extraction of retinoids, 333–334
  retinyl ester hydrolysis, 337
  separation of retinoids, 334–335, 340
 stability of compounds, 319
 standards, 314
 transgenic mouse mutants
  constant illumination response, 573
  dissection of eye, 567–568
  extraction, 568
  flash response, 572–573
  individual variation of retinas, 574
  retinal pigment epithelium p65 knockout mouse analysis, 722
 zebrafish larva eye retinoid analysis
  characterization of retinoids, 556–557
  chromatography, 556
  extraction, 555–556
  materials, 555
  principle, 552, 555
11-*cis*-Retinol
 formation from all-*trans* retinyl esters, see Isomerhydrolase
 high-performance liquid chromatography analysis, see Retinoids, high-performance liquid chromatography analysis
11-*cis*-Retinol dehydrogenase

epitope ectopic site studies of topology in mutants, principle, 347–348
glycosylation ectopic site studies of topology in mutants
  endoglycosidase H analysis, 355, 357
  mutant construction, 353–355
  principle, 346–347
high-performance liquid chromatography assay, 365–366, 368, 370–371
hydropathy analysis, 344–345
lumenal orientation of catalytic domain, 358
phase partion assay
  endogenous substrate assay, 364
  enzyme preparation, 363
  exogenous substrate assay, 364
  materials, 361
  membrane preparation, 363
  nicotinamide adenine dinucleotide tritiation, 361
  pH, 360
  principle, 360
  recombinant insect cell activity versus microsomes, 367–368
protease protection analysis
  principle of topology analysis, 345–346
  retinal pigment epithelium microsomes
    microsome preparation, 348–349
    proteolysis reaction, 349–350
    Western blot analysis, 350
  Triton X-100 studies, 356–357
  in vitro-translated enzymes
    comparison with retinal microsome enzyme topology, 350, 357
    microome preparation, 352
    proteinase K treatment, 352–353
    transcription, 351
    translation, 351–352
visual cycle role, 331, 344, 359, 371–372
Retinol isomerase, *see* Isomerhydrolase
Retinyl ester hydrolase, *see also specific enzymes*
  assay with radiolabeled substrates
    hydrolytic reaction confirmation, 386
    incubation conditions and quantification, 385
    overview, 384–385
    substrate synthesis, 385
  neutral, bile salt-independent activity in microsomes, 385–386

11-*cis*-retinyl ester hydrolase
  colocalization with 11-*cis*-retinyl esters in retinal pigment epithelium plasma membrane, 393–395, 398–400
  hydrolysis of 11-*cis*-retinyl esters *in situ*, 395–397
  isolation
    hydrophobic interaction chromatography, 398
    Percoll gradient separation of S2 membranes, 397–398
  simultaneous detection with lecithin:retinol acyltransferase and retinol isomerase activities
    all-*trans*-retinol preparation, 333
    cellular retinaldehyde-binding protein preparation and effects, 332–333, 340–341
    denaturing gel electrophoresis, 339
    high-performance liquid chromatography
      calculations, 339
      extraction of retinoids, 333–334
      retinyl ester hydrolysis, 337
      separation of retinoids, 334–335, 340
    incubation conditions, 333
    kinetic effects of phosphates, ATP, and alcohols, 341–343
    materials, 332
    stability of proteins, 339–340
    ultraviolet treatment of retinal pigment epithelium microsomes, 332, 340
  substrate specificity of retinal pigment epithelium microsome enzymes
    active site modification and inhibition analysis for different substrates, 390–391
    all-*trans*-retinyl ester hydrolase inhibitor studies
      bis(*p*-nitrophenyl) phosphate, 391–392
      phospholipase A_2, 392–393
    fatty acyl group preference, 388–389
    retinyl palmitate isomers, 386–388
    substrate competition studies, 389–390
  visual cycle role, 331, 384, 399–400
Reverse transcriptase–polymerase chain reaction
  cyclic GMP phosphodiesterase rescue

evaluation in mouse *rd* model of retinitis pigmentosa, 740
lecithin:retinol acyltransferase transcripts, 403–405
ribozyme cleavage products, 774–775
transgene quantification in mice, 176
RGR, *see* Retinal pigment epithelium retinal G-protein-coupled receptor
Rhodopsin
concentration determination in transgenic mouse mutants, 568
electroretinogram studies of mutants, 618–621
genotyping of mouse retinitis pigmentosa models, 524
intermediates, overview, 425
mutations in retinitis pigmentosa, 618–619
phosphorylation, *see* Phosphorylation, rhodopsin
photoaffinity labeling, *see* Photoaffinity labeling, rhodopsin
photobleaching recovery, *see* Photobleaching recovery, photoreceptors
retinal pigment epithelium p65 knockout mouse characterization, 722–723
Xenopus transgenesis and mutant expression, 64
Rhodopsin kinase
S-modulin cross-linking, 19–20
prenylated peptide analysis
high-performance liquid chromatography–mass spectrometry, 448–449, 451
preparation of peptide standards, 445–447
trypsinization, 447–448
Ribozyme
autosomal dominant retinitis pigmentosa, targeting of mutant protein transcripts in rat
challenges, 764, 775–776
cleavage reactions, incubation conditions, 771–772
cloning of ribozymes and targets, 767–768
expression level requirements for ribozymes, 776
materials, 767

multiple turnover kinetic analysis of cleavage, 772, 776
normal allele requirement for effective therapy, 775
preparative gel electrophoresis of RNA, 769–770
radiolabeling of RNA oligonucleotides, 769
reverse transcription–polymerase chain reaction of products, 774–775
RNA extraction from retina, 770
selectivity of cleavage, 775–776
single turnover kinetic analysis of cleavage, 772–774
specific radioactivity determination for labeled RNAs, 770–771
target site identification, 765–766
transcription, *in vitro*, 768–769
classification, 761–762
gene therapy
advantages over antisense targeting, 762
potential, 762
target specificity, 763–764
secondary structures of hammerhead and hairpin ribozymes, 763
Rod outer segment
cytoplasmic calcium, *see* Calcium flux
membrane fusion, *see also* Peripherin/rds
aggregation–adhesion analysis, 84–85
assay of peripherin/rds fragments
caveats, 80
contents mixing assay, 78–79
materials, 78
peptide effects on fusion, 79–80
quenching pair assay principle, 76
vesicle preparation, 78
assay of target membrane fusions
controls, 75
design principles, 66
disk membrane preparation, 69
disk rim-specific vesicle preparation, 69
fluorescence assay, 71–72
octadecylrhodamine B chloride labeling, 68–69
overview, 67
peripherin/rds purification from bovine retina and recombination with vesicles, 69–71

plasma membrane preparation, 67–68
bilayer destabilization analysis, 85–86
characteristics of target model membrane fusions
  disk membranes, 74
  disk rim-specific vesicles, 74–75
  overview, 73
  peripherin/rds recombined with vesicles, 74–75
fusion competency assay of canine kidney cells expressing mutant peripherin/rds
  advantages, 83
  cell culture, 80–81
  pitfalls, 83–84
  probe labeling of membranes, 82
  resonance energy transfer-based lipid mixing assay, 81–83
fusion protein mediation, 65
physiological functions, 65
steps, 84
phototransduction overview, 188–189
single-photon detection
  dialyzed outer segment recording, 195–196
  effective photon absorptions, estimation of average number, 196–197
  interpretation of electrophysiological recordings in terms of transduction cascade, 190, 199, 201–202
  number of responses for electrophysiological recordings, 190
  overview, 186–187
  single-photon response isolation, 197, 199
  size of current response, 189–190
  suction electrode recording, 190, 192–193
  tissue preparation, 187–188
  truncated outer segment recording, 193, 195
visual cycle role, 565–566
Rod worse than cone disease, electroretinogram studies, 624
Rom-1
  mutation in autosomal dominant retinitis pigmentosa, 671–672
  posttranslational modification, 673

sequence homology between species, 672–673
transient transfection with peripherin/rds in COS-1 cells
  analysis of protein complexes
    reagents, 679, 681
    retinal disease analysis, 684, 687
    velocity sedimentation analysis, 682–684
    Western blot analysis, 681–682
  cell growth and maintenance, 674–675
  expression vector construction for wild-type and mutant proteins, 675–677
  immunofluorescence assay of transfection efficiency, 677–679
  transfection, 677–678
ROS, *see* Rod outer segment
RPE, *see* Retinal pigment epithelium
RPE65, *see* Retinal pigment epithelium p65
RPGR, *see* Retinitis pigmentosa GTPase regulator
RT–PCR, *see* Reverse transcriptase–polymerase chain reaction

S

s26
  localization in retina, 14
  physical properties, 13
  purification
    anion-exchange chromatography, 11–12
    frog retina soluble protein preparation, 9–10
    hydrophobic affinity chromatography, 10–11
    recombinant protein from *Escherichia coli*, 13
S100A11
  antibody production, 109–110
  gel overlay assay of binding proteins, 116–117
  immunohistochemistry, 110
SCOP2
  function, 279
  gene cloning, 279–282

SDR, see Short-chain dehydrogenase/reductase
Short-chain dehydrogenase/reductase, see also specific enzymes
  activity of novel proteins, 382–383
  complementary DNA cloning of retinal enzymes, 376–377
  expressed sequence tags, database searching for cloning probes, 373–376
  expression in baculovirus–insect cell system, 380–382
  phase partition assay, 368, 382
  sequence analysis
    homology of superfamily, 372
    identification of family members, 377–378
    phylogenetic analysis, 378–379
  substrate specificity, 371–372
  tissue distribution, Northern blot analysis, 379–380
Single-photon detection, photoreceptors, see Rod outer segment
Single-strand conformation polymorphism
  green and green–red hybrid pigment gene expression in arrays, 668–670
  red/green pigment, ratio determination in retina, 667–668
  red pigment Ser180Ala polymorphism, 659–660
  retinal pigment epithelium p65
    labeled DNA, 708, 710
    primers, 709
    unlabeled DNA, 710
  X-linked red–green photopigment genes, 655, 658–659
in Situ hybridization, retinal neurons
  cryosection hybridization for double-label detection, 584–587
  cryosection hybridization for single-label detection
    applications, 581–582
    fixation of tissue, 582–583
    hybridization, 583
    immunohistochemical detection, 584
    posthybridization, 583–584
    prehybridization, 583
  detection techniques, 579, 584–585, 589–590
  digoxigenin–RNA probe synthesis, 580–581
  ribonuclease contamination prevention, 580
  whole mounts
    fixation of tissue, 587
    hybridization, 588
    immunohistochemical detection, 588–589
    posthybridization, 588
    prehybridization, 587–588
Southern blot
  transgenic mouse, sorting different lines from same founder, 174
  X-linked red–green photopigment genes, determination of number and ratio, 655
SPR, see Surface plasmon resonance
SSCP, see Single-strand conformation polymorphism
Sterol dehydrogenase, phase partition assay
  applications, 368, 382
  enzyme preparation, 363
  incubation conditions, 365
  materials, 361
  membrane preparation, 363
  nicotinamide adenine dinucleotide tritiation, 361
  pH, 360
  principle, 360
Surface plasmon resonance
  GRK1 binding kinetics analysis with recoverin or calmodulin, 30–31, 33
  ligand immobilization
    amine group coupling, 21–22
    biotinylated protein coupling to immobilized streptavidin, 24–25, 27–28
    histidine-tagged proteins, 28
    preservation of activity, 29
    sulfhydryl group coupling, 22
  overview, 20–21
  sensogram analysis, 29
  small molecule effects on protein–protein interactions in visual transduction cascade, 33–34, 36, 39
  structure–function analysis of mutant proteins, 29–30

T

α-Toxin, photoreceptor permeabilization cell death analysis, 277

confocal immunofluorescence localization
  antibody preparation, 272–273
  bovine retina sectioning and staining, 273
  data collection, 273, 275
  neurobiotin distribution following permeabilization, 277–278
  toxin distribution, 277
expression and purification of recombinant toxin, 271
mutant pore blocking with zinc, 271, 275
permeabilization mechanism and features, 270
rationale, 269–270
rhodopsin phosphorylation assay of permeabilization, 271–272, 275
Transducin
  abundance in rod outer segment, 253
  immobilization for surface plasmon resonance studies, 25, 27–28
  isoprenyl/methyl function studies of $\beta\gamma$
    ADP ribosylation by pertussis toxin, 459–460, 478–481
    effector enzyme activation assays, 459, 463, 465
    farnesylated cysteine analog probing of binding specificity, 462–464
    gel filtration analysis of subunit interactions, 481
    GTP exchange assay, 459, 462
    GTPγS binding to α catalyzed by metarhodopsin II, 476–478
    liposome binding assay, 460–461
    protein preparation for assays
      defarnesylated $\beta\gamma$, 457
      demethylated $\beta\gamma$, 457–459, 467, 469
      mutant $\beta\gamma$ expression and purification in baculovirus–insect cell system, 471–474
      reversed-phase high-performance liquid chromatography analysis of γ, 469, 471, 474
      rhodopsin, 456
      rod outer segment preparation, 467–468
      transducin, 456, 468–469
    rhodopsin binding, 462–463
    sites of modification, 466–467
    stripped rod outer segment binding assay, 475–476

promoter analysis of α subunit in transiently transfected *Xenopus* embryos, 57
subunits, functions and effectors, 454–456
visual cycle role, 453–454
Xenopus oocyte expression
  advantages, 41, 49
  α subunit expression, 47–49
  immunoprecipitation of radiolabeled proteins, 45–46
  materials and solutions, 41–43
  membrane preparation, 45
  nucleic acid injection, 43–44
  oocyte preparation, 43
  voltage clamp recording, 44–45
zebrafish larva eye, activation assays
  GTPγS *in situ* binding assay
    incubation conditions and radioactivity detection, 542
    materials, 541
    pob mutant analysis, 542, 544
    principle, 540–541
    tissue preparation, 541–542
  GTPγS *in vitro* binding assay
    incubation conditions and radioactivity detection, 544
    materials, 544
    principle, 544
    white light versus red light response, 544–545
Transgenic frog, *see Xenopus* transgenesis
Transgenic mouse
  GC-E knockout mouse
    electroretinogram recordings, 560–564
    generation, 559
    histological analysis, 560–561
    single-cell electrophysiology, 560–561, 563–564
  genotyping of retinitis pigmentosa models
    DNA isolation from tail biopsy, 522–523
    overview, 521–522
    polymerase chain reaction amplification, 523–524
    restriction digestion of amplified fragments, 524
    sequencing, 524–526
  opsin phosphorylation-site mutant expression

rationale, 168–169
constant illumination protocols
  excised eyes, 569
  living mice, 569
  retinoid responses, 573
DNA preparation for microinjection, 171
flash protocols
  excised eyes, 569
  living mice, 568
  retinoid responses, 572–573
founder screening, 173–174
gene dosage control with knockout crosses, 179–182, 184
housing of animals, 567
overexpression and retinal degradation, 178–179
overview, 565–567
retinal pigment epithelium p65 knockout mouse
  embryonic stem cell derivation, 714–716
  generation, 716
  phenotype, 722
  retinoids, high-performance liquid chromatography analysis, 722
  rhodopsin quantification and characterization, 722–723
visual cycle mutants
  mouse generation, 172–173
  photoreceptor membrane preparation, 569, 571
  retinoid analysis by high-performance liquid chromatography
    dissection of eye, 567–568
    extraction, 568
    individual variation of retinas, 574
  reverse transcriptase–polymerase chain reaction for transgene quantification, 176
  rhodopsin concentration determination, 568
  single-cell recordings from mutant-containing photoreceptors, 185
  Southern blot analysis for sorting different lines from same founder, 174
  trafficking of mutants, 184
  transgene constructs, 169–171
  Western blot analysis of protein expression, 574–575

Two-dimensional gel electrophoresis
  apparatus and isoelectric focusing tubes, 495–496
  bovine retina protein patterns, 500, 502
  buffers and solutions, 496–497
  genome database complementation, 493–494
  isoelectric focusing gel electrophoresis, 498–500
  mass spectrometry for peptide mass fingerprinting
    applications, 494–495, 510–511
    in-gel digestion
      protease selection, 502–503
      trypsin digestion, 503–504
    matrix-assisted laser desorption ionization–time of flight analysis
      criteria for identification using MS-Fit, 508–509
      database searching, 505–506
      equipment, 505
      large fragment fingerprinting, 509–510
      principle, 504–505
      spectra acquisition, 505
    prospects, 510–511
  overview, 492–493
  sample preparation, 497–498
  sodium dodecyl sulfate–polyacrylamide gel electrophoresis, 500

W

Western blot
  cyclic GMP phosphodiesterase, transgene expression, 740–741
  lecithin:retinol acyltransferase, 410
  peripherin/rds–rom-1 complex analysis, 681–682
  retinal pigment epithelium p65
    antibodies, 719–720
    bovine retinal pigment epithelium membranes, 720
    mouse eye, 720–722
  11-cis-retinol dehydrogenase, 350
  vision mutant protein expression analysis in transgenic mice, 574–575

X

Xenopus embryo
　advantages for photoreceptor studies, 50–51
　retinal development, 50
　transient transfection
　　embryo preparation, 54–55
　　lipofection, 56
　　luciferase reporter assay, 56–57
　　materials and solutions, 52–54
　　principle, 51–52
　　promoters, analysis of *cis*-acting elements, 57
　　trypsinization, 55–56
Xenopus melanophore, *see* Melanopsin
Xenopus oocyte, phototransduction protein expression
　advantages, 41, 49
　immunoprecipitation of radiolabeled proteins, 45–46
　materials and solutions, 41–43
　membrane preparation, 45
　nucleic acid injection, 43–44
　oocyte preparation, 43
　opsin
　　glycosylation forms, 46–47
　　Limulus opsin, 49
　　reconstitution, 49
　transducin α subunit, 47–49
　voltage clamp recording, 44–45
Xenopus transgenesis
　efficiency, 64
　egg quality, 58
　embryo selection, 63–64
　interphase egg extract, 60–61
　materials and solutions, 59–60
　nuclear transplantation, 62–63
　overview, 58–59
　restriction enzyme-mediated integration reaction, 62
　rhodopsin mutant expression, 64
　sperm nuclei preparation, 61–62
　Xenopus melanophore, melanopsin transgene expression
　　materials, 296–297
　　nuclear transplantation, 298–299
　　overview, 294–296
　　restriction-enzyme-mediated integration, 298
　　sperm and egg preparation, 297–298
　　transgene incorporation analysis, 302
　　video microscopy assay of light response, 304–305

Y

Yeast one-hybrid assay
　advantages and limitations, 592–593
　kits, 593
　principle, 591–592
　retinal transcription factor cloning
　　bait plasmid construction
　　　cloning target sequences into reporter vectors, 593–594
　　　target DNA sequence selection, 593
　　complementary DNA library from retina, construction and amplification, 594–596
　　dual-reporter selection
　　　false-positive minimization, 596
　　　separate reporter strains and diploid analysis, 597–598
　　　single dual-reporter strain, 597
　　positive clone characterization, 598–600
Yeast two-hybrid system, detection of interacting retinal proteins
　advantages and disadvantages, 689
　principle, 688–689
　retinitis pigmentosa GTPase regulator as bait
　bait
　　plasmid preparation, 691–693
　　plasmid, prescreen testing, 695
　　selection of region, 691
　complementary cDNA library characteristics, 690–691
　β-galactosidase filter lift assay, 696–698
　host strain, 694–695
　quantification of interaction, 700–703
　target plasmid preparation, 693–694
　transformation, 695–696
　verification of interaction, 698–700

Z

Zebrafish retina
　advantages as model system, 536–537

cyclic GMP phosphodiesterase assay
   incubation conditions and product separation, 546–547
   light sensitivity determination, 547
   materials, 546
   principle, 546
   time course of activation, 547, 549
guanylyl cyclase assay
   calcium dependence, 550–551
   incubation conditions and product separation, 550
   materials, 549–550
   principle, 549
homogenization of larval eye, 539–540
larval stocks and culture conditions, 538
nrb model, 537
retinoid analysis by high-performance liquid chromatography
   characterization of retinoids, 556–557
   chromatography, 556
   extraction, 555–556
   materials, 555
   principle, 552, 555
rhodopsin and cone opsin phosphorylation assay
   incubation conditions and gel autoradiography, 551–552
   materials, 551
   principle, 551
   white light versus red light response, 552
transducin activation assays
   GTPγS *in situ* binding assay
      incubation conditions and radioactivity detection, 542
      materials, 541
      pob mutant analysis, 542, 544
      principle, 540–541
      tissue preparation, 541–542
   GTPγS *in vitro* binding assay
      incubation conditions and radioactivity detection, 544
      materials, 544
      principle, 544
      white light versus red light response, 544–545

ISBN 0-12-182217-6

90038